D0216793

U.S. Customary Units and Their SI Equivalents

Quantity	U.S. Customary Units	SI Equivalent
Acceleration	ft/s^2	0.3048 m/s^2
	in./s^2	0.0254 m/s^2
Area	ft^2	0.0929 m^2
	in^2	645.2 mm^2
Energy	ft·lb	1.356 J
Force	kip	4.448 kN
	lb	4.448 N
	oz	0.2780 N
Impulse	lb·s	4.448 N·s
Length	ft	0.3048 m
	in.	25.40 mm
	mi	1.609 km
Mass	oz mass	28.35 g
	lb mass	0.4536 kg
	slug	14.59 kg
	ton	907.2 kg
Moment of a force	lb·ft	1.356 N·m
	lb·in.	0.1130 N·m
Moment of inertia		
Of an area	in^4	0.4162×10^6 mm^4
Of a mass	lb·ft·s^2	1.356 kg·m^2
Power	ft·lb/s	1.356 W
	hp	745.7 W
Pressure or stress	lb/ft^2	47.88 Pa
	lb/in^2 (psi)	6.895 kPa
Velocity	ft/s	0.3048 m/s
	in./s	0.0254 m/s
	mi/h (mph)	0.4470 m/s
	mi/h (mph)	1.609 km/h
Volume, solids	ft^3	0.02832 m^3
	in^3	16.39 cm^3
Liquids	gal	3.785 L
	qt	0.9464 L
Work	ft·lb	1.356 J

Seventh Edition

Mechanics of Materials

Ferdinand P. Beer

Late of Lehigh University

E. Russell Johnston, Jr.

Late of University of Connecticut

John T. DeWolf

University of Connecticut

David F. Mazurek

United States Coast Guard Academy

Mc
Graw
Hill
Education

MECHANICS OF MATERIALS, SEVENTH EDITION

Published by McGraw-Hill Education, 2 Penn Plaza, New York, NY 10121. Copyright © 2015 by McGraw-Hill Education. All rights reserved. Printed in the United States of America. Previous editions © 2012, 2009, 2006, and 2002. No part of this publication may be reproduced or distributed in any form or by any means, or stored in a database or retrieval system, without the prior written consent of McGraw-Hill Education, including, but not limited to, in any network or other electronic storage or transmission, or broadcast for distance learning.

Some ancillaries, including electronic and print components, may not be available to customers outside the United States.

This book is printed on acid-free paper.

1 2 3 4 5 6 7 8 9 0 QVS/QVS 1 0 9 8 7 6 5 4 3 2 1 0

ISBN 978-0-07-339823-5
MHID 0-07-339823-3

Senior Vice President, Products & Markets: *Kurt L. Strand*
Vice President, General Manager: *Marty Lange*
Vice President, Content Production & Technology Services: *Kimberly Meriwether David*
Editorial Director: *Thomas Timp*
Global Brand Manager: *Raghothaman Srinivasan*
Brand Manager: *Bill Stenquist*
Marketing Manager: *Heather Wagner*
Product Developer: *Robin Reed*
Director, Content Production: *Terri Schiesl*
Content Project Manager: *Jolynn Kilburg*
Buyer: *Nichole Birkenholz*
Media Project Manager: *Sandra Schnee*
Photo Research: *Carrie K. Burger*
In-House Designer: *Matthew Backhaus*
Cover Designer: *Matt Backhaus*
Cover Image Credit: *©Walter Bibikow*
Compositor: *RPK Editorial Services, Inc.*
Typeface: *9.5/12 Utopia Std*
Printer: *Quad/Graphics*

All credits appearing on page or at the end of the book are considered to be an extension of the copyright page.

The photo on the cover shows the steel sculpture "Venture" by Alex Liberman (1912-1999) in front of the Bank of America Building in Dallas, Texas. The building is supported by a combination of structural steel and reinforced concrete.

Library of Congress Cataloging-in-Publication

Data on File

The Internet addresses listed in the text were accurate at the time of publication. The inclusion of a website does not indicate an endorsement by the authors or McGraw-Hill Education, and McGraw-Hill Education does not guarantee the accuracy of the information presented at these sites.

www.mhhe.com

About the Authors

John T. DeWolf, Professor of Civil Engineering at the University of Connecticut, joined the Beer and Johnston team as an author on the second edition of *Mechanics of Materials*. John holds a B.S. degree in civil engineering from the University of Hawaii and M.E. and Ph.D. degrees in structural engineering from Cornell University. He is a Fellow of the American Society of Civil Engineers and a member of the Connecticut Academy of Science and Engineering. He is a registered Professional Engineer and a member of the Connecticut Board of Professional Engineers. He was selected as a University of Connecticut Teaching Fellow in 2006. Professional interests include elastic stability, bridge monitoring, and structural analysis and design.

David F. Mazurek, Professor of Civil Engineering at the United States Coast Guard Academy, joined the Beer and Johnston team as an author on the fifth edition. David holds a B.S. degree in ocean engineering and an M.S. degree in civil engineering from the Florida Institute of Technology, and a Ph.D. degree in civil engineering from the University of Connecticut. He is a registered Professional Engineer. He has served on the American Railway Engineering & Maintenance of Way Association's Committee 15—Steel Structures since 1991. He is a Fellow of the American Society of Civil Engineers, and was elected into the Connecticut Academy of Science and Engineering in 2013. Professional interests include bridge engineering, structural forensics, and blast-resistant design.

Contents

*Advanced or specialty topics

9 Deflection of Beams 599

10 Columns 691

11 Energy Methods 759

Appendices   A1

Preface

Objectives

The main objective of a basic mechanics course should be to develop in the engineering student the ability to analyze a given problem in a simple and logical manner and to apply to its solution a few fundamental and well-understood principles. This text is designed for the first course in mechanics of materials—or strength of materials—offered to engineering students in the sophomore or junior year. The authors hope that it will help instructors achieve this goal in that particular course in the same way that their other texts may have helped them in statics and dynamics. To assist in this goal, the seventh edition has undergone a complete edit of the language to make the book easier to read.

General Approach

In this text the study of the mechanics of materials is based on the understanding of a few basic concepts and on the use of simplified models. This approach makes it possible to develop all the necessary formulas in a rational and logical manner, and to indicate clearly the conditions under which they can be safely applied to the analysis and design of actual engineering structures and machine components.

Free-body Diagrams Are Used Extensively. Throughout the text free-body diagrams are used to determine external or internal forces. The use of "picture equations" will also help the students understand the superposition of loadings and the resulting stresses and deformations.

 The SMART Problem-Solving Methodology is Employed. New to this edition of the text, students are introduced to the SMART approach for solving engineering problems, whose acronym reflects the solution steps of *S*trategy, *M*odeling, *A*nalysis, and *R*eflect & *T*hink. This methodology is used in all Sample Problems, and it is intended that students will apply this approach in the solution of all assigned problems.

Design Concepts Are Discussed Throughout the Text Whenever Appropriate. A discussion of the application of the factor of safety to design can be found in Chap. 1, where the concepts of both allowable stress design and load and resistance factor design are presented.

A Careful Balance Between SI and U.S. Customary Units Is Consistently Maintained. Because it is essential that students be able to handle effectively both SI metric units and U.S. customary units, half the concept applications, sample problems, and problems to be assigned have been stated in SI units and half in U.S. customary units. Since a large number of problems are available, instructors can assign problems using each system of units in whatever proportion they find desirable for their class.

Optional Sections Offer Advanced or Specialty Topics. Topics such as residual stresses, torsion of noncircular and thin-walled members, bending of curved beams, shearing stresses in non-symmetrical members, and failure criteria have been included in optional sections for use in courses of varying emphases. To preserve the integrity of the subject, these topics are presented in the proper sequence, wherever they logically belong. Thus, even when not covered in the course, these sections are highly visible and can be easily referred to by the students if needed in a later course or in engineering practice. For convenience all optional sections have been indicated by asterisks.

Chapter Organization

It is expected that students using this text will have completed a course in statics. However, Chap. 1 is designed to provide them with an opportunity to review the concepts learned in that course, while shear and bending-moment diagrams are covered in detail in Secs. 5.1 and 5.2. The properties of moments and centroids of areas are described in Appendix A; this material can be used to reinforce the discussion of the determination of normal and shearing stresses in beams (Chaps. 4, 5, and 6).

The first four chapters of the text are devoted to the analysis of the stresses and of the corresponding deformations in various structural members, considering successively axial loading, torsion, and pure bending. Each analysis is based on a few basic concepts: namely, the conditions of equilibrium of the forces exerted on the member, the relations existing between stress and strain in the material, and the conditions imposed by the supports and loading of the member. The study of each type of loading is complemented by a large number of concept applications, sample problems, and problems to be assigned, all designed to strengthen the students' understanding of the subject.

The concept of stress at a point is introduced in Chap. 1, where it is shown that an axial load can produce shearing stresses as well as normal stresses, depending upon the section considered. The fact that stresses depend upon the orientation of the surface on which they are computed is emphasized again in Chaps. 3 and 4 in the cases of torsion and pure bending. However, the discussion of computational techniques—such as Mohr's circle—used for the transformation of stress at a point is delayed until Chap. 7, after students have had the opportunity to solve problems involving a combination of the basic loadings and have discovered for themselves the need for such techniques.

The discussion in Chap. 2 of the relation between stress and strain in various materials includes fiber-reinforced composite materials. Also, the study of beams under transverse loads is covered in two separate chapters. Chapter 5 is devoted to the determination of the normal stresses in a beam and to the design of beams based on the allowable normal stress in the material used (Sec. 5.3). The chapter begins with a discussion of the shear and bending-moment diagrams (Secs. 5.1 and 5.2) and includes an optional section on the use of singularity functions for the determination of the shear and bending moment in a beam (Sec. 5.4). The chapter ends with an optional section on nonprismatic beams (Sec. 5.5).

Chapter 6 is devoted to the determination of shearing stresses in beams and thin-walled members under transverse loadings. The formula for the shear flow, $q = VQ/I$, is derived in the traditional way. More advanced aspects of the design of beams, such as the determination of the principal stresses at the junction of the flange and web of a W-beam, are considered in Chap. 8, an optional chapter that may be covered after the transformations of stresses have been discussed in Chap. 7. The design of transmission shafts is in that chapter for the same reason, as well as the determination of stresses under combined loadings that can now include the determination of the principal stresses, principal planes, and maximum shearing stress at a given point.

Statically indeterminate problems are first discussed in Chap. 2 and considered throughout the text for the various loading conditions encountered. Thus, students are presented at an early stage with a method of solution that combines the analysis of deformations with the conventional analysis of forces used in statics. In this way, they will have become thoroughly familiar with this fundamental method by the end of the course. In addition, this approach helps the students realize that stresses themselves are statically indeterminate and can be computed only by considering the corresponding distribution of strains.

The concept of plastic deformation is introduced in Chap. 2, where it is applied to the analysis of members under axial loading. Problems involving the plastic deformation of circular shafts and of prismatic beams are also considered in optional sections of Chaps. 3, 4, and 6. While some of this material can be omitted at the choice of the instructor, its inclusion in the body of the text will help students realize the limitations of the assumption of a linear stress-strain relation and serve to caution them against the inappropriate use of the elastic torsion and flexure formulas.

The determination of the deflection of beams is discussed in Chap. 9. The first part of the chapter is devoted to the integration method and to the method of superposition, with an optional section (Sec. 9.3) based on the use of singularity functions. (This section should be used only if Sec. 5.4 was covered earlier.) The second part of Chap. 9 is optional. It presents the moment-area method in two lessons.

Chapter 10, which is devoted to columns, contains material on the design of steel, aluminum, and wood columns. Chapter 11 covers energy methods, including Castigliano's theorem.

Supplemental Resources for Instructors

Find the **Companion Website** for Mechanics of Materials at www.mhhe.com/beerjohnston. Included on the website are lecture PowerPoints, an image library, and animations. On the site you'll also find the **Instructor's Solutions Manual** (password-protected and available to instructors only) that accompanies the seventh edition. The manual continues the tradition of exceptional accuracy and normally keeps solutions contained to a single page for easier reference. The manual includes an in-depth review of the material in each chapter and houses tables designed to assist instructors in creating a schedule of assignments for their courses. The various topics covered in the text are listed in Table I, and a suggested number of periods to be spent on each topic is indicated. Table II provides a brief description of all groups of problems and a classification of the problems in each group according to the units used. A Course Organization Guide providing sample assignment schedules is also found on the website.

Via the website, instructors can also request access to **C.O.S.M.O.S.**, the Complete Online Solutions Manual Organization System that allows instructors to create custom homework, quizzes, and tests using end-of-chapter problems from the text.

 **McGraw-Hill Connect Engineering** provides online presentation, assignment, and assessment solutions. It connects your students with the tools and resources they'll need to achieve success. With Connect Engineering you can deliver assignments, quizzes, and tests online. A robust set of questions and activities are presented and aligned with the textbook's learning outcomes. As an instructor, you can edit existing questions and author entirely new problems. Integrate grade reports easily with Learning Management Systems (LMS), such as WebCT and Blackboard—and much more. ConnectPlus® Engineering provides students with all the advantages of Connect Engineering, plus 24/7 online access to a media-rich eBook, allowing seamless integration of text, media, and assessments. To learn more, visit www.mcgrawhillconnect.com.

LEARNSMART® **McGraw-Hill LearnSmart** is available as a standalone product or an integrated feature of McGraw-Hill Connect Engineering. It is an adaptive learning system designed to help students learn faster, study more efficiently, and retain more knowledge for greater success. LearnSmart assesses a student's knowledge of course content through a series of adaptive questions. It pinpoints concepts the student does not understand and maps out a personalized study plan for success. This innovative study tool also has features that allow instructors to see exactly what students have accomplished and a built-in assessment tool for graded assignments. Visit the following site for a demonstration. www.LearnSmartAdvantage.com

SMARTBOOK™ Powered by the intelligent and adaptive LearnSmart engine, **SmartBook** is the first and only continuously adaptive reading experience available today. Distinguishing what students know from what they don't, and honing in on concepts they are most likely to forget, SmartBook personalizes content for each student. Reading is no longer a passive and linear experience but an engaging and dynamic one, where students are more likely to master and retain important concepts, coming to class better prepared. SmartBook includes powerful reports that identify specific topics and learning objectives students need to study.

create™ Craft your teaching resources to match the way you teach! With **McGraw-Hill Create**, www.mcgrawhillcreate.com, you can easily rearrange chapters, combine material from other content sources, and quickly upload your original content, such as a course syllabus or teaching notes. Arrange your book to fit your teaching style. Create even allows you to personalize your book's appearance by selecting the cover and adding your name, school, and course information. Order a Create book and you'll receive a complimentary print review copy in 3–5 business days or a complimentary electronic review copy (eComp) via email in minutes. Go to www.mcgrawhillcreate.com today and register to experience how McGraw-Hill Create empowers you to teach *your* students *your* way.

Acknowledgments

The authors thank the many companies that provided photographs for this edition. We also wish to recognize the efforts of the staff of RPK Editorial Services, who diligently worked to edit, typeset, proofread, and generally scrutinize all of this edition's content. Our special thanks go to Amy Mazurek (B.S. degree in civil engineering from the Florida Institute of Technology, and a M.S. degree in civil engineering from the University of Connecticut) for her work in the checking and preparation of the solutions and answers of all the problems in this edition.

We also gratefully acknowledge the help, comments, and suggestions offered by the many reviewers and users of previous editions of *Mechanics of Materials*.

John T. DeWolf
David F. Mazurek

Guided Tour

Chapter Introduction. Each chapter begins with an introductory section that sets up the purpose and goals of the chapter, describing in simple terms the material that will be covered and its application to the solution of engineering problems. Chapter Objectives provide students with a preview of chapter topics.

Chapter Lessons. The body of the text is divided into units, each consisting of one or several theory sections, Concept Applications, one or several Sample Problems, and a large number of homework problems. The Companion Website contains a Course Organization Guide with suggestions on each chapter lesson.

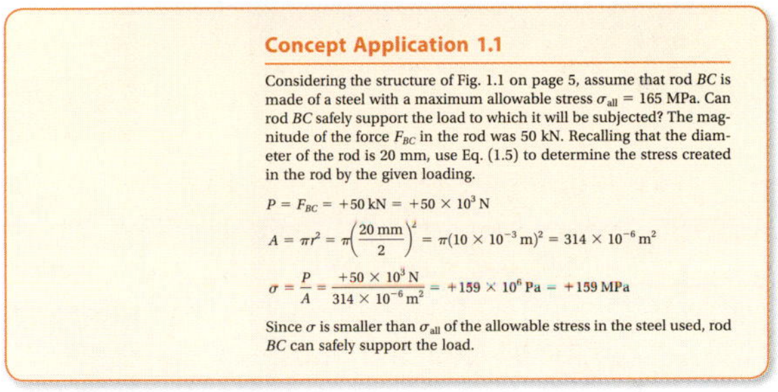

Concept Applications. Concept Applications are used extensively within individual theory sections to focus on specific topics, and they are designed to illustrate specific material being presented and facilitate its understanding.

Sample Problems. The Sample Problems are intended to show more comprehensive applications of the theory to the solution of engineering problems, and they employ the SMART problem-solving methodology that students are encouraged to use in the solution of their assigned problems. Since the sample problems have been set up in much the same form that students will use in solving the assigned problems, they serve the double purpose of amplifying the text and demonstrating the type of neat and orderly work that students should cultivate in their own solutions. In addition, in-problem references and captions have been added to the sample problem figures for contextual linkage to the step-by-step solution.

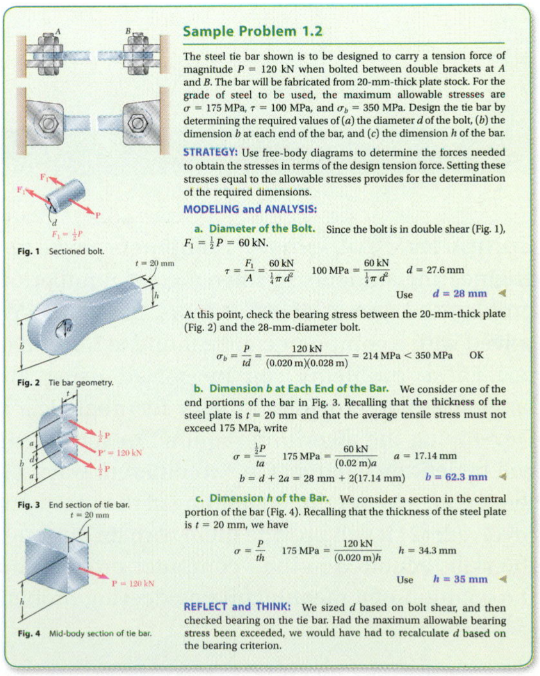

Homework Problem Sets. Over 25% of the nearly 1500 homework problems are new or updated. Most of the problems are of a practical nature and should appeal to engineering students. They are primarily designed, however, to illustrate the material presented in the text and to help students understand the principles used in mechanics of materials. The problems are grouped according to the portions of material they illustrate and are arranged in order of increasing difficulty. Answers to a majority of the problems are given at the end of the book. Problems for which the answers are given are set in blue type in the text, while problems for which no answer is given are set in red.

1

Introduction— Concept of Stress

Stresses occur in all structures subject to loads. This chapter will examine simple states of stress in elements, such as in the two-force members, bolts, and pins used in the structure shown.

Objectives
- **Review of statics** needed to determine forces in members of simple structures.
- **Introduce** concept of stress.
- **Define** different stress types: axial normal stress, shearing stress and bearing stress.
- **Discuss** engineer's two principal tasks, namely, the analysis and design of structures and machines.
- **Develop** problem solving approach.
- **Discuss** the components of stress on different planes and under different loading conditions.
- **Discuss** the many design considerations that an engineer should review before preparing a design.

Concept Application 1.1

Considering the structure of Fig. 1.1 on page 5, assume that rod BC is made of a steel with a maximum allowable stress $\sigma_{all} = 165$ MPa. Can rod BC safely support the load to which it will be subjected? The magnitude of the force F_{BC} in the rod was 50 kN. Recalling that the diameter of the rod is 20 mm, use Eq. (1.5) to determine the stress created in the rod by the given loading.

$$P = F_{BC} = +50 \text{ kN} = +50 \times 10^3 \text{ N}$$

$$A = \pi r^2 = \pi \left(\frac{20 \text{ mm}}{2}\right)^2 = \pi (10 \times 10^{-3} \text{ m})^2 = 314 \times 10^{-6} \text{ m}^2$$

$$\sigma = \frac{P}{A} = \frac{+50 \times 10^3 \text{ N}}{314 \times 10^{-6} \text{ m}^2} = +159 \times 10^6 \text{ Pa} = +159 \text{ MPa}$$

Since σ is smaller than σ_{all} of the allowable stress in the steel used, rod BC can safely support the load.

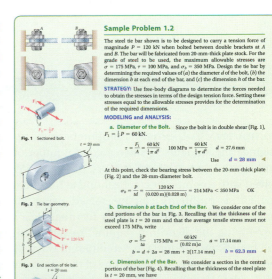

Sample Problem 1.2

The steel tie bar shown is to be designed to carry a tension force of magnitude $P = 120$ kN when bolted between double brackets at A and B. The bar will be fabricated from 20-mm-thick plate stock. For the grade of steel to be used, the maximum allowable stresses are $\sigma = 175$ MPa, $\tau = 100$ MPa, and $\sigma_b = 350$ MPa. Design the tie bar by determining the required values of (a) the diameter d of the bolt, (b) the dimension b at each end of the bar, and (c) the dimension h of the bar.

STRATEGY: Use free-body diagrams to determine the forces needed to obtain the stresses in terms of the design tension force. Setting these stresses equal to the allowable stresses provides for the determination of the required dimensions.

MODELING and ANALYSIS:

Fig. 1 Sectioned bolt.

a. Diameter of the Bolt. Since the bolt is in double shear (Fig. 1), $F_1 = \frac{1}{2}P = 60$ kN.

$$\tau = \frac{F_1}{A} = \frac{F_1}{\frac{1}{4}\pi d^2} \qquad 100 \text{ MPa} = \frac{60 \text{ kN}}{\frac{1}{4}\pi d^2} \qquad d = 27.6 \text{ mm}$$

$$\text{Use } \quad d = 28 \text{ mm} \ \blacktriangleleft$$

At this point, check the bearing stress between the 20-mm-thick plate (Fig. 2) and the 28-mm-diameter bolt.

$$\sigma_b = \frac{P}{td} = \frac{120 \text{ kN}}{(0.020 \text{ m})(0.028 \text{ m})} = 214 \text{ MPa} < 350 \text{ MPa} \quad \text{OK}$$

Fig. 2 Tie bar geometry.

b. Dimension b at Each End of the Bar. We consider one of the end portions of the bar in Fig. 3. Recalling that the thickness of the steel plate is $t = 20$ mm and that the average tensile stress must not exceed 175 MPa, write

$$\sigma = \frac{\frac{1}{2}P}{ta} \qquad 175 \text{ MPa} = \frac{60 \text{ kN}}{(0.02 \text{ m})a} \qquad a = 17.14 \text{ mm}$$

$$b = d + 2a = 28 \text{ mm} + 2(17.14 \text{ mm}) \qquad b = 62.3 \text{ mm} \ \blacktriangleleft$$

Fig. 3 End section of tie bar.
$t = 20$ mm

c. Dimension h of the Bar. We consider a section in the central portion of the bar (Fig. 4). Recalling that the thickness of the steel plate is $t = 20$ mm, we have

$$\sigma = \frac{P}{th} \qquad 175 \text{ MPa} = \frac{120 \text{ kN}}{(0.020 \text{ m})h} \qquad h = 34.3 \text{ mm}$$

$$\text{Use } \quad h = 35 \text{ mm} \ \blacktriangleleft$$

Fig. 4 Mid-body section of tie bar.

REFLECT and THINK: We sized d based on bolt shear, and then checked bearing on the tie bar. Had the maximum allowable bearing stress been exceeded, we would have had to recalculate d based on the bearing criterion.

Chapter Review and Summary. Each chapter ends with a review and summary of the material covered in that chapter. Subtitles are used to help students organize their review work, and cross-references have been included to help them find the portions of material requiring their special attention.

Review Problems. A set of review problems is included at the end of each chapter. These problems provide students further opportunity to apply the most important concepts introduced in the chapter.

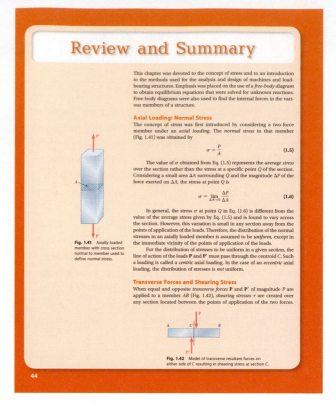

Review and Summary

This chapter was devoted to the concept of stress and to an introduction to the methods used for the analysis and design of machines and load-bearing structures. Emphasis was placed on the use of a *free-body diagram* to obtain equilibrium equations that were solved for unknown reactions. Free-body diagrams were also used to find the internal forces in the various members of a structure.

Axial Loading: Normal Stress
The concept of *stress* was first introduced by considering a two-force member under an *axial loading* (Fig. 1.41) was obtained by

$$\sigma = \frac{P}{A} \qquad (1.5)$$

The value of σ obtained from Eq. (1.5) represents the *average stress* over the section rather than the stress at a specific point Q of the section. Considering a small area ΔA surrounding Q and the magnitude ΔF of the force exerted on ΔA, the stress at point Q is

$$\sigma = \lim_{\Delta A \to 0} \frac{\Delta F}{\Delta A} \qquad (1.6)$$

In general, the stress σ at point Q in Eq. (1.6) is different from the value of the average stress given by Eq. (1.5) and is found to vary across the section. However, this variation is small in any section away from the points of application of the loads. Therefore, the distribution of the normal stresses in an axially loaded member is assumed to be *uniform*, except in the immediate vicinity of the points of application of the loads.

For the distribution of stresses to be uniform in a given section, the line of action of the loads $\mathbf{P}$ and $\mathbf{P'}$ must pass through the centroid C. Such a loading is called a *centric* axial loading. In the case of an *eccentric* axial loading, the distribution of stresses is *not* uniform.

Fig. 1.41 Axially loaded member with cross section normal to member used to define normal stress.

Transverse Forces and Shearing Stress
When equal and opposite *transverse forces* $\mathbf{P}$ and $\mathbf{P'}$ of magnitude P are applied to a member AB (Fig. 1.42), *shearing stresses* τ are created over any section located between the points of application of the two forces.

Fig. 1.42 Model of transverse resultant forces on either side of C resulting in shearing stress at section C.

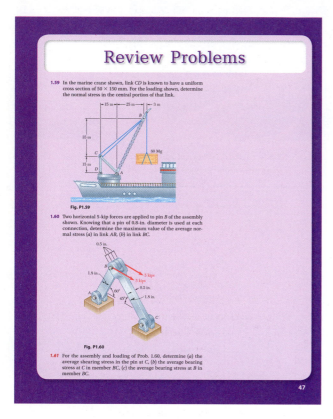

Review Problems

1.59 In the marine crane shown, link CD is known to have a uniform cross section of 50×150 mm. For the loading shown, determine the normal stress in the central portion of that link.

Fig. P1.59

1.60 Two horizontal 5-kip forces are applied to pin B of the assembly shown. Knowing that a pin of 0.8-in. diameter is used at each connection, determine the maximum value of the average normal stress (a) in link AB, (b) in link BC.

Fig. P1.60

1.61 For the assembly and loading of Prob. 1.60, determine (a) the average shearing stress in the pin at C, (b) the average bearing stress at C in member BC, (c) the average bearing stress at B in member BC.

Computer Problems. Computers make it possible for engineering students to solve a great number of challenging problems. A group of six or more problems designed to be solved with a computer can be found at the end of each chapter. These problems can be solved using any computer language that provides a basis for analytical calculations. Developing the algorithm required to solve a given problem will benefit the students in two different ways: (1) it will help them gain a better understanding of the mechanics principles involved; (2) it will provide them with an opportunity to apply the skills acquired in their computer programming course to the solution of a meaningful engineering problem.

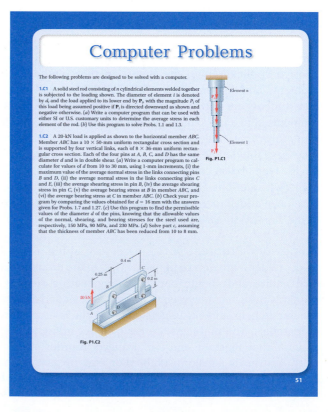

Computer Problems

The following problems are designed to be solved with a computer.

1.C1 A solid steel rod consisting of n cylindrical elements welded together is subjected to the loading shown. The diameter of element i is denoted by d_i and the load applied to its lower end by $\mathbf{P}_i$, with the magnitude P_i of this load being assumed positive if $\mathbf{P}_i$ is directed downward as shown and negative otherwise. (a) Write a computer program that can be used with either SI or U.S. customary units to determine the average stress in each element of the rod. (b) Use this program to solve Probs. 1.1 and 1.3.

Fig. P1.C1

1.C2 A 20-kN load is applied as shown to the horizontal member ABC. Member ABC has a 10×50-mm uniform rectangular cross section and is supported by four vertical links, each of 8×36-mm uniform rectangular cross section. Each of the four pins at A, B, C, and D has the same diameter d and is in double shear. (a) Write a computer program to calculate for values of d from 10 to 30 mm, using 1-mm increments, (i) the maximum value of the average normal stress in the links connecting pins B and D, (ii) the average normal stress in the links connecting pins C and E, (iii) the average shearing stress in pin B, (iv) the average shearing stress in pin C, (v) the average bearing stress at B in member ABC, and (vi) the average bearing stress at C in member ABC. (b) Check your program by comparing the values obtained for $d = 16$ mm with the answers given for Probs. 1.7 and 1.27. (c) Use this program to find the permissible values of the diameter d of the pins, knowing that the allowable values of the normal, shearing, and bearing stresses for the steel used are, respectively, 150 MPa, 90 MPa, and 230 MPa. (d) Solve part c, assuming that the thickness of member ABC has been reduced from 10 to 8 mm.

Fig. P1.C2

List of Symbols

a	Constant; distance
$\mathbf{A, B, C, \ldots}$	Forces; reactions
$A, B, C, \ldots$	Points
$A, \mathcal{a}$	Area
b	Distance; width
c	Constant; distance; radius
C	Centroid
$C_1, C_2, \ldots$	Constants of integration
C_P	Column stability factor
d	Distance; diameter; depth
D	Diameter
e	Distance; eccentricity; dilatation
E	Modulus of elasticity
f	Frequency; function
$\mathbf{F}$	Force
$F.S.$	Factor of safety
G	Modulus of rigidity; shear modulus
h	Distance; height
$\mathbf{H}$	Force
H, J, K	Points
$I, I_x, \ldots$	Moment of inertia
$I_{xy}, \ldots$	Product of inertia
J	Polar moment of inertia
k	Spring constant; shape factor; bulk modulus; constant
K	Stress concentration factor; torsional spring constant
l	Length; span
L	Length; span
L_e	Effective length
m	Mass
$\mathbf{M}$	Couple
$M, M_x, \ldots$	Bending moment
M_D	Bending moment, dead load (LRFD)
M_L	Bending moment, live load (LRFD)
M_U	Bending moment, ultimate load (LRFD)
n	Number; ratio of moduli of elasticity; normal direction
p	Pressure
$\mathbf{P}$	Force; concentrated load
P_D	Dead load (LRFD)
P_L	Live load (LRFD)

P_U	Ultimate load (LRFD)
q	Shearing force per unit length; shear flow
$\mathbf{Q}$	Force
Q	First moment of area
r	Radius; radius of gyration
$\mathbf{R}$	Force; reaction
R	Radius; modulus of rupture
s	Length
S	Elastic section modulus
t	Thickness; distance; tangential deviation
$\mathbf{T}$	Torque
T	Temperature
u, v	Rectangular coordinates
u	Strain-energy density
U	Strain energy; work
$\mathbf{v}$	Velocity
$\mathbf{V}$	Shearing force
V	Volume; shear
w	Width; distance; load per unit length
$\mathbf{W}, W$	Weight, load
x, y, z	Rectangular coordinates; distance; displacements; deflections
$\bar{x}, \bar{y}, \bar{z}$	Coordinates of centroid
Z	Plastic section modulus
α, β, γ	Angles
α	Coefficient of thermal expansion; influence coefficient
γ	Shearing strain; specific weight
γ_D	Load factor, dead load (LRFD)
γ_L	Load factor, live load (LRFD)
δ	Deformation; displacement
ϵ	Normal strain
θ	Angle; slope
λ	Direction cosine
ν	Poisson's ratio
ρ	Radius of curvature; distance; density
σ	Normal stress
τ	Shearing stress
ϕ	Angle; angle of twist; resistance factor
ω	Angular velocity

Seventh Edition

Mechanics of Materials

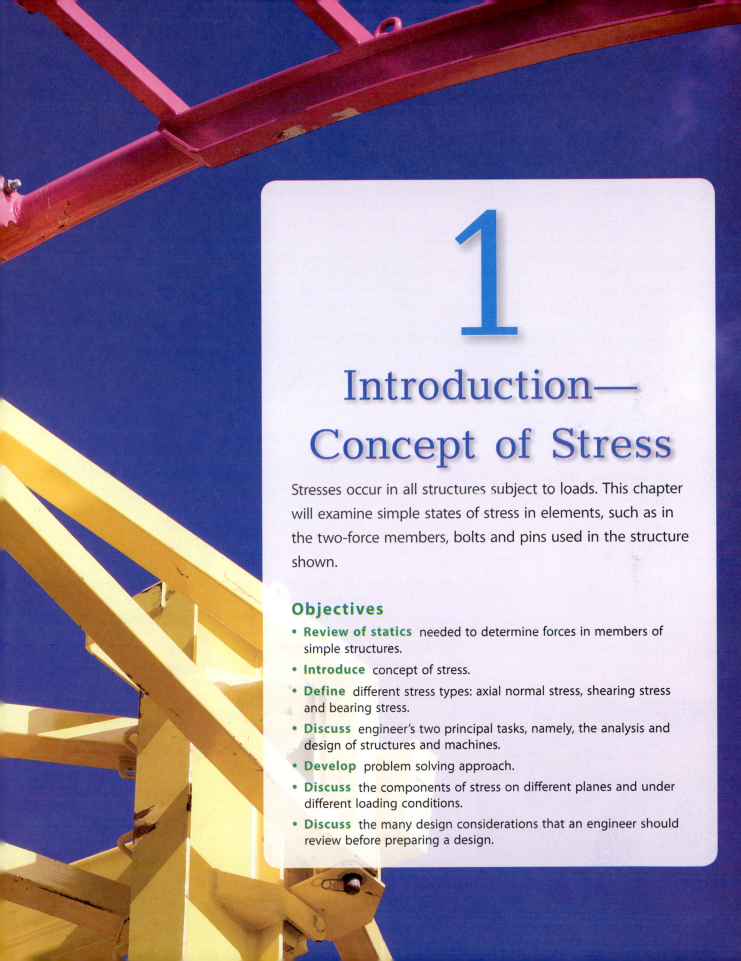

1

Introduction— Concept of Stress

Stresses occur in all structures subject to loads. This chapter will examine simple states of stress in elements, such as in the two-force members, bolts and pins used in the structure shown.

Objectives

- **Review of statics** needed to determine forces in members of simple structures.
- **Introduce** concept of stress.
- **Define** different stress types: axial normal stress, shearing stress and bearing stress.
- **Discuss** engineer's two principal tasks, namely, the analysis and design of structures and machines.
- **Develop** problem solving approach.
- **Discuss** the components of stress on different planes and under different loading conditions.
- **Discuss** the many design considerations that an engineer should review before preparing a design.

Introduction

The study of mechanics of materials provides future engineers with the means of analyzing and designing various machines and load-bearing structures involving the determination of *stresses* and *deformations*. This first chapter is devoted to the concept of *stress*.

Section 1.1 is a short review of the basic methods of statics and their application to determine the forces in the members of a simple structure consisting of pin-connected members. The concept of *stress* in a member of a structure and how that stress can be determined from the *force* in the member will be discussed in Sec. 1.2. You will consider the *normal stresses* in a member under axial loading, the *shearing stresses* caused by the application of equal and opposite transverse forces, and the *bearing stresses* created by bolts and pins in the members they connect.

Section 1.2 ends with a description of the method you should use in the solution of an assigned problem and a discussion of the numerical accuracy. These concepts will be applied in the analysis of the members of the simple structure considered earlier.

Again, a two-force member under axial loading is observed in Sec. 1.3 where the stresses on an *oblique* plane include both *normal* and *shearing* stresses, while Sec. 1.4 discusses that *six components* are required to describe the state of stress at a point in a body under the most general loading conditions.

Finally, Sec. 1.5 is devoted to the determination of the *ultimate strength* from test specimens and the use of a *factor of safety* to compute the *allowable load* for a structural component made of that material.

1.1   REVIEW OF THE METHODS OF STATICS

Consider the structure shown in Fig. 1.1, which was designed to support a 30-kN load. It consists of a boom *AB* with a 30 × 50-mm rectangular cross section and a rod *BC* with a 20-mm-diameter circular cross section. These are connected by a pin at *B* and are supported by pins and brackets at *A* and *C*, respectively. First draw a *free-body diagram* of the structure by detaching it from its supports at *A* and *C* and showing the reactions that these supports exert on the structure (Fig. 1.2). Note that the sketch of the structure has been simplified by omitting all unnecessary details. Many of you may have recognized at this point that *AB* and *BC* are *two-force members*. For those of you who have not, we will pursue our analysis, ignoring that fact and assuming that the directions of the reactions at *A* and *C* are unknown. Each of these reactions are represented by two components: $\mathbf{A}_x$ and $\mathbf{A}_y$ at *A*, and $\mathbf{C}_x$ and $\mathbf{C}_y$ at *C*. The equilibrium equations are.

$$+\!\!\curvearrowleft \Sigma M_C = 0: \qquad A_x(0.6\,\text{m}) - (30\,\text{kN})(0.8\,\text{m}) = 0$$

$$A_x = +40\,\text{kN} \tag{1.1}$$

$$\xrightarrow{+} \Sigma F_x = 0: \qquad A_x + C_x = 0$$

$$C_x = -A_x \qquad C_x = -40\,\text{kN} \tag{1.2}$$

$$+\!\!\uparrow \Sigma F_y = 0: \qquad A_y + C_y - 30\,\text{kN} = 0$$

$$A_y + C_y = +30\,\text{kN} \tag{1.3}$$

Photo 1.1   Crane booms used to load and unload ships.

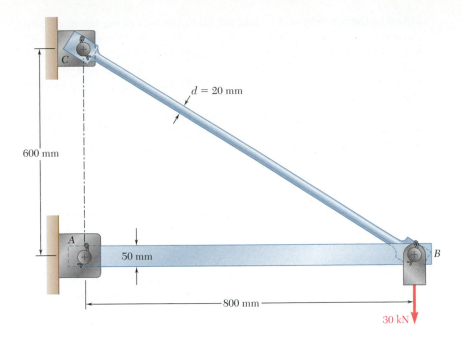

Fig. 1.1 Boom used to support a 30-kN load.

We have found two of the four unknowns, but cannot determine the other two from these equations, and no additional independent equation can be obtained from the free-body diagram of the structure. We must now dismember the structure. Considering the free-body diagram of the boom *AB* (Fig. 1.3), we write the following equilibrium equation:

$$+\uparrow \Sigma M_B = 0: \qquad -A_y(0.8 \text{ m}) = 0 \qquad A_y = 0 \qquad \textbf{(1.4)}$$

Substituting for A_y from Eq. (1.4) into Eq. (1.3), we obtain $C_y = +30$ kN. Expressing the results obtained for the reactions at *A* and *C* in vector form, we have

$$\mathbf{A} = 40 \text{ kN} \rightarrow \qquad \mathbf{C}_x = 40 \text{ kN} \leftarrow \qquad \mathbf{C}_y = 30 \text{ kN} \uparrow$$

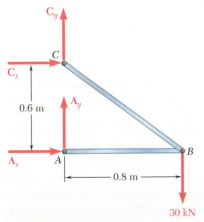

Fig. 1.2 Free-body diagram of boom showing applied load and reaction forces.

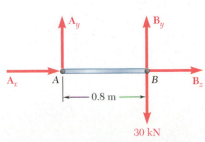

Fig. 1.3 Free-body diagram of member *AB* freed from structure.

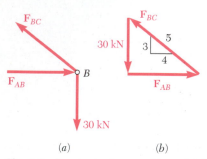

Fig. 1.4 Free-body diagram of boom's joint B and associated force triangle.

Note that the reaction at A is directed along the axis of the boom AB and causes compression in that member. Observe that the components C_x and C_y of the reaction at C are, respectively, proportional to the horizontal and vertical components of the distance from B to C and that the reaction at C is equal to 50 kN, is directed along the axis of the rod BC, and causes tension in that member.

These results could have been anticipated by recognizing that AB and BC are two-force members, i.e., members that are subjected to forces at only two points, these points being A and B for member AB, and B and C for member BC. Indeed, for a two-force member the lines of action of the resultants of the forces acting at each of the two points are equal and opposite and pass through both points. Using this property, we could have obtained a simpler solution by considering the free-body diagram of pin B. The forces on pin B, $\mathbf{F}_{AB}$ and $\mathbf{F}_{BC}$, are exerted, respectively, by members AB and BC and the 30-kN load (Fig. 1.4a). Pin B is shown to be in equilibrium by drawing the corresponding force triangle (Fig. 1.4b).

Since force $\mathbf{F}_{BC}$ is directed along member BC, its slope is the same as that of BC, namely, 3/4. We can, therefore, write the proportion

$$\frac{F_{AB}}{4} = \frac{F_{BC}}{5} = \frac{30 \text{ kN}}{3}$$

from which

$$F_{AB} = 40 \text{ kN} \qquad F_{BC} = 50 \text{ kN}$$

Forces $\mathbf{F}'_{AB}$ and $\mathbf{F}'_{BC}$ exerted by pin B on boom AB and rod BC are equal and opposite to $\mathbf{F}_{AB}$ and $\mathbf{F}_{BC}$ (Fig. 1.5).

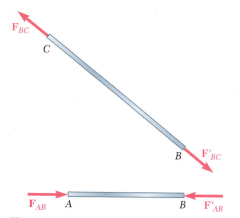

Fig. 1.5 Free-body diagrams of two-force members AB and BC.

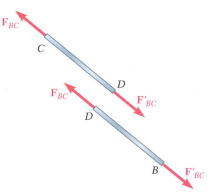

Fig. 1.6 Free-body diagrams of sections of rod BC.

Knowing the forces at the ends of each member, we can now determine the internal forces in these members. Passing a section at some arbitrary point D of rod BC, we obtain two portions BD and CD (Fig. 1.6). Since 50-kN forces must be applied at D to both portions of the rod to keep them in equilibrium, an internal force of 50 kN is produced in rod BC when a 30-kN load is applied at B. From the directions of the forces $\mathbf{F}_{BC}$ and $\mathbf{F}'_{BC}$ in Fig. 1.6 we see that the rod is in tension. A similar procedure enables us to determine that the internal force in boom AB is 40 kN and is in compression.

1.2 STRESSES IN THE MEMBERS OF A STRUCTURE

1.2A Axial Stress

In the preceding section, we found forces in individual members. This is the first and necessary step in the analysis of a structure. However it does not tell us whether the given load can be safely supported. Rod BC of the example considered in the preceding section is a two-force member and, therefore, the forces $\mathbf{F}_{BC}$ and $\mathbf{F}'_{BC}$ acting on its ends B and C (Fig. 1.5) are directed along the axis of the rod. Whether rod BC will break or not under this loading depends upon the value found for the internal force F_{BC}, the cross-sectional area of the rod, and the material of which the rod is made. Actually, the internal force F_{BC} represents the resultant of elementary forces distributed over the entire area A of the cross section (Fig. 1.7). The average

Photo 1.2 This bridge truss consists of two-force members that may be in tension or in compression.

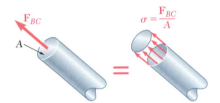

Fig. 1.7 Axial force represents the resultant of distributed elementary forces.

intensity of these distributed forces is equal to the force per unit area, F_{BC}/A, on the section. Whether or not the rod will break under the given loading depends upon the ability of the material to withstand the corresponding value F_{BC}/A of the intensity of the distributed internal forces.

Let us look at the uniformly distributed force using Fig. 1.8. The force per unit area, or intensity of the forces distributed over a given section, is called the *stress* and is denoted by the Greek letter σ (sigma). The stress in a member of cross-sectional area A subjected to an axial load $\mathbf{P}$ is obtained by dividing the magnitude P of the load by the area A:

$$\sigma = \frac{P}{A} \tag{1.5}$$

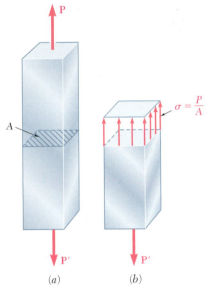

Fig. 1.8 (a) Member with an axial load. (b) Idealized uniform stress distribution at an arbitrary section.

A positive sign indicates a tensile stress (member in tension), and a negative sign indicates a compressive stress (member in compression).

As shown in Fig. 1.8, the section through the rod to determine the internal force in the rod and the corresponding stress is perpendicular to the axis of the rod. The corresponding stress is described as a *normal stress*. Thus, Eq. (1.5) gives the *normal stress in a member under axial loading*.

Note that in Eq. (1.5), σ represents the *average value* of the stress over the cross section, rather than the stress at a specific point of the cross section. To define the stress at a given point Q of the cross section, consider a small area ΔA (Fig. 1.9). Dividing the magnitude of $\Delta\mathbf{F}$ by ΔA, you obtain the average value of the stress over ΔA. Letting ΔA approach zero, the stress at point Q is

$$\sigma = \lim_{\Delta A \to 0} \frac{\Delta F}{\Delta A} \tag{1.6}$$

Fig. 1.9 Small area ΔA, at an arbitrary cross section point carries/axial ΔF in this axial member.

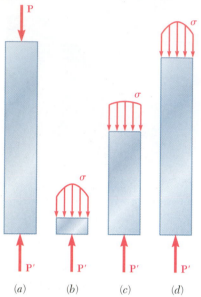

Fig. 1.10 Stress distributions at different sections along axially loaded member.

(a) (b) (c) (d)

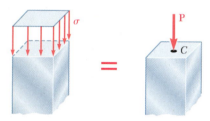

Fig. 1.11 Idealized uniform stress distribution implies the resultant force passes through the cross section's center.

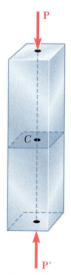

Fig. 1.12 Centric loading having resultant forces passing through the centroid of the section.

In general, the value for the stress σ at a given point Q of the section is different from that for the average stress given by Eq. (1.5), and σ is found to vary across the section. In a slender rod subjected to equal and opposite concentrated loads **P** and **P'** (Fig. 1.10a), this variation is small in a section away from the points of application of the concentrated loads (Fig. 1.10c), but it is quite noticeable in the neighborhood of these points (Fig. 1.10b and d).

It follows from Eq. (1.6) that the magnitude of the resultant of the distributed internal forces is

$$\int dF = \int_A \sigma \, dA$$

But the conditions of equilibrium of each of the portions of rod shown in Fig. 1.10 require that this magnitude be equal to the magnitude P of the concentrated loads. Therefore,

$$P = \int dF = \int_A \sigma \, dA \tag{1.7}$$

which means that the volume under each of the stress surfaces in Fig. 1.10 must be equal to the magnitude P of the loads. However, this is the only information derived from statics regarding the distribution of normal stresses in the various sections of the rod. The actual distribution of stresses in any given section is *statically indeterminate*. To learn more about this distribution, it is necessary to consider the deformations resulting from the particular mode of application of the loads at the ends of the rod. This will be discussed further in Chap. 2.

In practice, it is assumed that the distribution of normal stresses in an axially loaded member is uniform, except in the immediate vicinity of the points of application of the loads. The value σ of the stress is then equal to σ_{ave} and can be obtained from Eq. (1.5). However, realize that when we assume a uniform distribution of stresses in the section, it follows from elementary statics[†] that the resultant **P** of the internal forces must be applied at the centroid C of the section (Fig. 1.11). This means that *a uniform distribution of stress is possible only if the line of action of the concentrated loads* **P** *and* **P'** *passes through the centroid of the section considered* (Fig. 1.12). This type of loading is called *centric loading* and will take place in all straight two-force members found in trusses and pin-connected structures, such as the one considered in Fig. 1.1. However, if a two-force member is loaded axially, but *eccentrically*, as shown in Fig. 1.13a, the conditions of equilibrium of the portion of member in Fig. 1.13b show that the internal forces in a given section must be equivalent to a force **P** applied at the centroid of the section and a couple **M** of moment $M = Pd$. This distribution of forces—the corresponding distribution of stresses—*cannot be uniform*. Nor can the distribution of stresses be symmetric. This point will be discussed in detail in Chap. 4.

[†]See Ferdinand P. Beer and E. Russell Johnston, Jr., *Mechanics for Engineers,* 5th ed., McGraw-Hill, New York, 2008, or *Vector Mechanics for Engineers,* 10th ed., McGraw-Hill, New York, 2013, Secs. 5.2 and 5.3.

When SI metric units are used, P is expressed in newtons (N) and A in square meters (m^2), so the stress σ will be expressed in N/m^2. This unit is called a *pascal* (Pa). However, the pascal is an exceedingly small quantity and often multiples of this unit must be used: the kilopascal (kPa), the megapascal (MPa), and the gigapascal (GPa):

$$1\,\text{kPa} = 10^3\,\text{Pa} = 10^3\,\text{N/m}^2$$

$$1\,\text{MPa} = 10^6\,\text{Pa} = 10^6\,\text{N/m}^2$$

$$1\,\text{GPa} = 10^9\,\text{Pa} = 10^9\,\text{N/m}^2$$

When U.S. customary units are used, force P is usually expressed in pounds (lb) or kilopounds (kip), and the cross-sectional area A is given in square inches (in^2). The stress σ then is expressed in pounds per square inch (psi) or kilopounds per square inch (ksi).[†]

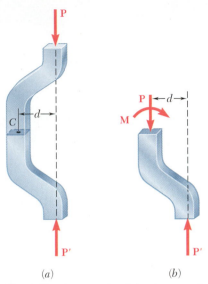

(a)   (b)

Fig. 1.13   An example of simple eccentric loading.

Concept Application 1.1

Considering the structure of Fig. 1.1 on page 5, assume that rod BC is made of a steel with a maximum allowable stress $\sigma_{\text{all}} = 165$ MPa. Can rod BC safely support the load to which it will be subjected? The magnitude of the force F_{BC} in the rod was 50 kN. Recalling that the diameter of the rod is 20 mm, use Eq. (1.5) to determine the stress created in the rod by the given loading.

$$P = F_{BC} = +50\,\text{kN} = +50 \times 10^3\,\text{N}$$

$$A = \pi r^2 = \pi \left(\frac{20\,\text{mm}}{2}\right)^2 = \pi (10 \times 10^{-3}\,\text{m})^2 = 314 \times 10^{-6}\,\text{m}^2$$

$$\sigma = \frac{P}{A} = \frac{+50 \times 10^3\,\text{N}}{314 \times 10^{-6}\,\text{m}^2} = +159 \times 10^6\,\text{Pa} = +159\,\text{MPa}$$

Since σ is smaller than σ_{all} of the allowable stress in the steel used, rod BC can safely support the load.

To be complete, our analysis of the given structure should also include the compressive stress in boom AB, as well as the stresses produced in the pins and their bearings. This will be discussed later in this chapter. You should also determine whether the deformations produced by the given loading are acceptable. The study of deformations under axial loads will be the subject of Chap. 2. For members in compression, the *stability* of the member (i.e., its ability to support a given load without experiencing a sudden change in configuration) will be discussed in Chap. 10.

[†]The principal SI and U.S. Customary units used in mechanics are listed in tables inside the front cover of this book. From the table on the right-hand side, 1 psi is approximately equal to 7 kPa and 1 ksi approximately equal to 7 MPa.

The engineer's role is not limited to the analysis of existing structures and machines subjected to given loading conditions. Of even greater importance is the *design* of new structures and machines, that is the selection of appropriate components to perform a given task.

Concept Application 1.2

As an example of design, let us return to the structure of Fig. 1.1 on page 5 and assume that aluminum with an allowable stress $\sigma_{all} = 100$ MPa is to be used. Since the force in rod BC is still $P = F_{BC} = 50$ kN under the given loading, from Eq. (1.5), we have

$$\sigma_{all} = \frac{P}{A} \qquad A = \frac{P}{\sigma_{all}} = \frac{50 \times 10^3 \text{ N}}{100 \times 10^6 \text{ Pa}} = 500 \times 10^{-6} \text{ m}^2$$

and since $A = \pi r^2$,

$$r = \sqrt{\frac{A}{\pi}} = \sqrt{\frac{500 \times 10^{-6} \text{ m}^2}{\pi}} = 12.62 \times 10^{-3} \text{ m} = 12.62 \text{ mm}$$

$$d = 2r = 25.2 \text{ mm}$$

Therefore, an aluminum rod 26 mm or more in diameter will be adequate.

1.2B Shearing Stress

The internal forces and the corresponding stresses discussed in Sec. 1.2A were normal to the section considered. A very different type of stress is obtained when transverse forces **P** and **P′** are applied to a member AB (Fig. 1.14). Passing a section at C between the points of application of the two forces (Fig. 1.15a), you obtain the diagram of portion AC shown in

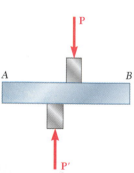

Fig. 1.14 Opposing transverse loads creating shear on member *AB*.

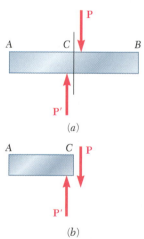

(a)

(b)

Fig. 1.15 This shows the resulting internal shear force on a section between transverse forces.

Fig. 1.15*b*. Internal forces must exist in the plane of the section, and their resultant is equal to **P**. These elementary internal forces are called *shearing forces,* and the magnitude P of their resultant is the *shear* in the section. Dividing the shear P by the area A of the cross section, you obtain the *average shearing stress* in the section. Denoting the shearing stress by the Greek letter τ (tau), write

$$\tau_{\text{ave}} = \frac{P}{A} \qquad (1.8)$$

The value obtained is an average value of the shearing stress over the entire section. Contrary to what was said earlier for normal stresses, the distribution of shearing stresses across the section *cannot* be assumed to be uniform. As you will see in Chap. 6, the actual value τ of the shearing stress varies from zero at the surface of the member to a maximum value τ_{max} that may be much larger than the average value τ_{ave}.

Photo 1.3   Cutaway view of a connection with a bolt in shear.

Shearing stresses are commonly found in bolts, pins, and rivets used to connect various structural members and machine components (Photo 1.3). Consider the two plates A and B, which are connected by a bolt CD (Fig. 1.16). If the plates are subjected to tension forces of magnitude F, stresses will develop in the section of bolt corresponding to the plane EE'. Drawing the diagrams of the bolt and of the portion located above the plane EE' (Fig. 1.17), the shear P in the section is equal to F. The average shearing stress in the section is obtained using Eq. (1.8) by dividing the shear $P = F$ by the area A of the cross section:

$$\tau_{\text{ave}} = \frac{P}{A} = \frac{F}{A} \qquad (1.9)$$

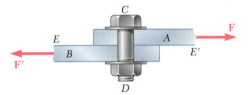

Fig. 1.16   Bolt subject to single shear.

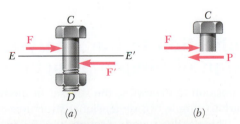

Fig. 1.17   (*a*) Diagram of bolt in single shear; (*b*) section E-E' of the bolt.

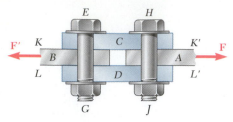

Fig. 1.18 Bolts subject to double shear.

The previous bolt is said to be in *single shear*. Different loading situations may arise, however. For example, if splice plates C and D are used to connect plates A and B (Fig. 1.18), shear will take place in bolt HJ in each of the two planes KK' and LL' (and similarly in bolt EG). The bolts are said to be in *double shear*. To determine the average shearing stress in each plane, draw free-body diagrams of bolt HJ and of the portion of the bolt located between the two planes (Fig. 1.19). Observing that the shear P in each of the sections is $P = F/2$, the average shearing stress is

$$\tau_{\text{ave}} = \frac{P}{A} = \frac{F/2}{A} = \frac{F}{2A} \tag{1.10}$$

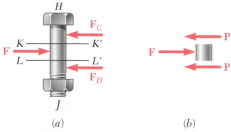

Fig. 1.19 (a) Diagram of bolt in double shear; (b) section K-K' and L-L' of the bolt.

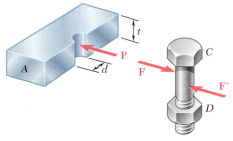

Fig. 1.20 Equal and opposite forces between plate and bolt, exerted over bearing surfaces.

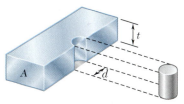

Fig. 1.21 Dimensions for calculating bearing stress area.

1.2C Bearing Stress in Connections

Bolts, pins, and rivets create stresses in the members they connect along the *bearing surface* or surface of contact. For example, consider again the two plates A and B connected by a bolt CD that were discussed in the preceding section (Fig. 1.16). The bolt exerts on plate A a force **P** equal and opposite to the force **F** exerted by the plate on the bolt (Fig. 1.20). The force **P** represents the resultant of elementary forces distributed on the inside surface of a half-cylinder of diameter d and of length t equal to the thickness of the plate. Since the distribution of these forces—and of the corresponding stresses—is quite complicated, in practice one uses an average nominal value σ_b of the stress, called the *bearing stress,* which is obtained by dividing the load P by the area of the rectangle representing the projection of the bolt on the plate section (Fig. 1.21). Since this area is equal to td, where t is the plate thickness and d the diameter of the bolt, we have

$$\sigma_b = \frac{P}{A} = \frac{P}{td} \tag{1.11}$$

1.2D Application to the Analysis and Design of Simple Structures

We are now in a position to determine the stresses in the members and connections of various simple two-dimensional structures and to design such structures. This is illustrated through the following Concept Application.

Concept Application 1.3

Returning to the structure of Fig. 1.1, we will determine the normal stresses, shearing stresses and bearing stresses. As shown in Fig. 1.22, the 20-mm-diameter rod *BC* has flat ends of 20 × 40-mm rectangular cross section, while boom *AB* has a 30 × 50-mm rectangular cross section and is fitted with a clevis at end *B*. Both members are connected at *B* by a pin from which the 30-kN load is suspended by means of a U-shaped bracket. Boom *AB* is supported at *A* by a pin fitted into a double bracket, while rod *BC* is connected at *C* to a single bracket. All pins are 25 mm in diameter.

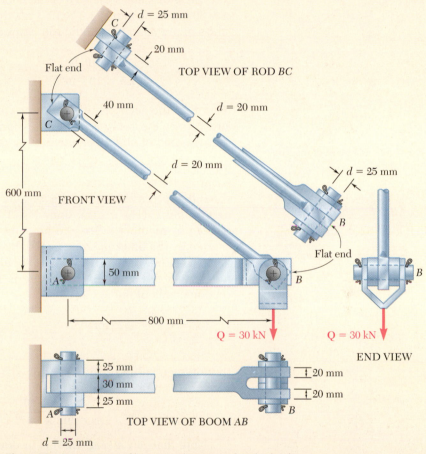

Fig. 1.22 Components of boom used to support 30 kN load.

Normal Stress in Boom *AB* and Rod *BC*. As found in Sec. 1.1A, the force in rod *BC* is $F_{BC} = 50$ kN (tension) and the area of its circular cross section is $A = 314 \times 10^{-6}$ m². The corresponding average normal stress is $\sigma_{BC} = +159$ MPa. However, the flat parts of the rod are also under tension and at the narrowest section. Where the hole is located, we have

$$A = (20 \text{ mm})(40 \text{ mm} - 25 \text{ mm}) = 300 \times 10^{-6} \text{ m}^2$$

(continued)

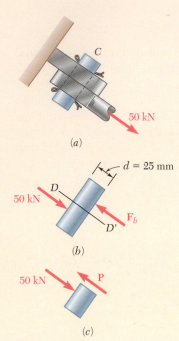

Fig. 1.23 Diagrams of the single shear pin at C.

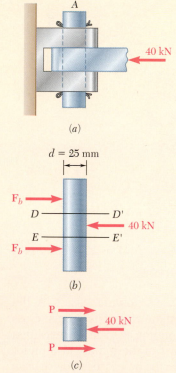

Fig. 1.24 Free-body diagrams of the double shear pin at A.

The corresponding average value of the stress is

$$(\sigma_{BC})_{\text{end}} = \frac{P}{A} = \frac{50 \times 10^3 \text{ N}}{300 \times 10^{-6} \text{ m}^2} = 167.0 \text{ MPa}$$

Note that this is an *average value*. Close to the hole the stress will actually reach a much larger value, as you will see in Sec. 2.11. Under an increasing load, the rod will fail near one of the holes rather than in its cylindrical portion; its design could be improved by increasing the width or the thickness of the flat ends of the rod.

Recall from Sec. 1.1A that the force in boom *AB* is $F_{AB} = 40$ kN (compression). Since the area of the boom's rectangular cross section is $A = 30 \text{ mm} \times 50 \text{ mm} = 1.5 \times 10^{-3} \text{ m}^2$, the average value of the normal stress in the main part of the rod between pins *A* and *B* is

$$\sigma_{AB} = -\frac{40 \times 10^3 \text{ N}}{1.5 \times 10^{-3} \text{ m}^2} = -26.7 \times 10^6 \text{ Pa} = -26.7 \text{ MPa}$$

Note that the sections of minimum area at *A* and *B* are not under stress, since the boom is in compression, and therefore *pushes* on the pins (instead of *pulling* on the pins as rod *BC* does).

Shearing Stress in Various Connections. To determine the shearing stress in a connection such as a bolt, pin, or rivet, you first show the forces exerted by the various members it connects. In the case of pin *C* (Fig. 1.23*a*), draw Fig. 1.23*b* to show the 50-kN force exerted by member *BC* on the pin, and the equal and opposite force exerted by the bracket. Drawing the diagram of the portion of the pin located below the plane *DD'* where shearing stresses occur (Fig. 1.23*c*), notice that the shear in that plane is $P = 50$ kN. Since the cross-sectional area of the pin is

$$A = \pi r^2 = \pi \left(\frac{25 \text{ mm}}{2}\right)^2 = \pi(12.5 \times 10^{-3} \text{ m})^2 = 491 \times 10^{-6} \text{ m}^2$$

the average value of the shearing stress in the pin at *C* is

$$\tau_{\text{ave}} = \frac{P}{A} = \frac{50 \times 10^3 \text{ N}}{491 \times 10^{-6} \text{ m}^2} = 102.0 \text{ MPa}$$

Note that pin *A* (Fig. 1.24) is in double shear. Drawing the free-body diagrams of the pin and the portion of pin located between the planes *DD'* and *EE'* where shearing stresses occur, we see that $P = 20$ kN and

$$\tau_{\text{ave}} = \frac{P}{A} = \frac{20 \text{ kN}}{491 \times 10^{-6} \text{ m}^2} = 40.7 \text{ MPa}$$

Pin *B* (Fig. 1.25*a*) can be divided into five portions that are acted upon by forces exerted by the boom, rod, and bracket. Portions *DE* (Fig. 1.25*b*) and *DG* (Fig. 1.25*c*) show that the shear in section *E* is $P_E = 15$ kN and the shear in section *G* is $P_G = 25$ kN. Since the loading

(continued)

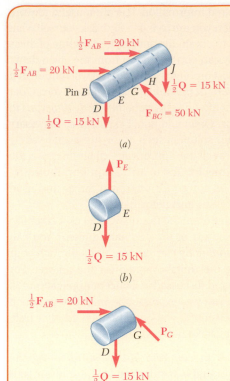

Fig. 1.25 Free-body diagrams for various sections at pin B.

of the pin is symmetric, the maximum value of the shear in pin B is $P_G = 25$ kN, and the largest the shearing stresses occur in sections G and H, where

$$\tau_{ave} = \frac{P_G}{A} = \frac{25 \text{ kN}}{491 \times 10^{-6} \text{ m}^2} = 50.9 \text{ MPa}$$

Bearing Stresses. Use Eq. (1.11) to determine the nominal bearing stress at A in member AB. From Fig. 1.22, $t = 30$ mm and $d = 25$ mm. Recalling that $P = F_{AB} = 40$ kN, we have

$$\sigma_b = \frac{P}{td} = \frac{40 \text{ kN}}{(30 \text{ mm})(25 \text{ mm})} = 53.3 \text{ MPa}$$

To obtain the bearing stress in the bracket at A, use $t = 2(25 \text{ mm}) = 50$ mm and $d = 25$ mm:

$$\sigma_b = \frac{P}{td} = \frac{40 \text{ kN}}{(50 \text{ mm})(25 \text{ mm})} = 32.0 \text{ MPa}$$

The bearing stresses at B in member AB, at B and C in member BC, and in the bracket at C are found in a similar way.

1.2E   Method of Problem Solution

You should approach a problem in mechanics as you would approach an actual engineering situation. By drawing on your own experience and intuition about physical behavior, you will find it easier to understand and formulate the problem. Your solution must be based on the fundamental principles of statics and on the principles you will learn in this text. Every step you take in the solution must be justified on this basis, leaving no room for your intuition or "feeling." After you have obtained an answer, you should check it. Here again, you may call upon your common sense and personal experience. If you are not completely satisfied with the result, you should carefully check your formulation of the problem, the validity of the methods used for its solution, and the accuracy of your computations.

In general, you can usually solve problems in several different ways; there is no one approach that works best for everybody. However, we have found that students often find it helpful to have a general set of guidelines to use for framing problems and planning solutions. In the Sample Problems throughout this text, we use a four-step approach for solving problems, which we refer to as the SMART methodology: Strategy, Modeling, Analysis, and Reflect & Think:

1. **Strategy.** The statement of a problem should be clear and precise, and should contain the given data and indicate what information is required. The first step in solving the problem is to decide what concepts you have learned that apply to the given situation and

connect the data to the required information. It is often useful to work backward from the information you are trying to find: ask yourself what quantities you need to know to obtain the answer, and if some of these quantities are unknown, how can you find them from the given data.

2. **Modeling.** The solution of most problems encountered will require that you first determine the *reactions at the supports* and *internal forces and couples*. It is important to include one or several *free-body diagrams* to support these determinations. Draw additional sketches as necessary to guide the remainder of your solution, such as for stress analyses.

3. **Analysis.** After you have drawn the appropriate diagrams, use the fundamental principles of mechanics to write equilibrium equations. These equations can be solved for unknown forces and used to compute the required stresses and deformations.

4. **Reflect & Think.** After you have obtained the answer, check it carefully. Does it make sense in the context of the original problem? You can often detect mistakes in *reasoning* by carrying the units through your computations and checking the units obtained for the answer. For example, in the design of the rod discussed in Concept Application 1.2, the required diameter of the rod was expressed in millimeters, which is the correct unit for a dimension; if you had obtained another unit, you would know that some mistake had been made.

You can often detect errors in *computation* by substituting the numerical answer into an equation that was not used in the solution and verifying that the equation is satisfied. The importance of correct computations in engineering cannot be overemphasized.

Numerical Accuracy.   The accuracy of the solution of a problem depends upon two items: (1) the accuracy of the given data and (2) the accuracy of the computations performed.

The solution cannot be more accurate than the less accurate of these two items. For example, if the loading of a beam is known to be 75,000 lb with a possible error of 100 lb either way, the relative error that measures the degree of accuracy of the data is

$$\frac{100 \text{ lb}}{75,000 \text{ lb}} = 0.0013 = 0.13\%$$

To compute the reaction at one of the beam supports, it would be meaningless to record it as 14,322 lb. The accuracy of the solution cannot be greater than 0.13%, no matter how accurate the computations are, and the possible error in the answer may be as large as $(0.13/100)(14,322 \text{ lb}) \approx 20$ lb. The answer should be properly recorded as $14,320 \pm 20$ lb.

In engineering problems, the data are seldom known with an accuracy greater than 0.2%. A practical rule is to use four figures to record numbers beginning with a "1" and three figures in all other cases. Unless otherwise indicated, the data given are assumed to be known with a comparable degree of accuracy. A force of 40 lb, for example, should be read 40.0 lb, and a force of 15 lb should be read 15.00 lb.

The speed and accuracy of calculators and computers makes the numerical computations in the solution of many problems much easier. However, students should not record more significant figures than can be justified merely because they are easily obtained. An accuracy greater than 0.2% is seldom necessary or meaningful in the solution of practical engineering problems.

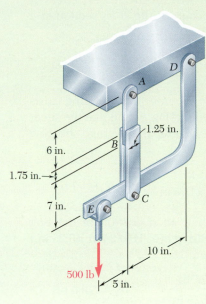

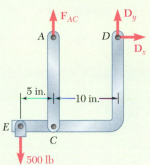

Fig. 1 Free-body diagram of hanger.

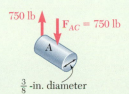

$\frac{3}{8}$-in. diameter

Fig. 2 Pin A.

Sample Problem 1.1

In the hanger shown, the upper portion of link *ABC* is $\frac{3}{8}$ in. thick and the lower portions are each $\frac{1}{4}$ in. thick. Epoxy resin is used to bond the upper and lower portions together at *B*. The pin at *A* has a $\frac{3}{8}$-in. diameter, while a $\frac{1}{4}$-in.-diameter pin is used at *C*. Determine (*a*) the shearing stress in pin *A*, (*b*) the shearing stress in pin *C*, (*c*) the largest normal stress in link *ABC*, (*d*) the average shearing stress on the bonded surfaces at *B*, and (*e*) the bearing stress in the link at *C*.

STRATEGY: Consider the free body of the hanger to determine the internal force for member *AB* and then proceed to determine the shearing and bearing forces applicable to the pins. These forces can then be used to determine the stresses.

MODELING: Draw the free-body diagram of the hanger to determine the support reactions (Fig. 1). Then draw the diagrams of the various components of interest showing the forces needed to determine the desired stresses (Figs. 2-6).

ANALYSIS:

Free Body: Entire Hanger. Since the link *ABC* is a two-force member (Fig. 1), the reaction at *A* is vertical; the reaction at *D* is represented by its components D_x and D_y. Thus,

$$+\!\uparrow\ \Sigma M_D = 0: \qquad (500\text{ lb})(15\text{ in.}) - F_{AC}(10\text{ in.}) = 0$$
$$F_{AC} = +750\text{ lb} \qquad F_{AC} = 750\text{ lb} \quad \textit{tension}$$

a. Shearing Stress in Pin A. Since this $\frac{3}{8}$-in.-diameter pin is in single shear (Fig. 2), write

$$\tau_A = \frac{F_{AC}}{A} = \frac{750\text{ lb}}{\frac{1}{4}\pi(0.375\text{ in.})^2} \qquad \tau_A = 6790\text{ psi} \ \blacktriangleleft$$

b. Shearing Stress in Pin C. Since this $\frac{1}{4}$-in.-diameter pin is in double shear (Fig. 3), write

$$\tau_C = \frac{\frac{1}{2}F_{AC}}{A} = \frac{375\text{ lb}}{\frac{1}{4}\pi(0.25\text{ in.})^2} \qquad \tau_C = 7640\text{ psi} \ \blacktriangleleft$$

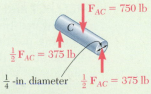

$\frac{1}{4}$-in. diameter

Fig. 3 Pin C.

(continued)

c. Largest Normal Stress in Link *ABC*. The largest stress is found where the area is smallest; this occurs at the cross section at *A* (Fig. 4) where the $\frac{3}{8}$-in. hole is located. We have

$$\sigma_A = \frac{F_{AC}}{A_{\text{net}}} = \frac{750 \text{ lb}}{(\frac{3}{8} \text{ in.})(1.25 \text{ in.} - 0.375 \text{ in.})} = \frac{750 \text{ lb}}{0.328 \text{ in}^2} \qquad \sigma_A = 2290 \text{ psi} \quad \blacktriangleleft$$

d. Average Shearing Stress at *B*. We note that bonding exists on both sides of the upper portion of the link (Fig. 5) and that the shear force on each side is $F_1 = (750 \text{ lb})/2 = 375 \text{ lb}$. The average shearing stress on each surface is

$$\tau_B = \frac{F_1}{A} = \frac{375 \text{ lb}}{(1.25 \text{ in.})(1.75 \text{ in.})} \qquad \tau_B = 171.4 \text{ psi} \quad \blacktriangleleft$$

e. Bearing Stress in Link at *C*. For each portion of the link (Fig. 6), $F_1 = 375 \text{ lb}$, and the nominal bearing area is $(0.25 \text{ in.})(0.25 \text{ in.}) = 0.0625 \text{ in}^2$.

$$\sigma_b = \frac{F_1}{A} = \frac{375 \text{ lb}}{0.0625 \text{ in}^2} \qquad \sigma_b = 6000 \text{ psi} \quad \blacktriangleleft$$

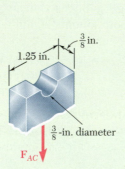

Fig. 4 Link *ABC* section at *A*.

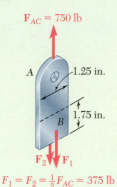

Fig. 5 Element *AB*.

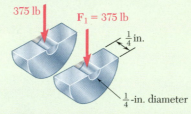

Fig. 6 Link *ABC* section at *C*.

REFLECT and THINK: This sample problem demonstrates the need to draw free-body diagrams of the separate components, carefully considering the behavior in each one. As an example, based on visual inspection of the hanger it is apparent that member *AC* should be in tension for the given load, and the analysis confirms this. Had a compression result been obtained instead, a thorough reexamination of the analysis would have been required.

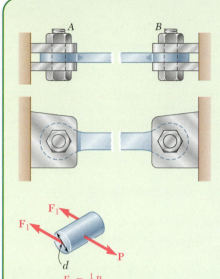

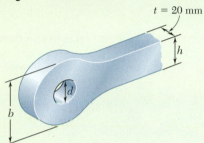

Fig. 1 Sectioned bolt.

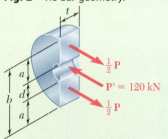

Fig. 2 Tie bar geometry.

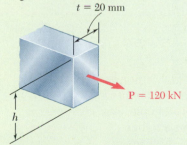

Fig. 4 Mid-body section of tie bar.

Sample Problem 1.2

The steel tie bar shown is to be designed to carry a tension force of magnitude $P = 120$ kN when bolted between double brackets at A and B. The bar will be fabricated from 20-mm-thick plate stock. For the grade of steel to be used, the maximum allowable stresses are $\sigma = 175$ MPa, $\tau = 100$ MPa, and $\sigma_b = 350$ MPa. Design the tie bar by determining the required values of (a) the diameter d of the bolt, (b) the dimension b at each end of the bar, and (c) the dimension h of the bar.

STRATEGY: Use free-body diagrams to determine the forces needed to obtain the stresses in terms of the design tension force. Setting these stresses equal to the allowable stresses provides for the determination of the required dimensions.

MODELING and ANALYSIS:

   a. Diameter of the Bolt.   Since the bolt is in double shear (Fig. 1), $F_1 = \frac{1}{2}P = 60$ kN.

$$\tau = \frac{F_1}{A} = \frac{60 \text{ kN}}{\frac{1}{4}\pi d^2} \qquad 100 \text{ MPa} = \frac{60 \text{ kN}}{\frac{1}{4}\pi d^2} \qquad d = 27.6 \text{ mm}$$

Use   **$d = 28$ mm**   ◄

At this point, check the bearing stress between the 20-mm-thick plate (Fig. 2) and the 28-mm-diameter bolt.

$$\sigma_b = \frac{P}{td} = \frac{120 \text{ kN}}{(0.020 \text{ m})(0.028 \text{ m})} = 214 \text{ MPa} < 350 \text{ MPa} \qquad \text{OK}$$

   b. Dimension b at Each End of the Bar.   We consider one of the end portions of the bar in Fig. 3. Recalling that the thickness of the steel plate is $t = 20$ mm and that the average tensile stress must not exceed 175 MPa, write

$$\sigma = \frac{\frac{1}{2}P}{ta} \qquad 175 \text{ MPa} = \frac{60 \text{ kN}}{(0.02 \text{ m})a} \qquad a = 17.14 \text{ mm}$$

$$b = d + 2a = 28 \text{ mm} + 2(17.14 \text{ mm}) \qquad \textbf{b = 62.3 mm} \quad ◄$$

   c. Dimension h of the Bar.   We consider a section in the central portion of the bar (Fig. 4). Recalling that the thickness of the steel plate is $t = 20$ mm, we have

$$\sigma = \frac{P}{th} \qquad 175 \text{ MPa} = \frac{120 \text{ kN}}{(0.020 \text{ m})h} \qquad h = 34.3 \text{ mm}$$

Use   **$h = 35$ mm**   ◄

REFLECT and THINK: We sized d based on bolt shear, and then checked bearing on the tie bar. Had the maximum allowable bearing stress been exceeded, we would have had to recalculate d based on the bearing criterion.

Problems

1.1 Two solid cylindrical rods AB and BC are welded together at B and loaded as shown. Knowing that $d_1 = 30$ mm and $d_2 = 50$ mm, find the average normal stress at the midsection of (a) rod AB, (b) rod BC.

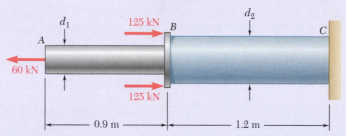

Fig. P1.1 and P1.2

1.2 Two solid cylindrical rods AB and BC are welded together at B and loaded as shown. Knowing that the average normal stress must not exceed 150 MPa in either rod, determine the smallest allowable values of the diameters d_1 and d_2.

1.3 Two solid cylindrical rods AB and BC are welded together at B and loaded as shown. Knowing that $P = 10$ kips, find the average normal stress at the midsection of (a) rod AB, (b) rod BC.

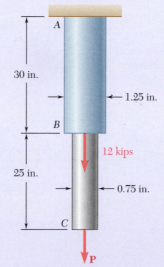

Fig. P1.3 and P1.4

1.4 Two solid cylindrical rods AB and BC are welded together at B and loaded as shown. Determine the magnitude of the force $\mathbf{P}$ for which the tensile stresses in rods AB and BC are equal.

1.5 A strain gage located at *C* on the surface of bone *AB* indicates that the average normal stress in the bone is 3.80 MPa when the bone is subjected to two 1200-N forces as shown. Assuming the cross section of the bone at *C* to be annular and knowing that its outer diameter is 25 mm, determine the inner diameter of the bone's cross section at *C*.

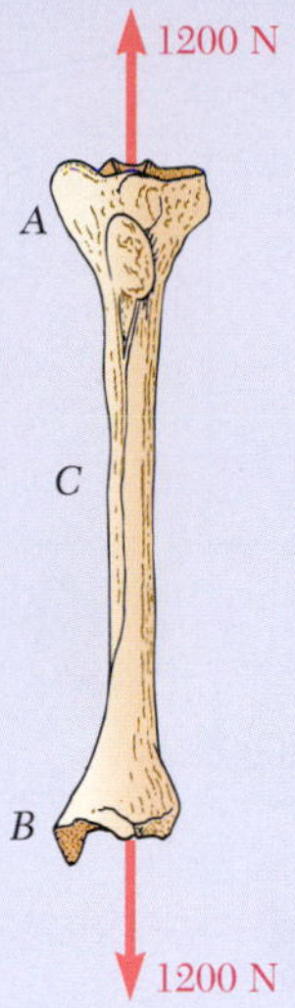

1.6 Two brass rods *AB* and *BC*, each of uniform diameter, will be brazed together at *B* to form a nonuniform rod of total length 100 m that will be suspended from a support at *A* as shown. Knowing that the density of brass is 8470 kg/m³, determine (*a*) the length of rod *AB* for which the maximum normal stress in *ABC* is minimum, (*b*) the corresponding value of the maximum normal stress.

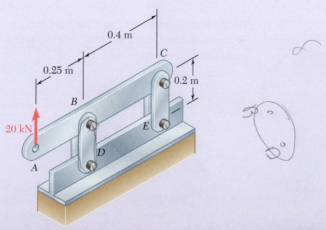

Fig. P1.5

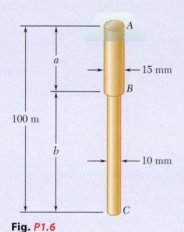

Fig. P1.6

1.7 Each of the four vertical links has an 8 × 36-mm uniform rectangular cross section, and each of the four pins has a 16-mm diameter. Determine the maximum value of the average normal stress in the links connecting (*a*) points *B* and *D*, (*b*) points *C* and *E*.

Fig. P1.7

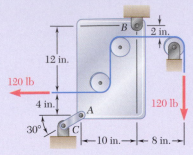

Fig. P1.8

1.8 Link *AC* has a uniform rectangular cross section $\frac{1}{8}$ in. thick and 1 in. wide. Determine the normal stress in the central portion of the link.

1.9 Three forces, each of magnitude P = 4 kN, are applied to the structure shown. Determine the cross-sectional area of the uniform portion of rod *BE* for which the normal stress in that portion is +100 MPa.

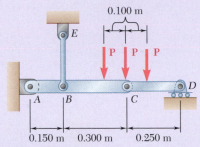

Fig. P1.9

1.10 Link BD consists of a single bar 1 in. wide and $\frac{1}{2}$ in. thick. Knowing that each pin has a $\frac{3}{8}$-in. diameter, determine the maximum value of the average normal stress in link *BD* if (*a*) $\theta = 0$, (*b*) $\theta = 90°$.

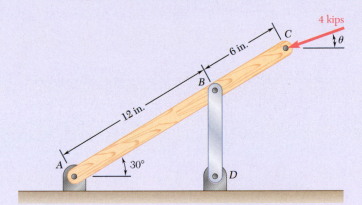

Fig. P1.10

1.11 For the Pratt bridge truss and loading shown, determine the average normal stress in member *BE*, knowing that the cross-sectional area of that member is 5.87 in².

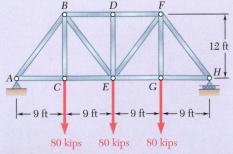

Fig. *P1.11*

1.12 The frame shown consists of *four* wooden members, *ABC*, *DEF*, *BE*, and *CF*. Knowing that each member has a 2 × 4-in. rectangular cross section and that each pin has a $\frac{1}{2}$-in. diameter, determine the maximum value of the average normal stress (*a*) in member *BE*, (*b*) in member *CF*.

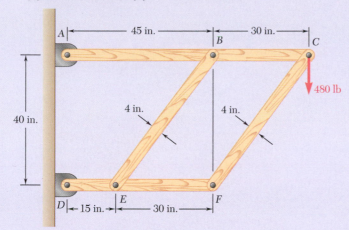

Fig. P1.12

1.13 An aircraft tow bar is positioned by means of a single hydraulic cylinder connected by a 25-mm-diameter steel rod to two identical arm-and-wheel units *DEF*. The mass of the entire tow bar is 200 kg, and its center of gravity is located at *G*. For the position shown, determine the normal stress in the rod.

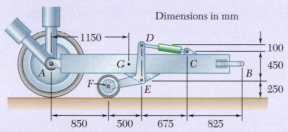

Fig. P1.13

1.14 Two hydraulic cylinders are used to control the position of the robotic arm *ABC*. Knowing that the control rods attached at *A* and *D* each have a 20-mm diameter and happen to be parallel in the position shown, determine the average normal stress in (*a*) member *AE*, (*b*) member *DG*.

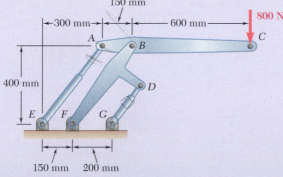

Fig. P1.14

1.15 Determine the diameter of the largest circular hole that can be punched into a sheet of polystyrene 6 mm thick, knowing that the force exerted by the punch is 45 kN and that a 55-MPa average shearing stress is required to cause the material to fail.

1.16 Two wooden planks, each $\frac{1}{2}$ in. thick and 9 in. wide, are joined by the dry mortise joint shown. Knowing that the wood used shears off along its grain when the average shearing stress reaches 1.20 ksi, determine the magnitude P of the axial load that will cause the joint to fail.

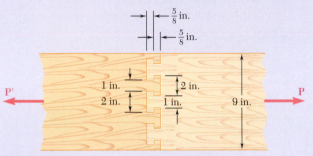

Fig. P1.16

1.17 When the force **P** reached 1600 lb, the wooden specimen shown failed in shear along the surface indicated by the dashed line. Determine the average shearing stress along that surface at the time of failure.

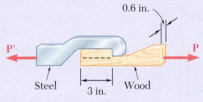

Fig. P1.17

1.18 A load **P** is applied to a steel rod supported as shown by an aluminum plate into which a 12-mm-diameter hole has been drilled. Knowing that the shearing stress must not exceed 180 MPa in the steel rod and 70 MPa in the aluminum plate, determine the largest load **P** that can be applied to the rod.

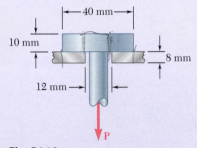

Fig. P1.18

1.19 The axial force in the column supporting the timber beam shown is $P = 20$ kips. Determine the smallest allowable length L of the bearing plate if the bearing stress in the timber is not to exceed 400 psi.

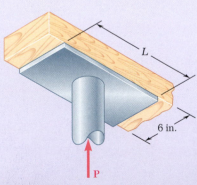

Fig. P1.19

1.20 Three wooden planks are fastened together by a series of bolts to form a column. The diameter of each bolt is 12 mm and the inner diameter of each washer is 16 mm, which is slightly larger than the diameter of the holes in the planks. Determine the smallest allowable outer diameter d of the washers, knowing that the average normal stress in the bolts is 36 MPa and that the bearing stress between the washers and the planks must not exceed 8.5 MPa.

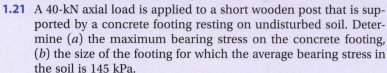

Fig. P1.20

1.21 A 40-kN axial load is applied to a short wooden post that is supported by a concrete footing resting on undisturbed soil. Determine (*a*) the maximum bearing stress on the concrete footing, (*b*) the size of the footing for which the average bearing stress in the soil is 145 kPa.

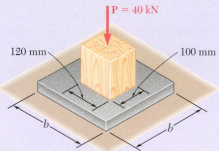

Fig. P1.21

1.22 An axial load **P** is supported by a short W8 × 40 column of cross-sectional area $A = 11.7$ in^2 and is distributed to a concrete foundation by a square plate as shown. Knowing that the average normal stress in the column must not exceed 30 ksi and that the bearing stress on the concrete foundation must not exceed 3.0 ksi, determine the side a of the plate that will provide the most economical and safe design.

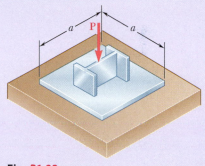

Fig. *P1.22*

1.23 Link *AB*, of width $b = 2$ in. and thickness $t = \frac{1}{4}$ in., is used to support the end of a horizontal beam. Knowing that the average normal stress in the link is −20 ksi and that the average shearing stress in each of the two pins is 12 ksi determine (*a*) the diameter d of the pins, (*b*) the average bearing stress in the link.

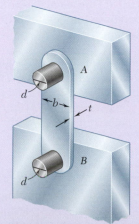

Fig. P1.23

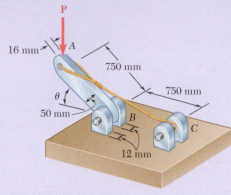

Fig. P1.24 and P1.25

1.24 Determine the largest load **P** that can be applied at A when $\theta = 60°$, knowing that the average shearing stress in the 10-mm-diameter pin at B must not exceed 120 MPa and that the average bearing stress in member AB and in the bracket at B must not exceed 90 MPa.

1.25 Knowing that $\theta = 40°$ and $P = 9$ kN, determine (*a*) the smallest allowable diameter of the pin at B if the average shearing stress in the pin is not to exceed 120 MPa, (*b*) the corresponding average bearing stress in member AB at B, (*c*) the corresponding average bearing stress in each of the support brackets at B.

1.26 The hydraulic cylinder CF, which partially controls the position of rod DE, has been locked in the position shown. Member BD is 15 mm thick and is connected at C to the vertical rod by a 9-mm-diameter bolt. Knowing that $P = 2$ kN and $\theta = 75°$, determine (*a*) the average shearing stress in the bolt, (*b*) the bearing stress at C in member BD.

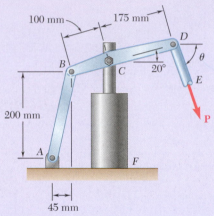

Fig. *P1.26*

1.27 For the assembly and loading of Prob. 1.7, determine (*a*) the average shearing stress in the pin at B, (*b*) the average bearing stress at B in member BD, (*c*) the average bearing stress at B in member ABC, knowing that this member has a 10×50-mm uniform rectangular cross section.

1.28 Two identical linkage-and-hydraulic-cylinder systems control the position of the forks of a fork-lift truck. The load supported by the one system shown is 1500 lb. Knowing that the thickness of member BD is $\frac{5}{8}$ in., determine (*a*) the average shearing stress in the $\frac{1}{2}$-in.-diameter pin at B, (*b*) the bearing stress at B in member BD.

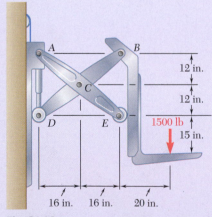

Fig. P1.28

1.3   STRESS ON AN OBLIQUE PLANE UNDER AXIAL LOADING

Previously, axial forces exerted on a two-force member (Fig. 1.26a) caused normal stresses in that member (Fig. 1.26b), while transverse forces exerted on bolts and pins (Fig. 1.27a) caused shearing stresses in those connections (Fig. 1.27b). Such a relation was observed between axial forces and normal stresses and transverse forces and shearing stresses, because stresses were being determined only on planes perpendicular to the axis of the member or connection. In this section, axial forces cause both normal and shearing stresses on planes that are not perpendicular to the axis of the member. Similarly, transverse forces exerted on a bolt or a pin cause both normal and shearing stresses on planes that are not perpendicular to the axis of the bolt or pin.

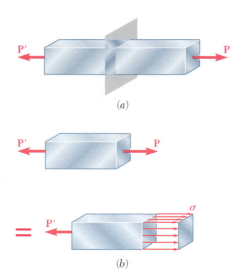

Fig. 1.26 Axial forces on a two-force member. (a) Section plane perpendicular to member away from load application. (b) Equivalent force diagram models of resultant force acting at centroid and uniform normal stress.

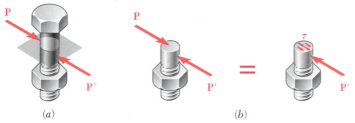

Fig. 1.27 (a) Diagram of a bolt from a single-shear joint with a section plane normal to the bolt. (b) Equivalent force diagram models of the resultant force acting at the section centroid and the uniform average shear stress.

Consider the two-force member of Fig. 1.26 that is subjected to axial forces **P** and **P′**. If we pass a section forming an angle θ with a normal plane (Fig. 1.28a) and draw the free-body diagram of the portion of member located to the left of that section (Fig. 1.28b), the equilibrium conditions of the free body show that the distributed forces acting on the section must be equivalent to the force **P**.

Resolving **P** into components **F** and **V**, respectively normal and tangential to the section (Fig. 1.28c),

$$F = P\cos\theta \qquad V = P\sin\theta \tag{1.12}$$

Force **F** represents the resultant of normal forces distributed over the section, and force **V** is the resultant of shearing forces (Fig. 1.28d). The average values of the corresponding normal and shearing stresses are obtained by dividing F and V by the area A_θ of the section:

$$\sigma = \frac{F}{A_\theta} \qquad \tau = \frac{V}{A_\theta} \tag{1.13}$$

Substituting for F and V from Eq. (1.12) into Eq. (1.13), and observing from Fig. 1.28c that $A_0 = A_\theta \cos\theta$ or $A_\theta = A_0/\cos\theta$, where A_0 is the area of a section perpendicular to the axis of the member, we obtain

$$\sigma = \frac{P\cos\theta}{A_0/\cos\theta} \qquad \tau = \frac{P\sin\theta}{A_0/\cos\theta}$$

or

$$\sigma = \frac{P}{A_0}\cos^2\theta \qquad \tau = \frac{P}{A_0}\sin\theta\cos\theta \tag{1.14}$$

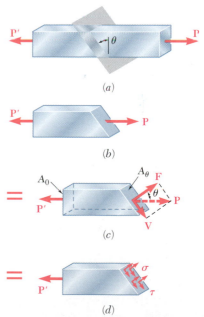

Fig. 1.28 Oblique section through a two-force member. (a) Section plane made at an angle θ to the member normal plane, (b) Free-body diagram of left section with internal resultant force **P**. (c) Free-body diagram of resultant force resolved into components **F** and **V** along the section plane's normal and tangential directions, respectively. (d) Free-body diagram with section forces **F** and **V** represented as normal stress, σ, and shearing stress, τ.

(a) Axial loading

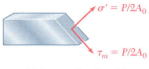

(b) Stresses for $\theta = 0$

(c) Stresses for $\theta = 45°$

(d) Stresses for $\theta = -45°$

Fig. 1.29 Selected stress results for axial loading.

Note from the first of Eqs. (1.14) that the normal stress σ is maximum when $\theta = 0$ (i.e., the plane of the section is perpendicular to the axis of the member). It approaches zero as θ approaches 90°. We check that the value of σ when $\theta = 0$ is

$$\sigma_m = \frac{P}{A_0} \tag{1.15}$$

The second of Eqs. (1.14) shows that the shearing stress τ is zero for $\theta = 0$ and $\theta = 90°$. For $\theta = 45°$, it reaches its maximum value

$$\tau_m = \frac{P}{A_0} \sin 45° \cos 45° = \frac{P}{2A_0} \tag{1.16}$$

The first of Eqs. (1.14) indicates that, when $\theta = 45°$, the normal stress σ' is also equal to $P/2A_0$:

$$\sigma' = \frac{P}{A_0} \cos^2 45° = \frac{P}{2A_0} \tag{1.17}$$

The results obtained in Eqs. (1.15), (1.16), and (1.17) are shown graphically in Fig. 1.29. The same loading may produce either a normal stress $\sigma_m = P/A_0$ and no shearing stress (Fig. 1.29b) or a normal and a shearing stress of the same magnitude $\sigma' = \tau_m = P/2A_0$ (Fig. 1.29c and d), depending upon the orientation of the section.

1.4 STRESS UNDER GENERAL LOADING CONDITIONS; COMPONENTS OF STRESS

The examples of the previous sections were limited to members under axial loading and connections under transverse loading. Most structural members and machine components are under more involved loading conditions.

Consider a body subjected to several loads $\mathbf{P}_1$, $\mathbf{P}_2$, etc. (Fig. 1.30). To understand the stress condition created by these loads at some point Q within the body, we shall first pass a section through Q, using a plane parallel to the yz plane. The portion of the body to the left of the section is subjected to some of the original loads, and to normal and shearing forces distributed over the section. We shall denote by $\Delta\mathbf{F}^x$ and $\Delta\mathbf{V}^x$, respectively, the normal and the shearing forces acting on a small area ΔA surrounding point Q (Fig. 1.31a). Note that the superscript x is used to indicate that the forces $\Delta\mathbf{F}^x$ and $\Delta\mathbf{V}^x$ act on a surface perpendicular to the x axis. While the normal force $\Delta\mathbf{F}^x$ has a well-defined direction, the shearing force $\Delta\mathbf{V}^x$ may have any direction in the plane of the section. We therefore resolve $\Delta\mathbf{V}^x$ into two component forces, $\Delta\mathbf{V}_y^x$ and $\Delta\mathbf{V}_z^x$, in directions parallel to the y and z axes, respectively (Fig. 1.31b). Dividing the magnitude of each force by the area ΔA and letting ΔA approach zero, we define the three stress components shown in Fig. 1.32:

$$\sigma_x = \lim_{\Delta A \to 0} \frac{\Delta F^x}{\Delta A}$$

$$\tau_{xy} = \lim_{\Delta A \to 0} \frac{\Delta V_y^x}{\Delta A} \qquad \tau_{xz} = \lim_{\Delta A \to 0} \frac{\Delta V_z^x}{\Delta A} \tag{1.18}$$

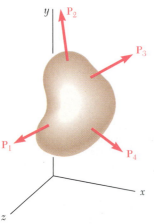

Fig. 1.30 Multiple loads on a general body.

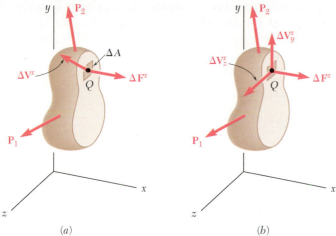

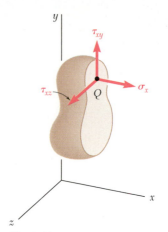

Fig. 1.31 (a) Resultant shear and normal forces, $\Delta \mathbf{V}^x$ and $\Delta \mathbf{F}^x$, acting on small area ΔA at point Q. (b) Forces on ΔA resolved into forces in coordinate directions.

Fig. 1.32 Stress components at point Q on the body to the left of the plane.

Note that the first subscript in σ_x, τ_{xy}, and τ_{xz} is used to indicate that the stresses are exerted *on a surface perpendicular to the x axis.* The second subscript in τ_{xy} and τ_{xz} identifies *the direction of the component.* The normal stress σ_x is positive if the corresponding arrow points in the positive x direction (i.e., if the body is in tension) and negative otherwise. Similarly, the shearing stress components τ_{xy} and τ_{xz} are positive if the corresponding arrows point, respectively, in the positive y and z directions.

This analysis also may be carried out by considering the portion of body located to the right of the vertical plane through Q (Fig. 1.33). The same magnitudes, but opposite directions, are obtained for the normal and shearing forces $\Delta \mathbf{F}^x$, $\Delta \mathbf{V}_y^x$, and $\Delta \mathbf{V}_z^x$. Therefore, the same values are obtained for the corresponding stress components. However as the section in Fig. 1.33, now faces the *negative x axis,* a positive sign for σ_x indicates that the corresponding arrow points *in the negative x direction.* Similarly, positive signs for τ_{xy} and τ_{xz} indicate that the corresponding arrows point in the negative y and z directions, as shown in Fig. 1.33.

Passing a section through Q parallel to the zx plane, we define the stress components, σ_y, τ_{yz}, and τ_{yx}. Then, a section through Q parallel to the xy plane yields the components σ_z, τ_{zx}, and τ_{zy}.

To visualize the stress condition at point Q, consider a small cube of side a centered at Q and the stresses exerted on each of the six faces of the cube (Fig. 1.34). The stress components shown are σ_x, σ_y, and σ_z, which represent the normal stress on faces respectively perpendicular to the x, y, and z axes, and the six shearing stress components τ_{xy}, τ_{xz}, etc. Recall that τ_{xy} represents the y component of the shearing stress exerted on the face perpendicular to the x axis, while τ_{yx} represents the x component of the shearing stress exerted on the face perpendicular to the y axis. Note that only three faces of the cube are actually visible in Fig. 1.34 and that equal and opposite stress components act on the hidden faces. While the stresses acting on the faces of the cube differ slightly from the stresses at Q, the error involved is small and vanishes as side a of the cube approaches zero.

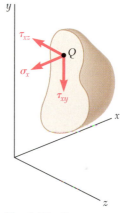

Fig. 1.33 Stress components at point Q on the body to the right of the plane.

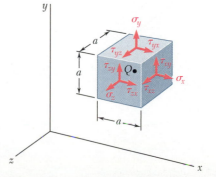

Fig. 1.34 Positive stress components at point Q.

Shearing stress components. Consider the free-body diagram of the small cube centered at point Q (Fig. 1.35). The normal and shearing forces acting on the various faces of the cube are obtained by multiplying the corresponding stress components by the area ΔA of each face. First write the following three equilibrium equations

$$\Sigma F_x = 0 \qquad \Sigma F_y = 0 \qquad \Sigma F_z = 0 \qquad\qquad \textbf{(1.19)}$$

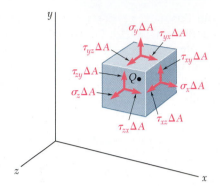

Fig. 1.35 Positive resultant forces on a small element at point Q resulting from a state of general stress.

Since forces equal and opposite to the forces actually shown in Fig. 1.35 are acting on the hidden faces of the cube, Eqs. (1.19) are satisfied. Considering the moments of the forces about axes x', y', and z' drawn from Q in directions respectively parallel to the x, y, and z axes, the three additional equations are

$$\Sigma M_{x'} = 0 \qquad \Sigma M_{y'} = 0 \qquad \Sigma M_{z'} = 0 \qquad\qquad \textbf{(1.20)}$$

Using a projection on the $x'y'$ plane (Fig. 1.36), note that the only forces with moments about the z axis different from zero are the shearing forces. These forces form two couples: a counterclockwise (positive) moment $(\tau_{xy} \Delta A)a$ and a clockwise (negative) moment $-(\tau_{yx} \Delta A)a$. The last of the three Eqs. (1.20) yields

$$+\!\curvearrowleft \Sigma M_z = 0: \qquad (\tau_{xy} \Delta A)a - (\tau_{yx} \Delta A)a = 0$$

from which

$$\boxed{\tau_{xy} = \tau_{yx}} \qquad\qquad \textbf{(1.21)}$$

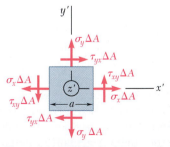

Fig. 1.36 Free-body diagram of small element at Q viewed on projected plane perpendicular to z'-axis. Resultant forces on positive and negative z' faces (not shown) act through the z'-axis, thus do not contribute to the moment about that axis.

This relationship shows that the y component of the shearing stress exerted on a face perpendicular to the x axis is equal to the x component of the shearing stress exerted on a face perpendicular to the y axis. From the remaining parts of Eqs. (1.20), we derive.

$$\tau_{yz} = \tau_{zy} \qquad \tau_{zx} = \tau_{xz} \qquad \textbf{(1.22)}$$

We conclude from Eqs. (1.21) and (1.22), only six stress components are required to define the condition of stress at a given point Q, instead of nine as originally assumed. These components are σ_x, σ_y, σ_z, τ_{xy}, τ_{yz}, and τ_{zx}. Also note that, at a given point, *shear cannot take place in one plane only;* an equal shearing stress must be exerted on another plane perpendicular to the first one. For example, considering the bolt of Fig. 1.29 and a small cube at the center Q (Fig. 1.37a), we see that shearing stresses of equal magnitude must be exerted on the two horizontal faces of the cube and on the two faces perpendicular to the forces **P** and **P′** (Fig. 1.37b).

Axial loading. Let us consider again a member under axial loading. If we consider a small cube with faces respectively parallel to the faces of the member and recall the results obtained in Sec. 1.3, the conditions of stress in the member may be described as shown in Fig. 1.38a; the only stresses are normal stresses σ_x exerted on the faces of the cube that are perpendicular to the x axis. However, if the small cube is rotated by 45° about the z axis so that its new orientation matches the orientation of the sections considered in Fig. 1.29c and d, normal and shearing stresses of equal magnitude are exerted on four faces of the cube (Fig. 1.38b). Thus, the same loading condition may lead to different interpretations of the stress situation at a given point, depending upon the orientation of the element considered. More will be said about this in Chap. 7: Transformation of Stress and Strain.

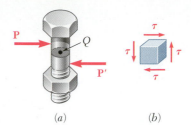

Fig. 1.37   Single-shear bolt with point Q chosen at the center. (b) Pure shear stress element at point Q.

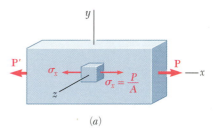

(a)

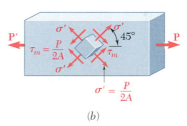

(b)

Fig. 1.38   Changing the orientation of the stress element produces different stress components for the same state of stress. This is studied in detail in Chapter 7.

1.5   DESIGN CONSIDERATIONS

In engineering applications, the determination of stresses is seldom an end in itself. Rather, the knowledge of stresses is used by engineers to assist in their most important task: the design of structures and machines that will safely and economically perform a specified function.

1.5A   Determination of the Ultimate Strength of a Material

An important element to be considered by a designer is how the material will behave under a load. This is determined by performing specific tests on prepared samples of the material. For example, a test specimen of steel may be prepared and placed in a laboratory testing machine to be subjected to a known centric axial tensile force, as described in Sec. 2.1B. As the magnitude of the force is increased, various dimensional changes such as length and diameter are measured. Eventually, the largest force that may be applied to the specimen is reached, and it either breaks or begins

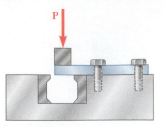

Fig. 1.39   Single shear test.

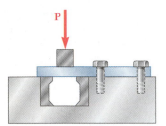

Fig. 1.40   Double shear test.

to carry less load. This largest force is called the *ultimate load* and is denoted by P_U. Since the applied load is centric, the ultimate load is divided by the original cross-sectional area of the rod to obtain the *ultimate normal stress* of the material. This stress, also known as the *ultimate strength in tension*, is

$$\sigma_U = \frac{P_U}{A} \tag{1.23}$$

Several test procedures are available to determine the *ultimate shearing stress* or *ultimate strength in shear*. The one most commonly used involves the twisting of a circular tube (Sec. 3.2). A more direct, if less accurate, procedure clamps a rectangular or round bar in a shear tool (Fig. 1.39) and applies an increasing load P until the ultimate load P_U for single shear is obtained. If the free end of the specimen rests on both of the hardened dies (Fig. 1.40), the ultimate load for double shear is obtained. In either case, the ultimate shearing stress τ_U is

$$\tau_U = \frac{P_U}{A} \tag{1.24}$$

In single shear, this area is the cross-sectional area A of the specimen, while in double shear it is equal to twice the cross-sectional area.

1.5B   Allowable Load and Allowable Stress: Factor of Safety

The maximum load that a structural member or a machine component will be allowed to carry under normal conditions is considerably smaller than the ultimate load. This smaller load is the *allowable load* (sometimes called the *working* or *design load*). Thus, only a fraction of the ultimate-load capacity of the member is used when the allowable load is applied. The remaining portion of the load-carrying capacity of the member is kept in reserve to assure its safe performance. The ratio of the ultimate load to the allowable load is used to define the *factor of safety*:[†]

$$\text{Factor of safety} = F.S. = \frac{\text{ultimate load}}{\text{allowable load}} \tag{1.25}$$

An alternative definition of the factor of safety is based on the use of stresses:

$$\text{Factor of safety} = F.S. = \frac{\text{ultimate stress}}{\text{allowable stress}} \tag{1.26}$$

These two expressions are identical when a linear relationship exists between the load and the stress. In most engineering applications,

[†]In some fields of engineering, notably aeronautical engineering, the *margin of safety* is used in place of the factor of safety. The margin of safety is defined as the factor of safety minus one; that is, margin of safety = *F.S.* − 1.00.

however, this relationship ceases to be linear as the load approaches its ultimate value, and the factor of safety obtained from Eq. (1.26) does not provide a true assessment of the safety of a given design. Nevertheless, the *allowable-stress method* of design, based on the use of Eq. (1.26), is widely used.

1.5C   Factor of Safety Selection

The selection of the factor of safety to be used is one of the most important engineering tasks. If a factor of safety is too small, the possibility of failure becomes unacceptably large. On the other hand, if a factor of safety is unnecessarily large, the result is an uneconomical or nonfunctional design. The choice of the factor of safety for a given design application requires engineering judgment based on many considerations.

1. *Variations that may occur in the properties of the member.* The composition, strength, and dimensions of the member are all subject to small variations during manufacture. In addition, material properties may be altered and residual stresses introduced through heating or deformation that may occur during manufacture, storage, transportation, or construction.

2. *The number of loadings expected during the life of the structure or machine.* For most materials, the ultimate stress decreases as the number of load cycles is increased. This phenomenon is known as *fatigue* and can result in sudden failure if ignored (see Sec. 2.1F).

3. *The type of loadings planned for in the design or that may occur in the future.* Very few loadings are known with complete accuracy—most design loadings are engineering estimates. In addition, future alterations or changes in usage may introduce changes in the actual loading. Larger factors of safety are also required for dynamic, cyclic, or impulsive loadings.

4. *Type of failure.* Brittle materials fail suddenly, usually with no prior indication that collapse is imminent. However, ductile materials, such as structural steel, normally undergo a substantial deformation called *yielding* before failing, providing a warning that overloading exists. Most buckling or stability failures are sudden, whether the material is brittle or not. When the possibility of sudden failure exists, a larger factor of safety should be used than when failure is preceded by obvious warning signs.

5. *Uncertainty due to methods of analysis.* All design methods are based on certain simplifying assumptions that result in calculated stresses being approximations of actual stresses.

6. *Deterioration that may occur in the future because of poor maintenance or unpreventable natural causes.* A larger factor of safety is necessary in locations where conditions such as corrosion and decay are difficult to control or even to discover.

7. *The importance of a given member to the integrity of the whole structure.* Bracing and secondary members in many cases can be designed with a factor of safety lower than that used for primary members.

In addition to these considerations, there is concern of the risk to life and property that a failure would produce. Where a failure would produce no risk to life and only minimal risk to property, the use of a smaller factor of safety can be acceptable. Finally, unless a careful design with a nonexcessive factor of safety is used, a structure or machine might not perform its design function. For example, high factors of safety may have an unacceptable effect on the weight of an aircraft.

For the majority of structural and machine applications, factors of safety are specified by design specifications or building codes written by committees of experienced engineers working with professional societies, industries, or federal, state, or city agencies. Examples of such design specifications and building codes are

1. *Steel:* American Institute of Steel Construction, Specification for Structural Steel Buildings
2. *Concrete:* American Concrete Institute, Building Code Requirement for Structural Concrete
3. *Timber:* American Forest and Paper Association, National Design Specification for Wood Construction
4. *Highway bridges:* American Association of State Highway Officials, Standard Specifications for Highway Bridges

1.5D Load and Resistance Factor Design

The allowable-stress method requires that all the uncertainties associated with the design of a structure or machine element be grouped into a single factor of safety. An alternative method of design makes it possible to distinguish between the uncertainties associated with the structure itself and those associated with the load it is designed to support. Called *Load and Resistance Factor Design (LRFD)*, this method allows the designer to distinguish between uncertainties associated with the *live load,* P_L (i.e., the active or time-varying load to be supported by the structure) and the *dead load,* P_D (i.e., the self weight of the structure contributing to the total load).

Using the LRFD method the *ultimate load,* P_U, of the structure (i.e., the load at which the structure ceases to be useful) should be determined. The proposed design is acceptable if the following inequality is satisfied:

$$\gamma_D P_D + \gamma_L P_L \le \phi P_U \tag{1.27}$$

The coefficient ϕ is the *resistance factor,* which accounts for the uncertainties associated with the structure itself and will normally be less than 1. The coefficients γ_D and γ_L are the *load factors;* they account for the uncertainties associated with the dead and live load and normally will be greater than 1, with γ_L generally larger than γ_D. While a few examples and assigned problems using LRFD are included in this chapter and in Chaps. 5 and 10, the allowable-stress method of design is primarily used in this text.

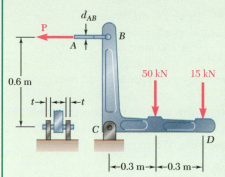

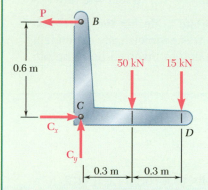

Fig. 1 Free-body diagram of bracket.

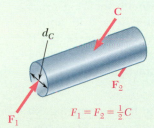

Fig. 2 Free-body diagram of pin at point *C*.

Sample Problem 1.3

Two loads are applied to the bracket *BCD* as shown. (*a*) Knowing that the control rod *AB* is to be made of a steel having an ultimate normal stress of 600 MPa, determine the diameter of the rod for which the factor of safety with respect to failure will be 3.3. (*b*) The pin at *C* is to be made of a steel having an ultimate shearing stress of 350 MPa. Determine the diameter of the pin *C* for which the factor of safety with respect to shear will also be 3.3. (*c*) Determine the required thickness of the bracket supports at *C*, knowing that the allowable bearing stress of the steel used is 300 MPa.

STRATEGY: Consider the free body of the bracket to determine the force **P** and the reaction at *C*. The resulting forces are then used with the allowable stresses, determined from the factor of safety, to obtain the required dimensions.

MODELING: Draw the free-body diagram of the hanger (Fig. 1), and the pin at *C* (Fig. 2).

ANALYSIS:

Free Body: Entire Bracket. Using Fig. 1, the reaction at *C* is represented by its components $\mathbf{C}_x$ and $\mathbf{C}_y$.

$$+\circlearrowleft \Sigma M_C = 0:\quad P(0.6\text{ m}) - (50\text{ kN})(0.3\text{ m}) - (15\text{ kN})(0.6\text{ m}) = 0 \quad P = 40\text{ kN}$$
$$\Sigma F_x = 0:\qquad\qquad\qquad C_x = 40\text{ kN}$$
$$\Sigma F_y = 0:\qquad\qquad\qquad C_y = 65\text{ kN}\qquad C = \sqrt{C_x^2 + C_y^2} = 76.3\text{ kN}$$

a. Control Rod *AB*. Since the factor of safety is 3.3, the allowable stress is

$$\sigma_{\text{all}} = \frac{\sigma_U}{F.S.} = \frac{600\text{ MPa}}{3.3} = 181.8\text{ MPa}$$

For *P* = 40 kN, the cross-sectional area required is

$$A_{\text{req}} = \frac{P}{\sigma_{\text{all}}} = \frac{40\text{ kN}}{181.8\text{ MPa}} = 220 \times 10^{-6}\text{ m}^2$$

$$A_{\text{req}} = \frac{\pi}{4} d_{AB}^2 = 220 \times 10^{-6}\text{ m}^2 \qquad d_{ab} = \textbf{16.74 mm} \blacktriangleleft$$

b. Shear in Pin *C*. For a factor of safety of 3.3, we have

$$\tau_{\text{all}} = \frac{\tau_U}{F.S.} = \frac{350\text{ MPa}}{3.3} = 106.1\text{ MPa}$$

(continued)

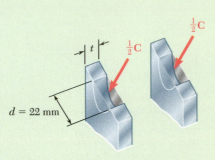

Fig. 3 Bearing loads at bracket support at point C.

As shown in Fig. 2 the pin is in double shear. We write

$$A_{\text{req}} = \frac{C/2}{\tau_{\text{all}}} = \frac{(76.3 \text{ kN})/2}{106.1 \text{ MPa}} = 360 \text{ mm}^2$$

$$A_{\text{req}} = \frac{\pi}{4} d_C^2 = 360 \text{ mm}^2 \qquad d_C = 21.4 \text{ mm} \qquad \textbf{Use: } d_C = 22 \text{ mm} \blacktriangleleft$$

c. Bearing at C. Using $d = 22$ mm, the nominal bearing area of each bracket is $22t$. From Fig. 3 the force carried by each bracket is $C/2$ and the allowable bearing stress is 300 MPa. We write

$$A_{\text{req}} = \frac{C/2}{\sigma_{\text{all}}} = \frac{(76.3 \text{ kN})/2}{300 \text{ MPa}} = 127.2 \text{ mm}^2$$

Thus, $22t = 127.2$ $\qquad t = 5.78$ mm $\qquad$ **Use: $t = 6$ mm** $\blacktriangleleft$

REFLECT and THINK: It was appropriate to design the pin C first and then its bracket, as the pin design was geometrically dependent upon diameter only, while the bracket design involved both the pin diameter and bracket thickness.

Sample Problem 1.4

The rigid beam BCD is attached by bolts to a control rod at B, to a hydraulic cylinder at C, and to a fixed support at D. The diameters of the bolts used are: $d_B = d_D = \frac{3}{8}$ in., $d_C = \frac{1}{2}$ in. Each bolt acts in double shear and is made from a steel for which the ultimate shearing stress is $\tau_U = 40$ ksi. The control rod AB has a diameter $d_A = \frac{7}{16}$ in. and is made of a steel for which the ultimate tensile stress is $\sigma_U = 60$ ksi. If the minimum factor of safety is to be 3.0 for the entire unit, determine the largest upward force that may be applied by the hydraulic cylinder at C.

STRATEGY: The factor of safety with respect to failure must be 3.0 or more in each of the three bolts and in the control rod. These four independent criteria need to be considered separately.

MODELING: Draw the free-body diagram of the bar (Fig. 1) and the bolts at B and C (Figs. 2 and 3). Determine the allowable value of the force **C** based on the required design criteria for each part.

ANALYSIS:

Free Body: Beam BCD. Using Fig. 1, first determine the force at C in terms of the force at B and in terms of the force at D.

$$+\!\!\curvearrowleft \Sigma M_D = 0: \qquad B(14 \text{ in.}) - C(8 \text{ in.}) = 0 \qquad C = 1.750B \quad \textbf{(1)}$$

$$+\!\!\curvearrowleft \Sigma M_B = 0: \qquad -D(14 \text{ in.}) + C(6 \text{ in.}) = 0 \qquad C = 2.33D \quad \textbf{(2)}$$

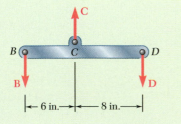

Fig. 1 Free-body diagram of beam BCD.

(continued)

[handwritten notes in left margin:]
bolt: $\sigma_U = 40$ ksi
rod $\sigma_U = 60$ kss $\frac{7}{16}$
AB
S.S = 3
C = ?

Control Rod.   For a factor of safety of 3.0

$$\sigma_{\text{all}} = \frac{\sigma_U}{F.S.} = \frac{60 \text{ ksi}}{3.0} = 20 \text{ ksi}$$

The allowable force in the control rod is

$$B = \sigma_{\text{all}}(A) = (20 \text{ ksi})\tfrac{1}{4}\pi\left(\tfrac{7}{16} \text{ in.}\right)^2 = 3.01 \text{ kips}$$

Using Eq. (1), the largest permitted value of C is

$$C = 1.750B = 1.750(3.01 \text{ kips}) \qquad C = \textbf{5.27 kips} \quad \blacktriangleleft$$

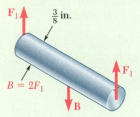

Fig. 2   Free-body diagram of pin at point B.

[figure labels: F_1, $\tfrac{3}{8}$ in., F_1, $B = 2F_1$, B]

Bolt at B. $\tau_{\text{all}} = \tau_U/F.S. = (40 \text{ ksi})/3 = 13.33 \text{ ksi}$. Since the bolt is in double shear (Fig. 2), the allowable magnitude of the force **B** exerted on the bolt is

$$B = 2F_1 = 2(\tau_{\text{all}}A) = 2(13.33 \text{ ksi})(\tfrac{1}{4}\pi)(\tfrac{3}{8} \text{ in.})^2 = 2.94 \text{ kips}$$

From Eq. (1), $C = 1.750B = 1.750(2.94 \text{ kips}) \qquad C = \textbf{5.15 kips} \quad \blacktriangleleft$

Bolt at D.   Since this bolt is the same as bolt B, the allowable force is $D = B = 2.94$ kips. From Eq. (2)

$$C = 2.33D = 2.33(2.94 \text{ kips}) \qquad C = \textbf{6.85 kips} \quad \blacktriangleleft$$

Bolt at C.   We again have $\tau_{\text{all}} = 13.33$ ksi. Using Fig. 3, we write

$$C = 2F_2 = 2(\tau_{\text{all}}A) = 2(13.33 \text{ ksi})(\tfrac{1}{4}\pi)(\tfrac{1}{2} \text{ in.})^2 \qquad C = \textbf{5.23 kips} \quad \blacktriangleleft$$

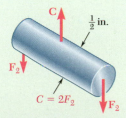

Fig. 3   Free-body diagram of pin at point C.

[figure labels: C, $\tfrac{1}{2}$ in., F_2, $C = 2F_2$, F_2]

Summary.   We have found separately four maximum allowable values of the force C. In order to satisfy all these criteria, choose the smallest value. $C = \textbf{5.15 kips} \quad \blacktriangleleft$

REFLECT and THINK: This example illustrates that all parts must satisfy the appropriate design criteria, and as a result, some parts have more capacity than needed.

Problems

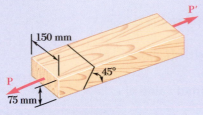

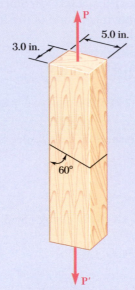

Fig. P1.29 and P1.30

Fig. P1.31 and P1.32

1.29 Two wooden members of uniform rectangular cross section are joined by the simple glued scarf splice shown. Knowing that $P = 11$ kN, determine the normal and shearing stresses in the glued splice.

1.30 Two wooden members of uniform rectangular cross section are joined by the simple glued scarf splice shown. Knowing that the maximum allowable shearing stress in the glued splice is 620 kPa, determine (*a*) the largest load **P** that can be safely applied, (*b*) the corresponding tensile stress in the splice.

1.31 The 1.4-kip load **P** is supported by two wooden members of uniform cross section that are joined by the simple glued scarf splice shown. Determine the normal and shearing stresses in the glued splice.

1.32 Two wooden members of uniform cross section are joined by the simple scarf splice shown. Knowing that the maximum allowable tensile stress in the glued splice is 75 psi, determine (*a*) the largest load **P** that can be safely supported, (*b*) the corresponding shearing stress in the splice.

1.33 A centric load **P** is applied to the granite block shown. Knowing that the resulting maximum value of the shearing stress in the block is 2.5 ksi, determine (*a*) the magnitude of **P**, (*b*) the orientation of the surface on which the maximum shearing stress occurs, (*c*) the normal stress exerted on that surface, (*d*) the maximum value of the normal stress in the block.

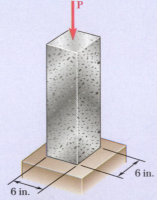

Fig. P1.33 and P1.34

1.34 A 240-kip load **P** is applied to the granite block shown. Determine the resulting maximum value of (*a*) the normal stress, (*b*) the shearing stress. Specify the orientation of the plane on which each of these maximum values occurs.

1.35 A steel pipe of 400-mm outer diameter is fabricated from 10-mm-thick plate by welding along a helix that forms an angle of 20° with a plane perpendicular to the axis of the pipe. Knowing that a 300-kN axial force **P** is applied to the pipe, determine the normal and shearing stresses in directions respectively normal and tangential to the weld.

1.36 A steel pipe of 400-mm outer diameter is fabricated from 10-mm-thick plate by welding along a helix that forms an angle of 20° with a plane perpendicular to the axis of the pipe. Knowing that the maximum allowable normal and shearing stresses in the directions respectively normal and tangential to the weld are $\sigma = 60$ MPa and $\tau = 36$ MPa, determine the magnitude P of the largest axial force that can be applied to the pipe.

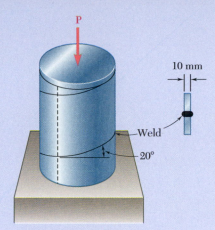

Fig. P1.35 and P1.36

1.37 A steel loop $ABCD$ of length 5 ft and of $\frac{3}{8}$-in. diameter is placed as shown around a 1-in.-diameter aluminum rod AC. Cables BE and DF, each of $\frac{1}{2}$-in. diameter, are used to apply the load **Q**. Knowing that the ultimate strength of the steel used for the loop and the cables is 70 ksi, and that the ultimate strength of the aluminum used for the rod is 38 ksi, determine the largest load **Q** that can be applied if an overall factor of safety of 3 is desired.

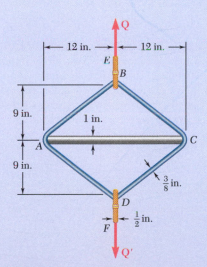

Fig. P1.37

1.38 Link BC is 6 mm thick, has a width $w = 25$ mm, and is made of a steel with a 480-MPa ultimate strength in tension. What is the factor of safety used if the structure shown was designed to support a 16-kN load **P**?

1.39 Link BC is 6 mm thick and is made of a steel with a 450-MPa ultimate strength in tension. What should be its width w if the structure shown is being designed to support a 20-kN load **P** with a factor of safety of 3?

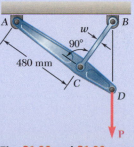

Fig. P1.38 and P1.39

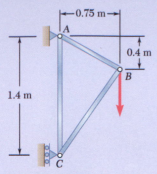

Fig. P1.40 and P1.41

1.40 Members *AB* and *BC* of the truss shown are made of the same alloy. It is known that a 20-mm-square bar of the same alloy was tested to failure and that an ultimate load of 120 kN was recorded. If a factor of safety of 3.2 is to be achieved for both bars, determine the required cross-sectional area of (*a*) bar *AB*, (*b*) bar *AC*.

1.41 Members *AB* and *BC* of the truss shown are made of the same alloy. It is known that a 20-mm-square bar of the same alloy was tested to failure and that an ultimate load of 120 kN was recorded. If bar *AB* has a cross-sectional area of 225 mm^2, determine (*a*) the factor of safety for bar *AB*, (*b*) the cross-sectional area of bar *AC* if it is to have the same factor of safety as bar *AB*.

1.42 Link *AB* is to be made of a steel for which the ultimate normal stress is 65 ksi. Determine the cross-sectional area of *AB* for which the factor of safety will be 3.20. Assume that the link will be adequately reinforced around the pins at *A* and *B*.

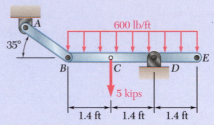

Fig. P1.42

1.43 Two wooden members are joined by plywood splice plates that are fully glued on the contact surfaces. Knowing that the clearance between the ends of the members is 6 mm and that the ultimate shearing stress in the glued joint is 2.5 MPa, determine the length *L* for which the factor of safety is 2.75 for the loading shown.

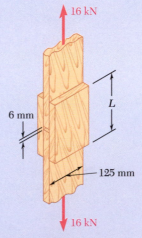

Fig. *P1.43*

1.44 For the joint and loading of Prob. 1.43, determine the factor of safety when *L* = 180 mm.

1.45 Three $\frac{3}{4}$-in.-diameter steel bolts are to be used to attach the steel plate shown to a wooden beam. Knowing that the plate will support a load $P = 24$ kips and that the ultimate shearing stress for the steel used is 52 ksi, determine the factor of safety for this design.

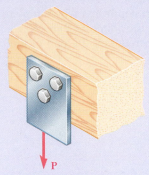

Fig. P1.45 and P1.46

1.46 Three steel bolts are to be used to attach the steel plate shown to a wooden beam. Knowing that the plate will support a load $P = 28$ kips, that the ultimate shearing stress for the steel used is 52 ksi, and that a factor of safety of 3.25 is desired, determine the required diameter of the bolts.

1.47 A load **P** is supported as shown by a steel pin that has been inserted in a short wooden member hanging from the ceiling. The ultimate strength of the wood used is 60 MPa in tension and 7.5 MPa in shear, while the ultimate strength of the steel is 145 MPa in shear. Knowing that $b = 40$ mm, $c = 55$ mm, and $d = 12$ mm, determine the load **P** if an overall factor of safety of 3.2 is desired.

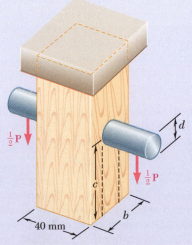

Fig. P1.47

1.48 For the support of Prob. 1.47, knowing that the diameter of the pin is $d = 16$ mm and that the magnitude of the load is $P = 20$ kN, determine (*a*) the factor of safety for the pin (*b*) the required values of *b* and *c* if the factor of safety for the wooden member is the same as that found in part *a* for the pin.

1.49 A steel plate $\frac{1}{4}$ in. thick is embedded in a concrete wall to anchor a high-strength cable as shown. The diameter of the hole in the plate is $\frac{3}{4}$ in., the ultimate strength of the steel used is 36 ksi. and the ultimate bonding stress between plate and concrete is 300 psi. Knowing that a factor of safety of 3.60 is desired when $P = 2.5$ kips, determine (a) the required width a of the plate, (b) the minimum depth b to which a plate of that width should be embedded in the concrete slab. (Neglect the normal stresses between the concrete and the end of the plate.)

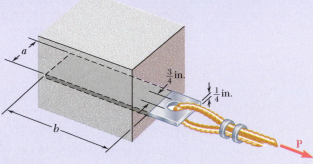

Fig. P1.49

1.50 Determine the factor of safety for the cable anchor in Prob. 1.49 when $P = 2.5$ kips, knowing that $a = 2$ in. and $b = 6$ in.

1.51 Link AC is made of a steel with a 65-ksi ultimate normal stress and has a $\frac{1}{4} \times \frac{1}{2}$-in. uniform rectangular cross section. It is connected to a support at A and to member BCD at C by $\frac{3}{4}$-in.-diameter pins, while member BCD is connected to its support at B by a $\frac{5}{16}$-in.-diameter pin. All of the pins are made of a steel with a 25-ksi ultimate shearing stress and are in single shear. Knowing that a factor of safety of 3.25 is desired, determine the largest load $\mathbf{P}$ that can be applied at D. Note that link AC is not reinforced around the pin holes.

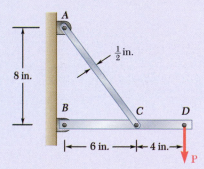

Fig. P1.51

1.52 Solve Prob. 1.51, assuming that the structure has been redesigned to use $\frac{5}{16}$-in.-diameter pins at A and C as well as at B and that no other changes have been made.

1.53 Each of the two vertical links CF connecting the two horizontal members AD and EG has a 10×40-mm uniform rectangular cross section and is made of a steel with an ultimate strength in tension of 400 MPa, while each of the pins at C and F has a 20-mm diameter and are made of a steel with an ultimate strength in shear of 150 MPa. Determine the overall factor of safety for the links CF and the pins connecting them to the horizontal members.

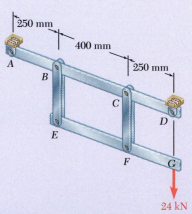

Fig. P1.53

1.54 Solve Prob. 1.53, assuming that the pins at C and F have been replaced by pins with a 30-mm diameter.

1.55 In the structure shown, an 8-mm-diameter pin is used at A, and 12-mm-diameter pins are used at B and D. Knowing that the ultimate shearing stress is 100 MPa at all connections and that the ultimate normal stress is 250 MPa in each of the two links joining B and D, determine the allowable load $\mathbf{P}$ if an overall factor of safety of 3.0 is desired.

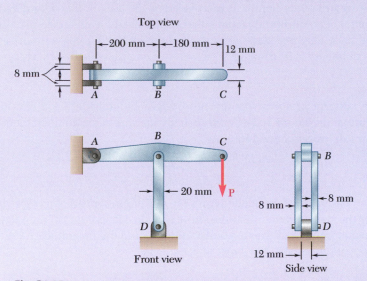

Fig. P1.55

1.56 In an alternative design for the structure of Prob. 1.55, a pin of 10-mm-diameter is to be used at A. Assuming that all other specifications remain unchanged, determine the allowable load $\mathbf{P}$ if an overall factor of safety of 3.0 is desired.

*1.57 A 40-kg platform is attached to the end B of a 50-kg wooden beam AB, which is supported as shown by a pin at A and by a slender steel rod BC with a 12-kN ultimate load. (*a*) Using the Load and Resistance Factor Design method with a resistance factor $\phi = 0.90$ and load factors $\gamma_D = 1.25$ and $\gamma_L = 1.6$, determine the largest load that can be safely placed on the platform. (*b*) What is the corresponding conventional factor of safety for rod BC?

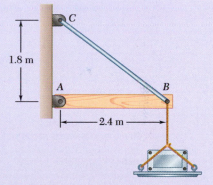

Fig. P1.57

*1.58 The Load and Resistance Factor Design method is to be used to select the two cables that will raise and lower a platform supporting two window washers. The platform weighs 160 lb and each of the window washers is assumed to weigh 195 lb with equipment. Since these workers are free to move on the platform, 75% of their total weight and the weight of their equipment will be used as the design live load of each cable. (*a*) Assuming a resistance factor $\phi = 0.85$ and load factors $\gamma_D = 1.2$ and $\gamma_L = 1.5$, determine the required minimum ultimate load of one cable. (*b*) What is the corresponding conventional factor of safety for the selected cables?

Fig. P1.58

Review and Summary

This chapter was devoted to the concept of stress and to an introduction to the methods used for the analysis and design of machines and load-bearing structures. Emphasis was placed on the use of a *free-body diagram* to obtain equilibrium equations that were solved for unknown reactions. Free-body diagrams were also used to find the internal forces in the various members of a structure.

Axial Loading: Normal Stress

The concept of *stress* was first introduced by considering a two-force member under an *axial loading*. The *normal stress* in that member (Fig. 1.41) was obtained by

$$\sigma = \frac{P}{A} \tag{1.5}$$

The value of σ obtained from Eq. (1.5) represents the *average stress* over the section rather than the stress at a specific point Q of the section. Considering a small area ΔA surrounding Q and the magnitude ΔF of the force exerted on ΔA, the stress at point Q is

$$\sigma = \lim_{\Delta A \to 0} \frac{\Delta F}{\Delta A} \tag{1.6}$$

In general, the stress σ at point Q in Eq. (1.6) is different from the value of the average stress given by Eq. (1.5) and is found to vary across the section. However, this variation is small in any section away from the points of application of the loads. Therefore, the distribution of the normal stresses in an axially loaded member is assumed to be *uniform,* except in the immediate vicinity of the points of application of the loads.

For the distribution of stresses to be uniform in a given section, the line of action of the loads **P** and **P'** must pass through the centroid C. Such a loading is called a *centric* axial loading. In the case of an *eccentric* axial loading, the distribution of stresses is *not* uniform.

Transverse Forces and Shearing Stress

When equal and opposite *transverse forces* **P** and **P'** of magnitude P are applied to a member AB (Fig. 1.42), *shearing stresses* τ are created over any section located between the points of application of the two forces.

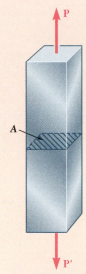

Fig. 1.41 Axially loaded member with cross section normal to member used to define normal stress.

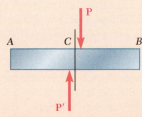

Fig. 1.42 Model of transverse resultant forces on either side of C resulting in shearing stress at section C.

These stresses vary greatly across the section and their distribution *cannot* be assumed to be uniform. However, dividing the magnitude *P*—referred to as the *shear* in the section—by the cross-sectional area *A*, the *average shearing stress* is:

$$\tau_{ave} = \frac{P}{A} \qquad (1.8)$$

Single and Double Shear

Shearing stresses are found in bolts, pins, or rivets connecting two structural members or machine components. For example, the shearing stress of bolt *CD* (Fig. 1.43), which is in *single shear,* is written as

$$\tau_{ave} = \frac{P}{A} = \frac{F}{A} \qquad (1.9)$$

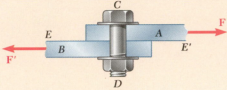

Fig. 1.43   Diagram of a single-shear joint.

The shearing stresses on bolts *EG* and *HJ* (Fig. 1.44), which are both in *double shear,* are written as

$$\tau_{ave} = \frac{P}{A} = \frac{F/2}{A} = \frac{F}{2A} \qquad (1.10)$$

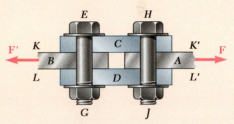

Fig. 1.44   Free-body diagram of a double-shear joint.

Bearing Stress

Bolts, pins, and rivets also create stresses in the members they connect along the *bearing surface* or surface of contact. Bolt *CD* of Fig. 1.43 creates stresses on the semicylindrical surface of plate *A* with which it is in contact (Fig. 1.45). Since the distribution of these stresses is quite complicated, one uses an average nominal value σ_b of the stress, called *bearing stress.*

$$\sigma_b = \frac{P}{A} = \frac{P}{td} \qquad (1.11)$$

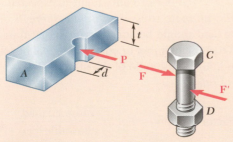

Fig. 1.45   Bearing stress from force *P* and the single-shear bolt associated with it.

Method of Solution

Your solution should begin with a clear and precise *statement* of the problem. Then draw one or several *free-body diagrams* that will be used to write *equilibrium equations.* These equations will be solved for *unknown forces,* from which the required *stresses* and *deformations* can be computed. Once the answer has been obtained, it should be *carefully checked.*

These guidelines are embodied by the SMART problem-solving methodology, where the steps of Strategy, Modeling, Analysis, and Reflect & Think are used. You are encouraged to apply this SMART methodology in the solution of all problems assigned from this text.

Stresses on an Oblique Section

When stresses are created on an *oblique section* in a two-force member under axial loading, both *normal* and *shearing* stresses occur. Denoting by θ the angle formed by the section with a normal plane (Fig. 1.46) and by A_0 the area of a section perpendicular to the axis of the member, the normal stress σ and the shearing stress τ on the oblique section are

Fig. 1.46 Axially loaded member with oblique section plane.

$$\sigma = \frac{P}{A_0} \cos^2 \theta \quad \tau = \frac{P}{A_0} \sin \theta \cos \theta \tag{1.14}$$

We observed from these formulas that the normal stress is maximum and equal to $\sigma_m = P/A_0$ for $\theta = 0$, while the shearing stress is maximum and equal to $\tau_m = P/2A_0$ for $\theta = 45°$. We also noted that $\tau = 0$ when $\theta = 0$, while $\sigma = P/2A_0$ when $\theta = 45°$.

Stress Under General Loading

Considering a small cube centered at Q (Fig. 1.47), σ_x is the normal stress exerted on a face of the cube perpendicular to the x axis, and τ_{xy} and τ_{xz} are the y and z components of the shearing stress exerted on the same face of the cube. Repeating this procedure for the other two faces of the cube and observing that $\tau_{xy} = \tau_{yx}$, $\tau_{yz} = \tau_{zy}$, and $\tau_{zx} = \tau_{xz}$, it was determined that *six stress components* are required to define the state of stress at a given point Q, being σ_x, σ_y, σ_z, τ_{xy}, τ_{yz}, and τ_{zx}.

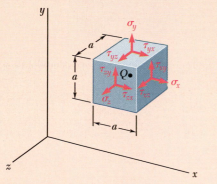

Fig. 1.47 Positive stress components at point Q.

Factor of Safety

The *ultimate load* of a given structural member or machine component is the load at which the member or component is expected to fail. This is computed from the *ultimate stress* or *ultimate strength* of the material used. The ultimate load should be considerably larger than the *allowable load* (i.e., the load that the member or component will be allowed to carry under normal conditions). The ratio of the ultimate load to the allowable load is the *factor of safety:*

$$\text{Factor of safety} = F.S. = \frac{\text{ultimate load}}{\text{allowable load}} \tag{1.25}$$

Load and Resistance Factor Design

Load and Resistance Factor Design (*LRFD*) allows the engineer to distinguish between the uncertainties associated with the structure and those associated with the load.

Review Problems

1.59 In the marine crane shown, link *CD* is known to have a uniform cross section of 50 × 150 mm. For the loading shown, determine the normal stress in the central portion of that link.

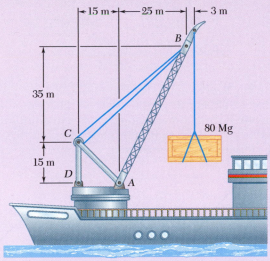

Fig. P1.59

1.60 Two horizontal 5-kip forces are applied to pin *B* of the assembly shown. Knowing that a pin of 0.8-in. diameter is used at each connection, determine the maximum value of the average normal stress (*a*) in link *AB*, (*b*) in link *BC*.

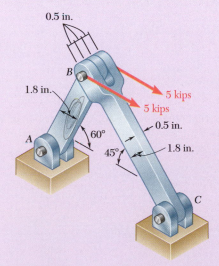

Fig. P1.60

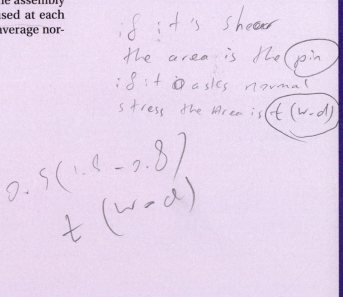

1.61 For the assembly and loading of Prob. 1.60, determine (*a*) the average shearing stress in the pin at *C*, (*b*) the average bearing stress at *C* in member *BC*, (*c*) the average bearing stress at *B* in member *BC*.

1.62 Two steel plates are to be held together by means of 16-mm-diameter high-strength steel bolts fitting snugly inside cylindrical brass spacers. Knowing that the average normal stress must not exceed 200 MPa in the bolts and 130 MPa in the spacers, determine the outer diameter of the spacers that yields the most economical and safe design.

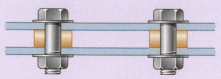

Fig. P1.62

1.63 A couple **M** of magnitude 1500 N · m is applied to the crank of an engine. For the position shown, determine (*a*) the force **P** required to hold the engine system in equilibrium, (*b*) the average normal stress in the connecting rod *BC*, which has a 450-mm² uniform cross section.

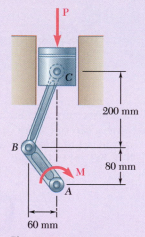

Fig. *P1.63*

1.64 Knowing that link *DE* is $\frac{1}{8}$ in. thick and 1 in. wide, determine the normal stress in the central portion of that link when (*a*) $\theta = 0$, (*b*) $\theta = 90°$.

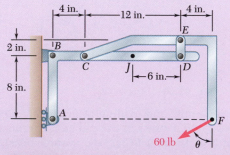

Fig. P1.64

1.65 A $\frac{5}{8}$-in.-diameter steel rod AB is fitted to a round hole near end C of the wooden member CD. For the loading shown, determine (a) the maximum average normal stress in the wood, (b) the distance b for which the average shearing stress is 100 psi on the surfaces indicated by the dashed lines, (c) the average bearing stress on the wood.

1.66 In the steel structure shown, a 6-mm-diameter pin is used at C and 10-mm-diameter pins are used at B and D. The ultimate shearing stress is 150 MPa at all connections, and the ultimate normal stress is 400 MPa in link BD. Knowing that a factor of safety of 3.0 is desired, determine the largest load **P** that can be applied at A. Note that link BD is not reinforced around the pin holes.

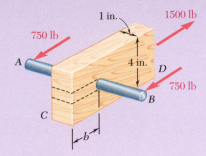

Fig. P1.65

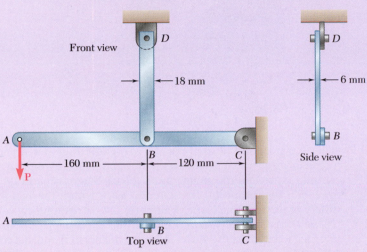

Fig. P1.66

1.67 Member ABC, which is supported by a pin and bracket at C and a cable BD, was designed to support the 16-kN load **P** as shown. Knowing that the ultimate load for cable BD is 100 kN, determine the factor of safety with respect to cable failure.

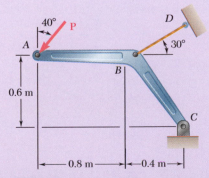

Fig. P1.67

1.68 A force **P** is applied as shown to a steel reinforcing bar that has been embedded in a block of concrete. Determine the smallest length L for which the full allowable normal stress in the bar can be developed. Express the result in terms of the diameter d of the bar, the allowable normal stress σ_{all} in the steel, and the average allowable bond stress τ_{all} between the concrete and the cylindrical surface of the bar. (Neglect the normal stresses between the concrete and the end of the bar.)

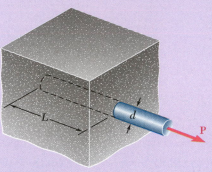

Fig. P1.68

1.69 The two portions of member AB are glued together along a plane forming an angle θ with the horizontal. Knowing that the ultimate stress for the glued joint is 2.5 ksi in tension and 1.3 ksi in shear, determine (a) the value of θ for which the factor of safety of the member is maximum, (b) the corresponding value of the factor of safety. (*Hint:* Equate the expressions obtained for the factors of safety with respect to the normal and shearing stresses.)

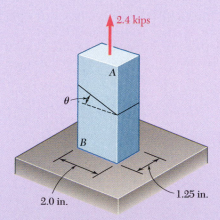

Fig. P1.69 and P1.70

1.70 The two portions of member AB are glued together along a plane forming an angle θ with the horizontal. Knowing that the ultimate stress for the glued joint is 2.5 ksi in tension and 1.3 ksi in shear, determine the range of values of θ for which the factor of safety of the members is at least 3.0.

Computer Problems

The following problems are designed to be solved with a computer.

1.C1 A solid steel rod consisting of n cylindrical elements welded together is subjected to the loading shown. The diameter of element i is denoted by d_i and the load applied to its lower end by $\mathbf{P}_i$, with the magnitude P_i of this load being assumed positive if $\mathbf{P}_i$ is directed downward as shown and negative otherwise. (*a*) Write a computer program that can be used with either SI or U.S. customary units to determine the average stress in each element of the rod. (*b*) Use this program to solve Probs. 1.1 and 1.3.

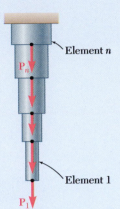

Fig. P1.C1

1.C2 A 20-kN load is applied as shown to the horizontal member ABC. Member ABC has a 10 × 50-mm uniform rectangular cross section and is supported by four vertical links, each of 8 × 36-mm uniform rectangular cross section. Each of the four pins at A, B, C, and D has the same diameter d and is in double shear. (*a*) Write a computer program to calculate for values of d from 10 to 30 mm, using 1-mm increments, (i) the maximum value of the average normal stress in the links connecting pins B and D, (ii) the average normal stress in the links connecting pins C and E, (iii) the average shearing stress in pin B, (iv) the average shearing stress in pin C, (v) the average bearing stress at B in member ABC, and (vi) the average bearing stress at C in member ABC. (*b*) Check your program by comparing the values obtained for $d = 16$ mm with the answers given for Probs. 1.7 and 1.27. (*c*) Use this program to find the permissible values of the diameter d of the pins, knowing that the allowable values of the normal, shearing, and bearing stresses for the steel used are, respectively, 150 MPa, 90 MPa, and 230 MPa. (*d*) Solve part c, assuming that the thickness of member ABC has been reduced from 10 to 8 mm.

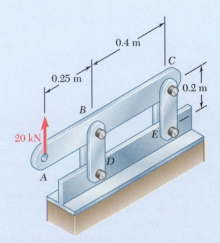

Fig. P1.C2

1.C3 Two horizontal 5-kip forces are applied to pin *B* of the assembly shown. Each of the three pins at *A*, *B*, and *C* has the same diameter *d* and is in double shear. (*a*) Write a computer program to calculate for values of *d* from 0.50 to 1.50 in., using 0.05-in. increments, (i) the maximum value of the average normal stress in member *AB*, (ii) the average normal stress in member *BC*, (iii) the average shearing stress in pin *A*, (iv) the average shearing stress in pin *C*, (v) the average bearing stress at *A* in member *AB*, (vi) the average bearing stress at *C* in member *BC*, and (vii) the average bearing stress at *B* in member *BC*. (*b*) Check your program by comparing the values obtained for *d* = 0.8 in. with the answers given for Probs. 1.60 and 1.61. (*c*) Use this program to find the permissible values of the diameter *d* of the pins, knowing that the allowable values of the normal, shearing, and bearing stresses for the steel used are, respectively, 22 ksi, 13 ksi, and 36 ksi. (*d*) Solve part *c*, assuming that a new design is being investigated in which the thickness and width of the two members are changed, respectively, from 0.5 to 0.3 in. and from 1.8 to 2.4 in.

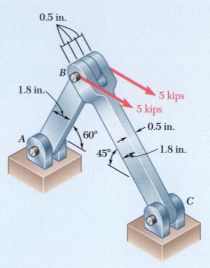

0.5 in.

B

5 kips

1.8 in.

5 kips

A   60°   0.5 in.

45°   1.8 in.

C

Fig. P1.C3

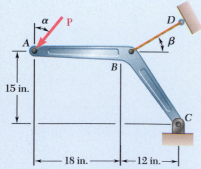

α   P

A

15 in.

B

D

β

C

18 in.   12 in.

Fig. P1.C4

1.C4 A 4-kip force **P** forming an angle α with the vertical is applied as shown to member *ABC*, which is supported by a pin and bracket at *C* and by a cable *BD* forming an angle β with the horizontal. (*a*) Knowing that the ultimate load of the cable is 25 kips, write a computer program to construct a table of the values of the factor of safety of the cable for values of α and β from 0 to 45°, using increments in α and β corresponding to 0.1 increments in tan α and tan β. (*b*) Check that for any given value of α, the maximum value of the factor of safety is obtained for β = 38.66° and explain why. (*c*) Determine the smallest possible value of the factor of safety for β = 38.66°, as well as the corresponding value of α, and explain the result obtained.

1.C5 A load **P** is supported as shown by two wooden members of uniform rectangular cross section that are joined by a simple glued scarf splice. (*a*) Denoting by σ_U and τ_U, respectively, the ultimate strength of the joint in tension and in shear, write a computer program which, for given values of *a*, *b*, *P*, σ_U and τ_U, expressed in either SI or U.S. customary units, and for values of α from 5 to 85° at 5° intervals, can calculate (i) the normal stress in the joint, (ii) the shearing stress in the joint, (iii) the factor of safety relative to failure in tension, (iv) the factor of safety relative to failure in shear, and (v) the overall factor of safety for the glued joint. (*b*) Apply this program, using the dimensions and loading of the members of Probs. 1.29 and 1.31, knowing that $\sigma_U = 150$ psi and $\tau_U = 214$ psi for the glue used in Prob. 1.29 and that $\sigma_U = 1.26$ MPa and $\tau_U = 1.50$ MPa for the glue used in Prob. 1.31. (*c*) Verify in each of these two cases that the shearing stress is maximum for $\alpha = 45°$.

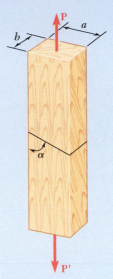

Fig. P1.C5

1.C6 Member *ABC* is supported by a pin and bracket at *A*, and by two links that are pin-connected to the member at *B* and to a fixed support at *D*. (*a*) Write a computer program to calculate the allowable load P_{all} for any given values of (i) the diameter d_1 of the pin at *A*, (ii) the common diameter d_2 of the pins at *B* and *D*, (iii) the ultimate normal stress σ_U in each of the two links, (iv) the ultimate shearing stress τ_U in each of the three pins, and (v) the desired overall factor of safety *F.S.* (*b*) Your program should also indicate which of the following three stresses is critical: the normal stress in the links, the shearing stress in the pin at *A*, or the shearing stress in the pins at *B* and *D*. (*c*) Check your program by using the data of Probs. 1.55 and 1.56, respectively, and comparing the answers obtained for P_{all} with those given in the text. (*d*) Use your program to determine the allowable load P_{all}, as well as which of the stresses is critical, when $d_1 = d_2 = 15$ mm, $\sigma_U = 110$ MPa for aluminum links, $\tau_U = 100$ MPa for steel pins, and *F.S.* = 3.2.

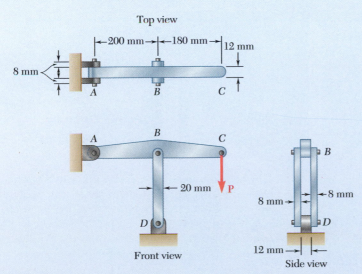

Fig. P1.C6

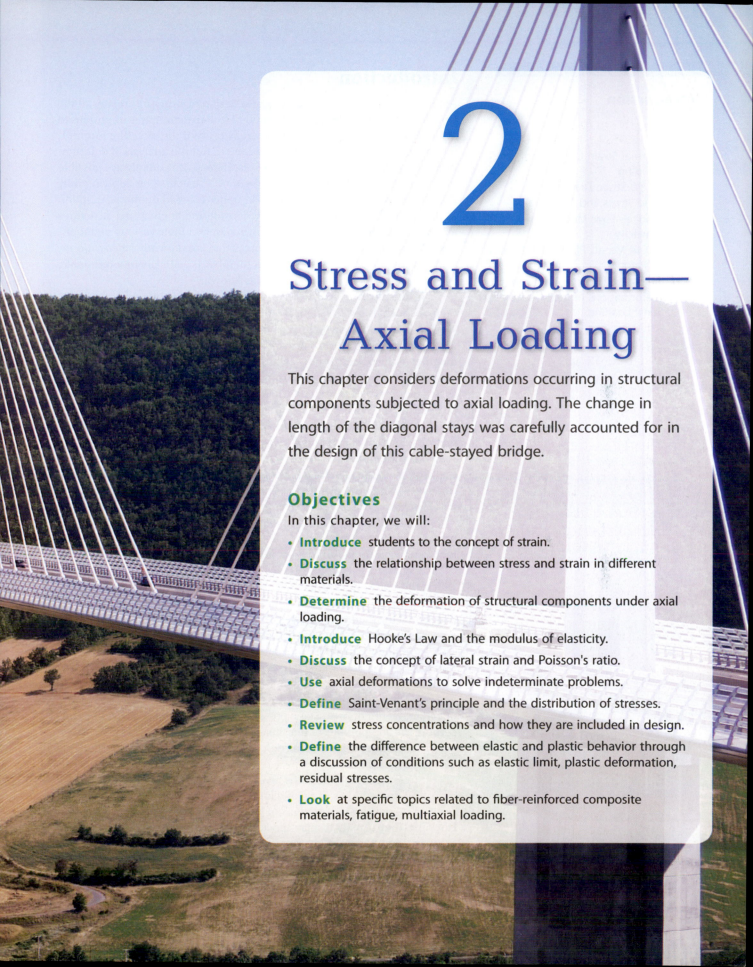

2

Stress and Strain—Axial Loading

This chapter considers deformations occurring in structural components subjected to axial loading. The change in length of the diagonal stays was carefully accounted for in the design of this cable-stayed bridge.

Objectives

In this chapter, we will:

- **Introduce** students to the concept of strain.

- **Discuss** the relationship between stress and strain in different materials.

- **Determine** the deformation of structural components under axial loading.

- **Introduce** Hooke's Law and the modulus of elasticity.

- **Discuss** the concept of lateral strain and Poisson's ratio.

- **Use** axial deformations to solve indeterminate problems.

- **Define** Saint-Venant's principle and the distribution of stresses.

- **Review** stress concentrations and how they are included in design.

- **Define** the difference between elastic and plastic behavior through a discussion of conditions such as elastic limit, plastic deformation, residual stresses.

- **Look** at specific topics related to fiber-reinforced composite materials, fatigue, multiaxial loading.

Introduction

An important aspect of the analysis and design of structures relates to the *deformations* caused by the loads applied to a structure. It is important to avoid deformations so large that they may prevent the structure from fulfilling the purpose for which it was intended. But the analysis of deformations also helps us to determine stresses. Indeed, it is not always possible to determine the forces in the members of a structure by applying only the principles of statics. This is because statics is based on the assumption of undeformable, rigid structures. By considering engineering structures as *deformable* and analyzing the deformations in their various members, it will be possible for us to compute forces that are *statically indeterminate*. The distribution of stresses in a given member is statically indeterminate, even when the force in that member is known.

In this chapter, you will consider the deformations of a structural member such as a rod, bar, or plate under *axial loading*. First, the *normal strain ϵ* in a member is defined as the *deformation of the member per unit length*. Plotting the stress σ versus the strain ε as the load applied to the member is increased produces a *stress-strain diagram* for the material used. From this diagram, some important properties of the material, such as its *modulus of elasticity*, and whether the material is *ductile* or *brittle* can be determined. While the behavior of most materials is independent of the direction of the load application, you will see that the response of fiber-reinforced composite materials depends upon the direction of the load.

From the stress-strain diagram, you also can determine whether the strains in the specimen will disappear after the load has been removed—when the material is said to behave *elastically*—or whether *a permanent set or plastic deformation* will result.

You will examine the phenomenon of *fatigue*, which causes structural or machine components to fail after a very large number of repeated loadings, even though the stresses remain in the elastic range.

Sections 2.2 and 2.3 discuss *statically indeterminate problems* in which the reactions and the internal forces *cannot* be determined from statics alone. Here the equilibrium equations derived from the free-body diagram of the member must be complemented by relationships involving deformations that are obtained from the geometry of the problem.

Additional constants associated with isotropic materials—i.e., materials with mechanical characteristics independent of direction—are introduced in Secs. 2.4 through 2.8. They include *Poisson's ratio*, relating lateral and axial strain, the *bulk modulus*, characterizing the change in volume of a material under hydrostatic pressure, and the *modulus of rigidity*, concerning the components of the shearing stress and shearing strain. Stress-strain relationships for an isotropic material under a multiaxial loading also are determined.

Stress-strain relationships involving modulus of elasticity, Poisson's ratio, and the modulus of rigidity are developed for fiber-reinforced composite materials under a multiaxial loading. While these materials are not isotropic, they usually display special *orthotropic* properties.

In Chap. 1, stresses were assumed uniformly distributed in any given cross section; they were also assumed to remain within the elastic range. The first assumption is discussed in Sec. 2.10, while *stress concentrations* near circular holes and fillets in flat bars are considered in Sec. 2.11.

Sections 2.12 and 2.13 discuss stresses and deformations in members made of a ductile material when the yield point of the material is exceeded, resulting in permanent *plastic deformations* and *residual stresses*.

2.1 AN INTRODUCTION TO STRESS AND STRAIN

2.1A Normal Strain Under Axial Loading

Consider a rod *BC* of length *L* and uniform cross-sectional area *A*, which is suspended from *B* (Fig. 2.1*a*). If you apply a load **P** to end *C*, the rod elongates (Fig. 2.1*b*). Plotting the magnitude *P* of the load against the deformation δ (Greek letter delta), you obtain a load-deformation diagram (Fig. 2.2). While this diagram contains information useful to the analysis of the rod under consideration, it cannot be used to predict the deformation of a rod of the same material but with different dimensions. Indeed, if a deformation δ is produced in rod *BC* by a load **P**, a load 2**P** is required to cause the same deformation in rod *B'C'* of the same length *L* but cross-sectional area 2*A* (Fig. 2.3). Note that in both cases the value of the stress is the same: $\sigma = P/A$. On the other hand, when load **P** is applied to a rod *B"C"* of the same cross-sectional area *A* but of length 2*L*, a deformation 2δ occurs in that rod (Fig. 2.4). This is a deformation twice as large as the deformation δ produced in rod *BC*. In both cases, the ratio of the deformation over the length of the rod is the same at δ/L. This introduces the concept of *strain*. We define the *normal strain* in a rod under axial loading as the *deformation per unit length* of that rod. The normal strain, ϵ (Greek letter epsilon), is

$$\epsilon = \frac{\delta}{L} \tag{2.1}$$

Plotting the stress $\sigma = P/A$ against the strain $\epsilon = \delta/L$ results in a curve that is characteristic of the properties of the material but does not depend upon the dimensions of the specimen used. This curve is called a *stress-strain diagram*.

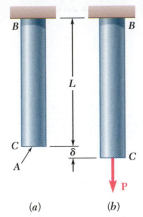

Fig. 2.1 Undeformed and deformed axially-loaded rod.

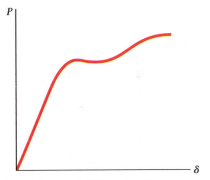

Fig. 2.2 Load-deformation diagram.

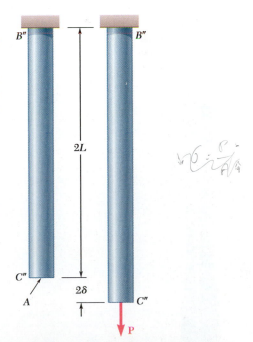

Fig. 2.4 The deformation is doubled when the rod length is doubled while keeping the load **P** and cross-sectional area *A* the same.

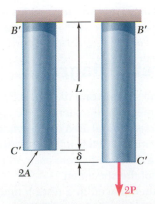

Fig. 2.3 Twice the load is required to obtain the same deformation δ when the cross-sectional area is doubled.

epsilon

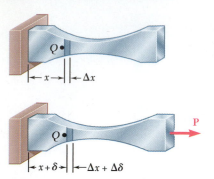

Fig. 2.5 Deformation of axially-loaded member of variable cross-sectional area.

Since rod *BC* in Fig. 2.1 has a uniform cross section of area *A*, the normal stress σ is assumed to have a constant value *P/A* throughout the rod. The strain ε is the ratio of the total deformation δ over the total length *L* of the rod. It too is consistent throughout the rod. However, for a member of variable cross-sectional area *A*, the normal stress $\sigma = P/A$ varies along the member, and it is necessary to define the strain at a given point *Q* by considering a small element of undeformed length Δx (Fig. 2.5). Denoting the deformation of the element under the given loading by $\Delta\delta$, the *normal strain at point Q* is defined as

$$\epsilon = \lim_{\Delta x \to 0} \frac{\Delta\delta}{\Delta x} = \frac{d\delta}{dx} \tag{2.2}$$

Since deformation and length are expressed in the same units, the normal strain ϵ obtained by dividing δ by *L* (or $d\delta$ by dx) is a *dimensionless quantity*. Thus, the same value is obtained for the normal strain, whether SI metric units or U.S. customary units are used. For instance, consider a bar of length *L* = 0.600 m and uniform cross section that undergoes a deformation $\delta = 150 \times 10^{-6}$ m. The corresponding strain is

$$\epsilon = \frac{\delta}{L} = \frac{150 \times 10^{-6}\,\text{m}}{0.600\,\text{m}} = 250 \times 10^{-6}\,\text{m/m} = 250 \times 10^{-6}$$

Note that the deformation also can be expressed in micrometers: $\delta = 150\,\mu$m and the answer written in micros (μ):

$$\epsilon = \frac{\delta}{L} = \frac{150\,\mu\text{m}}{0.600\,\text{m}} = 250\,\mu\text{m/m} = 250\,\mu$$

When U.S. customary units are used, the length and deformation of the same bar are *L* = 23.6 in. and $\delta = 5.91 \times 10^{-3}$ in. The corresponding strain is

$$\epsilon = \frac{\delta}{L} = \frac{5.91 \times 10^{-3}\,\text{in.}}{23.6\,\text{in.}} = 250 \times 10^{-6}\,\text{in./in.}$$

which is the same value found using SI units. However, when lengths and deformations are expressed in inches or microinches (μin.), keep the original units obtained for the strain. Thus, in the previous example, the strain would be recorded as either $\epsilon = 250 \times 10^{-6}$ in./in. or $\epsilon = 250\,\mu$in./in.

2.1B Stress-Strain Diagram

Tensile Test. To obtain the stress-strain diagram of a material, a *tensile test* is conducted on a specimen of the material. One type of specimen is shown in Photo 2.1. The cross-sectional area of the cylindrical central portion of the specimen is accurately determined and two gage marks are inscribed on that portion at a distance L_0 from each other. The distance L_0 is known as the *gage length* of the specimen.

The test specimen is then placed in a testing machine (Photo 2.2), which is used to apply a centric load **P**. As load **P** increases, the distance *L* between the two gage marks also increases (Photo 2.3). The distance *L* is measured with a dial gage, and the elongation $\delta = L - L_0$ is recorded

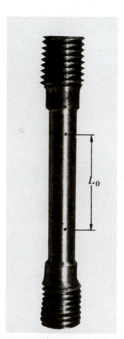

Photo 2.1 Typical tensile-test specimen. Undeformed gage length is L_0.

for each value of P. A second dial gage is often used simultaneously to measure and record the change in diameter of the specimen. From each pair of readings P and δ, the engineering stress σ is

$$\sigma = \frac{P}{A_0} \qquad (2.3)$$

and the engineering strain ε is

$$\epsilon = \frac{\delta}{L_0} \qquad (2.4)$$

The stress-strain diagram can be obtained by plotting ε as an abscissa and σ as an ordinate.

Stress-strain diagrams of materials vary widely, and different tensile tests conducted on the same material may yield different results, depending upon the temperature of the specimen and the speed of loading. However, some common characteristics can be distinguished from stress-strain diagrams to divide materials into two broad categories: *ductile* and *brittle* materials.

Ductile materials, including structural steel and many alloys of other materials are characterized by their ability to *yield* at normal temperatures. As the specimen is subjected to an increasing load, its length first increases linearly with the load and at a very slow rate. Thus, the initial portion of the stress-strain diagram is a straight line with a steep slope

Photo 2.2 Universal test machine used to test tensile specimens.

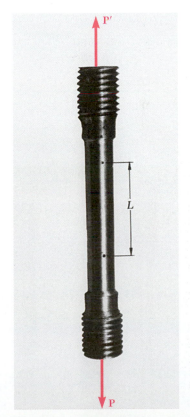

Photo 2.3 Elongated tensile test specimen having load P and deformed length $L > L_0$.

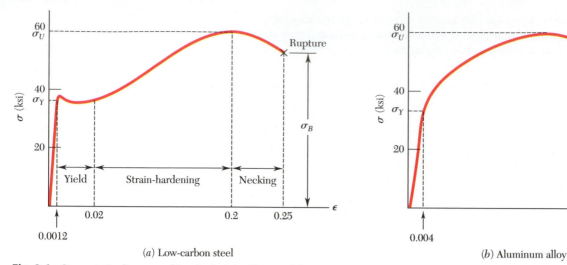

(a) Low-carbon steel

(b) Aluminum alloy

Fig. 2.6   Stress-strain diagrams of two typical ductile materials.

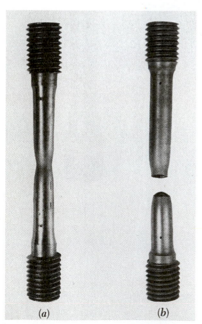

Photo 2.4   Ductile material tested specimens:
(a) with cross-section necking, (b) ruptured.

(Fig. 2.6). However, after a critical value σ_Y of the stress has been reached, the specimen undergoes a large deformation with a relatively small increase in the applied load. This deformation is caused by slippage along oblique surfaces and is due primarily to shearing stresses. After a maximum value of the load has been reached, the diameter of a portion of the specimen begins to decrease, due to local instability (Photo 2.4a). This phenomenon is known as *necking.* After necking has begun, lower loads are sufficient for specimen to elongate further, until it finally ruptures (Photo 2.4b). Note that rupture occurs along a cone-shaped surface that forms an angle of approximately 45° with the original surface of the specimen. This indicates that shear is primarily responsible for the failure of ductile materials, confirming the fact that shearing stresses under an axial load are largest on surfaces forming an angle of 45° with the load (see Sec. 1.3). Note from Fig. 2.6 that the elongation of a ductile specimen after it has ruptured can be 200 times as large as its deformation at yield. The stress σ_Y at which yield is initiated is called the *yield strength* of the material. The stress σ_U corresponding to the maximum load applied is known as the *ultimate strength.* The stress σ_B corresponding to rupture is called the *breaking strength.*

Brittle materials, comprising of cast iron, glass, and stone rupture without any noticeable prior change in the rate of elongation (Fig. 2.7). Thus, for brittle materials, there is no difference between the ultimate strength and the breaking strength. Also, the strain at the time of rupture is much smaller for brittle than for ductile materials. Note the absence of any necking of the specimen in the brittle material of Photo 2.5 and observe that rupture occurs along a surface perpendicular to the load. Thus, normal stresses are primarily responsible for the failure of brittle materials.[†]

[†]The tensile tests described in this section were assumed to be conducted at normal temperatures. However, a material that is ductile at normal temperatures may display the characteristics of a brittle material at very low temperatures, while a normally brittle material may behave in a ductile fashion at very high temperatures. At temperatures other than normal, therefore, one should refer to *a material in a ductile state* or to *a material in a brittle state,* rather than to a ductile or brittle material.

The stress-strain diagrams of Fig. 2.6 show that while structural steel and aluminum are both ductile, they have different yield characteristics. For structural steel (Fig. 2.6*a*), the stress remains constant over a large range of the strain after the onset of yield. Later, the stress must be increased to keep elongating the specimen until the maximum value σ_U has been reached. This is due to a property of the material known as *strain-hardening*. The *yield strength* of structural steel is determined during the tensile test by watching the load shown on the display of the testing machine. After increasing steadily, the load will suddenly drop to a slightly lower value, which is maintained for a certain period as the specimen keeps elongating. In a very carefully conducted test, one may be able to distinguish between the *upper yield point*, which corresponds to the load reached just before yield starts, and the *lower yield point*, which corresponds to the load required to maintain yield. Since the upper yield point is transient, the lower yield point is used to determine the yield strength of the material.

For aluminum (Fig. 2.6*b*) and of many other ductile materials, the stress keeps increasing—although not linearly—until the ultimate strength is reached. Necking then begins and eventually ruptures. For such materials, the yield strength σ_Y can be determined using the offset method. For example the yield strength at 0.2% offset is obtained by drawing through the point of the horizontal axis of abscissa $\epsilon = 0.2\%$ (or $\epsilon = 0.002$), which is a line parallel to the initial straight-line portion of the stress-strain diagram (Fig. 2.8). The stress σ_Y corresponding to the point Y is defined as the yield strength at 0.2% offset.

A standard measure of the ductility of a material is its *percent elongation*:

$$\text{Percent elongation} = 100\,\frac{L_B - L_0}{L_0}$$

where L_0 and L_B are the initial length of the tensile test specimen and its final length at rupture, respectively. The specified minimum elongation for a 2-in. gage length for commonly used steels with yield strengths up to 50 ksi is 21 percent. This means that the average strain at rupture should be at least 0.21 in./in.

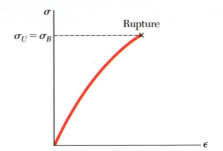

Fig. 2.7 Stress-strain diagram for a typical brittle material.

Photo 2.5 Ruptured brittle material specimen.

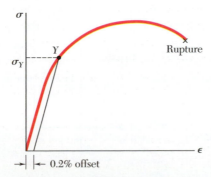

Fig. 2.8 Determination of yield strength by 0.2% offset method.

Another measure of ductility that is sometimes used is the *percent reduction in area*:

$$\text{Percent reduction in area} = 100 \frac{A_0 - A_B}{A_0}$$

where A_0 and A_B are the initial cross-sectional area of the specimen and its minimum cross-sectional area at rupture, respectively. For structural steel, percent reductions in area of 60 to 70 percent are common.

Compression Test.   If a specimen made of a ductile material is loaded in compression instead of tension, the stress-strain curve is essentially the same through its initial straight-line portion and through the beginning of the portion corresponding to yield and strain-hardening. Particularly noteworthy is the fact that for a given steel, the yield strength is the same in both tension and compression. For larger values of the strain, the tension and compression stress-strain curves diverge, and necking does not occur in compression. For most brittle materials, the ultimate strength in compression is much larger than in tension. This is due to the presence of flaws, such as microscopic cracks or cavities that tend to weaken the material in tension, while not appreciably affecting its resistance to compressive failure.

An example of brittle material with different properties in tension and compression is provided by *concrete*, whose stress-strain diagram is shown in Fig. 2.9. On the tension side of the diagram, we first observe a linear elastic range in which the strain is proportional to the stress. After the yield point has been reached, the strain increases faster than the stress until rupture occurs. The behavior of the material in compression is different. First, the linear elastic range is significantly larger. Second, rupture does not occur as the stress reaches its maximum value. Instead, the stress decreases in magnitude while the strain keeps increasing until rupture occurs. Note that the modulus of elasticity, which is represented by the slope of the stress-strain curve in its linear portion, is the same in tension and compression. This is true of most brittle materials.

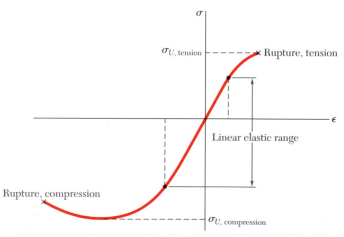

Fig. 2.9   Stress-strain diagram for concrete shows difference in tensile and compression response.

*2.1C   True Stress and True Strain

Recall that the stress plotted in Figs. 2.6 and 2.7 was obtained by dividing the load P by the cross-sectional area A_0 of the specimen measured before any deformation had taken place. Since the cross-sectional area of the specimen decreases as P increases, the stress plotted in these diagrams does not represent the actual stress in the specimen. The difference between the *engineering stress* $\sigma = P/A_0$ and the *true stress* $\sigma_t = P/A$ becomes apparent in ductile materials after yield has started. While the engineering stress σ, which is directly proportional to the load P, decreases with P during the necking phase, the true stress σ_t, which is proportional to P but also inversely proportional to A, keeps increasing until rupture of the specimen occurs.

For *engineering strain* $\epsilon = \delta/L_0$, instead of using the total elongation δ and the original value L_0 of the gage length, many scientists use all of the values of L that they have recorded. Dividing each increment ΔL of the distance between the gage marks by the corresponding value of L, the elementary strain $\Delta \epsilon = \Delta L/L$. Adding the successive values of $\Delta \epsilon$, the *true strain* ϵ_t is

$$\epsilon_t = \Sigma \Delta \epsilon = \Sigma (\Delta L/L)$$

With the summation replaced by an integral, the true strain can be expressed as:

$$\epsilon_t = \int_{L_0}^{L} \frac{dL}{L} = \ln \frac{L}{L_0} \qquad \textbf{(2.5)}$$

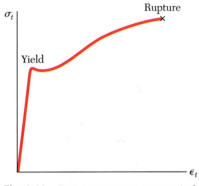

Fig. 2.10   True stress versus true strain for a typical ductile material.

Plotting true stress versus true strain (Fig. 2.10) more accurately reflects the behavior of the material. As already noted, there is no decrease in true stress during the necking phase. Also, the results obtained from either tensile or compressive tests yield essentially the same plot when true stress and true strain are used. This is not the case for large values of the strain when the engineering stress is plotted versus the engineering strain. However, in order to determine whether a load P will produce an acceptable stress and an acceptable deformation in a given member, engineers will use a diagram based on Eqs. (2.3) and (2.4) since these involve the cross-sectional area A_0 and the length L_0 of the member in its undeformed state, which are easily available.

2.1D   Hooke's Law; Modulus of Elasticity

Modulus of Elasticity.   Most engineering structures are designed to undergo relatively small deformations, involving only the straight-line portion of the corresponding stress-strain diagram. For that initial portion of the diagram (Fig. 2.6), the stress σ is directly proportional to the strain ϵ:

$$\sigma = E\epsilon \qquad \textbf{(2.6)}$$

This is known as *Hooke's law*, after Robert Hooke (1635–1703), an English scientist and one of the early founders of applied mechanics. The coefficient E of the material is the *modulus of elasticity* or *Young's modulus*, after the English scientist Thomas Young (1773–1829). Since the strain ϵ is a dimensionless quantity, E is expressed in the same units as stress σ—in pascals or one of its multiples for SI units and in psi or ksi for U.S. customary units.

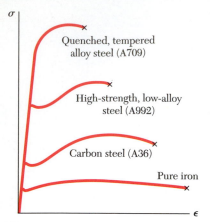

Fig. 2.11 Stress-strain diagrams for iron and different grades of steel.

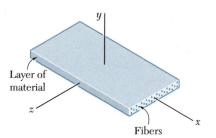

Fig. 2.12 Layer of fiber-reinforced composite material.

The largest value of stress for which Hooke's law can be used for a given material is the *proportional limit* of that material. For ductile materials possessing a well-defined yield point, as in Fig. 2.6a, the proportional limit almost coincides with the yield point. For other materials, the proportional limit cannot be determined as easily, since it is difficult to accurately determine the stress σ for which the relation between σ and ϵ ceases to be linear. For such materials, however, using Hooke's law for values of the stress slightly larger than the actual proportional limit will not result in any significant error.

Some physical properties of structural metals, such as strength, ductility, and corrosion resistance, can be greatly affected by alloying, heat treatment, and the manufacturing process used. For example, the stress-strain diagrams of pure iron and three different grades of steel (Fig. 2.11) show that large variations in the yield strength, ultimate strength, and final strain (ductility) exist. All of these metals possess the same modulus of elasticity—their "stiffness," or ability to resist a deformation within the linear range is the same. Therefore, if a high-strength steel is substituted for a lower-strength steel and if all dimensions are kept the same, the structure will have an increased load-carrying capacity, but its stiffness will remain unchanged.

For the materials considered so far, the relationship between normal stress and normal strain, $\sigma = E\epsilon$, is independent of the direction of loading. This is because the mechanical properties of each material, including its modulus of elasticity E, are independent of the direction considered. Such materials are said to be *isotropic*. Materials whose properties depend upon the direction considered are said to be *anisotropic*.

Fiber-Reinforced Composite Materials. An important class of anisotropic materials consists of *fiber-reinforced composite materials*. These are obtained by embedding fibers of a strong, stiff material into a weaker, softer material, called a *matrix*. Typical materials used as fibers are graphite, glass, and polymers, while various types of resins are used as a matrix. Figure 2.12 shows a layer, or *lamina*, of a composite material consisting of a large number of parallel fibers embedded in a matrix. An axial load applied to the lamina along the x axis, (in a direction parallel to the fibers) will create a normal stress σ_x in the lamina and a corresponding normal strain ϵ_x, satisfying Hooke's law as the load is increased and as long as the elastic limit of the lamina is not exceeded. Similarly, an axial load applied along the y axis, (in a direction perpendicular to the lamina) will create a normal stress σ_y and a normal strain ϵ_y, and an axial load applied along the z axis will create a normal stress σ_z and a normal strain ϵ_z, all satisfy Hooke's law. However, the moduli of elasticity E_x, E_y, and E_z corresponding, to each of these loadings will be different. Because the fibers are parallel to the x axis, the lamina will offer a much stronger resistance to a load directed along the x axis than to one directed along the y or z axis, and E_x will be much larger than either E_y or E_z.

A flat *laminate* is obtained by superposing a number of layers or laminas. If the laminate is subjected only to an axial load causing tension, the fibers in all layers should have the same orientation as the load in order to obtain the greatest possible strength. But if the laminate is in compression, the matrix material may not be strong enough to prevent the fibers from kinking or buckling. The lateral stability of the laminate can be increased by positioning some of the layers so that their fibers are perpendicular to the load. Positioning some layers so that their fibers are oriented at 30°, 45°, or 60° to the load also can be used to increase the resistance of the laminate to in-plane shear. Fiber-reinforced composite materials will be further discussed in Sec. 2.9, where their behavior under multiaxial loadings will be considered.

2.1E   Elastic Versus Plastic Behavior of a Material

Material behaves *elastically* if the strains in a test specimen from a given load disappear when the load is removed. The largest value of stress causing this elastic behavior is called the *elastic limit* of the material.

If the material has a well-defined yield point as in Fig. 2.6*a*, the elastic limit, the proportional limit, and the yield point are essentially equal. In other words, the material behaves elastically and linearly as long as the stress is kept below the yield point. However, if the yield point is reached, yield takes place as described in Sec. 2.1B. When the load is removed, the stress and strain decrease in a linear fashion along a line *CD* parallel to the straight-line portion *AB* of the loading curve (Fig. 2.13). The fact that ϵ does not return to zero after the load has been removed indicates that a *permanent set* or *plastic deformation* of the material has taken place. For most materials, the plastic deformation depends upon both the maximum value reached by the stress and the time elapsed before the load is removed. The stress-dependent part of the plastic deformation is called *slip*, and the time-dependent part—also influenced by the temperature—is *creep*.

When a material does not possess a well-defined yield point, the elastic limit cannot be determined with precision. However, assuming the elastic limit to be equal to the yield strength using the offset method (Sec. 2.1B) results in only a small error. Referring to Fig. 2.8, note that the straight line used to determine point *Y* also represents the unloading curve after a maximum stress σ_Y has been reached. While the material does not behave truly elastically, the resulting plastic strain is as small as the selected offset.

If, after being loaded and unloaded (Fig. 2.14), the test specimen is loaded again, the new loading curve will follow the earlier unloading curve until it almost reaches point *C*. Then it will bend to the right and connect with the curved portion of the original stress-strain diagram. This straight-line portion of the new loading curve is longer than the corresponding portion of the initial one. Thus, the proportional limit and the

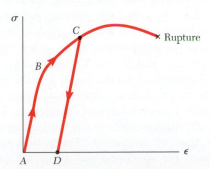

Fig. 2.13   Stress-strain response of ductile material loaded beyond yield and unloaded.

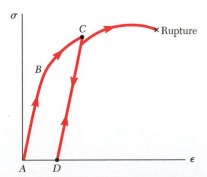

Fig. 2.14   Stress-strain response of ductile material reloaded after prior yielding and unloading.

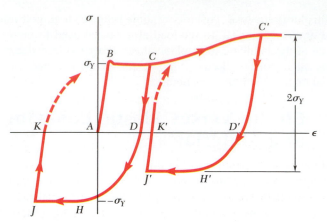

Fig. 2.15 Stress-strain response for mild steel subjected to two cases of reverse loading.

elastic limit have increased as a result of the strain-hardening that occurred during the earlier loading. However, since the point of rupture R remains unchanged, the ductility of the specimen, which should now be measured from point D, has decreased.

In previous discussions the specimen was loaded twice in the same direction (i.e., both loads were tensile loads). Now consider that the second load is applied in a direction opposite to that of the first one. Assume the material is mild steel where the yield strength is the same in tension and in compression. The initial load is tensile and is applied until point C is reached on the stress-strain diagram (Fig. 2.15). After unloading (point D), a compressive load is applied, causing the material to reach point H, where the stress is equal to $-\sigma_Y$. Note that portion DH of the stress-strain diagram is curved and does not show any clearly defined yield point. This is referred to as the *Bauschinger effect*. As the compressive load is maintained, the material yields along line HJ.

If the load is removed after point J has been reached, the stress returns to zero along line JK, and the slope of JK is equal to the modulus of elasticity E. The resulting permanent set AK may be positive, negative, or zero, depending upon the lengths of the segments BC and HJ. If a tensile load is applied again to the test specimen, the portion of the stress-strain diagram beginning at K (dashed line) will curve up and to the right until the yield stress σ_Y has been reached.

If the initial loading is large enough to cause strain-hardening of the material (point C'), unloading takes place along line $C'D'$. As the reverse load is applied, the stress becomes compressive, reaching its maximum value at H' and maintaining it as the material yields along line $H'J'$. While the maximum value of the compressive stress is less than σ_Y, the total change in stress between C' and H' is still equal to $2\sigma_Y$.

If point K or K' coincides with the origin A of the diagram, the permanent set is equal to zero, and the specimen may appear to have returned to its original condition. However, internal changes will have taken place and, the specimen will rupture without any warning after relatively few repetitions of the loading sequence. Thus, the excessive plastic deformations to which the specimen was subjected caused a radical change in the characteristics of the material. Therefore reverse loadings into the plastic range are seldom allowed, being permitted only under carefully controlled conditions such as in the straightening of damaged material and the final alignment of a structure or machine.

2.1F   Repeated Loadings and Fatigue

You might think that a given load may be repeated many times, provided that the stresses remain in the elastic range. Such a conclusion is correct for loadings repeated a few dozen or even a few hundred times. However, it is not correct when loadings are repeated thousands or millions of times. In such cases, rupture can occur at a stress much lower than the static breaking strength; this phenomenon is known as *fatigue*. A fatigue failure is of a brittle nature, even for materials that are normally ductile.

Fatigue must be considered in the design of all structural and machine components subjected to repeated or fluctuating loads. The number of loading cycles expected during the useful life of a component varies greatly. For example, a beam supporting an industrial crane can be loaded as many as two million times in 25 years (about 300 loadings per working day), an automobile crankshaft is loaded about half a billion times if the automobile is driven 200,000 miles, and an individual turbine blade can be loaded several hundred billion times during its lifetime.

Some loadings are of a fluctuating nature. For example, the passage of traffic over a bridge will cause stress levels that will fluctuate about the stress level due to the weight of the bridge. A more severe condition occurs when a complete reversal of the load occurs during the loading cycle. The stresses in the axle of a railroad car, for example, are completely reversed after each half-revolution of the wheel.

The number of loading cycles required to cause the failure of a specimen through repeated loadings and reverse loadings can be determined experimentally for any given maximum stress level. If a series of tests is conducted using different maximum stress levels, the resulting data is plotted as a σ-n curve. For each test, the maximum stress σ is plotted as an ordinate and the number of cycles n as an abscissa. Because of the large number of cycles required for rupture, the cycles n are plotted on a logarithmic scale.

A typical σ-n curve for steel is shown in Fig. 2.16. If the applied maximum stress is high, relatively few cycles are required to cause rupture. As the magnitude of the maximum stress is reduced, the number of cycles required to cause rupture increases, until the *endurance limit* is reached. The endurance limit is the stress for which failure does not occur, even for an indefinitely large number of loading cycles. For a low-carbon steel, such as structural steel, the endurance limit is about one-half of the ultimate strength of the steel.

For nonferrous metals, such as aluminum and copper, a typical σ-n curve (Fig. 2.16) shows that the stress at failure continues to decrease as the number of loading cycles is increased. For such metals, the *fatigue limit* is the stress corresponding to failure after a specified number of loading cycles.

Examination of test specimens, shafts, springs, and other components that have failed in fatigue shows that the failure initiated at a microscopic crack or some similar imperfection. At each loading, the crack was very slightly enlarged. During successive loading cycles, the crack propagated through the material until the amount of undamaged material was insufficient to carry the maximum load, and an abrupt, brittle failure

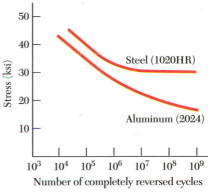

Fig. 2.16   Typical σ-n curves.

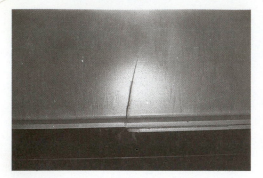

Photo 2.6 Fatigue crack in a steel girder of the Yellow Mill Pond Bridge, Connecticut, prior to repairs.

occurred. For example, Photo 2.6 shows a progressive fatigue crack in a highway bridge girder that initiated at the irregularity associated with the weld of a cover plate and then propagated through the flange and into the web. Because fatigue failure can be initiated at any crack or imperfection, the surface condition of a specimen has an important effect on the endurance limit obtained in testing. The endurance limit for machined and polished specimens is higher than for rolled or forged components or for components that are corroded. In applications in or near seawater or in other applications where corrosion is expected, a reduction of up to 50% in the endurance limit can be expected.

2.1G Deformations of Members Under Axial Loading

Consider a homogeneous rod BC of length L and uniform cross section of area A subjected to a centric axial load $\mathbf{P}$ (Fig. 2.17). If the resulting axial stress $\sigma = P/A$ does not exceed the proportional limit of the material, Hooke's law applies and

$$\sigma = E\epsilon \tag{2.6}$$

from which

$$\epsilon = \frac{\sigma}{E} = \frac{P}{AE} \tag{2.7}$$

Recalling that the strain ϵ in Sec. 2.1A is $\epsilon = \delta/L$

$$\delta = \epsilon L \tag{2.8}$$

and substituting for ϵ from Eq. (2.7) into Eq.(2.8):

$$\delta = \frac{PL}{AE} \tag{2.9}$$

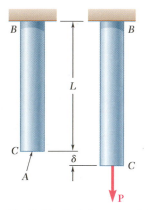

Fig. 2.17 Undeformed and deformed axially-loaded rod.

Equation (2.9) can be used only if the rod is homogeneous (constant E), has a uniform cross section of area A, and is loaded at its ends. If the rod is loaded at other points, or consists of several portions of various cross sections and possibly of different materials, it must be divided into component parts that satisfy the required conditions for the application of Eq. (2.9). Using the internal force P_i, length L_i, cross-sectional area A_i, and modulus of elasticity E_i, corresponding to part i, the deformation of the entire rod is

$$\delta = \sum_i \frac{P_i L_i}{A_i E_i} \tag{2.10}$$

In the case of a member of variable cross section (Fig. 2.18), the strain ϵ depends upon the position of the point Q, where it is computed as $\epsilon = d\delta/dx$ (Sec. 2.1A). Solving for $d\delta$ and substituting for ϵ from Eq. (2.7), the deformation of an element of length dx is

$$d\delta = \epsilon\, dx = \frac{P\, dx}{AE}$$

The total deformation δ of the member is obtained by integrating this expression over the length L of the member:

$$\delta = \int_0^L \frac{P\, dx}{AE} \tag{2.11}$$

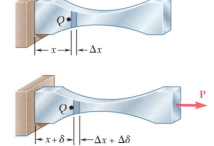

Fig. 2.18 Deformation of axially-loaded member of variable cross-sectional area.

Equation (2.11) should be used in place of (2.9) when both the cross-sectional area A is a function of x, or when the internal force P depends upon x, as is the case for a rod hanging under its own weight.

Concept Application 2.1

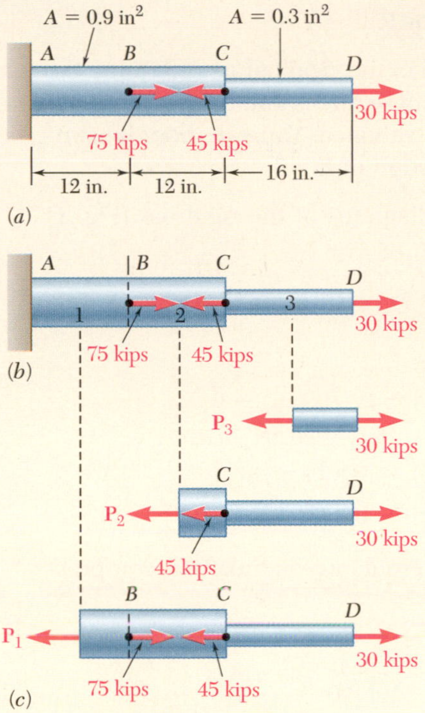

Determine the deformation of the steel rod shown in Fig. 2.19a under the given loads ($E = 29 \times 10^6$ psi).

The rod is divided into three component parts in Fig. 2.19b, so

$$L_1 = L_2 = 12\text{ in.} \qquad L_3 = 16\text{ in.}$$
$$A_1 = A_2 = 0.9\text{ in}^2 \qquad A_3 = 0.3\text{ in}^2$$

To find the internal forces P_1, P_2, and P_3, pass sections through each of the component parts, drawing each time the free-body diagram of the portion of rod located to the right of the section (Fig. 2.19c). Each of the free bodies is in equilibrium; thus

$$P_1 = 60\text{ kips} = 60 \times 10^3\text{ lb}$$
$$P_2 = -15\text{ kips} = -15 \times 10^3\text{ lb}$$
$$P_3 = 30\text{ kips} = 30 \times 10^3\text{ lb}$$

Using Eq. (2.10)

$$\delta = \sum_i \frac{P_i L_i}{A_i E_i} = \frac{1}{E}\left(\frac{P_1 L_1}{A_1} + \frac{P_2 L_2}{A_2} + \frac{P_3 L_3}{A_3}\right)$$
$$= \frac{1}{29 \times 10^6}\left[\frac{(60 \times 10^3)(12)}{0.9}\right.$$
$$\left. + \frac{(-15 \times 10^3)(12)}{0.9} + \frac{(30 \times 10^3)(16)}{0.3}\right]$$
$$\delta = \frac{2.20 \times 10^6}{29 \times 10^6} = 75.9 \times 10^{-3}\text{ in.}$$

Left figure labels:
$A = 0.9\text{ in}^2$ · $A = 0.3\text{ in}^2$ · A B C D · 30 kips · 75 kips · 45 kips · 12 in. · 12 in. · 16 in. · (a)

A $|B$ C D · 1 2 3 · 30 kips · 75 kips · 45 kips · (b)

P_3 · 30 kips · P_2 · C D · 30 kips · 45 kips · P_1 · B C D · 30 kips · 75 kips · 45 kips · (c)

Fig. 2.19 (a) Axially-loaded rod. (b) Rod divided into three sections. (c) Three sectioned free-body diagrams with internal resultant forces P_1, P_2, and P_3.

Rod BC of Fig. 2.17, used to derive Eq. (2.9), and rod AD of Fig. 2.19 have one end attached to a fixed support. In each case, the deformation δ of the rod was equal to the displacement of its free end. When both ends of a rod move, however, the deformation of the rod is measured by the *relative displacement* of one end of the rod with respect to the other. Consider the assembly shown in Fig. 2.20a, which consists of three elastic bars of length L connected by a rigid pin at A. If a load **P** is applied at B (Fig. 2.20b), each of the three bars will deform. Since the bars AC and AC' are attached to fixed supports at C and C', their common deformation is measured by the displacement δ_A of point A. On the other hand, since both ends of bar AB move, the deformation of AB is measured by the difference between the displacements δ_A and δ_B of points A and B, (i.e., by the relative displacement of B with respect to A). Denoting this relative displacement by $\delta_{B/A}$,

$$\delta_{B/A} = \delta_B - \delta_A = \frac{PL}{AE} \qquad (2.12)$$

where A is the cross-sectional area of AB and E is its modulus of elasticity.

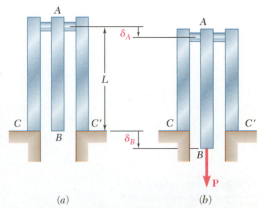

(a) (b)

Fig. 2.20 Example of relative end displacement, as exhibited by the middle bar. (a) Unloaded. (b) Loaded, with deformation.

α why not Bx ιαl By

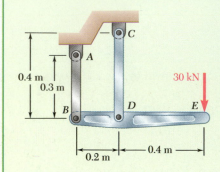

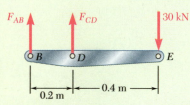

Fig. 1   Free-body diagram of rigid bar *BDE*.

Sample Problem 2.1

The rigid bar *BDE* is supported by two links *AB* and *CD*. Link *AB* is made of aluminum ($E = 70$ GPa) and has a cross-sectional area of 500 mm². Link *CD* is made of steel ($E = 200$ GPa) and has a cross-sectional area of 600 mm². For the 30-kN force shown, determine the deflection (*a*) of *B*, (*b*) of *D*, and (*c*) of *E*.

STRATEGY: Consider the free body of the rigid bar to determine the internal force of each link. Knowing these forces and the properties of the links, their deformations can be evaluated. You can then use simple geometry to determine the deflection of E.

MODELING: Draw the free body diagrams of the rigid bar (Fig. 1) and the two links (Fig. 2 and 3)

ANALYSIS:

Free Body: Bar BDE (Fig. 1)

$+\gamma \Sigma M_B = 0$: $-(30 \text{ kN})(0.6 \text{ m}) + F_{CD}(0.2 \text{ m}) = 0$

$\qquad\qquad\qquad F_{CD} = +90 \text{ kN} \qquad F_{CD} = 90 \text{ kN} \quad tension$

$+\gamma \Sigma M_D = 0$: $-(30 \text{ kN})(0.4 \text{ m}) - F_{AB}(0.2 \text{ m}) = 0$

$\qquad\qquad\qquad F_{AB} = -60 \text{ kN} \qquad F_{AB} = 60 \text{ kN} \quad compression$

a.   Deflection of B.   Since the internal force in link *AB* is compressive (Fig. 2), $P = -60$ kN and

$$\delta_B = \frac{PL}{AE} = \frac{(-60 \times 10^3 \text{ N})(0.3 \text{ m})}{(500 \times 10^{-6} \text{ m}^2)(70 \times 10^9 \text{ Pa})} = -514 \times 10^{-6} \text{ m}$$

The negative sign indicates a contraction of member *AB*. Thus, the deflection of end *B* is upward:

$$\delta_B = 0.514 \text{ mm} \uparrow \quad \blacktriangleleft$$

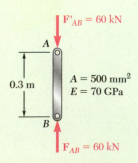

Fig. 2   Free-body diagram of two-force member *AB*.

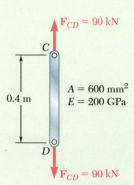

Fig. 3   Free-body diagram of two-force member *CD*.

(continued)

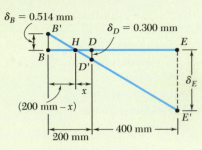

Fig. 4 Deflections at B and D of rigid bar are used to find δ_E.

b. Deflection of D. Since in rod CD (Fig. 3), $P = 90$ kN, write

$$\delta_D = \frac{PL}{AE} = \frac{(90 \times 10^3\,\text{N})(0.4\,\text{m})}{(600 \times 10^{-6}\,\text{m}^2)(200 \times 10^9\,\text{Pa})}$$
$$= 300 \times 10^{-6}\,\text{m} \qquad \delta_D = \mathbf{0.300\ mm\ \downarrow} \blacktriangleleft$$

c. Deflection of E. Referring to Fig. 4, we denote by B' and D' the displaced positions of points B and D. Since the bar BDE is rigid, points B', D', and E' lie in a straight line. Therefore,

$$\frac{BB'}{DD'} = \frac{BH}{HD} \qquad \frac{0.514\,\text{mm}}{0.300\,\text{mm}} = \frac{(200\,\text{mm}) - x}{x} \qquad x = 73.7\,\text{mm}$$

$$\frac{EE'}{DD'} = \frac{HE}{HD} \qquad \frac{\delta_E}{0.300\,\text{mm}} = \frac{(400\,\text{mm}) + (73.7\,\text{mm})}{73.7\,\text{mm}}$$

$$\delta_E = \mathbf{1.928\ mm\ \downarrow} \blacktriangleleft$$

REFLECT and THINK: Comparing the relative magnitude and direction of the resulting deflections, you can see that the answers obtained are consistent with the loading and the deflection diagram of Fig. 4.

Sample Problem 2.2

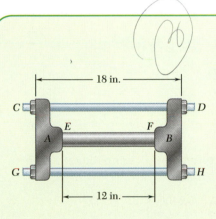

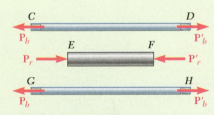

Fig. 1 Free-body diagrams of bolts and aluminum bar.

The rigid castings A and B are connected by two $\frac{3}{4}$-in.-diameter steel bolts CD and GH and are in contact with the ends of a 1.5-in.-diameter aluminum rod EF. Each bolt is single-threaded with a pitch of 0.1 in., and after being snugly fitted, the nuts at D and H are both tightened one-quarter of a turn. Knowing that E is 29×10^6 psi for steel and 10.6×10^6 psi for aluminum, determine the normal stress in the rod.

STRATEGY: The tightening of the nuts causes a displacement of the ends of the bolts relative to the rigid casting that is equal to the difference in displacements between the bolts and the rod. This will give a relation between the internal forces of the bolts and the rod that, when combined with a free body analysis of the rigid casting, will enable you to solve for these forces and determine the corresponding normal stress in the rod.

MODELING: Draw the free body diagrams of the bolts and rod (Fig. 1) and the rigid casting (Fig. 2).

ANALYSIS:

Deformations.

Bolts CD and GH. Tightening the nuts causes tension in the bolts (Fig. 1). Because of symmetry, both are subjected to the same

(continued)

internal force P_b and undergo the same deformation δ_b. Therefore,

$$\delta_b = +\frac{P_b L_b}{A_b E_b} = +\frac{P_b(18 \text{ in.})}{\frac{1}{4}\pi(0.75 \text{ in.})^2(29 \times 10^6 \text{ psi})} = +1.405 \times 10^{-6} P_b \quad (1)$$

Rod EF. The rod is in compression (Fig. 1), where the magnitude of the force is P_r and the deformation δ_r:

$$\delta_r = -\frac{P_r L_r}{A_r E_r} = -\frac{P_r(12 \text{ in.})}{\frac{1}{4}\pi(1.5 \text{ in.})^2(10.6 \times 10^6 \text{ psi})} = -0.6406 \times 10^{-6} P_r \quad (2)$$

Displacement of D Relative to B. Tightening the nuts one-quarter of a turn causes ends D and H of the bolts to undergo a displacement of $\frac{1}{4}(0.1 \text{ in.})$ *relative to casting B*. Considering end D,

$$\delta_{D/B} = \frac{1}{4}(0.1 \text{ in.}) = 0.025 \text{ in.} \quad (3)$$

But $\delta_{D/B} = \delta_D - \delta_B$, where δ_D and δ_B represent the displacements of D and B. If casting A is held in a fixed position while the nuts at D and H are being tightened, these displacements are equal to the deformations of the bolts and of the rod, respectively. Therefore,

$$\delta_{D/B} = \delta_b - \delta_r \quad (4)$$

Substituting from Eqs. (1), (2), and (3) into Eq. (4),

$$0.025 \text{ in.} = 1.405 \times 10^{-6} P_b + 0.6406 \times 10^{-6} P_r \quad (5)$$

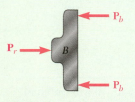

Fig. 2 Free-body diagram of rigid casting.

Free Body: Casting B (Fig. 2)

$$\xrightarrow{+} \Sigma F = 0: \qquad P_r - 2P_b = 0 \qquad P_r = 2P_b \quad (6)$$

Forces in Bolts and Rod

Substituting for P_r from Eq. (6) into Eq. (5), we have

$$0.025 \text{ in.} = 1.405 \times 10^{-6} P_b + 0.6406 \times 10^{-6}(2P_b)$$

$$P_b = 9.307 \times 10^3 \text{ lb} = 9.307 \text{ kips}$$

$$P_r = 2P_b = 2(9.307 \text{ kips}) = 18.61 \text{ kips}$$

Stress in Rod

$$\sigma_r = \frac{P_r}{A_r} = \frac{18.61 \text{ kips}}{\frac{1}{4}\pi(1.5 \text{ in.})^2} \qquad \sigma_r = 10.53 \text{ ksi} \blacktriangleleft$$

REFLECT and THINK: This is an example of a *statically indeterminate* problem, where the determination of the member forces could not be found by equilibrium alone. By considering the relative displacement characteristics of the members, you can obtain additional equations necessary to solve such problems. Situations like this will be examined in more detail in the following section.

Problems

2.1 A nylon thread is subjected to a 8.5-N tension force. Knowing that $E = 3.3$ GPa and that the length of the thread increases by 1.1%, determine (*a*) the diameter of the thread, (*b*) the stress in the thread.

2.2 A 4.8-ft-long steel wire of $\frac{1}{4}$-in.-diameter is subjected to a 750-lb tensile load. Knowing that $E = 29 \times 10^6$ psi, determine (*a*) the elongation of the wire, (*b*) the corresponding normal stress.

2.3 An 18-m-long steel wire of 5-mm diameter is to be used in the manufacture of a prestressed concrete beam. It is observed that the wire stretches 45 mm when a tensile force **P** is applied. Knowing that $E = 200$ GPa, determine (*a*) the magnitude of the force **P**, (*b*) the corresponding normal stress in the wire.

2.4 Two gage marks are placed exactly 250 mm apart on a 12-mm-diameter aluminum rod with $E = 73$ GPa and an ultimate strength of 140 MPa. Knowing that the distance between the gage marks is 250.28 mm after a load is applied, determine (*a*) the stress in the rod, (*b*) the factor of safety.

2.5 An aluminum pipe must not stretch more than 0.05 in. when it is subjected to a tensile load. Knowing that $E = 10.1 \times 10^6$ psi and that the maximum allowable normal stress is 14 ksi, determine (*a*) the maximum allowable length of the pipe, (*b*) the required area of the pipe if the tensile load is 127.5 kips.

2.6 A control rod made of yellow brass must not stretch more than 3 mm when the tension in the wire is 4 kN. Knowing that $E = 105$ GPa and that the maximum allowable normal stress is 180 MPa, determine (*a*) the smallest diameter rod that should be used, (*b*) the corresponding maximum length of the rod.

2.7 A steel control rod is 5.5 ft long and must not stretch more than 0.04 in. when a 2-kip tensile load is applied to it. Knowing that $E = 29 \times 10^6$ psi, determine (*a*) the smallest diameter rod that should be used, (*b*) the corresponding normal stress caused by the load.

2.8 A cast-iron tube is used to support a compressive load. Knowing that $E = 10 \times 10^6$ psi and that the maximum allowable change in length is 0.025%, determine (*a*) the maximum normal stress in the tube, (*b*) the minimum wall thickness for a load of 1600 lb if the outside diameter of the tube is 2.0 in.

2.9 A 4-m-long steel rod must not stretch more than 3 mm and the normal stress must not exceed 150 MPa when the rod is subjected to a 10-kN axial load. Knowing that $E = 200$ GPa, determine the required diameter of the rod.

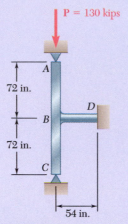

P = 130 kips

72 in.

A

D

B

72 in.

C

54 in.

Fig. P2.13

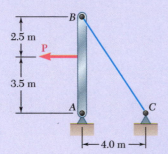

B

2.5 m

P

3.5 m

A

C

4.0 m

Fig. P2.14

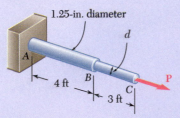

1.25-in. diameter

d

A

4 ft

B

3 ft

C

P

Fig. P2.15

2.10 A nylon thread is to be subjected to a 10-N tension. Knowing that $E = 3.2$ GPa, that the maximum allowable normal stress is 40 MPa, and that the length of the thread must not increase by more than 1%, determine the required diameter of the thread.

2.11 A block of 10-in. length and 1.8×1.6-in. cross section is to support a centric compressive load **P**. The material to be used is a bronze for which $E = 14 \times 10^6$ psi. Determine the largest load that can be applied, knowing that the normal stress must not exceed 18 ksi and that the decrease in length of the block should be at most 0.12% of its original length.

2.12 A square yellow-brass bar must not stretch more than 2.5 mm when it is subjected to a tensile load. Knowing that $E = 105$ GPa and that the allowable tensile strength is 180 MPa, determine (*a*) the maximum allowable length of the bar, (*b*) the required dimensions of the cross section if the tensile load is 40 kN.

2.13 Rod *BD* is made of steel ($E = 29 \times 10^6$ psi) and is used to brace the axially compressed member *ABC*. The maximum force that can be developed in member *BD* is 0.02*P*. If the stress must not exceed 18 ksi and the maximum change in length of *BD* must not exceed 0.001 times the length of *ABC*, determine the smallest-diameter rod that can be used for member *BD*.

2.14 The 4-mm-diameter cable *BC* is made of a steel with $E = 200$ GPa. Knowing that the maximum stress in the cable must not exceed 190 MPa and that the elongation of the cable must not exceed 6 mm, find the maximum load **P** that can be applied as shown.

2.15 A single axial load of magnitude $P = 15$ kips is applied at end *C* of the steel rod *ABC*. Knowing that $E = 30 \times 10^6$ psi, determine the diameter *d* of portion *BC* for which the deflection of point *C* will be 0.05 in.

2.16 A 250-mm-long aluminum tube ($E = 70$ GPa) of 36-mm outer diameter and 28-mm inner diameter can be closed at both ends by means of single-threaded screw-on covers of 1.5-mm pitch. With one cover screwed on tight, a solid brass rod ($E = 105$ GPa) of 25-mm diameter is placed inside the tube and the second cover is screwed on. Since the rod is slightly longer than the tube, it is observed that the cover must be forced against the rod by rotating it one-quarter of a turn before it can be tightly closed. Determine (*a*) the average normal stress in the tube and in the rod, (*b*) the deformations of the tube and of the rod.

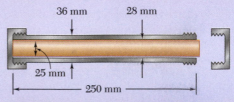

36 mm

28 mm

25 mm

250 mm

Fig. P2.16

2.17 The specimen shown has been cut from a $\frac{1}{4}$-in.-thick sheet of vinyl ($E = 0.45 \times 10^6$ psi) and is subjected to a 350-lb tensile load. Determine (*a*) the total deformation of the specimen, (*b*) the deformation of its central portion *BC*.

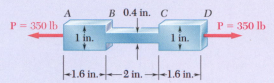

Fig. P2.17

2.18 The brass tube *AB* ($E = 105$ GPa) has a cross-sectional area of 140 mm² and is fitted with a plug at *A*. The tube is attached at *B* to a rigid plate that is itself attached at *C* to the bottom of an aluminum cylinder ($E = 72$ GPa) with a cross-sectional area of 250 mm². The cylinder is then hung from a support at *D*. In order to close the cylinder, the plug must move down through 1 mm. Determine the force **P** that must be applied to the cylinder.

2.19 Both portions of the rod *ABC* are made of an aluminum for which $E = 70$ GPa. Knowing that the magnitude of **P** is 4 kN, determine (*a*) the value of **Q** so that the deflection at *A* is zero, (*b*) the corresponding deflection of *B*.

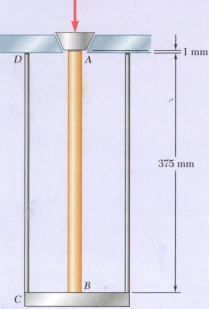

Fig. *P2.18*

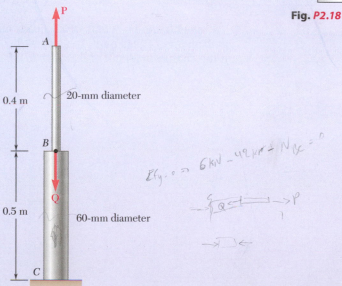

Fig. P2.19 and P2.20

2.20 The rod *ABC* is made of an aluminum for which $E = 70$ GPa. Knowing that $P = 6$ kN and $Q = 42$ kN, determine the deflection of (*a*) point *A*, (*b*) point *B*.

2.21 For the steel truss ($E = 200$ GPa) and loading shown, determine the deformations of members *AB* and *AD*, knowing that their cross-sectional areas are 2400 mm² and 1800 mm², respectively.

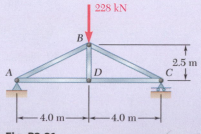

Fig. P2.21

2.22 For the steel truss ($E = 29 \times 10^6$ psi) and loading shown, determine the deformations of members BD and DE, knowing that their cross-sectional areas are 2 in² and 3 in², respectively.

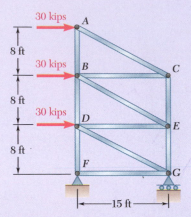

Fig. P2.22

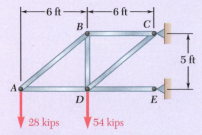

Fig. P2.23

2.23 Members AB and BC are made of steel ($E = 29 \times 10^6$ psi) with cross-sectional areas of 0.80 in² and 0.64 in², respectively. For the loading shown, determine the elongation of (a) member AB, (b) member BC.

2.24 The steel frame ($E = 200$ GPa) shown has a diagonal brace BD with an area of 1920 mm². Determine the largest allowable load **P** if the change in length of member BD is not to exceed 1.6 mm.

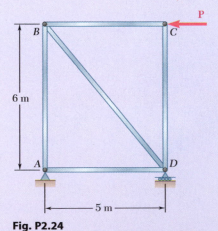

Fig. P2.24

2.25 Link BD is made of brass ($E = 105$ GPa) and has a cross-sectional area of 240 mm². Link CE is made of aluminum ($E = 72$ GPa) and has a cross-sectional area of 300 mm². Knowing that they support rigid member ABC, determine the maximum force **P** that can be applied vertically at point A if the deflection of A is not to exceed 0.35 mm.

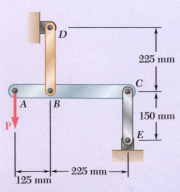

Fig. P2.25

2.26 Members *ABC* and *DEF* are joined with steel links ($E = 200$ GPa). Each of the links is made of a pair of 25×35-mm plates. Determine the change in length of (*a*) member *BE*, (*b*) member *CF*.

2.27 Each of the links *AB* and *CD* is made of aluminum ($E = 10.9 \times 10^6$ psi) and has a cross-sectional area of 0.2 in². Knowing that they support the rigid member *BC*, determine the deflection of point *E*.

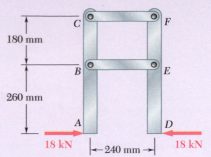

Fig. P2.26

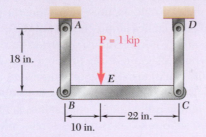

Fig. P2.27

2.28 The length of the $\frac{3}{32}$-in.-diameter steel wire *CD* has been adjusted so that with no load applied, a gap of $\frac{1}{16}$ in. exists between the end *B* of the rigid beam *ACB* and a contact point *E*. Knowing that $E = 29 \times 10^6$ psi, determine where a 50-lb block should be placed on the beam in order to cause contact between *B* and *E*.

2.29 A homogenous cable of length *L* and uniform cross section is suspended from one end. (*a*) Denoting by ρ the density (mass per unit volume) of the cable and by *E* its modulus of elasticity, determine the elongation of the cable due to its own weight. (*b*) Show that the same elongation would be obtained if the cable were horizontal and if a force equal to half of its weight were applied at each end.

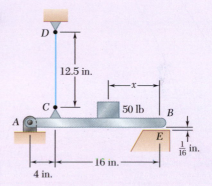

Fig. P2.28

2.30 The vertical load **P** is applied at the center *A* of the upper section of a homogeneous frustum of a circular cone of height *h*, minimum radius *a*, and maximum radius *b*. Denoting by *E* the modulus of elasticity of the material and neglecting the effect of its weight, determine the deflection of point *A*.

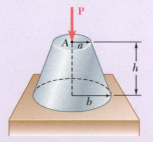

Fig. P2.30

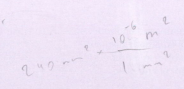

2.31 Denoting by ϵ the "engineering strain" in a tensile specimen, show that the true strain is $\epsilon_t = \ln(1 + \epsilon)$.

2.32 The volume of a tensile specimen is essentially constant while plastic deformation occurs. If the initial diameter of the specimen is d_1, show that when the diameter is d, the true strain is $\epsilon_t = 2\ln(d_1/d)$.

2.2 STATICALLY INDETERMINATE PROBLEMS

In the problems considered in the preceding section, we could always use free-body diagrams and equilibrium equations to determine the internal forces produced in the various portions of a member under given loading conditions. There are many problems, however, where the internal forces cannot be determined from statics alone. In most of these problems, the reactions themselves—the external forces—cannot be determined by simply drawing a free-body diagram of the member and writing the corresponding equilibrium equations. The equilibrium equations must be complemented by relationships involving deformations obtained by considering the geometry of the problem. Because statics is not sufficient to determine either the reactions or the internal forces, problems of this type are called *statically indeterminate*. The following concept applications show how to handle this type of problem.

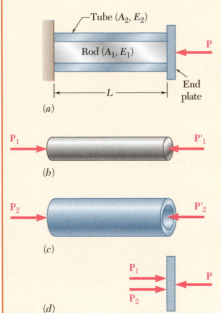

Tube (A_2, E_2)

Rod (A_1, E_1)

P

End plate

L

(a)

P_1 P'_1

(b)

P_2 P'_2

(c)

P_1

P_2 P

(d)

Fig. 2.21 (a) Concentric rod and tube, loaded by force P. (b) Free-body diagram of rod. (c) Free-body diagram of tube. (d) Free-body diagram of end plate.

Concept Application 2.2

A rod of length L, cross-sectional area A_1, and modulus of elasticity E_1, has been placed inside a tube of the same length L, but of cross-sectional area A_2 and modulus of elasticity E_2 (Fig. 2.21a). What is the deformation of the rod and tube when a force **P** is exerted on a rigid end plate as shown?

The axial forces in the rod and in the tube are P_1 and P_2, respectively. Draw free-body diagrams of all three elements (Fig. 2.21b, c, d). Only Fig. 2.21d yields any significant information, as:

$$P_1 + P_2 = P \tag{1}$$

Clearly, one equation is not sufficient to determine the two unknown internal forces P_1 and P_2. The problem is statically indeterminate.

However, the geometry of the problem shows that the deformations δ_1 and δ_2 of the rod and tube must be equal. Recalling Eq. (2.9), write

$$\delta_1 = \frac{P_1 L}{A_1 E_1} \qquad \delta_2 = \frac{P_2 L}{A_2 E_2} \tag{2}$$

Equating the deformations δ_1 and δ_2,

$$\frac{P_1}{A_1 E_1} = \frac{P_2}{A_2 E_2} \tag{3}$$

Equations (1) and (3) can be solved simultaneously for P_1 and P_2:

$$P_1 = \frac{A_1 E_1 P}{A_1 E_1 + A_2 E_2} \qquad P_2 = \frac{A_2 E_2 P}{A_1 E_1 + A_2 E_2}$$

Either of Eqs. (2) can be used to determine the common deformation of the rod and tube.

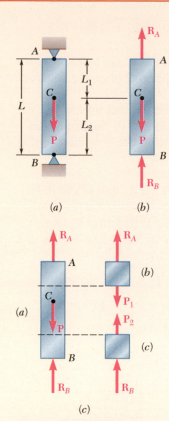

Fig. 2.22 (a) Restrained bar with axial load. (b) Free-body diagram of bar. (c) Free-body diagrams of sections above and below point C used to determine internal forces P_1 and P_2.

Concept Application 2.3

A bar AB of length L and uniform cross section is attached to rigid supports at A and B before being loaded. What are the stresses in portions AC and BC due to the application of a load P at point C (Fig. 2.22a)?

Drawing the free-body diagram of the bar (Fig. 2.22b), the equilibrium equation is

$$R_A + R_B = P \tag{1}$$

Since this equation is not sufficient to determine the two unknown reactions R_A and R_B, the problem is statically indeterminate.

However, the reactions can be determined if observed from the geometry that the total elongation δ of the bar must be zero. The elongations of the portions AC and BC are respectively δ_1 and δ_2, so

$$\delta = \delta_1 + \delta_2 = 0$$

Using Eq. (2.9), δ_1 and δ_2 can be expressed in terms of the corresponding internal forces P_1 and P_2,

$$\delta = \frac{P_1 L_1}{AE} + \frac{P_2 L_2}{AE} = 0 \tag{2}$$

Note from the free-body diagrams shown in parts b and c of Fig. 2.22c that $P_1 = R_A$ and $P_2 = -R_B$. Carrying these values into Equation (2),

$$R_A L_1 - R_B L_2 = 0 \tag{3}$$

Equations (1) and (3) can be solved simultaneously for R_A and R_B, as $R_A = PL_2/L$ and $R_B = PL_1/L$. The desired stresses σ_1 in AC and σ_2 in BC are obtained by dividing $P_1 = R_A$ and $P_2 = -R_B$ by the cross-sectional area of the bar:

$$\sigma_1 = \frac{PL_2}{AL} \qquad \sigma_2 = -\frac{PL_1}{AL}$$

Superposition Method.   A structure is statically indeterminate whenever it is held by more supports than are required to maintain its equilibrium. This results in more unknown reactions than available equilibrium equations. It is often convenient to designate one of the reactions as *redundant* and to eliminate the corresponding support. Since the stated conditions of the problem cannot be changed, the redundant reaction must be maintained in the solution. It will be treated as an *unknown load* that, together with the other loads, must produce deformations compatible with the original constraints. The actual solution of the problem considers separately the deformations caused by the given loads and the redundant reaction, and by adding—or *superposing*—the results obtained. The general conditions under which the combined effect of several loads can be obtained in this way are discussed in Sec. 2.5.

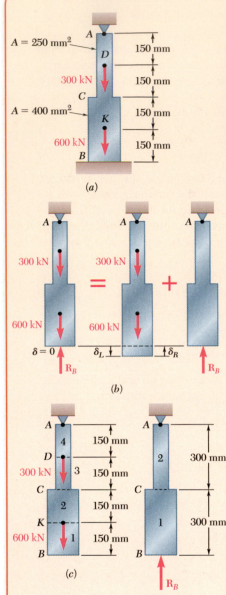

Fig. 2.23 (a) Restrained axially-loaded bar. (b) Reactions will be found by releasing constraint at point B and adding compressive force at point B to enforce zero deformation at point B. (c) Free-body diagram of released structure. (d) Free-body diagram of added reaction force at point B to enforce zero deformation at point B.

Concept Application 2.4

Determine the reactions at A and B for the steel bar and loading shown in Fig. 2.23a, assuming a close fit at both supports before the loads are applied.

We consider the reaction at B as redundant and release the bar from that support. The reaction R_B is considered to be an unknown load and is determined from the condition that the deformation δ of the bar equals zero.

The solution is carried out by considering the deformation $δ_L$ caused by the given loads and the deformation $δ_R$ due to the redundant reaction R_B (Fig. 2.23b).

The deformation $δ_L$ is obtained from Eq. (2.10) after the bar has been divided into four portions, as shown in Fig. 2.23c. Follow the same procedure as in Concept Application 2.1:

$$P_1 = 0 \quad P_2 = P_3 = 600 \times 10^3 \, N \quad P_4 = 900 \times 10^3 \, N$$

$$A_1 = A_2 = 400 \times 10^{-6} \, m^2 \quad A_3 = A_4 = 250 \times 10^{-6} \, m^2$$

$$L_1 = L_2 = L_3 = L_4 = 0.150 \, m$$

Substituting these values into Eq. (2.10),

$$δ_L = \sum_{i=1}^{4} \frac{P_i L_i}{A_i E} = \left(0 + \frac{600 \times 10^3 \, N}{400 \times 10^{-6} \, m^2} \right.$$

$$\left. + \frac{600 \times 10^3 \, N}{250 \times 10^{-6} \, m^2} + \frac{900 \times 10^3 \, N}{250 \times 10^{-6} \, m^2} \right) \frac{0.150 \, m}{E}$$

$$δ_L = \frac{1.125 \times 10^9}{E} \tag{1}$$

Considering now the deformation $δ_R$ due to the redundant reaction R_B, the bar is divided into two portions, as shown in Fig. 2.23d

$$P_1 = P_2 = -R_B$$

$$A_1 = 400 \times 10^{-6} \, m^2 \quad A_2 = 250 \times 10^{-6} \, m^2$$

$$L_1 = L_2 = 0.300 \, m$$

Substituting these values into Eq. (2.10),

$$δ_R = \frac{P_1 L_1}{A_1 E} + \frac{P_2 L_2}{A_2 E} = -\frac{(1.95 \times 10^3) R_B}{E} \tag{2}$$

Express the total deformation δ of the bar as zero:

$$δ = δ_L + δ_R = 0 \tag{3}$$

and, substituting for $δ_L$ and $δ_R$ from Eqs. (1) and (2) into Eqs. (3),

$$δ = \frac{1.125 \times 10^9}{E} - \frac{(1.95 \times 10^3) R_B}{E} = 0$$

(continued)

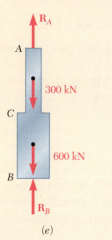

Fig. 2.23 (cont.) (e) Complete free-body diagram of *ACB*.

Solving for R_B,

$$R_B = 577 \times 10^3 \text{ N} = 577 \text{ kN}$$

The reaction R_A at the upper support is obtained from the free-body diagram of the bar (Fig. 2.23e),

$$+\uparrow \Sigma F_y = 0: \quad R_A - 300 \text{ kN} - 600 \text{ kN} + R_B = 0$$

$$R_A = 900 \text{ kN} - R_B = 900 \text{ kN} - 577 \text{ kN} = 323 \text{ kN}$$

Once the reactions have been determined, the stresses and strains in the bar can easily be obtained. Note that, while the total deformation of the bar is zero, each of its component parts *does deform* under the given loading and restraining conditions.

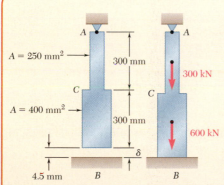

Fig. 2.24 Multi-section bar of Concept Application 2.4 with initial 4.5-mm gap at point *B*. Loading brings bar into contact with constraint.

Concept Application 2.5

Determine the reactions at A and B for the steel bar and loading of Concept Application 2.4, assuming now that a 4.5-mm clearance exists between the bar and the ground before the loads are applied (Fig. 2.24). Assume $E = 200$ GPa.

Considering the reaction at B to be redundant, compute the deformations δ_L and δ_R caused by the given loads and the redundant reaction $\mathbf{R}_B$. However, in this case, the total deformation is $\delta = 4.5$ mm. Therefore,

$$\delta = \delta_L + \delta_R = 4.5 \times 10^{-3} \text{ m} \tag{1}$$

Substituting for δ_L and δ_R into (Eq. 1), and recalling that $E = 200$ GPa $= 200 \times 10^9$ Pa,

$$\delta = \frac{1.125 \times 10^9}{200 \times 10^9} - \frac{(1.95 \times 10^3)R_B}{200 \times 10^9} = 4.5 \times 10^{-3} \text{ m}$$

Solving for R_B,

$$R_B = 115.4 \times 10^3 \text{ N} = 115.4 \text{ kN}$$

The reaction at A is obtained from the free-body diagram of the bar (Fig. 2.23e):

$$+\uparrow \Sigma F_y = 0: \quad R_A - 300 \text{ kN} - 600 \text{ kN} + R_B = 0$$
$$R_A = 900 \text{ kN} - R_B = 900 \text{ kN} - 115.4 \text{ kN} = 785 \text{ kN}$$

2.3 PROBLEMS INVOLVING TEMPERATURE CHANGES

Consider a homogeneous rod AB of uniform cross section that rests freely on a smooth horizontal surface (Fig. 2.25a). If the temperature of the rod is raised by ΔT, the rod elongates by an amount δ_T that is proportional to both the temperature change ΔT and the length L of the rod (Fig. 2.25b). Here

$$\delta_T = \alpha(\Delta T)L \tag{2.13}$$

where α is a constant characteristic of the material called the *coefficient of thermal expansion*. Since δ_T and L are both expressed in units of length, α represents a quantity *per degree C* or *per degree F*, depending whether the temperature change is expressed in degrees Celsius or Fahrenheit.

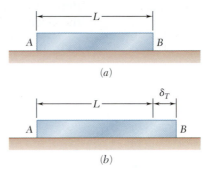

Fig. 2.25 Elongation of an unconstrained rod due to temperature increase.

Associated with deformation δ_T must be a strain $\epsilon_T = \delta_T/L$. Recalling Eq. (2.13),

$$\epsilon_T = \alpha\Delta T \tag{2.14}$$

The strain ϵ_T is called a *thermal strain*, as it is caused by the change in temperature of the rod. However, there is *no stress associated with the strain ϵ_T*.

Assume the same rod AB of length L is placed between two fixed supports at a distance L from each other (Fig. 2.26a). Again, there is neither stress nor strain in this initial condition. If we raise the temperature by ΔT, the rod cannot elongate because of the restraints imposed on its ends; the elongation δ_T of the rod is zero. Since the rod is homogeneous and of uniform cross section, the strain ϵ_T at any point is $\epsilon_T = \delta_T/L$ and thus is also zero. However, the supports will exert equal and opposite forces **P** and **P′** on the rod after the temperature has been raised, to keep it from elongating (Fig. 2.26b). It follows that a state of stress (with no corresponding strain) is created in the rod.

The problem created by the temperature change ΔT is statically indeterminate. Therefore, the magnitude P of the reactions at the supports is determined from the condition that the elongation of the rod is zero.

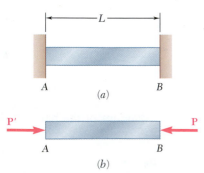

Fig. 2.26 Force P develops when the temperature of the rod increases while ends A and B are restrained.

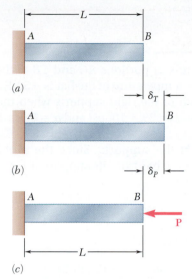

Fig. 2.27   Superposition method to find force at point B of restrained rod AB undergoing thermal expansion. (a) Initial rod length; (b) thermally expanded rod length; (c) force P pushes point B back to zero deformation.

Using the superposition method described in Sec. 2.2, the rod is detached from its support B (Fig. 2.27a) and elongates freely as it undergoes the temperature change ΔT (Fig. 2.27b). According to Eq. (2.13), the corresponding elongation is

$$\delta_T = \alpha(\Delta T)L$$

Applying now to end B the force **P** representing the redundant reaction, and recalling Eq. (2.9), a second deformation (Fig. 2.27c) is

$$\delta_P = \frac{PL}{AE}$$

Expressing that the total deformation δ must be zero,

$$\delta = \delta_T + \delta_P = \alpha(\Delta T)L + \frac{PL}{AE} = 0$$

from which

$$P = -AE\alpha(\Delta T)$$

The stress in the rod due to the temperature change ΔT is

$$\sigma = \frac{P}{A} = -E\alpha(\Delta T) \qquad \textbf{(2.15)}$$

The absence of any strain in the rod *applies only in the case of a homogeneous rod of uniform cross section.* Any other problem involving a restrained structure undergoing a change in temperature must be analyzed on its own merits. However, the same general approach can be used by considering the deformation due to the temperature change and the deformation due to the redundant reaction separately and superposing the two solutions obtained.

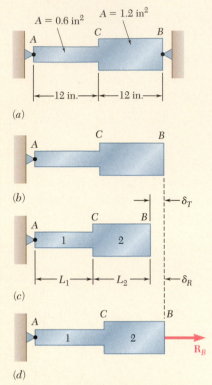

$A = 0.6 \text{ in}^2$ $A = 1.2 \text{ in}^2$

(a)

(b)

(c)

(d)

Fig. 2.28 (*a*) Restrained bar. (*b*) Bar at +75°F temperature. (*c*) Bar at lower temperature. (*d*) Force R_B needed to enforce zero deformation at point B.

Concept Application 2.6

Determine the values of the stress in portions *AC* and *CB* of the steel bar shown (Fig. 2.28*a*) when the temperature of the bar is −50°F, knowing that a close fit exists at both of the rigid supports when the temperature is +75°F. Use the values $E = 29 \times 10^6$ psi and $\alpha = 6.5 \times 10^{-6}/°F$ for steel.

Determine the reactions at the supports. Since the problem is statically indeterminate, detach the bar from its support at *B* and let it undergo the temperature change

$$\Delta T = (-50°F) - (75°F) = -125°F$$

The corresponding deformation (Fig. 2.28*c*) is

$$\delta_T = \alpha(\Delta T)L = (6.5 \times 10^{-6}/°F)(-125°F)(24 \text{ in.})$$
$$= -19.50 \times 10^{-3} \text{ in.}$$

Applying the unknown force $\mathbf{R}_B$ at end *B* (Fig. 2.28*d*), use Eq. (2.10) to express the corresponding deformation δ_R. Substituting

$$L_1 = L_2 = 12 \text{ in.}$$
$$A_1 = 0.6 \text{ in}^2 \quad A_2 = 1.2 \text{ in}^2$$
$$P_1 = P_2 = R_B \quad E = 29 \times 10^6 \text{ psi}$$

into Eq. (2.10), write

$$\delta_R = \frac{P_1 L_1}{A_1 E} + \frac{P_2 L_2}{A_2 E}$$

$$= \frac{R_B}{29 \times 10^6 \text{ psi}}\left(\frac{12 \text{ in.}}{0.6 \text{ in}^2} + \frac{12 \text{ in.}}{1.2 \text{ in}^2}\right)$$

$$= (1.0345 \times 10^{-6} \text{ in./lb})R_B$$

Expressing that the total deformation of the bar must be zero as a result of the imposed constraints, write

$$\delta = \delta_T + \delta_R = 0$$
$$= -19.50 \times 10^{-3} \text{ in.} + (1.0345 \times 10^{-6} \text{ in./lb})R_B = 0$$

from which

$$R_B = 18.85 \times 10^3 \text{ lb} = 18.85 \text{ kips}$$

The reaction at *A* is equal and opposite.

Noting that the forces in the two portions of the bar are $P_1 = P_2$ = 18.85 kips, obtain the following values of the stress in portions *AC* and *CB* of the bar:

(continued)

$$\sigma_1 = \frac{P_1}{A_1} = \frac{18.85 \text{ kips}}{0.6 \text{ in}^2} = +31.42 \text{ ksi}$$

$$\sigma_2 = \frac{P_2}{A_2} = \frac{18.85 \text{ kips}}{1.2 \text{ in}^2} = +15.71 \text{ ksi}$$

It cannot emphasized too strongly that, while the *total deformation* of the bar must be zero, the deformations of the portions *AC* and *CB are not zero*. A solution of the problem based on the assumption that these deformations are zero would therefore be wrong. Neither can the values of the strain in *AC* or *CB* be assumed equal to zero. To amplify this point, determine the strain ϵ_{AC} in portion *AC* of the bar. The strain ϵ_{AC} can be divided into two component parts; one is the thermal strain ϵ_T produced in the unrestrained bar by the temperature change ΔT (Fig. 2.28*c*). From Eq. (2.14),

$$\epsilon_T = \alpha \, \Delta T = (6.5 \times 10^{-6}/°\text{F})(-125°\text{F})$$

$$= -812.5 \times 10^{-6} \text{ in./in.}$$

The other component of ϵ_{AC} is associated with the stress σ_1 due to the force $\mathbf{R}_B$ applied to the bar (Fig. 2.28*d*). From Hooke's law, express this component of the strain as

$$\frac{\sigma_1}{E} = \frac{+31.42 \times 10^3 \text{ psi}}{29 \times 10^6 \text{ psi}} = +1083.4 \times 10^{-6} \text{ in./in.}$$

Add the two components of the strain in *AC* to obtain

$$\epsilon_{AC} = \epsilon_T + \frac{\sigma_1}{E} = -812.5 \times 10^{-6} + 1083.4 \times 10^{-6}$$

$$= +271 \times 10^{-6} \text{ in./in.}$$

A similar computation yields the strain in portion *CB* of the bar:

$$\epsilon_{CB} = \epsilon_T + \frac{\sigma_2}{E} = -812.5 \times 10^{-6} + 541.7 \times 10^{-6}$$

$$= -271 \times 10^{-6} \text{ in./in.}$$

The deformations δ_{AC} and δ_{CB} of the two portions of the bar are

$$\delta_{AC} = \epsilon_{AC}(AC) = (+271 \times 10^{-6})(12 \text{ in.})$$

$$= +3.25 \times 10^{-3} \text{ in.}$$

$$\delta_{CB} = \epsilon_{CB}(CB) = (-271 \times 10^{-6})(12 \text{ in.})$$

$$= -3.25 \times 10^{-3} \text{ in.}$$

Thus, while the sum $\delta = \delta_{AC} + \delta_{CB}$ of the two deformations is zero, neither of the deformations is zero.

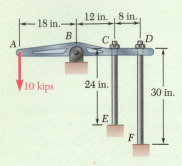

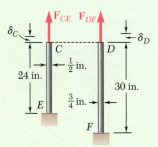

Fig. 1 Free-body diagram of rigid bar *ABCD*.

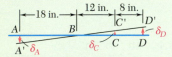

Fig. 2 Linearly proportional displacements along rigid bar *ABCD*.

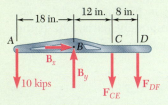

Fig. 3 Forces and deformations in *CE* and *DF*.

Sample Problem 2.3

The $\frac{1}{2}$-in.-diameter rod *CE* and the $\frac{3}{4}$-in.-diameter rod *DF* are attached to the rigid bar *ABCD* as shown. Knowing that the rods are made of aluminum and using $E = 10.6 \times 10^6$ psi, determine (*a*) the force in each rod caused by the loading shown and (*b*) the corresponding deflection of point *A*.

STRATEGY: To solve this statically indeterminate problem, you must supplement static equilibrium with a relative deflection analysis of the two rods.

MODELING: Draw the free body diagram of the bar (Fig. 1)

ANALYSIS:

Statics. Considering the free body of bar *ABCD* in Fig. 1, note that the reaction at *B* and the forces exerted by the rods are indeterminate. However, using statics,

$$+\circlearrowleft \Sigma M_B = 0: \qquad (10 \text{ kips})(18 \text{ in.}) - F_{CE}(12 \text{ in.}) - F_{DF}(20 \text{ in.}) = 0$$

$$12F_{CE} + 20F_{DF} = 180 \qquad \textbf{(1)}$$

Geometry. After application of the 10-kip load, the position of the bar is $A'BC'D'$ (Fig. 2). From the similar triangles BAA', BCC', and BDD',

$$\frac{\delta_C}{12 \text{ in.}} = \frac{\delta_D}{20 \text{ in.}} \qquad \delta_C = 0.6\delta_D \qquad \textbf{(2)}$$

$$\frac{\delta_A}{18 \text{ in.}} = \frac{\delta_D}{20 \text{ in.}} \qquad \delta_A = 0.9\delta_D \qquad \textbf{(3)}$$

Deformations. Using Eq. (2.9), and the data shown in Fig. 3, write

$$\delta_C = \frac{F_{CE}L_{CE}}{A_{CE}E} \qquad \delta_D = \frac{F_{DF}L_{DF}}{A_{DF}E}$$

Substituting for δ_C and δ_D into Eq. (2), write

$$\delta_C = 0.6\delta_D \qquad \frac{F_{CE}L_{CE}}{A_{CE}E} = 0.6\frac{F_{DF}L_{DF}}{A_{DF}E}$$

$$F_{CE} = 0.6\frac{L_{DF}}{L_{CE}}\frac{A_{CE}}{A_{DF}}F_{DF} = 0.6\left(\frac{30 \text{ in.}}{24 \text{ in.}}\right)\left[\frac{\frac{1}{4}\pi(\frac{1}{2} \text{ in.})^2}{\frac{1}{4}\pi(\frac{3}{4} \text{ in.})^2}\right]F_{DF} \qquad F_{CE} = 0.333F_{DF}$$

Force in Each Rod. Substituting for F_{CE} into Eq. (1) and recalling that all forces have been expressed in kips,

$$12(0.333F_{DF}) + 20F_{DF} = 180 \qquad \qquad F_{DF} = 7.50 \text{ kips} \ \blacktriangleleft$$

$$F_{CE} = 0.333F_{DF} = 0.333(7.50 \text{ kips}) \qquad F_{CE} = 2.50 \text{ kips} \ \blacktriangleleft$$

(continued)

Deflections.   The deflection of point D is

$$\delta_D = \frac{F_{DF}L_{DF}}{A_{DF}E} = \frac{(7.50 \times 10^3 \text{ lb})(30 \text{ in.})}{\frac{1}{4}\pi(\frac{3}{4} \text{ in.})^2(10.6 \times 10^6 \text{ psi})} \qquad \delta_D = 48.0 \times 10^{-3} \text{ in.}$$

Using Eq. (3),

$$\delta_A = 0.9\delta_D = 0.9(48.0 \times 10^{-3} \text{ in.}) \qquad \delta_A = 43.2 \times 10^{-3} \text{ in.} \quad \blacktriangleleft$$

REFLECT and THINK: You should note that as the rigid bar rotates about B, the deflections at C and D are proportional to their distance from the pivot point B, but *the forces exerted by the rods at these points are not.* Being statically indeterminate, these forces depend upon the deflection attributes of the rods as well as the equilibrium of the rigid bar.

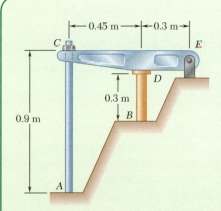

Fig. 1   Free-body diagram of bolt, cylinder and bar.

Sample Problem 2.4

The rigid bar CDE is attached to a pin support at E and rests on the 30-mm-diameter brass cylinder BD. A 22-mm-diameter steel rod AC passes through a hole in the bar and is secured by a nut that is snugly fitted when the temperature of the entire assembly is 20°C. The temperature of the brass cylinder is then raised to 50°C, while the steel rod remains at 20°C. Assuming that no stresses were present before the temperature change, determine the stress in the cylinder.

Rod AC: Steel	Cylinder BD: Brass
$E = 200$ GPa	$E = 105$ GPa
$\alpha = 11.7 \times 10^{-6}/°C$	$\alpha = 20.9 \times 10^{-6}/°C$

STRATEGY: You can use the method of superposition, considering $\mathbf{R}_B$ as redundant. With the support at B removed, the temperature rise of the cylinder causes point B to move down through δ_T. The reaction $\mathbf{R}_B$ must cause a deflection δ_1, equal to δ_T so that the final deflection of B will be zero (Fig. 2)

MODELING: Draw the free-body diagram of the entire assembly (Fig. 1).

ANALYSIS:

Statics.   Considering the free body of the entire assembly, write

$$+\!\!\uparrow\!\!\curvearrowleft \Sigma M_E = 0: \qquad R_A(0.75 \text{ m}) - R_B(0.3 \text{ m}) = 0 \qquad R_A = 0.4R_B \qquad \textbf{(1)}$$

(continued)

Deflection δ_T. Because of a temperature rise of $50° - 20° = 30°$C, the length of the brass cylinder increases by δ_T. (Fig. 2a).

$$\delta_T = L(\Delta T)\alpha = (0.3\text{ m})(30°\text{C})(20.9 \times 10^{-6}/°\text{C}) = 188.1 \times 10^{-6}\text{ m} \downarrow$$

Deflection δ_1. From Fig. 2b, note that $\delta_D = 0.4\delta_C$ and $\delta_1 = \delta_D + \delta_{B/D}$.

$$\delta_C = \frac{R_A L}{AE} = \frac{R_A(0.9\text{ m})}{\frac{1}{4}\pi(0.022\text{ m})^2(200\text{ GPa})} = 11.84 \times 10^{-9}R_A \uparrow$$

$$\delta_D = 0.40\delta_C = 0.4(11.84 \times 10^{-9}R_A) = 4.74 \times 10^{-9}R_A \uparrow$$

$$\delta_{B/D} = \frac{R_B L}{AE} = \frac{R_B(0.3\text{ m})}{\frac{1}{4}\pi(0.03\text{ m})^2(105\text{ GPa})} = 4.04 \times 10^{-9}R_B \uparrow$$

Recall from Eq. (1) that $R_A = 0.4R_B$, so

$$\delta_1 = \delta_D + \delta_{B/D} = [4.74(0.4R_B) + 4.04R_B]10^{-9} = 5.94 \times 10^{-9}R_B \uparrow$$

But $\delta_T = \delta_1$: $\qquad 188.1 \times 10^{-6}\text{ m} = 5.94 \times 10^{-9}R_B \qquad R_B = 31.7\text{ kN}$

Stress in Cylinder: $\quad \sigma_B = \dfrac{R_B}{A} = \dfrac{31.7\text{ kN}}{\frac{1}{4}\pi(0.03\text{ m})^2} \qquad \sigma_B = 44.8\text{ MPa}$ ◄

REFLECT and THINK: This example illustrates the large stresses that can develop in statically indeterminate systems due to even modest temperature changes. Note that if this assembly was statically determinate (i.e., the steel rod was removed), no stress at all would develop in the cylinder due to the temperature change.

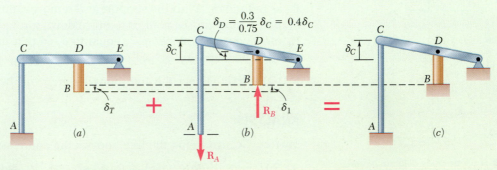

Fig. 2 Superposition of thermal and restraint force deformations (a) Support at B removed. (b) Reaction at B applied. (c) Final position.

Problems

2.33 An axial centric force of magnitude $P = 450$ kN is applied to the composite block shown by means of a rigid end plate. Knowing that $h = 10$ mm, determine the normal stress in (*a*) the brass core, (*b*) the aluminum plates.

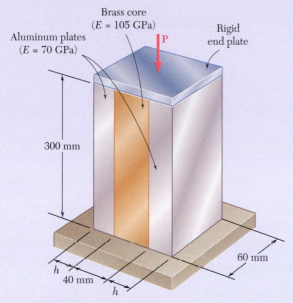

Brass core
($E = 105$ GPa)
Aluminum plates
($E = 70$ GPa)
Rigid end plate

300 mm

h
40 mm
h
60 mm

Fig. P2.33

2.34 For the composite block shown in Prob. 2.33, determine (*a*) the value of h if the portion of the load carried by the aluminum plates is half the portion of the load carried by the brass core, (*b*) the total load if the stress in the brass is 80 MPa.

2.35 The 4.5-ft concrete post is reinforced with six steel bars, each with a $1\frac{1}{8}$-in. diameter. Knowing that $E_s = 29 \times 10^6$ psi and $E_c = 4.2 \times 10^6$ psi, determine the normal stresses in the steel and in the concrete when a 350-kip axial centric force **P** is applied to the post.

2.36 For the post of Prob. 2.35, determine the maximum centric force that can be applied if the allowable normal stress is 20 ksi in the steel and 2.4 ksi in the concrete.

2.37 An axial force of 200 kN is applied to the assembly shown by means of rigid end plates. Determine (*a*) the normal stress in the aluminum shell, (*b*) the corresponding deformation of the assembly.

2.38 The length of the assembly shown decreases by 0.40 mm when an axial force is applied by means of rigid end plates. Determine (*a*) the magnitude of the applied force, (*b*) the corresponding stress in the brass core.

P

18 in.

4.5 ft

Fig. P2.35

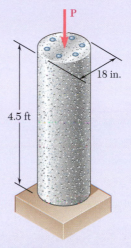

25 mm

Brass core
$E = 105$ GPa

300 mm

Aluminium shell
$E = 70$ GPa

60 mm

Fig. P2.37 and P2.38

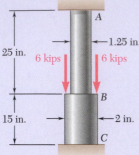

Fig. P2.39

2.39 A polystyrene rod consisting of two cylindrical portions AB and BC is restrained at both ends and supports two 6-kip loads as shown. Knowing that $E = 0.45 \times 10^6$ psi, determine (a) the reactions at A and C, (b) the normal stress in each portion of the rod.

2.40 Three steel rods ($E = 29 \times 10^6$ psi) support an 8.5-kip load **P**. Each of the rods AB and CD has a 0.32-in² cross-sectional area and rod EF has a 1-in² cross-sectional area. Neglecting the deformation of bar BED, determine (a) the change in length of rod EF, (b) the stress in each rod.

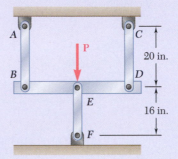

Fig. *P2.40*

2.41 Two cylindrical rods, one of steel and the other of brass, are joined at C and restrained by rigid supports at A and E. For the loading shown and knowing that $E_s = 200$ GPa and $E_b = 105$ GPa, determine (a) the reactions at A and E, (b) the deflection of point C.

2.42 Solve Prob. 2.41, assuming that rod AC is made of brass and rod CE is made of steel.

2.43 Each of the rods BD and CE is made of brass ($E = 105$ GPa) and has a cross-sectional area of 200 mm². Determine the deflection of end A of the rigid member ABC caused by the 2-kN load.

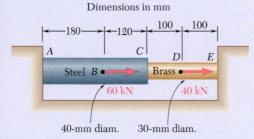

Fig. P2.41

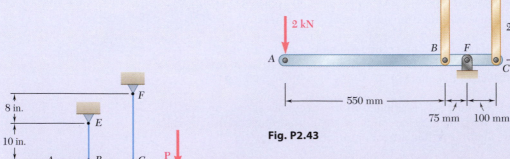

Fig. P2.43

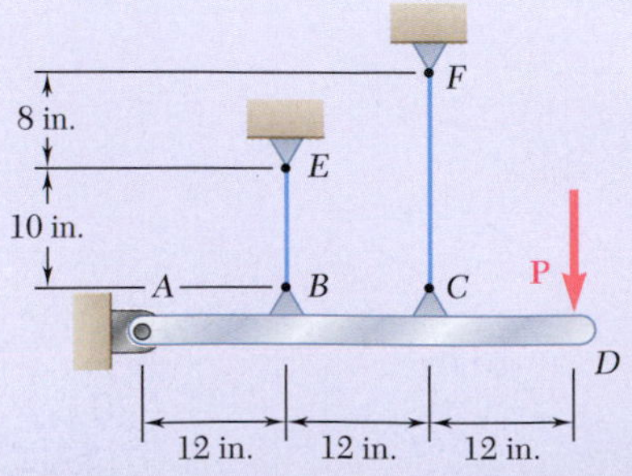

2.44 The rigid bar AD is supported by two steel wires of $\frac{1}{16}$-in. diameter ($E = 29 \times 10^6$ psi) and a pin and bracket at A. Knowing that the wires were initially taut, determine (a) the additional tension in each wire when a 220-lb load **P** is applied at D, (b) the corresponding deflection of point D.

Fig. P2.44

2.45 The rigid bar *ABC* is suspended from three wires of the same material. The cross-sectional area of the wire at *B* is equal to half of the cross-sectional area of the wires at *A* and *C*. Determine the tension in each wire caused by the load **P** shown.

2.46 The rigid bar *AD* is supported by two steel wires of $\frac{1}{16}$-in. diameter ($E = 29 \times 10^6$ psi) and a pin and bracket at *D*. Knowing that the wires were initially taut, determine (*a*) the additional tension in each wire when a 120-lb load **P** is applied at *B*, (*b*) the corresponding deflection of point *B*.

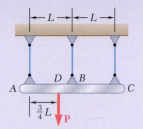

Fig. P2.45

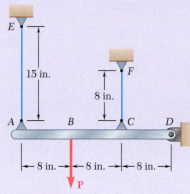

Fig. *P2.46*

2.47 The aluminum shell is fully bonded to the brass core and the assembly is unstressed at a temperature of 15°C. Considering only axial deformations, determine the stress in the aluminum when the temperature reaches 195°C.

2.48 Solve Prob. 2.47, assuming that the core is made of steel ($E_s = 200$ GPa, $\alpha_s = 11.7 \times 10^{-6}/°C$) instead of brass.

2.49 The brass shell ($\alpha_b = 11.6 \times 10^{-6}/°F$) is fully bonded to the steel core ($\alpha_s = 6.5 \times 10^{-6}/°F$). Determine the largest allowable increase in temperature if the stress in the steel core is not to exceed 8 ksi.

Brass core
E = 105 GPa
$\alpha = 20.9 \times 10^{-6}/°C$

Aluminum shell
E = 70 GPa
$\alpha = 23.6 \times 10^{-6}/°C$

Fig. P2.47

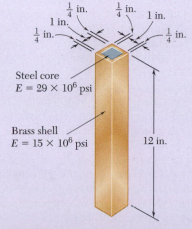

Fig. *P2.49*

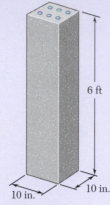

Fig. P2.50

2.50 The concrete post ($E_c = 3.6 \times 10^6$ psi and $\alpha_c = 5.5 \times 10^{-6}/°$F) is reinforced with six steel bars, each of $\frac{7}{8}$-in. diameter ($E_s = 29 \times 10^6$ psi and $\alpha_s = 6.5 \times 10^{-6}/°$F). Determine the normal stresses induced in the steel and in the concrete by a temperature rise of 65°F.

2.51 A rod consisting of two cylindrical portions AB and BC is restrained at both ends. Portion AB is made of steel ($E_s = 200$ GPa, $\alpha_s = 11.7 \times 10^{-6}/°$C) and portion BC is made of brass ($E_b = 105$ GPa, $\alpha_b = 20.9 \times 10^{-6}/°$C). Knowing that the rod is initially unstressed, determine the compressive force induced in ABC when there is a temperature rise of 50°C.

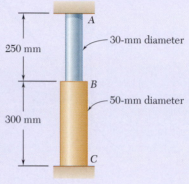

Fig. P2.51

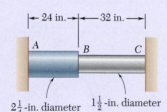

Fig. P2.52

2.52 A rod consisting of two cylindrical portions AB and BC is restrained at both ends. Portion AB is made of steel ($E_s = 29 \times 10^6$ psi, $\alpha_s = 6.5 \times 10^{-6}/°$F) and portion BC is made of aluminum ($E_a = 10.4 \times 10^6$ psi, $\alpha_a = 13.3 \times 10^{-6}/°$F). Knowing that the rod is initially unstressed, determine (a) the normal stresses induced in portions AB and BC by a temperature rise of 70°F, (b) the corresponding deflection of point B.

2.53 Solve Prob. 2.52, assuming that portion AB of the composite rod is made of aluminum and portion BC is made of steel.

2.54 The steel rails of a railroad track ($E_s = 200$ GPa, $\alpha_s = 11.7 \times 10^{-6}/°$C) were laid at a temperature of 6°C. Determine the normal stress in the rails when the temperature reaches 48°C, assuming that the rails (a) are welded to form a continuous track, (b) are 10 m long with 3-mm gaps between them.

2.55 Two steel bars ($E_s = 200$ GPa and $\alpha_s = 11.7 \times 10^{-6}/°$C) are used to reinforce a brass bar ($E_b = 105$ GPa, $\alpha_b = 20.9 \times 10^{-6}/°$C) that is subjected to a load $P = 25$ kN. When the steel bars were fabricated, the distance between the centers of the holes that were to fit on the pins was made 0.5 mm smaller than the 2 m needed. The steel bars were then placed in an oven to increase their length so that they would just fit on the pins. Following fabrication, the temperature in the steel bars dropped back to room temperature. Determine (a) the increase in temperature that was required to fit the steel bars on the pins, (b) the stress in the brass bar after the load is applied to it.

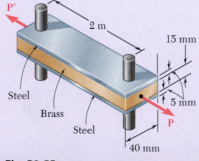

Fig. P2.55

2.56 Determine the maximum load P that can be applied to the brass bar of Prob. 2.55 if the allowable stress in the steel bars is 30 MPa and the allowable stress in the brass bar is 25 MPa.

2.57 An aluminum rod ($E_a = 70$ GPa, $\alpha_a = 23.6 \times 10^{-6}/°C$) and a steel link ($E_s = 200$ GPa, $\alpha_s = 11.7 \times 10^{-6}/°C$) have the dimensions shown at a temperature of 20°C. The steel link is heated until the aluminum rod can be fitted freely into the link. The temperature of the whole assembly is then raised to 150°C. Determine the final normal stress (a) in the rod, (b) in the link.

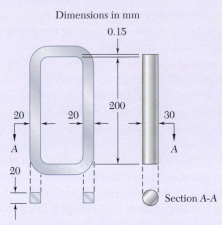

Dimensions in mm

Section A-A

Fig. P2.57

2.58 Knowing that a 0.02-in. gap exists when the temperature is 75°F, determine (a) the temperature at which the normal stress in the aluminum bar will be equal to -11 ksi, (b) the corresponding exact length of the aluminum bar.

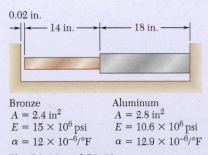

Bronze
$A = 2.4$ in^2
$E = 15 \times 10^6$ psi
$\alpha = 12 \times 10^{-6}/°F$

Aluminum
$A = 2.8$ in^2
$E = 10.6 \times 10^6$ psi
$\alpha = 12.9 \times 10^{-6}/°F$

Fig. P2.58 and P2.59

2.59 Determine (a) the compressive force in the bars shown after a temperature rise of 180°F, (b) the corresponding change in length of the bronze bar.

2.60 At room temperature (20°C) a 0.5-mm gap exists between the ends of the rods shown. At a later time when the temperature has reached 140°C, determine (a) the normal stress in the aluminum rod, (b) the change in length of the aluminum rod.

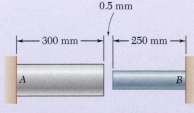

Aluminum
$A = 2000$ mm^2
$E = 75$ GPa
$\alpha = 23 \times 10^{-6}/°C$

Stainless steel
$A = 800$ mm^2
$E = 190$ GPa
$\alpha = 17.3 \times 10^{-6}/°C$

Fig. P2.60

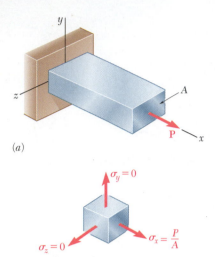

(a)

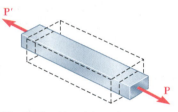

(b)

Fig. 2.29 A bar in uniaxial tension and a representative stress element.

Fig. 2.30 Materials undergo transverse contraction when elongated under axial load.

2.4 POISSON'S RATIO

When a homogeneous slender bar is axially loaded, the resulting stress and strain satisfy Hooke's law, as long as the elastic limit of the material is not exceeded. Assuming that the load **P** is directed along the x axis (Fig. 2.29a), $\sigma_x = P/A$, where A is the cross-sectional area of the bar, and from Hooke's law,

$$\epsilon_x = \sigma_x/E \tag{2.16}$$

where E is the modulus of elasticity of the material.

Also, the normal stresses on faces perpendicular to the y and z axes are zero: $\sigma_y = \sigma_z = 0$ (Fig. 2.29b). It would be tempting to conclude that the corresponding strains ϵ_y and ϵ_z are also zero. This *is not the case*. In all engineering materials, the elongation produced by an axial tensile force **P** in the direction of the force is accompanied by a contraction in any transverse direction (Fig. 2.30).[†] In this section and the following sections, all materials are assumed to be both *homogeneous* and *isotropic* (i.e., their mechanical properties are independent of both *position* and *direction*). It follows that the strain must have the same value for any transverse direction. Therefore, the loading shown in Fig. 2.29 must have $\epsilon_y = \epsilon_z$. This common value is the *lateral strain*. An important constant for a given material is its *Poisson's ratio*, named after the French mathematician Siméon Denis Poisson (1781–1840) and denoted by the Greek letter ν (nu).

$$\nu = -\frac{\text{lateral strain}}{\text{axial strain}} \tag{2.17}$$

or

$$\nu = -\frac{\epsilon_y}{\epsilon_x} = -\frac{\epsilon_z}{\epsilon_x} \tag{2.18}$$

for the loading condition represented in Fig. 2.29. Note the use of a minus sign in these equations to obtain a positive value for ν, as the axial and lateral strains have opposite signs for all engineering materials.[‡] Solving Eq. (2.18) for ϵ_y and ϵ_z, and recalling Eq. (2.16), write the following relationships, which fully describe the condition of strain under an axial load applied in a direction parallel to the x axis:

$$\epsilon_x = \frac{\sigma_x}{E} \qquad \epsilon_y = \epsilon_z = -\frac{\nu\sigma_x}{E} \tag{2.19}$$

[†]It also would be tempting, but equally wrong, to assume that the volume of the rod remains unchanged as a result of the combined effect of the axial elongation and transverse contraction (see Sec. 2.6).

[‡]However, some experimental materials, such as polymer foams, expand laterally when stretched. Since the axial and lateral strains have then the same sign, Poisson's ratio of these materials is negative. (*See* Roderic Lakes, "Foam Structures with a Negative Poisson's Ratio," *Science*, 27 February 1987, Volume 235, pp. 1038–1040.)

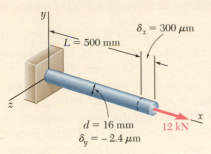

Fig. 2.31 Axially loaded rod.

Concept Application 2.7

A 500-mm-long, 16-mm-diameter rod made of a homogenous, isotropic material is observed to increase in length by 300 μm, and to decrease in diameter by 2.4 μm when subjected to an axial 12-kN load. Determine the modulus of elasticity and Poisson's ratio of the material.

The cross-sectional area of the rod is

$$A = \pi r^2 = \pi(8 \times 10^{-3}\,\text{m})^2 = 201 \times 10^{-6}\,\text{m}^2$$

Choosing the x axis along the axis of the rod (Fig. 2.31), write

$$\sigma_x = \frac{P}{A} = \frac{12 \times 10^3\,\text{N}}{201 \times 10^{-6}\,\text{m}^2} = 59.7\,\text{MPa}$$

$$\epsilon_x = \frac{\delta_x}{L} = \frac{300\,\mu\text{m}}{500\,\text{mm}} = 600 \times 10^{-6}$$

$$\epsilon_y = \frac{\delta_y}{d} = \frac{-2.4\,\mu\text{m}}{16\,\text{mm}} = -150 \times 10^{-6}$$

From Hooke's law, $\sigma_x = E\epsilon_x$,

$$E = \frac{\sigma_x}{\epsilon_x} = \frac{59.7\,\text{MPa}}{600 \times 10^{-6}} = 99.5\,\text{GPa}$$

and from Eq. (2.18),

$$\nu = -\frac{\epsilon_y}{\epsilon_x} = -\frac{-150 \times 10^{-6}}{600 \times 10^{-6}} = 0.25$$

2.5 MULTIAXIAL LOADING: GENERALIZED HOOKE'S LAW

All the examples considered so far in this chapter have dealt with slender members subjected to axial loads, i.e., to forces directed along a single axis. Consider now structural elements subjected to loads acting in the directions of the three coordinate axes and producing normal stresses σ_x, σ_y, and σ_z that are all different from zero (Fig. 2.32). This condition is a

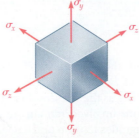

Fig. 2.32 State of stress for multiaxial loading.

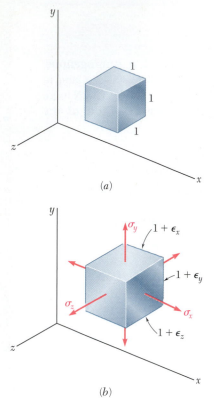

Fig. 2.33 Deformation of unit cube under multiaxial loading: (*a*) unloaded; (*b*) deformed.

multiaxial loading. Note that this is not the general stress condition described in Sec. 1.3, since no shearing stresses are included among the stresses shown in Fig. 2.32.

Consider an element of an isotropic material in the shape of a cube (Fig. 2.33*a*). Assume the side of the cube to be equal to unity, since it is always possible to select the side of the cube as a unit of length. Under the given multiaxial loading, the element will deform into a *rectangular parallelepiped* of sides equal to $1 + \epsilon_x$, $1 + \epsilon_y$, and $1 + \epsilon_z$, where ϵ_x, ϵ_y, and ϵ_z denote the values of the normal strain in the directions of the three coordinate axes (Fig. 2.33*b*). Note that, as a result of the deformations of the other elements of the material, the element under consideration could also undergo a translation, but the concern here is with the *actual deformation* of the element, not with any possible superimposed rigid-body displacement.

In order to express the strain components ϵ_x, ϵ_y, ϵ_z in terms of the stress components σ_x, σ_y, σ_z, consider the effect of each stress component and combine the results. This approach will be used repeatedly in this text, and is based on the *principle of superposition*. This principle states that the effect of a given combined loading on a structure can be obtained by *determining the effects of the various loads separately and combining the results*, provided that the following conditions are satisfied:

1. Each effect is linearly related to the load that produces it.
2. The deformation resulting from any given load is small and does not affect the conditions of application of the other loads.

For multiaxial loading, the first condition is satisfied if the stresses do not exceed the proportional limit of the material, and the second condition is also satisfied if the stress on any given face does not cause deformations of the other faces that are large enough to affect the computation of the stresses on those faces.

Considering the effect of the stress component σ_x, recall from Sec. 2.4 that σ_x causes a strain equal to σ_x/E in the x direction and strains equal to $-\nu\sigma_x/E$ in each of the y and z directions. Similarly, the stress component σ_y, if applied separately, will cause a strain σ_y/E in the y direction and strains $-\nu\sigma_y/E$ in the other two directions. Finally, the stress component σ_z causes a strain σ_z/E in the z direction and strains $-\nu\sigma_z/E$ in the x and y directions. Combining the results, the components of strain corresponding to the given multiaxial loading are

$$\begin{aligned} \epsilon_x &= +\frac{\sigma_x}{E} - \frac{\nu\sigma_y}{E} - \frac{\nu\sigma_z}{E} \\ \epsilon_y &= -\frac{\nu\sigma_x}{E} + \frac{\sigma_y}{E} - \frac{\nu\sigma_z}{E} \\ \epsilon_z &= -\frac{\nu\sigma_x}{E} - \frac{\nu\sigma_y}{E} + \frac{\sigma_z}{E} \end{aligned} \tag{2.20}$$

Equations (2.20) are the *generalized Hooke's law for the multiaxial loading of a homogeneous isotropic material.* As indicated earlier, these results are valid only as long as the stresses do not exceed the proportional limit and the deformations involved remain small. Also, a positive value for a stress component signifies tension and a negative value compression. Similarly, a positive value for a strain component indicates expansion in the corresponding direction and a negative value contraction.

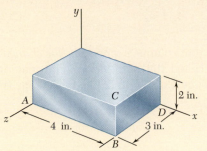

Fig. 2.34   Steel block under uniform pressure p.

Concept Application 2.8

The steel block shown (Fig. 2.34) is subjected to a uniform pressure on all its faces. Knowing that the change in length of edge AB is -1.2×10^{-3} in., determine (*a*) the change in length of the other two edges and (*b*) the pressure p applied to the faces of the block. Assume $E = 29 \times 10^{6}$ psi and $\nu = 0.29$.

a. Change in Length of Other Edges.   Substituting $\sigma_x = \sigma_y = \sigma_z = -p$ into Eqs. (2.20), the three strain components have the common value

$$\epsilon_x = \epsilon_y = \epsilon_z = -\frac{p}{E}(1 - 2\nu) \qquad (1)$$

Since

$$\epsilon_x = \delta_x/AB = (-1.2 \times 10^{-3}\,\text{in.})/(4\,\text{in.})$$
$$= -300 \times 10^{-6}\,\text{in./in.}$$

obtain

$$\epsilon_y = \epsilon_z = \epsilon_x = -300 \times 10^{-6}\,\text{in./in.}$$

from which

$$\delta_y = \epsilon_y(BC) = (-300 \times 10^{-6})(2\,\text{in.}) = -600 \times 10^{-6}\,\text{in.}$$
$$\delta_z = \epsilon_z(BD) = (-300 \times 10^{-6})(3\,\text{in.}) = -900 \times 10^{-6}\,\text{in.}$$

b. Pressure.   Solving Eq. (1) for p,

$$p = -\frac{E\epsilon_x}{1 - 2\nu} = -\frac{(29 \times 10^{6}\,\text{psi})(-300 \times 10^{-6})}{1 - 0.58}$$
$$p = 20.7\,\text{ksi}$$

*2.6   DILATATION AND BULK MODULUS

This section examines the effect of the normal stresses σ_x, σ_y, and σ_z on the volume of an element of isotropic material. Consider the element shown in Fig. 2.33. In its unstressed state, it is in the shape of a cube of unit volume. Under the stresses σ_x, σ_y, σ_z, it deforms into a rectangular parallelepiped of volume

$$v = (1 + \epsilon_x)(1 + \epsilon_y)(1 + \epsilon_z)$$

Since the strains ϵ_x, ϵ_y, ϵ_z are much smaller than unity, their products can be omitted in the expansion of the product. Therefore,

$$v = 1 + \epsilon_x + \epsilon_y + \epsilon_z$$

The change in volume e of the element is

$$e = v - 1 = 1 + \epsilon_x + \epsilon_y + \epsilon_z - 1$$

or

$$e = \epsilon_x + \epsilon_y + \epsilon_z \tag{2.21}$$

Since the element originally had a unit volume, e represents *the change in volume per unit volume* and is called the *dilatation* of the material. Substituting for ϵ_x, ϵ_y, and ϵ_z from Eqs. (2.20) into (2.21), the change is

$$e = \frac{\sigma_x + \sigma_y + \sigma_z}{E} - \frac{2\nu(\sigma_x + \sigma_y + \sigma_z)}{E}$$

$$e = \frac{1 - 2\nu}{E}(\sigma_x + \sigma_y + \sigma_z) \tag{2.22}†$$

When a body is subjected to a uniform hydrostatic pressure p, each of the stress components is equal to $-p$ and Eq. (2.22) yields

$$e = -\frac{3(1 - 2\nu)}{E}p \tag{2.23}$$

Introducing the constant

$$k = \frac{E}{3(1 - 2\nu)} \tag{2.24}$$

Eq. (2.23) is given in the form

$$e = -\frac{p}{k} \tag{2.25}$$

The constant k is known as the *bulk modulus* or *modulus of compression* of the material. It is expressed in pascals or in psi.

Because a stable material subjected to a hydrostatic pressure can only *decrease* in volume, the dilatation e in Eq. (2.25) is negative, and the bulk modulus k is a positive quantity. Referring to Eq. (2.24), $1 - 2\nu > 0$ or $\nu < \frac{1}{2}$. Recall from Sec. 2.4 that ν is positive for all engineering materials. Thus, for any engineering material,

$$0 < \nu < \tfrac{1}{2} \tag{2.26}$$

Note that an ideal material having ν equal to zero can be stretched in one direction without any lateral contraction. On the other hand, an ideal material for which $\nu = \frac{1}{2}$ and $k = \infty$ is perfectly incompressible ($e = 0$). Referring to Eq. (2.22) and noting that since $\nu < \frac{1}{2}$ in the elastic range, stretching an engineering material in one direction, for example in the x direction ($\sigma_x > 0$, $\sigma_y = \sigma_z = 0$), results in an increase of its volume ($e > 0$).‡

---

†Since the dilatation e represents a change in volume, it must be independent of the orientation of the element considered. It then follows from Eqs. (2.21) and (2.22) that the quantities $\epsilon_x + \epsilon_y + \epsilon_z$ and $\sigma_x + \sigma_y + \sigma_z$ are also independent of the orientation of the element. This property will be verified in Chap. 7.

‡However, in the plastic range, the volume of the material remains nearly constant.

Concept Application 2.9

Determine the change in volume ΔV of the steel block shown in Fig. 2.34, when it is subjected to the hydrostatic pressure $p = 180$ MPa. Use $E = 200$ GPa and $\nu = 0.29$.

From Eq. (2.24), the bulk modulus of steel is

$$k = \frac{E}{3(1 - 2\nu)} = \frac{200 \text{ GPa}}{3(1 - 0.58)} = 158.7 \text{ GPa}$$

and from Eq. (2.25), the dilatation is

$$e = -\frac{p}{k} = -\frac{180 \text{ MPa}}{158.7 \text{ GPa}} = -1.134 \times 10^{-3}$$

Since the volume V of the block in its unstressed state is

$$V = (80 \text{ mm})(40 \text{ mm})(60 \text{ mm}) = 192 \times 10^{3} \text{ mm}^{3}$$

and e represents the change in volume per unit volume, $e = \Delta V/V$,

$$\Delta V = eV = (-1.134 \times 10^{-3})(192 \times 10^{3} \text{ mm}^{3})$$

$$\Delta V = -218 \text{ mm}^{3}$$

2.7   SHEARING STRAIN

When we derived in Sec. 2.5 the relations (2.20) between normal stresses and normal strains in a homogeneous isotropic material, we assumed that no shearing stresses were involved. In the more general stress situation represented in Fig. 2.35, shearing stresses τ_{xy}, τ_{yz}, and τ_{zx} are present (as well as the corresponding shearing stresses τ_{yx}, τ_{zy}, and τ_{xz}). These stresses have no direct effect on the normal strains and, as long as all the deformations involved remain small, they will not affect the derivation nor the validity of Eqs. (2.20). The shearing stresses, however, tend to deform a cubic element of material into an *oblique* parallelepiped.

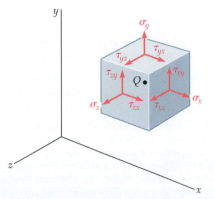

Fig. 2.35 Positive stress components at point Q for a general state of stress.

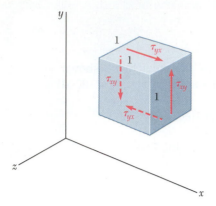

Fig. 2.36   Unit cubic element subjected to shearing stress.

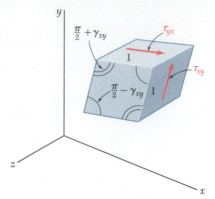

Fig. 2.37   Deformation of unit cubic element due to shearing stress.

Consider a cubic element (Fig. 2.36) subjected to only the shearing stresses τ_{xy} and τ_{yx} applied to faces of the element respectively perpendicular to the x and y axes. (Recall from Sec. 1.4 that $\tau_{xy} = \tau_{yx}$.) The cube is observed to deform into a rhomboid of sides equal to one (Fig. 2.37). Two of the angles formed by the four faces under stress are reduced from $\frac{\pi}{2}$ to $\frac{\pi}{2} - \gamma_{xy}$, while the other two are increased from $\frac{\pi}{2}$ to $\frac{\pi}{2} + \gamma_{xy}$. The small angle γ_{xy} (expressed in radians) defines the *shearing strain* corresponding to the x and y directions. When the deformation involves a *reduction* of the angle formed by the two faces oriented toward the positive x and y axes (as shown in Fig. 2.37), the shearing strain γ_{xy} is *positive;* otherwise, it is negative.

As a result of the deformations of the other elements of the material, the element under consideration also undergoes an overall rotation. The concern here is with the *actual deformation* of the element, not with any possible superimposed rigid-body displacement.[†]

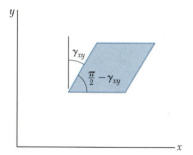

Fig. 2.38   Cubic element as viewed in *xy*-plane after rigid rotation.

Plotting successive values of τ_{xy} against the corresponding values of γ_{xy}, the shearing stress-strain diagram is obtained for the material. (This can be accomplished by carrying out a torsion test, as you will see in Chap. 3.) This diagram is similar to the normal stress-strain diagram from the tensile test described earlier; however, the values for the yield strength, ultimate strength, etc., are about half as large in shear as they are in tension. As for normal stresses and strains, the initial portion of the shearing stress-strain diagram is a straight line. For values of the shearing stress that do not exceed the proportional limit in shear, it can be written for any homogeneous isotropic material that

$$\tau_{xy} = G\gamma_{xy} \qquad (2.27)$$

This relationship is *Hooke's law for shearing stress and strain*, and the constant G is called the *modulus of rigidity* or *shear modulus* of the material.

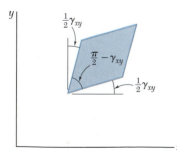

Fig. 2.39   Cubic element as viewed in *xy*-plane with equal rotation of *x* and *y* faces.

[†] In defining the strain γ_{xy}, some authors arbitrarily assume that the actual deformation of the element is accompanied by a rigid-body rotation where the horizontal faces of the element do not rotate. The strain γ_{xy} is then represented by the angle through which the other two faces have rotated (Fig. 2.38). Others assume a rigid-body rotates where the horizontal faces rotate through $\frac{1}{2}\gamma_{xy}$ counterclockwise and the vertical faces through $\frac{1}{2}\gamma_{xy}$ clockwise (Fig. 2.39). Since both assumptions are unnecessary and may lead to confusion, in this text you will associate the shearing strain γ_{xy} with the *change in the angle* formed by the two faces, rather than with the *rotation of a given face* under restrictive conditions.

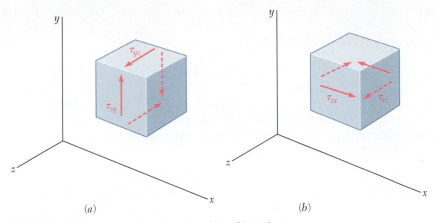

Fig. 2.40 States of pure shear in: (*a*) *yz*-plane; (*b*) *xz*-plane.

Since the strain γ_{xy} is defined as an angle in radians, it is dimensionless, and the modulus G is expressed in the same units as τ_{xy} in pascals or in psi. The modulus of rigidity G of any given material is less than one-half, but more than one-third of the modulus of elasticity E of that material.[†]

Now consider a small element of material subjected to shearing stresses τ_{yz} and τ_{zy} (Fig. 2.40*a*), where the shearing strain γ_{yz} is the change in the angle formed by the faces under stress. The shearing strain γ_{zx} is found in a similar way by considering an element subjected to shearing stresses τ_{zx} and τ_{xz} (Fig. 2.40*b*). For values of the stress that do not exceed the proportional limit, you can write two additional relationships:

$$\tau_{yz} = G\gamma_{yz} \qquad \tau_{zx} = G\gamma_{zx} \tag{2.28}$$

where the constant G is the same as in Eq. (2.27).

For the general stress condition represented in Fig. 2.35, and as long as none of the stresses involved exceeds the corresponding proportional limit, you can apply the principle of superposition and combine the results. The generalized Hooke's law for a homogeneous isotropic material under the most general stress condition is

$$\epsilon_x = +\frac{\sigma_x}{E} - \frac{\nu\sigma_y}{E} - \frac{\nu\sigma_z}{E}$$
$$\epsilon_y = -\frac{\nu\sigma_x}{E} + \frac{\sigma_y}{E} - \frac{\nu\sigma_z}{E}$$
$$\epsilon_z = -\frac{\nu\sigma_x}{E} - \frac{\nu\sigma_y}{E} + \frac{\sigma_z}{E} \tag{2.29}$$
$$\gamma_{xy} = \frac{\tau_{xy}}{G} \qquad \gamma_{yz} = \frac{\tau_{yz}}{G} \qquad \gamma_{zx} = \frac{\tau_{zx}}{G}$$

An examination of Eqs. (2.29) leads us to three distinct constants, E, ν, and G, which are used to predict the deformations caused in a given material by an arbitrary combination of stresses. Only two of these constants need be determined experimentally for any given material. The next section explains that the third constant can be obtained through a very simple computation.

[†]See Prob. 2.90.

Concept Application 2.10

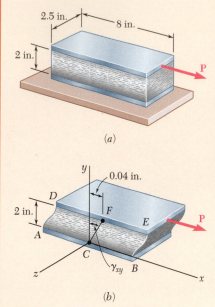

(a)

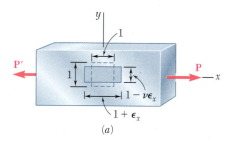

(b)

Fig. 2.41 *(a)* Rectangular block loaded in shear. *(b)* Deformed block showing the shearing strain.

A rectangular block of a material with a modulus of rigidity $G = 90$ ksi is bonded to two rigid horizontal plates. The lower plate is fixed, while the upper plate is subjected to a horizontal force **P** (Fig. 2.41*a*). Knowing that the upper plate moves through 0.04 in. under the action of the force, determine (*a*) the average shearing strain in the material and (*b*) the force **P** exerted on the upper plate.

a. Shearing Strain. The coordinate axes are centered at the midpoint C of edge AB and directed as shown (Fig. 2.41*b*). The shearing strain γ_{xy} is equal to the angle formed by the vertical and the line CF joining the midpoints of edges AB and DE. Noting that this is a very small angle and recalling that it should be expressed in radians, write

$$\gamma_{xy} \approx \tan \gamma_{xy} = \frac{0.04 \text{ in.}}{2 \text{ in.}} \qquad \gamma_{xy} = 0.020 \text{ rad}$$

b. Force Exerted on Upper Plate. Determine the shearing stress τ_{xy} in the material. Using Hooke's law for shearing stress and strain,

$$\tau_{xy} = G\gamma_{xy} = (90 \times 10^3 \text{ psi})(0.020 \text{ rad}) = 1800 \text{ psi}$$

The force exerted on the upper plate is

$$P = \tau_{xy}A = (1800 \text{ psi})(8 \text{ in.})(2.5 \text{ in.}) = 36.0 \times 10^3 \text{ lb}$$

$$P = 36.0 \text{ kips}$$

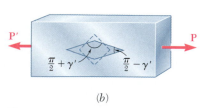

(a)

(b)

Fig. 2.42 Representations of strain in an axially-loaded bar: (*a*) cubic strain element faces aligned with coordinate axes; (*b*) cubic strain element faces rotated 45° about *z*-axis.

2.8 DEFORMATIONS UNDER AXIAL LOADING—RELATION BETWEEN E, ν, AND G

Section 2.4 showed that a slender bar subjected to an axial tensile load **P** directed along the *x* axis will elongate in the *x* direction and contract in both of the transverse *y* and *z* directions. If ϵ_x denotes the axial strain, the lateral strain is expressed as $\epsilon_y = \epsilon_z = -\nu\epsilon_x$, where ν is Poisson's ratio. Thus, an element in the shape of a cube of side equal to one and oriented as shown in Fig. 2.42*a* will deform into a rectangular parallelepiped of sides $1 + \epsilon_x$, $1 - \nu\epsilon_x$, and $1 - \nu\epsilon_x$. (Note that only one face of the element is shown in the figure.) On the other hand, if the element is oriented at 45° to the axis of the load (Fig. 2.42*b*), the face shown deforms into a rhombus. Therefore, the axial load **P** causes a shearing strain γ' equal to the amount by which each of the angles shown in Fig. 2.42*b* increases or decreases.[†]

The fact that shearing strains, as well as normal strains, result from an axial loading is not a surprise, since it was observed at the end of Sec. 1.4 that an axial load **P** causes normal and shearing stresses of equal magnitude on four of the faces of an element oriented at 45° to the axis of the member. This was illustrated in Fig. 1.38, which has been repeated

[†]Note that the load **P** also produces normal strains in the element shown in Fig. 2.42*b* (see Prob. 2.72).

here. It was also shown in Sec. 1.3 that the shearing stress is maximum on a plane forming an angle of 45° with the axis of the load. It follows from Hooke's law for shearing stress and strain that the shearing strain γ' associated with the element of Fig. 2.42b is also maximum: $\gamma' = \gamma_m$.

While a more detailed study of the transformations of strain is covered in Chap. 7, this section provides a relationship between the maximum shearing strain $\gamma' = \gamma_m$ associated with the element of Fig. 2.42b and the normal strain ϵ_x in the direction of the load. Consider the prismatic element obtained by intersecting the cubic element of Fig. 2.42a by a diagonal plane (Fig. 2.43a and b). Referring to Fig. 2.42a, this new element will deform into that shown in Fig. 2.43c, which has horizontal and vertical sides equal to $1 + \epsilon_x$ and $1 - \nu\epsilon_x$. But the angle formed by the oblique and horizontal faces of Fig. 2.43b is precisely half of one of the right angles of the cubic element in Fig. 2.42b. The angle β into which this angle deforms must be equal to half of $\pi/2 - \gamma_m$. Therefore,

$$\beta = \frac{\pi}{4} - \frac{\gamma_m}{2}$$

Applying the formula for the tangent of the difference of two angles,

$$\tan \beta = \frac{\tan \dfrac{\pi}{4} - \tan \dfrac{\gamma_m}{2}}{1 + \tan \dfrac{\pi}{4} \tan \dfrac{\gamma_m}{2}} = \frac{1 - \tan \dfrac{\gamma_m}{2}}{1 + \tan \dfrac{\gamma_m}{2}}$$

or since $\gamma_m/2$ is a very small angle,

$$\tan \beta = \frac{1 - \dfrac{\gamma_m}{2}}{1 + \dfrac{\gamma_m}{2}} \tag{2.30}$$

From Fig. 2.43c, observe that

$$\tan \beta = \frac{1 - \nu\epsilon_x}{1 + \epsilon_x} \tag{2.31}$$

Equating the right-hand members of Eqs. (2.30) and (2.31) and solving for γ_m, results in

$$\gamma_m = \frac{(1 + \nu)\epsilon_x}{1 + \dfrac{1 - \nu}{2}\epsilon_x}$$

Since $\epsilon_x \ll 1$, the denominator in the expression obtained can be assumed equal to one. Therefore,

$$\gamma_m = (1 + \nu)\epsilon_x \tag{2.32}$$

which is the desired relation between the maximum shearing strain γ_m and the axial strain ϵ_x.

To obtain a relation among the constants E, ν, and G, we recall that, by Hooke's law, $\gamma_m = \tau_m/G$, and for an axial loading, $\epsilon_x = \sigma_x/E$. Equation (2.32) can be written as

$$\frac{\tau_m}{G} = (1 + \nu)\frac{\sigma_x}{E}$$

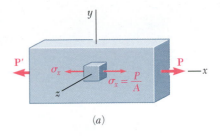

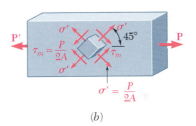

Fig. 1.38 (repeated)

(a)

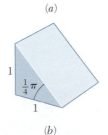

(b)

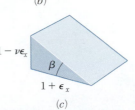

(c)

Fig. 2.43 (a) Cubic strain unit element, to be sectioned on a diagonal plane. (b) Undeformed section of unit element. (c) Deformed section of unit element.

or

$$\frac{E}{G} = (1 + \nu)\frac{\sigma_x}{\tau_m} \qquad (2.33)$$

Recall from Fig. 1.38 that $\sigma_x = P/A$ and $\tau_m = P/2A$, where A is the cross-sectional area of the member. Thus, $\sigma_x/\tau_m = 2$. Substituting this value into Eq. (2.33) and dividing both members by 2, the relationship is

$$\frac{E}{2G} = 1 + \nu \qquad (2.34)$$

which can be used to determine one of the constants E, ν, or G from the other two. For example, solving Eq. (2.34) for G,

$$G = \frac{E}{2(1 + \nu)} \qquad (2.35)$$

*2.9 STRESS-STRAIN RELATIONSHIPS FOR FIBER-REINFORCED COMPOSITE MATERIALS

Fiber-reinforced composite materials are fabricated by embedding fibers of a strong, stiff material into a weaker, softer material called a *matrix*. The relationship between the normal stress and the corresponding normal strain created in a lamina or layer of a composite material depends upon the direction in which the load is applied. Different moduli of elasticity, E_x, E_y, and E_z, are required to describe the relationship between normal stress and normal strain, according to whether the load is applied parallel to the fibers, perpendicular to the layer, or in a transverse direction.

Consider again the layer of composite material discussed in Sec. 2.1D and subject it to a uniaxial tensile load parallel to its fibers (Fig. 2.44a). It is assumed that the properties of the fibers and of the matrix have been combined or "smeared" into a fictitious, equivalent homogeneous material possessing these combined properties. In a small element of that layer of smeared material (Fig. 2.44b), the corresponding normal stress is σ_x and $\sigma_y = \sigma_z = 0$. As indicated in Sec. 2.1D, the corresponding normal strain in the x direction is $\epsilon_x = \sigma_x/E_x$, where E_x is the modulus of elasticity of the composite material in the x direction. As for isotropic materials, the elongation of the material in the x direction is accompanied by contractions in the y and z directions. These contractions depend upon the placement of the fibers in the matrix and generally will be different. Therefore, the lateral strains ϵ_y and ϵ_z also will be different, and the corresponding Poisson's ratios are

$$\nu_{xy} = -\frac{\epsilon_y}{\epsilon_x} \qquad \text{and} \qquad \nu_{xz} = -\frac{\epsilon_z}{\epsilon_x} \qquad (2.36)$$

Note that the first subscript in each of the Poisson's ratios ν_{xy} and ν_{xz} in Eqs. (2.36) refers to the direction of the load and the second to the direction of the contraction.

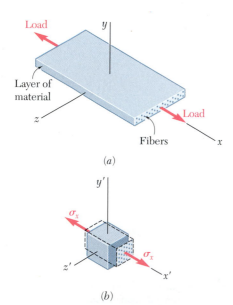

Fig. 2.44 Orthotropic fiber-reinforced composite material under uniaxial tensile load.

In the case of the *multiaxial loading* of a layer of a composite material, equations similar to Eqs. (2.20) of Sec. 2.5 can be used to describe the stress-strain relationship. In this case, three different values of the modulus of elasticity and six different values of Poisson's ratio are involved. We write

$$\epsilon_x = \frac{\sigma_x}{E_x} - \frac{\nu_{yx}\sigma_y}{E_y} - \frac{\nu_{zx}\sigma_z}{E_z}$$

$$\epsilon_y = -\frac{\nu_{xy}\sigma_x}{E_x} + \frac{\sigma_y}{E_y} - \frac{\nu_{zy}\sigma_z}{E_z} \qquad (2.37)$$

$$\epsilon_z = -\frac{\nu_{xz}\sigma_x}{E_x} - \frac{\nu_{yz}\sigma_y}{E_y} + \frac{\sigma_z}{E_z}$$

Equations (2.37) can be considered as defining the transformation of stress into strain for the given layer. It follows from a general property of such transformations that the coefficients of the stress components are symmetric:

$$\frac{\nu_{xy}}{E_x} = \frac{\nu_{yx}}{E_y} \qquad \frac{\nu_{yz}}{E_y} = \frac{\nu_{zy}}{E_z} \qquad \frac{\nu_{zx}}{E_z} = \frac{\nu_{xz}}{E_x} \qquad (2.38)$$

While different, these equations show that Poisson's ratios ν_{xy} and ν_{yx} are not independent; either of them can be obtained from the other if the corresponding values of the modulus of elasticity are known. The same is true of ν_{yz} and ν_{zy}, and of ν_{zx} and ν_{xz}.

Consider now the effect of shearing stresses on the faces of a small element of smeared layer. As discussed in Sec. 2.7 for isotropic materials, these stresses come in pairs of equal and opposite vectors applied to opposite sides of the given element and have no effect on the normal strains. Thus, Eqs. (2.37) remain valid. The shearing stresses, however, create shearing strains that are defined by equations similar to the last three of Eqs. (2.29) of Sec. 2.7, except that three different values of the modulus of rigidity, G_{xy}, G_{yz}, and G_{zx}, must be used:

$$\gamma_{xy} = \frac{\tau_{xy}}{G_{xy}} \qquad \gamma_{yz} = \frac{\tau_{yz}}{G_{yz}} \qquad \gamma_{zx} = \frac{\tau_{zx}}{G_{zx}} \qquad (2.39)$$

The fact that the three components of strain ϵ_x, ϵ_y, and ϵ_z can be expressed in terms of the normal stresses only and do not depend upon any shearing stresses characterizes *orthotropic materials* and distinguishes them from other anisotropic materials.

As in Sec. 2.1D, a flat *laminate* is obtained by superposing a number of layers or laminas. If the fibers in all layers are given the same orientation to withstand an axial tensile load, the laminate itself will be orthotropic. If the lateral stability of the laminate is increased by positioning some of its layers so that their fibers are at a right angle to the fibers of the other layers, the resulting laminate also will be orthotropic. On the other hand, if any of the layers of a laminate are positioned so that their fibers are neither parallel nor perpendicular to the fibers of other layers, the lamina generally will not be orthotropic.[†]

[†]For more information on fiber-reinforced composite materials, see Hyer, M. W., *Stress Analysis of Fiber-Reinforced Composite Materials*, DEStech Publications, Inc., Lancaster, PA, 2009.

Concept Application 2.11

A 60-mm cube is made from layers of graphite epoxy with fibers aligned in the x direction. The cube is subjected to a compressive load of 140 kN in the x direction. The properties of the composite material are: $E_x = 155.0$ GPa, $E_y = 12.10$ GPa, $E_z = 12.10$ GPa, $\nu_{xy} = 0.248$, $\nu_{xz} = 0.248$, and $\nu_{yz} = 0.458$. Determine the changes in the cube dimensions, knowing that (a) the cube is free to expand in the y and z directions (Fig. 2.45a); (b) the cube is free to expand in the z direction, but is restrained from expanding in the y direction by two fixed frictionless plates (Fig. 2.45b).

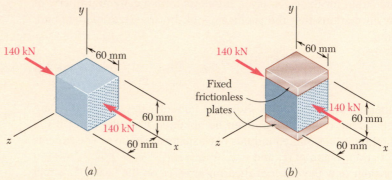

Fig. 2.45 Graphite-epoxy cube undergoing compression loading along the fiber direction; (a) unrestrained cube; (b) cube restrained in y direction.

a. Free in y and z Directions. Determine the stress σ_x in the direction of loading.

$$\sigma_x = \frac{P}{A} = \frac{-140 \times 10^3 \, \text{N}}{(0.060 \, \text{m})(0.060 \, \text{m})} = -38.89 \, \text{MPa}$$

Since the cube is not loaded or restrained in the y and z directions, we have $\sigma_y = \sigma_z = 0$. Thus, the right-hand members of Eqs. (2.37) reduce to their first terms. Substituting the given data into these equations,

$$\epsilon_x = \frac{\sigma_x}{E_x} = \frac{-38.89 \, \text{MPa}}{155.0 \, \text{GPa}} = -250.9 \times 10^{-6}$$

$$\epsilon_y = -\frac{\nu_{xy}\sigma_x}{E_x} = -\frac{(0.248)(-38.89 \, \text{MPa})}{155.0 \, \text{GPa}} = +62.22 \times 10^{-6}$$

$$\epsilon_z = -\frac{\nu_{xz}\sigma_x}{E_x} = -\frac{(0.248)(-38.69 \, \text{MPa})}{155.0 \, \text{GPa}} = +62.22 \times 10^{-6}$$

The changes in the cube dimensions are obtained by multiplying the corresponding strains by the length $L = 0.060$ m of the side of the cube:

$$\delta_x = \epsilon_x L = (-250.9 \times 10^{-6})(0.060 \, \text{m}) = -15.05 \, \mu\text{m}$$

$$\delta_y = \epsilon_y L = (+62.2 \times 10^{-6})(0.060 \, \text{m}) = +3.73 \, \mu\text{m}$$

$$\delta_z = \epsilon_z L = (+62.2 \times 10^{-6})(0.060 \, \text{m}) = +3.73 \, \mu\text{m}$$

(continued)

b. Free in *z* Direction, Restrained in *y* Direction. The stress in the *x* direction is the same as in part *a*, namely, $\sigma_x = 38.89$ MPa. Since the cube is free to expand in the *z* direction as in part *a*, $\sigma_z = 0$. But since the cube is now restrained in the *y* direction, the stress σ_y is not zero. On the other hand, since the cube cannot expand in the *y* direction, $\delta_y = 0$. Thus, $\epsilon_y = \delta_y/L = 0$. Set $\sigma_z = 0$ and $\epsilon_y = 0$ in the second of Eqs. (2.37) and solve that equation for σ_y:

$$\sigma_y = \left(\frac{E_y}{E_x}\right)\nu_{xy}\sigma_x = \left(\frac{12.10}{155.0}\right)(0.248)(-38.89 \text{ MPa})$$

$$= -752.9 \text{ kPa}$$

Now that the three components of stress have been determined, use the first and last of Eqs. (2.37) to compute the strain components ϵ_x and ϵ_z. But the first of these equations contains Poisson's ratio ν_{yx}, and as you saw earlier this ratio *is not equal* to the ratio ν_{xy} that was among the given data. To find ν_{yx}, use the first of Eqs. (2.38) and write

$$\nu_{yx} = \left(\frac{E_y}{E_x}\right)\nu_{xy} = \left(\frac{12.10}{155.0}\right)(0.248) = 0.01936$$

Now set $\sigma_z = 0$ in the first and third of Eqs. (2.37) and substitute the given values of E_x, E_y, ν_{xz}, and ν_{yz}, as well as the values obtained for σ_x, σ_y, and ν_{yx}, resulting in

$$\epsilon_x = \frac{\sigma_x}{E_x} - \frac{\nu_{yx}\sigma_y}{E_y} = \frac{-38.89 \text{ MPa}}{155.0 \text{ GPa}} - \frac{(0.01936)(-752.9 \text{ kPa})}{12.10 \text{ GPa}}$$

$$= -249.7 \times 10^{-6}$$

$$\epsilon_z = -\frac{\nu_{xz}\sigma_x}{E_x} - \frac{\nu_{yz}\sigma_y}{E_y} = -\frac{(0.248)(-38.89 \text{ MPa})}{155.0 \text{ GPa}} - \frac{(0.458)(-752.9 \text{ kPa})}{12.10 \text{ GPa}}$$

$$= +90.72 \times 10^{-6}$$

The changes in the cube dimensions are obtained by multiplying the corresponding strains by the length $L = 0.060$ m of the side of the cube:

$$\delta_x = \epsilon_x L = (-249.7 \times 10^{-6})(0.060 \text{ m}) = -14.98 \text{ } \mu\text{m}$$

$$\delta_y = \epsilon_y L = (0)(0.060 \text{ m}) = 0$$

$$\delta_z = \epsilon_z L = (+90.72 \times 10^{-6})(0.060 \text{ m}) = +5.44 \text{ } \mu\text{m}$$

Comparing the results of parts *a* and *b*, note that the difference between the values for the deformation δ_x in the direction of the fibers is negligible. However, the difference between the values for the lateral deformation δ_z is not negligible when the cube is restrained from deforming in the *y* direction.

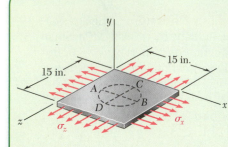

Sample Problem 2.5

A circle of diameter $d = 9$ in. is scribed on an unstressed aluminum plate of thickness $t = \frac{3}{4}$ in. Forces acting in the plane of the plate later cause normal stresses $\sigma_x = 12$ ksi and $\sigma_z = 20$ ksi. For $E = 10 \times 10^6$ psi and $\nu = \frac{1}{3}$, determine the change in (a) the length of diameter AB, (b) the length of diameter CD, (c) the thickness of the plate, and (d) the volume of the plate.

STRATEGY: You can use the generalized Hooke's Law to determine the components of strain. These strains can then be used to evaluate the various dimensional changes to the plate, and through the dilatation, also assess the volume change.

ANALYSIS:

Hooke's Law. Note that $\sigma_y = 0$. Using Eqs. (2.20), find the strain in each of the coordinate directions.

$$\epsilon_x = +\frac{\sigma_x}{E} - \frac{\nu\sigma_y}{E} - \frac{\nu\sigma_z}{E}$$

$$= \frac{1}{10 \times 10^6 \text{ psi}}\left[(12 \text{ ksi}) - 0 - \frac{1}{3}(20 \text{ ksi})\right] = +0.533 \times 10^{-3} \text{ in./in.}$$

$$\epsilon_y = -\frac{\nu\sigma_x}{E} + \frac{\sigma_y}{E} - \frac{\nu\sigma_z}{E}$$

$$= \frac{1}{10 \times 10^6 \text{ psi}}\left[-\frac{1}{3}(12 \text{ ksi}) + 0 - \frac{1}{3}(20 \text{ ksi})\right] = -1.067 \times 10^{-3} \text{ in./in.}$$

$$\epsilon_z = -\frac{\nu\sigma_x}{E} - \frac{\nu\sigma_y}{E} + \frac{\sigma_z}{E}$$

$$= \frac{1}{10 \times 10^6 \text{ psi}}\left[-\frac{1}{3}(12 \text{ ksi}) - 0 + (20 \text{ ksi})\right] = +1.600 \times 10^{-3} \text{ in./in.}$$

a. Diameter _AB_. The change in length is $\delta_{B/A} = \epsilon_x d$.

$$\delta_{B/A} = \epsilon_x d = (+0.533 \times 10^{-3} \text{ in./in.})(9 \text{ in.})$$

$$\delta_{B/A} = +4.8 \times 10^{-3} \text{ in.} \blacktriangleleft$$

b. Diameter _CD_.

$$\delta_{C/D} = \epsilon_z d = (+1.600 \times 10^{-3} \text{ in./in.})(9 \text{ in.})$$

$$\delta_{C/D} = +14.4 \times 10^{-3} \text{ in.} \blacktriangleleft$$

c. Thickness. Recalling that $t = \frac{3}{4}$ in.,

$$\delta_t = \epsilon_y t = (-1.067 \times 10^{-3} \text{ in./in.})(\tfrac{3}{4} \text{ in.})$$

$$\delta_t = -0.800 \times 10^{-3} \text{ in.} \blacktriangleleft$$

d. Volume of the Plate. Using Eq. (2.21),

$$e = \epsilon_x + \epsilon_y + \epsilon_z = (+0.533 - 1.067 + 1.600)10^{-3} = +1.067 \times 10^{-3}$$

$$\Delta V = eV = +1.067 \times 10^{-3}[(15 \text{ in.})(15 \text{ in.})(\tfrac{3}{4} \text{ in.})]$$

$$\Delta V = +0.180 \text{ in}^3 \blacktriangleleft$$

Problems

2.61 A standard tension test is used to determine the properties of an experimental plastic. The test specimen is a $\frac{5}{8}$-in.-diameter rod and it is subjected to an 800-lb tensile force. Knowing that an elongation of 0.45 in. and a decrease in diameter of 0.025 in. are observed in a 5-in. gage length, determine the modulus of elasticity, the modulus of rigidity, and Poisson's ratio for the material.

2.62 A 2-m length of an aluminum pipe of 240-mm outer diameter and 10-mm wall thickness is used as a short column to carry a 640-kN centric axial load. Knowing that $E = 73$ GPa and $\nu = 0.33$, determine (a) the change in length of the pipe, (b) the change in its outer diameter, (c) the change in its wall thickness.

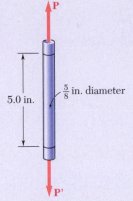

5.0 in.

$\frac{5}{8}$ in. diameter

Fig. P2.61

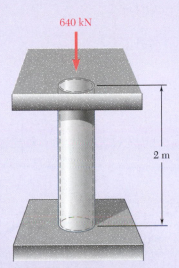

640 kN

2 m

Fig. *P2.62*

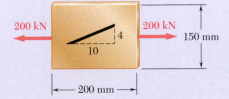

200 kN

200 kN

4

10

150 mm

200 mm

Fig. P2.63

2.63 A line of slope 4:10 has been scribed on a cold-rolled yellow-brass plate, 150 mm wide and 6 mm thick. Knowing that $E = 105$ GPa and $\nu = 0.34$, determine the slope of the line when the plate is subjected to a 200-kN centric axial load as shown.

2.64 A 2.75-kN tensile load is applied to a test coupon made from 1.6-mm flat steel plate ($E = 200$ GPa, $\nu = 0.30$). Determine the resulting change (a) in the 50-mm gage length, (b) in the width of portion AB of the test coupon, (c) in the thickness of portion AB, (d) in the cross-sectional area of portion AB.

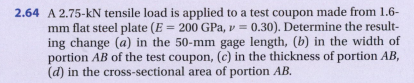

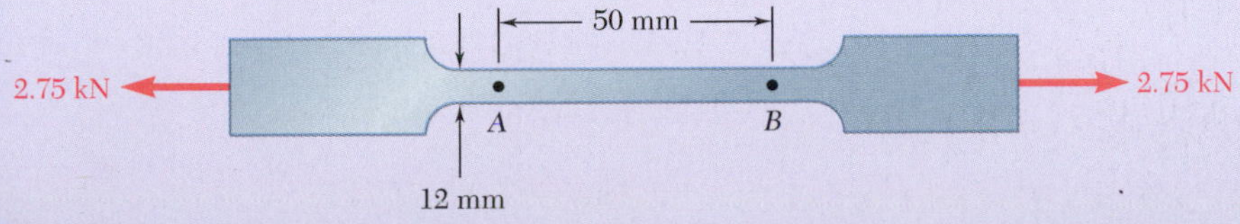

50 mm

2.75 kN

2.75 kN

A

B

12 mm

Fig. P2.64

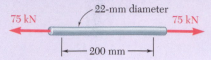

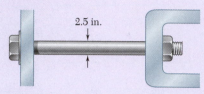

Fig. P2.65

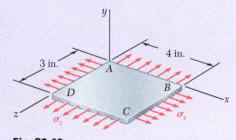

Fig. P2.66

2.65 In a standard tensile test a steel rod of 22-mm diameter is subjected to a tension force of 75 kN. Knowing that $\nu = 0.30$ and $E = 200$ GPa, determine (a) the elongation of the rod in a 200-mm gage length, (b) the change in diameter of the rod.

2.66 The change in diameter of a large steel bolt is carefully measured as the nut is tightened. Knowing that $E = 29 \times 10^6$ psi and $\nu = 0.30$, determine the internal force in the bolt if the diameter is observed to decrease by 0.5×10^{-3} in.

2.67 The brass rod AD is fitted with a jacket that is used to apply a hydrostatic pressure of 48 MPa to the 240-mm portion BC of the rod. Knowing that $E = 105$ GPa and $\nu = 0.33$, determine (a) the change in the total length AD, (b) the change in diameter at the middle of the rod.

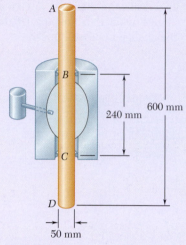

Fig. P2.67

2.68 A fabric used in air-inflated structures is subjected to a biaxial loading that results in normal stresses $\sigma_x = 18$ ksi and $\sigma_z = 24$ ksi. Knowing that the properties of the fabric can be approximated as $E = 12.6 \times 10^6$ psi and $\nu = 0.34$, determine the change in length of (a) side AB, (b) side BC, (c) diagonal AC.

2.69 A 1-in. square was scribed on the side of a large steel pressure vessel. After pressurization the biaxial stress condition at the square is as shown. Knowing that $E = 29 \times 10^6$ psi and $\nu = 0.30$, determine the change in length of (a) side AB, (b) side BC, (c) diagonal AC.

Fig. P2.68

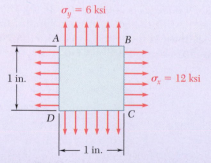

Fig. P2.69

2.70 The block shown is made of a magnesium alloy for which $E = 45$ GPa and $\nu = 0.35$. Knowing that $\sigma_x = -180$ MPa, determine (a) the magnitude of σ_y for which the change in the height of the block will be zero, (b) the corresponding change in the area of the face $ABCD$, (c) the corresponding change in the volume of the block.

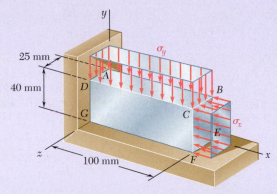

Fig. P2.70

2.71 The homogeneous plate $ABCD$ is subjected to a biaxial loading as shown. It is known that $\sigma_z = \sigma_0$ and that the change in length of the plate in the x direction must be zero, that is, $\epsilon_x = 0$. Denoting by E the modulus of elasticity and by ν Poisson's ratio, determine (a) the required magnitude of σ_x, (b) the ratio σ_0/ϵ_z.

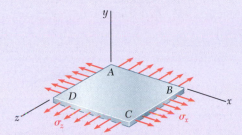

Fig. P2.71

2.72 For a member under axial loading, express the normal strain ϵ' in a direction forming an angle of 45° with the axis of the load in terms of the axial strain ϵ_x by (a) comparing the hypotenuses of the triangles shown in Fig. 2.43, which represent respectively an element before and after deformation, (b) using the values of the corresponding stresses σ' and σ_x shown in Fig. 1.38, and the generalized Hooke's law.

2.73 In many situations it is known that the normal stress in a given direction is zero. For example, $\sigma_z = 0$ in the case of the thin plate shown. For this case, which is known as *plane stress*, show that if the strains ϵ_x and ϵ_y have been determined experimentally, we can express σ_x, σ_y, and ϵ_z as follows:

$$\sigma_x = E\frac{\epsilon_x + \nu\epsilon_y}{1 - \nu^2}$$

$$\sigma_y = E\frac{\epsilon_y + \nu\epsilon_x}{1 - \nu^2}$$

$$\epsilon_z = -\frac{\nu}{1 - \nu}(\epsilon_x + \epsilon_y)$$

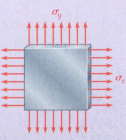

Fig. P2.73

2.74 In many situations physical constraints prevent strain from occurring in a given direction. For example, $\epsilon_z = 0$ in the case shown, where longitudinal movement of the long prism is prevented at every point. Plane sections perpendicular to the longitudinal axis remain plane and the same distance apart. Show that for this situation, which is known as *plane strain*, we can express σ_z, ϵ_x, and ϵ_y as follows:

$$\sigma_z = \nu(\sigma_x + \sigma_y)$$

$$\epsilon_x = \frac{1}{E}[(1 - \nu^2)\sigma_x - \nu(1 + \nu)\sigma_y]$$

$$\epsilon_y = \frac{1}{E}[(1 - \nu^2)\sigma_y - \nu(1 + \nu)\sigma_x]$$

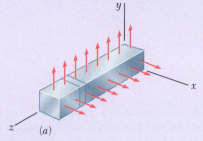

Fig. P2.74

2.75 The plastic block shown is bonded to a rigid support and to a vertical plate to which a 55-kip load **P** is applied. Knowing that for the plastic used $G = 150$ ksi, determine the deflection of the plate.

2.76 What load **P** should be applied to the plate of Prob. 2.75 to produce a $\frac{1}{16}$-in. deflection?

2.77 Two blocks of rubber with a modulus of rigidity $G = 12$ MPa are bonded to rigid supports and to a plate AB. Knowing that $c = 100$ mm and $P = 45$ kN, determine the smallest allowable dimensions a and b of the blocks if the shearing stress in the rubber is not to exceed 1.4 MPa and the deflection of the plate is to be at least 5 mm.

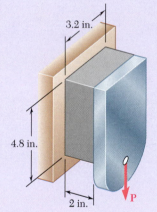

Fig. P2.75

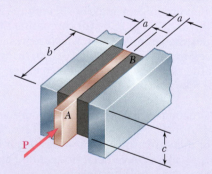

Fig. P2.77 and P2.78

2.78 Two blocks of rubber with a modulus of rigidity $G = 10$ MPa are bonded to rigid supports and to a plate AB. Knowing that $b = 200$ mm and $c = 125$ mm, determine the largest allowable load P and the smallest allowable thickness a of the blocks if the shearing stress in the rubber is not to exceed 1.5 MPa and the deflection of the plate is to be at least 6 mm.

2.79 An elastomeric bearing ($G = 130$ psi) is used to support a bridge girder as shown to provide flexibility during earthquakes. The beam must not displace more than $\frac{3}{8}$ in. when a 5-kip lateral load is applied as shown. Knowing that the maximum allowable shearing stress is 60 psi, determine (*a*) the smallest allowable dimension *b*, (*b*) the smallest required thickness *a*.

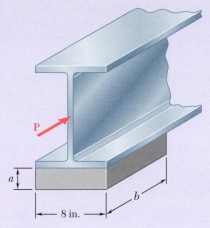

Fig. P2.79

2.80 For the elastomeric bearing in Prob. 2.79 with $b = 10$ in. and $a = 1$ in., determine the shearing modulus G and the shear stress τ for a maximum lateral load $P = 5$ kips and a maximum displacement $\delta = 0.4$ in.

2.81 A vibration isolation unit consists of two blocks of hard rubber bonded to a plate *AB* and to rigid supports as shown. Knowing that a force of magnitude $P = 25$ kN causes a deflection $\delta = 1.5$ mm of plate *AB*, determine the modulus of rigidity of the rubber used.

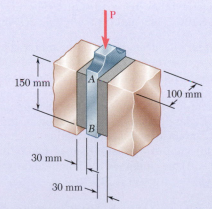

Fig. P2.81 and P2.82

2.82 A vibration isolation unit consists of two blocks of hard rubber with a modulus of rigidity $G = 19$ MPa bonded to a plate *AB* and to rigid supports as shown. Denoting by *P* the magnitude of the force applied to the plate and by δ the corresponding deflection, determine the effective spring constant, $k = P/\delta$, of the system.

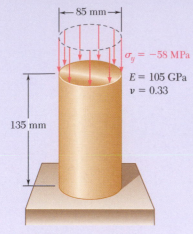

85 mm

$\sigma_y = -58$ MPa
$E = 105$ GPa
$\nu = 0.33$

135 mm

Fig. P2.84

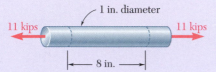

1 in. diameter

11 kips →   ← 11 kips

8 in.

Fig. P2.85

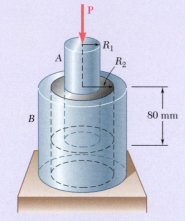

P

R_1
R_2

A

B

80 mm

Fig. *P2.87* and P2.88

$E_x = 50$ GPa $\nu_{xz} = 0.254$
$E_y = 15.2$ GPa $\nu_{xy} = 0.254$
$E_z = 15.2$ GPa $\nu_{zy} = 0.428$

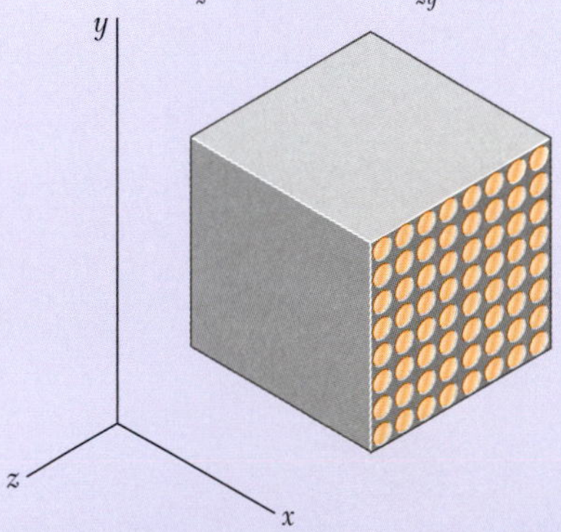

y

z   x

Fig. P2.91

*2.83 A 6-in.-diameter solid steel sphere is lowered into the ocean to a point where the pressure is 7.1 ksi (about 3 miles below the surface). Knowing that $E = 29 \times 10^6$ psi and $\nu = 0.30$, determine (a) the decrease in diameter of the sphere, (b) the decrease in volume of the sphere, (c) the percent increase in the density of the sphere.

*2.84 (a) For the axial loading shown, determine the change in height and the change in volume of the brass cylinder shown. (b) Solve part a, assuming that the loading is hydrostatic with $\sigma_x = \sigma_y = \sigma_z = -70$ MPa.

*2.85 Determine the dilatation e and the change in volume of the 8-in. length of the rod shown if (a) the rod is made of steel with $E = 29 \times 10^6$ psi and $\nu = 0.30$, (b) the rod is made of aluminum with $E = 10.6 \times 10^6$ psi and $\nu = 0.35$.

*2.86 Determine the change in volume of the 50-mm gage length segment AB in Prob. 2.64 (a) by computing the dilatation of the material, (b) by subtracting the original volume of portion AB from its final volume.

*2.87 A vibration isolation support consists of a rod A of radius $R_1 = 10$ mm and a tube B of inner radius $R_2 = 25$ mm bonded to an 80-mm-long hollow rubber cylinder with a modulus of rigidity $G = 12$ MPa. Determine the largest allowable force P that can be applied to rod A if its deflection is not to exceed 2.50 mm.

*2.88 A vibration isolation support consists of a rod A of radius R_1 and a tube B of inner radius R_2 bonded to an 80-mm-long hollow rubber cylinder with a modulus of rigidity $G = 10.93$ MPa. Determine the required value of the ratio R_2/R_1 if a 10-kN force P is to cause a 2-mm deflection of rod A.

*2.89 The material constants E, G, k, and ν are related by Eqs. (2.24) and (2.34). Show that any one of the constants may be expressed in terms of any other two constants. For example, show that (a) $k = GE/(9G - 3E)$ and (b) $\nu = (3k - 2G)/(6k + 2G)$.

*2.90 Show that for any given material, the ratio G/E of the modulus of rigidity over the modulus of elasticity is always less than $\frac{1}{2}$ but more than $\frac{1}{3}$. [Hint: Refer to Eq. (2.34) and to Sec. 2.1e.]

*2.91 A composite cube with 40-mm sides and the properties shown is made with glass polymer fibers aligned in the x direction. The cube is constrained against deformations in the y and z directions and is subjected to a tensile load of 65 kN in the x direction. Determine (a) the change in the length of the cube in the x direction and (b) the stresses σ_x, σ_y, and σ_z.

*2.92 The composite cube of Prob. 2.91 is constrained against deformation in the z direction and elongated in the x direction by 0.035 mm due to a tensile load in the x direction. Determine (a) the stresses σ_x, σ_y, and σ_z and (b) the change in the dimension in the y direction.

2.10   STRESS AND STRAIN DISTRIBUTION UNDER AXIAL LOADING: SAINT-VENANT'S PRINCIPLE

We have assumed so far that, in an axially loaded member, the normal stresses are uniformly distributed in any section perpendicular to the axis of the member. As we saw in Sec. 1.2A, such an assumption may be quite in error in the immediate vicinity of the points of application of the loads. However, the determination of the actual stresses in a given section of the member requires the solution of a statically indeterminate problem.

In Sec. 2.2, you saw that statically indeterminate problems involving the determination of *forces* can be solved by considering the *deformations* caused by these forces. It is thus reasonable to conclude that the determination of the *stresses* in a member requires the analysis of the strains produced by the stresses in the member. This is essentially the approach found in advanced textbooks, where the mathematical theory of elasticity is used to determine the distribution of stresses corresponding to various modes of application of the loads at the ends of the member. Given the more limited mathematical means at our disposal, our analysis of stresses will be restricted to the particular case when two rigid plates are used to transmit the loads to a member made of a homogeneous isotropic material (Fig. 2.46).

If the loads are applied at the center of each plate,[†] the plates will move toward each other without rotating, causing the member to get shorter, while increasing in width and thickness. It is assumed that the member will remain straight, plane sections will remain plane, and all elements of the member will deform in the same way, since this assumption is compatible with the given end conditions. Figure 2.47 shows a rubber model before and after loading.[‡] Now, if all elements deform in the same

Fig. 2.46   Axial load applied by rigid plates.

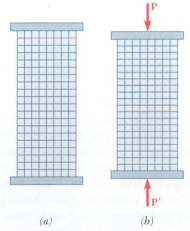

(a)   *(b)*

Fig. 2.47   Axial load applied by rigid plates to rubber model.

[†]More precisely, the common line of action of the loads should pass through the centroid of the cross section (cf. Sec. 1.2A).

[‡]Note that for long, slender members, another configuration is possible and will prevail if the load is sufficiently large; the member *buckles* and assumes a curved shape. This will be discussed in Chap. 10.

Fig. 2.48   Concentrated axial load applied to rubber model.

way, the distribution of strains throughout the member must be uniform. In other words, the axial strain ϵ_y and the lateral strain $\epsilon_x = -\nu\epsilon_y$ are constant. But, if the stresses do not exceed the proportional limit, Hooke's law applies, and $\sigma_y = E\epsilon_y$, so the normal stress σ_y is also constant. Thus, the distribution of stresses is uniform throughout the member, and at any point,

$$\sigma_y = (\sigma_y)_{\text{ave}} = \frac{P}{A}$$

If the loads are concentrated, as in Fig. 2.48, the elements in the immediate vicinity of the points of application of the loads are subjected to very large stresses, while other elements near the ends of the member are unaffected by the loading. This results in large deformations, strains, and stresses near the points of application of the loads, while no deformation takes place at the corners. Considering elements farther and farther from the ends, a progressive equalization of the deformations and a more uniform distribution of the strains and stresses are seen across a section of the member. Using the mathematical theory of elasticity found in advanced textbooks, Fig. 2.49 shows the resulting distribution of stresses across various sections of a thin rectangular plate subjected to concentrated loads. Note

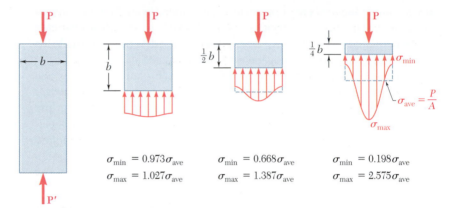

$$\sigma_{\min} = 0.973\sigma_{\text{ave}} \qquad \sigma_{\min} = 0.668\sigma_{\text{ave}} \qquad \sigma_{\min} = 0.198\sigma_{\text{ave}}$$
$$\sigma_{\max} = 1.027\sigma_{\text{ave}} \qquad \sigma_{\max} = 1.387\sigma_{\text{ave}} \qquad \sigma_{\max} = 2.575\sigma_{\text{ave}}$$

Fig. 2.49   Stress distributions in a plate under concentrated axial loads.

that at a distance b from either end, where b is the width of the plate, the stress distribution is nearly uniform across the section, and the value of the stress σ_y at any point of that section can be assumed to be equal to the average value P/A. Thus, at a distance equal to or greater than the width of the member, the distribution of stresses across a section is the same, whether the member is loaded as shown in Fig. 2.46 or Fig. 2.48. In other words, except in the immediate vicinity of the points of application of the loads, the stress distribution is assumed independent of the actual mode of application of the loads. This statement, which applies to axial loadings and to practically any type of load, is known as *Saint-Venant's principle*, after the French mathematician and engineer Adhémar Barré de Saint-Venant (1797–1886).

While Saint-Venant's principle makes it possible to replace a given loading by a simpler one to compute the stresses in a structural member, keep in mind two important points when applying this principle:

1. The actual loading and the loading used to compute the stresses must be *statically equivalent*.

2. Stresses cannot be computed in this manner in the immediate vicinity of the points of application of the loads. Advanced theoretical or experimental methods must be used to determine the distribution of stresses in these areas.

You should also observe that the plates used to obtain a uniform stress distribution in the member of Fig. 2.47 must allow the member to freely expand laterally. Thus, the plates cannot be rigidly attached to the member; assume them to be just in contact with the member and smooth enough not to impede lateral expansion. While such end conditions can be achieved for a member in compression, they cannot be physically realized in the case of a member in tension. It does not matter, whether or not an actual fixture can be realized and used to load a member so that the distribution of stresses in the member is uniform. The important thing is to *imagine a model* that will allow such a distribution of stresses and to keep this model in mind so that it can be compared with the actual loading conditions.

2.11   STRESS CONCENTRATIONS

As you saw in the preceding section, the stresses near the points of application of concentrated loads can reach values much larger than the average value of the stress in the member. When a structural member contains a discontinuity, such as a hole or a sudden change in cross section, high localized stresses can occur. Figures 2.50 and 2.51 show the distribution of stresses in critical sections corresponding to two situations. Figure 2.50 shows a flat bar with a *circular hole* and shows the stress distribution in a section passing through the center of the hole. Figure 2.51 shows a flat bar consisting of two portions of different widths connected by *fillets;* here the stress distribution is in the narrowest part of the connection, where the highest stresses occur.

These results were obtained experimentally through the use of a photoelastic method. Fortunately for the engineer, these results are independent of the size of the member and of the material used; they depend only upon the ratios of the geometric parameters involved (i.e., the ratio $2r/D$ for a circular hole, and the ratios r/d and D/d for fillets). Furthermore, the designer is more interested in the *maximum value* of the stress in a given section than the actual distribution of stresses. The main concern is to determine *whether* the allowable stress will be exceeded under a given loading, not *where* this value will be exceeded. Thus, the ratio

$$K = \frac{\sigma_{\max}}{\sigma_{\text{ave}}} \qquad (2.40)$$

is computed in the critical (narrowest) section of the discontinuity. This ratio is the *stress-concentration factor* of the discontinuity. Stress-concentration factors can be computed in terms of the ratios of the geometric parameters involved, and the results can be expressed in tables or graphs, as shown in Fig. 2.52. To determine the maximum stress occurring near a discontinuity in a given member subjected to a given axial load P, the designer needs to compute the average stress $\sigma_{\text{ave}} = P/A$ in the critical section and multiply the result obtained by the appropriate value of the stress-concentration factor K. Note that this procedure is valid only as long as $\sigma_{\max}$ does not exceed the proportional limit of the material, since the values of K plotted in Fig. 2.52 were obtained by assuming a linear relation between stress and strain.

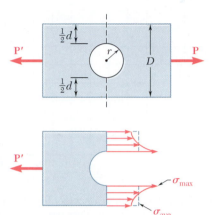

Fig. 2.50 Stress distribution near circular hole in flat bar under axial loading.

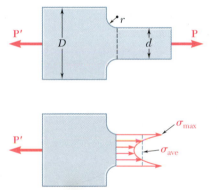

Fig. 2.51 Stress distribution near fillets in flat bar under axial loading.

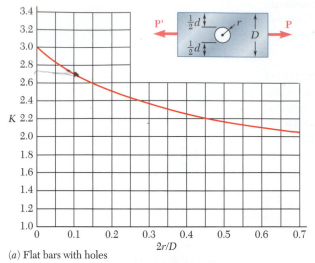

(a) Flat bars with holes

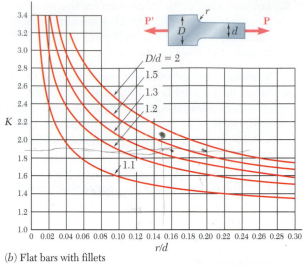

(b) Flat bars with fillets

Fig. 2.52 Stress concentration factors for flat bars under axial loading. Note that the average stress must be computed across the narrowest section: $\sigma_{ave} = P/td$, where t is the thickness of the bar. (Source: W. D. Pilkey and D.F. Pilkey, *Peterson's Stress Concentration Factors*, 3rd ed., John Wiley & Sons, New York, 2008.)

Concept Application 2.12

Determine the largest axial load **P** that can be safely supported by a flat steel bar consisting of two portions, both 10 mm thick and, respectively, 40 and 60 mm wide, connected by fillets of radius $r = 8$ mm. Assume an allowable normal stress of 165 MPa.

First compute the ratios

$$\frac{D}{d} = \frac{60 \text{ mm}}{40 \text{ mm}} = 1.50 \qquad \frac{r}{d} = \frac{8 \text{ mm}}{40 \text{ mm}} = 0.20$$

Using the curve in Fig. 2.52b corresponding to $D/d = 1.50$, the value of the stress-concentration factor corresponding to $r/d = 0.20$ is

$$K = 1.82$$

Then carrying this value into Eq. (2.40) and solving for σ_{ave},

$$\sigma_{ave} = \frac{\sigma_{max}}{1.82}$$

But σ_{max} cannot exceed the allowable stress $\sigma_{all} = 165$ MPa. Substituting this value for σ_{max}, the average stress in the narrower portion ($d = 40$ mm) of the bar should not exceed the value

$$\sigma_{ave} = \frac{165 \text{ MPa}}{1.82} = 90.7 \text{ MPa}$$

Recalling that $\sigma_{ave} = P/A$,

$$P = A\sigma_{ave} = (40 \text{ mm})(10 \text{ mm})(90.7 \text{ MPa}) = 36.3 \times 10^3 \text{ N}$$

$$P = 36.3 \text{ kN}$$

2.12 PLASTIC DEFORMATIONS

The results in the preceding sections were based on the assumption of a linear stress-strain relationship, where the proportional limit of the material was never exceeded. This is a reasonable assumption in the case of brittle materials, which rupture without yielding. For ductile materials, however, this implies that the yield strength of the material is not exceeded. The deformations will remain within the elastic range and the structural member will regain its original shape after all loads have been removed. However, if the stresses in any part of the member exceed the yield strength of the material, plastic deformations occur, and most of the results obtained in earlier sections cease to be valid. Then a more involved analysis, based on a nonlinear stress-strain relationship, must be carried out.

While an analysis taking into account the actual stress-strain relationship is beyond the scope of this text, we gain considerable insight into plastic behavior by considering an idealized *elastoplastic material* for which the stress-strain diagram consists of the two straight-line segments shown in Fig. 2.53. Note that the stress-strain diagram for mild steel in the elastic and plastic ranges is similar to this idealization. As long as the stress σ is less than the yield strength σ_Y, the material behaves elastically and obeys Hooke's law, $\sigma = E\epsilon$. When σ reaches the value σ_Y, the material starts yielding and keeps deforming plastically under a constant load. If the load is removed, unloading takes place along a straight-line segment CD parallel to the initial portion AY of the loading curve. The segment AD of the horizontal axis represents the strain corresponding to the permanent set or plastic deformation resulting from the loading and unloading of the specimen. While no actual material behaves exactly as shown in Fig. 2.53, this stress-strain diagram will prove useful in discussing the plastic deformations of ductile materials such as mild steel.

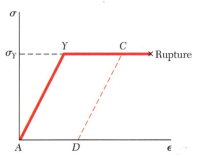

Fig. 2.53 Stress-strain diagram for an idealized elastoplastic material.

Concept Application 2.13

A rod of length $L = 500$ mm and cross-sectional area $A = 60$ mm^2 is made of an elastoplastic material having a modulus of elasticity $E = 200$ GPa in its elastic range and a yield point $\sigma_Y = 300$ MPa. The rod is subjected to an axial load until it is stretched 7 mm and the load is then removed. What is the resulting permanent set?

Referring to the diagram of Fig. 2.53, the maximum strain represented by the abscissa of point C is

$$\epsilon_C = \frac{\delta_C}{L} = \frac{7 \text{ mm}}{500 \text{ mm}} = 14 \times 10^{-3}$$

However, the yield strain, represented by the abscissa of point Y, is

$$\epsilon_Y = \frac{\sigma_Y}{E} = \frac{300 \times 10^6 \text{ Pa}}{200 \times 10^9 \text{ Pa}} = 1.5 \times 10^{-3}$$

The strain after unloading is represented by the abscissa ϵ_D of point D. Note from Fig. 2.53 that

$$\epsilon_D = AD = YC = \epsilon_C - \epsilon_Y$$
$$= 14 \times 10^{-3} - 1.5 \times 10^{-3} = 12.5 \times 10^{-3}$$

The permanent set is the deformation δ_D corresponding to the strain ϵ_D.

$$\delta_D = \epsilon_D L = (12.5 \times 10^{-3})(500 \text{ mm}) = 6.25 \text{ mm}$$

Concept Application 2.14

A 30-in.-long cylindrical rod of cross-sectional area $A_r = 0.075$ in^2 is placed inside a tube of the same length and of cross-sectional area $A_t = 0.100$ in^2. The ends of the rod and tube are attached to a rigid support on one side, and to a rigid plate on the other, as shown in the longitudinal section of Fig. 2.54a. The rod and tube are both assumed to be elastoplastic, with moduli of elasticity $E_r = 30 \times 10^6$ psi and $E_t = 15 \times 10^6$ psi, and yield strengths $(\sigma_r)_Y = 36$ ksi and $(\sigma_t)_Y = 45$ ksi. Draw the load-deflection diagram of the rod-tube assembly when a load P is applied to the plate as shown.

Determine the internal force and the elongation of the rod as it begins to yield

$$(P_r)_Y = (\sigma_r)_Y A_r = (36 \text{ ksi})(0.075 \text{ in}^2) = 2.7 \text{ kips}$$

$$(\delta_r)_Y = (\epsilon_r)_Y L = \frac{(\sigma_r)_Y}{E_r}L = \frac{36 \times 10^3 \text{ psi}}{30 \times 10^6 \text{ psi}}(30 \text{ in.})$$

$$= 36 \times 10^{-3} \text{ in.}$$

Since the material is elastoplastic, the force-elongation diagram *of the rod alone* consists of oblique and horizontal straight lines, as shown in Fig. 2.54b. Following the same procedure for the tube,

$$(P_t)_Y = (\sigma_t)_Y A_t = (45 \text{ ksi})(0.100 \text{ in}^2) = 4.5 \text{ kips}$$

$$(\delta_t)_Y = (\epsilon_t)_Y L = \frac{(\sigma_t)_Y}{E_t}L = \frac{45 \times 10^3 \text{ psi}}{15 \times 10^6 \text{ psi}}(30 \text{ in.})$$

$$= 90 \times 10^{-3} \text{ in.}$$

The load-deflection diagram of *the tube alone* is shown in Fig. 2.54c. Observing that the load and deflection of the rod-tube combination are

$$P = P_r + P_t \qquad \delta = \delta_r = \delta_t$$

we draw the required load-deflection diagram by adding the ordinates of the diagrams obtained for both the rod and the tube (Fig. 2.54d). Points Y_r and Y_t correspond to the onset of yield.

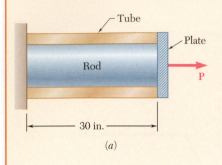

(a)

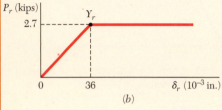

(b)

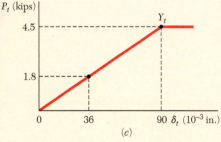

(c)

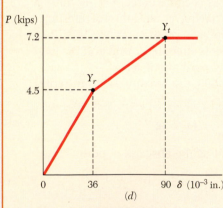

(d)

Fig. 2.54 (a) Concentric rod-tube assembly axially loaded by rigid plate. (b) Load-deflection response of the rod. (c) Load-deflection response of the tube. (d) Combined load-deflection response of the rod-tube assembly.

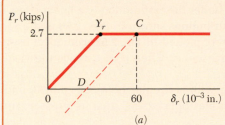

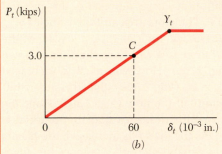

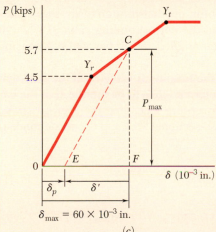

Fig. 2.55 (a) Rod load-deflection response with elastic unloading (red dashed line). (b) Tube load-deflection response; note that the given loading does not yield the tube, so unloading is along the original elastic loading line. (c) Combined rod-tube assembly load-deflection response with elastic unloading (red dashed line).

Concept Application 2.15

If the load **P** applied to the rod-tube assembly of Concept Application 2.14 is increased from zero to 5.7 kips and decreased back to zero, determine (a) the maximum elongation of the assembly and (b) the permanent set after the load has been removed.

a. Maximum Elongation. Referring to Fig. 2.54d, the load $P_{max} = 5.7$ kips corresponds to a point located on the segment $Y_r Y_t$ of the load-deflection diagram of the assembly. Thus, the rod has reached the plastic range with $P_r = (P_r)_Y = 2.7$ kips and $\sigma_r = (\sigma_r)_Y = 36$ ksi. However the tube is still in the elastic range with

$$P_t = P - P_r = 5.7 \text{ kips} - 2.7 \text{ kips} = 3.0 \text{ kips}$$

$$\sigma_t = \frac{P_t}{A_t} = \frac{3.0 \text{ kips}}{0.1 \text{ in}^2} = 30 \text{ ksi}$$

$$\delta_t = \epsilon_t L = \frac{\sigma_t}{E_t}L = \frac{30 \times 10^3 \text{ psi}}{15 \times 10^6 \text{ psi}}(30 \text{ in.}) = 60 \times 10^{-3} \text{ in.}$$

The maximum elongation of the assembly is

$$\delta_{max} = \delta_t = 60 \times 10^{-3} \text{ in.}$$

b. Permanent Set. As the load **P** decreases from 5.7 kips to zero, the internal forces P_r and P_t both decrease along a straight line, as shown in Fig. 2.55a and b. The force P_r decreases along line CD parallel to the initial portion of the loading curve, while the force P_t decreases along the original loading curve, since the yield stress was not exceeded in the tube. Their sum P will decrease along a line CE parallel to the portion $0Y_r$ of the load-deflection curve of the assembly (Fig. 2.55c). Referring to Fig. 2.55c, the slope of $0Y_r$ (and thus of CE) is

$$m = \frac{4.5 \text{ kips}}{36 \times 10^{-3} \text{ in.}} = 125 \text{ kips/in.}$$

The segment of line FE in Fig. 2.55c represents the deformation δ' of the assembly during the unloading phase, and the segment $0E$ is the permanent set δ_p after the load **P** has been removed. From triangle CEF,

$$\delta' = -\frac{P_{max}}{m} = -\frac{5.7 \text{ kips}}{125 \text{ kips/in.}} = -45.6 \times 10^{-3} \text{ in.}$$

The permanent set is

$$\delta_P = \delta_{max} + \delta' = 60 \times 10^{-3} - 45.6 \times 10^{-3}$$

$$= 14.4 \times 10^{-3} \text{ in.}$$

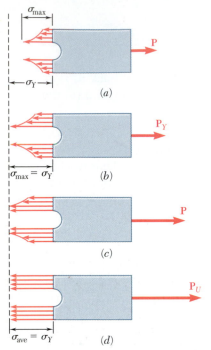

Fig. 2.56 Distribution of stresses in elastic-perfectly plastic material under increasing load.

Stress Concentrations. Recall that the discussion of stress concentrations of Sec. 2.11 was carried out under the assumption of a linear stress-strain relationship. The stress distributions shown in Figs. 2.50 and 2.51, and the stress-concentration factors plotted in Fig. 2.52 cannot be used when plastic deformations take place, i.e., when σ_{max} exceeds the yield strength σ_Y.

Consider again the flat bar with a circular hole of Fig. 2.50, and let us assume that the material is elastoplastic, i.e., that its stress-strain diagram is as shown in Fig. 2.53. As long as no plastic deformation takes place, the distribution of stresses is as indicated in Sec. 2.11 (Fig. 2.50a). The area under the stress-distribution curve represents the integral $\int \sigma \, dA$, which is equal to the load P. Thus this area and the value of σ_{max} must increase as the load P increases. As long as $\sigma_{max} \leq \sigma_Y$, all of the stress distributions obtained as P increases will have the shape shown in Fig. 2.50 and repeated in Fig. 2.56a. However, as P is increased beyond P_Y corresponding to $\sigma_{max} = \sigma_Y$ (Fig. 2.56b), the stress-distribution curve must flatten in the vicinity of the hole (Fig. 2.56c), since the stress cannot exceed the value σ_Y. This indicates that the material is yielding in the vicinity of the hole. As the load P is increased, the plastic zone where yield takes place keeps expanding until it reaches the edges of the plate (Fig. 2.56d). At that point, the distribution of stresses across the plate is uniform, $\sigma = \sigma_Y$, and the corresponding value $P = P_U$ of the load is the largest that can be applied to the bar without causing rupture.

It is interesting to compare the maximum value P_Y of the load that can be applied if no permanent deformation is to be produced in the bar with the value P_U that will cause rupture. Recalling the average stress, $\sigma_{ave} = P/A$, where A is the net cross-sectional area and the stress concentration factor, $K = \sigma_{max}/\sigma_{ave}$, write

$$P = \sigma_{ave} A = \frac{\sigma_{max} A}{K} \tag{2.41}$$

for any value of σ_{max} that does not exceed σ_Y. When $\sigma_{max} = \sigma_Y$ (Fig. 2.56b), $P = P_Y$, and Eq. (2.40) yields

$$P_Y = \frac{\sigma_Y A}{K} \tag{2.42}$$

On the other hand, when $P = P_U$ (Fig. 2.56d), $\sigma_{ave} = \sigma_Y$ and

$$P_U = \sigma_Y A \tag{2.43}$$

Comparing Eqs. (2.42) and (2.43),

$$P_Y = \frac{P_U}{K} \tag{2.44}$$

*2.13 RESIDUAL STRESSES

In Concept Application 2.13 of the preceding section, we considered a rod that was stretched beyond the yield point. As the load was removed, the rod did not regain its original length; it had been permanently deformed.

However, after the load was removed, all stresses disappeared. You should not assume that this will always be the case. Indeed, when only some of the parts of an indeterminate structure undergo plastic deformations, as in Concept Application 2.15, or when different parts of the structure undergo different plastic deformations, the stresses in the various parts of the structure will not return to zero after the load has been removed. Stresses called *residual stresses* will remain in various parts of the structure.

While computation of residual stresses in an actual structure can be quite involved, the following concept application provides a general understanding of the method to be used for their determination.

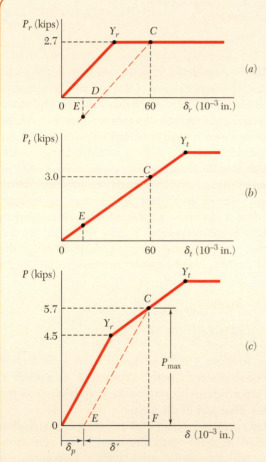

Fig. 2.57 (a) Rod load-deflection response with elastic unloading (red dashed line). (b) Tube load-deflection response; the given loading does not yield the tube, so unloading is along elastic loading line with residual tensile stress. (c) Combined rod-tube assembly load-deflection response with elastic unloading (red dashed line).

Concept Application 2.16

Determine the residual stresses in the rod and tube of Fig. 2.54a after the load **P** has been increased from zero to 5.7 kips and decreased back to zero.

Observe from the diagrams of Fig. 2.57 (similar to those in the previous concept application) that, after the load **P** has returned to zero, the internal forces P_r and P_t are *not* equal to zero. Their values have been indicated by point E in parts *a* and *b*. The corresponding stresses are not equal to zero either after the assembly has been unloaded. To determine these residual stresses, first determine the reverse stresses σ'_r and σ'_t caused by the unloading and add them to the maximum stresses $\sigma_r = 36$ ksi and $\sigma_t = 30$ ksi found in part *a* of Concept Application 2.15.

The strain caused by the unloading is the same in both the rod and the tube. It is equal to δ'/L, where δ' is the deformation of the assembly during unloading found in Concept Application 2.15:

$$\epsilon' = \frac{\delta'}{L} = \frac{-45.6 \times 10^{-3}\text{ in.}}{30\text{ in.}} = -1.52 \times 10^{-3}\text{ in./in.}$$

The corresponding reverse stresses in the rod and tube are

$$\sigma'_r = \epsilon' E_r = (-1.52 \times 10^{-3})(30 \times 10^6\text{ psi}) = -45.6\text{ ksi}$$

$$\sigma'_t = \epsilon' E_t = (-1.52 \times 10^{-3})(15 \times 10^6\text{ psi}) = -22.8\text{ ksi}$$

Then the residual stresses are found by superposing the stresses due to loading and the reverse stresses due to unloading.

$$(\sigma_r)_{\text{res}} = \sigma_r + \sigma'_r = 36\text{ ksi} - 45.6\text{ ksi} = -9.6\text{ ksi}$$

$$(\sigma_t)_{\text{res}} = \sigma_t + \sigma'_t = 30\text{ ksi} - 22.8\text{ ksi} = +7.2\text{ ksi}$$

Temperature Changes. Plastic deformations caused by temperature changes can also result in residual stresses. For example, consider a small plug that is to be welded to a large plate (Fig. 2.58). The plug can be

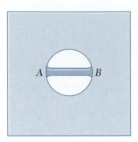

Fig. 2.58 Small rod welded to a large plate.

considered a small rod *AB* to be welded across a small hole in the plate. During the welding process, the temperature of the rod will be raised to over 1000°C, at which point its modulus of elasticity, stiffness, and stress will be almost zero. Since the plate is large, its temperature will not be increased significantly above room temperature (20°C). Thus, when the welding is completed, rod *AB* is at $T = 1000$°C with no stress and is attached to the plate, which is at 20°C.

As the rod cools, its modulus of elasticity increases. At about 500°C, it will approach its normal value of about 200 GPa. As the temperature of the rod decreases further, a situation similar to that considered in Sec. 2.3 and illustrated in Fig. 2.26 develops. Solving Eq. (2.15) for ΔT, making σ equal to the yield strength, assuming $\sigma_Y = 300$ MPa for the steel used, and $\alpha = 12 \times 10^{-6}$/°C, the temperature change that causes the rod to yield is

$$\Delta T = -\frac{\sigma}{E\alpha} = -\frac{300 \text{ MPa}}{(200 \text{ GPa})(12 \times 10^{-6}/°\text{C})} = -125°\text{C}$$

So the rod starts yielding at about 375°C and keeps yielding at a fairly constant stress level as it cools down to room temperature. As a result of welding, a residual stress (approximately equal to the yield strength of the steel used) is created in the plug and in the weld.

Residual stresses also occur as a result of the cooling of metals that have been cast or hot rolled. In these cases, the outer layers cool more rapidly than the inner core. This causes the outer layers to reacquire their stiffness (*E* returns to its normal value) faster than the inner core. When the entire specimen has returned to room temperature, the inner core will contract more than the outer layers. The result is residual longitudinal tensile stresses in the inner core and residual compressive stresses in the outer layers.

Residual stresses due to welding, casting, and hot rolling can be quite large (of the order of magnitude of the yield strength). These stresses can be removed by reheating the entire specimen to about 600°C and then allowing it to cool slowly over a period of 12 to 24 hours.

Sample Problem 2.6

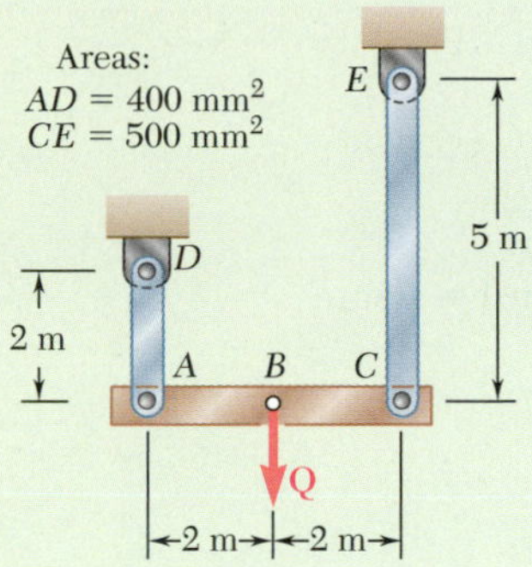

The rigid beam *ABC* is suspended from two steel rods as shown and is initially horizontal. The midpoint *B* of the beam is deflected 10 mm downward by the slow application of the force **Q**, after which the force is slowly removed. Knowing that the steel used for the rods is elasto-plastic with $E = 200$ GPa and $\sigma_Y = 300$ MPa, determine (*a*) the required maximum value of *Q* and the corresponding position of the beam and (*b*) the final position of the beam.

Areas:
$AD = 400$ mm^2
$CE = 500$ mm^2

5 m

2 m

D

E

A   *B*   *C*

Q

←2 m→←2 m→

STRATEGY: You can assume that plastic deformation would occur first in rod AD (which is a good assumption—*why?*), and then check this assumption.

MODELING AND ANALYSIS:

Statics. Since **Q** is applied at the midpoint of the beam (Fig. 1),

$$P_{AD} = P_{CE} \quad \text{and} \quad Q = 2P_{AD}$$

Elastic Action (Fig. 2). The maximum value of *Q* and the maximum elastic deflection of point *A* occur when $\sigma = \sigma_Y$ in rod *AD*.

$$(P_{AD})_{max} = \sigma_Y A = (300 \text{ MPa})(400 \text{ mm}^2) = 120 \text{ kN}$$

$$Q_{max} = 2(P_{AD})_{max} = 2(120 \text{ kN}) \qquad Q_{max} = 240 \text{ kN} \blacktriangleleft$$

$$\delta_{A_1} = \epsilon L = \frac{\sigma_Y}{E} L = \left(\frac{300 \text{ MPa}}{200 \text{ GPa}}\right)(2 \text{ m}) = 3 \text{ mm}$$

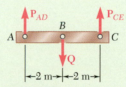

A P_{AD} B P_{CE} C

Q

←2 m→←2 m→

Fig. 1 Free-body diagram of rigid beam.

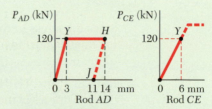

P_{AD} (kN) P_{CE} (kN)

120 — *Y*   *H*   120 — *Y*

J

0 3   11 14 mm   0   6 mm
  Rod *AD*   Rod *CE*

Fig. 2 Load-deflection diagrams for steel rods.

Since $P_{CE} = P_{AD} = 120$ kN, the stress in rod CE is

$$\sigma_{CE} = \frac{P_{CE}}{A} = \frac{120 \text{ kN}}{500 \text{ mm}^2} = 240 \text{ MPa}$$

The corresponding deflection of point C is

$$\delta_{C_1} = \epsilon L = \frac{\sigma_{CE}}{E} L = \left(\frac{240 \text{ MPa}}{200 \text{ GPa}}\right)(5 \text{ m}) = 6 \text{ mm}$$

The corresponding deflection of point B is

$$\delta_{B_1} = \tfrac{1}{2}(\delta_{A_1} + \delta_{C_1}) = \tfrac{1}{2}(3 \text{ mm} + 6 \text{ mm}) = 4.5 \text{ mm}$$

Since $\delta_B = 10$ mm, plastic deformation will occur.

Plastic Deformation.   For $Q = 240$ kN, plastic deformation occurs in rod AD, where $\sigma_{AD} = \sigma_Y = 300$ MPa. Since the stress in rod CE is within the elastic range, δ_C remains equal to 6 mm. From Fig. 3, the deflection δ_A for which $\delta_B = 10$ mm is obtained by writing

$$\delta_{B_2} = 10 \text{ mm} = \tfrac{1}{2}(\delta_{A_2} + 6 \text{ mm}) \qquad \delta_{A_2} = 14 \text{ mm}$$

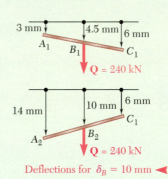

Fig. 3   Deflection of fully-loaded beam.

Unloading.   As force **Q** is slowly removed, the force P_{AD} decreases along line HJ parallel to the initial portion of the load-deflection diagram of rod AD. The final deflection of point A is

$$\delta_{A_3} = 14 \text{ mm} - 3 \text{ mm} = 11 \text{ mm}$$

Since the stress in rod CE remained within the elastic range, note that the final deflection of point C is zero. Fig. 4 illustrates the final position of the beam.

REFLECT and THINK: Due to symmetry in this determinate problem, the axial forces in the rods are equal. Given that the rods have identical material properties and that the cross-sectional area of rod AD is smaller than rod CE, you would therefore expect that rod AD would reach yield first (as assumed in the STRATEGY step).

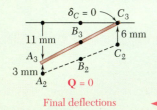

Final deflections

Fig. 4   Beam's final deflections with load removed.

Problems

2.93 Knowing that, for the plate shown, the allowable stress is 125 MPa, determine the maximum allowable value of P when (a) $r = 12$ mm, (b) $r = 18$ mm.

2.94 Knowing that $P = 38$ kN, determine the maximum stress when (a) $r = 10$ mm, (b) $r = 16$ mm, (c) $r = 18$ mm.

2.95 A hole is to be drilled in the plate at A. The diameters of the bits available to drill the hole range from $\frac{1}{2}$ to $1\frac{1}{2}$ in. in $\frac{1}{4}$-in. increments. If the allowable stress in the plate is 21 ksi, determine (a) the diameter d of the largest bit that can be used if the allowable load $\mathbf{P}$ at the hole is to exceed that at the fillets, (b) the corresponding allowable load $\mathbf{P}$.

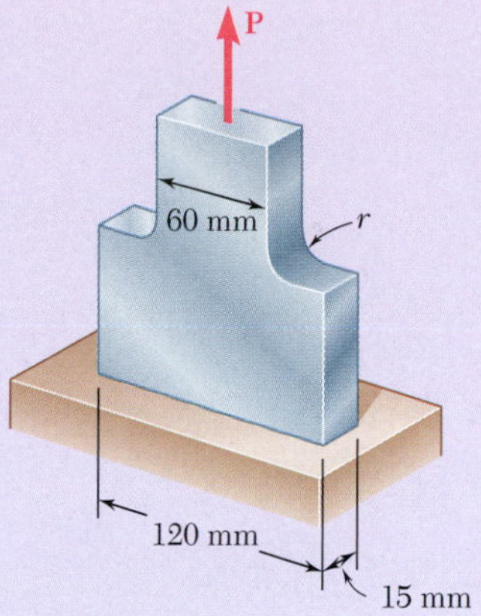

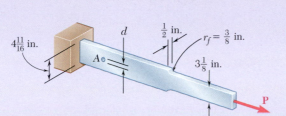

Fig. P2.95 and P2.96

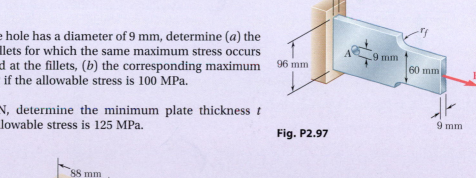

Fig. P2.93 and P2.94

2.96 (a) For $P = 13$ kips and $d = \frac{1}{2}$ in., determine the maximum stress in the plate shown. (b) Solve part a, assuming that the hole at A is not drilled.

2.97 Knowing that the hole has a diameter of 9 mm, determine (a) the radius r_f of the fillets for which the same maximum stress occurs at the hole A and at the fillets, (b) the corresponding maximum allowable load $\mathbf{P}$ if the allowable stress is 100 MPa.

2.98 For $P = 100$ kN, determine the minimum plate thickness t required if the allowable stress is 125 MPa.

Fig. P2.97

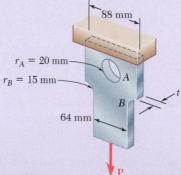

Fig. P2.98

2.99 (a) Knowing that the allowable stress is 20 ksi, determine the maximum allowable magnitude of the centric load **P**. (b) Determine the percent change in the maximum allowable magnitude of **P** if the raised portions are removed at the ends of the specimen.

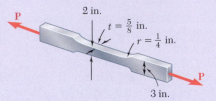

2 in.

$t = \frac{5}{8}$ in.

$r = \frac{1}{4}$ in.

3 in.

Fig. P2.99

2.100 A centric axial force is applied to the steel bar shown. Knowing that $\sigma_{all} = 20$ ksi, determine the maximum allowable load **P**.

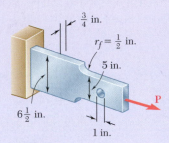

$\frac{3}{4}$ in.

$r_f = \frac{1}{2}$ in.

5 in.

$6\frac{1}{2}$ in.

1 in.

Fig. P2.100

2.101 The cylindrical rod AB has a length $L = 5$ ft and a 0.75-in. diameter; it is made of a mild steel that is assumed to be elastoplastic with $E = 29 \times 10^6$ psi and $\sigma_Y = 36$ ksi. A force **P** is applied to the bar and then removed to give it a permanent set δ_P. Determine the maximum value of the force **P** and the maximum amount δ_m by which the bar should be stretched if the desired value of δ_P is (a) 0.1 in., (b) 0.2 in.

B

L

A

P

Fig. P2.101 and P2.102

2.102 The cylindrical rod AB has a length $L = 6$ ft and a 1.25-in. diameter; it is made of a mild steel that is assumed to be elastoplastic with $E = 29 \times 10^6$ psi and $\sigma_Y = 36$ ksi. A force **P** is applied to the bar until end A has moved down by an amount δ_m. Determine the maximum value of the force **P** and the permanent set of the bar after the force has been removed, knowing (a) $\delta_m = 0.125$ in., (b) $\delta_m = 0.250$ in.

2.103 Rod AB is made of a mild steel that is assumed to be elastoplastic with $E = 200$ GPa and $\sigma_Y = 345$ MPa. After the rod has been attached to the rigid lever CD, it is found that end C is 6 mm too high. A vertical force $\mathbf{Q}$ is then applied at C until this point has moved to position C'. Determine the required magnitude of $\mathbf{Q}$ and the deflection δ_1 if the lever is to *snap* back to a horizontal position after $\mathbf{Q}$ is removed.

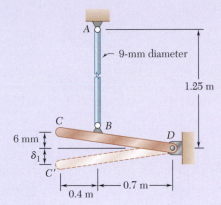

9-mm diameter

1.25 m

6 mm

δ_1

0.4 m

0.7 m

Fig. P2.103

2.104 Solve Prob. 2.103, assuming that the yield point of the mild steel is 250 MPa.

2.105 Rod ABC consists of two cylindrical portions AB and BC; it is made of a mild steel that is assumed to be elastoplastic with $E = 200$ GPa and $\sigma_Y = 250$ MPa. A force $\mathbf{P}$ is applied to the rod and then removed to give it a permanent set $\delta_P = 2$ mm. Determine the maximum value of the force $\mathbf{P}$ and the maximum amount δ_m by which the rod should be stretched to give it the desired permanent set.

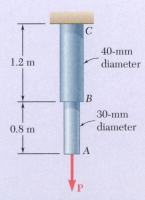

1.2 m

40-mm diameter

0.8 m

30-mm diameter

P

Fig. P2.105 and P2.106

2.106 Rod ABC consists of two cylindrical portions AB and BC; it is made of a mild steel that is assumed to be elastoplastic with $E = 200$ GPa and $\sigma_Y = 250$ MPa. A force $\mathbf{P}$ is applied to the rod until its end A has moved down by an amount $\delta_m = 5$ mm. Determine the maximum value of the force $\mathbf{P}$ and the permanent set of the rod after the force has been removed.

Fig. P2.107

2.107 Rod *AB* consists of two cylindrical portions *AC* and *BC*, each with a cross-sectional area of 1750 mm². Portion *AC* is made of a mild steel with $E = 200$ GPa and $\sigma_Y = 250$ MPa, and portion *BC* is made of a high-strength steel with $E = 200$ GPa and $\sigma_Y = 345$ MPa. A load **P** is applied at *C* as shown. Assuming both steels to be elastoplastic, determine (*a*) the maximum deflection of *C* if *P* is gradually increased from zero to 975 kN and then reduced back to zero, (*b*) the maximum stress in each portion of the rod, (*c*) the permanent deflection of *C*.

2.108 For the composite rod of Prob. 2.107, if *P* is gradually increased from zero until the deflection of point *C* reaches a maximum value of $\delta_m = 0.3$ mm and then decreased back to zero, determine, (*a*) the maximum value of *P*, (*b*) the maximum stress in each portion of the rod, (*c*) the permanent deflection of *C* after the load is removed.

2.109 Each cable has a cross-sectional area of 100 mm² and is made of an elastoplastic material for which $\sigma_Y = 345$ MPa and $E = 200$ GPa. A force **Q** is applied at *C* to the rigid bar *ABC* and is gradually increased from 0 to 50 kN and then reduced to zero. Knowing that the cables were initially taut, determine (*a*) the maximum stress that occurs in cable *BD*, (*b*) the maximum deflection of point *C*, (*c*) the final displacement of point *C*. (*Hint:* In part *c*, cable *CE* is not taut.)

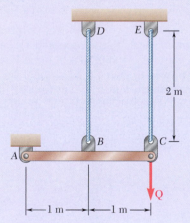

Fig. P2.109

2.110 Solve Prob. 2.109, assuming that the cables are replaced by rods of the same cross-sectional area and material. Further assume that the rods are braced so that they can carry compressive forces.

2.111 Two tempered-steel bars, each $\frac{3}{16}$ in. thick, are bonded to a $\frac{1}{2}$-in. mild-steel bar. This composite bar is subjected as shown to a centric axial load of magnitude *P*. Both steels are elastoplastic with $E = 29 \times 10^6$ psi and with yield strengths equal to 100 ksi and 50 ksi, respectively, for the tempered and mild steel. The load *P* is gradually increased from zero until the deformation of the bar reaches a maximum value $\delta_m = 0.04$ in. and then decreased back to zero. Determine (*a*) the maximum value of *P*, (*b*) the maximum stress in the tempered-steel bars, (*c*) the permanent set after the load is removed.

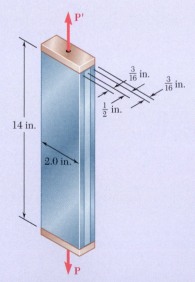

Fig. P2.111

2.112 For the composite bar of Prob. 2.111, if P is gradually increased from zero to 98 kips and then decreased back to zero, determine (a) the maximum deformation of the bar, (b) the maximum stress in the tempered-steel bars, (c) the permanent set after the load is removed.

2.113 The rigid bar ABC is supported by two links, AD and BE, of uniform 37.5×6-mm rectangular cross section and made of a mild steel that is assumed to be elastoplastic with $E = 200$ GPa and $\sigma_Y = 250$ MPa. The magnitude of the force **Q** applied at B is gradually increased from zero to 260 kN. Knowing that $a = 0.640$ m, determine (a) the value of the normal stress in each link, (b) the maximum deflection of point B.

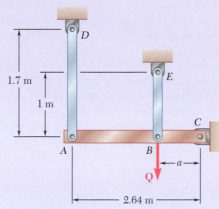

Fig. P2.113

2.114 Solve Prob. 2.113, knowing that $a = 1.76$ m and that the magnitude of the force **Q** applied at B is gradually increased from zero to 135 kN.

*2.115 Solve Prob. 2.113, assuming that the magnitude of the force **Q** applied at B is gradually increased from zero to 260 kN and then decreased back to zero. Knowing that $a = 0.640$ m, determine (a) the residual stress in each link, (b) the final deflection of point B. Assume that the links are braced so that they can carry compressive forces without buckling.

2.116 A uniform steel rod of cross-sectional area A is attached to rigid supports and is unstressed at a temperature of 45°F. The steel is assumed to be elastoplastic with $\sigma_Y = 36$ ksi and $E = 29 \times 10^6$ psi. Knowing that $\alpha = 6.5 \times 10^{-6}/°F$, determine the stress in the bar (a) when the temperature is raised to 320°F, (b) after the temperature has returned to 45°F.

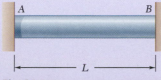

Fig. P2.116

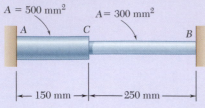

$A = 500 \text{ mm}^2$ $A = 300 \text{ mm}^2$

150 mm   250 mm

Fig. P2.117

2.117 The steel rod *ABC* is attached to rigid supports and is unstressed at a temperature of 25°C. The steel is assumed elastoplastic with $E = 200$ GPa and $\sigma_Y = 250$ MPa. The temperature of both portions of the rod is then raised to 150°C. Knowing that $\alpha = 11.7 \times 10^{-6}/°C$, determine (*a*) the stress in both portions of the rod, (*b*) the deflection of point *C*.

*2.118 Solve Prob. 2.117, assuming that the temperature of the rod is raised to 150°C and then returned to 25°C.

*2.119 For the composite bar of Prob. 2.111, determine the residual stresses in the tempered-steel bars if *P* is gradually increased from zero to 98 kips and then decreased back to zero.

*2.120 For the composite bar in Prob. 2.111, determine the residual stresses in the tempered-steel bars if *P* is gradually increased from zero until the deformation of the bar reaches a maximum value $\delta_m = 0.04$ in. and is then decreased back to zero.

*2.121 Narrow bars of aluminum are bonded to the two sides of a thick steel plate as shown. Initially, at $T_1 = 70°F$, all stresses are zero. Knowing that the temperature will be slowly raised to T_2 and then reduced to T_1, determine (*a*) the highest temperature T_2 that does *not* result in residual stresses, (*b*) the temperature T_2 that will result in a residual stress in the aluminum equal to 58 ksi. Assume $\alpha_a = 12.8 \times 10^{-6}/°F$ for the aluminum and $\alpha_s = 6.5 \times 10^{-6}/°F$ for the steel. Further assume that the aluminum is elastoplastic with $E = 10.9 \times 10^6$ psi and $\alpha_Y = 58$ ksi. (*Hint:* Neglect the small stresses in the plate.)

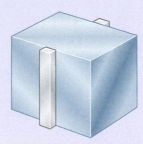

Fig. P2.121

*2.122 Bar *AB* has a cross-sectional area of 1200 mm² and is made of a steel that is assumed to be elastoplastic with $E = 200$ GPa and $\sigma_Y = 250$ MPa. Knowing that the force **F** increases from 0 to 520 kN and then decreases to zero, determine (*a*) the permanent deflection of point *C*, (*b*) the residual stress in the bar.

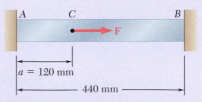

$a = 120$ mm

440 mm

Fig. P2.122

*2.123 Solve Prob. 2.122, assuming that $a = 180$ mm.

Review and Summary

Normal Strain

Consider a rod of length L and uniform cross section, and its deformation δ under an axial load **P** (Fig. 2.59). The *normal strain* ϵ in the rod is defined as the *deformation per unit length*:

$$\epsilon = \frac{\delta}{L} \qquad (2.1)$$

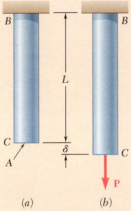

Fig. 2.59 Undeformed and deformed axially-loaded rod.

In the case of a rod of variable cross section, the normal strain at any given point Q is found by considering a small element of rod at Q:

$$\epsilon = \lim_{\Delta x \to 0} \frac{\Delta \delta}{\Delta x} = \frac{d\delta}{dx} \qquad (2.2)$$

Stress-Strain Diagram

A *stress-strain diagram* is obtained by plotting the stress σ versus the strain ϵ as the load increases. These diagrams can be used to distinguish between *brittle* and *ductile* materials. A brittle material ruptures without any noticeable prior change in the rate of elongation (Fig. 2.60), while a ductile material

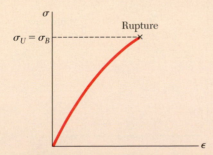

Fig. 2.60 Stress-strain diagram for a typical brittle material.

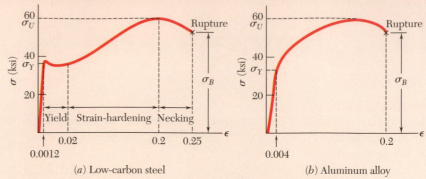

Fig. 2.61 Stress-strain diagrams of two typical ductile metal materials.

yields after a critical stress σ_Y (the *yield strength*) has been reached (Fig. 2.61). The specimen undergoes a large deformation before rupturing, with a relatively small increase in the applied load. An example of brittle material with different properties in tension and compression is *concrete*.

Hooke's Law and Modulus of Elasticity

The initial portion of the stress-strain diagram is a straight line. Thus, for small deformations, the stress is directly proportional to the strain:

$$\sigma = E\epsilon \qquad (2.6)$$

This relationship is *Hooke's law*, and the coefficient E is the *modulus of elasticity* of the material. The *proportional limit* is the largest stress for which Eq. (2.4) applies.

Properties of *isotropic* materials are independent of direction, while properties of *anisotropic* materials depend upon direction. *Fiber-reinforced composite materials* are made of fibers of a strong, stiff material embedded in layers of a weaker, softer material (Fig. 2.62).

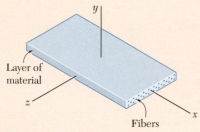

Fig. 2.62 Layer of fiber-reinforced composite material.

Elastic Limit and Plastic Deformation

If the strains caused in a test specimen by the application of a given load disappear when the load is removed, the material is said to behave *elastically*. The largest stress for which this occurs is called the *elastic limit* of the material. If the elastic limit is exceeded, the stress and strain decrease in a linear fashion when the load is removed, and the strain does not return to zero (Fig. 2.63), indicating that a *permanent set* or *plastic deformation* of the material has taken place.

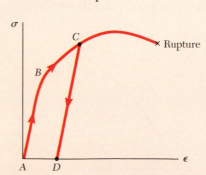

Fig. 2.63 Stress-strain response of ductile material loaded beyond yield and unloaded.

Fatigue and Endurance Limit

Fatigue causes the failure of structural or machine components after a very large number of repeated loadings, even though the stresses remain in the elastic range. A standard fatigue test determines the number n of successive loading-and-unloading cycles required to cause the failure of a specimen for any given maximum stress level σ and plots the resulting σ-n curve. The value of σ for which failure does not occur, even for an indefinitely large number of cycles, is known as the *endurance limit*.

Elastic Deformation Under Axial Loading

If a rod of length L and uniform cross section of area A is subjected at its end to a centric axial load **P** (Fig. 2.64), the corresponding deformation is

$$\delta = \frac{PL}{AE} \tag{2.9}$$

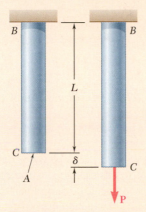

Fig. 2.64 Undeformed and deformed axially-loaded rod.

If the rod is loaded at several points or consists of several parts of various cross sections and possibly of different materials, the deformation δ of the rod must be expressed as the sum of the deformations of its component parts:

$$\delta = \sum_i \frac{P_i L_i}{A_i E_i} \tag{2.10}$$

Statically Indeterminate Problems

Statically indeterminate problems are those in which the reactions and the internal forces *cannot* be determined from statics alone. The equilibrium equations derived from the free-body diagram of the member under consideration were complemented by relations involving deformations and obtained from the geometry of the problem. The forces in the rod and in the tube of Fig. 2.65, for instance, were determined by observing that their sum is equal to P, and that they cause equal deformations in the rod and in the tube. Similarly, the reactions at the supports of the bar of

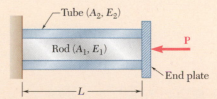

Fig. 2.65 Statically indeterminate problem where concentric rod and tube have same strain but different stresses.

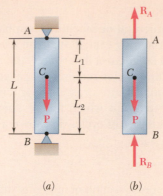

Fig. 2.66 (*a*) Axially-loaded statically-indeterminate member. (*b*) Free-body diagram.

Fig. 2.66 could not be obtained from the free-body diagram of the bar alone, but they could be determined by expressing that the total elongation of the bar must be equal to zero.

Problems with Temperature Changes

When the temperature of an *unrestrained rod AB* of length L is increased by ΔT, its elongation is

$$\delta_T = \alpha(\Delta T)L \tag{2.13}$$

where α is the *coefficient of thermal expansion* of the material. The corresponding strain, called *thermal strain*, is

$$\epsilon_T = \alpha \Delta T \tag{2.14}$$

and *no stress* is associated with this strain. However, if rod *AB* is *restrained* by fixed supports (Fig. 2.67), stresses develop in the rod as the temperature increases, because of the reactions at the supports. To determine the magnitude P of the reactions, the rod is first detached from its support at B (Fig. 2.68*a*).

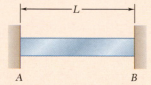

Fig. 2.67 Fully restrained bar of length *L*.

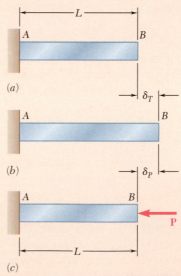

Fig. 2.68 Determination of reactions for bar of Fig. 2.67 subject to a temperature increase. (*a*) Support at *B* removed. (*b*) Thermal expansion. (*c*) Application of support reaction to counter thermal expansion.

The deformation δ_T of the rod occurs as it expands due to of the temperature change (Fig. 2.68b). The deformation δ_P caused by the force **P** is required to bring it back to its original length, so that it may be reattached to the support at B (Fig. 2.68c).

Lateral Strain and Poisson's Ratio

When an axial load **P** is applied to a homogeneous, slender bar (Fig. 2.69), it causes a strain, not only along the axis of the bar but in any transverse direction. This strain is the *lateral strain*, and the ratio of the lateral strain over the axial strain is called *Poisson's ratio*:

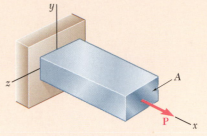

Fig. 2.69 A bar in uniaxial tension.

$$\nu = -\frac{\text{lateral strain}}{\text{axial strain}} \qquad (2.17)$$

Multiaxial Loading

The condition of strain under an axial loading in the x direction is

$$\epsilon_x = \frac{\sigma_x}{E} \qquad \epsilon_y = \epsilon_z = -\frac{\nu\sigma_x}{E} \qquad (2.19)$$

A *multiaxial loading* causes the state of stress shown in Fig. 2.70. The resulting strain condition was described by the *generalized Hooke's law* for a multiaxial loading.

$$\epsilon_x = +\frac{\sigma_x}{E} - \frac{\nu\sigma_y}{E} - \frac{\nu\sigma_z}{E}$$

$$\epsilon_y = -\frac{\nu\sigma_x}{E} + \frac{\sigma_y}{E} - \frac{\nu\sigma_z}{E} \qquad (2.20)$$

$$\epsilon_z = -\frac{\nu\sigma_x}{E} - \frac{\nu\sigma_y}{E} + \frac{\sigma_z}{E}$$

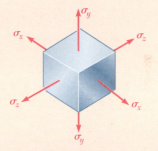

Fig. 2.70 State of stress for multiaxial loading.

Dilatation

If an element of material is subjected to the stresses σ_x, σ_y, σ_z, it will deform and a certain change of volume will result. The *change in volume per unit volume* is the *dilatation* of the material:

$$e = \frac{1 - 2\nu}{E}(\sigma_x + \sigma_y + \sigma_z) \qquad (2.22)$$

Bulk Modulus

When a material is subjected to a hydrostatic pressure p,

$$e = -\frac{p}{k} \qquad (2.25)$$

where k is the *bulk modulus* of the material:

$$k = \frac{E}{3(1 - 2\nu)} \qquad (2.24)$$

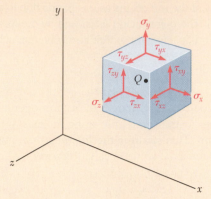

Fig. 2.71 Positive stress components at point Q for a general state of stress.

Shearing Strain: Modulus of Rigidity

The state of stress in a material under the most general loading condition involves shearing stresses, as well as normal stresses (Fig. 2.71). The shearing stresses tend to deform a cubic element of material into an oblique parallelepiped. The stresses τ_{xy} and τ_{yx} shown in Fig. 2.72 cause the angles formed by the faces on which they act to either increase or decrease by a small angle γ_{xy}. This angle defines the *shearing strain* corresponding to the x and y directions. Defining in a similar way the shearing strains γ_{yz} and γ_{zx}, the following relations were written:

$$\tau_{xy} = G\gamma_{xy} \qquad \tau_{yz} = G\gamma_{yz} \qquad \tau_{zx} = G\gamma_{zx} \qquad \textbf{(2.27, 28)}$$

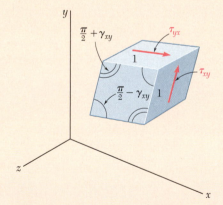

Fig. 2.72 Deformation of unit cubic element due to shearing stress.

which are valid for any homogeneous isotropic material within its proportional limit in shear. The constant G is the *modulus of rigidity* of the material, and the relationships obtained express *Hooke's law for shearing stress and strain*. Together with Eqs. (2.20), they form a group of equations representing the generalized Hooke's law for a homogeneous isotropic material under the most general stress condition.

While an axial load exerted on a slender bar produces only normal strains—both axial and transverse—on an element of material oriented along the axis of the bar, it will produce both normal and shearing strains on an element rotated through 45° (Fig. 2.73). The three constants E, ν, and G are not independent. They satisfy the relation

$$\frac{E}{2G} = 1 + \nu \qquad (2.34)$$

This equation can be used to determine any of the three constants in terms of the other two.

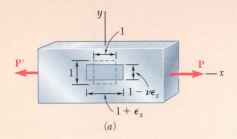

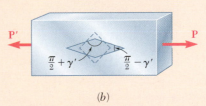

Fig. 2.73 Representations of strain in an axially-loaded bar: (a) cubic strain element with faces aligned with coordinate axes; (b) cubic strain element with faces rotated 45° about z-axis.

Saint-Venant's Principle

Saint-Venant's principle states that except in the immediate vicinity of the points of application of the loads, the distribution of stresses in a given member is independent of the actual mode of application of the loads. This principle makes it possible to assume a uniform distribution of stresses in a member subjected to concentrated axial loads, except close to the points of application of the loads, where stress concentrations will occur.

Stress Concentrations

Stress concentrations will also occur in structural members near a discontinuity, such as a hole or a sudden change in cross section. The ratio of the maximum value of the stress occurring near the discontinuity over the average stress computed in the critical section is referred to as the *stress-concentration factor* of the discontinuity:

$$K = \frac{\sigma_{max}}{\sigma_{ave}} \qquad (2.40)$$

Plastic Deformations

Plastic deformations occur in structural members made of a ductile material when the stresses in some part of the member exceed the yield strength of the material. An idealized *elastoplastic material* is characterized by the stress-strain diagram shown in Fig. 2.74. When an indeterminate structure

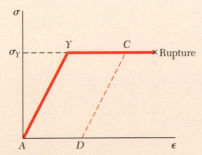

Fig. 2.74 Stress-strain diagram for an idealized elastoplastic material.

undergoes plastic deformations, the stresses do not, in general, return to zero after the load has been removed. The stresses remaining in the various parts of the structure are called *residual stresses* and can be determined by adding the maximum stresses reached during the loading phase and the reverse stresses corresponding to the unloading phase.

Review Problems

2.124 The uniform wire ABC, of unstretched length $2l$, is attached to the supports shown and a vertical load $\mathbf{P}$ is applied at the midpoint B. Denoting by A the cross-sectional area of the wire and by E the modulus of elasticity, show that, for $\delta \ll l$, the deflection at the midpoint B is

$$\delta = l \sqrt[3]{\frac{P}{AE}}$$

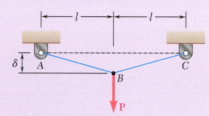

Fig. P2.124

2.125 The aluminum rod ABC ($E = 10.1 \times 10^6$ psi), which consists of two cylindrical portions AB and BC, is to be replaced with a cylindrical steel rod DE ($E = 29 \times 10^6$ psi) of the same overall length. Determine the minimum required diameter d of the steel rod if its vertical deformation is not to exceed the deformation of the aluminum rod under the same load and if the allowable stress in the steel rod is not to exceed 24 ksi.

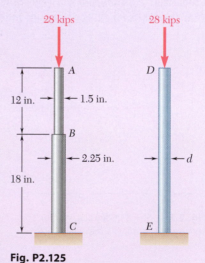

Fig. P2.125

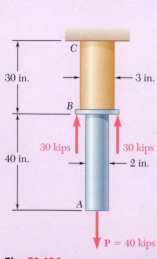

Fig. P2.126

2.126 Two solid cylindrical rods are joined at B and loaded as shown. Rod AB is made of steel ($E = 29 \times 10^6$ psi), and rod BC of brass ($E = 15 \times 10^6$ psi). Determine (*a*) the total deformation of the composite rod ABC, (*b*) the deflection of point B.

2.127 The brass strip *AB* has been attached to a fixed support at *A* and rests on a rough support at *B*. Knowing that the coefficient of friction is 0.60 between the strip and the support at *B*, determine the decrease in temperature for which slipping will impend.

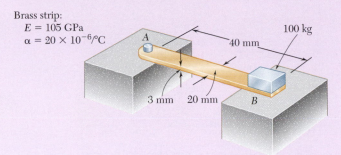

Brass strip:
$E = 105$ GPa
$\alpha = 20 \times 10^{-6}/°C$

100 kg

40 mm

3 mm   20 mm

Fig. P2.127

2.128 The specimen shown is made from a 1-in.-diameter cylindrical steel rod with two 1.5-in.-outer-diameter sleeves bonded to the rod as shown. Knowing that $E = 29 \times 10^6$ psi, determine (*a*) the load **P** so that the total deformation is 0.002 in., (*b*) the corresponding deformation of the central portion *BC*.

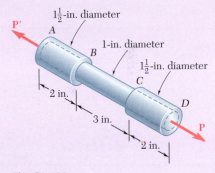

$1\frac{1}{2}$-in. diameter

1-in. diameter

$1\frac{1}{2}$-in. diameter

2 in.

3 in.

2 in.

Fig. P2.128

2.129 Each of the four vertical links connecting the two rigid horizontal members is made of aluminum ($E = 70$ GPa) and has a uniform rectangular cross section of 10×40 mm. For the loading shown, determine the deflection of (*a*) point *E*, (*b*) point *F*, (*c*) point *G*.

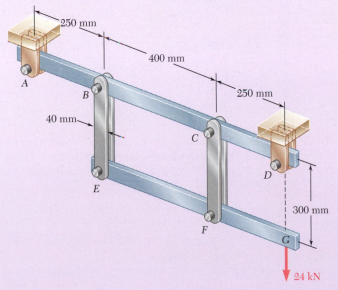

250 mm

400 mm

40 mm

250 mm

300 mm

24 kN

Fig. P2.129

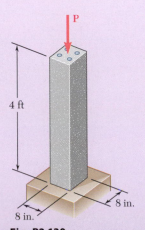

P

4 ft

8 in.

8 in.

Fig. P2.130

2.130 A 4-ft concrete post is reinforced with four steel bars, each with a $\frac{3}{4}$-in. diameter. Knowing that $E_s = 29 \times 10^6$ psi and $E_c = 3.6 \times 10^6$ psi, determine the normal stresses in the steel and in the concrete when a 150-kip axial centric force **P** is applied to the post.

2.131 The steel rods *BE* and *CD* each have a 16-mm diameter ($E = 200$ GPa); the ends of the rods are single-threaded with a pitch of 2.5 mm. Knowing that after being snugly fitted, the nut at *C* is tightened one full turn, determine (*a*) the tension in rod *CD*, (*b*) the deflection of point *C* of the rigid member *ABC*.

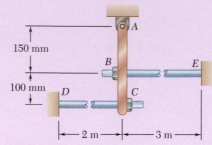

Fig. P2.131

2.132 The assembly shown consists of an aluminum shell ($E_a = 10.6 \times 10^6$ psi, $\alpha_a = 12.9 \times 10^{-6}/°F$) fully bonded to a steel core ($E_s = 29 \times 10^6$ psi, $\alpha_s = 6.5 \times 10^{-6}/°F$) and is unstressed. Determine (*a*) the largest allowable change in temperature if the stress in the aluminum shell is not to exceed 6 ksi, (*b*) the corresponding change in length of the assembly.

Fig. P2.132

2.133 The plastic block shown is bonded to a fixed base and to a horizontal rigid plate to which a force **P** is applied. Knowing that for the plastic used $G = 55$ ksi, determine the deflection of the plate when $P = 9$ kips.

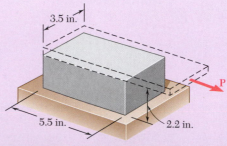

Fig. P2.133

2.134 The aluminum test specimen shown is subjected to two equal and opposite centric axial forces of magnitude P. (a) Knowing that $E = 70$ GPa and $\sigma_{all} = 200$ MPa, determine the maximum allowable value of P and the corresponding total elongation of the specimen. (b) Solve part a, assuming that the specimen has been replaced by an aluminum bar of the same length and a uniform 60×15-mm rectangular cross section.

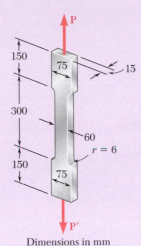

Dimensions in mm

Fig. P2.134

2.135 The uniform rod BC has cross-sectional area A and is made of a mild steel that can be assumed to be elastoplastic with a modulus of elasticity E and a yield strength σ_Y. Using the block-and-spring system shown, it is desired to simulate the deflection of end C of the rod as the axial force **P** is gradually applied and removed, that is, the deflection of points C and C' should be the same for all values of P. Denoting by μ the coefficient of friction between the block and the horizontal surface, derive an expression for (a) the required mass m of the block, (b) the required constant k of the spring.

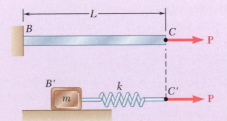

Fig. P2.135

Computer Problems

The following problems are designed to be solved with a computer. Write each program so that it can be used with either SI or U.S. customary units and in such a way that solid cylindrical elements may be defined by either their diameter or their cross-sectional area.

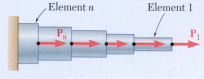

Fig. P2.C1

2.C1 A rod consisting of n elements, each of which is homogeneous and of uniform cross section, is subjected to the loading shown. The length of element i is denoted by L_i, its cross-sectional area by A_i, modulus of elasticity by E_i, and the load applied to its right end by $\mathbf{P}_i$, the magnitude P_i of this load being assumed to be positive if $\mathbf{P}_i$ is directed to the right and negative otherwise. (*a*) Write a computer program that can be used to determine the average normal stress in each element, the deformation of each element, and the total deformation of the rod. (*b*) Use this program to solve Probs. 2.20 and 2.126.

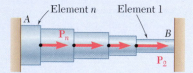

Fig. P2.C2

2.C2 Rod AB is horizontal with both ends fixed; it consists of n elements, each of which is homogeneous and of uniform cross section, and is subjected to the loading shown. The length of element i is denoted by L_i, its cross-sectional area by A_i, its modulus of elasticity by E_i, and the load applied to its right end by $\mathbf{P}_i$, the magnitude P_i of this load being assumed to be positive if $\mathbf{P}_i$ is directed to the right and negative otherwise. (Note that $P_1 = 0$.) (*a*) Write a computer program that can be used to determine the reactions at A and B, the average normal stress in each element, and the deformation of each element. (*b*) Use this program to solve Probs. 2.41 and 2.42.

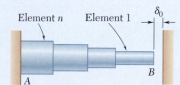

Fig. P2.C3

2.C3 Rod AB consists of n elements, each of which is homogeneous and of uniform cross section. End A is fixed, while initially there is a gap δ_0 between end B and the fixed vertical surface on the right. The length of element i is denoted by L_i, its cross-sectional area by A_i, its modulus of elasticity by E_i, and its coefficient of thermal expansion by α_i. After the temperature of the rod has been increased by ΔT, the gap at B is closed and the vertical surfaces exert equal and opposite forces on the rod. (*a*) Write a computer program that can be used to determine the magnitude of the reactions at A and B, the normal stress in each element, and the deformation of each element. (*b*) Use this program to solve Probs. 2.59 and 2.60.

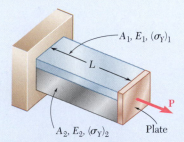

Fig. P2.C4

2.C4 Bar AB has a length L and is made of two different materials of given cross-sectional area, modulus of elasticity, and yield strength. The bar is subjected as shown to a load $\mathbf{P}$ that is gradually increased from zero until the deformation of the bar has reached a maximum value δ_m and then decreased back to zero. (*a*) Write a computer program that, for each of 25 values of δ_m equally spaced over a range extending from 0 to a value equal to 120% of the deformation causing both materials to yield, can be used to determine the maximum value P_m of the load, the maximum normal stress in each material, the permanent deformation δ_p of the bar, and the residual stress in each material. (*b*) Use this program to solve Probs. 2.111 and 2.112.

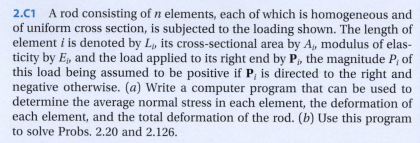

2.C5 The plate has a hole centered across the width. The stress concentration factor for a flat bar under axial loading with a centric hole is

$$K = 3.00 - 3.13\left(\frac{2r}{D}\right) + 3.66\left(\frac{2r}{D}\right)^2 - 1.53\left(\frac{2r}{D}\right)^3$$

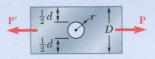

Fig. P2.C5

where r is the radius of the hole and D is the width of the bar. Write a computer program to determine the allowable load **P** for the given values of r, D, the thickness t of the bar, and the allowable stress σ_{all} of the material. Knowing that $t = \frac{1}{4}$ in., $D = 3.0$ in. and $\sigma_{all} = 16$ ksi, determine the allowable load **P** for values of r from 0.125 in. to 0.75 in., using 0.125 in. increments.

2.C6 A solid truncated cone is subjected to an axial force **P** as shown. The exact elongation is $(PL)/(2\pi c^2 E)$. By replacing the cone by n circular cylinders of equal thickness, write a computer program that can be used to calculate the elongation of the truncated cone. What is the percentage error in the answer obtained from the program using (*a*) $n = 6$, (*b*) $n = 12$, (*c*) $n = 60$?

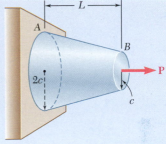

Fig. P2.C6

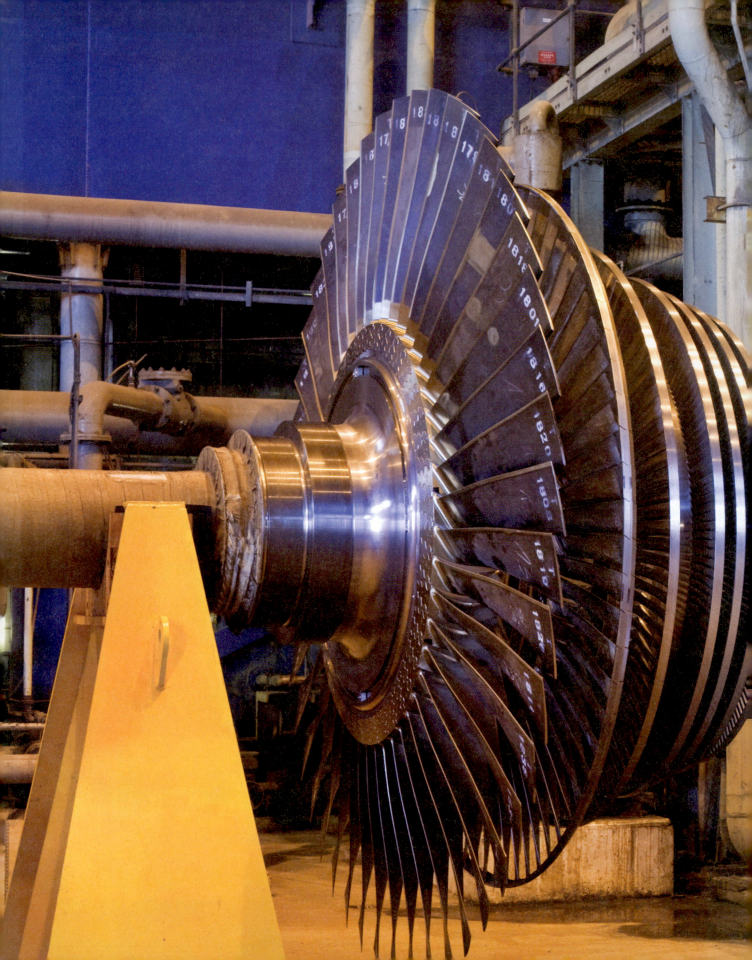

3

Torsion

In the part of the jet engine shown here, the central shaft links the components of the engine to develop the thrust that propels the aircraft.

Objectives

In this chapter, you will:

- **Introduce students** to the concept of torsion in structural members and machine parts
- **Define** shearing stresses and strains in a circular shaft subject to torsion
- **Define angle** of twist in terms of the applied torque, geometry of the shaft, and material
- **Use** torsional deformations to solve indeterminate problems
- **Design shafts** for power transmission
- **Review** stress concentrations and how they are included in torsion problems
- **Describe** the elastic-perfectly plastic response of circular shafts
- **Analyze** torsion for noncircular members
- **Define** the behavior of thin-walled hollow shafts

Introduction

In this chapter, structural members and machine parts that are in *torsion* will be analyzed, where the stresses and strains in members of circular cross section are subjected to twisting couples, or *torques*, **T** and **T'** (Fig. 3.1). These couples have a common magnitude T, and opposite senses. They are vector quantities and can be represented either by curved arrows (Fig. 3.1*a*) or by couple vectors (Fig. 3.1*b*).

Members in torsion are encountered in many engineering applications. The most common application is provided by *transmission shafts*, which are used to transmit power from one point to another (Photo 3.1). These shafts can be either solid, as shown in Fig. 3.1, or hollow.

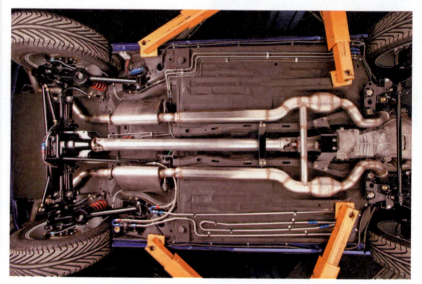

Photo 3.1   In this automotive power train, the shaft transmits power from the engine to the rear wheels.

The system shown in Fig. 3.2a consists of a turbine *A* and an electric generator *B* connected by a transmission shaft *AB*. Breaking the system into its three component parts (Fig. 3.2*b*), the turbine exerts a twisting couple or torque **T** on the shaft, which then exerts an equal torque on the generator. The generator reacts by exerting the equal and opposite torque **T'** on the shaft, and the shaft reacts by exerting the torque **T'** on the turbine.

First the stresses and deformations that take place in circular shafts will be analyzed. Then an important property of circular shafts is demonstrated: *When a circular shaft is subjected to torsion, every cross section remains plane and undistorted.* Therefore, while the various cross sections along the shaft rotate through different angles, each cross section rotates as a solid rigid slab. This property helps to determine the *distribution of shearing strains in a circular shaft and to conclude that the shearing strain varies linearly with the distance from the axis of the shaft.*

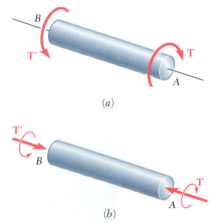

Fig. 3.1   Two equivalent ways to represent a torque in a free-body diagram.

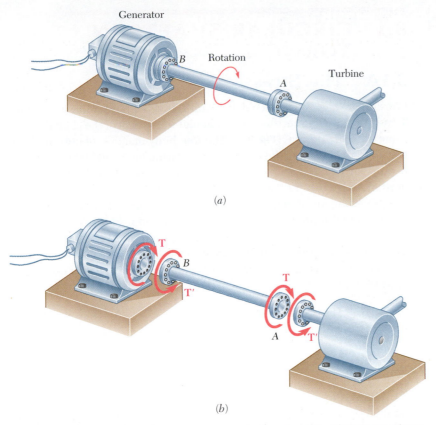

Generator

B

Rotation

A

Turbine

(a)

T

B

T′

T

A

T′

(b)

Fig. 3.2 *(a)* A generator receives power at a constant number of revolutions per minute from a turbine through shaft *AB*. *(b)* Free-body diagram of shaft *AB* along with the driving and reacting torques on the generator and turbine, respectively.

Deformations in the *elastic range* and Hooke's law for shearing stress and strain are used to determine the *distribution of shearing stresses* in a circular shaft and derive the *elastic torsion formulas*.

In Sec. 3.2, the *angle of twist* of a circular shaft is found when subjected to a given torque, assuming elastic deformations. The solution of problems involving *statically indeterminate shafts* is discussed in Sec. 3.3.

In Sec. 3.4, the *design of transmission shafts* is accomplished by determining the required physical characteristics of a shaft in terms of its speed of rotation and the power to be transmitted.

Section 3.5 accounts for stress concentrations where an abrupt change in diameter of the shaft occurs. In Secs. 3.6 to 3.8, stresses and deformations in circular shafts made of a ductile material are found when the yield point of the material is exceeded. You will then learn how to determine the permanent *plastic deformations* and *residual stresses* that remain in a shaft after it has been loaded beyond the yield point of the material.

The last sections of this chapter study the torsion of noncircular members (Sec. 3.9) and analyze the distribution of stresses in thin-walled hollow noncircular shafts (Sec. 3.10).

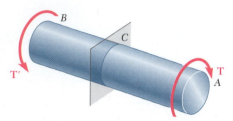

Fig. 3.3 Shaft subject to torques and a section plane at C.

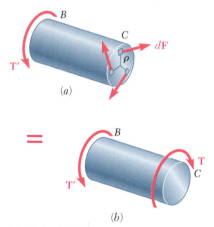

Fig. 3.4 (a) Free body diagram of section BC with torque at C represented by the contributions of small elements of area carrying forces dF a radius ρ from the section center. (b) Free-body diagram of section BC having all the small area elements summed resulting in torque T.

3.1 CIRCULAR SHAFTS IN TORSION

3.1A The Stresses in a Shaft

Consider a shaft AB subjected at A and B to equal and opposite torques **T** and **T′**. We pass a section perpendicular to the axis of the shaft through some arbitrary point C (Fig. 3.3). The free-body diagram of portion BC of the shaft must include the elementary shearing forces $d\mathbf{F}$, which are perpendicular to the radius of the shaft. These arise from the torque that portion AC exerts on BC as the shaft is twisted (Fig. 3.4a). The conditions of equilibrium for BC require that the system of these forces be equivalent to an internal torque **T**, as well as equal and opposite to **T′** (Fig. 3.4b). Denoting the perpendicular distance ρ from the force $d\mathbf{F}$ to the axis of the shaft and expressing that the sum of the moments of the shearing forces $d\mathbf{F}$ about the axis of the shaft is equal in magnitude to the torque **T**, write

$$\int \rho\, dF = T$$

Since $dF = \tau\, dA$, where τ is the shearing stress on the element of area dA, you also can write

$$\int \rho(\tau\, dA) = T \tag{3.1}$$

While these equations express an important condition that must be satisfied by the shearing stresses in any given cross section of the shaft, they do *not* tell us how these stresses are distributed in the cross section. Thus, the actual distribution of stresses under a given load is *statically indeterminate* (i.e., this distribution *cannot be determined by the methods of statics*). However, it was assumed in Sec. 1.2A that the normal stresses produced by an axial centric load were uniformly distributed, and this assumption was justified in Sec. 2.10, except in the neighborhood of concentrated loads. A similar assumption with respect to the distribution of shearing stresses in an elastic shaft *would be wrong*. Withhold any judgment until the *deformations* that are produced in the shaft have been analyzed. This will be done in the next section.

As indicated in Sec. 1.4, shear cannot take place in one plane only. Consider the very small element of shaft shown in Fig. 3.5. The torque applied to the shaft produces shearing stresses τ on the faces perpendicular to the axis of the shaft. However, the conditions of equilibrium (Sec. 1.4) require the existence of equal stresses on the faces formed by the two planes containing the axis of the shaft. That such shearing

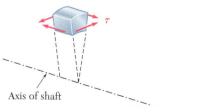

Axis of shaft

Fig. 3.5 Small element in shaft showing how shearing stress components act.

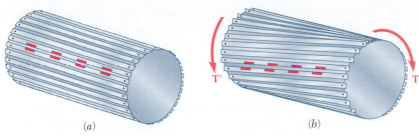

Fig. 3.6 Demonstration of shear in a shaft (a) undeformed; (b) loaded and deformed.

stresses actually occur in torsion can be demonstrated by considering a "shaft" made of separate slats pinned at both ends to disks, as shown in Fig. 3.6a. If markings have been painted on two adjoining slats, it is observed that the slats will slide with respect to each other when equal and opposite torques are applied to the ends of the "shaft" (Fig. 3.6b). While sliding will not actually take place in a shaft made of a homogeneous and cohesive material, the tendency for sliding will exist, showing that stresses occur on longitudinal planes as well as on planes perpendicular to the axis of the shaft.[†]

3.1B Deformations in a Circular Shaft

Deformation Characteristics. Consider a circular shaft attached to a fixed support at one end (Fig. 3.7a). If a torque **T** is applied to the other end, the shaft will twist, with its free end rotating through an angle ϕ called *the angle of twist* (Fig. 3.7b). Within a certain range of values of T, the angle of twist ϕ is proportional to T. Also, ϕ is proportional to the length L of the shaft. In other words, the angle of twist for a shaft of the same material and same cross section, but twice as long, will be twice as large under the same torque **T**.

When a circular shaft is subjected to torsion, *every cross section remains plane and undistorted.* In other words, while the various cross sections along the shaft rotate through different amounts, each cross section rotates as a solid rigid slab. This is illustrated in Fig. 3.8a, which shows the deformations in a rubber model subjected to torsion. This property is characteristic of circular shafts, whether solid or hollow—but not of members with noncircular cross section. For example, when a bar of square cross section is subjected to torsion, its various cross sections warp and do not remain plane (Fig. 3.8b).

The cross sections of a circular shaft remain plane and undistorted because a circular shaft is *axisymmetric* (i.e., its appearance remains the same when it is viewed from a fixed position and rotated about its axis through an arbitrary angle). Square bars, on the other hand, retain the same appearance only if they are rotated through 90° or 180°. Theoretically the axisymmetry of circular shafts can be used to prove that their cross sections remain plane and undistorted.

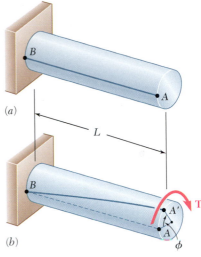

Fig. 3.7 Shaft with fixed support and line AB drawn showing deformation under torsion loading: (a) unloaded; (b) loaded

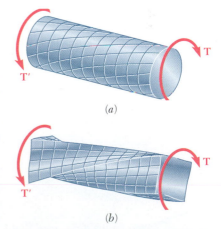

Fig. 3.8 Comparison of deformations in (a) circular and (b) square shafts.

[†]The twisting of a cardboard tube that has been slit lengthwise provides another demonstration of the existence of shearing stresses on longitudinal planes.

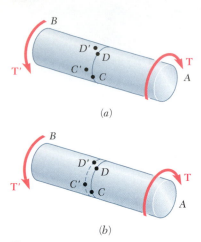

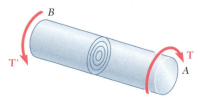

Fig. 3.9 Shaft subject to twisting.

Fig. 3.10 Concentric circles at a cross section.

Consider points C and D located on the circumference of a given cross section, and let C' and D' be the positions after the shaft has been twisted (Fig. 3.9*a*). The axisymmetry requires that the rotation that would have brought D into D' will bring C into C'. Thus, C' and D' must lie on the circumference of a circle, and the arc $C'D'$ must be equal to the arc CD (Fig. 3.9*b*).

Assume that C' and D' lie on a different circle, and the new circle is located to the left of the original circle, as shown in Fig. 3.9*b*. The same situation will prevail for any other cross section, since all cross sections of the shaft are subjected to the same internal torque T, and looking at the shaft from its end A shows that the loading causes any given circle drawn on the shaft to move *away*. But viewed from B, the given load looks the same (a clockwise couple in the foreground and a counterclockwise couple in the background), where the circle moves *toward* you. This contradiction proves that C' and D' lie on the same circle as C and D. Thus, as the shaft is twisted, the original circle just rotates in its own plane. Since the same reasoning can be applied to any smaller, concentric circle located in the cross section, the entire cross section remains plane (Fig. 3.10).

This argument does not preclude the possibility for the various concentric circles of Fig. 3.10 to rotate by different amounts when the shaft is twisted. But if that were so, a given diameter of the cross section would be distorted into a curve, as shown in Fig. 3.11*a*. Looking at this curve from A, the outer layers of the shaft get more twisted than the inner ones, while looking from B reveals the opposite (Fig. 3.11*b*). This inconsistency indicates that any diameter of a given cross section remains straight (Fig. 3.11*c*); therefore, any given cross section of a circular shaft remains plane and undistorted.

Now consider the mode of application of the twisting couples **T** and **T'**. If *all* sections of the shaft, from one end to the other, are to remain plane and undistorted, the couples are applied so the ends of the shaft remain plane and undistorted. This can be accomplished by applying the couples **T** and **T'** to rigid plates that are solidly attached to the ends of the shaft (Fig. 3.12*a*). All sections will remain plane and undistorted when the loading is applied, and the resulting deformations will be uniform throughout the entire length of the shaft. All of the equally spaced circles shown in Fig. 3.12*a* will rotate by the same amount relative to their neighbors, and each of the straight lines will be transformed into a curve (helix) intersecting the various circles at the same angle (Fig. 3.12*b*).

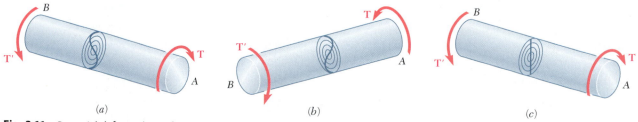

Fig. 3.11 Potential deformations of diameter lines if section's concentric circles rotate different amounts (*a*, *b*) or the same amount (*c*).

Shearing Strains. The examples given in this and the following sections are based on the assumption of rigid end plates. However, loading conditions may differ from those corresponding to the model of Fig. 3.12. This model helps to define a torsion problem for which we can obtain an exact solution. By use of Saint-Venant's principle, the results obtained for this idealized model may be extended to most engineering applications.

Now we will determine the distribution of *shearing strains* in a circular shaft of length L and radius c that has been twisted through an angle ϕ (Fig. 3.13a). Detaching from the shaft a cylinder of radius ρ, consider the small square element formed by two adjacent circles and two adjacent straight lines traced on the surface before any load is applied (Fig. 3.13b). As the shaft is subjected to a torsional load, the element deforms into a rhombus (Fig. 3.13c). Here the shearing strain γ in a given element is measured by the change in the angles formed by the sides of that element (Sec. 2.7). Since the circles defining two of the sides remain unchanged, the shearing strain γ must be equal to the angle between lines AB and $A'B$.

Figure 3.13c shows that, for small values of γ, the arc length AA' is expressed as $AA' = L\gamma$. But since $AA' = \rho\phi$, it follows that $L\gamma = \rho\phi$, or

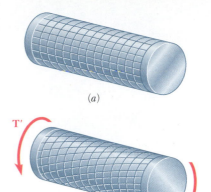

(a)

(b)

Fig. 3.12 Visualization of deformation resulting from twisting couples: (a) undeformed, (b) deformed.

$$\gamma = \frac{\rho\phi}{L} \tag{3.2}$$

where γ and ϕ are in radians. This equation shows that the shearing strain γ at a given point of a shaft in torsion is proportional to the angle of twist ϕ. It also shows that γ is proportional to the distance ρ from the axis of the shaft to that point. Thus, *the shearing strain in a circular shaft varies linearly with the distance from the axis of the shaft.*

From Eq. (3.2), the shearing strain is maximum on the surface of the shaft, where $\rho = c$.

$$\gamma_{\max} = \frac{c\phi}{L} \tag{3.3}$$

Eliminating ϕ from Eqs. (3.2) and (3.3), the shearing strain γ at a distance ρ from the axis of the shaft is

$$\gamma = \frac{\rho}{c}\gamma_{\max} \tag{3.4}$$

3.1C Stresses in the Elastic Range

When the torque **T** is such that all shearing stresses in the shaft remain below the yield strength τ_Y, the stresses in the shaft will remain below both the proportional limit and the elastic limit. Thus, Hooke's law will apply, and there will be no permanent deformation.

Recalling Hooke's law for shearing stress and strain from Sec. 2.7, write

$$\tau = G\gamma \tag{3.5}$$

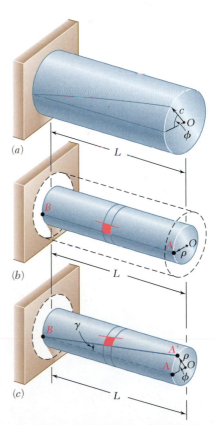

(a)

(b)

(c)

Fig. 3.13 Shearing strain deformation. (a) The angle of twist ϕ. (b) Undeformed portion of shaft of radius ρ. (c) Deformed portion of shaft; angle of twist ϕ and shearing strain γ share the same arc length AA'.

where G is the modulus of rigidity or shear modulus of the material. Multiplying both members of Eq. (3.4) by G, write

$$G\gamma = \frac{\rho}{c} G\gamma_{max}$$

or, making use of Eq. (3.5),

$$\tau = \frac{\rho}{c} \tau_{max} \qquad (3.6)$$

This equation shows that, as long as the yield strength (or proportional limit) is not exceeded in any part of a circular shaft, *the shearing stress in the shaft varies linearly with the distance ρ from the axis of the shaft.* Figure 3.14a shows the stress distribution in a solid circular shaft of radius c. A hollow circular shaft of inner radius c_1 and outer radius c_2 is shown in Fig. 3.14b. From Eq. (3.6),

$$\tau_{min} = \frac{c_1}{c_2} \tau_{max} \qquad (3.7)$$

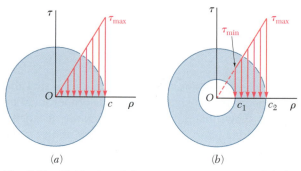

(a)   (b)

Fig. 3.14 Distribution of shearing stresses in a torqued shaft: (a) Solid shaft, (b) Hollow shaft.

Recall from Sec. 3.1A that the sum of the moments of the elementary forces exerted on any cross section of the shaft must be equal to the magnitude T of the torque exerted on the shaft:

$$\int \rho(\tau\, dA) = T \qquad (3.1)$$

Substituting for τ from Eq. (3.6) into Eq. (3.1),

$$T = \int \rho\tau\, dA = \frac{\tau_{max}}{c} \int \rho^2\, dA$$

The integral in the last part represents the polar moment of inertia J of the cross section with respect to its center O. Therefore,

$$T = \frac{\tau_{max} J}{c} \qquad (3.8)$$

or solving for τ_{max},

$$\tau_{max} = \frac{Tc}{J} \qquad (3.9)$$

Substituting for τ_{max} from Eq. (3.9) into Eq. (3.6), the shearing stress at any distance ρ from the axis of the shaft is

$$\tau = \frac{T\rho}{J} \qquad (3.10)$$

Equations (3.9) and (3.10) are known as the *elastic torsion formulas*. Recall from statics that the polar moment of inertia of a circle of radius c is $J = \frac{1}{2}\pi c^4$. For a hollow circular shaft of inner radius c_1 and outer radius c_2, the polar moment of inertia is

$$J = \tfrac{1}{2}\pi c_2^4 - \tfrac{1}{2}\pi c_1^4 = \tfrac{1}{2}\pi(c_2^4 - c_1^4) \qquad (3.11)$$

When SI metric units are used in Eq. (3.9) or (3.10), T is given in N·m, c or ρ in meters, and J in m^4. The resulting shearing stress is given in N/m^2, that is, pascals (Pa). When U.S. customary units are used, T is given in lb·in., c or ρ in inches, and J in in^4. The resulting shearing stress is given in psi.

Concept Application 3.1

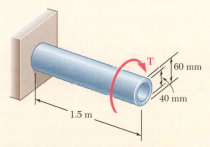

Fig. 3.15   Hollow, fixed-end shaft having torque T applied at end.

A hollow cylindrical steel shaft is 1.5 m long and has inner and outer diameters respectively equal to 40 and 60 mm (Fig. 3.15). (*a*) What is the largest torque that can be applied to the shaft if the shearing stress is not to exceed 120 MPa? (*b*) What is the corresponding minimum value of the shearing stress in the shaft?

The largest torque **T** that can be applied to the shaft is the torque for which $\tau_{max} = 120$ MPa. Since this is less than the yield strength for any steel, use Eq. (3.9). Solving this equation for T,

$$T = \frac{J\tau_{max}}{c} \qquad (1)$$

Recalling that the polar moment of inertia J of the cross section is given by Eq. (3.11), where $c_1 = \frac{1}{2}(40\text{ mm}) = 0.02\text{ m}$ and $c_2 = \frac{1}{2}(60\text{ mm}) = 0.03\text{ m}$, write

$$J = \tfrac{1}{2}\pi (c_2^4 - c_1^4) = \tfrac{1}{2}\pi(0.03^4 - 0.02^4) = 1.021 \times 10^{-6}\text{ m}^4$$

Substituting for J and τ_{max} into Eq. (1) and letting $c = c_2 = 0.03$ m,

$$T = \frac{J\tau_{max}}{c} = \frac{(1.021 \times 10^{-6}\text{ m}^4)(120 \times 10^6\text{ Pa})}{0.03\text{ m}} = 4.08\text{ kN·m}$$

The minimum shearing stress occurs on the inner surface of the shaft. Equation (3.7) expresses that τ_{min} and τ_{max} are respectively proportional to c_1 and c_2:

$$\tau_{min} = \frac{c_1}{c_2}\tau_{max} = \frac{0.02\text{ m}}{0.03\text{ m}}(120\text{ MPa}) = 80\text{ MPa}$$

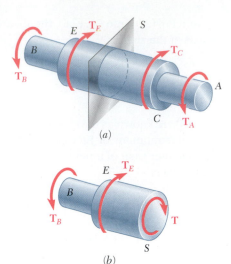

Fig. 3.16 Shaft with variable cross section. (*a*) With applied torques and section *S*. (*b*) Free-body diagram of sectioned shaft.

The torsion formulas of Eqs. (3.9) and (3.10) were derived for a shaft of uniform circular cross section subjected to torques at its ends. However, they also can be used for a shaft of variable cross section or for a shaft subjected to torques at locations other than its ends (Fig. 3.16*a*). The distribution of shearing stresses in a given cross section *S* of the shaft is obtained from Eq. (3.9), where *J* is the polar moment of inertia of that section and *T* represents the *internal torque* in that section. *T* is obtained by drawing the free-body diagram of the portion of shaft located on one side of the section (Fig. 3.16*b*) and writing that the sum of the torques applied (including the internal torque **T**) is zero (see Sample Prob. 3.1).

Our analysis of stresses in a shaft has been limited to shearing stresses due to the fact that the element selected was oriented so that its faces were either parallel or perpendicular to the axis of the shaft (Fig. 3.5). Now consider two elements *a* and *b* located on the surface of a circular shaft subjected to torsion (Fig. 3.17). Since the faces of element *a* are respectively parallel and perpendicular to the axis of the shaft, the only stresses on the element are the shearing stresses

$$\tau_{max} = \frac{Tc}{J} \tag{3.9}$$

On the other hand, the faces of element *b*, which form arbitrary angles with the axis of the shaft, are subjected to a combination of normal and shearing stresses. Consider the stresses and resulting forces on faces that

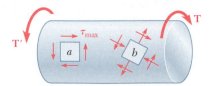

Fig. 3.17 Circular shaft with stress elements at different orientations.

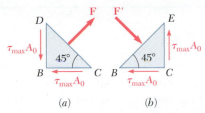

Fig. 3.18   Forces on faces at 45° to shaft axis.

are at 45° to the axis of the shaft. The free-body diagrams of the two tri-angular elements are shown in Fig. 3.18. From Fig. 3.18a, the stresses exerted on the faces BC and BD are the shearing stresses $\tau_{max} = Tc/J$. The magnitude of the corresponding shear forces is $\tau_{max}A_0$, where A_0 is the area of the face. Observing that the components along DC of the two shear forces are equal and opposite, the force **F** exerted on DC must be perpendicular to that face and is a tensile force. Its magnitude is

$$F = 2(\tau_{max}A_0)\cos 45° = \tau_{max}A_0\sqrt{2} \tag{3.12}$$

The corresponding stress is obtained by dividing the force F by the area A of face DC. Observing that $A = A_0\sqrt{2}$,

$$\sigma = \frac{F}{A} = \frac{\tau_{max}A_0\sqrt{2}}{A_0\sqrt{2}} = \tau_{max} \tag{3.13}$$

A similar analysis of the element of Figure 3.18b shows that the stress on the face BE is $\sigma = -\tau_{max}$. Therefore, the stresses exerted on the faces of an element c at 45° to the axis of the shaft (Fig. 3.19) are normal stresses equal to $\pm\tau_{max}$. Thus, while element a in Fig. 3.19 is in pure shear, element c in the same figure is subjected to a tensile stress on two of its faces and a compressive stress on the other two. Also note that all of the stresses involved have the same magnitude, Tc/J.[†]

Because ductile materials generally fail in shear, a specimen subjected to torsion breaks along a plane perpendicular to its longitudinal axis (Photo 3.2a). On the other hand, brittle materials are weaker in tension than in shear. Thus, when subjected to torsion, a brittle material tends to break along surfaces perpendicular to the direction in which tension is maximum, forming a 45° angle with the longitudinal axis of the specimen (Photo 3.2b).

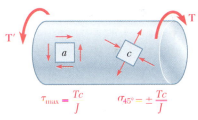

Fig. 3.19   Shaft elements with only shearing stresses or normal stresses.

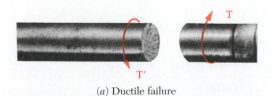

(a) Ductile failure

(b) Brittle failure

Photo 3.2   Shear failure of shaft subject to torque.

[†]Stresses on elements of arbitrary orientation, such as in Fig. 3.18b, will be discussed in Chap. 7.

Sample Problem 3.1

Shaft BC is hollow with inner and outer diameters of 90 mm and 120 mm, respectively. Shafts AB and CD are solid and of diameter d. For the loading shown, determine (a) the maximum and minimum shearing stress in shaft BC, (b) the required diameter d of shafts AB and CD if the allowable shearing stress in these shafts is 65 MPa.

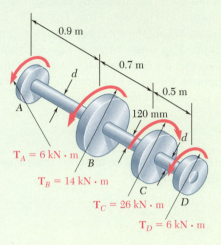

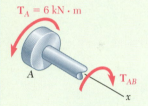

Fig. 1 Free-body diagram for section to left of cut between A and B.

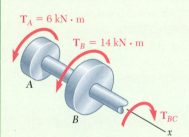

Fig. 2 Free-body diagram for section to left of cut between B and C.

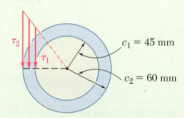

Fig. 3 Shearing stress distribution on cross section.

STRATEGY: Use free-body diagrams to determine the torque in each shaft. The torques can then be used to find the stresses for shaft BC and the required diameters for shafts AB and CD.

MODELING: Denoting by $\mathbf{T}_{AB}$ the torque in shaft AB (Fig. 1), we pass a section through shaft AB and, for the free body shown, we write

$$\Sigma M_x = 0: \qquad (6 \text{ kN·m}) - T_{AB} = 0 \qquad T_{AB} = 6 \text{ kN·m}$$

We now pass a section through shaft BC (Fig. 2) and, for the free body shown, we have

$$\Sigma M_x = 0: \quad (6 \text{ kN·m}) + (14 \text{ kN·m}) - T_{BC} = 0 \qquad T_{BC} = 20 \text{ kN·m}$$

ANALYSIS:

a. Shaft BC.   For this hollow shaft we have

$$J = \frac{\pi}{2}(c_2^4 - c_1^4) = \frac{\pi}{2}[(0.060)^4 - (0.045)^4] = 13.92 \times 10^{-6} \text{ m}^4$$

Maximum Shearing Stress.   On the outer surface, we have

$$\tau_{\max} = \tau_2 = \frac{T_{BC}c_2}{J} = \frac{(20 \text{ kN·m})(0.060 \text{ m})}{13.92 \times 10^{-6} \text{ m}^4} \qquad \tau_{\max} = 86.2 \text{ MPa} \blacktriangleleft$$

(continued)

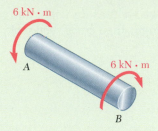

Fig. 4 Free-body diagram of shaft portion *AB*.

Minimum Shearing Stress. As shown in Fig. 3 the stresses are proportional to the distance from the axis of the shaft.

$$\frac{\tau_{min}}{\tau_{max}} = \frac{c_1}{c_2} \qquad \frac{\tau_{min}}{86.2 \text{ MPa}} = \frac{45 \text{ mm}}{60 \text{ mm}} \qquad \tau_{min} = 64.7 \text{ MPa} \blacktriangleleft$$

b. Shafts AB and CD. We note that both shafts have the same torque $T = 6$ kN·m (Fig. 4). Denoting the radius of the shafts by c and knowing that $\tau_{all} = 65$ MPa, we write

$$\tau = \frac{Tc}{J} \qquad 65 \text{ MPa} = \frac{(6 \text{ kN·m})c}{\frac{\pi}{2}c^4}$$

$$c^3 = 58.8 \times 10^{-6} \text{ m}^3 \qquad c = 38.9 \times 10^{-3} \text{ m}$$

$$d = 2c = 2(38.9 \text{ mm}) \qquad d = 77.8 \text{ mm} \blacktriangleleft$$

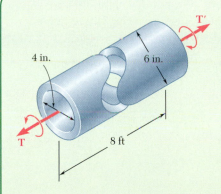

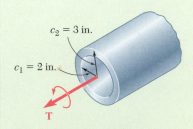

Fig. 1 Shaft as designed.

Sample Problem 3.2

The preliminary design of a motor to generator connection calls for the use of a large hollow shaft with inner and outer diameters of 4 in. and 6 in., respectively. Knowing that the allowable shearing stress is 12 ksi, determine the maximum torque that can be transmitted by (*a*) the shaft as designed, (*b*) a solid shaft of the same weight, and (*c*) a hollow shaft of the same weight and an 8-in. outer diameter.

STRATEGY: Use Eq. (3.9) to determine the maximum torque using the allowable stress.

MODELING and ANALYSIS:

a. Hollow Shaft as Designed. Using Fig. 1 and setting $\tau_{all} = 12$ ksi, we write

$$J = \frac{\pi}{2}(c_2^4 - c_1^4) = \frac{\pi}{2}[(3 \text{ in.})^4 - (2 \text{ in.})^4] = 102.1 \text{ in}^4$$

Using Eq. (3.9), we write

$$\tau_{max} = \frac{Tc_2}{J} \qquad 12 \text{ ksi} = \frac{T(3 \text{ in.})}{102.1 \text{ in}^4} \qquad T = 408 \text{ kip·in.} \blacktriangleleft$$

(continued)

b. Solid Shaft of Equal Weight. For the shaft as designed and this solid shaft to have the same weight and length, their cross-sectional areas must be equal, i.e. $A_{(a)} = A_{(b)}$.

$$\pi[(3\text{ in.})^2 - (2\text{ in.})^2] = \pi c_3^2 \qquad c_3 = 2.24\text{ in.}$$

Using Fig. 2 and setting $\tau_{\text{all}} = 12$ ksi, we write

$$\tau_{\text{max}} = \frac{Tc_3}{J} \qquad 12\text{ ksi} = \frac{T(2.24\text{ in.})}{\dfrac{\pi}{2}(2.24\text{ in.})^4} \qquad T = 211\text{ kip·in.} \blacktriangleleft$$

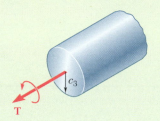

Fig. 2 Solid shaft having equal weight.

c. Hollow Shaft of 8-in. Diameter. For equal weight, the cross-sectional areas again must be equal, i.e., $A_{(a)} = A_{(c)}$ (Fig. 3). We determine the inside diameter of the shaft by writing

$$\pi[(3\text{ in.})^2 - (2\text{ in.})^2] = \pi[(4\text{ in.})^2 - c_5^2] \qquad c_5 = 3.317\text{ in.}$$

For $c_5 = 3.317$ in. and $c_4 = 4$ in.,

$$J = \frac{\pi}{2}[(4\text{ in.})^4 - (3.317\text{ in.})^4] = 212\text{ in}^4$$

With $\tau_{\text{all}} = 12$ ksi and $c_4 = 4$ in.,

$$\tau_{\text{max}} = \frac{Tc_4}{J} \qquad 12\text{ ksi} = \frac{T(4\text{ in.})}{212\text{ in}^4} \qquad T = 636\text{ kip·in.} \blacktriangleleft$$

REFLECT and THINK: This example illustrates the advantage obtained when the shaft material is further from the centroidal axis.

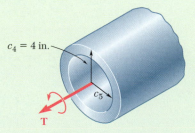

Fig. 3 Hollow shaft with an 8-in. outer diameter, having equal weight.

Problems

3.1 Determine the torque **T** that causes a maximum shearing stress of 70 MPa in the steel cylindrical shaft shown.

3.2 For the cylindrical shaft shown, determine the maximum shearing stress caused by a torque of magnitude $T = 800$ N·m.

3.3 (*a*) Determine the torque **T** that causes a maximum shearing stress of 45 MPa in the hollow cylindrical steel shaft shown. (*b*) Determine the maximum shearing stress caused by the same torque **T** in a solid cylindrical shaft of the same cross-sectional area.

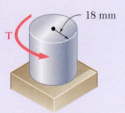

Fig. P3.1 and P3.2

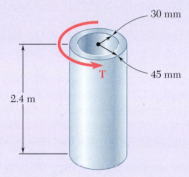

Fig. P3.3

3.4 (*a*) Determine the maximum shearing stress caused by a 40-kip·in. torque **T** in the 3-in.-diameter solid aluminum shaft shown. (*b*) Solve part *a*, assuming that the solid shaft has been replaced by a hollow shaft of the same outer diameter and of 1-in. inner diameter.

3.5 (*a*) For the 3-in.-diameter solid cylinder and loading shown, determine the maximum shearing stress. (*b*) Determine the inner diameter of the 4-in.-diameter hollow cylinder shown, for which the maximum stress is the same as in part *a*.

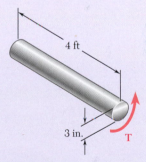

Fig. P3.4

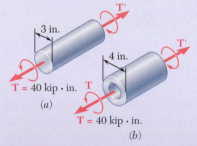

Fig. P3.5

Fig. P3.6

3.6 A torque $T = 3$ kN·m is applied to the solid bronze cylinder shown. Determine (*a*) the maximum shearing stress, (*b*) the shearing stress at point D, which lies on a 15-mm-radius circle drawn on the end of the cylinder, (*c*) the percent of the torque carried by the portion of the cylinder within the 15-mm radius.

3.7 The solid spindle AB is made of a steel with an allowable shearing stress of 12 ksi, and sleeve CD is made of a brass with an allowable shearing stress of 7 ksi. Determine (*a*) the largest torque **T** that can be applied at A if the allowable shearing stress is not to be exceeded in sleeve CD, (*b*) the corresponding required value of the diameter d_s of spindle AB.

3.8 The solid spindle AB has a diameter $d_s = 1.5$ in. and is made of a steel with an allowable shearing stress of 12 ksi, while sleeve CD is made of a brass with an allowable shearing stress of 7 ksi. Determine the largest torque **T** that can be applied at A.

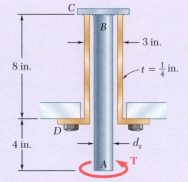

Fig. P3.7 and *P3.8*

3.9 The torques shown are exerted on pulleys A, B, and C. Knowing that both shafts are solid, determine the maximum shearing stress in (*a*) shaft AB, (*b*) shaft BC.

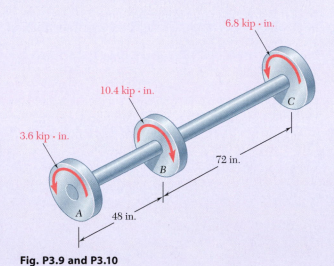

Fig. P3.9 and P3.10

3.10 The shafts of the pulley assembly shown are to be redesigned. Knowing that the allowable shearing stress in each shaft is 8.5 ksi, determine the smallest allowable diameter of (*a*) shaft AB, (*b*) shaft BC.

3.11 Knowing that each of the shafts *AB*, *BC*, and *CD* consist of a solid circular rod, determine (*a*) the shaft in which the maximum shearing stress occurs, (*b*) the magnitude of that stress.

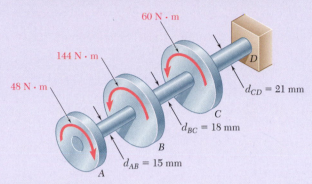

Fig. P3.11 and *P3.12*

3.12 Knowing that an 8-mm-diameter hole has been drilled through each of the shafts *AB*, *BC*, and *CD*, determine (*a*) the shaft in which the maximum shearing stress occurs, (*b*) the magnitude of that stress.

3.13 Under normal operating conditions, the electric motor exerts a torque of 2.4 kN·m on shaft *AB*. Knowing that each shaft is solid, determine the maximum shearing stress in (*a*) shaft *AB*, (*b*) shaft *BC*, (*c*) shaft *CD*.

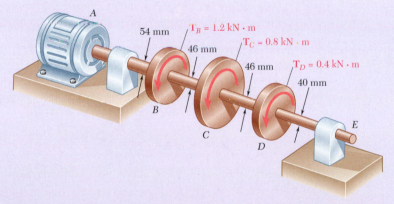

Fig. P3.13

3.14 In order to reduce the total mass of the assembly of Prob. 3.13, a new design is being considered in which the diameter of shaft *BC* will be smaller. Determine the smallest diameter of shaft *BC* for which the maximum value of the shearing stress in the assembly will not be increased.

3.15 The allowable shearing stress is 15 ksi in the 1.5-in.-diameter steel rod *AB* and 8 ksi in the 1.8-in.-diameter brass rod *BC*. Neglecting the effect of stress concentrations, determine the largest torque **T** that can be applied at *A*.

3.16 The allowable shearing stress is 15 ksi in the steel rod *AB* and 8 ksi in the brass rod *BC*. Knowing that a torque of magnitude $T = 10$ kip·in. is applied at *A*, determine the required diameter of (*a*) rod *AB*, (*b*) rod *BC*.

Fig. P3.15 and P3.16

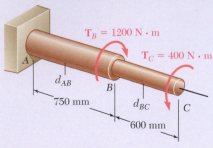

Fig. P3.17 and P3.18

3.17 The solid shaft shown is formed of a brass for which the allowable shearing stress is 55 MPa. Neglecting the effect of stress concentrations, determine the smallest diameters d_{AB} and d_{BC} for which the allowable shearing stress is not exceeded.

3.18 Solve Prob. 3.17 assuming that the direction of $\mathbf{T}_C$ is reversed.

3.19 The solid rod AB has a diameter $d_{AB} = 60$ mm and is made of a steel for which the allowable shearing stress is 85 MPa. The pipe CD, which has an outer diameter of 90 mm and a wall thickness of 6 mm, is made of an aluminum for which the allowable shearing stress is 54 MPa. Determine the largest torque $\mathbf{T}$ that can be applied at A.

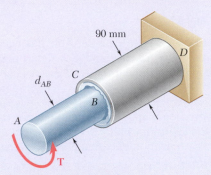

Fig. P3.19 and P3.20

3.20 The solid rod AB has a diameter $d_{AB} = 60$ mm. The pipe CD has an outer diameter of 90 mm and a wall thickness of 6 mm. Knowing that both the rod and the pipe are made of steel for which the allowable shearing stress is 75 MPa, determine the largest torque $\mathbf{T}$ that can be applied at A.

3.21 A torque of magnitude $T = 1000$ N·m is applied at D as shown. Knowing that the allowable shearing stress is 60 MPa in each shaft, determine the required diameter of (*a*) shaft AB, (*b*) shaft CD.

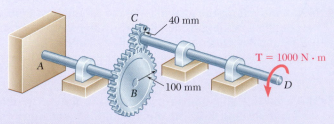

Fig. P3.21 and P3.22

3.22 A torque of magnitude $T = 1000$ N·m is applied at D as shown. Knowing that the diameter of shaft AB is 56 mm and that the diameter of shaft CD is 42 mm, determine the maximum shearing stress in (*a*) shaft AB, (*b*) shaft CD.

3.23 Under normal operating conditions a motor exerts a torque of magnitude T_F at F. The shafts are made of a steel for which the allowable shearing stress is 12 ksi and have diameters $d_{CDE} = 0.900$ in. and $d_{FGH} = 0.800$ in. Knowing that $r_D = 6.5$ in. and $r_G = 4.5$ in., determine the largest allowable value of T_F.

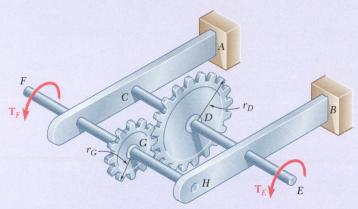

Fig. P3.23 and *P3.24*

3.24 Under normal operating conditions a motor exerts a torque of magnitude $T_F = 1200$ lb·in. at F. Knowing that $r_D = 8$ in., $r_G = 3$ in., and the allowable shearing stress is 10.5 ksi in each shaft, determine the required diameter of (*a*) shaft *CDE*, (*b*) shaft *FGH*.

3.25 The two solid shafts are connected by gears as shown and are made of a steel for which the allowable shearing stress is 7000 psi. Knowing the diameters of the two shafts are, respectively, $d_{BC} = 1.6$ in. and $d_{EF} = 1.25$ in. determine the largest torque $\mathbf{T}_C$ that can be applied at C.

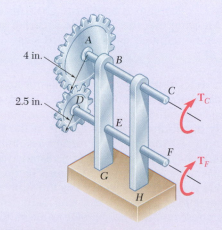

Fig. P3.25 and *P3.26*

3.26 The two solid shafts are connected by gears as shown and are made of a steel for which the allowable shearing stress is 8500 psi. Knowing that a torque of magnitude $T_C = 5$ kip·in. is applied at C and that the assembly is in equilibrium, determine the required diameter of (*a*) shaft *BC*, (*b*) shaft *EF*.

3.27 For the gear train shown, the diameters of the three solid shafts are:

$$d_{AB} = 20 \text{ mm} \qquad d_{CD} = 25 \text{ mm} \qquad d_{EF} = 40 \text{ mm}$$

Knowing that for each shaft the allowable shearing stress is 60 MPa, determine the largest torque **T** that can be applied.

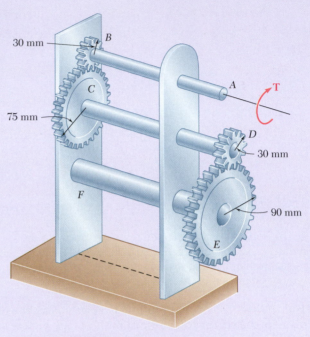

Fig. P3.27 and P3.28

3.28 A torque $T = 900$ N·m is applied to shaft AB of the gear train shown. Knowing that the allowable shearing stress is 80 MPa, determine the required diameter of (*a*) shaft AB, (*b*) shaft CD, (*c*) shaft EF.

3.29 While the exact distribution of the shearing stresses in a hollow-cylindrical shaft is as shown in Fig. P3.29*a*, an approximate value can be obtained for τ_{max} by assuming that the stresses are uniformly distributed over the area A of the cross section, as shown in Fig. P3.29*b*, and then further assuming that all of the elementary shearing forces act at a distance from O equal to the mean radius $\frac{1}{2}(c_1 + c_2)$ of the cross section. This approximate value is $\tau_0 = T/Ar_m$, where T is the applied torque. Determine the ratio τ_{max}/τ_0 of the true value of the maximum shearing stress and its approximate value τ_0 for values of c_1/c_2 respectively equal to 1.00, 0.95, 0.75, 0.50, and 0.

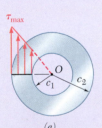

Fig. P3.29

3.30 (*a*) For a given allowable shearing stress, determine the ratio T/w of the maximum allowable torque T and the weight per unit length w for the hollow shaft shown. (*b*) Denoting by $(T/w)_0$ the value of this ratio for a solid shaft of the same radius c_2, express the ratio T/w for the hollow shaft in terms of $(T/w)_0$ and c_1/c_2.

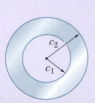

Fig. *P3.30*

3.2 ANGLE OF TWIST IN THE ELASTIC RANGE

In this section, a relationship will be determined between the angle of twist ϕ of a circular shaft and the torque **T** exerted on the shaft. The entire shaft is assumed to remain elastic. Considering first the case of a shaft of length L with a uniform cross section of radius c subjected to a torque **T** at its free end (Fig. 3.20), recall that the angle of twist ϕ and the maximum shearing strain γ_{max} are related as

$$\gamma_{max} = \frac{c\phi}{L} \qquad (3.3)$$

But in the elastic range, the yield stress is not exceeded anywhere in the shaft. Hooke's law applies, and $\gamma_{max} = \tau_{max}/G$. Recalling Eq. (3.9),

$$\gamma_{max} = \frac{\tau_{max}}{G} = \frac{Tc}{JG} \qquad (3.14)$$

Equating the right-hand members of Eqs. (3.3) and (3.14) and solving for ϕ, write

$$\phi = \frac{TL}{JG} \qquad (3.15)$$

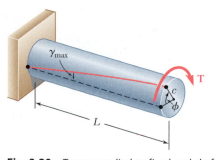

Fig. 3.20 Torque applied to fixed end shaft resulting in angle of twist ϕ.

where ϕ is in radians. The relationship obtained shows that, within the elastic range, *the angle of twist ϕ is proportional to the torque T applied to the shaft*. This agrees with the discussion at the beginning of Sec. 3.1B.

   Equation (3.15) provides a convenient method to determine the modulus of rigidity. A cylindrical rod of a material is placed in a *torsion testing machine* (Photo 3.3). Torques of increasing magnitude T are applied to the specimen, and the corresponding values of the angle of twist ϕ in a length L of the specimen are recorded. As long as the yield stress of the material is not exceeded, the points obtained by plotting ϕ against T fall on a straight line. The slope of this line represents the quantity JG/L, from which the modulus of rigidity G can be computed.

Photo 3.3   Tabletop torsion testing machine.

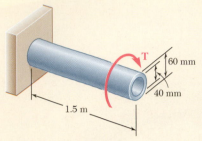

Fig. 3.15 (repeated) Hollow, fixed-end shaft having torque T applied at end.

Concept Application 3.2

What torque should be applied to the end of the shaft of Concept Application 3.1 to produce a twist of 2°? Use the value $G = 77$ GPa for the modulus of rigidity of steel.

Solving Eq. (3.15) for T, write

$$T = \frac{JG}{L}\phi$$

Substituting the given values

$$G = 77 \times 10^9 \, \text{Pa} \qquad L = 1.5 \, \text{m}$$

$$\phi = 2°\left(\frac{2\pi \, \text{rad}}{360°}\right) = 34.9 \times 10^{-3} \, \text{rad}$$

and recalling that, for the given cross section,

$$J = 1.021 \times 10^{-6} \, \text{m}^4$$

we have

$$T = \frac{JG}{L}\phi = \frac{(1.021 \times 10^{-6} \, \text{m}^4)(77 \times 10^9 \, \text{Pa})}{1.5 \, \text{m}}(34.9 \times 10^{-3} \, \text{rad})$$

$$T = 1.829 \times 10^3 \, \text{N·m} = 1.829 \, \text{kN·m}$$

Concept Application 3.3

What angle of twist will create a shearing stress of 70 MPa on the inner surface of the hollow steel shaft of Concept Applications 3.1 and 3.2?

One method for solving this problem is to use Eq. (3.10) to find the torque T corresponding to the given value of τ and Eq. (3.15) to determine the angle of twist ϕ corresponding to the value of T just found.

A more direct solution is to use Hooke's law to compute the shearing strain on the inner surface of the shaft:

$$\gamma_{\text{min}} = \frac{\tau_{\text{min}}}{G} = \frac{70 \times 10^6 \, \text{Pa}}{77 \times 10^9 \, \text{Pa}} = 909 \times 10^{-6}$$

Recalling Eq. (3.2), which was obtained by expressing the length of arc AA' in Fig. 3.13c in terms of both γ and ϕ, we have

$$\phi = \frac{L\gamma_{\text{min}}}{c_1} = \frac{1500 \, \text{mm}}{20 \, \text{mm}}(909 \times 10^{-6}) = 68.2 \times 10^{-3} \, \text{rad}$$

To obtain the angle of twist in degrees, write

$$\phi = (68.2 \times 10^{-3} \, \text{rad})\left(\frac{360°}{2\pi \, \text{rad}}\right) = 3.91°$$

Equation (3.15) can be used for the angle of twist only if the shaft is homogeneous (constant G), has a uniform cross section, and is loaded only at its ends. If the shaft is subjected to torques at locations other than its ends or if it has several portions with various cross sections and possibly of different materials, it must be divided into parts that satisfy the required conditions for Eq. (3.15). For shaft AB shown in Fig. 3.21, four different parts should be considered: AC, CD, DE, and EB. The total angle of twist of the shaft (i.e., the angle through which end A rotates with respect to end B) is obtained by *algebraically* adding the angles of twist of each component part. Using the internal torque T_i, length L_i, cross-sectional polar moment of inertia J_i, and modulus of rigidity G_i, corresponding to part i, the total angle of twist of the shaft is

$$\phi = \sum_i \frac{T_i L_i}{J_i G_i} \tag{3.16}$$

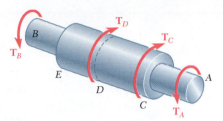

Fig. 3.21 Shaft with multiple cross-section dimensions and multiple loads.

The internal torque T_i in any given part of the shaft is obtained by passing a section through that part and drawing the free-body diagram of the portion of shaft located on one side of the section. This procedure is applied in Sample Prob. 3.3.

For a shaft with a variable circular cross section, as shown in Fig. 3.22, Eq. (3.15) is applied to a disk of thickness dx. The angle by which one face of the disk rotates with respect to the other is

$$d\phi = \frac{T\,dx}{JG}$$

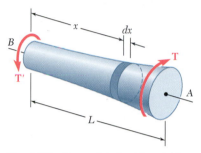

Fig. 3.22 Torqued shaft with variable cross section.

where J is a function of x. Integrating in x from 0 to L, the total angle of twist of the shaft is

$$\phi = \int_0^L \frac{T\,dx}{JG} \tag{3.17}$$

The shafts shown in Figs. 3.15 and 3.20 both had one end attached to a fixed support. In each case, the angle of twist ϕ was equal to the angle of rotation of its free end. When both ends of a shaft rotate, however, the angle of twist of the shaft is equal to the angle through which one end of the shaft rotates *with respect to the other*. For example, consider the assembly shown in Fig. 3.23a, consisting of two elastic shafts AD and BE, each of length L, radius c, modulus of rigidity G, and attached to gears meshed at C. If a torque $\mathbf{T}$ is applied at E (Fig. 3.23b), both shafts will be twisted. Since the end D of shaft AD is fixed, the angle of twist of AD is measured by the angle of rotation ϕ_A of end A. On the other hand, since both ends of shaft BE rotate, the angle of twist of BE is equal to the difference between the angles of rotation ϕ_B and ϕ_E (i.e., the angle of twist is equal to the angle through which end E rotates with respect to end B). This relative angle of rotation, $\phi_{E/B}$, is

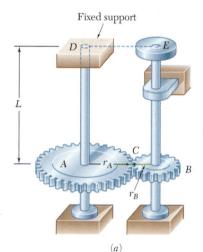

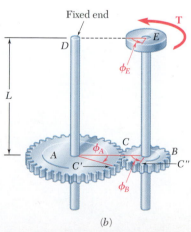

Fig. 3.23 (a) Gear assembly for transmitting torque from point E to point D. (b) Angles of twist at disk E, gear B, and gear A.

$$\phi_{E/B} = \phi_E - \phi_B = \frac{TL}{JG}$$

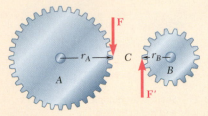

Fig. 3.24 Gear teeth forces for gears *A* and *B*.

Concept Application 3.4

For the assembly of Fig. 3.23, knowing that $r_A = 2r_B$, determine the angle of rotation of end *E* of shaft *BE* when the torque **T** is applied at *E*.

First determine the torque $\mathbf{T}_{AD}$ exerted on shaft *AD*. Observing that equal and opposite forces **F** and **F′** are applied on the two gears at *C* (Fig. 3.24) and recalling that $r_A = 2r_B$, the torque exerted on shaft *AD* is twice as large as the torque exerted on shaft *BE*. Thus, $T_{AD} = 2T$.

Since the end *D* of shaft *AD* is fixed, the angle of rotation ϕ_A of gear *A* is equal to the angle of twist of the shaft and is

$$\phi_A = \frac{T_{AD}L}{JG} = \frac{2TL}{JG}$$

Since the arcs *CC′* and *CC″* in Fig. 3.23*b* must be equal, $r_A\phi_A = r_B\phi_B$. So,

$$\phi_B = (r_A/r_B)\phi_A = 2\phi_A$$

Therefore,

$$\phi_B = 2\phi_A = \frac{4TL}{JG}$$

Next, consider shaft *BE*. The angle of twist of the shaft is equal to the angle $\phi_{E/B}$ through which end *E* rotates with respect to end *B*. Thus,

$$\phi_{E/B} = \frac{T_{BE}L}{JG} = \frac{TL}{JG}$$

The angle of rotation of end *E* is obtained by

$$\phi_E = \phi_B + \phi_{E/B}$$

$$= \frac{4TL}{JG} + \frac{TL}{JG} = \frac{5TL}{JG}$$

3.3 STATICALLY INDETERMINATE SHAFTS

There are situations where the internal torques cannot be determined from statics alone. In such cases, the external torques (i.e., those exerted on the shaft by the supports and connections) cannot be determined from the free-body diagram of the entire shaft. The equilibrium equations must be complemented by relations involving the deformations of the shaft and obtained by the geometry of the problem. Because statics is not sufficient to determine external and internal torques, the shafts are *statically inde-terminate*. The following Concept Application as well as Sample Prob. 3.5 show how to analyze statically indeterminate shafts.

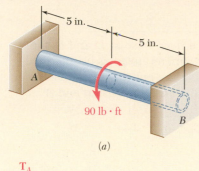

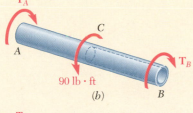

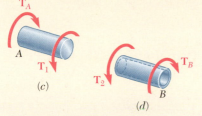

Fig. 3.25 (a) Shaft with central applied torque and fixed ends. (b) Free-body diagram of shaft AB. (c) Free-body diagrams for solid and hollow segments.

Concept Application 3.5

A circular shaft AB consists of a 10-in.-long, $\frac{7}{8}$-in.-diameter steel cylinder, in which a 5-in.-long, $\frac{5}{8}$-in.-diameter cavity has been drilled from end B. The shaft is attached to fixed supports at both ends, and a 90 lb·ft torque is applied at its midsection (Fig. 3.25a). Determine the torque exerted on the shaft by each of the supports.

Drawing the free-body diagram of the shaft and denoting by $\mathbf{T}_A$ and $\mathbf{T}_B$ the torques exerted by the supports (Fig. 3.25b), the equilibrium equation is

$$T_A + T_B = 90 \text{ lb·ft}$$

Since this equation is not sufficient to determine the two unknown torques $\mathbf{T}_A$ and $\mathbf{T}_B$, the shaft is statically indeterminate.

However, $\mathbf{T}_A$ and $\mathbf{T}_B$ can be determined if we observe that the total angle of twist of shaft AB must be zero, since both of its ends are restrained. Denoting by ϕ_1 and ϕ_2, respectively, the angles of twist of portions AC and CB, we write

$$\phi = \phi_1 + \phi_2 = 0$$

From the free-body diagram of a small portion of shaft including end A (Fig. 3.25c), we note that the internal torque T_1 in AC is equal to T_A; from the free-body diagram of a small portion of shaft including end B (Fig. 3.25d), we note that the internal torque T_2 in CB is equal to T_B. Recalling Eq. (3.15) and observing that portions AC and CB of the shaft are twisted in opposite senses, write

$$\phi = \phi_1 + \phi_2 = \frac{T_A L_1}{J_1 G} - \frac{T_B L_2}{J_2 G} = 0$$

Solving for T_B,

$$T_B = \frac{L_1 J_2}{L_2 J_1} T_A$$

Substituting the numerical data gives

$$L_1 = L_2 = 5 \text{ in.}$$

$$J_1 = \tfrac{1}{2}\pi\left(\tfrac{7}{16}\text{ in.}\right)^4 = 57.6 \times 10^{-3} \text{ in}^4$$

$$J_2 = \tfrac{1}{2}\pi\left[\left(\tfrac{7}{16}\text{ in.}\right)^4 - \left(\tfrac{5}{16}\text{ in.}\right)^4\right] = 42.6 \times 10^{-3} \text{ in}^4$$

Therefore,

$$T_B = 0.740\, T_A$$

Substitute this expression into the original equilibrium equation:

$$1.740\, T_A = 90 \text{ lb·ft}$$

$$T_A = 51.7 \text{ lb·ft} \qquad T_B = 38.3 \text{ lb·ft}$$

Sample Problem 3.3

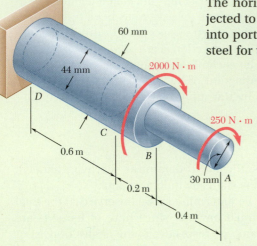

The horizontal shaft AD is attached to a fixed base at D and is subjected to the torques shown. A 44-mm-diameter hole has been drilled into portion CD of the shaft. Knowing that the entire shaft is made of steel for which $G = 77$ GPa, determine the angle of twist at end A.

STRATEGY: Use free-body diagrams to determine the torque in each shaft segment AB, BC, and CD. Then use Eq. (3.16) to determine the angle of twist at end A.

MODELING:

Passing a section through the shaft between A and B (Fig. 1), we find

$$\Sigma M_x = 0: \qquad (250 \text{ N·m}) - T_{AB} = 0 \qquad T_{AB} = 250 \text{ N·m}$$

Passing now a section between B and C (Fig. 2) we have

$$\Sigma M_x = 0: (250 \text{ N·m}) + (2000 \text{ N·m}) - T_{BC} = 0 \qquad T_{BC} = 2250 \text{ N·m}$$

Since no torque is applied at C,

$$T_{CD} = T_{BC} = 2250 \text{ N·m}$$

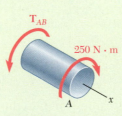

Fig. 1 Free-body diagram for finding internal torque in segment AB.

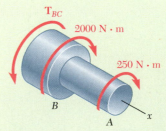

Fig. 2 Free-body diagram for finding internal torque in segment BC.

(continued)

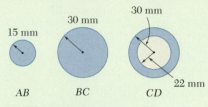

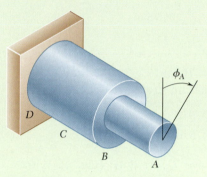

Fig. 3   Dimensions for three cross sections of shaft.

Fig. 4   Representation of angle of twist at end A.

ANALYSIS:

Polar Moments of Inertia

Using Fig. 3

$$J_{AB} = \frac{\pi}{2}c^4 = \frac{\pi}{2}(0.015\ \text{m})^4 = 0.0795 \times 10^{-6}\ \text{m}^4$$

$$J_{BC} = \frac{\pi}{2}c^4 = \frac{\pi}{2}(0.030\ \text{m})^4 = 1.272 \times 10^{-6}\ \text{m}^4$$

$$J_{CD} = \frac{\pi}{2}(c_2^4 - c_1^4) = \frac{\pi}{2}[(0.030\ \text{m})^4 - (0.022\ \text{m})^4] = 0.904 \times 10^{-6}\ \text{m}^4$$

Angle of Twist.   From Fig. 4, using Eq. (3.16) and recalling that $G = 77$ GPa for the entire shaft, we have

$$\phi_A = \sum_i \frac{T_i L_i}{J_i G} = \frac{1}{G}\left(\frac{T_{AB}L_{AB}}{J_{AB}} + \frac{T_{BC}L_{BC}}{J_{BC}} + \frac{T_{CD}L_{CD}}{J_{CD}}\right)$$

$$\phi_A = \frac{1}{77\ \text{GPa}}\left[\frac{(250\ \text{N·m})(0.4\ \text{m})}{0.0795 \times 10^{-6}\ \text{m}^4} + \frac{(2250)(0.2)}{1.272 \times 10^{-6}} + \frac{(2250)(0.6)}{0.904 \times 10^{-6}}\right]$$

$$= 0.01634 + 0.00459 + 0.01939 = 0.0403\ \text{rad}$$

$$\phi_A = (0.0403\ \text{rad})\frac{360°}{2\pi\ \text{rad}} \qquad\qquad \phi_A = 2.31° \blacktriangleleft$$

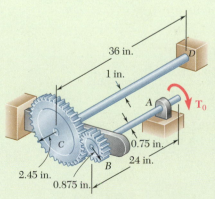

Sample Problem 3.4

Two solid steel shafts are connected by the gears shown. Knowing that for each shaft $G = 11.2 \times 10^6$ psi and the allowable shearing stress is 8 ksi, determine (*a*) the largest torque $\mathbf{T}_0$ that may be applied to end A of shaft AB and (*b*) the corresponding angle through which end A of shaft AB rotates.

STRATEGY: Use the free-body diagrams and kinematics to determine the relation between the torques and twist in each shaft segment, AB and CD. Then use the allowable stress to determine the torque that can be applied and Eq. (3.15) to determine the angle of twist at end A.

(continued)

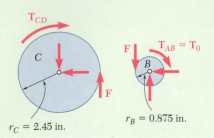

Fig. 1   Free-body diagrams of gears B and C.

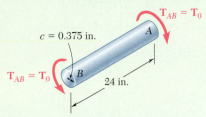

Fig. 2   Angle of twists for gears B and C.

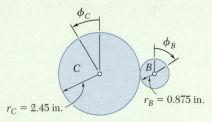

Fig. 3   Free-body diagram of shaft AB.

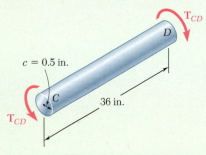

Fig. 4   Free-body diagram of shaft CD.

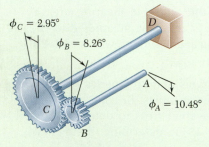

Fig. 5   Angle of twist results.

MODELING: Denoting by F the magnitude of the tangential force between gear teeth (Fig. 1), we have

Gear B. $\Sigma M_B = 0$: $F(0.875 \text{ in.}) - T_0 = 0$ $T_{CD} = 2.8 T_0$ **(1)**

Gear C. $\Sigma M_C = 0$: $F(2.45 \text{ in.}) - T_{CD} = 0$

Using kinematics with Fig. 2, we see that the peripheral motions of the gears are equal and write

$$r_B \phi_B = r_C \phi_C \qquad \phi_B = \phi_C \frac{r_C}{r_B} = \phi_C \frac{2.45 \text{ in.}}{0.875 \text{ in.}} = 2.8 \phi_C \qquad \textbf{(2)}$$

ANALYSIS:

a. Torque T_0.   For shaft AB, $T_{AB} = T_0$ and $c = 0.375$ in. (Fig. 3); considering maximum permissible shearing stress, we write

$$\tau = \frac{T_{AB} c}{J} \qquad 8000 \text{ psi} = \frac{T_0 (0.375 \text{ in.})}{\frac{1}{2}\pi(0.375 \text{ in.})^4} \qquad T_0 = 663 \text{ lb·in.} \quad \blacktriangleleft$$

For shaft CD using Eq. (1) we have $T_{CD} = 2.8 T_0$ (Fig. 4). With $c = 0.5$ in. and $\tau_{all} = 8000$ psi, we write

$$\tau = \frac{T_{CD} c}{J} \qquad 8000 \text{ psi} = \frac{2.8 T_0 (0.5 \text{ in.})}{\frac{1}{2}\pi(0.5 \text{ in.})^4} \qquad T_0 = 561 \text{ lb·in.} \quad \blacktriangleleft$$

The maximum permissible torque is the smaller value obtained for T_0.

$$T_0 = 561 \text{ lb·in.} \quad \blacktriangleleft$$

b. Angle of Rotation at End A.   We first compute the angle of twist for each shaft.

Shaft AB.   For $T_{AB} = T_0 = 561$ lb·in., we have

$$\phi_{A/B} = \frac{T_{AB} L}{JG} = \frac{(561 \text{ lb·in.})(24 \text{ in.})}{\frac{1}{2}\pi(0.375 \text{ in.})^4 (11.2 \times 10^6 \text{ psi})} = 0.0387 \text{ rad} = 2.22°$$

Shaft CD. $T_{CD} = 2.8 T_0 = 2.8(561 \text{ lb·in.})$

$$\phi_{C/D} = \frac{T_{CD} L}{JG} = \frac{2.8(561 \text{ lb·in.})(36 \text{ in.})}{\frac{1}{2}\pi(0.5 \text{ in.})^4 (11.2 \times 10^6 \text{ psi})} = 0.0514 \text{ rad} = 2.95°$$

Since end D of shaft CD is fixed, we have $\phi_C = \phi_{C/D} = 2.95°$. Using Eq. (2) with Fig. 5, we find the angle of rotation of gear B is

$$\phi_B = 2.8 \phi_C = 2.8(2.95°) = 8.26°$$

For end A of shaft AB, we have

$$\phi_A = \phi_B + \phi_{A/B} = 8.26° + 2.22° \qquad \phi_A = 10.48° \quad \blacktriangleleft$$

Sample Problem 3.5

A steel shaft and an aluminum tube are connected to a fixed support and to a rigid disk as shown in the cross section. Knowing that the initial stresses are zero, determine the maximum torque T_0 that can be applied to the disk if the allowable stresses are 120 MPa in the steel shaft and 70 MPa in the aluminum tube. Use $G = 77$ GPa for steel and $G = 27$ GPa for aluminum.

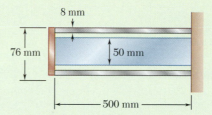

STRATEGY: We know that the applied load is resisted by both the shaft and the tube, but we do not know the portion carried by each part. Thus we need to look at the deformations. We know that both the shaft and tube are connected to the rigid disk and that the angle of twist is therefore the same for each. Once we know the portion of the torque carried by each part, we can use the allowable stress for each to determine which one governs and use this to determine the maximum torque.

MODELING:

We first draw a free-body diagram of the disk (Fig. 1) and find

$$T_0 = T_1 + T_2 \tag{1}$$

Knowing that the angle of twist is the same for the shaft and tube, we write

$$\phi_1 = \phi_2: \quad \frac{T_1 L_1}{J_1 G_1} = \frac{T_2 L_2}{J_2 G_2}$$

$$\frac{T_1(0.5\,\text{m})}{(2.003 \times 10^{-6}\,\text{m}^4)(27\,\text{GPa})} = \frac{T_2(0.5\,\text{m})}{(0.614 \times 10^{-6}\,\text{m}^4)(77\,\text{GPa})}$$

$$T_2 = 0.874 T_1 \tag{2}$$

Fig. 1 Free-body diagram of end cap.

(continued)

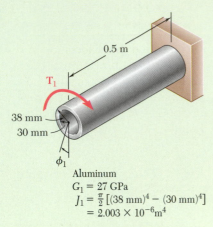

0.5 m

T_1

38 mm
30 mm

ϕ_1

Aluminum
$G_1 = 27$ GPa
$J_1 = \frac{\pi}{2}\left[(38 \text{ mm})^4 - (30 \text{ mm})^4\right]$
$\quad = 2.003 \times 10^{-6} \text{m}^4$

Fig. 2 Torque and angle of twist for hollow shaft.

ANALYSIS: We need to determine which part reaches its allowable stress first, and so we arbitrarily assume that the requirement $\tau_{alum} \leq$ 70 MPa is critical. For the aluminum tube in Fig. 2, we have

$$T_1 = \frac{\tau_{alum} J_1}{c_1} = \frac{(70 \text{ MPa})(2.003 \times 10^{-6} \text{ m}^4)}{0.038 \text{ m}} = 3690 \text{ N·m}$$

Using Eq. (2), compute the corresponding value T_2 and then find the maximum shearing stress in the steel shaft of Fig. 3.

$$T_2 = 0.874 T_1 = 0.874(3690) = 3225 \text{ N·m}$$

$$\tau_{steel} = \frac{T_2 c_2}{J_2} = \frac{(3225 \text{ N·m})(0.025 \text{ m})}{0.614 \times 10^{-6} \text{ m}^4} = 131.3 \text{ MPa}$$

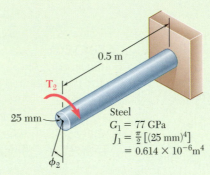

0.5 m

T_2

25 mm

Steel
$G_1 = 77$ GPa
$J_1 = \frac{\pi}{2}\left[(25 \text{ mm})^4\right]$
$\quad = 0.614 \times 10^{-6} \text{m}^4$

ϕ_2

Fig. 3 Torque and angle of twist for solid shaft.

Note that the allowable steel stress of 120 MPa is exceeded; the assumption was *wrong*. Thus, the maximum torque $\mathbf{T}_0$ will be obtained by making $\tau_{steel} = 120$ MPa. Determine the torque $\mathbf{T}_2$:

$$T_2 = \frac{\tau_{steel} J_2}{c_2} = \frac{(120 \text{ MPa})(0.614 \times 10^{-6} \text{ m}^4)}{0.025 \text{ m}} = 2950 \text{ N·m}$$

From Eq. (2), we have

$$2950 \text{ N·m} = 0.874 T_1 \qquad T_1 = 3375 \text{ N·m}$$

Using Eq. (1), we obtain the maximum permissible torque:

$$T_0 = T_1 + T_2 = 3375 \text{ N·m} + 2950 \text{ N·m}$$

$$T_0 = 6.325 \text{ kN·m} \quad \blacktriangleleft$$

REFLECT and THINK: This example illustrates that each part must not exceed its maximum allowable stress. Since the steel shaft reaches its allowable stress level first, the maximum stress in the aluminum shaft is below its maximum.

Problems

3.31 Determine the largest allowable diameter of a 3-m-long steel rod ($G = 77.2$ GPa) if the rod is to be twisted through 30° without exceeding a shearing stress of 80 MPa.

3.32 The ship at A has just started to drill for oil on the ocean floor at a depth of 5000 ft. Knowing that the top of the 8-in.-diameter steel drill pipe ($G = 11.2 \times 10^6$ psi) rotates through two complete revolutions before the drill bit at B starts to operate, determine the maximum shearing stress caused in the pipe by torsion.

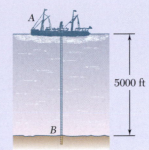

Fig. P3.32

3.33 (*a*) For the solid steel shaft shown, determine the angle of twist at A. Use $G = 11.2 \times 10^6$ psi. (*b*) Solve part *a*, assuming that the steel shaft is hollow with a 1.5-in. outer radius and a 0.75-in. inner radius.

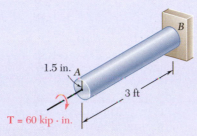

Fig. P3.33

3.34 (*a*) For the aluminum pipe shown ($G = 27$ GPa), determine the torque $\mathbf{T}_0$ causing an angle of twist of 2°. (*b*) Determine the angle of twist if the same torque $\mathbf{T}_0$ is applied to a solid cylindrical shaft of the same length and cross-sectional area.

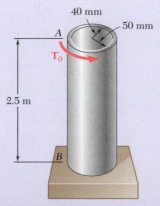

Fig. P3.34

3.35 The electric motor exerts a 500 N·m-torque on the aluminum shaft *ABCD* when it is rotating at a constant speed. Knowing that *G* = 27 GPa and that the torques exerted on pulleys *B* and *C* are as shown, determine the angle of twist between (*a*) *B* and *C*, (*b*) *B* and *D*.

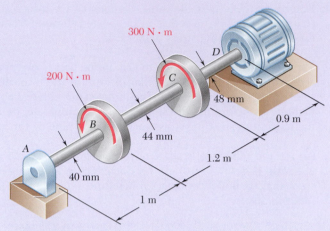
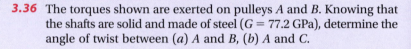
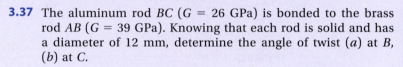

Fig. P3.35

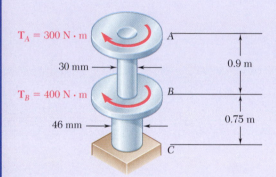

Fig. P3.36

3.36 The torques shown are exerted on pulleys *A* and *B*. Knowing that the shafts are solid and made of steel (*G* = 77.2 GPa), determine the angle of twist between (*a*) *A* and *B*, (*b*) *A* and *C*.

3.37 The aluminum rod *BC* (*G* = 26 GPa) is bonded to the brass rod *AB* (*G* = 39 GPa). Knowing that each rod is solid and has a diameter of 12 mm, determine the angle of twist (*a*) at *B*, (*b*) at *C*.

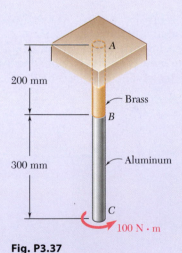

Fig. P3.37

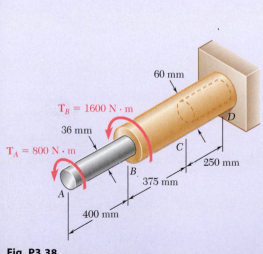

Fig. P3.38

3.38 The aluminum rod *AB* (*G* = 27 GPa) is bonded to the brass rod *BD* (*G* = 39 GPa). Knowing that portion *CD* of the brass rod is hollow and has an inner diameter of 40 mm, determine the angle of twist at *A*.

3.39 The solid spindle AB has a diameter $d_s = 1.75$ in. and is made of a steel with $G = 11.2 \times 10^6$ psi and $\tau_{all} = 12$ ksi, while sleeve CD is made of a brass with $G = 5.6 \times 10^6$ psi and $\tau_{all} = 7$ ksi. Determine (a) the largest torque **T** that can be applied at A if the given allowable stresses are not to be exceeded and if the angle of twist of sleeve CD is not to exceed $0.375°$, (b) the corresponding angle through which end A rotates.

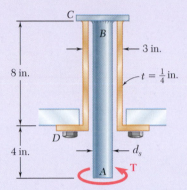

Fig. P3.39 and P3.40

3.40 The solid spindle AB has a diameter $d_s = 1.5$ in. and is made of a steel with $G = 11.2 \times 10^6$ psi and $\tau_{all} = 12$ ksi, while sleeve CD is made of a brass with $G = 5.6 \times 10^6$ psi and $\tau_{all} = 7$ ksi. Determine the largest angle through which end A can be rotated.

3.41 Two shafts, each of $\frac{7}{8}$-in. diameter, are connected by the gears shown. Knowing that $G = 11.2 \times 10^6$ psi and that the shaft at F is fixed, determine the angle through which end A rotates when a 1.2 kip·in. torque is applied at A.

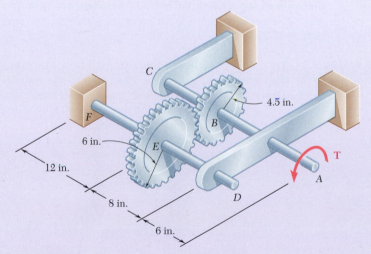

Fig. P3.41

3.42 Two solid steel shafts, each of 30-mm diameter, are connected by the gears shown. Knowing that $G = 77.2$ GPa, determine the angle through which end A rotates when a torque of magnitude $T = 200$ N·m is applied at A.

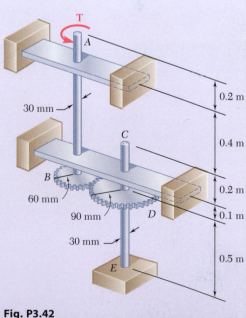

Fig. P3.42

3.43 A coder F, used to record in digital form the rotation of shaft A, is connected to the shaft by means of the gear train shown, which consists of four gears and three solid steel shafts each of diameter d. Two of the gears have a radius r and the other two a radius nr. If the rotation of the coder F is prevented, determine in terms of T, l, G, J, and n the angle through which end A rotates.

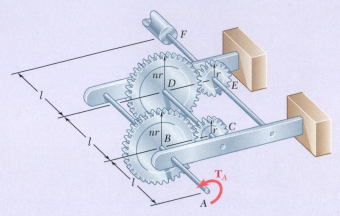

Fig. *P3.43*

3.44 For the gear train described in Prob. 3.43, determine the angle through which end A rotates when $T = 5$ lb·in., $l = 2.4$ in., $d = \frac{1}{16}$ in., $G = 11.2 \times 10^6$ psi, and $n = 2$.

3.45 The design specifications of a 1.2-m-long solid circular transmission shaft require that the angle of twist of the shaft not exceed 4° when a torque of 750 N·m is applied. Determine the required diameter of the shaft, knowing that the shaft is made of a steel with an allowable shearing stress of 90 MPa and a modulus of rigidity of 77.2 GPa.

3.46 and 3.47 The solid cylindrical rod BC of length $L = 24$ in. is attached to the rigid lever AB of length $a = 15$ in. and to the support at C. Design specifications require that the displacement of A not exceed 1 in. when a 100-lb force $\mathbf{P}$ is applied at A. For the material indicated, determine the required diameter of the rod.

      3.46 *Steel*: $\tau_{\text{all}} = 15$ ksi, $G = 11.2 \times 10^6$ psi.
      3.47 *Aluminum*: $\tau_{\text{all}} = 10$ ksi, $G = 3.9 \times 10^6$ psi.

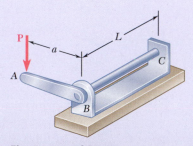

Fig. P3.46 and P3.47

3.48 The design of the gear-and-shaft system shown requires that steel shafts of the same diameter be used for both *AB* and *CD*. It is further required that $\tau_{max} \leq 60$ MPa and that the angle ϕ_D through which end *D* of shaft *CD* rotates not exceed 1.5°. Knowing that $G = 77.2$ GPa, determine the required diameter of the shafts.

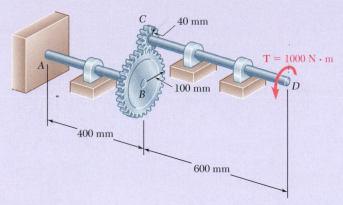

Fig. P3.48

3.49 The electric motor exerts a torque of 800 N·m on the steel shaft *ABCD* when it is rotating at a constant speed. Design specifications require that the diameter of the shaft be uniform from *A* to *D* and that the angle of twist between *A* and *D* not exceed 1.5°. Knowing that $\tau_{max} \leq 60$ MPa and $G = 77.2$ GPa, determine the minimum diameter shaft that can be used.

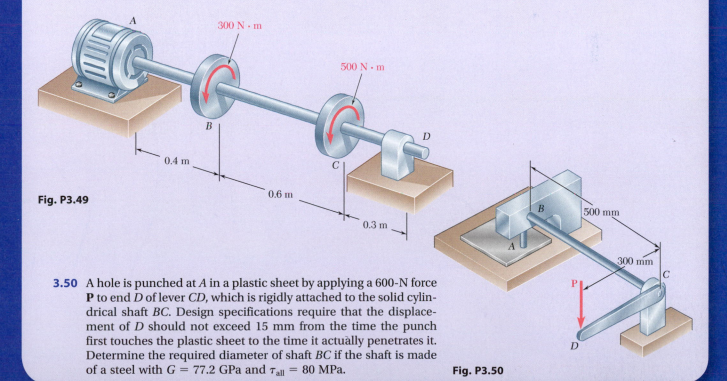

Fig. P3.49

3.50 A hole is punched at *A* in a plastic sheet by applying a 600-N force **P** to end *D* of lever *CD*, which is rigidly attached to the solid cylindrical shaft *BC*. Design specifications require that the displacement of *D* should not exceed 15 mm from the time the punch first touches the plastic sheet to the time it actually penetrates it. Determine the required diameter of shaft *BC* if the shaft is made of a steel with $G = 77.2$ GPa and $\tau_{all} = 80$ MPa.

Fig. P3.50

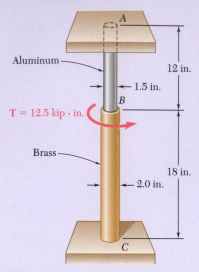

Fig. P3.51

3.51 The solid cylinders AB and BC are bonded together at B and are attached to fixed supports at A and C. Knowing that the modulus of rigidity is 3.7×10^6 psi for aluminum and 5.6×10^6 psi for brass, determine the maximum shearing stress (*a*) in cylinder AB, (*b*) in cylinder BC.

3.52 Solve Prob. 3.51, assuming that cylinder AB is made of steel, for which $G = 11.2 \times 10^6$ psi.

3.53 The composite shaft shown consists of a 0.2-in.-thick brass jacket ($G = 5.6 \times 10^6$ psi) bonded to a 1.2-in.-diameter steel core ($G_{\text{steel}} = 11.2 \times 10^6$ psi). Knowing that the shaft is subjected to 5 kip·in. torques, determine (*a*) the maximum shearing stress in the brass jacket, (*b*) the maximum shearing stress in the steel core, (*c*) the angle of twist of end B relative to end A.

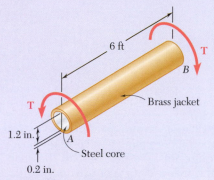

Fig. P3.53 and P3.54

3.54 The composite shaft shown consists of a 0.2-in.-thick brass jacket ($G = 5.6 \times 10^6$ psi) bonded to a 1.2-in.-diameter steel core ($G_{\text{steel}} = 11.2 \times 10^6$ psi). Knowing that the shaft is being subjected to the torques shown, determine the largest angle through which it can be twisted if the following allowable stresses are not to be exceeded: $\tau_{\text{steel}} = 15$ ksi and $\tau_{\text{brass}} = 8$ ksi.

3.55 Two solid steel shafts ($G = 77.2$ GPa) are connected to a coupling disk B and to fixed supports at A and C. For the loading shown, determine (*a*) the reaction at each support, (*b*) the maximum shearing stress in shaft AB, (*c*) the maximum shearing stress in shaft BC.

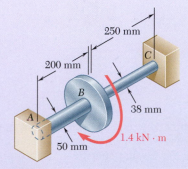

Fig. P3.55

3.56 Solve Prob. 3.55, assuming that the shaft AB is replaced by a hollow shaft of the same outer diameter and 25-mm inner diameter.

3.57 and 3.58 Two solid steel shafts are fitted with flanges that are then connected by bolts as shown. The bolts are slightly undersized and permit a 1.5° rotation of one flange with respect to the other before the flanges begin to rotate as a single unit. Knowing that $G = 77.2$ GPa, determine the maximum shearing stress in each shaft when a torque of **T** of magnitude 500 N·m is applied to the flange indicated.

3.57 The torque **T** is applied to flange B.
3.58 The torque **T** is applied to flange C.

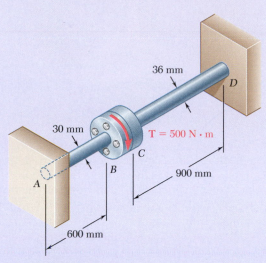

Fig. P3.57 and P3.58

3.59 The steel jacket CD has been attached to the 40-mm-diameter steel shaft AE by means of *rigid* flanges welded to the jacket and to the rod. The outer diameter of the jacket is 80 mm and its wall thickness is 4 mm. If 500-N·m torques are applied as shown, determine the maximum shearing stress in the jacket.

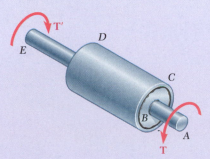

Fig. P3.59

3.60 A torque **T** is applied as shown to a solid tapered shaft AB. Show by integration that the angle of twist at A is

$$\phi = \frac{7TL}{12\pi Gc^4}$$

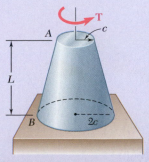

Fig. P3.60

183

3.61 The mass moment of inertia of a gear is to be determined experimentally by using a torsional pendulum consisting of a 6-ft steel wire. Knowing that $G = 11.2 \times 10^6$ psi, determine the diameter of the wire for which the torsional spring constant will be 4.27 lb·ft/rad.

Fig. P3.61

3.62 A solid shaft and a hollow shaft are made of the same material and are of the same weight and length. Denoting by n the ratio c_1/c_2, show that the ratio T_s/T_h of the torque T_s in the solid shaft to the torque T_h in the hollow shaft is (a) $\sqrt{(1 - n^2)/(1 + n^2)}$ if the maximum shearing stress is the same in each shaft, (b) $(1 - n^2)/(1 + n^2)$ if the angle of twist is the same for each shaft.

3.63 An annular plate of thickness t and modulus G is used to connect shaft AB of radius r_1 to tube CD of radius r_2. Knowing that a torque **T** is applied to end A of shaft AB and that end D of tube CD is fixed, (a) determine the magnitude and location of the maximum shearing stress in the annular plate, (b) show that the angle through which end B of the shaft rotates with respect to end C of the tube is

$$\phi_{BC} = \frac{T}{4\pi Gt}\left(\frac{1}{r_1^2} - \frac{1}{r_2^2}\right)$$

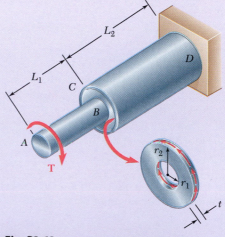

Fig. P3.63

3.4 DESIGN OF TRANSMISSION SHAFTS

The principal specifications to be met in the design of a transmission shaft are the *power* to be transmitted and the *speed of rotation* of the shaft. The role of the designer is to select the material and the dimensions of the cross section of the shaft so that the maximum shearing stress allowable will not be exceeded when the shaft is transmitting the required power at the specified speed.

To determine the torque exerted on the shaft, the power P associated with the rotation of a rigid body subjected to a torque $\mathbf{T}$ is

$$P = T\omega \tag{3.18}$$

where ω is the angular velocity of the body in radians per second (rad/s). But $\omega = 2\pi f$, where f is the frequency of the rotation, (i.e., the number of revolutions per second). The unit of frequency is $1\ \text{s}^{-1}$ and is called a *hertz* (Hz). Substituting for ω into Eq. (3.18),

$$P = 2\pi fT \tag{3.19}$$

When SI units are used with f expressed in Hz and T in N·m, the power will be in N·m/s—that is, in *watts* (W). Solving Eq. (3.19) for T, the torque exerted on a shaft transmitting the power P at a frequency of rotation f is

$$T = \frac{P}{2\pi f} \tag{3.20}$$

After determining the torque $\mathbf{T}$ to be applied to the shaft and selecting the material to be used, the designer carries the values of T and the maximum allowable stress into Eq. (3.9).

$$\frac{J}{c} = \frac{T}{\tau_{\max}} \tag{3.21}$$

This also provides the minimum allowable parameter J/c. When SI units are used, T is expressed in N·m, $\tau_{\max}$ in Pa (or N/m^2), and J/c in m^3. For a solid circular shaft, $J = \frac{1}{2}\pi c^4$, and $J/c = \frac{1}{2}\pi c^3$; substituting this value for J/c into Eq. (3.21) and solving for c yields the minimum allowable value for the radius of the shaft. For a hollow circular shaft, the critical parameter is J/c_2, where c_2 is the outer radius of the shaft; the value of this parameter may be computed from Eq. (3.11) to determine whether a given cross section will be acceptable.

When U.S. customary units are used, the frequency is usually expressed in rpm and the power in horsepower (hp). Before applying Eq. (3.20), it is then necessary to convert the frequency into revolutions per second (i.e., hertz) and the power into ft·lb/s or in·lb/s using:

$$1\ \text{rpm} = \frac{1}{60}\,\text{s}^{-1} = \frac{1}{60}\text{Hz}$$

$$1\ \text{hp} = 550\ \text{ft·lb/s} = 6600\ \text{in·lb/s}$$

When the power is given in in·lb/s, Eq. (3.20) yields the value of the torque T in lb·in. Carrying this value of T into Eq. (3.21), and expressing τ_{max} in psi, the parameter J/c is given in in^3.

Photo 3.4 In a complex gear train, the maximum allowable shearing stress of the weakest member must not be exceeded.

Concept Application 3.6

What size of shaft should be used for the rotor of a 5-hp motor operating at 3600 rpm if the shearing stress is not to exceed 8500 psi in the shaft?

The power of the motor in in·lb/s and its frequency in cycles per second (or hertz)

$$P = (5 \text{ hp})\left(\frac{6600 \text{ in·lb/s}}{1 \text{ hp}}\right) = 33{,}000 \text{ in·lb/s}$$

$$f = (3600 \text{ rpm})\frac{1 \text{ Hz}}{60 \text{ rpm}} = 60 \text{ Hz} = 60 \text{ s}^{-1}$$

The torque exerted on the shaft is given by Eq. (3.20):

$$T = \frac{P}{2\pi f} = \frac{33{,}000 \text{ in·lb/s}}{2\pi (60 \text{ s}^{-1})} = 87.54 \text{ lb·in.}$$

Substituting for T and τ_{max} into Eq. (3.21),

$$\frac{J}{c} = \frac{T}{\tau_{max}} = \frac{87.54 \text{ lb·in.}}{8500 \text{ psi}} = 10.30 \times 10^{-3} \text{ in}^3$$

But $J/c = \frac{1}{2}\pi c^3$ for a solid shaft. Therefore,

$$\frac{1}{2}\pi c^3 = 10.30 \times 10^{-3} \text{ in}^3$$
$$c = 0.1872 \text{ in.}$$
$$d = 2c = 0.374 \text{ in.}$$

A $\frac{3}{8}$-in. shaft should be used.

Concept Application 3.7

A shaft consisting of a steel tube of 50-mm outer diameter is to transmit 100 kW of power while rotating at a frequency of 20 Hz. Determine the tube thickness that should be used if the shearing stress is not to exceed 60 MPa.

The torque exerted on the shaft is given by Eq. (3.20):

$$T = \frac{P}{2\pi f} = \frac{100 \times 10^3\,\text{W}}{2\pi\,(20\,\text{Hz})} = 795.8\,\text{N·m}$$

From Eq. (3.21), the parameter J/c_2 must be at least equal to

$$\frac{J}{c_2} = \frac{T}{\tau_{\max}} = \frac{795.8\,\text{N·m}}{60 \times 10^6\,\text{N/m}^2} = 13.26 \times 10^{-6}\,\text{m}^3 \tag{1}$$

But, from Eq. (3.10),

$$\frac{J}{c_2} = \frac{\pi}{2c_2}(c_2^4 - c_1^4) = \frac{\pi}{0.050}[(0.025)^4 - c_1^4] \tag{2}$$

Equating the right-hand members of Eqs. (1) and (2),

$$(0.025)^4 - c_1^4 = \frac{0.050}{\pi}(13.26 \times 10^{-6})$$

$$c_1^4 = 390.6 \times 10^{-9} - 211.0 \times 10^{-9} = 179.6 \times 10^{-9}\,\text{m}^4$$

$$c_1 = 20.6 \times 10^{-3}\,\text{m} = 20.6\,\text{mm}$$

The corresponding tube thickness is

$$c_2 - c_1 = 25\,\text{mm} - 20.6\,\text{mm} = 4.4\,\text{mm}$$

A tube thickness of 5 mm should be used.

3.5   STRESS CONCENTRATIONS IN CIRCULAR SHAFTS

The torsion formula $\tau_{\max} = Tc/J$ was derived in Sec. 3.1C for a circular shaft of uniform cross section. Moreover, the shaft in Sec. 3.1B was loaded at its ends through rigid end plates solidly attached to it. However, torques are usually applied to the shaft through either flange couplings (Fig. 3.26a) or gears connected to the shaft by keys fitted into keyways (Fig. 3.26b). In both cases, the distribution of stresses in and near the section where the torques are applied should be different from that given by the torsion formula. For example, high concentrations of stresses occur in the neighborhood of the keyway shown in Fig. 3.26b. These localized stresses can be determined through experimental stress analysis methods or through the use of the mathematical theory of elasticity.

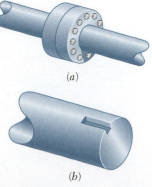

Fig. 3.26   Coupling of shafts using (a) bolted flange, (b) slot for keyway.

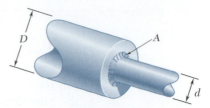

Fig. 3.27 Shafts having two different diameters with a fillet at the junction.

The torsion formula also can be used for a shaft of variable circular cross section. For a shaft with an abrupt change in the diameter of its cross section, stress concentrations occur near the discontinuity, with the highest stresses occurring at A (Fig. 3.27). These stresses can be reduced using a fillet, and the maximum value of the shearing stress at the fillet is

$$\tau_{max} = K\frac{Tc}{J} \tag{3.22}$$

where the stress Tc/J is the stress computed for the smaller-diameter shaft and K is a stress concentration factor. Since K depends upon the ratio of the two diameters and the ratio of the radius of the fillet to the diameter of the smaller shaft, it can be computed and recorded in the form of a table or a graph, as shown in Fig. 3.28. However, this procedure for determining localized shearing stresses is valid only as long as the value of τ_{max} given by Eq. (3.22) does not exceed the proportional limit of the material, since the values of K plotted in Fig. 3.28 were obtained under the assumption of a linear relation between shearing stress and shearing strain. If plastic deformations occur, the result is a maximum stress lower than those indicated by Eq. (3.22).

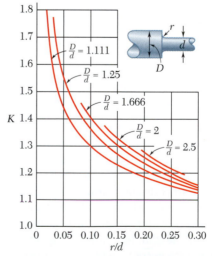

Fig. 3.28 Plot of stress concentration factors for fillets in circular shafts. (Source: W. D. Pilkey and D. F. Pilkey, *Peterson's Stress Concentration Factors*, 3rd ed., John Wiley & Sons, New York, 2008.)

Sample Problem 3.6

The stepped shaft shown is to rotate at 900 rpm as it transmits power from a turbine to a generator. The grade of steel specified in the design has an allowable shearing stress of 8 ksi. (*a*) For the preliminary design shown, determine the maximum power that can be transmitted. (*b*) If in the final design the radius of the fillet is increased so that $r = \frac{15}{16}$ in., what will be the percent change, relative to the preliminary design, in the power that can be transmitted?

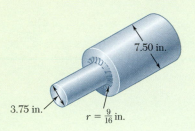

7.50 in.

3.75 in.

$r = \frac{9}{16}$ in.

STRATEGY: Use Fig. 3.28 to account for the influence of stress concentrations on the torque and Eq. (3.20) to determine the maximum power that can be transmitted.

MODELING and ANALYSIS:

a. Preliminary Design. Using the notation of Fig. 3.28, we have: $D = 7.50$ in., $d = 3.75$ in., $r = \frac{9}{16}$ in. $= 0.5625$ in.

$$\frac{D}{d} = \frac{7.50 \text{ in.}}{3.75 \text{ in.}} = 2 \qquad \frac{r}{d} = \frac{0.5625 \text{ in.}}{3.75 \text{ in.}} = 0.15$$

A stress concentration factor $K = 1.33$ is found from Fig. 3.28.

Torque. Recalling Eq. (3.22), we write

$$\tau_{max} = K\frac{Tc}{J} \qquad T = \frac{J}{c}\frac{\tau_{max}}{K} \qquad (1)$$

where J/c refers to the smaller-diameter shaft:

$$J/c = \tfrac{1}{2}\pi c^3 = \tfrac{1}{2}\pi(1.875 \text{ in.})^3 = 10.35 \text{ in}^3$$

and where

$$\frac{\tau_{max}}{K} = \frac{8 \text{ ksi}}{1.33} = 6.02 \text{ ksi}$$

(continued)

Substituting into Eq. (1), we find (Fig. 1) $T = (10.35 \text{ in}^3)(6.02 \text{ ksi}) = 62.3 \text{ kip·in.}$

Power. Since $f = (900 \text{ rpm})\dfrac{1 \text{ Hz}}{60 \text{ rpm}} = 15 \text{ Hz} = 15 \text{ s}^{-1}$, we write

$$P_a = 2\pi f T = 2\pi(15 \text{ s}^{-1})(62.3 \text{ kip·in.}) = 5.87 \times 10^6 \text{ in·lb/s}$$
$$P_a = (5.87 \times 10^6 \text{ in·lb/s})(1 \text{ hp}/6600 \text{ in·lb/s}) \quad P_a = 890 \text{ hp} \quad \blacktriangleleft$$

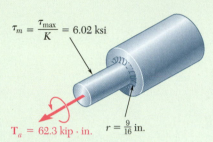

$\tau_m = \dfrac{\tau_{\max}}{K} = 6.02 \text{ ksi}$

$\mathbf{T}_a = 62.3 \text{ kip · in.}$ $r = \frac{9}{16} \text{ in.}$

Fig. 1 Allowable torque for design having $r = 9/16$ in.

b. Final Design. For $r = \frac{15}{16}$ in. = 0.9375 in.,

$$\frac{D}{d} = 2 \qquad \frac{r}{d} = \frac{0.9375 \text{ in.}}{3.75 \text{ in.}} = 0.250 \qquad K = 1.20$$

Following the procedure used previously, we write (Fig. 2)

$$\frac{\tau_{\max}}{K} = \frac{8 \text{ ksi}}{1.20} = 6.67 \text{ ksi}$$

$$T = \frac{J}{c}\frac{\tau_{\max}}{K} = (10.35 \text{ in}^3)(6.67 \text{ ksi}) = 69.0 \text{ kip·in.}$$

$$P_b = 2\pi f T = 2\pi(15 \text{ s}^{-1})(69.0 \text{ kip·in.}) = 6.50 \times 10^6 \text{ in·lb/s}$$

$$P_b = (6.50 \times 10^6 \text{ in·lb/s})(1 \text{ hp}/6600 \text{ in·lb/s}) = 985 \text{ hp}$$

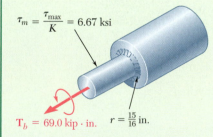

$\tau_m = \dfrac{\tau_{\max}}{K} = 6.67 \text{ ksi}$

$\mathbf{T}_b = 69.0 \text{ kip · in.}$ $r = \frac{15}{16} \text{ in.}$

Fig. 2 Allowable torque for design having $r = 15/16$ in.

Percent Change in Power

$$\text{Percent change} = 100\frac{P_b - P_a}{P_a} = 100\frac{985 - 890}{890} = +11\% \quad \blacktriangleleft$$

REFLECT and THINK: As demonstrated, a small increase in radius of the fillet at the transition in the shaft produces a significant change in the maximum power transmitted.

Problems

3.64 Determine the maximum shearing stress in a solid shaft of 1.5-in. diameter as it transmits 75 hp at a speed of (*a*) 750 rpm, (*b*) 1500 rpm.

3.65 Determine the maximum shearing stress in a solid shaft of 12-mm diameter as it transmits 2.5 kW at a frequency of (*a*) 25 Hz, (*b*) 50 Hz.

3.66 Using an allowable shearing stress of 4.5 ksi, design a solid steel shaft to transmit 12 hp at a speed of (*a*) 1200 rpm, (*b*) 2400 rpm.

3.67 Using an allowable shearing stress of 50 MPa, design a solid steel shaft to transmit 15 kW at a frequency of (*a*) 30 Hz, (*b*) 60 Hz.

3.68 While a steel shaft of the cross section shown rotates at 120 rpm, a stroboscopic measurement indicates that the angle of twist is 2° in a 4-m length. Using $G = 77.2$ GPa, determine the power being transmitted.

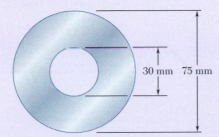

30 mm   75 mm

Fig. P3.68

3.69 Determine the required thickness of the 50-mm tubular shaft of Concept Application 3.7, if it is to transmit the same power while rotating at a frequency of 30 Hz.

3.70 A steel drive shaft is 6 ft long and its outer and inner diameters are respectively equal to 2.25 in. and 1.75 in. Knowing that the shaft transmits 240 hp while rotating at 1800 rpm, determine (*a*) the maximum shearing stress, (*b*) the angle of twist of the shaft ($G = 11.2 \times 10^6$ psi).

3.71 The hollow steel shaft shown ($G = 77.2$ GPa, $\tau_{\text{all}} = 50$ MPa) rotates at 240 rpm. Determine (*a*) the maximum power that can be transmitted, (*b*) the corresponding angle of twist of the shaft.

3.72 A steel pipe of 3.5-in. outer diameter is to be used to transmit a torque of 3000 lb·ft without exceeding an allowable shearing stress of 8 ksi. A series of 3.5-in.-outer-diameter pipes is available for use. Knowing that the wall thickness of the available pipes varies from 0.25 in. to 0.50 in. in 0.0625-in. increments, choose the lightest pipe that can be used.

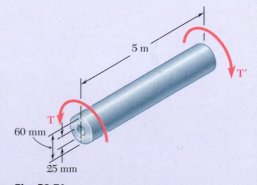

5 m

T′

60 mm

T

25 mm

Fig. P3.71

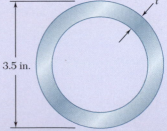

t

3.5 in.

Fig. *P3.72*

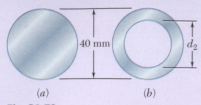

(a)   *(b)*

Fig. P3.73

3.73 The design of a machine element calls for a 40-mm-outer-diameter shaft to transmit 45 kW. (*a*) If the speed of rotation is 720 rpm, determine the maximum shearing stress in shaft *a*. (*b*) If the speed of rotation can be increased 50% to 1080 rpm, determine the largest inner diameter of shaft *b* for which the maximum shearing stress will be the same in each shaft.

3.74 Three shafts and four gears are used to form a gear train that will transmit power from the motor at *A* to a machine tool at *F*. (Bearings for the shafts are omitted in the sketch.) The diameter of each shaft is as follows: $d_{AB} = 16\text{mm}$, $d_{CD} = 20$ mm, $d_{EF} = 28$ mm. Knowing that the frequency of the motor is 24 Hz and that the allowable shearing stress for each shaft is 75 MPa, determine the maximum power that can be transmitted.

3.75 Three shafts and four gears are used to form a gear train that will transmit 7.5 kW from the motor at *A* to a machine tool at *F*. (Bearings for the shafts are omitted in the sketch.) Knowing that the frequency of the motor is 30 Hz and that the allowable stress for each shaft is 60 MPa, determine the required diameter of each shaft.

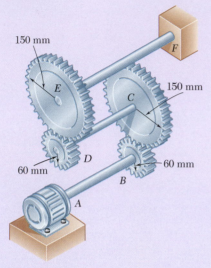

Fig. P3.74 and P3.75

3.76 The two solid shafts and gears shown are used to transmit 16 hp from the motor at *A* operating at a speed of 1260 rpm, to a machine tool at *D*. Knowing that each shaft has a diameter of 1 in., determine the maximum shearing stress (*a*) in shaft *AB*, (*b*) in shaft *CD*.

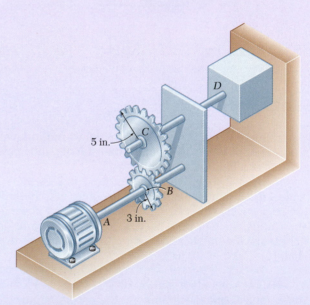

Fig. P3.76 and P3.77

3.77 The two solid shafts and gears shown are used to transmit 16 hp from the motor at *A* operating at a speed of 1260 rpm to a machine tool at *D*. Knowing that the maximum allowable shearing stress is 8 ksi, determine the required diameter (*a*) of shaft *AB*, (*b*) of shaft *CD*.

3.78 The shaft-disk-belt arrangement shown is used to transmit 3 hp from point A to point D. (a) Using an allowable shearing stress of 9500 psi, determine the required speed of shaft AB. (b) Solve part a, assuming that the diameters of shafts AB and CD are, respectively, 0.75 in. and 0.625 in.

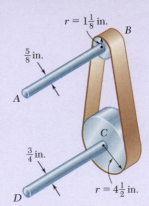

Fig. P3.78

3.79 A 5-ft-long solid steel shaft of 0.875-in. diameter is to transmit 18 hp between a motor and a machine tool. Determine the lowest speed at which the shaft can rotate, knowing that $G = 11.2 \times 10^6$ psi, that the maximum shearing stress must not exceed 4.5 ksi, and the angle of twist must not exceed 3.5°.

3.80 A 2.5-m-long steel shaft of 30-mm diameter rotates at a frequency of 30 Hz. Determine the maximum power that the shaft can transmit, knowing that $G = 77.2$ GPa, that the allowable shearing stress is 50 MPa, and that the angle of twist must not exceed 7.5°.

3.81 The design specifications of a 1.2-m-long solid transmission shaft require that the angle of twist of the shaft not exceed 4° when a torque of 750 N·m is applied. Determine the required diameter of the shaft, knowing that the shaft is made of a steel with an allowable shearing stress of 90 MPa and a modulus of rigidity of 77.2 GPa.

3.82 A 1.5-m-long tubular steel shaft ($G = 77.2$ GPa) of 38-mm outer diameter d_1 and 30-mm inner diameter d_2 is to transmit 100 kW between a turbine and a generator. Knowing that the allowable shearing stress is 60 MPa and that the angle of twist must not exceed 3°, determine the minimum frequency at which the shaft can rotate.

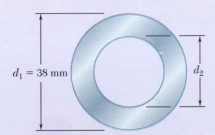

Fig. *P3.82* and *P3.83*

3.83 A 1.5-m-long tubular steel shaft of 38-mm outer diameter d_1 is to be made of a steel for which $\tau_{all} = 65$ MPa and $G = 77.2$ GPa. Knowing that the angle of twist must not exceed 4° when the shaft is subjected to a torque of 600 N·m, determine the largest inner diameter d_2 that can be specified in the design.

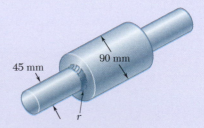

Fig. P3.84

3.84 The stepped shaft shown must transmit 40 kW at a speed of 720 rpm. Determine the minimum radius r of the fillet if an allowable stress of 36 MPa is not to be exceeded.

3.85 The stepped shaft shown rotates at 450 rpm. Knowing that $r = 0.5$ in., determine the maximum power that can be transmitted without exceeding an allowable shearing stress of 7500 psi.

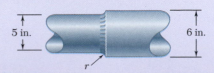

Fig. *P3.85*

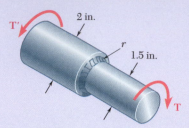

Fig. P3.86

3.86 Knowing that the stepped shaft shown transmits a torque of magnitude $T = 2.50$ kip·in., determine the maximum shearing stress in the shaft when the radius of the fillet is (a) $r = \frac{1}{8}$ in., (b) $r = \frac{3}{16}$ in.

3.87 The stepped shaft shown must rotate at a frequency of 50 Hz. Knowing that the radius of the fillet is $r = 8$ mm and the allowable shearing stress is 45 MPa, determine the maximum power that can be transmitted.

Fig. P3.87 and P3.88

3.88 The stepped shaft shown must transmit 45 kW. Knowing that the allowable shearing stress in the shaft is 40 MPa and that the radius of the fillet is $r = 6$ mm, determine the smallest permissible speed of the shaft.

3.89 A torque of magnitude $T = 200$ lb·in. is applied to the stepped shaft shown, which has a full quarter-circular fillet. Knowing that $D = 1$ in., determine the maximum shearing stress in the shaft when (a) $d = 0.8$ in., (b) $d = 0.9$ in.

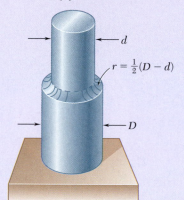

Full quarter-circular fillet extends to edge of larger shaft.

Fig. P3.89, P3.90 and *P3.91*

3.90 In the stepped shaft shown, which has a full quarter-circular fillet, the allowable shearing stress is 80 MPa. Knowing that $D = 30$ mm, determine the largest allowable torque that can be applied to the shaft if (a) $d = 26$ mm, (b) $d = 24$ mm.

3.91 In the stepped shaft shown, which has a full quarter-circular fillet, $D = 1.25$ in. and $d = 1$ in. Knowing that the speed of the shaft is 2400 rpm and that the allowable shearing stress is 7500 psi, determine the maximum power that can be transmitted by the shaft.

*3.6   PLASTIC DEFORMATIONS IN CIRCULAR SHAFTS

Equations (3.10) and (3.15) for the stress distribution and the angle of twist for a circular shaft subjected to a torque **T** assume that Hooke's law applied throughout the shaft. If the yield strength is exceeded in some portion of the shaft, or the material involved is a brittle material with a nonlinear shearing-stress-strain diagram, these relationships cease to be valid. This section will develop a more general method—used when Hooke's law does not apply—to determine the distribution of stresses in a solid circular shaft and compute the torque required to produce a given angle of twist.

No specific stress-strain relationship was assumed in Sec. 3.1B, when the shearing strain γ varied linearly with the distance ρ from the axis of the shaft (Fig. 3.29). Thus,

$$\gamma = \frac{\rho}{c}\gamma_{max} \tag{3.4}$$

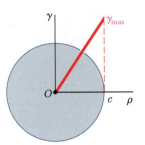

Fig. 3.29   Distribution of shearing strain for torsion of a circular shaft.

where c is the radius of the shaft.

Assuming that the maximum value τ_{max} of the shearing stress τ has been specified, the plot of τ versus ρ may be obtained as follows. We first determine from the shearing-stress-strain diagram the value of γ_{max} corresponding to τ_{max} (Fig. 3.30), and carry this value into Eq. (3.4). Then, for each value of ρ, we determine the corresponding value of γ from Eq. (3.4) or Fig. 3.29 and obtain from the stress-strain diagram of Fig. 3.30 the shearing stress τ corresponding to this value of γ. Plotting τ against ρ yields the desired distribution of stresses (Fig. 3.31).

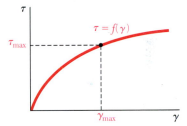

Fig. 3.30   Nonlinear shearing-stress-strain diagram.

We now recall that, when we derived Eq. (3.1) in Sec. 3.1A, we assumed no particular relation between shearing stress and strain. We may therefore use Eq. (3.1) to determine the torque **T** corresponding to the shearing-stress distribution obtained in Fig. 3.31. Considering an annular element of radius ρ and thickness $d\rho$, we express the element of area in Eq. (3.1) as $dA = 2\pi\rho\,d\rho$ and write

$$T = \int_0^c \rho\tau(2\pi\rho\,d\rho)$$

or

$$T = 2\pi\int_0^c \rho^2\tau\,d\rho \tag{3.23}$$

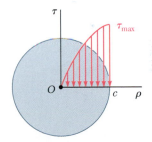

Fig. 3.31   Shearing strain distribution for shaft with nonlinear stress-strain response.

where τ is the function of ρ plotted in Fig. 3.31.

If τ is a known analytical function of γ, Eq. (3.4) can be used to express τ as a function of ρ, and the integral in Eq. (3.23) can be determined analytically. Otherwise, the torque **T** can be obtained through numerical integration. This computation becomes more meaningful if we observe that the integral in Eq. (3.23) represents the second moment, or the moment of inertia, with respect to the vertical axis of the area in Fig. 3.31 located above the horizontal axis and bounded by the stress-distribution curve.

The ultimate torque T_U, associated with the failure of the shaft, can be determined from the ultimate shearing stress τ_U by choosing $\tau_{max} = \tau_U$ and carrying out the computations indicated earlier. However, it is often

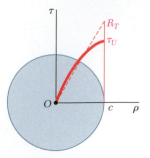

Fig. 3.32 Stress distribution in circular shaft at failure.

more convenient to determine T_U experimentally by twisting a specimen until it breaks. Assuming a fictitious linear distribution of stresses, Eq. (3.9) can thus be used to determine the corresponding maximum shearing stress R_T:

$$R_T = \frac{T_U c}{J} \qquad (3.24)$$

The fictitious stress R_T is called the *modulus of rupture in torsion*. It can be used to determine the ultimate torque T_U of a shaft made of the same material but of different dimensions by solving Eq. (3.24) for T_U. Since the actual and the fictitious linear stress distributions shown in Fig. 3.32 must yield the same value for the ultimate torque T_U, the areas must also have the same moment of inertia with respect to the vertical axis. Thus, the modulus of rupture R_T is always larger than the actual ultimate shearing stress τ_U.

In some cases, the stress distribution and the torque **T** corresponding to a given angle of twist ϕ can be determined from the equation of Sec. 3.1B for shearing strain γ in terms of ϕ, ρ, and the length L of the shaft:

$$\gamma = \frac{\rho \phi}{L} \qquad (3.2)$$

With ϕ and L given, Eq. (3.2) provides the value of γ corresponding to any given value of ρ. Using the stress-strain diagram of the material, obtain the corresponding value of the shearing stress τ and plot τ against ρ. Once the shearing-stress distribution is obtained, the torque **T** can be determined analytically or numerically.

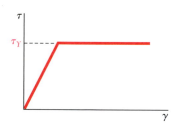

Fig. 3.33 Elastoplastic stress-strain diagram.

*3.7 CIRCULAR SHAFTS MADE OF AN ELASTOPLASTIC MATERIAL

Consider the idealized case of a *solid circular shaft made of an elastoplastic material* having the shearing-stress-strain diagram shown in Fig. 3.33. Using this diagram, we can proceed as indicated earlier and find the stress distribution across a section of the shaft for any value of the torque **T**.

As long as the shearing stress τ does not exceed the yield strength τ_Y, Hooke's law applies, and the stress distribution across the section is linear (Fig. 3.34a) with $\tau_{\max}$ given as:

$$\tau_{\max} = \frac{Tc}{J} \qquad (3.9)$$

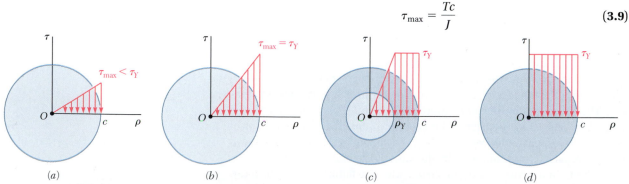

(a) (b) (c) (d)

Fig. 3.34 Stress distributions for elastoplastic shaft at different stages of loading: (a) elastic, (b) impending yield, (c) partially yielded, and (d) fully yielded.

As the torque increases, τ_{max} eventually reaches the value τ_Y (Fig. 3.34b). Substituting into Eq. (3.9) and solving for the corresponding value of the torque T_Y at the onset of yield

$$T_Y = \frac{J}{c}\tau_Y \tag{3.25}$$

This value is the *maximum elastic torque*, since it is the largest torque for which the deformation remains fully elastic. For a solid circular shaft $J/c = \frac{1}{2}\pi c^3$, we have

$$T_Y = \frac{1}{2}\pi c^3 \tau_Y \tag{3.26}$$

As the torque is increased, a plastic region develops in the shaft around an elastic core of radius ρ_Y (Fig. 3.34c). In this plastic region, the stress is uniformly equal to τ_Y, while in the elastic core, the stress varies linearly with ρ and can be expressed as

$$\tau = \frac{\tau_Y}{\rho_Y}\rho \tag{3.27}$$

As T is increased, the plastic region expands until, at the limit, the deformation is fully plastic (Fig. 3.34d).

Equation (3.23) is used to determine the torque T corresponding to a given radius ρ_Y of the elastic core. Recalling that τ is given by Eq. (3.27) for $0 \le \rho \le \rho_Y$ and is equal to τ_Y for $\rho_Y \le \rho \le c$,

$$T = 2\pi \int_0^{\rho_Y} \rho^2 \left(\frac{\tau_Y}{\rho_Y}\rho\right) d\rho + 2\pi \int_{\rho_Y}^c \rho^2 \tau_Y \, d\rho$$

$$= \frac{1}{2}\pi \rho_Y^3 \tau_Y + \frac{2}{3}\pi c^3 \tau_Y - \frac{2}{3}\pi \rho_Y^3 \tau_Y$$

$$T = \frac{2}{3}\pi c^3 \tau_Y \left(1 - \frac{1}{4}\frac{\rho_Y^3}{c^3}\right) \tag{3.28}$$

or in view of Eq. (3.26),

$$T = \frac{4}{3}T_Y\left(1 - \frac{1}{4}\frac{\rho_Y^3}{c^3}\right) \tag{3.29}$$

where T_Y is the maximum elastic torque. As ρ_Y approaches zero, the torque approaches the limiting value

$$T_p = \frac{4}{3}T_Y \tag{3.30}$$

This value, which corresponds to a fully plastic deformation (Fig. 3.34d), is the *plastic torque* of the shaft. Note that Eq. (3.30) is valid only for a *solid circular shaft made of an elastoplastic material.*

Since the distribution of *strain* across the section remains linear after the onset of yield, Eq. (3.2) remains valid and can be used to express the radius ρ_Y of the elastic core in terms of the angle of twist ϕ. If ϕ is large enough to cause a plastic deformation, the radius ρ_Y of the elastic core is obtained by making γ equal to the yield strain γ_Y in Eq. (3.2) and solving for the corresponding value ρ_Y of the distance ρ.

$$\rho_Y = \frac{L\gamma_Y}{\phi} \tag{3.31}$$

Using the angle of twist at the onset of yield ϕ_Y (i.e., when $\rho_Y = c$) and making $\phi = \phi_Y$ and $\rho_Y = c$ in Eq. (3.31), we have

$$c = \frac{L\gamma_Y}{\phi_Y} \tag{3.32}$$

Dividing Eq. (3.31) by (3.32)—member by member—provides the relationship:[†]

$$\frac{\rho_Y}{c} = \frac{\phi_Y}{\phi} \tag{3.33}$$

If we carry the expression obtained for ρ_Y/c into Eq. (3.29), the torque T as a function of the angle of twist ϕ is

$$T = \frac{4}{3}T_Y\left(1 - \frac{1}{4}\frac{\phi_Y^3}{\phi^3}\right) \tag{3.34}$$

where T_Y and ϕ_Y are the torque and the angle of twist at the onset of yield. Note that Eq. (3.34) can be used only for values of ϕ larger than ϕ_Y. For $\phi < \phi_Y$, the relation between T and ϕ is linear and given by Eq. (3.15). Combining both equations, the plot of T against ϕ is as represented in Fig. 3.35. As ϕ increases indefinitely, T approaches the limiting value $T_p = \frac{4}{3}T_Y$ corresponding to the case of a fully developed plastic zone (Fig. 3.34d). While the value T_p cannot actually be reached, Eq. (3.34) indicates that it is rapidly approached as ϕ increases. For $\phi = 2\phi_Y$, T is within about 3% of T_p, and for $\phi = 3\phi_Y$, it is within about 1%.

Since the plot of T against ϕ for an idealized elastoplastic material (Fig. 3.35) differs greatly from the shearing-stress-strain diagram (Fig. 3.33), it is clear that the shearing-stress-strain diagram of an actual material cannot be obtained directly from a torsion test carried out on a solid circular rod made of that material. However, a fairly accurate diagram can be obtained from a torsion test if a portion of the specimen consists of a thin circular tube.[‡] Indeed, the shearing stress will have a constant value τ in that portion. Thus, Eq. (3.1) reduces to

$$T = \rho A\tau$$

where ρ is the average radius of the tube and A is its cross-sectional area. The shearing stress is proportional to the torque, and τ easily can be computed from the corresponding values of T. The corresponding shearing strain γ can be obtained from Eq. (3.2) and from the values of ϕ and L measured on the tubular portion of the specimen.

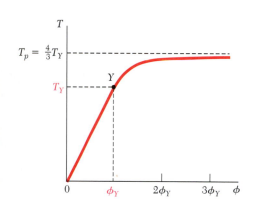

Fig. 3.35 Load-displacement relation for elastoplastic material.

[†]Equation (3.33) applies to any ductile material with a well-defined yield point, since its derivation is independent of the shape of the stress-strain diagram beyond the yield point.

[‡]In order to minimize the possibility of failure by buckling, the specimen should be made so that the length of the tubular portion is no longer than its diameter.

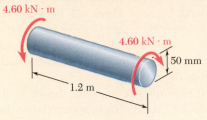

4.60 kN · m

4.60 kN · m

50 mm

1.2 m

Fig. 3.36   Loaded circular shaft.

Concept Application 3.8

A solid circular shaft, 1.2 m long and 50 mm in diameter, is subjected to a 4.60-kN·m torque at each end (Fig. 3.36). Assuming the shaft to be made of an elastoplastic material with a yield strength in shear of 150 MPa and a modulus of rigidity of 77 GPa, determine (a) the radius of the elastic core, (b) the angle of twist of the shaft.

a. Radius of Elastic Core.   Determine the torque T_Y at the onset of yield. Using Eq. (3.25) with $\tau_Y = 150$ MPa, $c = 25$ mm, and

$$J = \tfrac{1}{2}\pi c^4 = \tfrac{1}{2}\pi(25 \times 10^{-3}\,\text{m})^4 = 614 \times 10^{-9}\,\text{m}^4$$

write

$$T_Y = \frac{J\tau_Y}{c} = \frac{(614 \times 10^{-9}\,\text{m}^4)(150 \times 10^6\,\text{Pa})}{25 \times 10^{-3}\,\text{m}} = 3.68\,\text{kN·m}$$

Solving Eq. (3.29) for $(\rho_Y/c)^3$ and substituting the values of T and T_Y, we have

$$\left(\frac{\rho_Y}{c}\right)^3 = 4 - \frac{3T}{T_Y} = 4 - \frac{3(4.60\,\text{kN·m})}{3.68\,\text{kN·m}} = 0.250$$

$$\frac{\rho_Y}{c} = 0.630 \qquad \rho_Y = 0.630(25\,\text{mm}) = 15.8\,\text{mm}$$

b. Angle of Twist.   The angle of twist ϕ_Y is determined at the onset of yield from Eq. (3.15) as

$$\phi_Y = \frac{T_Y L}{JG} = \frac{(3.68 \times 10^3\,\text{N·m})(1.2\,\text{m})}{(614 \times 10^{-9}\,\text{m}^4)(77 \times 10^9\,\text{Pa})} = 93.4 \times 10^{-3}\,\text{rad}$$

Solving Eq. (3.33) for ϕ and substituting the values obtained for ϕ_Y and ρ_Y/c, write

$$\phi = \frac{\phi_Y}{\rho_Y/c} = \frac{93.4 \times 10^{-3}\,\text{rad}}{0.630} = 148.3 \times 10^{-3}\,\text{rad}$$

or

$$\phi = (148.3 \times 10^{-3}\,\text{rad})\left(\frac{360°}{2\pi\,\text{rad}}\right) = 8.50°$$

*3.8   RESIDUAL STRESSES IN CIRCULAR SHAFTS

In the two preceding sections, we saw that a plastic region will develop in a shaft subjected to a large enough torque, and that the shearing stress τ at any given point in the plastic region may be obtained from the shearing-stress-strain diagram of Fig. 3.30. If the torque is removed, the resulting

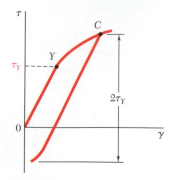

Fig. 3.37 Shear stress-strain diagram for loading past yield, followed by unloading until compressive yield occurs.

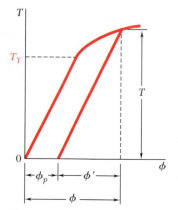

Fig. 3.38 Torque-angle of twist response for loading past yield, followed by unloading.

reduction of stress and strain at the point considered will take place along a straight line (Fig. 3.37). As you will see further in this section, the final value of the stress will not, in general, be zero. There will be a residual stress at most points, and that stress may be either positive or negative. We note that, as was the case for the normal stress, the shearing stress will keep decreasing until it has reached a value equal to its maximum value at C minus twice the yield strength of the material.

Consider again the idealized elastoplastic material shown in the shearing-stress-strain diagram of Fig. 3.33. Assuming that the relationship between τ and γ at any point of the shaft remains linear as long as the stress does not decrease by more than $2\tau_Y$, we can use Eq. (3.15) to obtain the angle through which the shaft untwists as the torque decreases back to zero. As a result, the unloading of the shaft is represented by a straight line on the T-ϕ diagram (Fig. 3.38). Note that the angle of twist does not return to zero after the torque has been removed. Indeed, the loading and unloading of the shaft result in a permanent deformation characterized by

$$\phi_p = \phi - \phi' \tag{3.35}$$

where ϕ corresponds to the loading phase and can be obtained from T by solving Eq. (3.34) with ϕ' corresponding to the unloading phase obtained from Eq. (3.15).

The residual stresses in an elastoplastic material are obtained by applying the principle of superposition (Sec. 2.13). We consider, on one hand, the stresses due to the application of the given torque **T** and, on the other, the stresses due to the equal and opposite torque which is applied to unload the shaft. The first group of stresses reflects the elastoplastic behavior of the material during the loading phase (Fig. 3.39a). The second group has the linear behavior of the same material during the unloading phase (Fig. 3.39b). Adding the two groups of stresses provides the distribution of the residual stresses in the shaft (Fig. 3.39c).

Figure 3.39c shows that some residual stresses have the same sense as the original stresses, while others have the opposite sense. This was to be expected since, according to Eq. (3.1), the relationship

$$\int \rho(\tau \, dA) = 0 \tag{3.36}$$

must be verified after the torque has been removed.

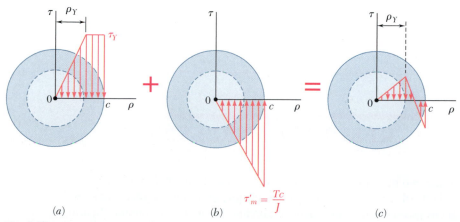

Fig. 3.39 Stress distributions for unloading of shaft with elastoplastic material.

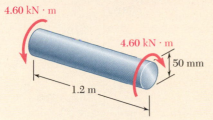

4.60 kN · m

4.60 kN · m

50 mm

1.2 m

Fig. 3.36 (repeated) Loaded circular shaft.

Concept Application 3.9

For the shaft of Concept Application 3.8, shown in Fig. 3.36, determine (*a*) the permanent twist and (*b*) the distribution of residual stresses after the 4.60-kN·m torque has been removed.

a. Permanent Twist. Recall from Concept Application 3.8 that the angle of twist corresponding to the given torque is $\phi = 8.50°$. The angle ϕ' through which the shaft untwists as the torque is removed is obtained from Eq. (3.15). Substituting the given data,

$$T = 4.60 \times 10^3 \, \text{N·m}$$
$$L = 1.2 \, \text{m}$$
$$G = 77 \times 10^9 \, \text{Pa}$$

and $J = 614 \times 10^{-9} \, \text{m}^4$, we have

$$\phi' = \frac{TL}{JG} = \frac{(4.60 \times 10^3 \, \text{N·m})(1.2 \, \text{m})}{(614 \times 10^{-9} \, \text{m}^4)(77 \times 10^9 \, \text{Pa})}$$
$$= 116.8 \times 10^{-3} \, \text{rad}$$

or

$$\phi' = (116.8 \times 10^{-3} \, \text{rad})\frac{360°}{2\pi \, \text{rad}} = 6.69°$$

The permanent twist is

$$\phi_p = \phi - \phi' = 8.50° - 6.69° = 1.81°$$

b. Residual Stresses. Recall from Concept Application 3.8 that the yield strength is $\tau_Y = 150$ MPa and the radius of the elastic core corresponding to the torque is $\rho_Y = 15.8$ mm. The distribution of the stresses in the loaded shaft is as shown in Fig. 3.40*a*.

The distribution of stresses due to the opposite 4.60-kN·m torque required to unload the shaft is linear, as shown in Fig. 3.40*b*. The maximum stress in the distribution of the reverse stresses is obtained from Eq. (3.9):

$$\tau'_{\text{max}} = \frac{Tc}{J} = \frac{(4.60 \times 10^3 \, \text{N·m})(25 \times 10^{-3} \, \text{m})}{614 \times 10^{-9} \, \text{m}^4}$$
$$= 187.3 \, \text{MPa}$$

Superposing the two distributions of stresses gives the residual stresses shown in Fig. 3.40*c*. Even though the reverse stresses exceed the yield strength τ_Y, the assumption of a linear distribution of these stresses is valid, since they do not exceed $2\tau_Y$.

(continued)

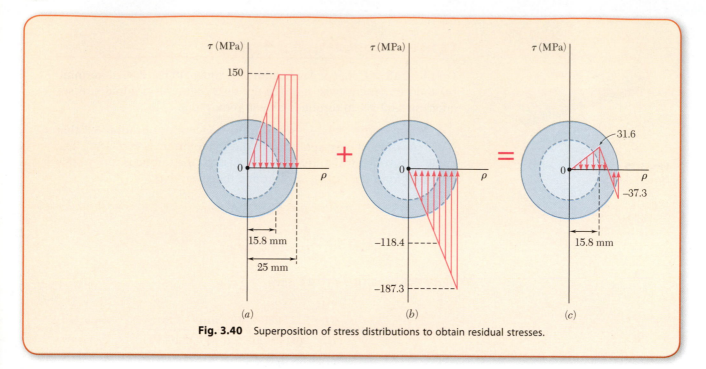

Fig. 3.40 Superposition of stress distributions to obtain residual stresses.

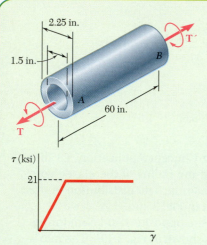

Fig. 1 Elastoplastic stress-strain diagram.

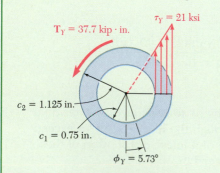

Fig. 2 Shearing stress distribution at impending yield.

Sample Problem 3.7

Shaft AB is made of a mild steel that is assumed to be elastoplastic with $G = 11.2 \times 10^6$ psi and $\tau_Y = 21$ ksi. A torque **T** is applied and gradually increased in magnitude. Determine the magnitude of **T** and the corresponding angle of twist when (*a*) yield first occurs and (*b*) the deformation has become fully plastic.

STRATEGY: We use the geometric properties and the resulting stress distribution on the cross section to determine the torque. The angle of twist is then determined using Eq. (3.2), applied to the portion of the cross section that is still elastic.

MODELING and ANALYSIS:

The geometric properties of the cross section are

$$c_1 = \tfrac{1}{2}(1.5 \text{ in.}) = 0.75 \text{ in.} \qquad c_2 = \tfrac{1}{2}(2.25 \text{ in.}) = 1.125 \text{ in.}$$

$$J = \tfrac{1}{2}\pi(c_2^4 - c_1^4) = \tfrac{1}{2}\pi[(1.125 \text{ in.})^4 - (0.75 \text{ in.})^4] = 2.02 \text{ in}^4$$

a. Onset of Yield. For $\tau_{\max} = \tau_Y = 21$ ksi (Figs. 1 and 2), we find

$$T_Y = \frac{\tau_Y J}{c_2} = \frac{(21 \text{ ksi})(2.02 \text{ in}^4)}{1.125 \text{ in.}}$$

$$T_Y = 37.7 \text{ kip·in.} \quad \blacktriangleleft$$

(continued)

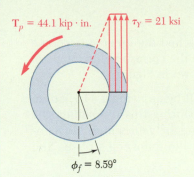

Fig. 3 Shearing stress distribution at fully plastic state.

Making $\rho = c_2$ and $\gamma = \gamma_Y$ in Eq. (3.2) and solving for ϕ, we obtain the value of ϕ_Y:

$$\phi_Y = \frac{\gamma_Y L}{c_2} = \frac{\tau_Y L}{c_2 G} = \frac{(21 \times 10^3 \text{ psi})(60 \text{ in.})}{(1.125 \text{ in.})(11.2 \times 10^6 \text{ psi})} = 0.100 \text{ rad}$$

$$\phi_Y = 5.73° \blacktriangleleft$$

b. Fully Plastic Deformation. When the plastic zone reaches the inner surface (Fig. 3), the stresses are uniformly distributed. Using Eq. (3.23), we write

$$T_p = 2\pi\tau_Y \int_{c_1}^{c_2} \rho^2 \, d\rho = \tfrac{2}{3}\pi\tau_Y(c_2^3 - c_1^3)$$

$$= \tfrac{2}{3}\pi(21 \text{ ksi})[(1.125 \text{ in.})^3 - (0.75 \text{ in.})^3]$$

$$T_p = 44.1 \text{ kip·in.} \blacktriangleleft$$

When yield first occurs on the inner surface, the deformation is fully plastic; we have from Eq. (3.2),

$$\phi_f = \frac{\gamma_Y L}{c_1} = \frac{\tau_Y L}{c_1 G} = \frac{(21 \times 10^3 \text{ psi})(60 \text{ in.})}{(0.75 \text{ in.})(11.2 \times 10^6 \text{ psi})} = 0.150 \text{ rad}$$

$$\phi_f = 8.59° \blacktriangleleft$$

REFLECT and THINK: For larger angles of twist, the torque remains constant; the T-ϕ diagram of the shaft is shown (Fig. 4).

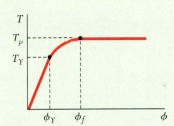

Fig. 4 Torque-angle of twist diagram for hollow shaft.

Sample Problem 3.8

For the shaft of Sample Problem 3.7 determine the residual stresses and the permanent angle of twist after the torque $T_p = 44.1$ kip·in. has been removed.

STRATEGY: We begin with the tube loaded by the fully plastic torque in Sample Problem 3.7. We apply an equal and opposite torque, knowing that the stresses induced from this unloading are elastic. Combining the stresses gives the residual stresses, and the change in the angle of twist is fully elastic.

(continued)

MODELING and ANALYSIS:

Recall that when the plastic zone first reached the inner surface, the applied torque was $T_p = 44.1$ kip·in. and the corresponding angle of twist was $\phi_f = 8.59°$. These values are shown in Figure 1a.

Elastic Unloading. We unload the shaft by applying a 44.1 kip·in. torque in the sense shown in Fig. 1b. During this unloading, the behavior of the material is linear. Recalling the values found in Sample Prob. 3.7 for c_1, c_2, and J, we obtain the following stresses and angle of twist:

$$\tau_{max} = \frac{Tc_2}{J} = \frac{(44.1 \text{ kip·in.})(1.125 \text{ in.})}{2.02 \text{ in}^4} = 24.56 \text{ ksi}$$

$$\tau_{min} = \tau_{max}\frac{c_1}{c_2} = (24.56 \text{ ksi})\frac{0.75 \text{ in.}}{1.125 \text{ in.}} = 16.37 \text{ ksi}$$

$$\phi' = \frac{TL}{JG} = \frac{(44.1 \times 10^3 \text{ psi})(60 \text{ in.})}{(2.02 \text{ in}^4)(11.2 \times 10^6 \text{ psi})} = 0.1170 \text{ rad} = 6.70°$$

Residual Stresses and Permanent Twist. The results of the loading (Fig. 1a) and the unloading (Fig. 1b) are superposed (Fig. 1c) to obtain the residual stresses and the permanent angle of twist ϕ_p.

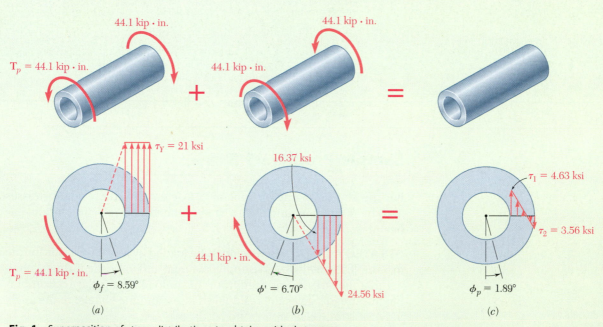

Fig. 1 Superposition of stress distributions to obtain residual stresses.

Problems

3.92 The solid circular shaft shown is made of a steel that is assumed to be elastoplastic with $\tau_Y = 145$ MPa. Determine the magnitude T of the applied torques when the plastic zone is (*a*) 16 mm deep, (*b*) 24 mm deep.

3.93 A 1.25-in. diameter solid rod is made of an elastoplastic material with $\tau_Y = 5$ ksi. Knowing that the elastic core of the rod is 1 in. in diameter, determine the magnitude of the applied torque **T**.

Fig. P3.92

3.94 The solid shaft shown is made of a mild steel that is assumed to be elastoplastic with $G = 11.2 \times 10^6$ psi and $\tau_Y = 21$ ksi. Determine the maximum shearing stress and the radius of the elastic core caused by the application of a torque of magnitude (*a*) $T = 100$ kip·in., (*b*) $T = 140$ kip·in.

3.95 The solid shaft shown is made of a mild steel that is assumed to be elastoplastic with $G = 77.2$ GPa and $\tau_Y = 145$ MPa. Determine the maximum shearing stress and the radius of the elastic core caused by the application of a torque of magnitude (*a*) $T = 600$ N·m, (*b*) $T = 1000$ N·m.

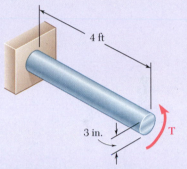

Fig. P3.94

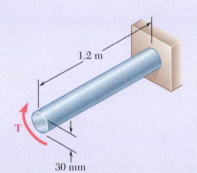

Fig. P3.95 and *P3.96*

3.96 The solid shaft shown is made of a mild steel that is assumed to be elastoplastic with $\tau_Y = 145$ MPa. Determine the radius of the elastic core caused by the application of a torque equal to $1.1\ T_Y$, where T_Y is the magnitude of the torque at the onset of yield.

3.97 It is observed that a straightened paper clip can be twisted through several revolutions by the application of a torque of approximately 60 N·m. Knowing that the diameter of the wire in the paper clip is 0.9 mm, determine the approximate value of the yield stress of the steel.

3.98 The solid shaft shown is made of a mild steel that is assumed to be elastoplastic with $G = 77.2$ GPa and $\tau_Y = 145$ MPa. Determine the angle of twist caused by the application of a torque of magnitude (*a*) $T = 600$ N·m, (*b*) $T = 1000$ N·m.

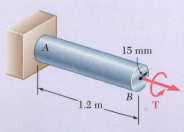

Fig. P3.98

3.99 For the solid circular shaft of Prob. 3.94, determine the angle of twist caused by the application of a torque of magnitude (*a*) $T = 80$ kip·in., (*b*) $T = 130$ kip·in.

3.100 For the solid shaft of Prob. 3.98, determine (*a*) the magnitude of the torque **T** required to twist the shaft through an angle of 15°, (*b*) the radius of the corresponding elastic core.

3.101 A 3-ft-long solid shaft has a diameter of 2.5 in. and is made of a mild steel that is assumed to be elastoplastic with $\tau_Y = 21$ ksi and $G = 11.2 \times 10^6$ psi. Determine the torque required to twist the shaft through an angle of (*a*) 2.5°, (*b*) 5°.

3.102 An 18-mm-diameter solid circular shaft is made of a material that is assumed to be elastoplastic with $\tau_Y = 145$ MPa and $G = 77.2$ GPa. For a 1.2-m length of the shaft, determine the maximum shearing stress and the angle of twist caused by a 200-N·m torque.

3.103 A 0.75-in.-diameter solid circular shaft is made of a material that is assumed to be elastoplastic with $\tau_Y = 20$ ksi and $G = 11.2 \times 10^6$ psi. For a 4-ft length of the shaft, determine the maximum shearing stress and the angle of twist caused by a 1800-lb·in. torque.

3.104 The shaft *AB* is made of a material that is elastoplastic with $\tau_Y = 90$ MPa and $G = 30$ GPa. For the loading shown, determine (*a*) the radius of the elastic core of the shaft, (*b*) the angle of twist at end *B*.

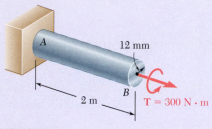

Fig. P3.104

3.105 A solid circular rod is made of a material that is assumed to be elastoplastic. Denoting by T_Y and ϕ_Y, respectively, the torque and the angle of twist at the onset of yield, determine the angle of twist if the torque is increased to (*a*) $T = 1.1\ T_Y$, (*b*) $T = 1.25\ T_Y$, (*c*) $T = 1.3\ T_Y$.

3.106 A hollow shaft is 0.9 m long and has the cross section shown. The steel is assumed to be elastoplastic with $\tau_Y = 180$ MPa and $G = 77.2$ GPa. Determine (*a*) the angle of twist at which the section first becomes fully plastic, (*b*) the corresponding magnitude of the applied torque.

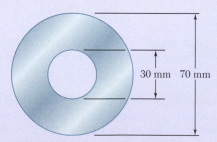

Fig. P3.106 and P3.107

3.107 A hollow shaft is 0.9 m long and has the cross section shown. The steel is assumed to be elastoplastic with $\tau_Y = 180$ MPa and $G = 77.2$ GPa. Determine the applied torque and the corresponding angle of twist (*a*) at the onset of yield, (*b*) when the plastic zone is 10 mm deep.

3.108 A steel rod is machined to the shape shown to form a tapered solid shaft to which a torque is of magnitude $T = 75$ kip·in. is applied. Assuming the steel to be elastoplastic with $\tau_Y = 21$ ksi and $G = 11.2 \times 10^6$ psi, determine (a) the radius of the elastic core in portion AB of the shaft, (b) the length of portion CD that remains fully elastic.

3.109 If the torque applied to the tapered shaft of Prob. 3.108 is slowly increased, determine (a) the magnitude T of the largest torque that can be applied to the shaft, (b) the length of the portion CD that remains fully elastic.

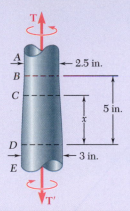

Fig. P3.108 and **P3.109**

3.110 A solid brass rod of 1.2-in. diameter is subjected to a torque that causes a maximum shearing stress of 13.5 ksi in the rod. Using the τ-γ diagram shown for the brass rod used, determine (a) the magnitude of the torque, (b) the angle of twist in a 24-in. length of the rod.

3.111 A solid brass rod of 0.8-in. diameter and 30-in. length is twisted through an angle of 10°. Using the τ-γ diagram shown for the brass rod used, determine (a) the magnitude of the torque applied to the rod, (b) the maximum shearing stress in the rod.

3.112 A 50-mm diameter cylinder is made of a brass for which the stress-strain diagram is as shown. Knowing that the angle of twist is 5° in a 725-mm length, determine by approximate means the magnitude T of torque applied to the shaft.

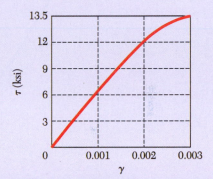

Fig. P3.110 and P3.111

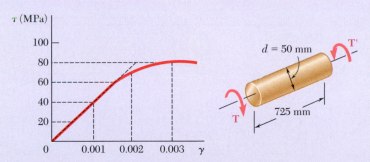

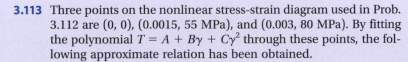

Fig. P3.112

3.113 Three points on the nonlinear stress-strain diagram used in Prob. 3.112 are (0, 0), (0.0015, 55 MPa), and (0.003, 80 MPa). By fitting the polynomial $T = A + B\gamma + C\gamma^2$ through these points, the following approximate relation has been obtained.

$$T = 46.7 \times 10^9 \gamma - 6.67 \times 10^{12} \gamma^2$$

Solve Prob. 3.112 using this relation, Eq. (3.2), and Eq. (3.23).

3.114 The solid circular drill rod AB is made of a steel that is assumed to be elastoplastic with $\tau_Y = 22$ ksi and $G = 11.2 \times 10^6$ psi. Knowing that a torque $T = 75$ kip·in. is applied to the rod and then removed, determine the maximum residual shearing stress in the rod.

3.115 In Prob. 3.114, determine the permanent angle of twist of the rod.

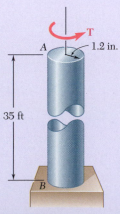

Fig. P3.114

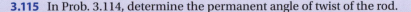

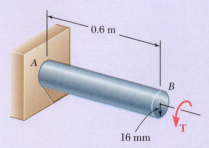

Fig. P3.116

3.116 The solid shaft shown is made of a steel that is assumed to be elastoplastic with $\tau_Y = 145$ MPa and $G = 77.2$ GPa. The torque is increased in magnitude until the shaft has been twisted through 6°; the torque is then removed. Determine (*a*) the magnitude and location of the maximum residual shearing stress, (*b*) the permanent angle of twist.

3.117 After the solid shaft of Prob. 3.116 has been loaded and unloaded as described in that problem, a torque $\mathbf{T}_1$ of sense opposite to the original torque $\mathbf{T}$ is applied to the shaft. Assuming no change in the value of ϕ_Y, determine the angle of twist ϕ_1 for which yield is initiated in this second loading and compare it with the angle ϕ_Y for which the shaft started to yield in the original loading.

3.118 The hollow shaft shown is made of a steel that is assumed to be elastoplastic with $\tau_Y = 145$ MPa and $G = 77.2$ GPa. The magnitude T of the torques is slowly increased until the plastic zone first reaches the inner surface of the shaft; the torques are then removed. Determine the magnitude and location of the maximum residual shearing stress in the rod.

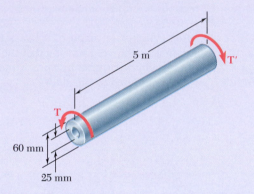

Fig. P3.118

3.119 In Prob. 3.118, determine the permanent angle of twist of the rod.

3.120 A torque $\mathbf{T}$ applied to a solid rod made of an elastoplastic material is increased until the rod is fully plastic and then removed. (*a*) Show that the distribution of residual shearing stresses is as represented in the figure. (*b*) Determine the magnitude of the torque due to the stresses acting on the portion of the rod located within a circle of radius c_0.

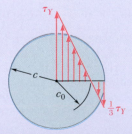

Fig. P3.120

*3.9 TORSION OF NONCIRCULAR MEMBERS

The formulas obtained for the distributions of strain and stress under a torsional loading in Sec. 3.1 apply only to members with a circular cross section. They were derived based on the assumption that the cross section of the member remained plane and undistorted. This assumption depends upon the *axisymmetry* of the member (i.e., the fact that its appearance remains the same when viewed from a fixed position and rotated about its axis through an arbitrary angle).

A square bar, on the other hand, retains the same appearance only when it is rotated through 90° or 180°. Following a line of reasoning similar to that used in Sec. 3.1B, one could show that the diagonals of the square cross section of the bar and the lines joining the midpoints of the sides of that section remain straight (Fig. 3.41). However, because of the lack of axisymmetry of the bar, any other line drawn in its cross section will deform when it is twisted, and the cross section will be warped out of its original plane.

Equations (3.4) and (3.6) define the distributions of strain and stress in an elastic circular shaft but cannot be used for noncircular members. For example, it would be wrong to assume that the shearing stress in the cross section of a square bar varies linearly with the distance from the axis of the bar and is therefore largest at the corners of the cross section. The shearing stress is actually zero at these points.

Consider a small cubic element located at a corner of the cross section of a square bar in torsion and select coordinate axes parallel to the edges (Fig. 3.42*a*). Since the face perpendicular to the *y* axis is part of the free surface of the bar, all stresses on this face must be zero. Referring to Fig. 3.42*b*, we write

$$\tau_{yx} = 0 \qquad \tau_{yz} = 0 \qquad \text{(3.37)}$$

For the same reason, all stresses on the face perpendicular to the *z* axis must be zero, and

$$\tau_{zx} = 0 \qquad \tau_{zy} = 0 \qquad \text{(3.38)}$$

It follows from the first of Eqs. (3.37) and the first of Eqs. (3.38) that

$$\tau_{xy} = 0 \qquad \tau_{xz} = 0 \qquad \text{(3.39)}$$

Thus, both components of the shearing stress on the face perpendicular to the axis of the bar are zero. Thus, there is no shearing stress at the corners of the cross section of the bar.

By twisting a rubber model of a square bar, one finds no deformations—and no stresses—occur along the edges of the bar, while the largest deformations—and the largest stresses—occur along the center line of each of the faces of the bar (Fig. 3.43).

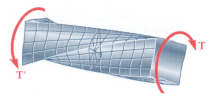
Fig. 3.41 Twisting a shaft of square cross section.

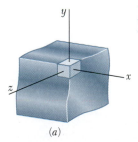
(a)

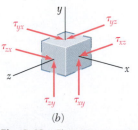
(b)

Fig. 3.42 Element at corner of square bar in torsion: (*a*) location of element in shaft and (*b*) potential shearing stress components on element.

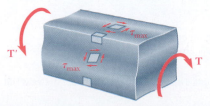
Fig. 3.43 Stress elements in a torsionally loaded, deformed square bar.

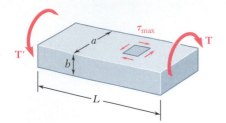

Fig. 3.44 Shaft with rectangular cross section, showing the location of maximum shearing stress.

The determination of the stresses in noncircular members subjected to a torsional loading is beyond the scope of this text. However, results obtained from the mathematical theory of elasticity for straight bars with a *uniform rectangular cross section* are given here for our use.[†] Denoting by L the length of the bar, by a and b, respectively, the wider and narrower side of its cross section, and by T the magnitude of the torque applied to the bar (Fig. 3.44), the maximum shearing stress occurs along the center line of the *wider* face and is equal to

$$\tau_{max} = \frac{T}{c_1 ab^2} \tag{3.40}$$

The angle of twist can be expressed as

$$\phi = \frac{TL}{c_2 ab^3 G} \tag{3.41}$$

Coefficients c_1 and c_2 depend only upon the ratio a/b and are given in Table 3.1 for a number of values of that ratio. Note that Eqs. (3.40) and (3.41) are valid only within the elastic range.

Table 3.1 shows that for $a/b \geq 5$, the coefficients c_1 and c_2 are equal. It may be shown that for such values of a/b, we have

$$c_1 = c_2 = \tfrac{1}{3}(1 - 0.630 b/a) \quad \text{(for } a/b \geq 5 \text{ only)} \tag{3.42}$$

The distribution of shearing stresses in a noncircular member may be visualized by using the *membrane analogy*. A homogeneous elastic membrane attached to a fixed frame and subjected to a uniform pressure on one of its sides constitutes an *analog* of the bar in torsion, (i.e., the determination of the deformation of the membrane depends upon the solution of the same partial differential equation as the determination of the shearing stresses in the bar.)[‡] More specifically, if Q is a point of the cross section of the bar and Q' the corresponding point of the membrane (Fig. 3.45), the

Table 3.1. Coefficients for Rectangular Bars in Torsion

a/b	c_1	c_2
1.0	0.208	0.1406
1.2	0.219	0.1661
1.5	0.231	0.1958
2.0	0.246	0.229
2.5	0.258	0.249
3.0	0.267	0.263
4.0	0.282	0.281
5.0	0.291	0.291
10.0	0.312	0.312
∞	0.333	0.333

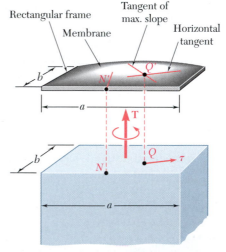

Fig. 3.45 Application of membrane analogy to shaft with rectangular cross section.

[†]See S. P. Timoshenko and J. N. Goodier, *Theory of Elasticity*, 3d ed., McGraw-Hill, New York, 1969, sec. 109.

[‡]Ibid. Sec. 107.

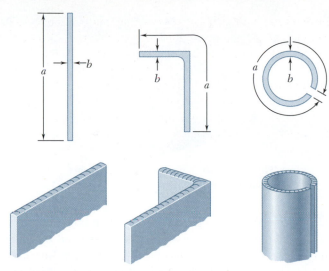

Fig. 3.46   Membrane analogy for various thin-walled members.

shearing stress τ at Q has the same direction as the horizontal tangent to the membrane at Q', and its magnitude is proportional to the maximum slope of the membrane at Q'.[†] Furthermore, the applied torque is proportional to the volume between the membrane and the plane of the fixed frame. For the membrane of Fig. 3.45, which is attached to a rectangular frame, the steepest slope occurs at the midpoint N' of the larger side of the frame. Thus, the maximum shearing stress in a bar of rectangular cross section occurs at the midpoint N of the larger side of that section.

The membrane analogy can be used just as effectively to visualize the shearing stresses in any straight bar of uniform, noncircular cross section. In particular, consider several thin-walled members with the cross sections shown in Fig. 3.46 that are subjected to the same torque. Using the membrane analogy to help us visualize the shearing stresses, we note that since the same torque is applied to each member, the same volume is located under each membrane, and the maximum slope is about the same in each case. Thus, for a thin-walled member of uniform thickness and arbitrary shape, the maximum shearing stress is the same as for a rectangular bar with a very large value of a/b and can be determined from Eq. (3.40) with $c_1 = 0.333$.[‡]

*3.10   THIN-WALLED HOLLOW SHAFTS

In the preceding section we saw that the determination of stresses in noncircular members generally requires the use of advanced mathematical methods. In thin-walled hollow noncircular shafts, a good approximation of the distribution of stresses in the shaft can be obtained by a simple computation. Consider a hollow cylindrical member of *noncircular* section

---

[†]This is the slope measured in a direction perpendicular to the horizontal tangent at Q'.

[‡]It also could be shown that the angle of twist can be determined from Eq. (3.41) with $c_2 = 0.333$.

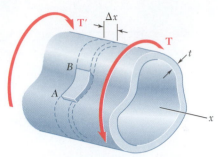

Fig. 3.47 Thin-walled hollow shaft subject to torsional loading.

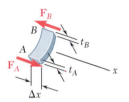

Fig. 3.48 Segment of thin-walled hollow shaft.

Fig. 3.49 Small stress element from segment.

Fig. 3.50 Direction of shearing stress on cross section.

subjected to a torsional loading (Fig. 3.47).[†] While the thickness t of the wall may vary within a transverse section, it is assumed that it remains small compared to the other dimensions of the member. Now detach the colored portion of wall AB bounded by two transverse planes at a distance Δx from each other and by two longitudinal planes perpendicular to the wall. Since the portion AB is in equilibrium, the sum of the forces exerted on it in the longitudinal x direction must be zero (Fig. 3.48). The only forces involved in this direction are the shearing forces $\mathbf{F}_A$ and $\mathbf{F}_B$ exerted on the ends of portion AB. Therefore,

$$\Sigma F_x = 0: \qquad\qquad F_A - F_B = 0 \qquad\qquad \textbf{(3.43)}$$

Now express F_A as the product of the longitudinal shearing stress τ_A on the small face at A and of the area $t_A \, \Delta x$ of that face:

$$F_A = \tau_A(t_A \, \Delta x)$$

While the shearing stress is independent of the x coordinate of the point considered, it may vary across the wall. Thus, τ_A represents the average value of the stress computed across the wall. Expressing F_B in a similar way and substituting for F_A and F_B into (3.43), write

$$\tau_A(t_A \, \Delta x) - \tau_B(t_B \, \Delta x) = 0$$

or $$\qquad\qquad\qquad \tau_A t_A = \tau_B t_B \qquad\qquad \textbf{(3.44)}$$

Since A and B were chosen arbitrarily, Eq. (3.44) shows that the product τt of the longitudinal shearing stress τ and the wall thickness t is constant throughout the member. Denoting this product by q, we have

$$q = \tau t = \text{constant} \qquad\qquad \textbf{(3.45)}$$

Now detach a small element from the wall portion AB (Fig. 3.49). Since the outer and inner faces are part of the free surface of the hollow member, the stresses are equal to zero. Recalling Eqs. (1.21) and (1.22) of Sec. 1.4, the stress components indicated on the other faces by dashed arrows are also zero, while those represented by solid arrows are equal. Thus, the shearing stress at any point of a transverse section of the hollow member is parallel to the wall surface (Fig. 3.50), and its average value computed across the wall satisfies Eq. (3.45).

[†]The wall of the member must enclose a single cavity and must not be slit open. In other words, the member should be topologically equivalent to a hollow circular shaft.

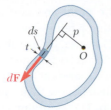

Fig. 3.51   Shear force in the wall.

At this point, an analogy can be made between the distribution of the shearing stresses τ in the transverse section of a thin-walled hollow shaft and the distributions of the velocities v in water flowing through a closed channel of unit depth and variable width. While the velocity v of the water varies from point to point on account of the variation in the width t of the channel, the rate of flow, $q = vt$, remains constant throughout the channel, just as τt in Eq. (3.45). Because of this, the product $q = \tau t$ is called the *shear flow* in the wall of the hollow shaft.

We will now derive a relation between the torque T applied to a hollow member and the shear flow q in its wall. Consider a small element of the wall section, of length ds (Fig. 3.51). The area of the element is $dA = t\,ds$, and the magnitude of the shearing force $d\mathbf{F}$ exerted on the element is

$$dF = \tau\,dA = \tau(t\,ds) = (\tau t)\,ds = q\,ds \qquad (3.46)$$

The moment dM_O of this force about an arbitrary point O within the cavity of the member can be obtained by multiplying dF by the perpendicular distance p from O to the line of action of $d\mathbf{F}$.

$$dM_O = p\,dF = p(q\,ds) = q(p\,ds) \qquad (3.47)$$

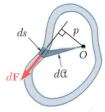

Fig. 3.52   Infinitesimal area used in finding the resultant torque.

But the product $p\,ds$ is equal to twice the area $d\mathcal{Q}$ of the colored triangle in Fig. 3.52. Thus,

$$dM_O = q(2d\mathcal{Q}) \qquad (3.48)$$

Since the integral around the wall section of the left-hand member of Eq. (3.48) represents the sum of the moments of all the elementary shearing forces exerted on the wall section and this sum is equal to the torque T applied to the hollow member,

$$T = \oint dM_O = \oint q(2d\mathcal{Q})$$

The shear flow q being a constant, write

$$T = 2q\mathcal{Q} \qquad (3.49)$$

where $\mathcal{Q}$ is the area bounded by the center line of the wall cross section (Fig. 3.53).

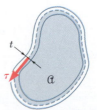

Fig. 3.53   Area for shear flow.

The shearing stress τ at any given point of the wall can be expressed in terms of the torque T if q is substituted from Eq. (3.45) into Eq. (3.49). Solving for τ:

$$\tau = \frac{T}{2t\mathcal{Q}} \qquad (3.50)$$

where t is the wall thickness at the point considered and α the area bounded by the center line. Recall that τ represents the average value of the shearing stress across the wall. However, for elastic deformations, the distribution of stresses across the wall can be assumed to be uniform, and thus Eq. (3.50) yields the actual shearing stress at a given point of the wall.

The angle of twist of a thin-walled hollow shaft can be obtained also by using the method of energy (Chap. 11). Assuming an elastic deformation, it is shown[†] that the angle of twist of a thin-walled shaft of length L and modulus of rigidity G is

$$\phi = \frac{TL}{4\alpha^2 G} \oint \frac{ds}{t} \tag{3.51}$$

where the integral is computed along the center line of the wall section.

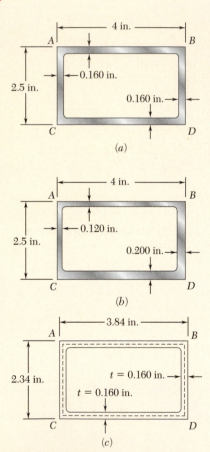

(a)

(b)

(c)

Fig. 3.54 Thin-walled aluminum tube: (a) with uniform thickness, (b) with non-uniform thickness, (c) area bounded by center line of wall thickness.

Concept Application 3.10

Structural aluminum tubing of 2.5×4-in. rectangular cross section was fabricated by extrusion. Determine the shearing stress in each of the four walls of a portion of such tubing when it is subjected to a torque of 24 kip·in., assuming (a) a uniform 0.160-in. wall thickness (Fig. 3.54a) and (b) that as a result of defective fabrication, walls AB and AC are 0.120-in. thick and walls BD and CD are 0.200-in. thick (Fig. 3.54b).

a. Tubing of Uniform Wall Thickness. The area bounded by the center line (Fig. 3.54c) is

$$\alpha = (3.84 \text{ in.})(2.34 \text{ in.}) = 8.986 \text{ in}^2$$

Since the thickness of each of the four walls is $t = 0.160$ in., from Eq. (3.50), the shearing stress in each wall is

$$\tau = \frac{T}{2t\alpha} = \frac{24 \text{ kip·in.}}{2(0.160 \text{ in.})(8.986 \text{ in}^2)} = 8.35 \text{ ksi}$$

b. Tubing with Variable Wall Thickness. Observing that the area α bounded by the center line is the same as in part a, and substituting successively $t = 0.120$ in. and $t = 0.200$ in. into Eq. (3.50), we have

$$\tau_{AB} = \tau_{AC} = \frac{24 \text{ kip·in.}}{2(0.120 \text{ in.})(8.986 \text{ in}^2)} = 11.13 \text{ ksi}$$

and

$$\tau_{BD} = \tau_{CD} = \frac{24 \text{ kip·in.}}{2(0.200 \text{ in.})(8.986 \text{ in}^2)} = 6.68 \text{ ksi}$$

Note that the stress in a given wall depends only upon its thickness.

[†]See Prob. 11.70.

Sample Problem 3.9

Using $\tau_{\text{all}} = 40$ MPa, determine the largest torque that may be applied to each of the brass bars and to the brass tube shown in the figure below. Note that the two solid bars have the same cross-sectional area, and that the square bar and square tube have the same outside dimensions.

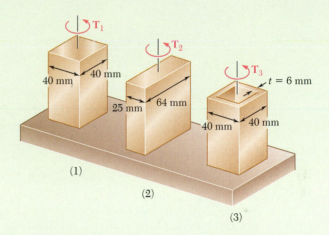

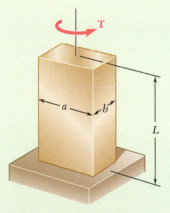

Fig. 1 General dimensions of solid rectangular bar in torsion.

STRATEGY: We obtain the torque using Eq. (3.40) for the solid cross sections and Eq. (3.50) for the hollow cross section.

MODELING and ANALYSIS:

1. Bar with Square Cross Section. For a solid bar of rectangular cross section (Fig. 1), the maximum shearing stress is given by Eq. (3.40)

$$\tau_{\text{max}} = \frac{T}{c_1 a b^2}$$

where the coefficient c_1 is obtained from Table 3.1.

$$a = b = 0.040 \text{ m} \qquad \frac{a}{b} = 1.00 \qquad c_1 = 0.208$$

For $\tau_{\text{max}} = \tau_{\text{all}} = 40$ MPa, we have

$$\tau_{\text{max}} = \frac{T_1}{c_1 a b^2} \qquad 40 \text{ MPa} = \frac{T_1}{0.208(0.040 \text{ m})^3} \qquad \textcolor{blue}{T_1 = 532 \text{ N·m}} \ \blacktriangleleft$$

2. Bar with Rectangular Cross Section. We now have

$$a = 0.064 \text{ m} \qquad b = 0.025 \text{ m} \qquad \frac{a}{b} = 2.56$$

(continued)

Interpolating in Table 3.1: $c_1 = 0.259$

$$\tau_{max} = \frac{T_2}{c_1 ab^2} \qquad 40\,\text{MPa} = \frac{T_2}{0.259(0.064\,\text{m})(0.025\,\text{m})^2} \qquad T_2 = 414\,\text{N·m} \quad \blacktriangleleft$$

3. Square Tube. For a tube of thickness t (Fig. 2), the shearing stress is given by Eq. (3.50)

$$\tau = \frac{T}{2t\alpha}$$

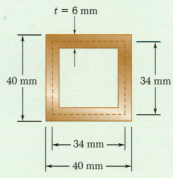

$t = 6$ mm

40 mm   34 mm

├── 34 mm ──┤

├── 40 mm ──┤

Fig. 2 Hollow, square brass bar section dimensions.

where α is the area bounded by the center line of the cross section. We have

$$\alpha = (0.034\,\text{m})(0.034\,\text{m}) = 1.156 \times 10^{-3}\,\text{m}^2$$

We substitute $\tau = \tau_{all} = 40$ MPa and $t = 0.006$ m and solve for the allowable torque:

$$\tau = \frac{T}{2t\alpha} \qquad 40\,\text{MPa} = \frac{T_3}{2(0.006\,\text{m})(1.156 \times 10^{-3}\,\text{m}^2)} \qquad T_3 = 555\,\text{N·m} \quad \blacktriangleleft$$

REFLECT and THINK: Comparing the capacity of the bar of solid square cross section with that of the tube with the same outer dimensions demonstrates the ability of the tube to carry a larger torque.

Problems

3.121 Determine the smallest allowable square cross section of a steel shaft of length 20 ft if the maximum shearing stress is not to exceed 10 ksi when the shaft is twisted through one complete revolution. Use $G = 11.2 \times 10^6$ psi.

3.122 Determine the smallest allowable length of a stainless steel shaft of $\frac{3}{8} \times \frac{3}{4}$-in. cross section if the shearing stress is not to exceed 15 ksi when the shaft is twisted through 15°. Use $G = 11.2 \times 10^6$ psi.

3.123 Using $\tau_{all} = 70$ MPa and $G = 27$ GPa, determine for each of the aluminum bars shown the largest torque **T** that can be applied and the corresponding angle of twist at end B.

3.124 Knowing that the magnitude of the torque **T** is 200 N·m and that $G = 27$ GPa, determine for each of the aluminum bars shown the maximum shearing stress and the angle of twist at end B.

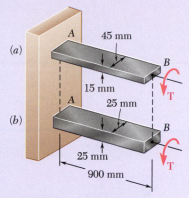

Fig. P3.123 and P3.124

3.125 Determine the largest torque **T** that can be applied to each of the two brass bars shown and the corresponding angle of twist at B, knowing that $\tau_{all} = 12$ ksi and $G = 5.6 \times 10^6$ psi.

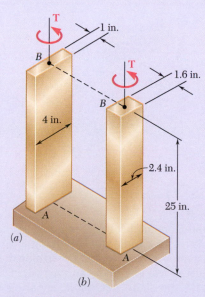

Fig. P3.125 and P3.126

3.126 Each of the two brass bars shown is subjected to a torque of magnitude $T = 12.5$ kip·in. Knowing that $G = 5.6 \times 10^6$ psi, determine for each bar the maximum shearing stress and the angle of twist at B.

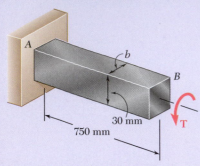

Fig. P3.127 and P3.128

3.127 The torque **T** causes a rotation of 0.6° at end B of the aluminum bar shown. Knowing that $b = 15$ mm and $G = 26$ GPa, determine the maximum shearing stress in the bar.

3.128 The torque **T** causes a rotation of 2° at end B of the stainless steel bar shown. Knowing that $b = 20$ mm and $G = 75$ GPa, determine the maximum shearing stress in the bar.

3.129 Two shafts are made of the same material. The cross section of shaft A is a square of side b and that of shaft B is a circle of diameter b. Knowing that the shafts are subjected to the same torque, determine the ratio τ_A/τ_B of maximum shearing stresses occurring in the shafts.

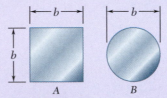

Fig. P3.129

3.130 Shafts A and B are made of the same material and have the same cross-sectional area, but A has a circular cross section and B has a square cross section. Determine the ratio of the maximum torques T_A and T_B when the two shafts are subjected to the same maximum shearing stress ($\tau_A = \tau_B$). Assume both deformations to be elastic.

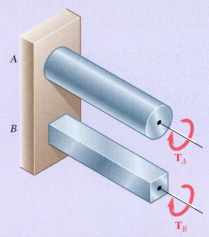

Fig. *P3.130*, P3.131 and P3.132

3.131 Shafts A and B are made of the same material and have the same length and cross-sectional area, but A has a circular cross section and B has a square cross section. Determine the ratio of the maximum values of the angles ϕ_A and ϕ_B when the two shafts are subjected to the same maximum shearing stress ($\tau_A = \tau_B$). Assume both deformations to be elastic.

3.132 Shafts A and B are made of the same material and have the same cross-sectional area, but A has a circular cross section and B has a square cross section. Determine the ratio of the angles ϕ_A and ϕ_B through which shafts A and B are respectively twisted when the two shafts are subjected to the same torque ($T_A = T_B$). Assume both deformations to be elastic.

3.133 A torque of magnitude $T = 2$ kip·in. is applied to each of the steel bars shown. Knowing that $\tau_{all} = 6$ ksi, determine the required dimension b for each bar.

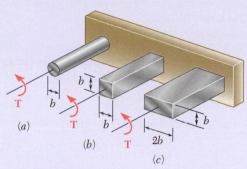

Fig. P3.133 and *P3.134*

3.134 A torque of magnitude $T = 300$ N·m is applied to each of the aluminum bars shown. Knowing that $\tau_{all} = 60$ MPa, determine the required dimension b for each bar.

3.135 A 1.25-m-long steel angle has an L127 × 76 × 6.4 cross section. From Appendix C we find that the thickness of the section is 6.4 mm and that its area is 1250 mm². Knowing that $\tau_{all} = 60$ MPa and that $G = 77.2$ GPa, and ignoring the effect of stress concentrations, determine (*a*) the largest torque **T** that can be applied, (*b*) the corresponding angle of twist.

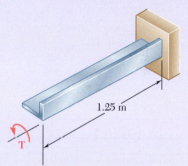

Fig. P3.135

3.136 A 36-kip·in. torque is applied to a 10-ft-long steel angle with an L8 × 8 × 1 cross section. From Appendix C we find that the thickness of the section is 1 in. and that its area is 15 in². Knowing that $G = 11.2 \times 10^6$ psi, determine (*a*) the maximum shearing stress along line *a-a*, (*b*) the angle of twist.

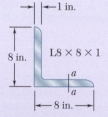

Fig. P3.136

3.137 A 4-m-long steel member has a W310 × 60 cross section. Knowing that $G = 77.2$ GPa and that the allowable shearing stress is 40 MPa, determine (*a*) the largest torque **T** that can be applied, (*b*) the corresponding angle of twist. Refer to Appendix C for the dimensions of the cross section and neglect the effect of stress concentrations. (*Hint:* consider the web and flanges separately and obtain a relation between the torques exerted on the web and a flange, respectively, by expressing that the resulting angles of twist are equal.)

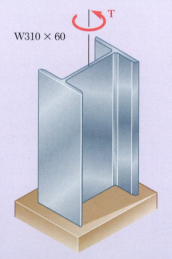

Fig. P3.137

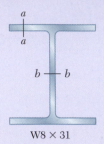

W8 × 31
Fig. P3.138

3.138 An 8-ft-long steel member with a W8 × 31 cross section is subjected to a 5-kip·in. torque. The properties of the rolled-steel section are given in Appendix C. Knowing that $G = 11.2 \times 10^6$ psi, determine (*a*) the maximum shearing stress along line *a-a*, (*b*) the maximum shearing stress along line *b-b*, (*c*) the angle of twist. (See hint of Prob. 3.137.)

3.139 A 5-kip·ft torque is applied to a hollow aluminum shaft having the cross section shown. Neglecting the effect of stress concentrations, determine the shearing stress at points *a* and *b*.

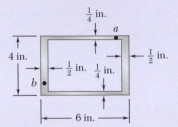

Fig. P3.139

3.140 A torque $T = 750$ kN·m is applied to the hollow shaft shown that has a uniform 8-mm wall thickness. Neglecting the effect of stress concentrations, determine the shearing stress at points *a* and *b*.

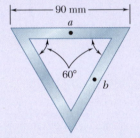

Fig. *P3.140*

3.141 A 750-N·m torque is applied to a hollow shaft having the cross section shown and a uniform 6-mm wall thickness. Neglecting the effect of stress concentrations, determine the shearing stress at points *a* and *b*.

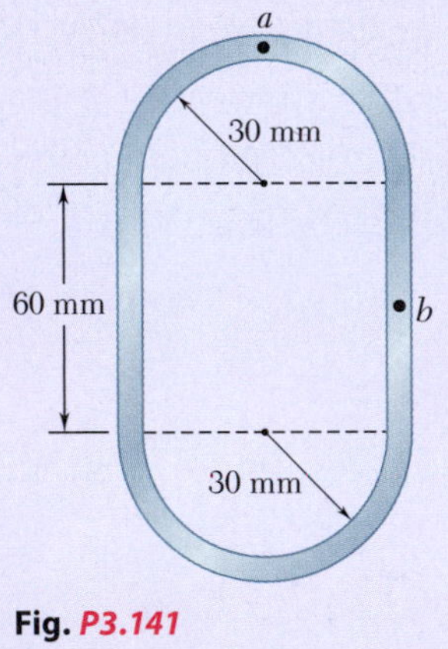

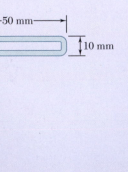

Fig. *P3.141*

3.142 and 3.143 A hollow member having the cross section shown is formed from sheet metal of 2-mm thickness. Knowing that the shearing stress must not exceed 3 MPa, determine the largest torque that can be applied to the member.

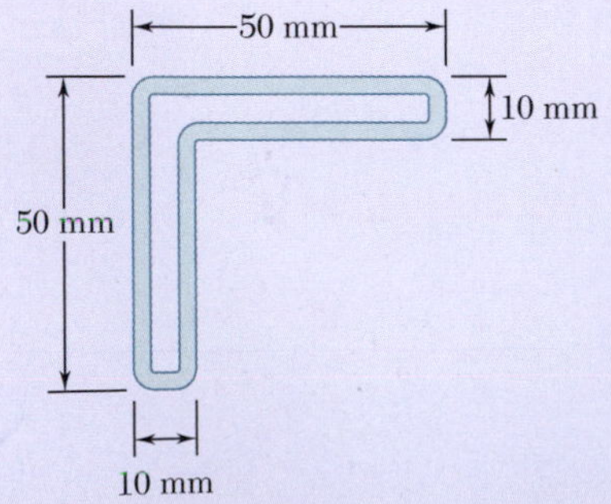

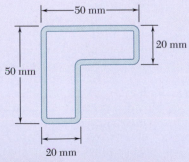

Fig. P3.142

10 mm
Fig. P3.143

3.144 A 90-N·m torque is applied to a hollow shaft having the cross section shown. Neglecting the effect of stress concentrations, determine the shearing stress at points a and b.

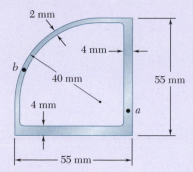

Fig. P3.144

3.145 and 3.146 A hollow member having the cross section shown is to be formed from sheet metal of 0.06-in. thickness. Knowing that a 1250-lb·in. torque will be applied to the member, determine the smallest dimension d that can be used if the shearing stress is not to exceed 750 psi.

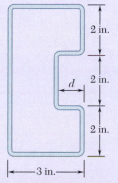

Fig. P3.145

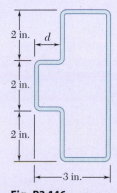

Fig. P3.146

3.147 A cooling tube having the cross section shown is formed from a sheet of stainless steel of 3-mm thickness. The radii $c_1 = 150$ mm and $c_2 = 100$ mm are measured to the center line of the sheet metal. Knowing that a torque of magnitude $T = 3$ kN·m is applied to the tube, determine (a) the maximum shearing stress in the tube, (b) the magnitude of the torque carried by the outer circular shell. Neglect the dimension of the small opening where the outer and inner shells are connected.

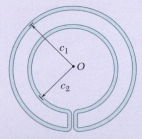

Fig. P3.147

3.148 A hollow cylindrical shaft was designed to have a uniform wall thickness of 0.1 in. Defective fabrication, however, resulted in the shaft having the cross section shown. Knowing that a 15-kip·in. torque is applied to the shaft, determine the shearing stresses at points a and b.

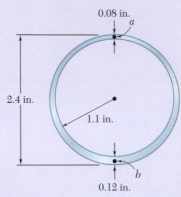

0.08 in.

a

2.4 in.

1.1 in.

b

0.12 in.

Fig. *P3.148*

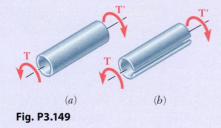

T′   T′

T   T

(a)   (b)

Fig. P3.149

3.149 Equal torques are applied to thin-walled tubes of the same length L, same thickness t, and same radius c. One of the tubes has been slit lengthwise as shown. Determine (a) the ratio τ_b / τ_a of the maximum shearing stresses in the tubes, (b) the ratio ϕ_b / ϕ_a of the angles of twist of the tubes.

3.150 A hollow cylindrical shaft of length L, mean radius c_m, and uniform thickness t is subjected to a torque of magnitude T. Consider, on the one hand, the values of the average shearing stress τ_{ave} and the angle of twist ϕ obtained from the elastic torsion formulas developed in Secs. 3.1C and 3.2 and, on the other hand, the corresponding values obtained from the formulas developed in Sec. 3.10 for thin-walled shafts. (a) Show that the relative error introduced by using the thin-walled-shaft formulas rather than the elastic torsion formulas is the same for τ_{ave} and ϕ and that the relative error is positive and proportional to the ratio t / c_m. (b) Compare the percent error corresponding to values of the ratio t / c_m of 0.1, 0.2, and 0.4.

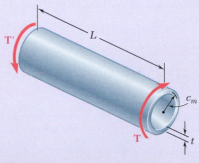

T′ L

c_m

T t

Fig. P3.150

Review and Summary

This chapter was devoted to the analysis and design of *shafts* subjected to twisting couples, or *torques*. Except for the last two sections of the chapter, our discussion was limited to *circular shafts*.

Deformations in Circular Shafts

The distribution of stresses in the cross section of a circular shaft is *statically indeterminate*. The determination of these stresses requires a prior analysis of the *deformations* occurring in the shaft [Sec. 3.1B]. In a circular shaft subjected to torsion, *every cross section remains plane and undistorted*. The *shearing strain* in a small element with sides parallel and perpendicular to the axis of the shaft and at a distance ρ from that axis is

$$\gamma = \frac{\rho\phi}{L} \tag{3.2}$$

where ϕ is the angle of twist for a length L of the shaft (Fig. 3.55). Equation (3.2) shows that the *shearing strain in a circular shaft varies linearly with the distance from the axis of the shaft*. It follows that the strain is maximum at the surface of the shaft, where ρ is equal to the radius c of the shaft:

$$\gamma_{max} = \frac{c\phi}{L} \qquad \gamma = \frac{\rho}{c}\gamma_{max} \tag{3.3, 4}$$

Shearing Stresses in Elastic Range

The relationship between *shearing stresses* in a circular shaft within the elastic range [Sec. 3.1C] and Hooke's law for shearing stress and strain, $\tau = G\gamma$, is

$$\tau = \frac{\rho}{c}\tau_{max} \tag{3.6}$$

which shows that within the elastic range, the *shearing stress τ in a circular shaft also varies linearly with the distance from the axis of the shaft*. Equating the sum of the moments of the elementary forces exerted on any section of the shaft to the magnitude T of the torque applied to the shaft, the *elastic torsion formulas* are

$$\tau_{max} = \frac{Tc}{J} \qquad \tau = \frac{T\rho}{J} \tag{3.9, 10}$$

where c is the radius of the cross section and J its centroidal polar moment of inertia. $J = \frac{1}{2}\pi c^4$ for a solid shaft, and $J = \frac{1}{2}\pi(c_2^4 - c_1^4)$ for a hollow shaft of inner radius c_1 and outer radius c_2.

We noted that while the element a in Fig. 3.56 is in pure shear, the element c in the same figure is subjected to normal stresses of the same magnitude,

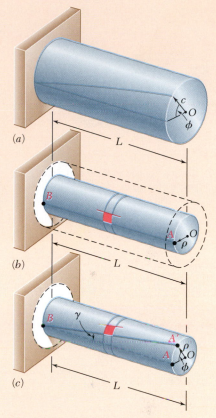

Fig. 3.55 Torsional deformations. (*a*) The angle of twist ϕ. (*b*) Undeformed portion of shaft of radius ρ. (*c*) Deformed portion of shaft; angle of twist ϕ and shearing strain γ share same arc length AA'.

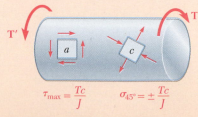

$$\tau_{max} = \frac{Tc}{J} \qquad \sigma_{45°} = \pm\frac{Tc}{J}$$

Fig. 3.56 Shaft elements with only shearing stresses or normal stresses.

Tc/J, with two of the normal stresses being tensile and two compressive. This explains why in a torsion test ductile materials, which generally fail in shear, will break along a plane perpendicular to the axis of the specimen, while brittle materials, which are weaker in tension than in shear, will break along surfaces forming a 45° angle with that axis.

Angle of Twist

Within the elastic range, the angle of twist ϕ of a circular shaft is proportional to the torque T applied to it (Fig. 3.57).

$$\phi = \frac{TL}{JG} \text{ (units of radians)} \tag{3.15}$$

where L = length of shaft
  J = polar moment of inertia of cross section
  G = modulus of rigidity of material
  ϕ is in *radians*

If the shaft is subjected to torques at locations other than its ends or consists of several parts of various cross sections and possibly of different materials, the angle of twist of the shaft must be expressed as the *algebraic sum* of the angles of twist of its component parts:

$$\phi = \sum_i \frac{T_i L_i}{J_i G_i} \tag{3.16}$$

When both ends of a shaft BE rotate (Fig. 3.58), the angle of twist is equal to the *difference* between the angles of rotation ϕ_B and ϕ_E of its ends. When two shafts AD and BE are connected by gears A and B, the torques applied by gear A on shaft AD and gear B on shaft BE are *directly proportional* to the radii r_A and r_B of the two gears—since the forces applied on each other by the gear teeth at C are equal and opposite. On the other hand, the angles ϕ_A and ϕ_B are *inversely proportional* to r_A and r_B—since the arcs CC' and CC'' described by the gear teeth are equal.

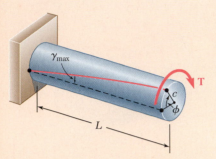

Fig. 3.57 Torque applied to fixed end shaft resulting in angle of twist ϕ.

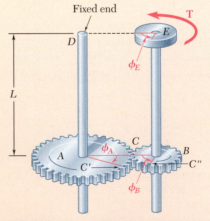

Fig. 3.58 Angles of twist at E, gear B, and gear A for a meshed-gear system.

Statically Indeterminate Shafts

If the reactions at the supports of a shaft or the internal torques cannot be determined from statics alone, the shaft is said to be *statically indeterminate*. The equilibrium equations obtained from free-body diagrams must be complemented by relationships involving deformations of the shaft and obtained from the geometry of the problem.

Transmission Shafts

For the *design of transmission shafts*, the power P transmitted is

$$P = 2\pi fT \tag{3.19}$$

where T is the torque exerted at each end of the shaft and f the *frequency* or speed of rotation of the shaft. The unit of frequency is the revolution per second (s^{-1}) or hertz (Hz). If SI units are used, T is expressed in newton-meters (N·m) and P in *watts* (W). If U.S. customary units are used, T is expressed in lb·ft or lb·in., and P in ft·lb/s or in·lb/s; the power can be converted into *horsepower* (hp) through

$$1 \text{ hp} = 550 \text{ ft·lb/s} = 6600 \text{ in·lb/s}$$

To design a shaft to transmit a given power P at a frequency f, solve Eq. (3.19) for T. This value and the maximum allowable value of τ for the material can be used with Eq. (3.9) to determine the required shaft diameter.

Stress Concentrations

Stress concentrations in circular shafts result from an abrupt change in the diameter of a shaft and can be reduced through the use of a *fillet* (Fig. 3.59). The maximum value of the shearing stress at the fillet is

$$\tau_{\max} = K\frac{Tc}{J} \tag{3.22}$$

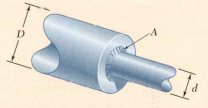

Fig. 3.59 Shafts having two different diameters with a fillet at the junction.

where the stress Tc/J is computed for the smaller-diameter shaft and K is a stress concentration factor.

Plastic Deformations

Even when Hooke's law does not apply, the distribution of *strains* in a circular shaft is always linear. If the shearing-stress-strain diagram for the material is known, it is possible to plot the shearing stress τ against the distance ρ from the axis of the shaft for any given value of $\tau_{\max}$ (Fig. 3.60). Summing the torque of annular elements of radius ρ and thickness $d\rho$, the torque T is

$$T = \int_0^c \rho\tau(2\pi\rho \, d\rho) = 2\pi\int_0^c \rho^2\tau \, d\rho \tag{3.23}$$

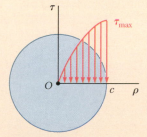

Fig. 3.60 Shearing stress distribution for shaft with nonlinear stress-strain response.

where τ is the function of ρ plotted in Fig. 3.60.

Modulus of Rupture

An important value of the torque is the ultimate torque T_U, which causes failure of the shaft. This can be determined either experimentally, or by Eq. (3.22) with $\tau_{\max}$ chosen equal to the ultimate shearing stress τ_U of the

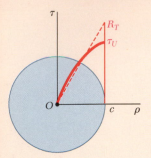

Fig. 3.61 Stress distribution in circular shaft at failure.

material. From T_U, and assuming a linear stress distribution (Fig 3.61), we determined the corresponding fictitious stress $R_T = T_U c/J$, known as the *modulus of rupture in torsion*.

Solid Shaft of Elastoplastic Material

In a *solid circular shaft* made of an *elastoplastic material*, as long as τ_{max} does not exceed the yield strength τ_Y of the material, the stress distribution across a section of the shaft is linear (Fig. 3.62a). The torque T_Y corresponding to $\tau_{max} = \tau_Y$ (Fig. 3.62b) is the *maximum elastic torque*. For a solid circular shaft of radius c,

$$T_Y = \tfrac{1}{2}\pi c^3 \tau_Y \tag{3.26}$$

As the torque increases, a plastic region develops in the shaft around an elastic core of radius ρ_Y. The torque T corresponding to a given value of ρ_Y is

$$T = \frac{4}{3}T_Y\left(1 - \frac{1}{4}\frac{\rho_Y^3}{c^3}\right) \tag{3.29}$$

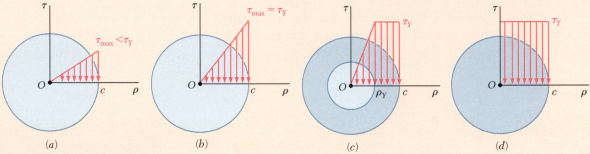

Fig. 3.62 Stress distributions for elastoplastic shaft at different stages of loading: (*a*) elastic, (*b*) impending yield, (*c*) partially yielded, and (*d*) fully yielded.

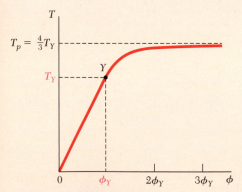

Fig. 3.63 Load-displacement relation for elastoplastic material.

As ρ_Y approaches zero, the torque approaches a limiting value T_p, called the *plastic torque*:

$$T_p = \frac{4}{3}T_Y \tag{3.30}$$

Plotting the torque T against the angle of twist ϕ of a solid circular shaft (Fig. 3.63), the segment of straight line $0Y$ defined by Eq. (3.15) and followed by a curve approaching the straight line $T = T_p$ is

$$T = \frac{4}{3}T_Y\left(1 - \frac{1}{4}\frac{\phi_Y^3}{\phi^3}\right) \tag{3.34}$$

Permanent Deformation and Residual Stresses

Loading a circular shaft beyond the onset of yield and unloading it results in a *permanent deformation* characterized by the angle of twist $\phi_p = \phi - \phi'$, where ϕ corresponds to the loading phase described in the previous paragraph, and ϕ' to the unloading phase represented by a straight line in

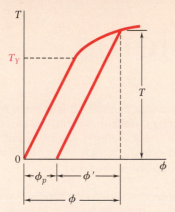

Fig. 3.64 Torque-angle of twist response for loading past yield and, followed by unloading.

Fig. 3.64. *Residual stresses* in the shaft can be determined by adding the maximum stresses reached during the loading phase and the reverse stresses corresponding to the unloading phase.

Torsion of Noncircular Members

The equations for the distribution of strain and stress in circular shafts are based on the fact that due to the axisymmetry of these members, cross sections remain plane and undistorted. This property does not hold for noncircular members, such as the square bar of Fig. 3.65.

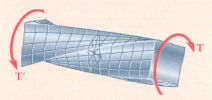

Fig. 3.65 Twisting a shaft of square cross section.

Bars of Rectangular Cross Section

For straight bars with a *uniform rectangular cross section* (Fig. 3.66), the maximum shearing stress occurs along the center line of the *wider* face of the bar. The *membrane analogy* can be used to visualize the distribution of stresses in a noncircular member.

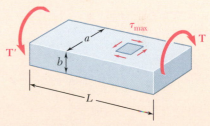

Fig. 3.66 Shaft with rectangular cross section, showing the location of maximum shearing stress.

Thin-Walled Hollow Shafts

The shearing stress in *noncircular thin-walled hollow shafts* is parallel to the wall surface and varies both across and along the wall cross section. Denoting the average value of the shearing stress τ, computed across the wall at a given point of the cross section, and by t the thickness of the wall at that point (Fig. 3.67), we demonstrated that the product $q = \tau t$, called the *shear flow*, is constant along the cross section.

The average shearing stress τ at any given point of the cross section is

$$\tau = \frac{T}{2t\mathfrak{a}} \tag{3.50}$$

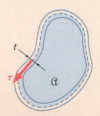

Fig. 3.67 Area for shear flow.

Review Problems

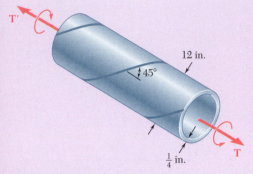

Fig. P3.151

3.151 A steel pipe of 12-in. outer diameter is fabricated from $\frac{1}{4}$-in.-thick plate by welding along a helix that forms an angle of 45° with a plane parallel to the axis of the pipe. Knowing that the maximum allowable tensile stress in the weld is 12 ksi, determine the largest torque that can be applied to the pipe.

3.152 A torque of magnitude $T = 120$ N·m is applied to shaft AB of the gear train shown. Knowing that the allowable shearing stress is 75 MPa in each of the three solid shafts, determine the required diameter of (*a*) shaft AB, (*b*) shaft CD, (*c*) shaft EF.

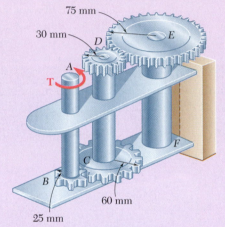

Fig. *P3.152*

3.153 Two solid shafts are connected by gears as shown. Knowing that $G = 77.2$ GPa for each shaft, determine the angle through which end A rotates when $T_A = 1200$ N·m.

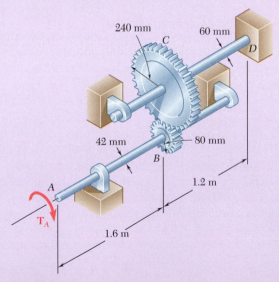

Fig. P3.153

3.154 In the bevel-gear system shown, $\alpha = 18.43°$. Knowing that the allowable shearing stress is 8 ksi in each shaft and that the system is in equilibrium, determine the largest torque $\mathbf{T}_A$ that can be applied at A.

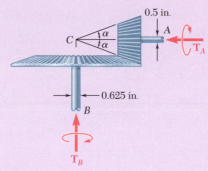

Fig. P3.154

3.155 The design specifications for the gear-and-shaft system shown require that the same diameter be used for both shafts and that the angle through which pulley A will rotate when subjected to a 2-kip·in. torque $\mathbf{T}_A$ while pulley D is held fixed will not exceed 7.5°. Determine the required diameter of the shafts if both shafts are made of a steel with $G = 11.2 \times 10^6$ psi and $\tau_{all} = 12$ ksi.

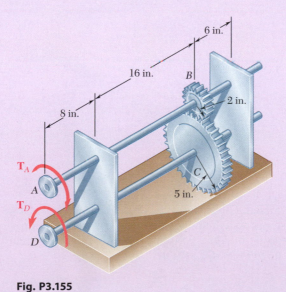

Fig. P3.155

3.156 A torque of magnitude $T = 4$ kN·m is applied at end A of the composite shaft shown. Knowing that the modulus of rigidity is 77.2 GPa for the steel and 27 GPa for the aluminum, determine (*a*) the maximum shearing stress in the steel core, (*b*) the maximum shearing stress in the aluminum jacket, (*c*) the angle of twist at A.

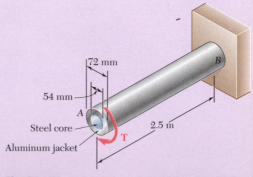

Fig. P3.156

3.157 Ends A and D of the two solid steel shafts AB and CD are fixed, while ends B and C are connected to gears as shown. Knowing that the allowable shearing stress is 50 MPa in each shaft, determine the largest torque $\mathbf{T}$ that can be applied to gear B.

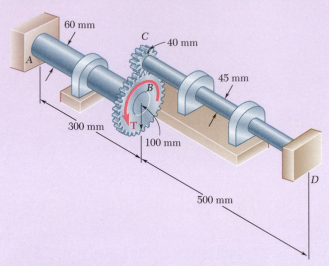

Fig. P3.157

3.158 As the hollow steel shaft shown rotates at 180 rpm, a stroboscopic measurement indicates that the angle of twist of the shaft is 3°. Knowing that $G = 77.2$ GPa, determine (a) the power being transmitted, (b) the maximum shearing stress in the shaft.

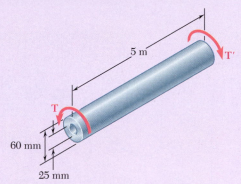

Fig. P3.158

3.159 Knowing that the allowable shearing stress is 8 ksi for the stepped shaft shown, determine the magnitude T of the largest torque that can be transmitted by the shaft when the radius of the fillet is (a) $r = \frac{3}{16}$ in., (b) $r = \frac{1}{4}$ in.

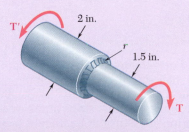

Fig. P3.159

3.160 A hollow brass shaft has the cross section shown. Knowing that the shearing stress must not exceed 12 ksi and neglecting the effect of stress concentrations, determine the largest torque that can be applied to the shaft.

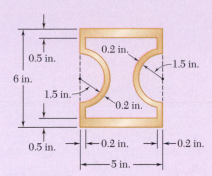

Fig. P3.160

3.161 Two solid brass rods AB and CD are brazed to a brass sleeve EF. Determine the ratio d_2/d_1 for which the same maximum shearing stress occurs in the rods and in the sleeve.

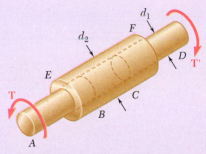

Fig. P3.161

3.162 The shaft AB is made of a material that is elastoplastic with $\tau_Y = 12.5$ ksi and $G = 4 \times 10^6$ psi. For the loading shown, determine (a) the radius of the elastic core of the shaft, (b) the angle of twist of the shaft.

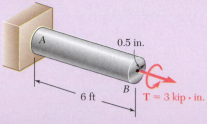

Fig. P3.162

Computer Problems

The following problems are designed to be solved with a computer. Write each program so that it can be used with either SI or U.S. Customary units.

3.C1 Shaft AB consists of n homogeneous cylindrical elements, which can be solid or hollow. Its end A is fixed, while its end B is free, and it is subjected to the loading shown. The length of element i is denoted by L_i, its outer diameter by OD_i, its inner diameter by ID_i, its modulus of rigidity by G_i, and the torque applied to its right end by $\mathbf{T}_i$, the magnitude T_i of this torque being assumed to be positive if $\mathbf{T}_i$ is counterclockwise from end B and negative otherwise. (Note that $ID_i = 0$ if the element is solid.) (*a*) Write a computer program that can be used to determine the maximum shearing stress in each element, the angle of twist of each element, and the angle of twist of the entire shaft. (*b*) Use this program to solve Probs. 3.35, 3.36, and 3.38.

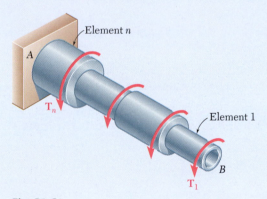

Fig. P3.C1

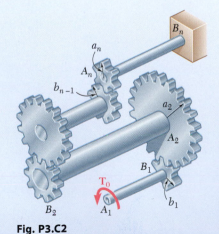

Fig. P3.C2

3.C2 The assembly shown consists of n cylindrical shafts, which can be solid or hollow, connected by gears and supported by brackets (not shown). End A_1 of the first shaft is free and is subjected to a torque $\mathbf{T}_0$, while end B_n of the last shaft is fixed. The length of shaft A_iB_i is L_i, its outer diameter OD_i, its inner diameter ID_i, and its modulus of rigidity G_i. (Note that $ID_i = 0$ if the element is solid.) The radius of gear A_i is a_i, and the radius of gear B_i is b_i. (*a*) Write a computer program that can be used to determine the maximum shearing stress in each shaft, the angle of twist of each shaft, and the angle through which end A_i rotates. (*b*) Use this program to solve Probs. 3.41 and 3.44.

3.C3 Shaft AB consists of n homogeneous cylindrical elements, which can be solid or hollow. Both of its ends are fixed, and it is subjected to the loading shown. The length of element i is denoted by L_i, its outer diameter by OD_i, its inner diameter by ID_i, its modulus of rigidity by G_i, and the torque applied to its right end by $\mathbf{T}_i$, the magnitude T_i of this torque being assumed to be positive if $\mathbf{T}_i$ is observed as counterclockwise from end B and negative otherwise. Note that $ID_i = 0$ if the element is solid and also that $T_1 = 0$. Write a computer program that can be used to determine the reactions at A and B, the maximum shearing stress in each element, and the angle of twist of each element. Use this program (a) to solve Prob. 3.55 and (b) to determine the maximum shearing stress in the shaft of Sample Problem 3.7.

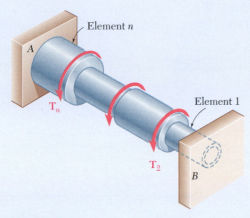

Fig. P3.C3

3.C4 The homogeneous, solid cylindrical shaft AB has a length L, a diameter d, a modulus of rigidity G, and a yield strength τ_Y. It is subjected to a torque $\mathbf{T}$ that is gradually increased from zero until the angle of twist of the shaft has reached a maximum value ϕ_m and then decreased back to zero. (a) Write a computer program that, for each of 16 values of ϕ_m equally spaced over a range extending from 0 to a value 3 times as large as the angle of twist at the onset of yield, can be used to determine the maximum value T_m of the torque, the radius of the elastic core, the maximum shearing stress, the permanent twist, and the residual shearing stress both at the surface of the shaft and at the interface of the elastic core and the plastic region. (b) Use this program to obtain approximate answers to Probs. 3.114, 3.115, 3.116.

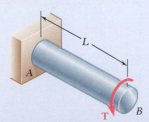

Fig. P3.C4

3.C5 The exact expression is given in Prob. 3.64 for the angle of twist of the solid tapered shaft AB when a torque $\mathbf{T}$ is applied as shown. Derive an approximate expression for the angle of twist by replacing the tapered shaft by n cylindrical shafts of equal length and of radius $r_i = (n + i - \frac{1}{2})(c/n)$, where $i = 1, 2, \ldots, n$. Using for T, L, G, and c values of your choice, determine the percentage error in the approximate expression when (a) $n = 4$, (b) $n = 8$, (c) $n = 20$, and (d) $n = 100$.

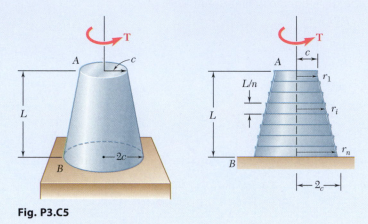

Fig. P3.C5

3.C6 A torque $\mathbf{T}$ is applied as shown to the long, hollow, tapered shaft AB of uniform thickness t. Derive an approximate expression for the angle of twist by replacing the tapered shaft by n cylindrical rings of equal length and of radius $r_i = (n + i - \frac{1}{2})(c/n)$, where $i = 1, 2, \ldots, n$. Using for T, L, G, c, and t values of your choice, determine the percentage error in the approximate expression when (a) $n = 4$, (b) $n = 8$, (c) $n = 20$, and (d) $n = 100$.

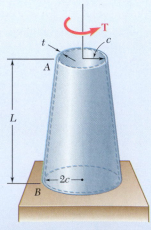

Fig. P3.C6

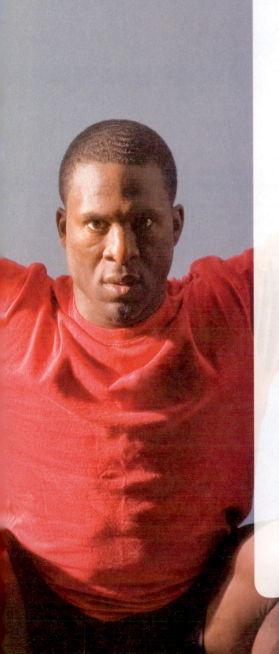

4

Pure Bending

The normal stresses and the curvature resulting from pure bending, such as those developed in the center portion of the barbell shown, will be studied in this chapter.

Objectives

In this chapter, you will:

- **Introduce** students to bending behavior
- **Define** the deformations, strains, and normal stresses in beams subject to pure bending
- **Describe** the behavior of composite beams made of more than one material
- **Review** stress concentrations and how they are included in the design of beams
- **Study** plastic deformations to determine how to evaluate beams made of elastoplastic materials
- **Analyze** members subject to eccentric axial loading, involving both axial stresses and bending stresses
- **Review** beams subject to unsymmetric bending, i.e., where bending does not occur in a plane of symmetry
- **Study** bending of curved members

Introduction

This chapter and the following two analyze the stresses and strains in prismatic members subjected to *bending*. Bending is a major concept used in the design of many machine and structural components, such as beams and girders.

This chapter is devoted to the analysis of prismatic members subjected to equal and opposite couples **M** and **M′** acting in the same longitudinal plane. Such members are said to be in *pure bending*. The members are assumed to possess a plane of symmetry with the couples **M** and **M′** acting in that plane (Fig. 4.1).

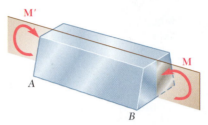

Fig. 4.1   Member in pure bending

An example of pure bending is provided by the bar of a typical barbell as it is held overhead by a weight lifter as shown in the opening photo for this chapter. The bar carries equal weights at equal distances from the hands of the weight lifter. Because of the symmetry of the free-body diagram of the bar (Fig. 4.2a), the reactions at the hands must be equal and opposite to the weights. Therefore, as far as the middle portion *CD* of the bar is concerned, the weights and the reactions can be replaced by two equal and opposite 960-lb·in. couples (Fig. 4.2b), showing that the middle portion of the bar is in pure bending. A similar analysis of a small sport buggy (Photo 4.1) shows that the axle is in pure bending between the two points where it is attached to the frame.

The results obtained from the direct applications of pure bending will be used in the analysis of other types of loadings, such as *eccentric axial loadings* and *transverse loadings*.

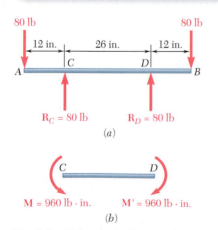

Fig. 4.2   (a) Free-body diagram of the barbell pictured in the chapter opening photo and (b) free-body diagram of the center portion of the bar, which is in pure bending.

Photo 4.1   The center portion of the rear axle of the sport buggy is in pure bending.

Photo 4.2 shows a 12-in. steel bar clamp used to exert 150-lb forces on two pieces of lumber as they are being glued together. Figure 4.3*a* shows the equal and opposite forces exerted by the lumber on the clamp. These forces result in an *eccentric loading* of the straight portion of the clamp. In Fig. 4.3*b*, a section *CC'* has been passed through the clamp and a free-body diagram has been drawn of the upper half of the clamp. The internal forces in the section are equivalent to a 150-lb axial tensile force **P** and a 750-lb·in. couple **M**. By combining our knowledge of the stresses under a *centric* load and the results of an analysis of stresses in pure bending, the distribution of stresses under an *eccentric* load is obtained. This is discussed in Sec. 4.8.

Photo 4.2 Clamp used to glue lumber pieces together.

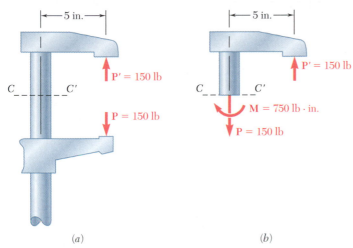

Fig. 4.3 (*a*) Free-body diagram of a clamp, (*b*) free-body diagram of the upper portion of the clamp.

The study of pure bending plays an essential role in the study of beams (i.e., prismatic members) subjected to various types of *transverse loads*. Consider a cantilever beam *AB* supporting a concentrated load **P** at its free end (Fig. 4.4*a*). If a section is passed through *C* at a distance *x* from *A*, the free-body diagram of *AC* (Fig. 4.4*b*) shows that the internal forces in the section consist of a force **P'** equal and opposite to **P** and a couple **M** of magnitude $M = Px$. The distribution of normal stresses in the section can be obtained from the couple **M** as if the beam were in pure bending. The shearing stresses in the section depend on the force **P'**, and their distribution over a given section is discussed in Chap. 6.

The first part of this chapter covers the analysis of stresses and deformations caused by pure bending in a homogeneous member possessing a plane of symmetry and made of a material following Hooke's law. The methods of statics are used in Sec. 4.1A to derive three fundamental equations which must be satisfied by the normal stresses in any given cross section of the member. In Sec. 4.1B, it will be proved that *transverse sections remain plane* in a member subjected to pure bending, while in Sec. 4.2, formulas are developed to determine the *normal stresses* and *radius of curvature* for that member within the elastic range.

Sec. 4.4 covers the stresses and deformations in *composite members* made of more than one material, such as reinforced-concrete beams, which utilize the best features of steel and concrete and are extensively used in the

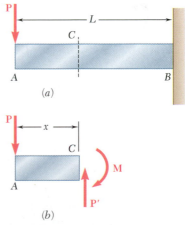

Fig. 4.4 (*a*) Cantilevered beam with end loading. (*b*) As portion *AC* shows, beam is not in pure bending.

construction of buildings and bridges. You will learn to draw a *transformed section* representing a member made of a homogeneous material that undergoes the same deformations as the composite member under the same loading. The transformed section is used to find the stresses and deformations in the original composite member. Section 4.5 is devoted to the determination of *stress concentrations* occurring where the cross section of a member undergoes a sudden change.

Section 4.6 covers *plastic deformations*, where the members are made of a material that does not follow Hooke's law and are subjected to bending. The stresses and deformations in members made of an *elastoplastic material* are discussed in Sec. 4.6A. Starting with the *maximum elastic moment M_Y*, which corresponds to the onset of yield, you will consider the effects of increasingly larger moments until the *plastic moment M_p* is reached. You will also determine the *permanent deformations* and *residual stresses* that result from such loadings (Sec. 4.6C).

In Sec. 4.7, you will analyze an *eccentric axial loading* in a plane of symmetry (Fig. 4.3) by superposing the stresses due to pure bending and a centric axial loading.

The study of the bending of prismatic members concludes with the analysis of *unsymmetric bending* (Sec. 4.8), and the study of the general case of *eccentric axial loading* (Sec. 4.9). The final section of this chapter is devoted to the determination of the stresses in *curved members* (Sec. 4.10).

4.1 SYMMETRIC MEMBERS IN PURE BENDING

4.1A Internal Moment and Stress Relations

Consider a prismatic member AB possessing a plane of symmetry and subjected to equal and opposite couples **M** and **M′** acting in that plane (Fig. 4.5a). If a section is passed through the member AB at some arbitrary point C, the conditions of equilibrium of the portion AC of the member require the internal forces in the section to be equivalent to the couple **M** (Fig. 4.5b). The moment M of that couple is the *bending moment* in the section. Following the usual convention, a positive sign is assigned to M when the member is bent as shown in Fig. 4.5a (i.e., when the concavity of the beam faces upward) and a negative sign otherwise.

Denoting by σ_x the normal stress at a given point of the cross section and by τ_{xy} and τ_{xz} the components of the shearing stress, we express that

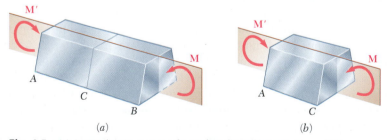

Fig. 4.5 (*a*) A member in a state of pure bending. (*b*) Any intermediate portion of *AB* will also be in pure bending.

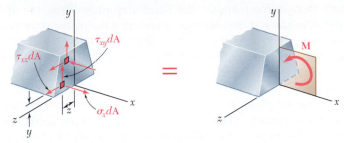

Fig. 4.6 Stresses resulting from pure bending moment M.

the system of the elementary internal forces exerted on the section is equivalent to the couple **M** (Fig. 4.6).

Recall from statics that a couple **M** actually consists of two equal and opposite forces. The sum of the components of these forces in any direction is therefore equal to zero. Moreover, the moment of the couple is the same about *any* axis perpendicular to its plane and is zero about any axis contained in that plane. Selecting arbitrarily the *z* axis shown in Fig. 4.6, the equivalence of the elementary internal forces and the couple **M** is expressed by writing that the sums of the components and moments of the forces are equal to the corresponding components and moments of the couple **M**:

x components:	$\int \sigma_x \, dA = 0$	**(4.1)**
Moments about *y* axis:	$\int z\sigma_x \, dA = 0$	**(4.2)**
Moments about *z* axis:	$\int (-y\sigma_x \, dA) = M$	**(4.3)**

Three additional equations could be obtained by setting equal to zero the sums of the *y* components, *z* components, and moments about the *x* axis, but these equations would involve only the components of the shearing stress and, as you will see in the next section, the components of the shearing stress are both equal to zero.

Two remarks should be made at this point:

1. The minus sign in Eq. (4.3) is due to the fact that a tensile stress ($\sigma_x > 0$) leads to a negative moment (clockwise) of the normal force $\sigma_x \, dA$ about the *z* axis.
2. Equation (4.2) could have been anticipated, since the application of couples in the plane of symmetry of member *AB* result in a distribution of normal stresses symmetric about the *y* axis.

Once more, note that the actual distribution of stresses in a given cross section cannot be determined from statics alone. It is *statically indeterminate* and may be obtained only by analyzing the *deformations* produced in the member.

4.1B Deformations

We will now analyze the deformations of a prismatic member possessing a plane of symmetry. Its ends are subjected to equal and opposite couples **M** and **M**′ acting in the plane of symmetry. The member will bend under the action of the couples, but will remain symmetric with respect to that plane (Fig. 4.7). Moreover, since the bending moment *M* is the same in

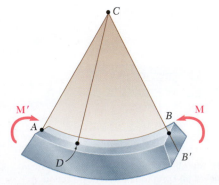

Fig. 4.7 Initially straight members in pure bending deform into a circular arc.

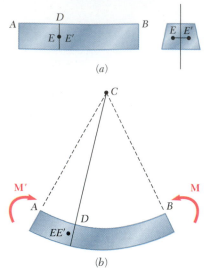

Fig. 4.8 (*a*) Two points in a cross section at *D* that is perpendicular to the member's axis. (*b*) Considering the possibility that these points do not remain in the cross section after bending.

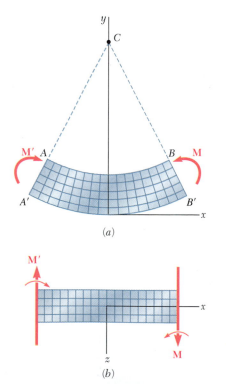

Fig. 4.9 Member subject to pure bending shown in two views. (*a*) Longitudinal, vertical section (plane of symmetry). (*b*) Longitudinal, horizontal section.

any cross section, the member will bend uniformly. Thus, the line *AB* along the upper face of the member intersecting the plane of the couples will have a constant curvature. In other words, the line *AB* will be transformed into a circle of center *C*, as will the line *A'B'* along the lower face of the member. Note that the line *AB* will decrease in length when the member is bent (i.e., when *M* > 0), while *A'B'* will become longer.

Next we will prove that any cross section perpendicular to the axis of the member remains plane, and that the plane of the section passes through *C*. If this were not the case, we could find a point *E* of the original section through *D* (Fig. 4.8*a*) which, after the member has been bent, would *not* lie in the plane perpendicular to the plane of symmetry that contains line *CD* (Fig. 4.8*b*). But, because of the symmetry of the member, there would be another point *E'* that would be transformed exactly in the same way. Let us assume that, after the beam has been bent, both points would be located to the left of the plane defined by *CD,* as shown in Fig. 4.8*b*. Since the bending moment *M* is the same throughout the member, a similar situation would prevail in any other cross section, and the points corresponding to *E* and *E'* would also move to the left. Thus, an observer at *A* would conclude that the loading causes the points *E* and *E'* in the various cross sections to move forward (toward the observer). But an observer at *B*, to whom the loading looks the same, and who observes the points *E* and *E'* in the same positions (except that they are now inverted) would reach the opposite conclusion. This inconsistency leads us to conclude that *E* and *E'* will lie in the plane defined by *CD* and, therefore, that the section remains plane and passes through *C*. We should note, however, that this discussion does not rule out the possibility of deformations within the plane of the section (see Sec. 4.3).

Suppose that the member is divided into a large number of small cubic elements with faces respectively parallel to the three coordinate planes. The property we have established requires that these elements be transformed as shown in Fig. 4.9 when the member is subjected to the couples **M** and **M'**. Since all the faces represented in the two projections of Fig. 4.9 are at 90° to each other, we conclude that $\gamma_{xy} = \gamma_{zx} = 0$ and, thus, that $\tau_{xy} = \tau_{xz} = 0$. Regarding the three stress components that we have not yet discussed, namely, σ_y, σ_z, and τ_{yz}, we note that they must be zero on the surface of the member. Since, on the other hand, the deformations involved do not require any interaction between the elements of a given transverse cross section, we can assume that these three stress components are equal to zero throughout the member. This assumption is verified, both from experimental evidence and from the theory of elasticity, for slender members undergoing small deformations.[†] We conclude that the only nonzero stress component exerted on any of the small cubic elements considered here is the normal component σ_x. Thus, at any point of a slender member in pure bending, we have a state of *uniaxial stress*. Recalling that, for *M* > 0, lines *AB* and *A'B'* are observed, respectively, to decrease and increase in length, we note that the strain ϵ_x and the stress σ_x are negative in the upper portion of the member (*compression*) and positive in the lower portion (*tension*).

It follows from above that a surface parallel to the upper and lower faces of the member must exist where ϵ_x and σ_x are zero. This surface is

[†]Also see Prob. 4.32.

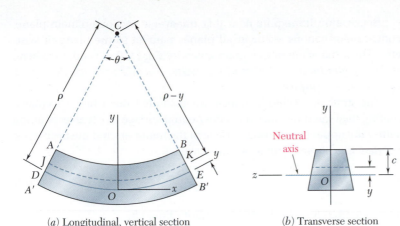

(a) Longitudinal, vertical section
(plane of symmetry)

(b) Transverse section

Fig. 4.10 Establishment of neutral axis. (a) Longitudinal-vertical view. (b) Transverse section at origin.

called the *neutral surface.* The neutral surface intersects the plane of symmetry along an arc of circle *DE* (Fig. 4.10a), and it intersects a transverse section along a straight line called the *neutral axis* of the section (Fig. 4.10b). The origin of coordinates is now selected on the neutral surface—rather than on the lower face of the member—so that the distance from any point to the neutral surface is measured by its coordinate *y.*

Denoting by ρ the radius of arc *DE* (Fig. 4.10a), by θ the central angle corresponding to *DE,* and observing that the length of *DE* is equal to the length *L* of the undeformed member, we write

$$L = \rho\theta \qquad (4.4)$$

Considering the arc *JK* located at a distance *y* above the neutral surface, its length L' is

$$L' = (\rho - y)\theta \qquad (4.5)$$

Since the original length of arc *JK* was equal to *L,* the deformation of *JK* is

$$\delta = L' - L \qquad (4.6)$$

or, substituting from Eqs. (4.4) and (4.5) into Eq. (4.6),

$$\delta = (\rho - y)\theta - \rho\theta = -y\theta \qquad (4.7)$$

The longitudinal strain ϵ_x in the elements of *JK* is obtained by dividing δ by the original length *L* of *JK.* Write

$$\epsilon_x = \frac{\delta}{L} = \frac{-y\theta}{\rho\theta}$$

or

$$\epsilon_x = -\frac{y}{\rho} \qquad (4.8)$$

The minus sign is due to the fact that it is assumed the bending moment is positive, and thus the beam is concave upward.

Because of the requirement that transverse sections remain plane, identical deformations occur in all planes parallel to the plane of symmetry. Thus, the value of the strain given by Eq. (4.8) is valid anywhere, and the *longitudinal normal strain ϵ_x varies linearly with the distance y from the neutral surface.*

The strain ϵ_x reaches its maximum absolute value when y is largest. Denoting the largest distance from the neutral surface as c (corresponding to either the upper or the lower surface of the member) and the *maximum absolute value* of the strain as ϵ_m, we have

$$\epsilon_m = \frac{c}{\rho} \tag{4.9}$$

Solving Eq. (4.9) for ρ and substituting into Eq. (4.8),

$$\epsilon_x = -\frac{y}{c}\epsilon_m \tag{4.10}$$

To compute the strain or stress at a given point of the member, we must first locate the neutral surface in the member. To do this, we must specify the stress-strain relation of the material used, as will be considered in the next section.[†]

4.2   STRESSES AND DEFORMATIONS IN THE ELASTIC RANGE

We now consider the case when the bending moment M is such that the normal stresses in the member remain below the yield strength σ_Y. This means that the stresses in the member remain below the proportional limit and the elastic limit as well. There will be no permanent deformation, and Hooke's law for uniaxial stress applies. Assuming the material to be homogeneous and denoting its modulus of elasticity by E, the normal stress in the longitudinal x direction is

$$\sigma_x = E\epsilon_x \tag{4.11}$$

Recalling Eq. (4.10) and multiplying both members by E, we write

$$E\epsilon_x = -\frac{y}{c}(E\epsilon_m)$$

or using Eq. (4.11),

$$\sigma_x = -\frac{y}{c}\sigma_m \tag{4.12}$$

where σ_m denotes the *maximum absolute value* of the stress. This result shows that, *in the elastic range, the normal stress varies linearly with the distance from the neutral surface* (Fig. 4.11).

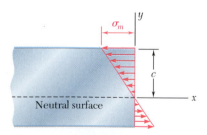

Fig. 4.11   Bending stresses vary linearly with distance from the neutral axis.

[†]Let us note that, if the member possesses both a vertical and a horizontal plane of symmetry (e.g., a member with a rectangular cross section) and the stress-strain curve is the same in tension and compression, the neutral surface will coincide with the plane of symmetry (see Sec. 4.6).

Note that neither the location of the neutral surface nor the maximum value σ_m of the stress have yet to be determined. Both can be found using Eqs. (4.1) and (4.3). Substituting for σ_x from Eq. (4.12) into Eq. (4.1), write

$$\int \sigma_x \, dA = \int \left(-\frac{y}{c}\sigma_m \right) dA = -\frac{\sigma_m}{c} \int y \, dA = 0$$

from which

$$\int y \, dA = 0 \qquad\qquad \textbf{(4.13)}$$

This equation shows that the first moment of the cross section about its neutral axis must be zero.[†] Thus, for a member subjected to pure bending and *as long as the stresses remain in the elastic range, the neutral axis passes through the centroid of the section.*

Recall Eq. (4.3), which was developed with respect to an *arbitrary* horizontal z axis:

$$\int (-y\sigma_x \, dA) = M \qquad\qquad \textbf{(4.3)}$$

Specifying that the z axis coincides with the neutral axis of the cross section, substitute σ_x from Eq. (4.12) into Eq. (4.3):

$$\int (-y) \left(-\frac{y}{c}\sigma_m \right) dA = M$$

or

$$\frac{\sigma_m}{c} \int y^2 \, dA = M \qquad\qquad \textbf{(4.14)}$$

Recall that for pure bending the neutral axis passes through the centroid of the cross section and I is the moment of inertia or second moment of area of the cross section with respect to a centroidal axis perpendicular to the plane of the couple **M**. Solving Eq. (4.14) for σ_m,[‡]

$$\sigma_m = \frac{Mc}{I} \qquad\qquad \textbf{(4.15)}$$

Substituting for σ_m from Eq. (4.15) into Eq. (4.12), we obtain the normal stress σ_x at any distance y from the neutral axis:

$$\sigma_x = -\frac{My}{I} \qquad\qquad \textbf{(4.16)}$$

Equations (4.15) and (4.16) are called the *elastic flexure formulas*, and the normal stress σ_x caused by the bending or "flexing" of the member is often referred to as the *flexural stress*. The stress is compressive ($\sigma_x < 0$) above the neutral axis ($y > 0$) when the bending moment M is positive and tensile ($\sigma_x > 0$) when M is negative.

[†]See Appendix A for a discussion of the moments of areas.

[‡]Recall that the bending moment is assumed to be positive. If the bending moment is negative, M should be replaced in Eq. (4.15) by its absolute value $|M|$.

Returning to Eq. (4.15), the ratio I/c depends only on the geometry of the cross section. This ratio is defined as the *elastic section modulus S*, where

$$\text{Elastic section modulus} = S = \frac{I}{c} \qquad (4.17)$$

Substituting S for I/c into Eq. (4.15), this equation in alternative form is

$$\sigma_m = \frac{M}{S} \qquad (4.18)$$

Since the maximum stress σ_m is inversely proportional to the elastic section modulus S, beams should be designed with as large a value of S as is practical. For example, a wooden beam with a rectangular cross section of width b and depth h has

$$S = \frac{I}{c} = \frac{\frac{1}{12}bh^3}{h/2} = \tfrac{1}{6}bh^2 = \tfrac{1}{6}Ah \qquad (4.19)$$

where A is the cross-sectional area of the beam. For two beams with the same cross-sectional area A (Fig. 4.12), the beam with the larger depth h will have the larger section modulus and will be the more effective in resisting bending.[†]

In the case of structural steel (Photo 4.3), American standard beams (S-beams) and wide-flange beams (W-beams) are preferred to other

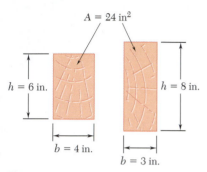

$A = 24$ in^2

$h = 6$ in.

$b = 4$ in.

$h = 8$ in.

$b = 3$ in.

Fig. 4.12 Wood beam cross sections.

Photo 4.3 Wide-flange steel beams are used in the frame of this building.

[†]However, large values of the ratio h/b could result in lateral instability of the beam.

shapes because a large portion of their cross section is located far from the neutral axis (Fig. 4.13). Thus, for a given cross-sectional area and a given depth, their design provides large values of I and S. Values of the elastic section modulus of commonly manufactured beams can be obtained from tables listing the various geometric properties of such beams. To determine the maximum stress σ_m in a given section of a standard beam, the engineer needs only to read the value of the elastic section modulus S in such a table and divide the bending moment M in the section by S.

The deformation of the member caused by the bending moment M is measured by the *curvature* of the neutral surface. The curvature is defined as the reciprocal of the radius of curvature ρ and can be obtained by solving Eq. (4.9) for $1/\rho$:

$$\frac{1}{\rho} = \frac{\epsilon_m}{c} \tag{4.20}$$

In the elastic range, $\epsilon_m = \sigma_m/E$. Substituting for ϵ_m into Eq. (4.20) and recalling Eq. (4.15), write

$$\frac{1}{\rho} = \frac{\sigma_m}{Ec} = \frac{1}{Ec}\frac{Mc}{I}$$

or

$$\frac{1}{\rho} = \frac{M}{EI} \tag{4.21}$$

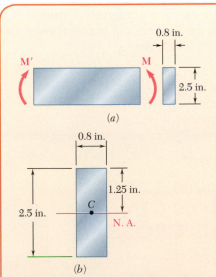

Fig. 4.13 Two types of steel beam cross sections: (a) American Standard beam (S) (b) wide-flange beam (W).

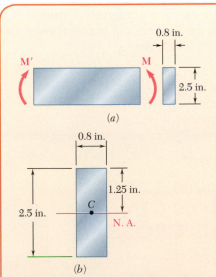

Fig. 4.14 (a) Bar of rectangular cross-section in pure bending. (b) Centroid and dimensions of cross section.

Concept Application 4.1

A steel bar of 0.8×2.5-in. rectangular cross section is subjected to two equal and opposite couples acting in the vertical plane of symmetry of the bar (Fig. 4.14a). Determine the value of the bending moment M that causes the bar to yield. Assume $\sigma_Y = 36$ ksi.

Since the neutral axis must pass through the centroid C of the cross section, $c = 1.25$ in. (Fig. 4.14b). On the other hand, the centroidal moment of inertia of the rectangular cross section is

$$I = \tfrac{1}{12}bh^3 = \tfrac{1}{12}(0.8 \text{ in.})(2.5 \text{ in.})^3 = 1.042 \text{ in}^4$$

Solving Eq. (4.15) for M, and substituting the above data,

$$M = \frac{I}{c}\sigma_m = \frac{1.042 \text{ in}^4}{1.25 \text{ in.}}(36 \text{ ksi})$$

$$M = 30 \text{ kip·in.}$$

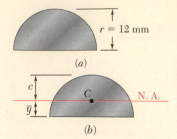

Fig. 4.15 (a) Semi-circular section of rod in pure bending. (b) Centroid and neutral axis of cross section.

Concept Application 4.2

An aluminum rod with a semicircular cross section of radius $r = 12$ mm (Fig. 4.15a) is bent into the shape of a circular arc of mean radius $\rho = 2.5$ m. Knowing that the flat face of the rod is turned toward the center of curvature of the arc, determine the maximum tensile and compressive stress in the rod. Use $E = 70$ GPa.

We can use Equation (4.21) to determine the bending moment M corresponding to the given radius of curvature ρ and then Eq. (4.15) to determine σ_m. However, it is simpler to use Eq. (4.9) to determine ϵ_m and Hooke's law to obtain σ_m.

The ordinate $\bar{y}$ of the centroid C of the semicircular cross section is

$$\bar{y} = \frac{4r}{3\pi} = \frac{4(12 \text{ mm})}{3\pi} = 5.093 \text{ mm}$$

The neutral axis passes through C (Fig. 4.15b), and the distance c to the point of the cross section farthest away from the neutral axis is

$$c = r - \bar{y} = 12 \text{ mm} - 5.093 \text{ mm} = 6.907 \text{ mm}$$

Using Eq. (4.9),

$$\epsilon_m = \frac{c}{\rho} = \frac{6.907 \times 10^{-3} \text{ m}}{2.5 \text{ m}} = 2.763 \times 10^{-3}$$

and applying Hooke's law,

$$\sigma_m = E\epsilon_m = (70 \times 10^9 \text{ Pa})(2.763 \times 10^{-3}) = 193.4 \text{ MPa}$$

Since this side of the rod faces away from the center of curvature, the stress obtained is a tensile stress. The maximum compressive stress occurs on the flat side of the rod. Using the fact that the stress is proportional to the distance from the neutral axis, write

$$\sigma_{\text{comp}} = -\frac{\bar{y}}{c}\sigma_m = -\frac{5.093 \text{ mm}}{6.907 \text{ mm}}(193.4 \text{ MPa})$$

$$= -142.6 \text{ MPa}$$

4.3 DEFORMATIONS IN A TRANSVERSE CROSS SECTION

While Sec. 4.1b showed that the transverse cross section of a member in pure bending remains plane, there is the possibility of deformations within the plane of the section. Recall from Sec. 2.4 that elements in a state of uniaxial stress, $\sigma_x \neq 0$, $\sigma_y = \sigma_z = 0$, are deformed in the transverse y

and z directions, as well as in the axial x direction. The normal strains ϵ_y and ϵ_z depend upon Poisson's ratio ν for the material used and are expressed as

$$\epsilon_y = -\nu\epsilon_x \qquad \epsilon_z = -\nu\epsilon_x$$

or recalling Eq. (4.8),

$$\epsilon_y = \frac{\nu y}{\rho} \qquad \epsilon_z = \frac{\nu y}{\rho} \qquad \text{(4.22)}$$

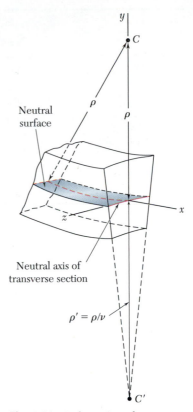

Fig. 4.16 Deformation of a transverse cross section.

These relationships show that the elements located above the neutral surface ($y > 0$) expand in both the y and z directions, while the elements located below the neutral surface ($y < 0$) contract. In a member of rectangular cross section, the expansion and contraction of the various elements in the vertical direction will compensate, and no change in the vertical dimension of the cross section will be observed. As far as the deformations in the horizontal transverse z direction are concerned, however, the expansion of the elements located above the neutral surface and the corresponding contraction of the elements located below that surface will result in the various horizontal lines in the section being bent into arcs of circle (Fig. 4.16). This situation is similar to that in a longitudinal cross section. Comparing the second of Eqs. (4.22) with Eq. (4.8), the neutral axis of the transverse section is bent into a circle of radius $\rho' = \rho/\nu$. The center C' of this circle is located below the neutral surface (assuming $M > 0$) (i.e., on the side opposite to the center of curvature C). The reciprocal of the radius of curvature ρ' represents the curvature of the transverse cross section and is called the *anticlastic curvature*.

$$\text{Anticlastic curvature} = \frac{1}{\rho'} = \frac{\nu}{\rho} \qquad \text{(4.23)}$$

In this section we will now discuss the manner in which the couples **M** and **M**′ are applied to the member. If *all* transverse sections of the member, from one end to the other, are to remain plane and free of shearing stresses, the couples must be applied so that the ends remain plane and free of shearing stresses. This can be accomplished by applying the couples **M** and **M**′ to the member through the use of rigid and smooth plates (Fig. 4.17). The forces exerted by the plates will be normal to the end sections, and these sections, while remaining plane, will be free to deform, as described earlier in this section.

Note that these loading conditions cannot be actually realized, since they require each plate to exert tensile forces on the corresponding end section below its neutral axis, while allowing the section to freely deform in its own plane. The fact that the rigid-end-plates model of Fig. 4.17 cannot be physically realized, however, does not detract from its importance, which is to allow us to *visualize* the loading conditions corresponding to the relationships in the preceding sections. Actual loading conditions may differ appreciably from this idealized model. Using Saint-Venant's principle, however, these relationships can be used to compute stresses in engineering situations, as long as the section considered is not too close to the points where the couples are applied.

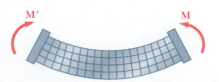

Fig. 4.17 Pure bending with end plates to insure plane sections remain plane.

Sample Problem 4.1

The rectangular tube shown is extruded from an aluminum alloy for which $\sigma_Y = 40$ ksi, $\sigma_U = 60$ ksi, and $E = 10.6 \times 10^6$ psi. Neglecting the effect of fillets, determine (*a*) the bending moment M for which the factor of safety will be 3.00 and (*b*) the corresponding radius of curvature of the tube.

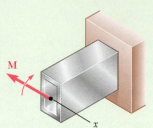

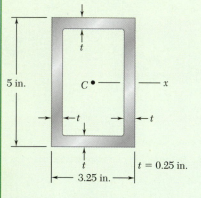

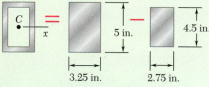

Fig. 1 Superposition for calculating moment of inertia.

STRATEGY: Use the factor of safety to determine the allowable stress. Then calculate the bending moment and radius of curvature using Eqs. (4.15) and (4.21).

MODELING and ANALYSIS:

Moment of Inertia. Considering the cross-sectional area of the tube as the difference between the two rectangles shown in Fig. 1 and recalling the formula for the centroidal moment of inertia of a rectangle, write

$$I = \tfrac{1}{12}(3.25)(5)^3 - \tfrac{1}{12}(2.75)(4.5)^3 \qquad I = 12.97 \text{ in}^4$$

Allowable Stress. For a factor of safety of 3.00 and an ultimate stress of 60 ksi, we have

$$\sigma_{all} = \frac{\sigma_U}{F.S.} = \frac{60 \text{ ksi}}{3.00} = 20 \text{ ksi}$$

Since $\sigma_{all} < \sigma_Y$, the tube remains in the elastic range and we can apply the results of Sec. 4.2.

a. Bending Moment. With $c = \tfrac{1}{2}(5 \text{ in.}) = 2.5$ in., we write

$$\sigma_{all} = \frac{Mc}{I} \qquad M = \frac{I}{c}\sigma_{all} = \frac{12.97 \text{ in}^4}{2.5 \text{ in.}}(20 \text{ ksi}) \qquad M = 103.8 \text{ kip·in.} \blacktriangleleft$$

b. Radius of Curvature. Using Fig. 2 and recalling that $E = 10.6 \times 10^6$ psi, we substitute this value and the values obtained for I and M into Eq. (4.21) and find

$$\frac{1}{\rho} = \frac{M}{EI} = \frac{103.8 \times 10^3 \text{ lb·in.}}{(10.6 \times 10^6 \text{ psi})(12.97 \text{ in}^4)} = 0.755 \times 10^{-3} \text{ in}^{-1}$$

$$\rho = 1325 \text{ in.} \qquad\qquad \rho = 110.4 \text{ ft} \blacktriangleleft$$

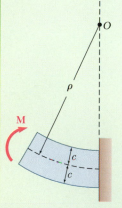

Fig. 2 Deformed shape of beam.

(continued)

REFLECT and THINK: Alternatively, we can calculate the radius of curvature using Eq. (4.9). Since we know that the maximum stress is $\sigma_{all} = 20$ ksi, the maximum strain ϵ_m can be determined, and Eq. (4.9) gives

$$\epsilon_m = \frac{\sigma_{all}}{E} = \frac{20 \text{ ksi}}{10.6 \times 10^6 \text{ psi}} = 1.887 \times 10^{-3} \text{ in./in.}$$

$$\epsilon_m = \frac{c}{\rho} \qquad \rho = \frac{c}{\epsilon_m} = \frac{2.5 \text{ in.}}{1.887 \times 10^{-3} \text{ in./in.}}$$

$$\rho = 1325 \text{ in.} \qquad\qquad \rho = 110.4 \text{ ft} \blacktriangleleft$$

Sample Problem 4.2

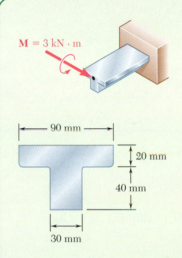

A cast-iron machine part is acted upon by the 3 kN·m couple shown. Knowing that $E = 165$ GPa and neglecting the effect of fillets, determine (*a*) the maximum tensile and compressive stresses in the casting and (*b*) the radius of curvature of the casting.

STRATEGY: The moment of inertia is determined, recognizing that it is first necessary to determine the location of the neutral axis. Then Eqs. (4.15) and (4.21) are used to determine the stresses and radius of curvature.

MODELING and ANALYSIS:

Centroid. Divide the T-shaped cross section into two rectangles as shown in Fig. 1 and write

	Area, mm²	$\bar{y}$, mm	$\bar{y}A$, mm³	
1	$(20)(90) = 1800$	50	90×10^3	$\bar{Y}\Sigma A = \Sigma \bar{y}A$
2	$\underline{(40)(30) = 1200}$	20	$\underline{24 \times 10^3}$	$\bar{Y}(3000) = 114 \times 10^6$
	$\Sigma A = 3000$		$\Sigma \bar{y}A = 114 \times 10^3$	$\bar{Y} = 38$ mm

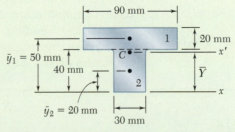

Fig. 1 Composite areas for calculating centroid.

(continued)

Fig. 2 Composite areas for calculating moment of inertia.

Centroidal Moment of Inertia. The parallel-axis theorem is used to determine the moment of inertia of each rectangle (Fig. 2) with respect to the axis x' that passes through the centroid of the composite section. Adding the moments of inertia of the rectangles, write

$$I_{x'} = \Sigma(\bar{I} + Ad^2) = \Sigma(\tfrac{1}{12}bh^3 + Ad^2)$$

$$= \tfrac{1}{12}(90)(20)^3 + (90 \times 20)(12)^2 + \tfrac{1}{12}(30)(40)^3 + (30 \times 40)(18)^2$$

$$= 868 \times 10^3 \, \text{mm}^4$$

$$I = 868 \times 10^{-9} \, \text{m}^4$$

a. Maximum Tensile Stress. Since the applied couple bends the casting downward, the center of curvature is located below the cross section. The maximum tensile stress occurs at point A (Fig. 3), which is farthest from the center of curvature.

$$\sigma_A = \frac{Mc_A}{I} = \frac{(3 \, \text{kN·m})(0.022 \, \text{m})}{868 \times 10^{-9} \, \text{m}^4} \qquad \sigma_A = +76.0 \, \text{MPa} \blacktriangleleft$$

Maximum Compressive Stress. This occurs at point B (Fig. 3):

$$\sigma_B = -\frac{Mc_B}{I} = -\frac{(3 \, \text{kN·m})(0.038 \, \text{m})}{868 \times 10^{-9} \, \text{m}^4} \qquad \sigma_B = -131.3 \, \text{MPa} \blacktriangleleft$$

b. Radius of Curvature. From Eq. (4.21), using Fig. 3, we have

$$\frac{1}{\rho} = \frac{M}{EI} = \frac{3 \, \text{kN·m}}{(165 \, \text{GPa})(868 \times 10^{-9} \, \text{m}^4)}$$

$$= 20.95 \times 10^{-3} \, \text{m}^{-1} \qquad \rho = 47.7 \, \text{m} \blacktriangleleft$$

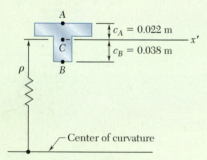

Fig. 3 Radius of curvature is measured to the centroid of the cross section.

REFLECT and THINK: Note the T-section has a vertical plane of symmetry, with the applied moment in that plane. Thus the couple of this applied moment lies in the plane of symmetry, resulting in symmetrical bending. Had the couple been in another plane, we would have unsymmetric bending and thus would need to apply the principles of Sec. 4.8.

Problems

4.1 and 4.2 Knowing that the couple shown acts in a vertical plane, determine the stress at (*a*) point *A*, (*b*) point *B*.

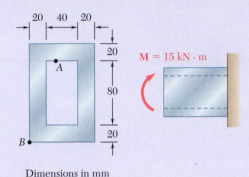

Dimensions in mm

Fig. P4.1

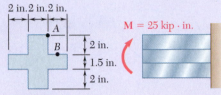

Fig. P4.2

4.3 Using an allowable stress of 155 MPa, determine the largest bending moment **M** that can be applied to the wide-flange beam shown. Neglect the effect of fillets.

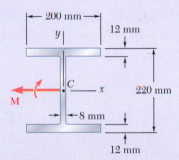

Fig. P4.3

4.4 Solve Prob. 4.3, assuming that the wide-flange beam is bent about the *y* axis by a couple of moment M_y.

4.5 Using an allowable stress of 16 ksi, determine the largest couple that can be applied to each pipe.

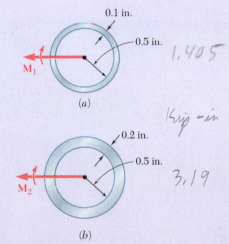

(*a*)

(*b*)

Fig. P4.5

4.6 Knowing that the couple shown acts in a vertical plane, determine the stress at (*a*) point *A*, (*b*) point *B*.

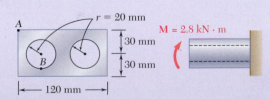

Fig. P4.6

4.7 and 4.8 Two W4 × 13 rolled sections are welded together as shown. Knowing that for the steel alloy used $\sigma_U = 58$ ksi and using a factor of safety of 3.0, determine the largest couple that can be applied when the assembly is bent about the z axis.

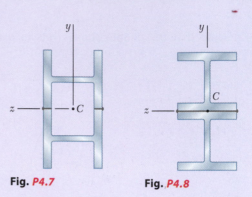

Fig. *P4.7*   Fig. *P4.8*

4.9 through 4.11 Two vertical forces are applied to a beam of the cross section shown. Determine the maximum tensile and compressive stresses in portion *BC* of the beam.

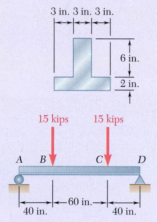

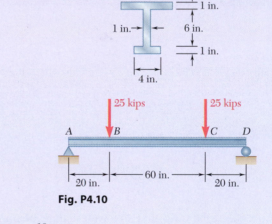

Fig. P4.9   Fig. P4.10

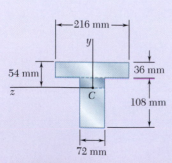

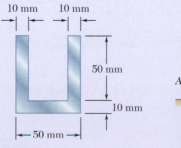

Fig. P4.11

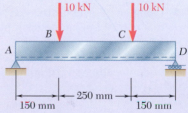

Fig. P4.12

4.12 Knowing that a beam of the cross section shown is bent about a horizontal axis and that the bending moment is 6 kN·m, determine the total force acting on the shaded portion of the web.

4.13 Knowing that a beam of the cross section shown is bent about a horizontal axis and that the bending moment is 4 kN·m, determine the total force acting on the shaded portion of the beam.

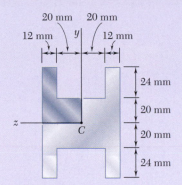

Fig. P4.13

4.14 Solve Prob. 4.13, assuming that the beam is bent about a vertical axis by a couple of moment 4 kN·m.

4.15 Knowing that for the extruded beam shown the allowable stress is 12 ksi in tension and 16 ksi in compression, determine the largest couple **M** that can be applied.

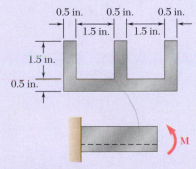

Fig. P4.15

4.16 The beam shown is made of a nylon for which the allowable stress is 24 MPa in tension and 30 MPa in compression. Determine the largest couple **M** that can be applied to the beam.

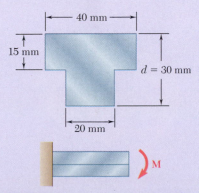

Fig. P4.16

4.17 Solve Prob. 4.16, assuming that $d = 40$ mm.

4.18 Knowing that for the beam shown the allowable stress is 12 ksi in tension and 16 ksi in compression, determine the largest couple **M** that can be applied.

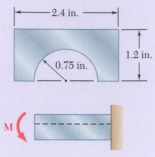

Fig. P4.18

4.19 and *4.20* Knowing that for the extruded beam shown the allowable stress is 120 MPa in tension and 150 MPa in compression, determine the largest couple **M** that can be applied.

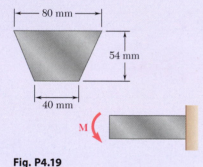

Fig. P4.19

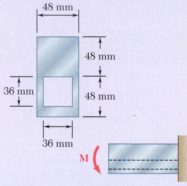

Fig. *P4.20*

4.21 Straight rods of 6-mm diameter and 30-m length are stored by coiling the rods inside a drum of 1.25-m inside diameter. Assuming that the yield strength is not exceeded, determine (*a*) the maximum stress in a coiled rod, (*b*) the corresponding bending moment in the rod. Use $E = 200$ GPa.

Fig. P4.21

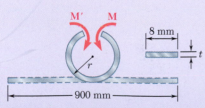

Fig. P4.22

4.22 A 900-mm strip of steel is bent into a full circle by two couples applied as shown. Determine (*a*) the maximum thickness *t* of the strip if the allowable stress of the steel is 420 MPa, (*b*) the corresponding moment *M* of the couples. Use $E = 200$ GPa.

4.23 Straight rods of 0.30-in. diameter and 200-ft length are some-times used to clear underground conduits of obstructions or to thread wires through a new conduit. The rods are made of high-strength steel and, for storage and transportation, are wrapped on spools of 5-ft diameter. Assuming that the yield strength is not exceeded, determine (a) the maximum stress in a rod, when the rod, which is initially straight, is wrapped on a spool, (b) the cor-responding bending moment in the rod. Use $E = 29 \times 10^6$ psi.

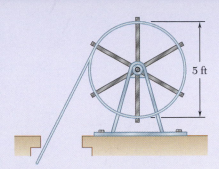

Fig. P4.23

4.24 A 60-N·m couple is applied to the steel bar shown. (a) Assuming that the couple is applied about the z axis as shown, determine the maximum stress and the radius of curvature of the bar. (b) Solve part a, assuming that the couple is applied about the y axis. Use $E = 200$ GPa.

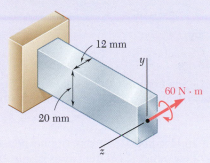

Fig. P4.24

4.25 (a) Using an allowable stress of 120 MPa, determine the largest couple **M** that can be applied to a beam of the cross section shown. (b) Solve part a, assuming that the cross section of the beam is an 80-mm square.

4.26 A thick-walled pipe is bent about a horizontal axis by a couple **M**. The pipe may be designed with or without four fins. (a) Using an allowable stress of 20 ksi, determine the largest couple that may be applied if the pipe is designed with four fins as shown. (b) Solve part a, assuming that the pipe is designed with no fins.

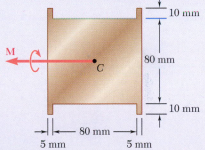

Fig. P4.25

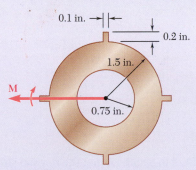

Fig. P4.26

4.27 A couple **M** will be applied to a beam of rectangular cross section that is to be sawed from a log of circular cross section. Determine the ratio d/b for which (a) the maximum stress σ_m will be as small as possible, (b) the radius of curvature of the beam will be maximum.

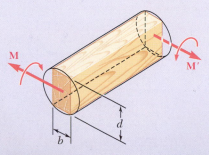

Fig. P4.27

4.28 A portion of a square bar is removed by milling, so that its cross section is as shown. The bar is then bent about its horizontal axis by a couple **M**. Considering the case where $h = 0.9h_0$, express the maximum stress in the bar in the form $\sigma_m = k\sigma_0$ where σ_0 is the maximum stress that would have occurred if the original square bar had been bent by the same couple **M**, and determine the value of k.

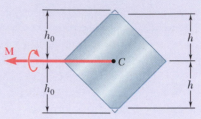

Fig. *P4.28*

4.29 In Prob. 4.28, determine (*a*) the value of h for which the maximum stress σ_m is as small as possible, (*b*) the corresponding value of k.

4.30 For the bar and loading of Concept Application 4.1, determine (*a*) the radius of curvature ρ, (*b*) the radius of curvature ρ' of a transverse cross section, (*c*) the angle between the sides of the bar that were originally vertical. Use $E = 29 \times 10^6$ psi and $\nu = 0.29$.

4.31 A W200 × 31.3 rolled-steel beam is subjected to a couple **M** of moment 45 kN·m. Knowing that $E = 200$ GPa and $\nu = 0.29$, determine (*a*) the radius of curvature ρ, (*b*) the radius of curvature ρ' of a transverse cross section.

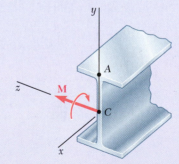

Fig. P4.31

4.32 It was assumed in Sec. 4.1B that the normal stresses σ_y in a member in pure bending are negligible. For an initially straight elastic member of rectangular cross section, (*a*) derive an approximate expression for σ_y as a function of y, (*b*) show that $(\sigma_y)_{max} = -(c/2\rho)(\sigma_x)_{max}$ and, thus, that σ_y can be neglected in all practical situations. (*Hint*: Consider the free-body diagram of the portion of beam located below the surface of ordinate y and assume that the distribution of the stress σ_x is still linear.)

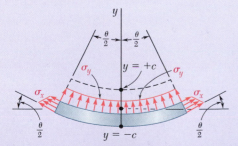

Fig. P4.32

4.4 MEMBERS MADE OF COMPOSITE MATERIALS

The derivations given in Sec. 4.2 are based on the assumption of a homogeneous material with a given modulus of elasticity E. If the member is made of two or more materials with different moduli of elasticity, the member is a composite member.

Consider a bar consisting of two portions of different materials bonded together as shown in Fig. 4.18. This composite bar will deform as described in Sec. 4.1B, since its cross section remains the same throughout its entire length, and since no assumption was made in Sec. 4.1B regarding the stress-strain relationship of the material or materials involved. Thus, the normal strain ϵ_x still varies linearly with the distance y from the neutral axis of the section (Fig. 4.21a and b), and formula (4.8) holds:

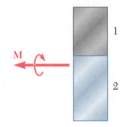

Fig. 4.18 Cross section made with different materials

$$\epsilon_x = -\frac{y}{\rho} \tag{4.8}$$

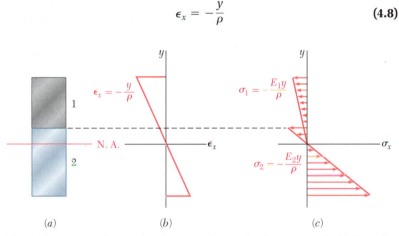

Fig. 4.19 Stress and strain distributions in bar Made of two materials. (a) Neutral axis shifted from centroid. (b) Strain distribution. (c) Corresponding stress distribution.

However, it cannot be assumed that the neutral axis passes through the centroid of the composite section, and one of the goals of this analysis is to determine the location of this axis.

Since the moduli of elasticity E_1 and E_2 of the two materials are different, the equations for the normal stress in each material are

$$\sigma_1 = E_1\epsilon_x = -\frac{E_1 y}{\rho}$$

$$\sigma_2 = E_2\epsilon_x = -\frac{E_2 y}{\rho} \tag{4.24}$$

A stress-distribution curve is obtained that consists of two segments with straight lines as shown in Fig. 4.19c. It follows from Eqs. (4.24) that the force dF_1 exerted on an element of area dA of the upper portion of the cross section is

$$dF_1 = \sigma_1\, dA = -\frac{E_1 y}{\rho}\, dA \tag{4.25}$$

while the force dF_2 exerted on an element of the same area dA of the lower portion is

$$dF_2 = \sigma_2 \, dA = -\frac{E_2 y}{\rho} \, dA \qquad \textbf{(4.26)}$$

Denoting the ratio E_2/E_1 of the two moduli of elasticity by n, we can write

$$dF_2 = -\frac{(nE_1)y}{\rho} \, dA = -\frac{E_1 y}{\rho}(n \, dA) \qquad \textbf{(4.27)}$$

Comparing Eqs. (4.25) and (4.27), we note that the same force dF_2 would be exerted on an element of area $n \, dA$ of the first material. Thus, the resistance to bending of the bar would remain the same if both portions were made of the first material, provided that the width of each element of the lower portion were multiplied by the factor n. Note that this widening (if $n > 1$) or narrowing (if $n < 1$) must be *in a direction parallel to the neutral axis of the section*, since it is essential that the distance y of each element from the neutral axis remain the same. This new cross section is called the *transformed section* of the member (Fig. 4.20).

Since the transformed section represents the cross section of a member made of a *homogeneous material* with a modulus of elasticity E_1, the method described in Sec. 4.2 can be used to determine the neutral axis of the section and the normal stress at various points. The neutral axis is drawn *through the centroid of the transformed section* (Fig. 4.21), and the stress σ_x at any point of the corresponding homogeneous member obtained from Eq. (4.16) is

$$\sigma_x = -\frac{My}{I} \qquad \textbf{(4.16)}$$

where y is the distance from the neutral surface and I is *the moment of inertia of the transformed section* with respect to its centroidal axis.

To obtain the stress σ_1 at a point located in the upper portion of the cross section of the original composite bar, compute the stress σ_x at the corresponding point of the transformed section. However, to obtain the stress σ_2 at a point in the lower portion of the cross section, we must *multiply by n* the stress σ_x computed at the corresponding point of the transformed section. Indeed, the same elementary force dF_2 is applied to an element of area $n \, dA$ of the transformed section and to an element of area dA of the original section. Thus, the stress σ_2 at a point of the original section must be n times larger than the stress at the corresponding point of the transformed section.

The deformations of a composite member can also be determined by using the transformed section. We recall that the transformed section represents the cross section of a member, made of a homogeneous material of modulus E_1, which deforms in the same manner as the composite member. Therefore, using Eq. (4.21), we write that the curvature of the composite member is

$$\frac{1}{\rho} = \frac{M}{E_1 I}$$

where I is the moment of inertia of the transformed section with respect to its neutral axis.

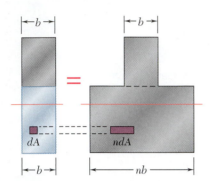

Fig. 4.20 Transformed section based on replacing lower material with that used on top.

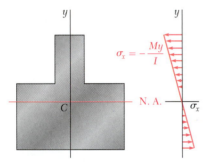

Fig. 4.21 Distribution of stresses in transformed section.

Concept Application 4.3

A bar obtained by bonding together pieces of steel ($E_s = 29 \times 10^6$ psi) and brass ($E_b = 15 \times 10^6$ psi) has the cross section shown (Fig. 4.22a). Determine the maximum stress in the steel and in the brass when the bar is in pure bending with a bending moment $M = 40$ kip·in.

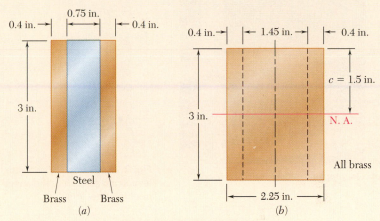

Fig. 4.22 (a) Composite bar. (b) Transformed section.

The transformed section corresponding to an equivalent bar made entirely of brass is shown in Fig. 4.22b. Since

$$n = \frac{E_s}{E_b} = \frac{29 \times 10^6 \text{ psi}}{15 \times 10^6 \text{ psi}} = 1.933$$

the width of the central portion of brass, which replaces the original steel portion, is obtained by multiplying the original width by 1.933:

$$(0.75 \text{ in.})(1.933) = 1.45 \text{ in.}$$

Note that this change in dimension occurs in a direction parallel to the neutral axis. The moment of inertia of the transformed section about its centroidal axis is

$$I = \tfrac{1}{12} bh^3 = \tfrac{1}{12}(2.25 \text{ in.})(3 \text{ in.})^3 = 5.063 \text{ in}^4$$

and the maximum distance from the neutral axis is $c = 1.5$ in. Using Eq. (4.15), the maximum stress in the transformed section is

$$\sigma_m = \frac{Mc}{I} = \frac{(40 \text{ kip·in.})(1.5 \text{ in.})}{5.063 \text{ in}^4} = 11.85 \text{ ksi}$$

This value also represents the maximum stress in the brass portion of the original composite bar. The maximum stress in the steel portion, however, will be larger than for the transformed section, since the area of the central portion must be reduced by the factor $n = 1.933$. Thus,

$$(\sigma_{\text{brass}})_{\text{max}} = 11.85 \text{ ksi}$$

$$(\sigma_{\text{steel}})_{\text{max}} = (1.933)(11.85 \text{ ksi}) = 22.9 \text{ ksi}$$

Photo 4.4 Reinforced concrete building frame.

An important example of structural members made of two different materials is furnished by *reinforced concrete beams* (Photo 4.4). These beams, when subjected to positive bending moments, are reinforced by steel rods placed a short distance above their lower face (Fig. 4.23a). Since concrete is very weak in tension, it cracks below the neutral surface, and the steel rods carry the entire tensile load, while the upper part of the concrete beam carries the compressive load.

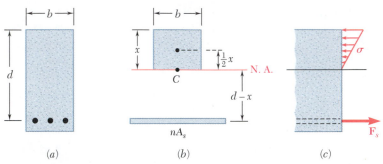

Fig. 4.23 Reinforced concrete beam: (a) Cross section showing location of reinforcing steel. (b) Transformed section of all concrete. (c) Concrete stresses and resulting steel force.

To obtain the transformed section of a reinforced concrete beam, we replace the total cross-sectional area A_s of the steel bars by an equivalent area nA_s, where n is the ratio E_s/E_c of the moduli of elasticity of steel and concrete (Fig. 4.23b). Since the concrete in the beam acts effectively only in compression, only the portion located above the neutral axis should be used in the transformed section.

The position of the neutral axis is obtained by determining the distance x from the upper face of the beam to the centroid C of the transformed section. Using the width of the beam b and the distance d from the upper face to the center line of the steel rods, the first moment of the transformed section with respect to the neutral axis must be zero. Since the first moment of each portion of the transformed section is obtained by multiplying its area by the distance of its own centroid from the neutral axis,

$$(bx)\frac{x}{2} - nA_s(d - x) = 0$$

or

$$\frac{1}{2}bx^2 + nA_sx - nA_sd = 0 \qquad \textbf{(4.28)}$$

Solving this quadratic equation for x, both the position of the neutral axis in the beam and the portion of the cross section of the concrete beam that is effectively used are obtained.

The stresses in the transformed section are determined as explained earlier in this section (see Sample Prob. 4.4). The distribution of the compressive stresses in the concrete and the resultant $\mathbf{F}_s$ of the tensile forces in the steel rods are shown in Fig. 4.23c.

4.5   STRESS CONCENTRATIONS

The formula $\sigma_m = Mc/I$ for a member with a plane of symmetry and a uniform cross section is accurate throughout the entire length of the member only if the couples **M** and **M**′ are applied through the use of rigid and smooth plates. Under other conditions of application of the loads, stress concentrations exist near the points where the loads are applied.

Higher stresses also occur if the cross section of the member undergoes a sudden change. Two particular cases are a flat bar with a sudden change in width and a flat bar with grooves. Since the distribution of stresses in the critical cross sections depends only upon the geometry of the members, stress-concentration factors can be determined for various ratios of the parameters involved and recorded, as shown in Figs. 4.24 and 4.25. The value of the maximum stress in the critical cross section is expressed as

$$\sigma_m = K\frac{Mc}{I} \qquad (4.29)$$

where K is the stress-concentration factor and c and I refer to the critical section (i.e., the section of width d). Figures 4.24 and 4.25 clearly show the importance of using fillets and grooves of radius r as large as practical.

Finally, as for axial loading and torsion, the values of the factors K are computed under the assumption of a linear relation between stress and strain. In many applications, plastic deformations occur and result in values of the maximum stress lower than those indicated by Eq. (4.29).

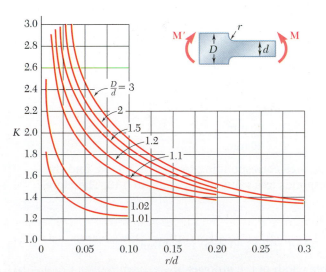

Fig. 4.24 Stress-concentration factors for *flat bars* with fillets under pure bending. (Source: W. D. Pilkey and D. F. Pilkey, *Peterson's Stress Concentration Factors*, 3rd ed., John Wiley & Sons, New York, 2008.)

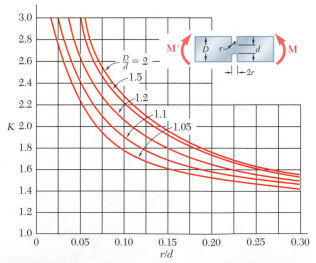

Fig. 4.25 Stress-concentration factors for *flat bars* with grooves (notches) under pure bending. (Source: W. D. Pilkey and D. F. Pilkey, *Peterson's Stress Concentration Factors*, 3rd ed., John Wiley & Sons, New York, 2008.)

Concept Application 4.4

Grooves 10 mm deep are to be cut in a steel bar which is 60 mm wide and 9 mm thick (Fig. 4.26). Determine the smallest allowable width of the grooves if the stress in the bar is not to exceed 150 MPa when the bending moment is equal to 180 N·m.

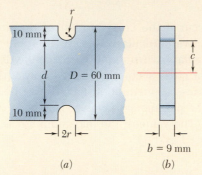

Fig. 4.26   (a) Notched bar dimensions. (b) Cross section.

Note from Fig. 4.26a that

$$d = 60 \text{ mm} - 2(10 \text{ mm}) = 40 \text{ mm}$$
$$c = \tfrac{1}{2}d = 20 \text{ mm} \qquad b = 9 \text{ mm}$$

The moment of inertia of the critical cross section about its neutral axis is

$$I = \tfrac{1}{12}bd^3 = \tfrac{1}{12}(9 \times 10^{-3}\,\text{m})(40 \times 10^{-3}\,\text{m})^3$$
$$= 48 \times 10^{-9}\,\text{m}^4$$

The value of the stress Mc/I is

$$\frac{Mc}{I} = \frac{(180 \text{ N·m})(20 \times 10^{-3}\,\text{m})}{48 \times 10^{-9}\,\text{m}^4} = 75 \text{ MPa}$$

Substituting this value for Mc/I into Eq. (4.29) and making $\sigma_m = 150$ MPa, write

$$150 \text{ MPa} = K(75 \text{ MPa})$$
$$K = 2$$

On the other hand,

$$\frac{D}{d} = \frac{60 \text{ mm}}{40 \text{ mm}} = 1.5$$

Using the curve of Fig. 4.25 corresponding to $D/d = 1.5$, we find that the value $K = 2$ corresponds to a value of r/d equal to 0.13. Therefore,

$$\frac{r}{d} = 0.13$$

$$r = 0.13d = 0.13(40 \text{ mm}) = 5.2 \text{ mm}$$

The smallest allowable width of the grooves is

$$2r = 2(5.2 \text{ mm}) = 10.4 \text{ mm}$$

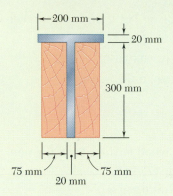

Sample Problem 4.3

Two steel plates have been welded together to form a beam in the shape of a T that has been strengthened by securely bolting to it the two oak timbers shown in the figure. The modulus of elasticity is 12.5 GPa for the wood and 200 GPa for the steel. Knowing that a bending moment $M = 50$ kN·m is applied to the composite beam, determine (*a*) the maximum stress in the wood and (*b*) the stress in the steel along the top edge.

STRATEGY: The beam is first transformed to a beam made of a single material (either steel or wood). The moment of inertia is then determined for the transformed section, and this is used to determine the required stresses, remembering that the actual stresses must be based on the original material.

MODELING:

Transformed Section.   First compute the ratio

$$n = \frac{E_s}{E_w} = \frac{200\ \text{GPa}}{12.5\ \text{GPa}} = 16$$

Multiplying the horizontal dimensions of the steel portion of the section by $n = 16$, a transformed section made entirely of wood is obtained.

Neutral Axis.   Fig. 1 shows the transformed section. The neutral axis passes through the centroid of the transformed section. Since the section consists of two rectangles,

$$\overline{Y} = \frac{\Sigma \overline{y}A}{\Sigma A} = \frac{(0.160\ \text{m})(3.2\ \text{m} \times 0.020\ \text{m}) + 0}{3.2\ \text{m} \times 0.020\ \text{m} + 0.470\ \text{m} \times 0.300\ \text{m}} = 0.050\ \text{m}$$

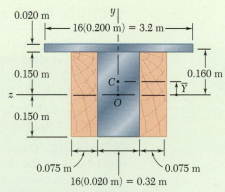

Fig. 1 Transformed cross section.

(continued)

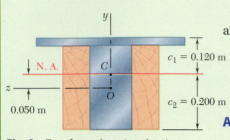

Fig. 2   Transformed section showing neutral axis and distances to extreme fibers.

Centroidal Moment of Inertia.   Using Fig. 2 and the parallel-axis theorem,

$$I = \tfrac{1}{12}(0.470)(0.300)^3 + (0.470 \times 0.300)(0.050)^2$$
$$+ \tfrac{1}{12}(3.2)(0.020)^3 + (3.2 \times 0.020)(0.160 - 0.050)^2$$
$$I = 2.19 \times 10^{-3}\,\text{m}^4$$

ANALYSIS:

a. Maximum Stress in Wood.   The wood farthest from the neutral axis is located along the bottom edge, where $c_2 = 0.200$ m.

$$\sigma_w = \frac{Mc_2}{I} = \frac{(50 \times 10^3\,\text{N·m})(0.200\,\text{m})}{2.19 \times 10^{-3}\,\text{m}^4}$$

$$\sigma_w = 4.57\,\text{MPa} \;\blacktriangleleft$$

b. Stress in Steel.   Along the top edge, $c_1 = 0.120$ m. From the transformed section we obtain an equivalent stress in wood, which must be multiplied by n to obtain the stress in steel.

$$\sigma_s = n\frac{Mc_1}{I} = (16)\frac{(50 \times 10^3\,\text{N·m})(0.120\,\text{m})}{2.19 \times 10^{-3}\,\text{m}^4}$$

$$\sigma_s = 43.8\,\text{MPa} \;\blacktriangleleft$$

REFLECT and THINK: Since the transformed section was based on a beam made entirely of wood, it was necessary to use n to get the actual stress in the steel. Furthermore, at any common distance from the neutral axis, the stress in the steel will be substantially greater than that in the wood, reflective of the much larger modulus of elasticity for the steel.

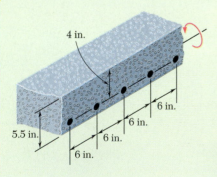

Sample Problem 4.4

A concrete floor slab is reinforced by $\tfrac{5}{8}$-in.-diameter steel rods placed 1.5 in. above the lower face of the slab and spaced 6 in. on centers, as shown in the figure. The modulus of elasticity is 3.6×10^6 psi for the concrete used and 29×10^6 psi for the steel. Knowing that a bending moment of 40 kip·in. is applied to each 1-ft width of the slab, determine (*a*) the maximum stress in the concrete and (*b*) the stress in the steel.

STRATEGY: Transform the section to a single material, concrete, and then calculate the moment of inertia for the transformed section. Continue by calculating the required stresses, remembering that the actual stresses must be based on the original material.

(continued)

MODELING:

Transformed Section.   Consider a portion of the slab 12 in. wide, in which there are two $\frac{5}{8}$-in.-diameter rods having a total cross-sectional area

$$A_s = 2\left[\frac{\pi}{4}\left(\frac{5}{8}\text{ in.}\right)^2\right] = 0.614\text{ in}^2$$

Since concrete acts only in compression, all the tensile forces are carried by the steel rods, and the transformed section (Fig. 1) consists of the two areas shown. One is the portion of concrete in compression (located above the neutral axis), and the other is the transformed steel area nA_s. We have

$$n = \frac{E_s}{E_c} = \frac{29 \times 10^6\text{ psi}}{3.6 \times 10^6\text{ psi}} = 8.06$$

$$nA_s = 8.06(0.614\text{ in}^2) = 4.95\text{ in}^2$$

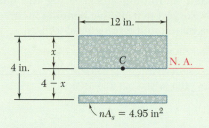

Fig. 1   Transformed section.

Neutral Axis.   The neutral axis of the slab passes through the centroid of the transformed section. Summing moments of the transformed area about the neutral axis, write

$$12x\left(\frac{x}{2}\right) - 4.95(4 - x) = 0 \qquad x = 1.450\text{ in.}$$

Moment of Inertia.   Using Fig. 2, the centroidal moment of inertia of the transformed area is

$$I = \tfrac{1}{3}(12)(1.450)^3 + 4.95(4 - 1.450)^2 = 44.4\text{ in}^4$$

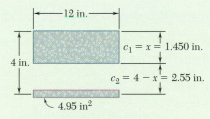

Fig. 2   Dimensions of transformed section used to calculate moment of inertia.

ANALYSIS:

a. Maximum Stress in Concrete.   Fig. 3 shows the stresses on the cross section. At the top of the slab, we have $c_1 = 1.450$ in. and

$$\sigma_c = \frac{Mc_1}{I} = \frac{(40\text{ kip·in.})(1.450\text{ in.})}{44.4\text{ in}^4} \qquad \sigma_c = 1.306\text{ ksi} \blacktriangleleft$$

b. Stress in Steel.   For the steel, we have $c_2 = 2.55$ in., $n = 8.06$ and

$$\sigma_s = n\frac{Mc_2}{I} = 8.06\frac{(40\text{ kip·in.})(2.55\text{ in.})}{44.4\text{ in}^4} \qquad \sigma_s = 18.52\text{ ksi} \blacktriangleleft$$

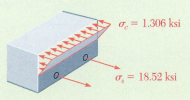

Fig. 3   Stress diagram.

REFLECT and THINK: Since the transformed section was based on a beam made entirely of concrete, it was necessary to use n to get the actual stress in the steel. The difference in the resulting stresses reflects the large differences in the moduli of elasticity.

Problems

4.33 and 4.34 A bar having the cross section shown has been formed by securely bonding brass and aluminum stock. Using the data given below, determine the largest permissible bending moment when the composite bar is bent about a horizontal axis.

	Aluminum	Brass
Modulus of elasticity	70 GPa	105 GPa
Allowable stress	100 MPa	160 MPa

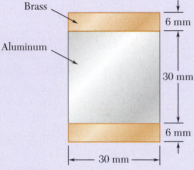

Fig. P4.33

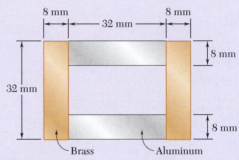

Fig. P4.34

4.35 and 4.36 For the composite bar indicated, determine the largest permissible bending moment when the bar is bent about a vertical axis.

> **4.35** Bar of Prob. 4.33.
> **4.36** Bar of Prob. 4.34.

4.37 and 4.38 Wooden beams and steel plates are securely bolted together to form the composite member shown. Using the data given below, determine the largest permissible bending moment when the member is bent about a horizontal axis.

	Wood	Steel
Modulus of elasticity	2×10^6 psi	29×10^6 psi
Allowable stress	2000 psi	22 ksi

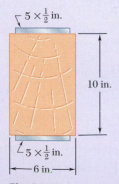

Fig. P4.37

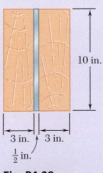

Fig. P4.38

4.39 and 4.40 A copper strip ($E_c = 105$ GPa) and an aluminum strip ($E_a = 75$ GPa) are bonded together to form the composite beam shown. Knowing that the beam is bent about a horizontal axis by a couple of moment $M = 35$ N·m, determine the maximum stress in (*a*) the aluminum strip, (*b*) the copper strip.

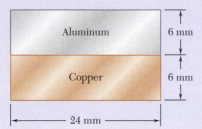

Fig. P4.39

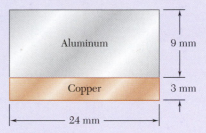

Fig. P4.40

4.41 and 4.42 The 6 × 12-in. timber beam has been strengthened by bolting to it the steel reinforcement shown. The modulus of elasticity for wood is 1.8×10^6 psi and for steel is 29×10^6 psi. Knowing that the beam is bent about a horizontal axis by a couple of moment $M = 450$ kip·in., determine the maximum stress in (*a*) the wood, (*b*) the steel.

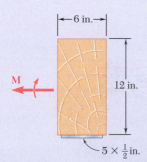

Fig. P4.41

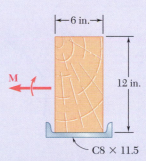

Fig. *P4.42*

4.43 and 4.44 For the composite beam indicated, determine the radius of curvature caused by the couple of moment 35 N·m.
    4.43 Beam of Prob. 4.39.
    4.44 Beam of Prob. 4.40.

4.45 and 4.46 For the composite beam indicated, determine the radius of curvature caused by the couple of moment 450 kip·in.
    4.45 Beam of Prob. 4.41.
    4.46 Beam of Prob. 4.42.

4.47 A concrete slab is reinforced by $\frac{5}{8}$-in.-diameter steel rods placed on 5.5-in. centers as shown. The modulus of elasticity is 3×10^6 psi for the concrete and 29×10^6 psi for the steel. Using an allowable stress of 1400 psi for the concrete and 20 ksi for the steel, determine the largest bending moment in a portion of slab 1 ft wide.

4.48 Solve Prob. 4.47, assuming that the spacing of the $\frac{5}{8}$-in.-diameter steel rods is increased to 7.5 in.

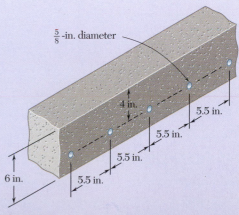

Fig. P4.47

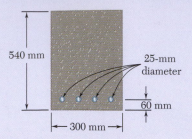

Fig. P4.49

4.49 The reinforced concrete beam shown is subjected to a positive bending moment of 175 kN·m. Knowing that the modulus of elasticity is 25 GPa for the concrete and 200 GPa for the steel, determine (a) the stress in the steel, (b) the maximum stress in the concrete.

4.50 Solve Prob. 4.49, assuming that the 300-mm width is increased to 350 mm.

4.51 Knowing that the bending moment in the reinforced concrete beam is +100 kip·ft and that the modulus of elasticity is 3.625×10^6 psi for the concrete and 29×10^6 psi for the steel, determine (a) the stress in the steel, (b) the maximum stress in the concrete.

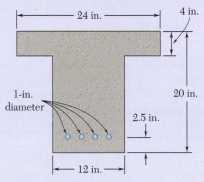

Fig. P4.51

4.52 A concrete beam is reinforced by three steel rods placed as shown. The modulus of elasticity is 3×10^6 psi for the concrete and 29×10^6 psi for the steel. Using an allowable stress of 1350 psi for the concrete and 20 ksi for the steel, determine the largest allowable positive bending moment in the beam.

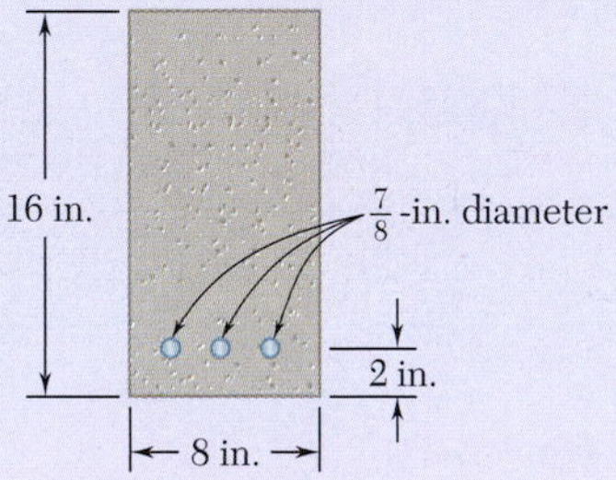

Fig. P4.52

4.53 The design of a reinforced concrete beam is said to be *balanced* if the maximum stresses in the steel and concrete are equal, respectively, to the allowable stresses σ_s and σ_c. Show that to achieve a balanced design the distance x from the top of the beam to the neutral axis must be

$$x = \frac{d}{1 + \dfrac{\sigma_s E_c}{\sigma_c E_s}}$$

where E_c and E_s are the moduli of elasticity of concrete and steel, respectively, and d is the distance from the top of the beam to the reinforcing steel.

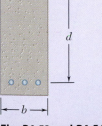

Fig. P4.53 and P4.54

4.54 For the concrete beam shown, the modulus of elasticity is 25 GPa for the concrete and 200 GPa for the steel. Knowing that $b = 200$ mm and $d = 450$ mm, and using an allowable stress of 12.5 MPa for the concrete and 140 MPa for the steel, determine (a) the required area A_s of the steel reinforcement if the beam is to be balanced, (b) the largest allowable bending moment. (See Prob. 4.53 for definition of a balanced beam.)

4.55 and 4.56 Five metal strips, each 0.5 × 1.5-in. cross section, are bonded together to form the composite beam shown. The modulus of elasticity is 30×10^6 psi for the steel, 15×10^6 psi for the brass, and 10×10^6 psi for the aluminum. Knowing that the beam is bent about a horizontal axis by a couple of moment 12 kip·in., determine (*a*) the maximum stress in each of the three metals, (*b*) the radius of curvature of the composite beam.

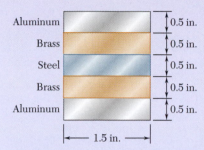

Fig. P4.55

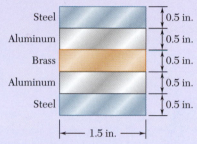

Fig. P4.56

4.57 The composite beam shown is formed by bonding together a brass rod and an aluminum rod of semicircular cross sections. The modulus of elasticity is 15×10^6 psi for the brass and 10×10^6 psi for the aluminum. Knowing that the composite beam is bent about a horizontal axis by couples of moment 8 kip·in., determine the maximum stress (*a*) in the brass, (*b*) in the aluminum.

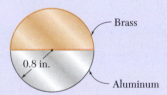

Fig. P4.57

4.58 A steel pipe and an aluminum pipe are securely bonded together to form the composite beam shown. The modulus of elasticity is 200 GPa for the steel and 70 GPa for the aluminum. Knowing that the composite beam is bent by a couple of moment 500 N·m, determine the maximum stress (*a*) in the aluminum, (*b*) in the steel.

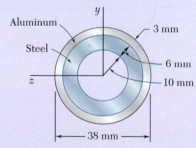

Fig. P4.58

4.59 The rectangular beam shown is made of a plastic for which the value of the modulus of elasticity in tension is one-half of its value in compression. For a bending moment $M = 600$ N·m, determine the maximum (*a*) tensile stress, (*b*) compressive stress.

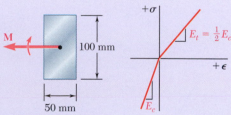

Fig. P4.59

*4.60 A rectangular beam is made of material for which the modulus of elasticity is E_t in tension and E_c in compression. Show that the curvature of the beam in pure bending is

$$\frac{1}{\rho} = \frac{M}{E_r I}$$

where

$$E_r = \frac{4 E_t E_c}{(\sqrt{E_t} + \sqrt{E_c})^2}$$

4.61 Knowing that $M = 250$ N·m, determine the maximum stress in the beam shown when the radius r of the fillets is (a) 4 mm, (b) 8 mm.

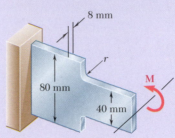

Fig. P4.61 and *P4.62*

4.62 Knowing that the allowable stress for the beam shown is 90 MPa, determine the allowable bending moment M when the radius r of the fillets is (a) 8 mm, (b) 12 mm.

4.63 Semicircular grooves of radius r must be milled as shown in the sides of a steel member. Using an allowable stress of 8 ksi, determine the largest bending moment that can be applied to the member when (a) $r = \frac{3}{8}$ in, (b) $r = \frac{3}{4}$ in.

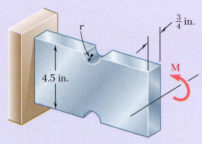

Fig. P4.63 and P4.64

4.64 Semicircular grooves of radius r must be milled as shown in the sides of a steel member. Knowing that $M = 4$ kip·in., determine the maximum stress in the member when the radius r of the semicircular grooves is (a) $r = \frac{3}{8}$ in, (b) $r = \frac{3}{4}$ in.

4.65 A couple of moment $M = 2$ kN·m is to be applied to the end of a steel bar. Determine the maximum stress in the bar (a) if the bar is designed with grooves having semicircular portions of radius $r = 10$ mm, as shown in Fig. *a*, (b) if the bar is redesigned by removing the material to the left and right of the dashed lines as shown in Fig. *b*.

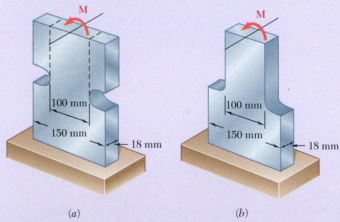

(a)   (b)

Fig. P4.65 and *P4.66*

4.66 The allowable stress used in the design of a steel bar is 80 MPa. Determine the largest couple **M** that can be applied to the bar (a) if the bar is designed with grooves having semicircular portions of radius $r = 15$ mm, as shown in Fig. *a*, (b) if the bar is redesigned by removing the material to the left and right of the dashed lines as shown in Fig. *b*.

*4.6 PLASTIC DEFORMATIONS

In the fundamental relation $\sigma_x = -My/I$ in Sec. 4.2, Hooke's law was applied throughout the member. If the yield strength is exceeded in some portion of the member or the material involved is a brittle material with a nonlinear stress-strain diagram, this relationship ceases to be valid. This section develops a more general method for the determination of the distribution of stresses in a member in pure bending that can be used when Hooke's law does not apply.

Recall that no specific stress-strain relationship was assumed in Sec. 4.1B, when it was proved that the normal strain ϵ_x varies linearly with the distance y from the neutral surface. This property can be used now to write

$$\epsilon_x = -\frac{y}{c}\epsilon_m \qquad (4.10)$$

where y represents the distance of the point considered from the neutral surface, and c is the maximum value of y.

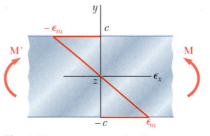

Fig. 4.27 Linear strain distribution in member under pure bending.

However, we cannot assume that the neutral axis passes through the centroid of a given section, since this property was derived in Sec. 4.2 under the assumption of elastic deformations. The neutral axis must be located by trial and error until a distribution of stresses has been found that satisfies Eqs. (4.1) and (4.3) of Sec. 4.1. However, in a member possessing both a vertical and a horizontal plane of symmetry and made of a material characterized by the same stress-strain relationship in tension and compression, the neutral axis coincides with the horizontal axis of symmetry of that section. The properties of the material require that the stresses be symmetric with respect to the neutral axis (i.e., with respect to *some* horizontal axis) and this condition is met (and Eq. (4.1) satisfied) only if that axis is the horizontal axis of symmetry.

The distance y in Eq. (4.10) is measured from the horizontal axis of symmetry z of the cross section, and the distribution of strain ϵ_x is linear and symmetric with respect to that axis (Fig. 4.27). On the other hand, the stress-strain curve is symmetric with respect to the origin of coordinates (Fig. 4.28).

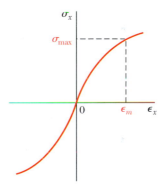

Fig. 4.28 Material with nonlinear stress-strain diagram.

The distribution of stresses in the cross section of the member (i.e., the plot of σ_x versus y) is obtained as follows. Assuming that σ_{max} has been specified, we first determine the value of ϵ_m from the stress-strain diagram and carry it into Eq. (4.10). Then for each value of y, determine the corresponding value of ϵ_x from Eq. (4.10) or Fig. 4.27, and obtain from the stress-strain diagram of Fig. 4.28 the stress σ_x corresponding to ϵ_x. Plotting σ_x against y yields the desired distribution of stresses (Fig. 4.29).

Recall that Eq. (4.3) assumed no particular relation between stress and strain. Therefore, Eq. (4.3) can be used to determine the bending moment M corresponding to the stress distribution obtained in Fig. 4.29. Considering a member with a rectangular cross section of width b, the element of area in Eq. (4.3) is expressed as $dA = b\,dy$, so

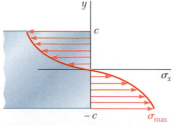

Fig. 4.29 Nonlinear stress distribution in member under pure bending.

$$M = -b\int_{-c}^{c} y\sigma_x\,dy \qquad (4.30)$$

where σ_x is the function of y plotted in Fig. 4.29. Since σ_x is an odd function of y, Eq. (4.30) in the alternative form is

$$M = -2b \int_0^c y\sigma_x \, dy \qquad (4.31)$$

If σ_x is a known analytical function of ϵ_x, Eq. (4.10) can be used to express σ_x as a function of y, and the integral in Eq. (4.31) can be determined analytically. Otherwise, the bending moment M can be obtained through a numerical integration. This computation becomes more meaningful if it is noted that the integral in Eq. (4.31) represents the first moment with respect to the horizontal axis of the area in Fig. 4.29 that is located above the horizontal axis and is bounded by the stress-distribution curve and the vertical axis.

An important value is the ultimate bending moment M_U, which causes failure of the member. This can be determined from the ultimate strength σ_U of the material by choosing $\sigma_{max} = \sigma_U$. However, it is found more convenient in practice to determine M_U experimentally for a specimen of a given material. Assuming a fictitious linear distribution of stresses, Eq. (4.15) is used to determine the corresponding maximum stress R_B:

$$R_B = \frac{M_U c}{I} \qquad (4.32)$$

The fictitious stress R_B is called the *modulus of rupture in bending* of the material. It can be used to determine the ultimate bending moment M_U of a member made of the same material and having a cross section of the same shape, but of different dimensions, by solving Eq. (4.32) for M_U. Since, in the case of a member with a rectangular cross section, the actual and the fictitious linear stress distributions shown in Fig. 4.30 must yield the same value M_U for the ultimate bending moment, the areas they define must have the same first moment with respect to the horizontal axis. Thus, the modulus of rupture R_B will always be larger than the actual ultimate strength σ_U.

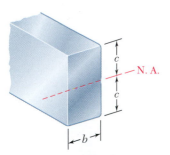

Fig. 4.30 Member stress distribution at ultimate moment M_U.

*4.6A Members Made of Elastoplastic Material

To gain a better insight into the plastic behavior of a member in bending, consider a member made of an *elastoplastic material* and assume the member to have a *rectangular cross section* of width b and depth $2c$ (Fig. 4.31). Recall from Sec. 2.12 the stress-strain diagram for an idealized elastoplastic material is as shown in Fig. 4.32.

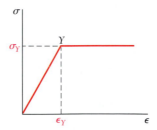

Fig. 4.31 Member with rectangular cross section.

Fig. 4.32 Idealized elastoplastic stress-strain diagram.

As long as the normal stress σ_x does not exceed the yield strength σ_Y, Hooke's law applies, and the stress distribution across the section is linear (Fig. 4.33a). The maximum value of the stress is

$$\sigma_m = \frac{Mc}{I} \qquad (4.15)$$

As the bending moment increases, σ_m eventually reaches σ_Y (Fig. 4.33b). Substituting this value into Eq. (4.15) and solving for M, the value M_Y of the bending moment at the onset of yield is

$$M_Y = \frac{I}{c}\sigma_Y \qquad (4.33)$$

The moment M_Y is called the *maximum elastic moment*, since it is the largest moment for which the deformation remains fully elastic. Recalling that, for the rectangular cross section,

$$\frac{I}{c} = \frac{b(2c)^3}{12c} = \frac{2}{3}bc^2 \qquad (4.34)$$

so

$$M_Y = \frac{2}{3}bc^2\sigma_Y \qquad (4.35)$$

As the bending moment increases further, plastic zones develop in the member. The stress is uniformly equal to $-\sigma_Y$ in the upper zone and to $+\sigma_Y$ in the lower zone (Fig. 4.33c). Between the plastic zones, an elastic core subsists in which the stress σ_x varies linearly with y:

$$\sigma_x = -\frac{\sigma_Y}{y_Y}y \qquad (4.36)$$

Here y_Y represents half the thickness of the elastic core. As M increases, the plastic zones expand, and at the limit, the deformation is fully plastic (Fig. 4.33d).

Equation (4.31) is used to determine the value of the bending moment M corresponding to a given thickness $2y_Y$ of the elastic core. Recalling that σ_x is given by Eq. (4.36) for $0 \le y \le y_Y$ and is equal to $-\sigma_Y$ for $y_Y \le y \le c$,

$$M = -2b\int_0^{y_Y} y\left(-\frac{\sigma_Y}{y_Y}y\right)dy - 2b\int_{y_Y}^{c} y(-\sigma_Y)\,dy$$

$$= \frac{2}{3}by_Y^2\sigma_Y + bc^2\sigma_Y - by_Y^2\sigma_Y$$

$$M = bc^2\sigma_Y\left(1 - \frac{1}{3}\frac{y_Y^2}{c^2}\right) \qquad (4.37)$$

or in view of Eq. (4.35),

$$M = \frac{3}{2}M_Y\left(1 - \frac{1}{3}\frac{y_Y^2}{c^2}\right) \qquad (4.38)$$

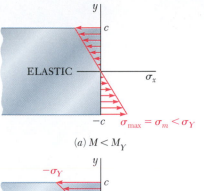

(a) $M < M_Y$

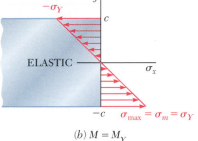

(b) $M = M_Y$

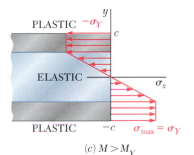

(c) $M > M_Y$

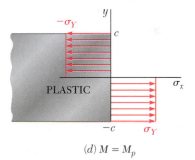

(d) $M = M_p$

Fig. 4.33 Bending stress distribution in a member for : (a) elastic, $M < M_Y$ (b) yield impending, $M = M_Y$, (c) partially yielded, $M > M_Y$, and (d) fully plastic, $M = M_p$.

where M_Y is the maximum elastic moment. Note that as y_Y approaches zero, the bending moment approaches the limiting value

$$M_p = \frac{3}{2}M_Y \qquad (4.39)$$

This value of the bending moment corresponds to fully plastic deformation (Fig. 4.33*d*) and is called the *plastic moment* of the member. Note that Eq. (4.39) is valid only for a *rectangular member made of an elastoplastic material*.

The distribution of *strain* across the section remains linear after the onset of yield. Therefore, Eq. (4.8) remains valid and can be used to determine the half-thickness y_Y of the elastic core:

$$y_Y = \epsilon_Y \rho \qquad (4.40)$$

where ϵ_Y is the yield strain and ρ is the radius of curvature corresponding to a bending moment $M \geq M_Y$. When the bending moment is equal to M_Y, $y_Y = c$ and Eq. (4.40) yields

$$c = \epsilon_Y \rho_Y \qquad (4.41)$$

where ρ_Y is the radius of curvature corresponding to M_Y. Dividing Eq. (4.40) by Eq. (4.41) member by member, the relationship is[†]

$$\frac{y_Y}{c} = \frac{\rho}{\rho_Y} \qquad (4.42)$$

Substituting for y_Y/c from Eq. (4.42) into Eq. (4.38), the bending moment M is a function of the radius of curvature ρ of the neutral surface:

$$M = \frac{3}{2}M_Y\left(1 - \frac{1}{3}\frac{\rho^2}{\rho_Y^2}\right) \qquad (4.43)$$

Note that Eq. (4.43) is valid only after the onset of yield for values of M larger than M_Y. For $M < M_Y$, Eq. (4.21) should be used.

Observe from Eq. (4.43) that the bending moment reaches $M_p = \frac{3}{2}M_Y$ only when $\rho = 0$. Since we clearly cannot have a zero radius of curvature at every point of the neutral surface, a fully plastic deformation cannot develop in pure bending. However, in Chap. 6 it will be shown that such a situation may occur at one point in a beam under a transverse loading.

The stress distributions in a rectangular member corresponding to the maximum elastic moment M_Y and to the limiting case of the plastic moment M_p are represented in Fig. 4.34. Since, the resultants of the tensile and compressive forces must pass through the centroids of and be equal in magnitude to the volumes representing the stress distributions, then

$$R_Y = \tfrac{1}{2}bc\sigma_Y$$

and

$$R_p = bc\sigma_Y$$

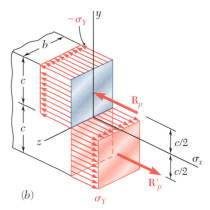

Fig. 4.34 Stress distributions in member at (*a*) maximum elastic moment and at (*b*) plastic moment.

[†]Equation (4.42) applies to any member made of any ductile material with a well-defined yield point, since its derivation is independent of both the shape of the cross section and the shape of stress-strain diagram beyond the yield point.

The moments of the corresponding couples are, respectively,

$$M_Y = (\tfrac{4}{3}c)R_Y = \tfrac{2}{3}bc^2\sigma_Y \qquad (4.44)$$

and

$$M_p = cR_p = bc^2\sigma_Y \qquad (4.45)$$

Thus for a rectangular member $M_p = \tfrac{3}{2}M_Y$ as required by Eq. (4.39).

For beams of *nonrectangular cross section*, the computation of the maximum elastic moment M_Y and of the plastic moment M_p is usually simplified if a graphical method of analysis is used, as shown in Sample Prob. 4.5. In this case, the ratio $k = M_p/M_Y$ is generally not equal to $\tfrac{3}{2}$. For structural shapes such as wide-flange beams, this ratio varies approximately from 1.08 to 1.14. Because it depends only upon the shape of the cross section, the ratio $k = M_p/M_Y$ is called the *shape factor* of the cross section. Note that if the shape factor k and the maximum elastic moment M_Y of a beam are known, the plastic moment M_p of the beam can be obtained by

$$M_p = kM_Y \qquad (4.46)$$

The ratio M_p/σ_Y is called the *plastic section modulus* of the member and is denoted by Z. When the plastic section modulus Z and the yield strength σ_Y of a beam are known, the plastic moment M_p of the beam can be obtained by

$$M_p = Z\sigma_Y \qquad (4.47)$$

Recalling from Eq. (4.18) that $M_Y = S\sigma_Y$ and comparing this relationship with Eq. (4.47), the shape factor $k = M_p/M_Y$ of a given cross section is the ratio of the plastic and elastic section moduli:

$$k = \frac{M_p}{M_Y} = \frac{Z\sigma_Y}{S\sigma_Y} = \frac{Z}{S} \qquad (4.48)$$

Considering a rectangular beam of width b and depth h, note from Eqs. (4.45) and (4.47) that the *plastic section modulus* of a rectangular beam is

$$Z = \frac{M_p}{\sigma_Y} = \frac{bc^2\sigma_Y}{\sigma_Y} = bc^2 = \tfrac{1}{4}bh^2$$

However, recall from Eq. (4.19) that the *elastic section modulus* of the same beam is

$$S = \tfrac{1}{6}bh^2$$

Substituting the values obtained for Z and S into Eq. (4.48), the shape factor of a rectangular beam is

$$k = \frac{Z}{S} = \frac{\tfrac{1}{4}bh^2}{\tfrac{1}{6}bh^2} = \frac{3}{2}$$

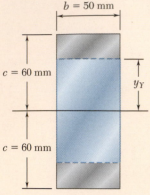

Fig. 4.35 Rectangular cross section with load $M_Y < M < M_p$.

Concept Application 4.5

A member of uniform rectangular cross section 50×120 mm (Fig. 4.35) is subjected to a bending moment $M = 36.8$ kN·m. Assuming that the member is made of an elastoplastic material with a yield strength of 240 MPa and a modulus of elasticity of 200 GPa, determine (a) the thickness of the elastic core and (b) the radius of curvature of the neutral surface.

a. Thickness of Elastic Core.   Determine the maximum elastic moment M_Y. Substituting the given data into Eq. (4.34),

$$\frac{I}{c} = \frac{2}{3}bc^2 = \frac{2}{3}(50 \times 10^{-3}\,\text{m})(60 \times 10^{-3}\,\text{m})^2$$

$$= 120 \times 10^{-6}\,\text{m}^3$$

Then carrying this value and $\sigma_Y = 240$ MPa into Eq. (4.33),

$$M_Y = \frac{I}{c}\sigma_Y = (120 \times 10^{-6}\,\text{m}^3)(240\,\text{MPa}) = 28.8\,\text{kN·m}$$

Substituting the values of M and M_Y into Eq. (4.38),

$$36.8\,\text{kN·m} = \frac{3}{2}(28.8\,\text{kN·m})\left(1 - \frac{1}{3}\frac{y_Y^2}{c^2}\right)$$

$$\left(\frac{y_Y}{c}\right)^2 = 0.444 \qquad \frac{y_Y}{c} = 0.666$$

Since $c = 60$ mm,

$$y_Y = 0.666(60\,\text{mm}) = 40\,\text{mm}$$

Thus, the thickness $2y_Y$ of the elastic core is 80 mm.

b. Radius of Curvature.   The yield strain is

$$\epsilon_Y = \frac{\sigma_Y}{E} = \frac{240 \times 10^6\,\text{Pa}}{200 \times 10^9\,\text{Pa}} = 1.2 \times 10^{-3}$$

Solving Eq. (4.40) for ρ and substituting the values obtained for y_Y and ϵ_Y,

$$\rho = \frac{y_Y}{\epsilon_Y} = \frac{40 \times 10^{-3}\,\text{m}}{1.2 \times 10^{-3}} = 33.3\,\text{m}$$

*4.6B Members with a Single Plane of Symmetry

So far the member in bending has had two planes of symmetry: one containing the couples **M** and **M′** and one perpendicular to that plane. Now consider when the member possesses only one plane of symmetry containing the couples **M** and **M′**. Our analysis will be limited to the situation where the deformation is fully plastic, with the normal stress uniformly equal to $-\sigma_Y$ above the neutral surface and $+\sigma_Y$ below that surface (Fig. 4.36a).

As indicated in Sec. 4.6, the neutral axis cannot be assumed to coincide with the centroidal axis of the cross section when the cross section is not symmetric to that axis. To locate the neutral axis, we consider that the resultant $\mathbf{R}_1$ of the elementary compressive forces is exerted on the portion A_1 of the cross section located above the neutral axis, and the resultant $\mathbf{R}_2$ of the tensile forces is exerted on the portion A_2 located below the neutral axis (Fig. 4.36b). Since the forces $\mathbf{R}_1$ and $\mathbf{R}_2$ form a couple equivalent to the one applied to the member, they must have the same magnitude. Therefore $R_1 = R_2$, or $A_1\sigma_Y = A_2\sigma_Y$, from which we conclude that $A_1 = A_2$. Therefore, *the neutral axis divides the cross section into portions of equal areas*. Note that the axis obtained in this way is *not* a centroidal axis of the section.

The lines of action of the resultants $\mathbf{R}_1$ and $\mathbf{R}_2$ pass through the centroids C_1 and C_2 of the two portions just defined. Denoting by d the distance between C_1 and C_2 and by A the total area of the cross section, the plastic moment of the member is

$$M_p = \left(\frac{1}{2}A\sigma_Y\right)d$$

The actual computation of the plastic moment of a member with only one plane of symmetry is given in Sample Prob. 4.6.

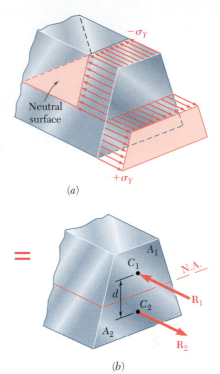

(a)

(b)

Fig. 4.36 Nonsymmetrical beam subject to plastic moment. (a) Stress distributions and (b) resultant forces acting at tension/compression centroids.

*4.6C Residual Stresses

We have just seen that plastic zones develop in a member made of an elastoplastic material if the bending moment is large enough. When the bending moment is decreased back to zero, the corresponding reduction in stress and strain at any given point is represented by a straight line on the stress-strain diagram, as shown in Fig. 4.37. The final value of the stress at a point will not (in general) be zero. There is a residual stress at most points, and that stress may or may not have the same sign as the maximum stress reached at the end of the loading phase.

Since the linear relation between σ_x and ϵ_x applies at all points of the member during the unloading phase, Eq. (4.16) can be is used to obtain the change in stress at any given point. The unloading phase can be handled by assuming the member to be fully elastic.

The residual stresses are obtained by applying the principle of superposition in a manner similar to that described in Sec. 2.13 for an axial centric loading and used again in Sec. 3.8 for torsion. We consider, on one hand, the stresses due to the application of the given bending moment M, and on the other, the reverse stresses due to the equal and opposite bending moment $-M$ that is applied to unload the member. The first group of stresses reflect the *elastoplastic* behavior of the material during the loading phase, and the second group the *linear* behavior of the same material during the unloading phase. Adding the two groups of stresses provides the distribution of residual stresses in the member.

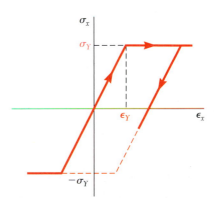

Fig. 4.37 Elastoplastic material stress-strain diagram with load reversal.

Concept Application 4.6

For the member of Fig. 4.35, determine (*a*) the distribution of the residual stresses, (*b*) the radius of curvature, after the bending moment has been decreased from its maximum value of 36.8 kN·m back to zero.

a. Distribution of Residual Stresses.

Recall from Concept Application 4.5 that the yield strength is $\sigma_Y = 240$ MPa and the thickness of the elastic core is $2y_Y = 80$ mm. The distribution of the stresses in the loaded member is as shown in Fig. 4.38*a*.

The distribution of the reverse stresses due to the opposite 36.8 kN·m bending moment required to unload the member is linear and is shown in Fig. 4.38*b*. The maximum stress σ'_m in that distribution is obtained from Eq. (4.15). Recalling that $I/c = 120 \times 10^{-6}$ m³,

$$\sigma'_m = \frac{Mc}{I} = \frac{36.8 \text{ kN·m}}{120 \times 10^{-6} \text{ m}^3} = 306.7 \text{ MPa}$$

Superposing the two distributions of stresses, obtain the residual stresses shown in Fig. 4.38*c*. We note that even though the reverse stresses are larger than the yield strength σ_Y, the assumption of a linear distribution of the reverse stresses is valid, since they do not exceed $2\sigma_Y$.

b. Radius of Curvature after Unloading.

We apply Hooke's law at any point of the core $|y| < 40$ mm, since no plastic deformation has occurred in that portion of the member. Thus, the residual strain at the distance $y = 40$ mm is

$$\epsilon_x = \frac{\sigma_x}{E} = \frac{-35.5 \times 10^6 \text{ Pa}}{200 \times 10^9 \text{ Pa}} = -177.5 \times 10^{-6}$$

Solving Eq. (4.8) for ρ and substituting the appropriate values of y and ϵ_x gives

$$\rho = -\frac{y}{\epsilon_x} = \frac{40 \times 10^{-3} \text{ m}}{177.5 \times 10^{-6}} = 225 \text{ m}$$

The value obtained for ρ after the load has been removed represents a permanent deformation of the member.

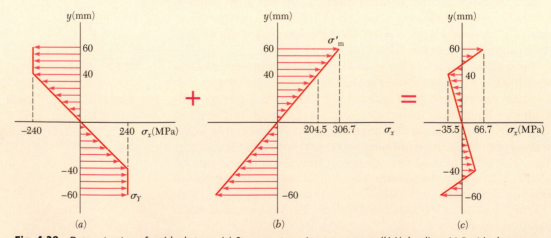

Fig. 4.38 Determination of residual stress: (*a*) Stresses at maximum moment. (*b*) Unloading. (*c*) Residual stresses.

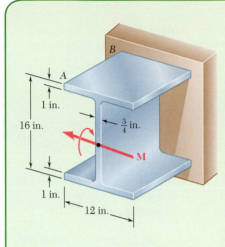

Sample Problem 4.5

Beam AB has been fabricated from a high-strength low-alloy steel that is assumed to be elastoplastic with $E = 29 \times 10^6$ psi and $\sigma_Y = 50$ ksi. Neglecting the effect of fillets, determine the bending moment M and the corresponding radius of curvature (a) when yield first occurs, (b) when the flanges have just become fully plastic.

STRATEGY: Up to the point that yielding first occurs at the top and bottom of this symmetrical section, the stresses and radius of curvature are calculated assuming elastic behavior. A further increase in load causes plastic behavior over parts of the cross section, and it is then necessary to work with the resulting stress distribution on the cross section to obtain the corresponding moment and radius of curvature.

MODELING and ANALYSIS:

a. Onset of Yield. The centroidal moment of inertia of the section is

$$I = \tfrac{1}{12}(12 \text{ in.})(16 \text{ in.})^3 - \tfrac{1}{12}(12 \text{ in.} - 0.75 \text{ in.})(14 \text{ in.})^3 = 1524 \text{ in}^4$$

Bending Moment. For $\sigma_{max} = \sigma_Y = 50$ ksi and $c = 8$ in., we have

$$M_Y = \frac{\sigma_Y I}{c} = \frac{(50 \text{ ksi})(1524 \text{ in}^4)}{8 \text{ in.}} \qquad M_Y = 9525 \text{ kip·in.} \quad \blacktriangleleft$$

Radius of Curvature. As shown in Fig. 1, the strain at the top and bottom is the strain at initial yielding, $\epsilon_Y = \sigma_Y/E = (50 \text{ ksi})/(29 \times 10^6 \text{ psi}) = 0.001724$. Noting that $c = 8$ in., we have from Eq. (4.41)

$$c = \epsilon_Y \rho_Y \qquad 8 \text{ in.} = 0.001724 \rho_Y \qquad \rho_Y = 4640 \text{ in.} \quad \blacktriangleleft$$

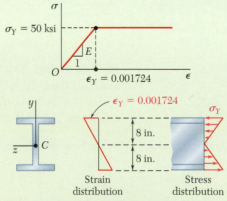

Fig. 1 Elastoplastic material response and elastic strain and stress distributions.

(continued)

b. Flanges Fully Plastic. When the flanges have just become fully plastic, the strains and stresses in the section are as shown in Fig. 2.

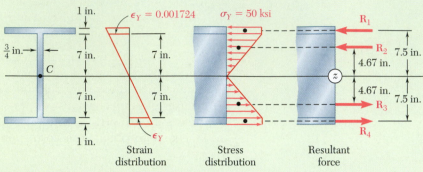

Fig. 2 Strain and stress distributions with flanges fully plastic.

The compressive forces exerted on the top flange and on the top half of the web are replaced by their resultants R_1 and R_2. Similarly, replace the tensile stresses by R_3 and R_4.

$$R_1 = R_4 = (50 \text{ ksi})(12 \text{ in.})(1 \text{ in.}) = 600 \text{ kips}$$
$$R_2 = R_3 = \tfrac{1}{2}(50 \text{ ksi})(7 \text{ in.})(0.75 \text{ in.}) = 131.3 \text{ kips}$$

Bending Moment. Summing the moments of R_1, R_2, R_3, and R_4 about the z axis, write

$$M = 2[R_1(7.5 \text{ in.}) + R_2(4.67 \text{ in.})]$$
$$= 2[(600)(7.5) + (131.3)(4.67)] \qquad M = 10{,}230 \text{ kip·in.} \quad \blacktriangleleft$$

Radius of Curvature. Since $y_Y = 7$ in. for this loading, we have from Eq. (4.40)

$$y_Y = \epsilon_Y \rho \qquad 7 \text{ in.} = (0.001724)\rho \qquad \rho = 4060 \text{ in.} = 338 \text{ ft} \quad \blacktriangleleft$$

REFLECT and THINK: Once the load is increased beyond that which causes initial yielding, it is necessary to work with the actual stress distribution to determine the applied moment. The radius of curvature is based on the elastic portion of the beam.

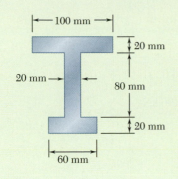

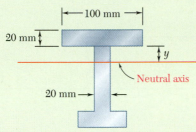

Fig. 1 For fully plastic deformation, neutral axis divides the cross section into two equal areas.

Sample Problem 4.6

Determine the plastic moment M_p of a beam with the cross section shown when the beam is bent about a horizontal axis. Assume that the material is elastoplastic with a yield strength of 240 MPa.

STRATEGY: All portions of the cross section are yielding, and the resulting stress distribution must be used to determine the moment. Since the beam is not symmetrical, it is first necessary to determine the location of the neutral axis.

MODELING:

Neutral Axis. When the deformation is fully plastic, the neutral axis divides the cross section into two portions of equal areas (Fig. 1). Since the total area is

$$A = (100)(20) + (80)(20) + (60)(20) = 4800 \text{ mm}^2$$

the area located above the neutral axis must be 2400 mm². Write

$$(20)(100) + 20y = 2400 \qquad y = 20 \text{ mm}$$

Note that the neutral axis does *not* pass through the centroid of the cross section.

ANALYSIS:

Plastic Moment. Using Fig. 2, the resultant R_i of the elementary forces exerted on the partial area A_i is equal to

$$R_i = A_i\sigma_Y$$

and passes through the centroid of that area. We have

$$R_1 = A_1\sigma_Y = [(0.100 \text{ m})(0.020 \text{ m})]240 \text{ MPa} = 480 \text{ kN}$$

$$R_2 = A_2\sigma_Y = [(0.020 \text{ m})(0.020 \text{ m})]240 \text{ MPa} = 96 \text{ kN}$$

$$R_3 = A_3\sigma_Y = [(0.020 \text{ m})(0.060 \text{ m})]240 \text{ MPa} = 288 \text{ kN}$$

$$R_4 = A_4\sigma_Y = [(0.060 \text{ m})(0.020 \text{ m})]240 \text{ MPa} = 288 \text{ kN}$$

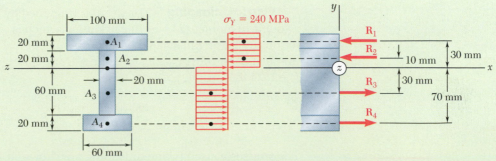

Fig. 2 Fully plastic stress distributions and resultant forces for finding the plastic moment.

(continued)

The plastic moment M_p is obtained by summing the moments of the forces about the z axis.

$$M_p = (0.030\ \text{m})R_1 + (0.010\ \text{m})R_2 + (0.030\ \text{m})R_3 + (0.070\ \text{m})R_4$$

$$= (0.030\ \text{m})(480\ \text{kN}) + (0.010\ \text{m})(96\ \text{kN})$$

$$+ (0.030\ \text{m})(288\ \text{kN}) + (0.070\ \text{m})(288\ \text{kN})$$

$$= 44.16\ \text{kN·m} \qquad\qquad M_p = 44.2\ \text{kN·m} \quad \blacktriangleleft$$

REFLECT and THINK: Since the cross section is *not* symmetric about the z axis, the sum of the moments of $\mathbf{R}_1$ and $\mathbf{R}_2$ is *not* equal to the sum of the moments of $\mathbf{R}_3$ and $\mathbf{R}_4$.

Sample Problem 4.7

For the beam of Sample Prob. 4.5, determine the residual stresses and the permanent radius of curvature after the 10,230-kip·in. couple **M** has been removed.

STRATEGY: Start with the moment and stress distribution when the flanges have just become plastic. The beam is then unloaded by a couple that is equal and opposite to the couple originally applied. During the unloading, the action of the beam is fully elastic. The stresses due to the original loading and those due to the unloading are superposed to obtain the residual stress distribution.

MODELING and ANALYSIS:

Loading. In Sample Prob. 4.5, a couple of moment $M = 10{,}230$ kip·in. was applied and the stresses shown in Fig. 1*a* were obtained.

Elastic Unloading. The beam is unloaded by the application of a couple of moment $M = -10{,}230$ kip·in. (which is equal and opposite to the couple originally applied). During this unloading, the action of the beam is fully elastic; recalling from Sample Prob. 4.5 that $I = 1524\ \text{in}^4$

$$\sigma'_m = \frac{Mc}{I} = \frac{(10{,}230\ \text{kip·in.})(8\ \text{in.})}{1524\ \text{in}^4} = 53.70\ \text{ksi}$$

The stresses caused by the unloading are shown in Fig. 1*b*.

(continued)

Residual Stresses. We superpose the stresses due to the loading (Fig. 1*a*) and to the unloading (Fig. 1*b*) and obtain the residual stresses in the beam (Fig. 1*c*).

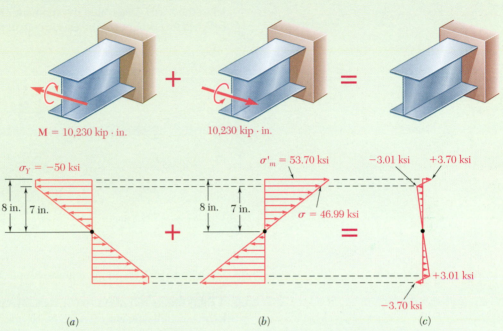

Fig. 1 Superposition of plastic loading and elastic unloading to obtain residual stresses.

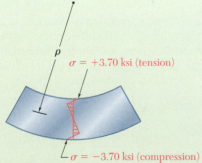

Fig. 2 Representation of the permanent radius of curvature.

Permanent Radius of Curvature. At $y = 7$ in. the residual stress is $\sigma = -3.01$ ksi. Since no plastic deformation occurred at this point, Hooke's law can be used, and $\epsilon_x = \sigma/E$. Recalling Eq. (4.8), we write

$$\rho = -\frac{y}{\epsilon_x} = -\frac{yE}{\sigma} = -\frac{(7 \text{ in.})(29 \times 10^6 \text{ psi})}{-3.01 \text{ ksi}} = +67{,}400 \text{ in.} \quad \rho = 5620 \text{ ft} \blacktriangleleft$$

REFLECT and THINK: From Fig. 2, note that the residual stress is tensile on the upper face of the beam and compressive on the lower face, even though the beam is concave upward.

Problems

4.67 The prismatic bar shown is made of a steel that is assumed to be elastoplastic with $\sigma_Y = 300$ MPa and is subjected to a couple **M** parallel to the x axis. Determine the moment M of the couple for which (a) yield first occurs, (b) the elastic core of the bar is 4 mm thick.

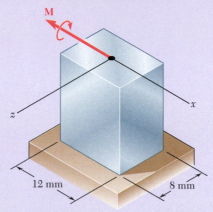

Fig. P4.67

4.68 Solve Prob. 4.67, assuming that the couple **M** is parallel to the z axis.

4.69 A solid square rod of side 0.6 in. is made of a steel that is assumed to be elastoplastic with $E = 29 \times 10^6$ psi and $\sigma_Y = 48$ ksi. Knowing that a couple **M** is applied and maintained about an axis parallel to a side of the cross section, determine the moment M of the couple for which the radius of curvature is 6 ft.

4.70 For the solid square rod of Prob. 4.69, determine the moment M for which the radius of curvature is 3 ft.

4.71 The prismatic rod shown is made of a steel that is assumed to be elastoplastic with $E = 200$ GPa and $\sigma_Y = 280$ MPa. Knowing that couples **M** and **M′** of moment 525 N·m are applied and maintained about axes parallel to the y axis, determine (a) the thickness of the elastic core, (b) the radius of curvature of the bar.

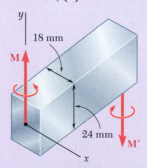

Fig. P4.71

4.72 Solve Prob. 4.71, assuming that the couples **M** and **M′** are applied and maintained about axes parallel to the x axis.

4.73 and 4.74   A beam of the cross section shown is made of a steel that is assumed to be elastoplastic with $E = 200$ GPa and $\sigma_Y = 240$ MPa. For bending about the z axis, determine the bending moment at which (*a*) yield first occurs, (*b*) the plastic zones at the top and bottom of the bar are 30 mm thick.

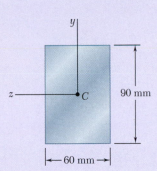

Fig. *P4.73*

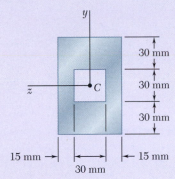

Fig. *P4.74*

4.75 and 4.76   A beam of the cross section shown is made of a steel that is assumed to be elastoplastic with $E = 29 \times 10^6$ psi and $\sigma_Y = 42$ ksi. For bending about the z axis, determine the bending moment at which (*a*) yield first occurs, (*b*) the plastic zones at the top and bottom of the bar are 3 in. thick.

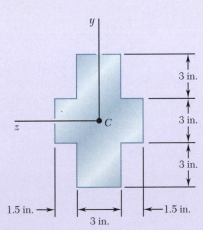

Fig. **P4.75**

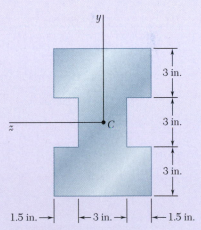

Fig. **P4.76**

4.77 through 4.80   For the beam indicated, determine (*a*) the plastic moment M_p, (*b*) the shape factor of the cross section.

      4.77 Beam of Prob. **4.73.**
      4.78 Beam of Prob. **4.74.**
      4.79 Beam of Prob. 4.75.
      4.80 Beam of Prob. **4.76.**

4.81 through 4.83 Determine the plastic moment M_p of a steel beam of the cross section shown, assuming the steel to be elastoplastic with a yield strength of 240 MPa.

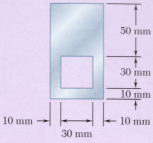

50 mm
30 mm
10 mm
10 mm → ← → ← 10 mm
30 mm

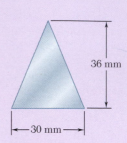

36 mm
30 mm

$r = 18$ mm

Fig. P4.81   Fig. P4.82   Fig. *P4.83*

4.84 Determine the plastic moment M_p of a steel beam of the cross section shown, assuming the steel to be elastoplastic with a yield strength of 42 ksi.

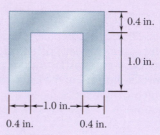

0.4 in.
1.0 in.
←→ ←1.0 in.→ ←→
0.4 in.   0.4 in.

Fig. P4.84

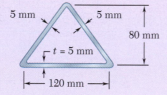

5 mm   5 mm
$t = 5$ mm
80 mm
120 mm

Fig. *P4.85*

4.85 Determine the plastic moment M_p of the cross section shown when the beam is bent about a horizontal axis. Assume the material to be elastoplastic with a yield strength of 175 MPa.

4.86 Determine the plastic moment M_p of a steel beam of the cross section shown, assuming the steel to be elastoplastic with a yield strength of 36 ksi.

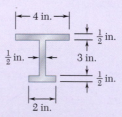

← 4 in. →
$\frac{1}{2}$ in.
$\frac{1}{2}$ in. → ←   3 in.
$\frac{1}{2}$ in.
2 in.

Fig. P4.86

4.87 and 4.88 For the beam indicated, a couple of moment equal to the full plastic moment M_p is applied and then removed. Using a yield strength of 240 MPa, determine the residual stress at $y = 45$ mm.

4.87 Beam of Prob. 4.73.
4.88 Beam of Prob. 4.74.

4.89 and 4.90 A bending couple is applied to the bar indicated, causing plastic zones 3 in. thick to develop at the top and bottom of the bar. After the couple has been removed, determine (a) the residual stress at $y = 4.5$ in., (b) the points where the residual stress is zero, (c) the radius of curvature corresponding to the permanent deformation of the bar.

      4.89 Beam of Prob. 4.75.
      4.90 Beam of Prob. 4.76.

4.91 A bending couple is applied to the beam of Prob. 4.73, causing plastic zones 30 mm thick to develop at the top and bottom of the beam. After the couple has been removed, determine (a) the residual stress at $y = 45$ mm, (b) the points where the residual stress is zero, (c) the radius of curvature corresponding to the permanent deformation of the beam.

4.92 A beam of the cross section shown is made of a steel that is assumed to be elastoplastic with $E = 29 \times 10^6$ psi and $\sigma_Y = 42$ ksi. A bending couple is applied to the beam about the z axis, causing plastic zones 2 in. thick to develop at the top and bottom of the beam. After the couple has been removed, determine (a) the residual stress at $y = 2$ in., (b) the points where the residual stress is zero, (c) the radius of curvature corresponding to the permanent deformation of the beam.

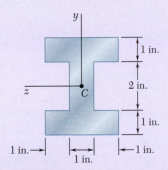

Fig. P4.92

4.93 A rectangular bar that is straight and unstressed is bent into an arc of circle of radius ρ by two couples of moment M. After the couples are removed, it is observed that the radius of curvature of the bar is ρ_R. Denoting by ρ_Y the radius of curvature of the bar at the onset of yield, show that the radii of curvature satisfy the following relation:

$$\frac{1}{\rho_R} = \frac{1}{\rho}\left\{1 - \frac{3}{2}\frac{\rho}{\rho_Y}\left[1 - \frac{1}{3}\left(\frac{\rho}{\rho_Y}\right)^2\right]\right\}$$

4.94 A solid bar of rectangular cross section is made of a material that is assumed to be elastoplastic. Denoting by M_Y and ρ_Y, respectively, the bending moment and radius of curvature at the onset of yield, determine (a) the radius of curvature when a couple of moment $M = 1.25\,M_Y$ is applied to the bar, (b) the radius of curvature after the couple is removed. Check the results obtained by using the relation derived in Prob. 4.93.

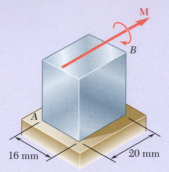

Fig. *P4.95*

16 mm   20 mm

4.95 The prismatic bar *AB* is made of a steel that is assumed to be elastoplastic and for which $E = 200$ GPa. Knowing that the radius of curvature of the bar is 2.4 m when a couple of moment $M = 350$ N·m is applied as shown, determine (*a*) the yield strength of the steel, (*b*) the thickness of the elastic core of the bar.

4.96 The prismatic bar *AB* is made of an aluminum alloy for which the tensile stress-strain diagram is as shown. Assuming that the σ-ϵ diagram is the same in compression as in tension, determine (*a*) the radius of curvature of the bar when the maximum stress is 250 MPa, (*b*) the corresponding value of the bending moment. (*Hint*: For part *b*, plot σ versus y and use an approximate method of integration.)

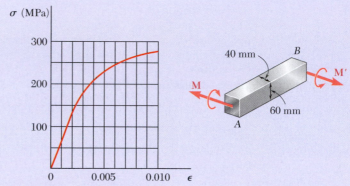

Fig. P4.96

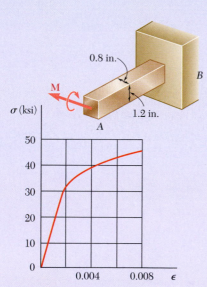

Fig. *P4.97*

4.97 The prismatic bar *AB* is made of a bronze alloy for which the tensile stress-strain diagram is as shown. Assuming that the σ-ϵ diagram is the same in compression as in tension, determine (*a*) the maximum stress in the bar when the radius of curvature of the bar is 100 in., (*b*) the corresponding value of the bending moment. (See hint given in Prob. 4.96.)

4.98 A prismatic bar of rectangular cross section is made of an alloy for which the stress-strain diagram can be represented by the relation $\epsilon = k\sigma^n$ for $\sigma > 0$ and $\epsilon = -|k\sigma^n|$ for $\sigma < 0$. If a couple **M** is applied to the bar, show that the maximum stress is

$$\sigma_m = \frac{1 + 2n}{3n}\frac{Mc}{I}$$

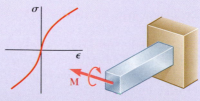

Fig. P4.98

4.7 ECCENTRIC AXIAL LOADING IN A PLANE OF SYMMETRY

We saw in Sec. 1.2A that the distribution of stresses in the cross section of a member under axial loading can be assumed uniform only if the line of action of the loads **P** and **P'** passes through the centroid of the cross section. Such a loading is said to be *centric*. Let us now analyze the distribution of stresses when the line of action of the loads does *not* pass through the centroid of the cross section, i.e., when the loading is *eccentric*.

Two examples of an eccentric loading are shown in Photos 4.5 and 4.6. In Photo 4.5, the weight of the lamp causes an eccentric loading on the post. Likewise, the vertical forces exerted on the press in Photo. 4.6 cause an eccentric loading on the back column of the press.

Photo 4.5 Walkway light.

Photo 4.6 Bench press.

In this section, our analysis will be limited to members that possess a plane of symmetry, and it will be assumed that the loads are applied in the plane of symmetry of the member (Fig. 4.39a). The internal forces acting on a given cross section may then be represented by a force **F** applied at the centroid *C* of the section and a couple **M** acting in the plane of symmetry of the member (Fig. 4.39b). The conditions of equilibrium of the free body *AC* require that the force **F** be equal and opposite to **P'** and that the moment of the couple **M** be equal and opposite to the moment of **P'** about *C*. Denoting by *d* the distance from the centroid *C* to the line of action *AB* of the forces **P** and **P'**, we have

$$F = P \quad \text{and} \quad M = Pd \tag{4.49}$$

We now observe that the internal forces in the section would have been represented by the same force and couple if the straight portion *DE* of member *AB* had been detached from *AB* and subjected simultaneously to the centric loads **P** and **P'** and to the bending couples **M** and **M'** (Fig. 4.40). Thus, the stress distribution due to the original eccentric

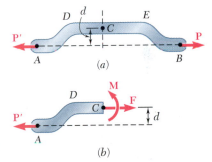

Fig. 4.39 (a) Member with eccentric loading. (b) Free-body diagram of the member with internal loads at section *C*.

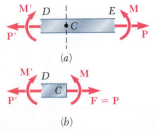

Fig. 4.40 (a) Free-body diagram of straight portion *DE*. (b) Free-body diagram of portion *CD*.

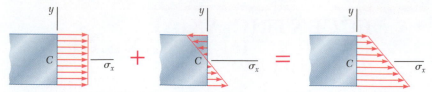

Fig. 4.41   Stress distribution for eccentric loading is obtained by superposing the axial and pure bending distributions.

loading can be obtained by superposing the uniform stress distribution corresponding to the centric loads **P** and **P′** and the linear distribution corresponding to the bending couples **M** and **M′** (Fig. 4.41). Write

$$\sigma_x = (\sigma_x)_{\text{centric}} + (\sigma_x)_{\text{bending}}$$

or recalling Eqs. (1.5) and (4.16),

$$\sigma_x = \frac{P}{A} - \frac{My}{I} \tag{4.50}$$

where A is the area of the cross section and I its centroidal moment of inertia and y is measured from the centroidal axis of the cross section. This relationship shows that the distribution of stresses across the section is *linear but not uniform*. Depending upon the geometry of the cross section and the eccentricity of the load, the combined stresses may all have the same sign, as shown in Fig. 4.41, or some may be positive and others negative, as shown in Fig. 4.42. In the latter case, there will be a line in the section, along which $\sigma_x = 0$. This line represents the *neutral axis* of the section. We note that the neutral axis does *not* coincide with the centroidal axis of the section, since $\sigma_x \neq 0$ for $y = 0$.

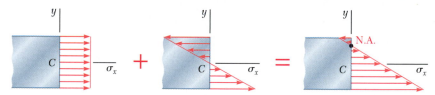

Fig. 4.42   Alternative stress distribution for eccentric loading that results in zones of tension and compression.

The results obtained are valid only to the extent that the conditions of applicability of the superposition principle (Sec. 2.5) and of Saint-Venant's principle (Sec. 2.10) are met. This means that the stresses involved must not exceed the proportional limit of the material. The deformations due to bending must not appreciably affect the distance d in Fig. 4.39*a*, and the cross section where the stresses are computed must not be too close to points D or E. The first of these requirements clearly shows that the superposition method cannot be applied to plastic deformations.

Concept Application 4.7

An open-link chain is obtained by bending low-carbon steel rods of 0.5-in. diameter into the shape shown (Fig. 4.43a). Knowing that the chain carries a load of 160 lb, determine (a) the largest tensile and compressive stresses in the straight portion of a link, (b) the distance between the centroidal and the neutral axis of a cross section.

a. Largest Tensile and Compressive Stresses. The internal forces in the cross section are equivalent to a centric force **P** and a bending couple **M** (Fig. 4.43b) of magnitudes

$$P = 160 \text{ lb}$$

$$M = Pd = (160 \text{ lb})(0.65 \text{ in.}) = 104 \text{ lb·in.}$$

The corresponding stress distributions are shown in Fig. 4.43c and d. The distribution due to the centric force P is uniform and equal to $\sigma_0 = P/A$. We have

$$A = \pi c^2 = \pi (0.25 \text{ in.})^2 = 0.1963 \text{ in}^2$$

$$\sigma_0 = \frac{P}{A} = \frac{160 \text{ lb}}{0.1963 \text{ in}^2} = 815 \text{ psi}$$

The distribution due to the bending couple **M** is linear with a maximum stress $\sigma_m = Mc/I$. We write

$$I = \tfrac{1}{4}\pi c^4 = \tfrac{1}{4}\pi (0.25 \text{ in.})^4 = 3.068 \times 10^{-3} \text{ in}^4$$

$$\sigma_m = \frac{Mc}{I} = \frac{(104 \text{ lb·in.})(0.25 \text{ in.})}{3.068 \times 10^{-3} \text{ in}^4} = 8475 \text{ psi}$$

Superposing the two distributions, we obtain the stress distribution corresponding to the given eccentric loading (Fig. 4.43e). The largest

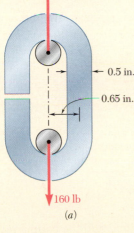

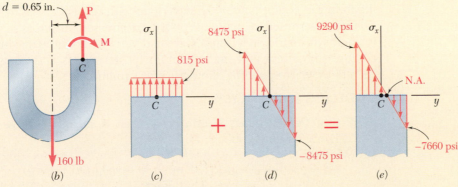

Fig. 4.43 (a) Open chain link under loading. (b) Free-body diagram for section at C. (c) Axial stress at section C. (d) Bending stress at C. (e) Superposition of stresses.

(continued)

tensile and compressive stresses in the section are found to be, respectively,

$$\sigma_t = \sigma_0 + \sigma_m = 815 + 8475 = 9290 \text{ psi}$$
$$\sigma_c = \sigma_0 - \sigma_m = 815 - 8475 = -7660 \text{ psi}$$

b. Distance Between Centroidal and Neutral Axes. The distance y_0 from the centroidal to the neutral axis of the section is obtained by setting $\sigma_x = 0$ in Eq. (4.50) and solving for y_0:

$$0 = \frac{P}{A} - \frac{My_0}{I}$$

$$y_0 = \left(\frac{P}{A}\right)\left(\frac{I}{M}\right) = (815 \text{ psi})\frac{3.068 \times 10^{-3}\text{ in}^4}{104 \text{ lb·in.}}$$

$$y_0 = 0.0240 \text{ in.}$$

Sample Problem 4.8

Knowing that for the cast iron link shown the allowable stresses are 30 MPa in tension and 120 MPa in compression, determine the largest force **P** which can be applied to the link. (*Note*: The T-shaped cross section of the link has previously been considered in Sample Prob. 4.2.)

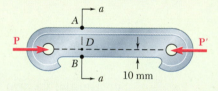

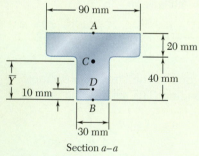

Fig. 1 Section geometry to find centroid location.

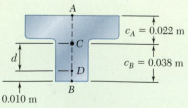

Fig. 2 Dimensions for finding d.

STRATEGY: The stresses due to the axial load and the couple resulting from the eccentricity of the axial load with respect to the neutral axis are superposed to obtain the maximum stresses. The cross section is singly symmetric, so it is necessary to determine both the maximum compression stress and the maximum tension stress and compare each to the corresponding allowable stress to find **P**.

MODELING and ANALYSIS:

Properties of Cross Section. The cross section is shown in Fig. 1. From Sample Prob. 4.2, we have

$$A = 3000 \text{ mm}^2 = 3 \times 10^{-3}\text{ m}^2 \qquad \overline{Y} = 38 \text{ mm} = 0.038 \text{ m}$$
$$I = 868 \times 10^{-9}\text{ m}^4$$

We now write (Fig. 2): $d = (0.038 \text{ m}) - (0.010 \text{ m}) = 0.028 \text{ m}$

(continued)

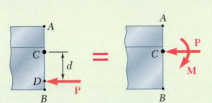

Fig. 3 Equivalent force-couple system at centroid *C*.

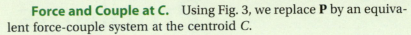

Force and Couple at C. Using Fig. 3, we replace **P** by an equivalent force-couple system at the centroid *C*.

$$P = P \qquad M = P(d) = P(0.028 \text{ m}) = 0.028P$$

The force **P** acting at the centroid causes a uniform stress distribution (Fig. 4*a*). The bending couple **M** causes a linear stress distribution (Fig. 4*b*).

$$\sigma_0 = \frac{P}{A} = \frac{P}{3 \times 10^{-3}} = 333P \qquad \text{(Compression)}$$

$$\sigma_1 = \frac{Mc_A}{I} = \frac{(0.028P)(0.022)}{868 \times 10^{-9}} = 710P \qquad \text{(Tension)}$$

$$\sigma_2 = \frac{Mc_B}{I} = \frac{(0.028P)(0.038)}{868 \times 10^{-9}} = 1226P \qquad \text{(Compression)}$$

Superposition. The total stress distribution (Fig. 4*c*) is found by superposing the stress distributions caused by the centric force **P** and by the couple **M**. Since tension is positive, and compression negative, we have

$$\sigma_A = -\frac{P}{A} + \frac{Mc_A}{I} = -333P + 710P = +377P \qquad \text{(Tension)}$$

$$\sigma_B = -\frac{P}{A} - \frac{Mc_B}{I} = -333P - 1226P = -1559P \qquad \text{(Compression)}$$

Largest Allowable Force. The magnitude of **P** for which the tensile stress at point *A* is equal to the allowable tensile stress of 30 MPa is found by writing

$$\sigma_A = 377P = 30 \text{ MPa} \qquad\qquad P = 79.6 \text{ kN} \quad \blacktriangleleft$$

We also determine the magnitude of **P** for which the stress at *B* is equal to the allowable compressive stress of 120 MPa.

$$\sigma_B = -1559P = -120 \text{ MPa} \qquad\qquad P = 77.0 \text{ kN} \quad \blacktriangleleft$$

The magnitude of the largest force **P** that can be applied without exceeding either of the allowable stresses is the smaller of the two values we have found.

$$P = 77.0 \text{ kN} \quad \blacktriangleleft$$

Fig. 4 Stress distribution at section *C* is superposition of axial and bending distributions.

Problems

4.99 Knowing that the magnitude of the horizontal force **P** is 8 kN, determine the stress at (*a*) point *A*, (*b*) point *B*.

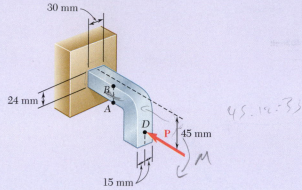

45.12.33

M

Fig. P4.99

4.100 A short wooden post supports a 6-kip axial load as shown. Determine the stress at point *A* when (*a*) *b* = 0, (*b*) *b* = 1.5 in., (*c*) *b* = 3 in.

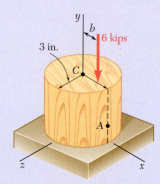

Fig. P4.100

4.101 Two forces **P** can be applied separately or at the same time to a plate that is welded to a solid circular bar of radius *r*. Determine the largest compressive stress in the circular bar, (*a*) when both forces are applied, (*b*) when only one of the forces is applied.

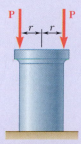

Fig. P4.101

4.102 A short 120 × 180-mm column supports the three axial loads shown. Knowing that section *ABD* is sufficiently far from the loads to remain plane, determine the stress at (*a*) corner *A*, (*b*) corner *B*.

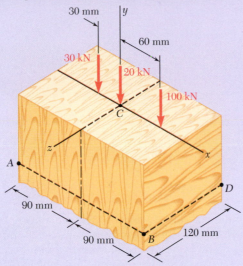

Fig. *P4.102*

4.103 As many as three axial loads, each of magnitude *P* = 50 kN, can be applied to the end of a W200 × 31.1 rolled-steel shape. Determine the stress at point *A*, (*a*) for the loading shown, (*b*) if loads are applied at points 1 and 2 only.

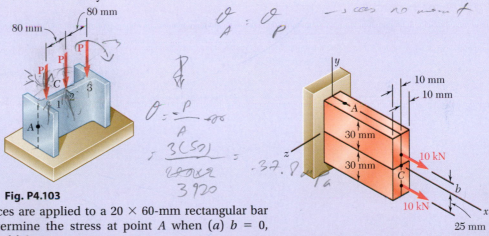

Fig. P4.103

4.104 Two 10-kN forces are applied to a 20 × 60-mm rectangular bar as shown. Determine the stress at point *A* when (*a*) *b* = 0, (*b*) *b* = 15 mm, (*c*) *b* = 25 mm.

Fig. *P4.104*

4.105 Portions of a $\frac{1}{2} \times \frac{1}{2}$-in. square bar have been bent to form the two machine components shown. Knowing that the allowable stress is 15 ksi, determine the maximum load that can be applied to each component.

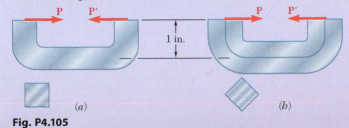

Fig. P4.105

4.106 Knowing that the allowable stress in section *ABD* is 80 MPa, determine the largest force **P** that can be applied to the bracket shown.

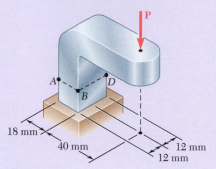

Fig. P4.106

4.107 A milling operation was used to remove a portion of a solid bar of square cross section. Knowing that $a = 30$ mm, $d = 20$ mm, and $\sigma_{all} = 60$ MPa, determine the magnitude P of the largest forces that can be safely applied at the centers of the ends of the bar.

4.108 A milling operation was used to remove a portion of a solid bar of square cross section. Forces of magnitude $P = 18$ kN are applied at the centers of the ends of the bar. Knowing that $a = 30$ mm and $\sigma_{all} = 135$ MPa, determine the smallest allowable depth d of the milled portion of the bar.

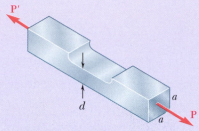

Fig. P4.107 and P4.108

4.109 The two forces shown are applied to a rigid plate supported by a steel pipe of 8-in. outer diameter and 7-in. inner diameter. Determine the value of **P** for which the maximum compressive stress in the pipe is 15 ksi.

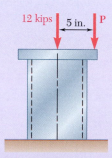

Fig. P4.109

4.110 An offset h must be introduced into a solid circular rod of diameter d. Knowing that the maximum stress after the offset is introduced must not exceed 5 times the stress in the rod when it is straight, determine the largest offset that can be used.

4.111 An offset h must be introduced into a metal tube of 0.75-in. outer diameter and 0.08-in. wall thickness. Knowing that the maximum stress after the offset is introduced must not exceed 4 times the stress in the tube when it is straight, determine the largest offset that can be used.

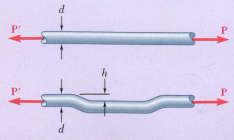

Fig. P4.110 and *P4.111*

4.112 A short column is made by nailing four 1 × 4-in. planks to a 4 × 4-in. timber. Using an allowable stress of 600 psi, determine the largest compressive load **P** that can be applied at the center of the top section of the timber column as shown if (*a*) the column is as described, (*b*) plank 1 is removed, (*c*) planks 1 and 2 are removed, (*d*) planks 1, 2, and 3 are removed, (*e*) all planks are removed.

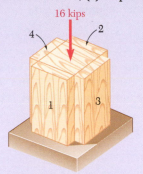

16 kips

Fig. P4.112

4.113 A vertical rod is attached at point *A* to the cast iron hanger shown. Knowing that the allowable stresses in the hanger are $\sigma_{all} = +5$ ksi and $\sigma_{all} = -12$ ksi, determine the largest downward force and the largest upward force that can be exerted by the rod.

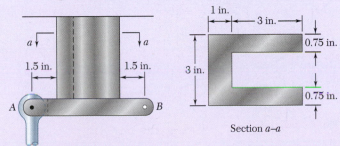

Fig. P4.113

4.114 Solve Prob. 4.113, assuming that the vertical rod is attached at point *B* instead of point *A*.

4.115 Knowing that the clamp shown has been tightened until $P = 400$ N, determine (*a*) the stress at point *A*, (*b*) the stress at point *B*, (*c*) the location of the neutral axis of section *a-a*.

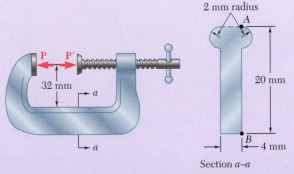

Fig. P4.115

4.116 The shape shown was formed by bending a thin steel plate. Assuming that the thickness *t* is small compared to the length *a* of each side of the shape, determine the stress (*a*) at *A*, (*b*) at *B*, (*c*) at *C*.

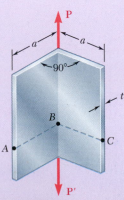

Fig. P4.116

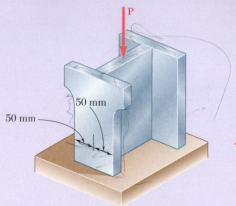

Fig. P4.117

4.117 Three steel plates, each of 25 × 150-mm cross section, are welded together to form a short H-shaped column. Later, for architectural reasons, a 25-mm strip is removed from each side of one of the flanges. Knowing that the load remains centric with respect to the original cross section, and that the allowable stress is 100 MPa, determine the largest force **P** (a) that could be applied to the original column, (b) that can be applied to the modified column.

4.118 A vertical force **P** of magnitude 20 kips is applied at point C located on the axis of symmetry of the cross section of a short column. Knowing that $y = 5$ in., determine (a) the stress at point A, (b) the stress at point B, (c) the location of the neutral axis.

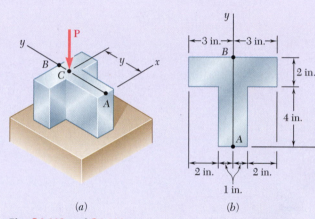

(a)   (b)

Fig. *P4.118* and *P4.119*

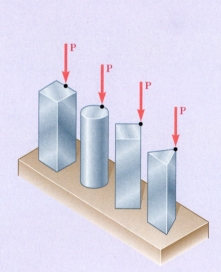

Fig. P4.120

4.119 A vertical force **P** is applied at point C located on the axis of symmetry of the cross section of a short column. Determine the range of values of y for which tensile stresses do not occur in the column.

4.120 The four bars shown have the same cross-sectional area. For the given loadings, show that (a) the maximum compressive stresses are in the ratio 4:5:7:9, (b) the maximum tensile stresses are in the ratio 2:3:5:3. (*Note:* the cross section of the triangular bar is an equilateral triangle.)

4.121 An eccentric force **P** is applied as shown to a steel bar of 25 × 90-mm cross section. The strains at A and B have been measured and found to be

$$\epsilon_A = +350\ \mu \qquad \epsilon_B = -70\ \mu$$

Knowing that $E = 200$ GPa, determine (a) the distance d, (b) the magnitude of the force **P**.

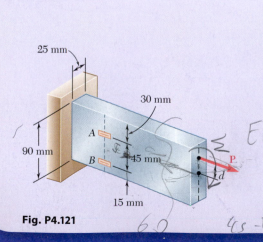

Fig. P4.121

4.122 Solve Prob. 4.121, assuming that the measured strains are

$$\epsilon_A = +600\ \mu \qquad \epsilon_B = +420\ \mu$$

4.123 The C-shaped steel bar is used as a dynamometer to determine the magnitude P of the forces shown. Knowing that the cross section of the bar is a square of side 40 mm and that the strain on the inner edge was measured and found to be 450 μ, determine the magnitude P of the forces. Use $E = 200$ GPa.

4.124 A short length of a rolled-steel column supports a rigid plate on which two loads $\mathbf{P}$ and $\mathbf{Q}$ are applied as shown. The strains at two points A and B on the centerline of the outer faces of the flanges have been measured and found to be

$$\epsilon_A = -400 \times 10^{-6} \text{ in./in.} \qquad \epsilon_B = -300 \times 10^{-6} \text{ in./in.}$$

Knowing that $E = 29 \times 10^6$ psi, determine the magnitude of each load.

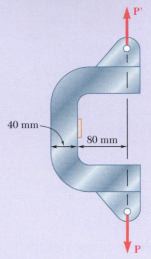

Fig. P4.123

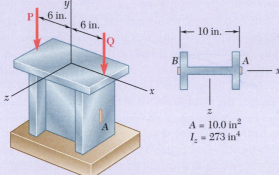
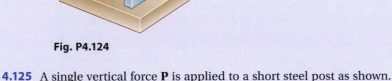

$A = 10.0 \text{ in}^2$
$I_z = 273 \text{ in}^4$

Fig. P4.124

4.125 A single vertical force $\mathbf{P}$ is applied to a short steel post as shown. Gages located at A, B, and C indicate the following strains:

$$\epsilon_A = -500 \ \mu \qquad \epsilon_B = -1000 \ \mu \qquad \epsilon_C = -200 \ \mu$$

Knowing that $E = 29 \times 10^6$ psi, determine (a) the magnitude of $\mathbf{P}$, (b) the line of action of $\mathbf{P}$, (c) the corresponding strain at the hidden edge of the post, where $x = -2.5$ in. and $z = -1.5$ in.

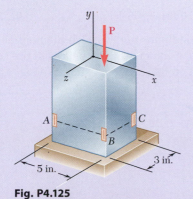

Fig. P4.125

4.126 The eccentric axial force $\mathbf{P}$ acts at point D, which must be located 25 mm below the top surface of the steel bar shown. For $P = 60$ kN, (a) determine the depth d of the bar for which the tensile stress at point A is maximum, (b) the corresponding stress at A.

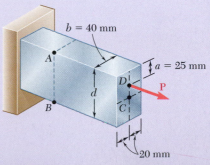

Fig. P4.126

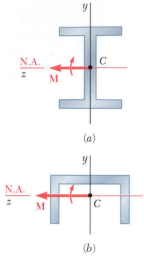

Fig. 4.44 Moment in plane of symmetry.

4.8 UNSYMMETRIC BENDING ANALYSIS

Our analysis of pure bending has been limited so far to members possessing at least one plane of symmetry and subjected to couples acting in that plane. Because of the symmetry of such members and of their loadings, the members remain symmetric with respect to the plane of the couples and thus bend in that plane (Sec. 4.1B). This is illustrated in Fig. 4.44; part *a* shows the cross section of a member possessing two planes of symmetry, one vertical and one horizontal, and part *b* the cross section of a member with a single, vertical plane of symmetry. In both cases the couple exerted on the section acts in the vertical plane of symmetry of the member and is represented by the horizontal couple vector **M**, and in both cases the neutral axis of the cross section is found to coincide with the axis of the couple.

Let us now consider situations where the bending couples do *not* act in a plane of symmetry of the member, either because they act in a different plane, or because the member does not possess any plane of symmetry. In such situations, we cannot assume that the member will bend in the plane of the couples. This is illustrated in Fig. 4.45. In each part of the figure, the couple exerted on the section has again been assumed to act in a vertical plane and has been represented by a horizontal couple vector **M**. However, since the vertical plane is not a plane of symmetry, *we cannot expect the member to bend in that plane or the neutral axis of the section to coincide with the axis of the couple.*

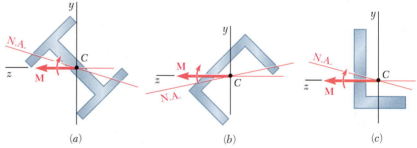

Fig. 4.45 Moment not in plane of symmetry.

The precise conditions under which the neutral axis of a cross section of arbitrary shape coincides with the axis of the couple **M** representing the forces acting on that section is shown in Fig. 4.46. Both the couple

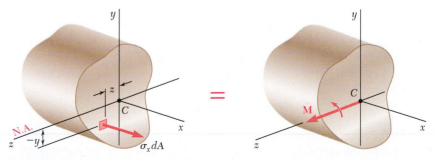

Fig. 4.46 Section of arbitrary shape where the neutral axis coincides with the axis of couple **M**.

vector **M** and the neutral axis are assumed to be directed along the z axis. Recall from Sec. 4.1A that the elementary internal forces $\sigma_x dA$ form a system equivalent to the couple **M**. Thus,

x components:	$\int \sigma_x dA = 0$	**(4.1)**
moments about y axis:	$\int z\sigma_x dA = 0$	**(4.2)**
moments about z axis:	$\int(-y\sigma_x dA) = M$	**(4.3)**

When all of the stresses are within the proportional limit, the first of these equations leads to the requirement that the neutral axis be a centroidal axis, and the last to the fundamental relation $\sigma_x = -My/I$. Since we had assumed in Sec. 4.1A that the cross section was symmetric with respect to the y axis, Eq. (4.2) was dismissed as trivial at that time. Now that we are considering a cross section of arbitrary shape, Eq. (4.2) becomes highly significant. Assuming the stresses to remain within the proportional limit of the material, $\sigma_x = -\sigma_m y/c$ is substituted into Eq. (4.2) for

$$\int z\left(-\frac{\sigma_m y}{c}\right) dA = 0 \quad \text{or} \quad \int yz\, dA = 0 \qquad \textbf{(4.51)}$$

The integral $\int yz\, dA$ represents the product of inertia I_{yz} of the cross section with respect to the y and z axes, and will be zero if these axes are the *principal centroidal axes of the cross section.*[†] Thus the neutral axis of the cross section coincides with the axis of the couple **M** representing the forces acting on that section *if, and only if, the couple vector* **M** *is directed along one of the principal centroidal axes of the cross section.*

Note that the cross sections shown in Fig. 4.44 are symmetric with respect to at least one of the coordinate axes. In each case, the y and z axes are the principal centroidal axes of the section. Since the couple vector **M** is directed along one of the principal centroidal axes, the neutral axis coincides with the axis of the couple. Also, if the cross sections are rotated through 90° (Fig. 4.47), the couple vector **M** is still directed along a principal centroidal axis, and the neutral axis again coincides with the axis of the couple, even though in case b the couple does *not* act in a plane of symmetry of the member.

In Fig. 4.45, neither of the coordinate axes is an axis of symmetry for the sections shown, and the coordinate axes are not principal axes. Thus, the couple vector **M** is not directed along a principal centroidal axis, and the neutral axis does not coincide with the axis of the couple. However, any given section possesses principal centroidal axes, even if it is unsymmetric, as the section shown in Fig. 4.45c, and these axes may be determined analytically or by using Mohr's circle.[†] If the couple vector **M** is directed along one of the principal centroidal axes of the section, the neutral axis will coincide with the axis of the couple (Fig. 4.48), and the equations derived for symmetric members can be used to determine the stresses.

As you will see presently, the principle of superposition can be used to determine stresses in the most general case of unsymmetric bending. Consider first a member with a vertical plane of symmetry subjected to

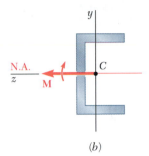

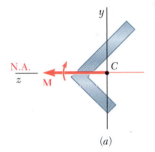

Fig. 4.47 Moment aligned with principal centroidal axis.

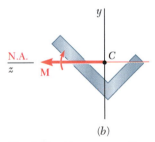

Fig. 4.48 Moment not aligned with principal centroidal axis.

[†]See Ferdinand P. Beer and E. Russell Johnston, Jr., *Mechanics for Engineers*, 5th ed., McGraw-Hill, New York, 2008, or *Vector Mechanics for Engineers*, 10th ed., McGraw-Hill, New York, 2013, Secs. 9.8–9.10.

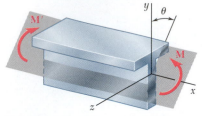

Fig. 4.49 Unsymmetric bending, with bending moment not in a plane of symmetry.

bending couples **M** and **M′** acting in a plane forming an angle θ with the vertical plane (Fig. 4.49). The couple vector **M** representing the forces acting on a given cross section forms the same angle θ with the horizontal z axis (Fig. 4.50). Resolving the vector **M** into component vectors **M**$_z$ and **M**$_y$ along the z and y axes, respectively, gives

$$M_z = M \cos \theta \qquad M_y = M \sin \theta \tag{4.52}$$

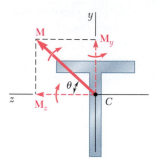

Fig. 4.50 Applied moment is resolved into y and z components.

Since the y and z axes are the principal centroidal axes of the cross section, Eq. (4.16) determines the stresses resulting from the application of either of the couples represented by **M**$_z$ and **M**$_y$. The couple **M**$_z$ acts in a vertical plane and bends the member in that plane (Fig. 4.51). The resulting stresses are

$$\sigma_x = -\frac{M_z y}{I_z} \tag{4.53}$$

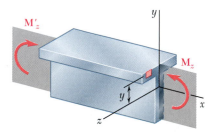

Fig. 4.51 M_z acts in a plane that includes a principal centroidal axis, bending the member in the vertical plane.

where I_z is the moment of inertia of the section about the principal centroidal z axis. The negative sign is due to the compression above the xz plane ($y > 0$) and tension below ($y < 0$). The couple **M**$_y$ acts in a horizontal plane and bends the member in that plane (Fig. 4.52). The resulting stresses are

$$\sigma_x = +\frac{M_y z}{I_y} \tag{4.54}$$

where I_y is the moment of inertia of the section about the principal centroidal y axis, and where the positive sign is due to the fact that we have

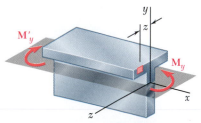

Fig. 4.52 M_y acts in a plane that includes a principal centroidal axis, bending the member in the horizontal plane.

tension to the left of the vertical xy plane ($z > 0$) and compression to its right ($z < 0$). The distribution of the stresses caused by the original couple **M** is obtained by superposing the stress distributions defined by Eqs. (4.53) and (4.54), respectively. We have

$$\sigma_x = -\frac{M_z y}{I_z} + \frac{M_y z}{I_y} \qquad \textbf{(4.55)}$$

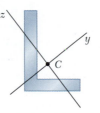

Fig. 4.53 Unsymmetric cross section with principal axes.

Note that the expression obtained can also be used to compute the stresses in an unsymmetric section, as shown in Fig. 4.53, once the principal centroidal y and z axes have been determined. However, Eq. (4.55) is valid only if the conditions of applicability of the principle of superposition are met. It should not be used if the combined stresses exceed the proportional limit of the material or if the deformations caused by one of the couples appreciably affect the distribution of the stresses due to the other.

Equation (4.55) shows that the distribution of stresses caused by unsymmetric bending is linear. However, the neutral axis of the cross section will not, in general, coincide with the axis of the bending couple. Since the normal stress is zero at any point of the neutral axis, the equation defining that axis is obtained by setting $\sigma_x = 0$ in Eq. (4.55).

$$-\frac{M_z y}{I_z} + \frac{M_y z}{I_y} = 0$$

Solving for y and substituting for M_z and M_y from Eqs. (4.52) gives

$$y = \left(\frac{I_z}{I_y} \tan \theta\right) z \qquad \textbf{(4.56)}$$

This equation is for a straight line of slope $m = (I_z/I_y) \tan \theta$. Thus, the angle ϕ that the neutral axis forms with the z axis (Fig. 4.54) is defined by the relation

$$\tan \phi = \frac{I_z}{I_y} \tan \theta \qquad \textbf{(4.57)}$$

where θ is the angle that the couple vector **M** forms with the same axis. Since I_z and I_y are both positive, ϕ and θ have the same sign. Furthermore, $\phi > \theta$ when $I_z > I_y$, and $\phi < \theta$ when $I_z < I_y$. Thus, the neutral axis is always located between the couple vector **M** and the principal axis corresponding to the minimum moment of inertia.

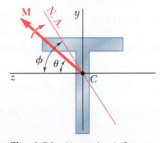

Fig. 4.54 Neutral axis for unsymmetric bending.

Concept Application 4.8

A 1600-lb·in. couple is applied to a wooden beam, of rectangular cross section 1.5 by 3.5 in., in a plane forming an angle of 30° with the vertical (Fig. 4.55a). Determine (a) the maximum stress in the beam and (b) the angle that the neutral surface forms with the horizontal plane.

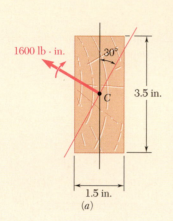
(a)

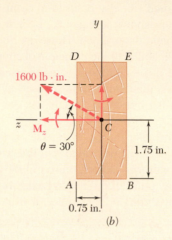
(b)

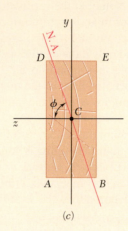
(c)

a. Maximum Stress. The components $\mathbf{M}_z$ and $\mathbf{M}_y$ of the couple vector are first determined (Fig. 4.55b):

$$M_z = (1600 \text{ lb·in.}) \cos 30° = 1386 \text{ lb·in.}$$

$$M_y = (1600 \text{ lb·in.}) \sin 30° = 800 \text{ lb·in.}$$

Compute the moments of inertia of the cross section with respect to the z and y axes:

$$I_z = \tfrac{1}{12}(1.5 \text{ in.})(3.5 \text{ in.})^3 = 5.359 \text{ in}^4$$

$$I_y = \tfrac{1}{12}(3.5 \text{ in.})(1.5 \text{ in.})^3 = 0.9844 \text{ in}^4$$

The largest tensile stress due to $\mathbf{M}_z$ occurs along AB and is

$$\sigma_1 = \frac{M_z y}{I_z} = \frac{(1386 \text{ lb·in.})(1.75 \text{ in.})}{5.359 \text{ in}^4} = 452.6 \text{ psi}$$

The largest tensile stress due to $\mathbf{M}_y$ occurs along AD and is

$$\sigma_2 = \frac{M_y z}{I_y} = \frac{(800 \text{ lb·in.})(0.75 \text{ in.})}{0.9844 \text{ in}^4} = 609.5 \text{ psi}$$

The largest tensile stress due to the combined loading, therefore, occurs at A and is

$$\sigma_{\max} = \sigma_1 + \sigma_2 = 452.6 + 609.5 = 1062 \text{ psi}$$

The largest compressive stress has the same magnitude and occurs at E.

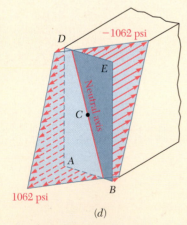

Fig. 4.55 (a) Rectangular wood beam subject to unsymmetric bending. (b) Bending moment resolved into components. (c) Cross section with neutral axis. (d) Stress distribution.

(continued)

b. Angle of Neutral Surface with Horizontal Plane. The angle ϕ that the neutral surface forms with the horizontal plane (Fig. 4.55c) is obtained from Eq. (4.57):

$$\tan \phi = \frac{I_z}{I_y} \tan \theta = \frac{5.359 \text{ in}^4}{0.9844 \text{ in}^4} \tan 30° = 3.143$$

$$\phi = 72.4°$$

The distribution of the stresses across the section is shown in Fig. 4.55d.

4.9 GENERAL CASE OF ECCENTRIC AXIAL LOADING ANALYSIS

In Sec. 4.7 we analyzed the stresses produced in a member by an eccentric axial load applied in a plane of symmetry of the member. We will now study the more general case when the axial load is not applied in a plane of symmetry.

Consider a straight member AB subjected to equal and opposite eccentric axial forces **P** and **P**′ (Fig. 4.56a), and let a and b be the distances from the line of action of the forces to the principal centroidal axes of the cross section of the member. The eccentric force **P** is statically equivalent to the system consisting of a centric force **P** and of the two couples **M**$_y$ and **M**$_z$ of moments $M_y = Pa$ and $M_z = Pb$ in Fig. 4.56b. Similarly, the eccentric force **P**′ is equivalent to the centric force **P**′ and the couples **M**′$_y$ and **M**′$_z$.

By virtue of Saint-Venant's principle (Sec. 2.10), replace the original loading of Fig. 4.56a by the statically equivalent loading of Fig. 4.56b to determine the distribution of stresses in section S of the member (as long as that section is not too close to either end). The stresses due to the loading of Fig. 4.56b can be obtained by superposing the stresses corresponding to the centric axial load **P** and to the bending couples **M**$_y$ and **M**$_z$, as long as the conditions of the principle of superposition are satisfied (Sec. 2.5). The stresses due to the centric load **P** are given by Eq. (1.5), and the stresses due to the bending couples by Eq. (4.55). Therefore,

$$\sigma_x = \frac{P}{A} - \frac{M_z y}{I_z} + \frac{M_y z}{I_y} \quad \text{(4.58)}$$

where y and z are measured from the principal centroidal axes of the section. This relationship shows that the distribution of stresses across the section is *linear*.

In computing the combined stress σ_x from Eq. (4.58), be sure to correctly determine the sign of each of the three terms in the right-hand member, since each can be positive or negative, depending upon the

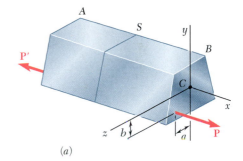

(a)

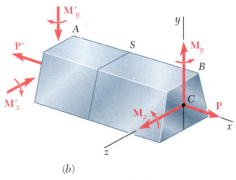

(b)

Fig. 4.56 Eccentric axial loading. (a) Axial force applied away from section centroid. (b) Equivalent force-couple system acting at centroid.

sense of the loads **P** and **P**′ and the location of their line of action with respect to the principal centroidal axes of the cross section. The combined stresses σ_x obtained from Eq. (4.58) at various points of the section may all have the same sign, or some may be positive and others negative. In the latter case, there will be a line in the section along which the stresses are zero. Setting $\sigma_x = 0$ in Eq. (4.58), the equation of a straight line representing the *neutral axis* of the section is

$$\frac{M_z}{I_z}y - \frac{M_y}{I_y}z = \frac{P}{A}$$

Concept Application 4.9

A vertical 4.80-kN load is applied as shown on a wooden post of rectangular cross section, 80 by 120 mm (Fig. 4.57*a*). (*a*) Determine the stress at points *A*, *B*, *C*, and *D*. (*b*) Locate the neutral axis of the cross section.

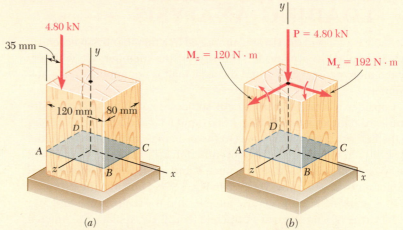

Fig. 4.57 (*a*) Eccentric load on a rectangular wood column. (*b*) Equivalent force-couple system for eccentric load.

a. Stresses. The given eccentric load is replaced by an equivalent system consisting of a centric load **P** and two couples **M**$_x$ and **M**$_z$ represented by vectors directed along the principal centroidal axes of the section (Fig. 4.57*b*). Thus

$$M_x = (4.80\ \text{kN})(40\ \text{mm}) = 192\ \text{N·m}$$

$$M_z = (4.80\ \text{kN})(60\ \text{mm} - 35\ \text{mm}) = 120\ \text{N·m}$$

Compute the area and the centroidal moments of inertia of the cross section:

$$A = (0.080\ \text{m})(0.120\ \text{m}) = 9.60 \times 10^{-3}\ \text{m}^2$$

$$I_x = \tfrac{1}{12}(0.120\ \text{m})(0.080\ \text{m})^3 = 5.12 \times 10^{-6}\ \text{m}^4$$

$$I_z = \tfrac{1}{12}(0.080\ \text{m})(0.120\ \text{m})^3 = 11.52 \times 10^{-6}\ \text{m}^4$$

(continued)

The stress σ_0 due to the centric load **P** is negative and uniform across the section:

$$\sigma_0 = \frac{P}{A} = \frac{-4.80 \text{ kN}}{9.60 \times 10^{-3} \text{ m}^2} = -0.5 \text{ MPa}$$

The stresses due to the bending couples $\mathbf{M}_x$ and $\mathbf{M}_z$ are linearly distributed across the section with maximum values equal to

$$\sigma_1 = \frac{M_x z_{\max}}{I_x} = \frac{(192 \text{ N·m})(40 \text{ mm})}{5.12 \times 10^{-6} \text{ m}^4} = 1.5 \text{ MPa}$$

$$\sigma_2 = \frac{M_z x_{\max}}{I_z} = \frac{(120 \text{ N·m})(60 \text{ mm})}{11.52 \times 10^{-6} \text{ m}^4} = 0.625 \text{ MPa}$$

The stresses at the corners of the section are

$$\sigma_y = \sigma_0 \pm \sigma_1 \pm \sigma_2$$

where the signs must be determined from Fig. 4.57*b*. Noting that the stresses due to $\mathbf{M}_x$ are positive at *C* and *D* and negative at *A* and *B*, and the stresses due to $\mathbf{M}_z$ are positive at *B* and *C* and negative at *A* and *D*, we obtain

$$\sigma_A = -0.5 - 1.5 - 0.625 = -2.625 \text{ MPa}$$

$$\sigma_B = -0.5 - 1.5 + 0.625 = -1.375 \text{ MPa}$$

$$\sigma_C = -0.5 + 1.5 + 0.625 = +1.625 \text{ MPa}$$

$$\sigma_D = -0.5 + 1.5 - 0.625 = +0.375 \text{ MPa}$$

b. Neutral Axis.   The stress will be zero at a point *G* between *B* and *C*, and at a point *H* between *D* and *A* (Fig. 4.57*c*). Since the stress distribution is linear,

$$\frac{BG}{80 \text{ mm}} = \frac{1.375}{1.625 + 1.375} \qquad BG = 36.7 \text{ mm}$$

$$\frac{HA}{80 \text{ mm}} = \frac{2.625}{2.625 + 0.375} \qquad HA = 70 \text{ mm}$$

The neutral axis can be drawn through points *G* and *H* (Fig. 4.57*d*). The distribution of the stresses across the section is shown in Fig. 4.57*e*.

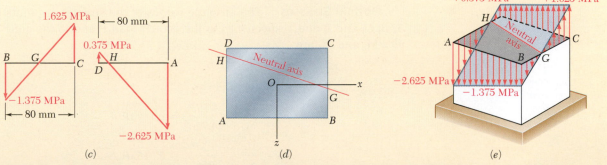

Fig. 4.57   (*cont.*) (*c*) Stress distributions along edges *BC* and *AD*. (*d*) Neutral axis is line through points *G* and *H*. (*e*) Stress distribution for eccentric load.

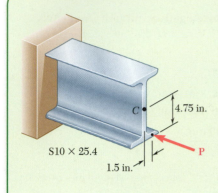

S10 × 25.4

4.75 in.

1.5 in.

C

P

Sample Problem 4.9

A horizontal load **P** is applied as shown to a short section of an S10 × 25.4 rolled-steel member. Knowing that the compressive stress in the member is not to exceed 12 ksi, determine the largest permissible load **P**.

STRATEGY: The load is applied eccentrically with respect to both centroidal axes of the cross section. The load is replaced with an equivalent force-couple system at the centroid of the cross section. The stresses due to the axial load and the two couples are then superposed to determine the maximum stresses on the cross section.

MODELING and ANALYSIS:

Properties of Cross Section. The cross section is shown in Fig. 1, and the following data are taken from Appendix C.

> *Area:* $A = 7.46$ in^2
> *Section moduli:* $S_x = 24.7$ in^3 $S_y = 2.91$ in^3

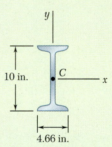

10 in.

4.66 in.

C

y

x

Fig. 1 Rolled-steel member

Force and Couple at C. Using Fig. 2, we replace **P** by an equivalent force-couple system at the centroid C of the cross section.

$$M_x = (4.75 \text{ in.})P \qquad M_y = (1.5 \text{ in.})P$$

Note that the couple vectors $\mathbf{M}_x$ and $\mathbf{M}_y$ are directed along the principal axes of the cross section.

Normal Stresses. The absolute values of the stresses at points A, B, D, and E due, respectively, to the centric load **P** and to the couples $\mathbf{M}_x$ and $\mathbf{M}_y$ are

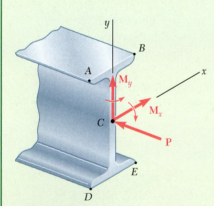

A

B

M_y

M_x

C

P

E

D

y

x

Fig. 2 Equivalent force-couple system at section centroid.

$$\sigma_1 = \frac{P}{A} = \frac{P}{7.46 \text{ in}^2} = 0.1340P$$

$$\sigma_2 = \frac{M_x}{S_x} = \frac{4.75P}{24.7 \text{ in}^3} = 0.1923P$$

$$\sigma_3 = \frac{M_y}{S_y} = \frac{1.5P}{2.91 \text{ in}^3} = 0.5155P$$

(continued)

Superposition. The total stress at each point is found by superposing the stresses due to **P**, $\mathbf{M}_x$, and $\mathbf{M}_y$. We determine the sign of each stress by carefully examining the sketch of the force-couple system.

$$\sigma_A = -\sigma_1 + \sigma_2 + \sigma_3 = -0.1340P + 0.1923P + 0.5155P = +0.574P$$

$$\sigma_B = -\sigma_1 + \sigma_2 - \sigma_3 = -0.1340P + 0.1923P - 0.5155P = -0.457P$$

$$\sigma_D = -\sigma_1 - \sigma_2 + \sigma_3 = -0.1340P - 0.1923P + 0.5155P = +0.189P$$

$$\sigma_E = -\sigma_1 - \sigma_2 - \sigma_3 = -0.1340P - 0.1923P - 0.5155P = -0.842P$$

Largest Permissible Load. The maximum compressive stress occurs at point E. Recalling that $\sigma_{\text{all}} = -12$ ksi, we write

$$\sigma_{\text{all}} = \sigma_E \qquad -12 \text{ ksi} = -0.842P \qquad P = 14.3 \text{ kips} \quad \blacktriangleleft$$

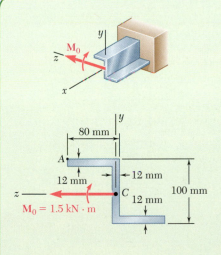

*Sample Problem 4.10

A couple of magnitude $M_0 = 1.5$ kN·m acting in a vertical plane is applied to a beam having the Z-shaped cross section shown. Determine (a) the stress at point A and (b) the angle that the neutral axis forms with the horizontal plane. The moments and product of inertia of the section with respect to the y and z axes have been computed and are

$$I_y = 3.25 \times 10^{-6} \, \text{m}^4$$

$$I_z = 4.18 \times 10^{-6} \, \text{m}^4$$

$$I_{yz} = 2.87 \times 10^{-6} \, \text{m}^4$$

STRATEGY: The Z-shaped cross section does not have an axis of symmetry, so it is first necessary to determine the orientation of the principal axes and the corresponding moments of inertia. The applied load is then resolved into components along the principal axes. The stresses due to the axial load and the two couples are then superposed to determine the stress at point A. The angle between the neutral axis and horizontal plane is then found using Eq. (4.57).

(continued)

MODELING and ANALYSIS:

Principal Axes. We draw Mohr's circle and determine the orientation of the principal axes and the corresponding principal moments of inertia. (Fig. 1)[†]

$$\tan 2\theta_m = \frac{FZ}{EF} = \frac{2.87}{0.465} \qquad 2\theta_m = 80.8° \qquad \theta_m = 40.4°$$

$$R^2 = (EF)^2 + (FZ)^2 = (0.465)^2 + (2.87)^2 \quad R = 2.91 \times 10^{-6}\,\text{m}^4$$

$$I_u = I_{\min} = OU = I_{\text{ave}} - R = 3.72 - 2.91 = 0.810 \times 10^{-6}\,\text{m}^4$$

$$I_v = I_{\max} = OV = I_{\text{ave}} + R = 3.72 + 2.91 = 6.63 \times 10^{-6}\,\text{m}^4$$

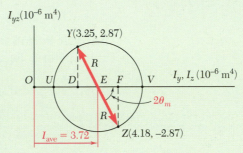

Fig. 1 Mohr's circle analysis.

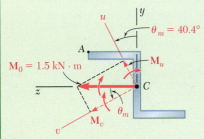

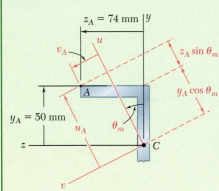

Fig. 2 Bending moment resolved along principal axes.

Loading. As shown in Fig. 2, the applied couple $\mathbf{M}_0$ is resolved into components parallel to the principal axes.

$$M_u = M_0 \sin\theta_m = 1500 \sin 40.4° = 972\,\text{N·m}$$

$$M_v = M_0 \cos\theta_m = 1500 \cos 40.4° = 1142\,\text{N·m}$$

a. Stress at A. The perpendicular distances from each principal axis to point A shown in Fig. 3 and are

$$u_A = y_A \cos\theta_m + z_A \sin\theta_m = 50 \cos 40.4° + 74 \sin 40.4° = 86.0\,\text{mm}$$

$$v_A = -y_A \sin\theta_m + z_A \cos\theta_m = -50 \sin 40.4° + 74 \cos 40.4° = 23.9\,\text{mm}$$

Considering separately the bending about each principal axis, note that $\mathbf{M}_u$ produces a tensile stress at point A while $\mathbf{M}_v$ produces a compressive stress at the same point.

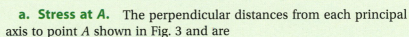

$$\sigma_A = +\frac{M_u v_A}{I_u} - \frac{M_v u_A}{I_v} = +\frac{(972\,\text{N·m})(0.0239\,\text{m})}{0.810 \times 10^{-6}\,\text{m}^4} - \frac{(1142\,\text{N·m})(0.0860\,\text{m})}{6.63 \times 10^{-6}\,\text{m}^4}$$

$$= +(28.68\,\text{MPa}) - (14.81\,\text{MPa}) \qquad \sigma_A = +13.87\,\text{MPa} \ \blacktriangleleft$$

Fig. 3 Location of A relative to principal axis.

b. Neutral Axis. As shown in Fig. 4, we find the angle ϕ that the neutral axis forms with the v axis.

$$\tan\phi = \frac{I_v}{I_u} \tan\theta_m = \frac{6.63}{0.810} \tan 40.4° \qquad \phi = 81.8°$$

The angle β formed by the neutral axis and the horizontal is

$$\beta = \phi - \theta_m = 81.8° - 40.4° = 41.4° \qquad \beta = 41.4° \ \blacktriangleleft$$

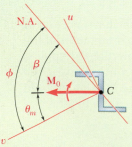

Fig. 4 Cross section with neutral axis.

[†]See Ferdinand P. Beer and E. Russell Johnston, Jr., *Mechanics for Engineers*, 5th ed., McGraw-Hill, New York, 2008, or *Vector Mechanics for Engineers*–10th ed., McGraw-Hill, New York, 2013, Secs. 9.8–9.10.

Problems

4.127 through 4.134 The couple **M** is applied to a beam of the cross section shown in a plane forming an angle β with the vertical. Determine the stress at (*a*) point *A*, (*b*) point *B*, (*c*) point *D*.

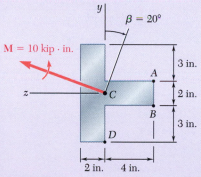

Fig. P4.127

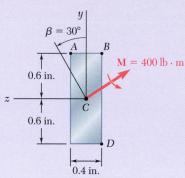

Fig. P4.128

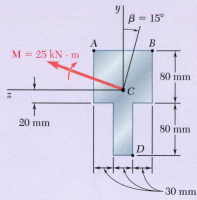

Fig. P4.129

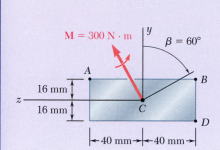

Fig. P4.130

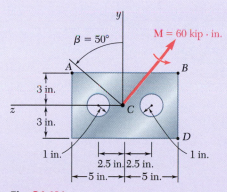

Fig. *P4.131*

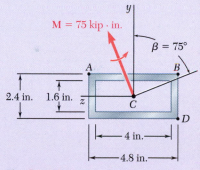

Fig. *P4.132*

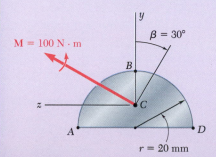

Fig. P4.133

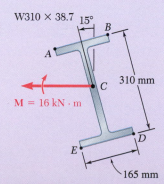

Fig. P4.134

4.135 through 4.140 The couple **M** acts in a vertical plane and is applied to a beam oriented as shown. Determine (*a*) the angle that the neutral axis forms with the horizontal, (*b*) the maximum tensile stress in the beam.

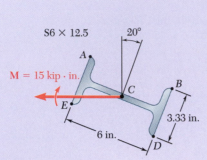

S6 × 12.5

M = 15 kip · in.

20°

3.33 in.

6 in.

Fig. P4.135

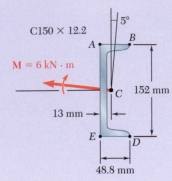

C150 × 12.2

5°

M = 6 kN · m

152 mm

13 mm

48.8 mm

Fig. *P4.136*

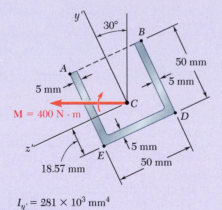

y'

30°

50 mm

5 mm

5 mm

M = 400 N · m

z'

5 mm

50 mm

18.57 mm

$I_{y'} = 281 \times 10^3 \text{ mm}^4$
$I_{z'} = 176.9 \times 10^3 \text{ mm}^4$

Fig. P4.137

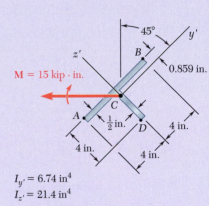

45°

y'

z'

0.859 in.

M = 15 kip · in.

$\frac{1}{2}$ in.

4 in.

4 in.

4 in.

$I_{y'} = 6.74 \text{ in}^4$
$I_{z'} = 21.4 \text{ in}^4$

Fig. P4.138

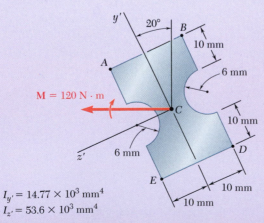

y'

20°

10 mm

6 mm

M = 120 N · m

10 mm

z'

6 mm

10 mm

10 mm

$I_{y'} = 14.77 \times 10^3 \text{ mm}^4$
$I_{z'} = 53.6 \times 10^3 \text{ mm}^4$

Fig. P4.139

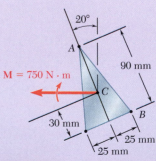

20°

90 mm

M = 750 N · m

30 mm

25 mm

25 mm

Fig. *P4.140*

*4.141 through *4.143 The couple **M** acts in a vertical plane and is applied to a beam oriented as shown. Determine the stress at point *A*.

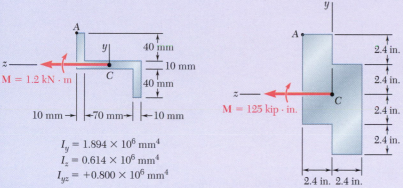

$I_y = 1.894 \times 10^6 \text{ mm}^4$
$I_z = 0.614 \times 10^6 \text{ mm}^4$
$I_{yz} = +0.800 \times 10^6 \text{ mm}^4$

Fig. P4.141

$I_y = 8.7 \text{ in}^4$
$I_z = 24.5 \text{ in}^4$
$I_{yz} = +8.3 \text{ in}^4$

Fig. P4.142

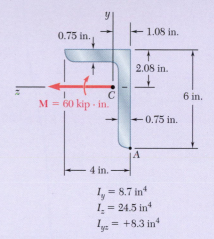

Fig. P4.143

4.144 The tube shown has a uniform wall thickness of 12 mm. For the loading given, determine (*a*) the stress at points *A* and *B*, (*b*) the point where the neutral axis intersects line *ABD*.

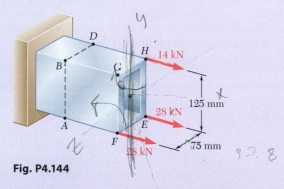

Fig. P4.144

4.145 A horizontal load **P** of magnitude 100 kN is applied to the beam shown. Determine the largest distance *a* for which the maximum tensile stress in the beam does not exceed 75 MPa.

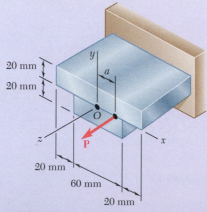

Fig. P4.145

315

4.146 Knowing that $P = 90$ kips, determine the largest distance a for which the maximum compressive stress does not exceed 18 ksi.

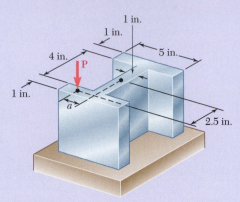

Fig. P4.146 and P4.147

4.147 Knowing that $a = 1.25$ in., determine the largest value of $\mathbf{P}$ that can be applied without exceeding either of the following allowable stresses:

$$\sigma_{ten} = 10 \text{ ksi} \qquad \sigma_{comp} = 18 \text{ ksi}$$

4.148 A rigid circular plate of 125-mm radius is attached to a solid 150×200-mm rectangular post, with the center of the plate directly above the center of the post. If a 4-kN force $\mathbf{P}$ is applied at E with $\theta = 30°$, determine (a) the stress at point A, (b) the stress at point B, (c) the point where the neutral axis intersects line ABD.

4.149 In Prob. 4.148, determine (a) the value of θ for which the stress at D reaches its largest value, (b) the corresponding values of the stress at A, B, C, and D.

4.150 A beam having the cross section shown is subjected to a couple $\mathbf{M}_0$ that acts in a vertical plane. Determine the largest permissible value of the moment M_0 of the couple if the maximum stress in the beam is not to exceed 12 ksi. *Given*: $I_y = I_z = 11.3$ in.^{4}, $A = 4.75$ in.^{2}, $k_{min} = 0.983$ in. (*Hint*: By reason of symmetry, the principal axes form an angle of 45° with the coordinate axes. Use the relations $I_{min} = Ak^2_{min}$ and $I_{min} + I_{max} = I_y + I_z$.)

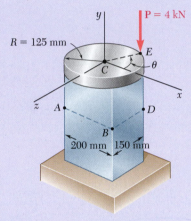

Fig. P4.148

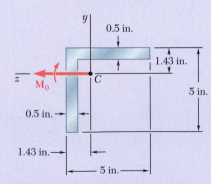

Fig. P4.150

4.151 Solve Prob. 4.150, assuming that the couple $\mathbf{M}_0$ acts in a horizontal plane.

4.152 The Z section shown is subjected to a couple M_0 acting in a vertical plane. Determine the largest permissible value of the moment M_0 of the couple if the maximum stress is not to exceed 80 MPa. *Given:* $I_{max} = 2.28 \times 10^{-6}$ m^4, $I_{min} = 0.23 \times 10^{-6}$ m^4, principal axes 25.7° ⬎ and 64.3° ⬏.

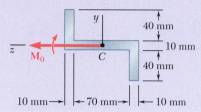

Fig. P4.152

4.153 Solve Prob. 4.152 assuming that the couple M_0 acts in a horizontal plane.

4.154 An extruded aluminum member having the cross section shown is subjected to a couple acting in a vertical plane. Determine the largest permissible value of the moment M_0 of the couple if the maximum stress is not to exceed 12 ksi. *Given:* $I_{max} = 0.957$ in^4, $I_{min} = 0.427$ in^4, principal axes 29.4° ⬏ and 60.6° ⬎.

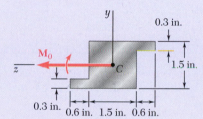

Fig. P4.154

4.155 A beam having the cross section shown is subjected to a couple M_0 acting in a vertical plane. Determine the largest permissible value of the moment M_0 of the couple if the maximum stress is not to exceed 100 MPa. *Given:* $I_y = I_z = b^4/36$ and $I_{yz} = b^4/72$.

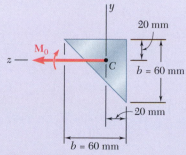

Fig. P4.155

4.156 Show that, if a solid rectangular beam is bent by a couple applied in a plane containing one diagonal of a rectangular cross section, the neutral axis will lie along the other diagonal.

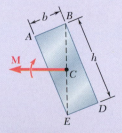

Fig. P4.156

4.157 (*a*) Show that the stress at corner *A* of the prismatic member shown in Fig. *a* will be zero if the vertical force **P** is applied at a point located on the line

$$\frac{x}{b/6} + \frac{z}{h/6} = 1$$

(*b*) Further show that, if no tensile stress is to occur in the member, the force **P** must be applied at a point located within the area bounded by the line found in part *a* and three similar lines corresponding to the condition of zero stress at *B*, *C*, and *D*, respectively. This area, shown in Fig. *b*, is known as the *kern* of the cross section.

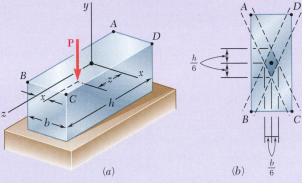

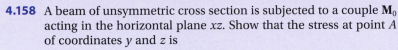

(*a*)   (*b*)

Fig. P4.157

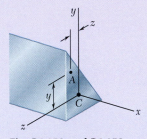

Fig. P4.158 and P4.159

4.158 A beam of unsymmetric cross section is subjected to a couple **M**$_0$ acting in the horizontal plane *xz*. Show that the stress at point *A* of coordinates *y* and *z* is

$$\sigma_A = \frac{zI_z - yI_{yz}}{I_yI_z - I_{yz}^2}M_y$$

where I_y, I_z, and I_{yz} denote the moments and product of inertia of the cross section with respect to the coordinate axes, and M_y the moment of the couple.

4.159 A beam of unsymmetric cross section is subjected to a couple **M**$_0$ acting in the vertical plane *xy*. Show that the stress at point *A* of coordinates *y* and *z* is

$$\sigma_A = -\frac{yI_y - zI_{yz}}{I_yI_z - I_{yz}^2}M_z$$

where I_y, I_z, and I_{yz} denote the moments and product of inertia of the cross section with respect to the coordinate axes, and M_z the moment of the couple.

4.160 (*a*) Show that, if a vertical force **P** is applied at point *A* of the section shown, the equation of the neutral axis *BD* is

$$\left(\frac{x_A}{r_z^2}\right)x + \left(\frac{z_A}{r_x^2}\right)z = -1$$

where r_z and r_x denote the radius of gyration of the cross section with respect to the *z* axis and the *x* axis, respectively. (*b*) Further show that, if a vertical force **Q** is applied at any point located on line *BD*, the stress at point *A* will be zero.

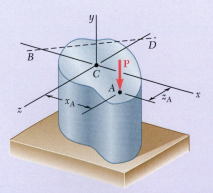

Fig. P4.160

*4.10 CURVED MEMBERS

Our analysis of stresses due to bending has been restricted so far to straight members. In this section, the stresses are caused by the application of equal and opposite couples to members that are initially curved. Our discussion is limited to curved members with uniform cross sections possessing a plane of symmetry in which the bending couples are applied. It is assumed that all stresses remain below the proportional limit.

If the initial curvature of the member is small (i.e., the radius of curvature is large compared to the depth of its cross section) an approximation can be obtained for the distribution of stresses by assuming the member to be straight and using the formulas derived in Secs. 4.1B and 4.2.[†]

However, when the radius of curvature and the dimensions of the cross section of the member are of the same order of magnitude, it is necessary to use a different method of analysis, which was first introduced by the German engineer E. Winkler (1835–1888).

Consider the curved member of uniform cross section shown in Fig. 4.58. Its transverse section is symmetric with respect to the y axis (Fig. 4.58b) and, in its unstressed state, its upper and lower surfaces intersect the vertical xy plane along arcs of circle AB and FG centered

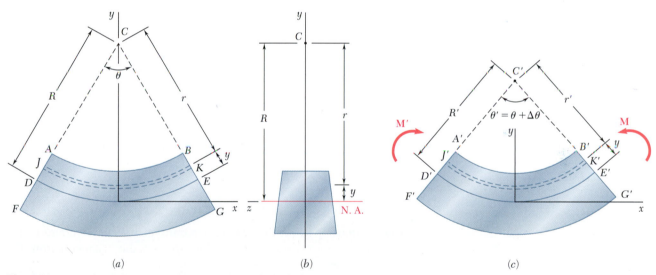

        (a)   (b)   (c)

Fig. 4.58 Curved member in pure bending: (a) undeformed, (b) cross section, and (c) deformed.

at C (Fig. 4.58a). Now apply two equal and opposite couples **M** and **M**′ in the plane of symmetry of the member (Fig. 4.58c). A reasoning similar to that of Sec. 4.1B would show that any transverse plane section containing C remains plane, and the various arcs of circle indicated in Fig. 4.58a are transformed into circular and concentric arcs with a center C' different from C. If the couples **M** and **M**′ are directed as shown, the curvature of the various arcs of circle increases; that is $A'C' < AC$. Also, the couples **M** and **M**′ cause the length of the upper surface of the member to decrease ($A'B' < AB$) and the length of the lower

---

[†]See Prob. 4.166.

surface to increase ($F'G' > FG$). Therefore, we conclude that a *neutral surface* must exist in the member, the length of which remains constant. The intersection of the neutral surface with the xy plane is shown in Fig. 4.58a by the arc DE of radius R, and in Fig. 4.58c by the arc $D'E'$ of radius R'. The central angles θ and θ' corresponding respectively to DE and $D'E'$ express the fact that the length of the neutral surface remains constant by

$$R\theta = R'\theta' \tag{4.59}$$

Considering the arc of circle JK located at a distance y above the neutral surface and denoting respectively by r and r' the radius of this arc before and after the bending couples have been applied, the deformation of JK is

$$\delta = r'\theta' - r\theta \tag{4.60}$$

Observing from Fig. 4.58 that

$$r = R - y \qquad r' = R' - y \tag{4.61}$$

and substituting these expressions into Eq. (4.60),

$$\delta = (R' - y)\theta' - (R - y)\theta$$

or recalling Eq. (4.59) and setting $\theta' - \theta = \Delta\theta$,

$$\delta = -y\,\Delta\theta \tag{4.62}$$

The normal strain ϵ_x in the elements of JK is obtained by dividing the deformation δ by the original length $r\theta$ of arc JK:

$$\epsilon_x = \frac{\delta}{r\theta} = -\frac{y\,\Delta\theta}{r\theta}$$

Recalling the first of the relationships in Eq. (4.61),

$$\epsilon_x = -\frac{\Delta\theta}{\theta}\frac{y}{R - y} \tag{4.63}$$

This relationship shows that, while each transverse section remains plane, the normal strain ϵ_x *does not vary linearly* with the distance y from the neutral surface.

The normal stress σ_x can be obtained from Hooke's law, $\sigma_x = E\epsilon_x$, by substituting for ϵ_x from Eq. (4.63):

$$\sigma_x = -\frac{E\,\Delta\theta}{\theta}\frac{y}{R - y} \tag{4.64}$$

or alternatively, recalling the first of Eqs. (4.61),

$$\sigma_x = -\frac{E\,\Delta\theta}{\theta}\frac{R - r}{r} \tag{4.65}$$

Equation (4.64) shows that, like ϵ_x, the normal stress σ_x *does not vary linearly* with the distance y from the neutral surface. Plotting σ_x versus y, an arc of hyperbola is obtained (Fig. 4.59).

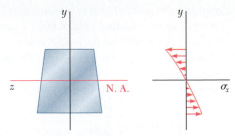

Fig. 4.59   Stress distribution in curved beam.

In order to determine the location of the neutral surface in the member and the value of the coefficient $E\,\Delta\theta/\theta$ used in Eqs. (4.64) and (4.65), we recall that the elementary forces acting on any transverse section must be statically equivalent to the bending couple **M**. Expressing that the sum of the elementary forces acting on the section must be zero and that the sum of their moments about the transverse z axis must be equal to the bending moment M, write the equations

$$\int \sigma_x \, dA = 0 \qquad\qquad\text{(4.1)}$$

and

$$\int (-y\sigma_x \, dA) = M \qquad\qquad\text{(4.3)}$$

Substituting for σ_x from Eq. (4.65) into Eq. (4.1), write

$$-\int \frac{E\,\Delta\theta}{\theta} \frac{R - r}{r} dA = 0$$

$$\int \frac{R - r}{r} dA = 0$$

$$R \int \frac{dA}{r} - \int dA = 0$$

from which it follows that the distance R from the center of curvature C to the neutral surface is defined by

$$R = \frac{A}{\displaystyle\int \frac{dA}{r}} \qquad\qquad\text{(4.66)}$$

Note that the value obtained for R is not equal to the distance $\bar{r}$ from C to the centroid of the cross section, since $\bar{r}$ is defined by a different relationship, namely,

$$\bar{r} = \frac{1}{A} \int r \, dA \qquad\qquad\text{(4.67)}$$

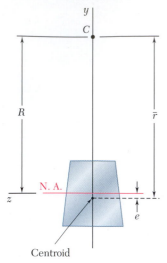

Fig. 4.60 Parameter *e* locates neutral axis relative to the centroid of a curved member section.

Thus, in a curved member *the neutral axis of a transverse section does not pass through the centroid of that section* (Fig. 4.60).[†] Expressions for the radius R of the neutral surface will be derived for some specific cross-sectional shapes in Concept Application 4.10 and in Probs. 4.187 through 4.189. These expressions are shown in Fig. 4.61.

Substituting now for σ_x from (4.65) into Eq. (4.3), write

$$\int \frac{E\,\Delta\theta}{\theta}\frac{R-r}{r}y\,dA = M$$

or since $y = R - r$,

$$\frac{E\,\Delta\theta}{\theta}\int \frac{(R-r)^2}{r}dA = M$$

Expanding the square in the integrand, we obtain after reductions

$$\frac{E\,\Delta\theta}{\theta}\left[R^2\int \frac{dA}{r} - 2RA + \int r\,dA\right] = M$$

Recalling Eqs. (4.66) and (4.67), we note that the first term in the brackets is equal to RA, while the last term is equal to $\bar{r}A$. Therefore,

$$\frac{E\,\Delta\theta}{\theta}(RA - 2RA + \bar{r}A) = M$$

and solving for $E\,\Delta\theta/\theta$,

$$\frac{E\,\Delta\theta}{\theta} = \frac{M}{A(\bar{r} - R)} \tag{4.68}$$

Referring to Fig. 4.58, $\Delta\theta > 0$ for $M > 0$. It follows that $\bar{r} - R > 0$, or $R < \bar{r}$, regardless of the shape of the section. Thus, the neutral axis of a transverse section is always located between the centroid of the section and the center of curvature of the member (Fig. 4.60). Setting $\bar{r} - R = e$, Eq. (4.68) is written in the form

$$\frac{E\,\Delta\theta}{\theta} = \frac{M}{Ae} \tag{4.69}$$

[†]However, an interesting property of the neutral surface is noted if Eq. (4.66) is written in the alternative form

$$\frac{1}{R} = \frac{1}{A}\int \frac{1}{r}dA \tag{4.66a}$$

Equation (4.66a) shows that, if the member is divided into a large number of fibers of cross-sectional area dA, the curvature $1/R$ of the neutral surface is equal to the average value of the curvature $1/r$ of the various fibers.

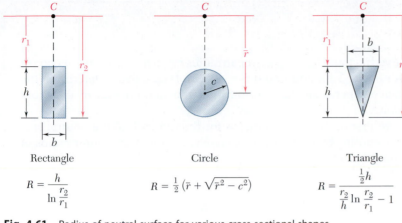
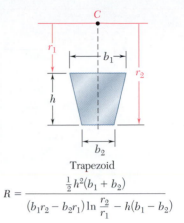

Rectangle	Circle	Triangle	Trapezoid
$R = \dfrac{h}{\ln \dfrac{r_2}{r_1}}$	$R = \tfrac{1}{2}\left(\bar{r} + \sqrt{\bar{r}^2 - c^2}\right)$	$R = \dfrac{\tfrac{1}{2}h}{\dfrac{r_2}{h}\ln \dfrac{r_2}{r_1} - 1}$	$R = \dfrac{\tfrac{1}{2}h^2(b_1 + b_2)}{(b_1 r_2 - b_2 r_1)\ln \dfrac{r_2}{r_1} - h(b_1 - b_2)}$

Fig. 4.61 Radius of neutral surface for various cross-sectional shapes.

Substituting for $E\,\Delta\theta/\theta$ from Eq. (4.69) into Eqs. (4.64) and (4.65), the alternative expressions for the normal stress σ_x in a curved beam are

$$\sigma_x = -\frac{My}{Ae(R - y)} \tag{4.70}$$

and

$$\sigma_x = \frac{M(r - R)}{Aer} \tag{4.71}$$

Note that the parameter e in the previous equations is a small quantity obtained by subtracting two lengths of comparable size, R and $\bar{r}$. In order to determine σ_x with a reasonable degree of accuracy, it is necessary to compute R and $\bar{r}$ very accurately, particularly when both of these quantities are large (i.e., when the curvature of the member is small). However, it is possible in such a case to obtain a good approximation for σ_x by using the formula $\sigma_x = -My/I$ developed for straight members.

We will now determine the change in curvature of the neutral surface caused by the bending moment M. Solving Eq. (4.59) for the curvature $1/R'$ of the neutral surface in the deformed member,

$$\frac{1}{R'} = \frac{1}{R}\frac{\theta'}{\theta}$$

or setting $\theta' = \theta + \Delta\theta$ and recalling Eq. (4.69),

$$\frac{1}{R'} = \frac{1}{R}\left(1 + \frac{\Delta\theta}{\theta}\right) = \frac{1}{R}\left(1 + \frac{M}{EAe}\right)$$

the change in curvature of the neutral surface is

$$\frac{1}{R'} - \frac{1}{R} = \frac{M}{EAeR} \tag{4.72}$$

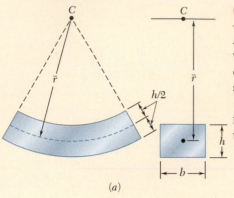

(a)

Concept Application 4.10

A curved rectangular bar has a mean radius $\bar{r} = 6$ in. and a cross section of width $b = 2.5$ in. and depth $h = 1.5$ in. (Fig. 4.62a). Determine the distance e between the centroid and the neutral axis of the cross section.

We first derive the expression for the radius R of the neutral surface. Denoting by r_1 and r_2, respectively, the inner and outer radius of the bar (Fig. 4.62b), use Eq. (4.66) to write

$$R = \frac{A}{\int_{r_1}^{r_2} \dfrac{dA}{r}} = \frac{bh}{\int_{r_1}^{r_2} \dfrac{b\,dr}{r}} = \frac{h}{\int_{r_1}^{r_2} \dfrac{dr}{r}}$$

$$R = \frac{h}{\ln\dfrac{r_2}{r_1}} \tag{4.73}$$

For the given data,

$$r_1 = \bar{r} - \tfrac{1}{2}h = 6 - 0.75 = 5.25 \text{ in.}$$

$$r_2 = \bar{r} + \tfrac{1}{2}h = 6 + 0.75 = 6.75 \text{ in.}$$

Substituting for h, r_1, and r_2 into Eq. (4.73),

$$R = \frac{h}{\ln\dfrac{r_2}{r_1}} = \frac{1.5 \text{ in.}}{\ln\dfrac{6.75}{5.25}} = 5.9686 \text{ in.}$$

The distance between the centroid and the neutral axis of the cross section (Fig. 4.62c) is thus

$$e = \bar{r} - R = 6 - 5.9686 = 0.0314 \text{ in.}$$

Note that it was necessary to calculate R with five significant figures in order to obtain e with the usual degree of accuracy.

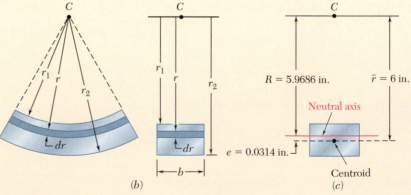

Fig. 4.62 (a) Curved rectangular bar. (b) Dimensions for curved bar. (c) Location of the neutral axis.

Concept Application 4.11

For the bar of Concept Application 4.10, determine the largest tensile and compressive stresses, knowing that the bending moment in the bar is $M = 8$ kip·in.

Use Eq. (4.71) with the given data

$$M = 8 \text{ kip·in.} \qquad A = bh = (2.5 \text{ in.})(1.5 \text{ in.}) = 3.75 \text{ in}^2$$

and the values obtained in Concept Application 4.10 for R and e:

$$R = 5.969 \qquad e = 0.0314 \text{ in.}$$

First using $r = r_2 = 6.75$ in. in Eq. (4.71), write

$$\sigma_{max} = \frac{M(r_2 - R)}{Aer_2}$$

$$= \frac{(8 \text{ kip·in.})(6.75 \text{ in.} - 5.969 \text{ in.})}{(3.75 \text{ in}^2)(0.0314 \text{ in.})(6.75 \text{ in.})}$$

$$\sigma_{max} = 7.86 \text{ ksi}$$

Now using $r = r_1 = 5.25$ in. in Eq. (4.71),

$$\sigma_{min} = \frac{M(r_1 - R)}{Aer_1}$$

$$= \frac{(8 \text{ kip·in.})(5.25 \text{ in.} - 5.969 \text{ in.})}{(3.75 \text{ in}^2)(0.0314 \text{ in.})(5.25 \text{ in.})}$$

$$\sigma_{min} = -9.30 \text{ ksi}$$

Remark. Compare the values obtained for σ_{max} and σ_{min} with the result for a straight bar. Using Eq. (4.15) of Sec. 4.2,

$$\sigma_{max, min} = \pm \frac{Mc}{I}$$

$$= \pm \frac{(8 \text{ kip·in.})(0.75 \text{ in.})}{\frac{1}{12}(2.5 \text{ in.})(1.5 \text{ in.})^3} = \pm 8.53 \text{ ksi}$$

Sample Problem 4.11

A machine component has a T-shaped cross section and is loaded as shown. Knowing that the allowable compressive stress is 50 MPa, determine the largest force **P** that can be applied to the component.

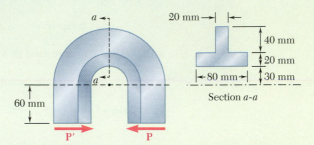

Section *a-a*

STRATEGY: The properties are first determined for the singly-symmetric cross section. The force and couple at the critical section are used to calculate the maximum compressive stress, which is obtained by superposing the axial stress and the bending stress determined from Eqs. (4.66) and (4.71). This stress is then equated to the allowable compressive stress to determine the force **P**.

MODELING and ANALYSIS:

Centroid of the Cross Section.   Locate the centroid *D* of the cross section (Fig. 1)

	A_i, mm^2	$\bar{r}_i$, mm	$\bar{r}_i A_i$, mm^3	
1	$(20)(80) = 1600$	40	64×10^3	$\bar{r}\Sigma A_i = \Sigma \bar{r}_i A_i$
2	$(40)(20) = 800$	70	56×10^3	$\bar{r}(2400) = 120 \times 10^3$
	$\Sigma A_i = 2400$		$\Sigma \bar{r}_i A_i = 120 \times 10^3$	$\bar{r} = 50\text{ mm} = 0.050\text{ m}$

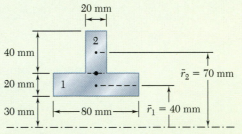

Fig. 1   Composite areas to calculate centroid location.

Force and Couple at D.   The internal forces in section *a–a* are equivalent to a force **P** acting at *D* and a couple **M** of moment (Fig. 2)

$$M = P(50\text{ mm} + 60\text{ mm}) = (0.110\text{ m})P$$

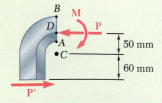

Fig. 2   Free-body diagram of left side.

(continued)

Superposition. The centric force **P** causes a uniform compressive stress on section *a–a*, shown in Fig. 3a. The bending couple **M** causes a varying stress distribution [Eq. (4.71)], shown in Fig. 3b. We note that the couple **M** tends to increase the curvature of the member and is therefore positive (see Fig. 4.58). The total stress at a point of section *a–a* located at distance *r* from the center of curvature *C* is

$$\sigma = -\frac{P}{A} + \frac{M(r - R)}{Aer} \qquad (1)$$

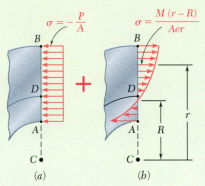

Figs. 3 Stress distribution is the superposition of (a) axial stress and (b) bending stress.

Radius of Neutral Surface. Using Fig. 4, we now determine the radius *R* of the neutral surface by using Eq. (4.66).

$$R = \frac{A}{\displaystyle\int \frac{dA}{r}} = \frac{2400 \text{ mm}^2}{\displaystyle\int_{r_1}^{r_2} \frac{(80 \text{ mm}) \, dr}{r} + \int_{r_2}^{r_3} \frac{(20 \text{ mm}) \, dr}{r}}$$

$$= \frac{2400}{80 \ln \dfrac{50}{30} + 20 \ln \dfrac{90}{50}} = \frac{2400}{40.866 + 11.756} = 45.61 \text{ mm}$$

$$= 0.04561 \text{ m}$$

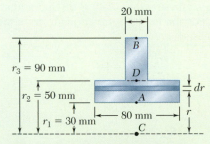

Fig. 4 Geometry of cross section.

We also compute: $e = \bar{r} - R = 0.05000 \text{ m} - 0.04561 \text{ m} = 0.00439 \text{ m}$

Allowable Load. We observe that the largest compressive stress will occur at point *A* where *r* = 0.030 m. Recalling that $\sigma_{\text{all}} = 50$ MPa and using Eq. (1), write

$$-50 \times 10^6 \text{ Pa} = -\frac{P}{2.4 \times 10^{-3} \text{ m}^2} + \frac{(0.110 \, P)(0.030 \text{ m} - 0.04561 \text{ m})}{(2.4 \times 10^{-3} \text{ m}^2)(0.00439 \text{ m})(0.030 \text{ m})}$$

$$-50 \times 10^6 = -417P - 5432P \qquad\qquad P = 8.55 \text{ kN} \blacktriangleleft$$

Problems

4.161 For the curved bar shown, determine the stress at point A when (*a*) $h = 50$ mm, (*b*) $h = 60$ mm.

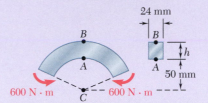

Fig. P4.161 and P4.162

4.162 For the curved bar shown, determine the stress at points A and B when $h = 55$ mm.

4.163 For the machine component and loading shown, determine the stress at point A when (*a*) $h = 2$ in., (*b*) $h = 2.6$ in.

4.164 For the machine component and loading shown, determine the stress at points A and B when $h = 2.5$ in.

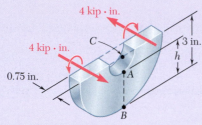

Fig. P4.163 and P4.164

4.165 The curved bar shown has a cross section of 40×60 mm and an inner radius $r_1 = 15$ mm. For the loading shown, determine the largest tensile and compressive stresses.

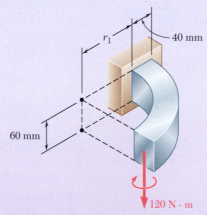

Fig. *P4.165* and *P4.166*

4.166 For the curved bar and loading shown, determine the percent error introduced in the computation of the maximum stress by assuming that the bar is straight. Consider the case when (*a*) $r_1 = 20$ mm, (*b*) $r_1 = 200$ mm, (*c*) $r_1 = 2$ m.

4.167 Steel links having the cross section shown are available with different central angles β. Knowing that the allowable stress is 12 ksi, determine the largest force **P** that can be applied to a link for which $\beta = 90°$.

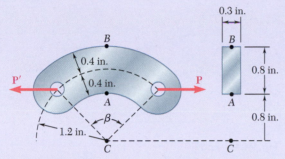

Fig. P4.167

4.168 Solve Prob. 4.167, assuming that $\beta = 60°$.

4.169 The curved bar shown has a cross section of 30×30 mm. Knowing that the allowable compressive stress is 175 MPa, determine the largest allowable distance a.

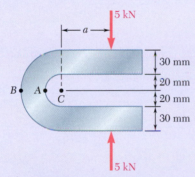

Fig. P4.169

4.170 For the split ring shown, determine the stress at (a) point A, (b) point B.

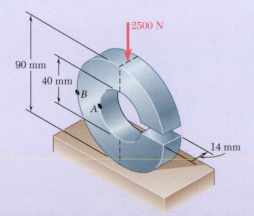

Fig. P4.170

4.171 Three plates are welded together to form the curved beam shown. For $M = 8$ kip·in., determine the stress at (a) point A, (b) point B, (c) the centroid of the cross section.

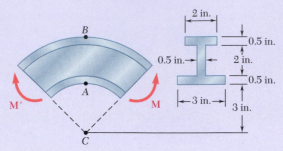

Fig. P4.171 and *P4.172*

4.172 Three plates are welded together to form the curved beam shown. For the given loading, determine the distance e between the neutral axis and the centroid of the cross section.

4.173 and 4.174 Knowing that the maximum allowable stress is 45 MPa, determine the magnitude of the largest moment M that can be applied to the components shown.

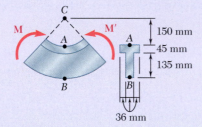

Fig. P4.173

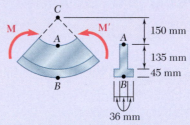

Fig. P4.174

4.175 The split ring shown has an inner radius $r_1 = 0.8$ in. and a *circular* cross section of diameter $d = 0.6$ in. Knowing that each of the 120-lb forces is applied at the centroid of the cross section, determine the stress (a) at point A, (b) at point B.

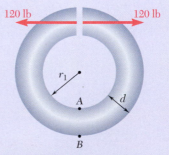

Fig. P4.175

4.176 Solve Prob. 4.175, assuming that the ring has an inner radius $r_1 = 0.6$ in. and a cross-sectional diameter $d = 0.8$ in.

4.177 The bar shown has a circular cross section of 14-mm diameter. Knowing that $a = 32$ mm, determine the stress at (a) point A, (b) point B.

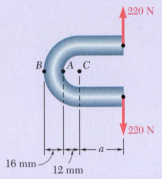

220 N

B · A · C

220 N

16 mm — 12 mm — a

Fig. P4.177 and P4.178

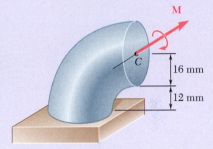

Fig. P4.179

M

C

16 mm

12 mm

4.178 The bar shown has a circular cross section of 14-mm diameter. Knowing that the allowable stress is 38 MPa, determine the largest permissible distance a from the line of action of the 220-N forces to the plane containing the center of curvature of the bar.

4.179 The curved bar shown has a circular cross section of 32-mm diameter. Determine the largest couple **M** that can be applied to the bar about a horizontal axis if the maximum stress is not to exceed 60 MPa.

4.180 Knowing that $P = 10$ kN, determine the stress at (a) point A, (b) point B.

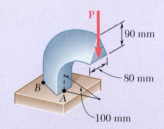

P

90 mm

80 mm

B

A

100 mm

Fig. P4.180

4.181 and 4.182 Knowing that $M = 5$ kip·in., determine the stress at (a) point A, (b) point B.

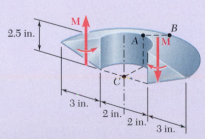

2.5 in.

M

A

B

M

C

3 in.

2 in.

2 in.

3 in.

Fig. P4.181

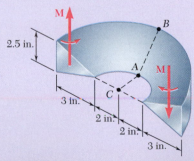

M

2.5 in.

B

A

M

C

3 in.

2 in.

2 in.

3 in.

Fig. P4.182

4.183 Knowing that the machine component shown has a trapezoidal cross section with $a = 3.5$ in. and $b = 2.5$ in., determine the stress at (*a*) point *A*, (*b*) point *B*.

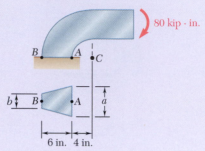

Fig. P4.183 and P4.184

4.184 Knowing that the machine component shown has a trapezoidal cross section with $a = 2.5$ in. and $b = 3.5$ in., determine the stress at (*a*) point *A*, (*b*) point *B*.

4.185 For the curved beam and loading shown, determine the stress at (*a*) point *A*, (*b*) point *B*.

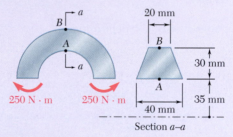

Fig. P4.185

4.186 For the crane hook shown, determine the largest tensile stress in section *a-a*.

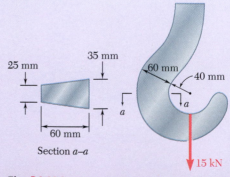

Fig. *P4.186*

*4.187 through 4.189 Using Eq. (4.66), derive the expression for R given in Fig. 4.61 for

      *4.187 A circular cross section.

      4.188 A trapezoidal cross section.

      4.189 A triangular cross section.

4.190 Show that if the cross section of a curved beam consists of two or more rectangles, the radius R of the neutral surface can be expressed as

$$R = \frac{A}{\ln\left[\left(\dfrac{r_2}{r_1}\right)^{b_1}\left(\dfrac{r_3}{r_2}\right)^{b_2}\left(\dfrac{r_4}{r_3}\right)^{b_3}\right]}$$

where A is the total area of the cross section.

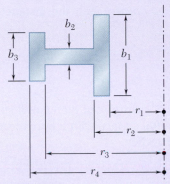

Fig. P4.190

*4.191 For a curved bar of rectangular cross section subjected to a bending couple **M**, show that the radial stress at the neutral surface is

$$\sigma_r = \frac{M}{Ae}\left(1 - \frac{r_1}{R} - \ln\frac{R}{r_1}\right)$$

and compute the value of σ_r for the curved bar of Concept Applications 4.10 and 4.11. (*Hint*: consider the free-body diagram of the portion of the beam located above the neutral surface.)

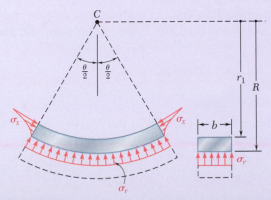

Fig. P4.191

Review and Summary

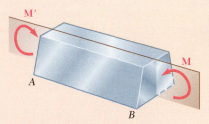

Fig. 4.63 Member in pure-bending.

This chapter was devoted to the analysis of members in *pure bending*. The stresses and deformation in members subjected to equal and opposite couples **M** and **M′** acting in the same longitudinal plane (Fig. 4.63) were studied.

Normal Strain in Bending

In members possessing a plane of symmetry and subjected to couples acting in that plane, it was proven that *transverse sections remain plane* as a member is deformed. A member in pure bending also has a *neutral surface* along which normal strains and stresses are zero. The longitudinal *normal strain* ϵ_x *varies linearly* with the distance y from the neutral surface:

$$\epsilon_x = -\frac{y}{\rho} \qquad (4.8)$$

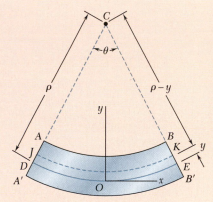

Fig. 4.64 Deformation with respect to neutral axis

where ρ is the *radius of curvature* of the neutral surface (Fig. 4.64). The intersection of the neutral surface with a transverse section is known as the *neutral axis* of the section.

Normal Stress in Elastic Range

For members made of a material that follows Hooke's law, the *normal stress* σ_x *varies linearly* with the distance from the neutral axis (Fig. 4.65). Using the maximum stress σ_m, the normal stress is

$$\sigma_x = -\frac{y}{c}\sigma_m \qquad (4.12)$$

where c is the largest distance from the neutral axis to a point in the section.

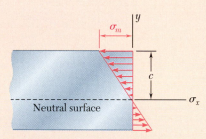

Fig. 4.65 Stress distribution for the elastic flexure formula.

Elastic Flexure Formula

By setting the sum of the elementary forces $\sigma_x\, dA$ equal to zero, we proved that the *neutral axis passes through the centroid* of the cross section of a member in pure bending. Then by setting the sum of the moments of the elementary forces equal to the bending moment, the *elastic flexure formula* is

$$\sigma_m = \frac{Mc}{I} \qquad (4.15)$$

where I is the moment of inertia of the cross section with respect to the neutral axis. The normal stress at any distance y from the neutral axis is

$$\sigma_x = -\frac{My}{I} \qquad (4.16)$$

Elastic Section Modulus

Noting that I and c depend only on the geometry of the cross section we introduced the *elastic section modulus*

$$S = \frac{I}{c} \tag{4.17}$$

Use the section modulus to write an alternative expression for the maximum normal stress:

$$\sigma_m = \frac{M}{S} \tag{4.18}$$

Curvature of Member

The *curvature* of a member is the reciprocal of its radius of curvature, and may be found by

$$\frac{1}{\rho} = \frac{M}{EI} \tag{4.21}$$

Anticlastic Curvature

In the bending of homogeneous members possessing a plane of symmetry, deformations occur in the plane of a transverse cross section and result in *anticlastic curvature* of the members.

Members Made of Several Materials

We considered the bending of members made of several materials with *different moduli of elasticity*. While transverse sections remain plane, the *neutral axis does not pass through the centroid* of the composite cross section (Fig. 4.66). Using the ratio of the moduli of elasticity of the materials, we obtained a *transformed section* corresponding to an equivalent member made entirely of one material. The methods previously developed are used to determine the stresses in this equivalent homogeneous member (Fig. 4.67), and the ratio of the moduli of elasticity is used to determine the stresses in the composite beam.

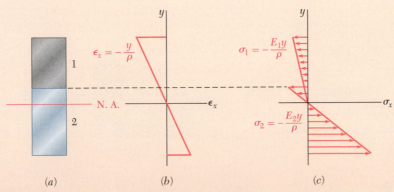

Fig. 4.66 (a) Composite section. (b) Strain distribution. (c) Stress distribution.

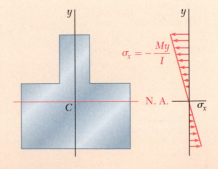

Fig. 4.67 Transformed section.

Stress Concentrations

Stress concentrations occur in members in pure bending and were discussed; charts giving stress-concentration factors for flat bars with fillets and grooves also were presented in Figs. 4.24 and 4.25.

Plastic Deformations

A rectangular beam made of an *elastoplastic material* (Fig. 4.68) was analyzed as the magnitude of the bending moment was increased (Fig. 4.69). The *maximum elastic moment* M_Y occurs when yielding is initiated in the beam (Fig. 4.69b). As the bending moment is increased, plastic zones develop (Fig. 4.69c), and the size of the elastic core of the member is decreased. When the beam becomes fully plastic (Fig. 4.69d), the maximum or *plastic moment* M_p is obtained. *Permanent deformations* and *residual stresses* remain in a member after the loads that caused yielding have been removed.

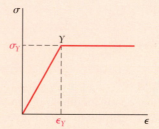

Fig. 4.68 Elastoplastic stress-strain diagram.

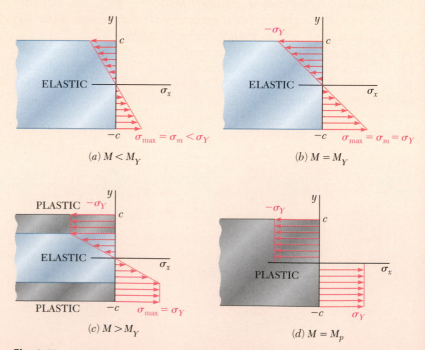

(a) $M < M_Y$

(b) $M = M_Y$

(c) $M > M_Y$

(d) $M = M_p$

Fig. 4.69 Bending stress distribution in a member for : (a) elastic, $M < M_Y$ (b) yield impending, $M = M_Y$, (c) partially yielded, $M > M_Y$, and (d) fully plastic, $M = M_p$.

Eccentric Axial Loading

When a member is loaded *eccentrically in a plane of symmetry*, the *eccentric load* is replaced with a force-couple system located at the centroid of the cross section (Fig. 4.70). The stresses from the centric load and the bending couple are superposed (Fig. 4.71):

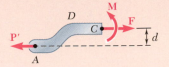

Fig. 4.70 Section of an eccentrically loaded member.

$$\sigma_x = \frac{P}{A} - \frac{My}{I} \qquad (4.50)$$

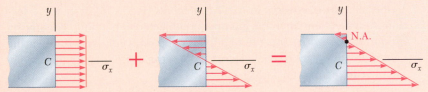

Fig. 4.71 Stress distribution for eccentric loading is obtained by superposing the axial and pure bending distributions.

Unsymmetric Bending

For bending of members of *unsymmetric cross section*, the flexure formula may be used, provided that the couple vector **M** is directed along one of the principal centroidal axes of the cross section. When necessary, **M** can be resolved into components along the principal axes, and the stresses superposed due to the component couples (Figs. 4.72 and 4.73).

$$\sigma_x = -\frac{M_z y}{I_z} + \frac{M_y z}{I_y} \qquad (4.55)$$

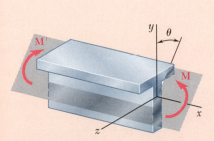

Fig. 4.72 Unsymmetric bending with bending moment not in a plane of symmetry.

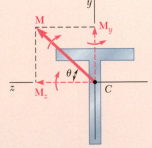

Fig. 4.73 Applied moment resolved into y and z components.

For the couple **M** shown in Fig. 4.74, the orientation of the neutral axis is defined by

$$\tan \phi = \frac{I_z}{I_y} \tan \theta \qquad (4.57)$$

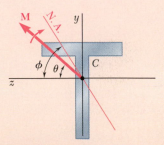

Fig. 4.74 Neutral axis for unsymmetric bending.

General Eccentric Axial Loading

For the general case of *eccentric axial loading*, the load is replaced by a force-couple system located at the centroid. The stresses are superposed due to the centric load and the two component couples directed along the principal axes:

$$\sigma_x = \frac{P}{A} - \frac{M_z y}{I_z} + \frac{M_y z}{I_y} \tag{4.58}$$

Curved Members

In the analysis of stresses in *curved members* (Fig. 4.75), transverse sections remain plane when the member is subjected to bending. The *stresses do not vary linearly*, and the neutral surface does not pass through the centroid of the section. The distance R from the center of curvature of the member to the neutral surface is

$$R = \frac{A}{\displaystyle\int \frac{dA}{r}} \tag{4.66}$$

where A is the area of the cross section. The normal stress at a distance y from the neutral surface is

$$\sigma_x = -\frac{My}{Ae(R - y)} \tag{4.70}$$

where M is the bending moment and e is the distance from the centroid of the section to the neutral surface.

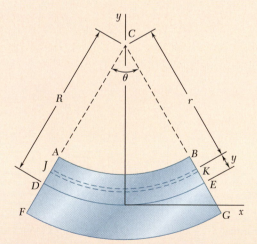

Fig. 4.75 Curved member geometry.

Review Problems

4.192 Two vertical forces are applied to a beam of the cross section shown. Determine the maximum tensile and compressive stresses in portion *BC* of the beam.

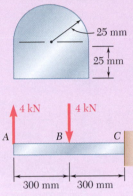

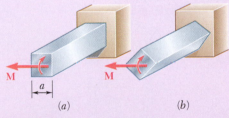

Fig. P4.192

4.193 A steel band saw blade that was originally straight passes over 8-in.-diameter pulleys when mounted on a band saw. Determine the maximum stress in the blade, knowing that it is 0.018 in. thick and 0.625 in. wide. Use $E = 29 \times 10^6$ psi.

Fig. P4.193

4.194 A couple of magnitude *M* is applied to a square bar of side *a*. For each of the orientations shown, determine the maximum stress and the curvature of the bar.

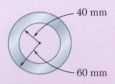

(a) (b)

Fig. P4.194

4.195 Determine the plastic moment M_p of a steel beam of the cross section shown, assuming the steel to be elastoplastic with a yield strength of 240 MPa.

40 mm

60 mm

Fig. P4.195

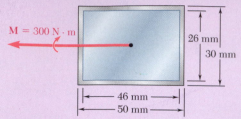

Fig. P4.196

4.196 In order to increase corrosion resistance, a 2-mm-thick cladding of aluminum has been added to a steel bar as shown. The modulus of elasticity is 200 GPa for steel and 70 GPa for aluminum. For a bending moment of 300 N·m, determine (*a*) the maximum stress in the steel, (*b*) the maximum stress in the aluminum, (*c*) the radius of curvature of the bar.

4.197 The vertical portion of the press shown consists of a rectangular tube of wall thickness $t = 10$ mm. Knowing that the press has been tightened on wooden planks being glued together until $P = 20$ kN, determine the stress at (*a*) point A, (*b*) point B.

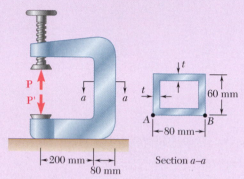

Fig. P4.197

4.198 The four forces shown are applied to a rigid plate supported by a solid steel post of radius *a*. Knowing that $P = 24$ kips and $a = 1.6$ in., determine the maximum stress in the post when (*a*) the force at D is removed, (*b*) the forces at C and D are removed.

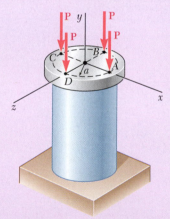

Fig. P4.198

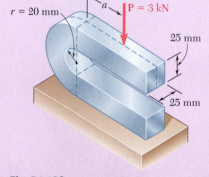

Fig. P4.199

4.199 The curved portion of the bar shown has an inner radius of 20 mm. Knowing that the allowable stress in the bar is 150 MPa, determine the largest permissible distance *a* from the line of action of the 3-kN force to the vertical plane containing the center of curvature of the bar.

4.200 Determine the maximum stress in each of the two machine elements shown.

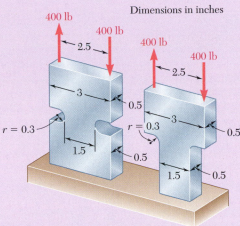

Dimensions in inches

400 lb
400 lb
2.5
3
0.5
r = 0.3
1.5
0.5

400 lb
400 lb
2.5
3
r = 0.3
0.5
1.5
0.5

Fig. P4.200

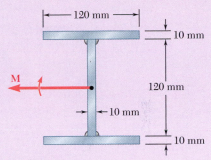

120 mm
10 mm
M
120 mm
10 mm
10 mm

Fig. P4.201

4.201 Three 120 × 10-mm steel plates have been welded together to form the beam shown. Assuming that the steel is elastoplastic with $E = 200$ GPa and $\sigma_Y = 300$ MPa, determine (a) the bending moment for which the plastic zones at the top and bottom of the beam are 40 mm thick, (b) the corresponding radius of curvature of the beam.

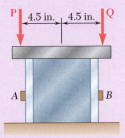

P   4.5 in.   4.5 in.   Q

A   B

Fig. P4.202

4.202 A short length of a W8 × 31 rolled-steel shape supports a rigid plate on which two loads **P** and **Q** are applied as shown. The strains at two points A and B on the centerline of the outer faces of the flanges have been measured and found to be

$$\epsilon_A = -550 \times 10^{-6} \text{ in./in.} \qquad \epsilon_B = -680 \times 10^{-6} \text{ in./in.}$$

Knowing that $E = 29 \times 10^6$ psi, determine the magnitude of each load.

4.203 Two thin strips of the same material and same cross section are bent by couples of the same magnitude and glued together. After the two surfaces of contact have been securely bonded, the couples are removed. Denoting by σ_1 the maximum stress and by ρ_1 the radius of curvature of each strip while the couples were applied, determine (a) the final stresses at points A, B, C, and D, (b) the final radius of curvature.

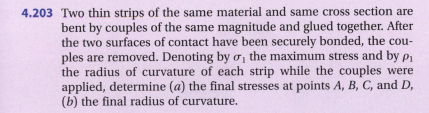

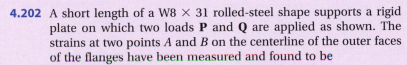

σ_1
A
B
σ_1
C
σ_1
D
σ_1

M₁   M'₁
M₁   M'₁

Fig. P4.203

Computer Problems

The following problems are designed to be solved with a computer.

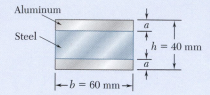

Fig. P4.C1

4.C1 Two aluminum strips and a steel strip are to be bonded together to form a composite member of width $b = 60$ mm and depth $h = 40$ mm. The modulus of elasticity is 200 GPa for the steel and 75 GPa for the aluminum. Knowing that $M = 1500$ N·m, write a computer program to calculate the maximum stress in the aluminum and in the steel for values of a from 0 to 20 mm using 2-mm increments. Using appropriate smaller increments, determine (a) the largest stress that can occur in the steel and (b) the corresponding value of a.

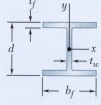

Fig. P4.C2

4.C2 A beam of the cross section shown, made of a steel that is assumed to be elastoplastic with a yield strength σ_Y and a modulus of elasticity E is bent about the x axis. (a) Denoting by y_Y the half thickness of the elastic core, write a computer program to calculate the bending moment M and the radius of curvature ρ for values of y_Y from $\frac{1}{2}d$ to $\frac{1}{6}d$ using decrements equal to $\frac{1}{2}t_f$. Neglect the effect of fillets. (b) Use this program to solve Prob. 4.201.

4.C3 An 8-kip·in. couple **M** is applied to a beam of the cross section shown in a plane forming an angle β with the vertical. Noting that the centroid of the cross section is located at C and that the y and z axes are principal axes, write a computer program to calculate the stress at A, B, C, and D for values of β from 0 to 180° using 10° increments. (*Given:* $I_y = 6.23$ in^4 and $I_z = 1.481$ in^4.)

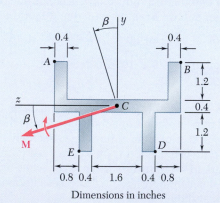

Dimensions in inches

Fig. P4.C3

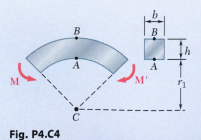

Fig. P4.C4

4.C4 Couples of moment $M = 2$ kN·m are applied as shown to a curved bar having a rectangular cross section with $h = 100$ mm and $b = 25$ mm. Write a computer program and use it to calculate the stresses at points A and B for values of the ratio r_1/h from 10 to 1 using decrements of 1, and from 1 to 0.1 using decrements of 0.1. Using appropriate smaller increments, determine the ratio r_1/h for which the maximum stress in the curved bar is 50% larger than the maximum stress in a straight bar of the same cross section.

4.C5 The couple **M** is applied to a beam of the cross section shown. (*a*) Write a computer program that, for loads expressed in either SI or U.S. customary units, can be used to calculate the maximum tensile and compressive stresses in the beam. (*b*) Use this program to solve Probs. 4.9, 4.10, and 4.11.

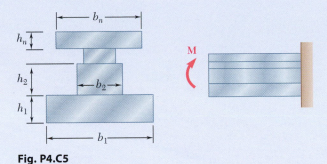

Fig. P4.C5

4.C6 A solid rod of radius $c = 1.2$ in. is made of a steel that is assumed to be elastoplastic with $E = 29{,}000$ ksi and $\sigma_Y = 42$ ksi. The rod is subjected to a couple of moment M that increases from zero to the maximum elastic moment M_Y and then to the plastic moment M_p. Denoting by y_Y the half thickness of the elastic core, write a computer program and use it to calculate the bending moment M and the radius of curvature ρ for values of y_Y from 1.2 in. to 0 using 0.2-in. decrements. (*Hint:* Divide the cross section into 80 horizontal elements of 0.03-in. height.)

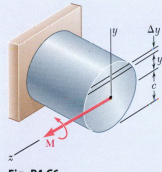

Fig. P4.C6

4.C7 The machine element of Prob. 4.182 is to be redesigned by removing part of the triangular cross section. It is believed that the removal of a small triangular area of width a will lower the maximum stress in the element. In order to verify this design concept, write a computer program to calculate the maximum stress in the element for values of a from 0 to 1 in. using 0.1-in. increments. Using appropriate smaller increments, determine the distance a for which the maximum stress is as small as possible and the corresponding value of the maximum stress.

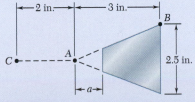

Fig. P4.C7

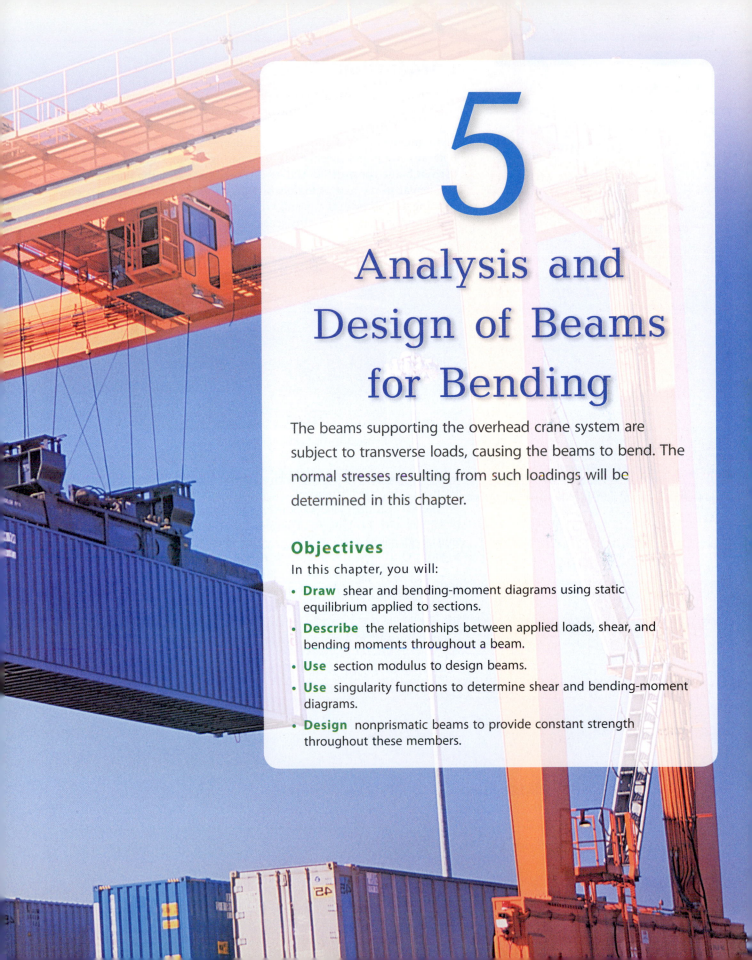

5

Analysis and Design of Beams for Bending

The beams supporting the overhead crane system are subject to transverse loads, causing the beams to bend. The normal stresses resulting from such loadings will be determined in this chapter.

Objectives

In this chapter, you will:

- **Draw** shear and bending-moment diagrams using static equilibrium applied to sections.
- **Describe** the relationships between applied loads, shear, and bending moments throughout a beam.
- **Use** section modulus to design beams.
- **Use** singularity functions to determine shear and bending-moment diagrams.
- **Design** nonprismatic beams to provide constant strength throughout these members.

Introduction

This chapter and most of the next one are devoted to the analysis and the design of *beams,* which are structural members supporting loads applied at various points along the member. Beams are usually long, straight prismatic members. Steel and aluminum beams play an important part in both structural and mechanical engineering. Timber beams are widely used in home construction (Photo 5.1). In most cases, the loads are perpendicular to the axis of the beam. This *transverse loading* causes only bending and shear in the beam. When the loads are not at a right angle to the beam, they also produce axial forces in the beam.

Photo 5.1   Timber beams used in a residential dwelling.

The transverse loading of a beam may consist of *concentrated loads* P_1, P_2, . . . expressed in newtons, pounds, or their multiples of kilonewtons and kips (Fig. 5.1a); of a *distributed load* w expressed in N/m, kN/m, lb/ft, or kips/ft (Fig. 5.1b); or of a combination of both. When the load w per unit length has a constant value over part of the beam (as between A and B in Fig. 5.1b), the load is *uniformly distributed.*

Beams are classified according to the way they are supported, as shown in Fig. 5.2. The distance L is called the *span.* Note that the reactions at the supports of the beams in Fig. 5.2 *a, b,* and *c* involve a total of only three unknowns and can be determined by the methods of statics. Such beams are said to be *statically determinate.* On the other hand, the

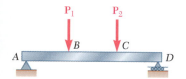

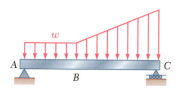

(a) Concentrated loads

(b) Distributed loads

Fig. 5.1   Transversely loaded beams.

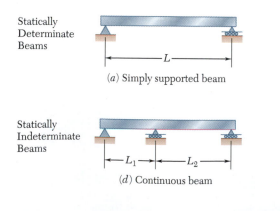

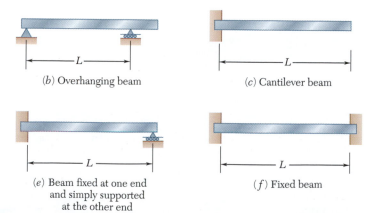

Statically Determinate Beams

(a) Simply supported beam

(b) Overhanging beam

(c) Cantilever beam

Statically Indeterminate Beams

(d) Continuous beam

(e) Beam fixed at one end and simply supported at the other end

(f) Fixed beam

Fig. 5.2   Common beam support configurations.

reactions at the supports of the beams in Fig. 5.2 *d, e,* and *f* involve more than three unknowns and cannot be determined by the methods of statics alone. The properties of the beams with regard to their resistance to deformations must be taken into consideration. Such beams are said to be *statically indeterminate,* and their analysis will be discussed in Chap. 9.

Sometimes two or more beams are connected by hinges to form a single continuous structure. Two examples of beams hinged at a point *H* are shown in Fig. 5.3. Note that the reactions at the supports involve four unknowns and cannot be determined from the free-body diagram of the two-beam system. They can be determined by recognizing that the internal moment at the hinge is zero. Then, after considering the free-body diagram of each beam separately, six unknowns are involved (including two force components at the hinge), and six equations are available.

When a beam is subjected to transverse loads, the internal forces in any section of the beam consist of a shear force **V** and a bending couple **M**. For example, a simply supported beam *AB* is carrying two concentrated loads and a uniformly distributed load (Fig. 5.4*a*). To determine the internal forces in a section through point *C,* draw the free-body diagram of the entire beam to obtain the reactions at the supports (Fig. 5.4*b*). Passing a section through *C,* then draw the free-body diagram of *AC* (Fig. 5.4*c*), from which the shear force **V** and the bending couple **M** are found.

The bending couple **M** creates *normal stresses* in the cross section, while the shear force **V** creates *shearing stresses.* In most cases, the dominant criterion in the design of a beam for strength is the maximum value of the normal stress in the beam. The normal stresses in a beam are the subject of this chapter, while shearing stresses are discussed in Chap. 6.

Since the distribution of the normal stresses in a given section depends only upon the bending moment *M* and the geometry of the section,[†] the elastic flexure formulas derived in Sec. 4.2 are used to determine the maximum stress, as well as the stress at any given point;[‡]

$$\sigma_m = \frac{|M|c}{I} \tag{5.1}$$

and

$$\sigma_x = -\frac{My}{I} \tag{5.2}$$

where *I* is the moment of inertia of the cross section with respect to a centroidal axis perpendicular to the plane of the couple, *y* is the distance from the neutral surface, and *c* is the maximum value of that distance (Fig. 4.11). Also recall from Sec. 4.2 that the maximum value σ_m of the normal stress can be expressed in terms of the section modulus *S.* Thus

$$\sigma_m = \frac{|M|}{S} \tag{5.3}$$

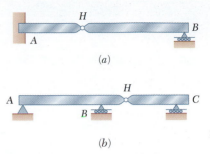

(a)

(b)

Fig. 5.3 Beams connected by hinges.

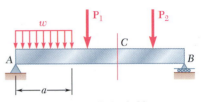

(a) Transversely-loaded beam

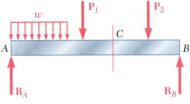

(b) Free-body diagram to find support reactions

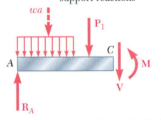

(c) Free-body diagram to find internal forces at *C*

Fig. 5.4 Analysis of a simply supported beam.

[†]It is assumed that the distribution of the normal stresses in a given cross section is not affected by the deformations caused by the shearing stresses. This assumption will be verified in Sec. 6.2.

[‡]Recall from Sec. 4.1 that *M* can be positive or negative, depending upon whether the concavity of the beam at the point considered faces upward or downward. Thus, in a transverse loading the sign of *M* can vary along the beam. On the other hand, since σ_m is a positive quantity, the absolute value of *M* is used in Eq. (5.1).

The fact that σ_m is inversely proportional to S underlines the importance of selecting beams with a large section modulus. Section moduli of various rolled-steel shapes are given in Appendix C, while the section modulus of a rectangular shape is

$$S = \tfrac{1}{6}bh^2 \tag{5.4}$$

where b and h are, respectively, the width and the depth of the cross section.

Equation (5.3) also shows that for a beam of uniform cross section, σ_m is proportional to $|M|$. Thus, the maximum value of the normal stress in the beam occurs in the section where $|M|$ is largest. One of the most important parts of the design of a beam for a given loading condition is the determination of the location and magnitude of the largest bending moment.

This task is made easier if a *bending-moment diagram* is drawn, where the bending moment M is determined at various points of the beam and plotted against the distance x measured from one end. It is also easier if a *shear diagram* is drawn by plotting the shear V against x. The sign convention used to record the values of the shear and bending moment is discussed in Sec. 5.1.

In Sec. 5.2 relationships between load, shear, and bending moments are derived and used to obtain the shear and bending-moment diagrams. This approach facilitates the determination of the largest absolute value of the bending moment and the maximum normal stress in the beam.

In Sec. 5.3 beams are designed for bending such that the maximum normal stress in these beams will not exceed their allowable values.

Another method to determine the maximum values of the shear and bending moment is based on expressing V and M in terms of *singularity functions.* This is discussed in Sec. 5.4. This approach lends itself well to the use of computers and will be expanded in Chap. 9 for the determination of the slope and deflection of beams.

Finally, the design of *nonprismatic beams* (i.e., beams with a variable cross section) is discussed in Sec. 5.5. By selecting the shape and size of the variable cross section so that its elastic section modulus $S = I/c$ varies along the length of the beam in the same way as $|M|$, it is possible to design beams where the maximum normal stress in each section is equal to the allowable stress of the material. Such beams are said to be of *constant strength.*

5.1   SHEAR AND BENDING-MOMENT DIAGRAMS

The maximum absolute values of the shear and bending moment in a beam are easily found if V and M are plotted against the distance x measured from one end of the beam. Besides, as you will see in Chap. 9, the knowledge of M as a function of x is essential to determine the deflection of a beam.

In this section, the shear and bending-moment diagrams are obtained by determining the values of V and M at selected points of the beam. These values are found by passing a section through the point to

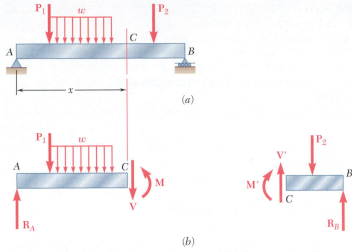

(a)

(b)

Fig. 5.5   Determination of shear force, *V*, and bending moment, *M*, at a given section. (*a*) Loaded beam with section indicated at arbitrary position *x*. (*b*) Free-body diagrams drawn to the left and right of the section at *C*.

be determined (Fig. 5.5*a*) and considering the equilibrium of the portion of beam located on either side of the section (Fig. 5.5*b*). Since the shear forces **V** and **V**′ have opposite senses, recording the shear at point *C* with an up or down arrow is meaningless, unless it is indicated at the same time which of the free bodies *AC* and *CB* is being considered. For this reason, the shear *V* is recorded with a *plus sign* if the shear forces are directed as in Fig. 5.5*b* and a *minus sign* otherwise. A similar convention is applied for the bending moment *M*.[†] Summarizing the sign conventions:

The shear V and the bending moment M at a given point of a beam are positive when the internal forces and couples acting on each portion of the beam are directed as shown in Fig. 5.6a.

1. *The shear at any given point of a beam is positive when the* **external** *forces (loads and reactions) acting on the beam tend to shear off the beam at that point as indicated in Fig. 5.6b.*
2. *The bending moment at any given point of a beam is positive when the* **external** *forces acting on the beam tend to bend the beam at that point as indicated in Fig. 5.6c.*

It is helpful to note that the values of the shear and of the bending moment are positive in the left half of a simply supported beam carrying a single concentrated load at its midpoint, as is discussed in the following Concept Application.

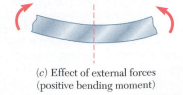

(*a*) Internal forces
(positive shear and positive bending moment)

(*b*) Effect of external forces
(positive shear)

(*c*) Effect of external forces
(positive bending moment)

Fig. 5.6   Sign convention for shear and bending moment.

[†]This convention is the same that we used earlier in Sec. 4.1.

Concept Application 5.1

Draw the shear and bending-moment diagrams for a simply supported beam *AB* of span *L* subjected to a single concentrated load **P** at its midpoint *C* (Fig. 5.7a).

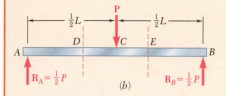

(b)

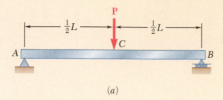

(a)

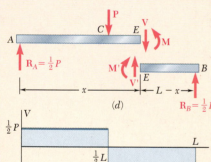

(c)

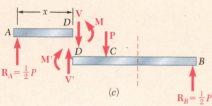

(d)

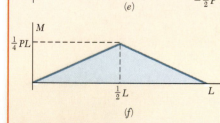

(e)

(f)

Fig. 5.7 (a) Simply supported beam with midpoint load, **P.** (b) Free-body diagram of entire beam. (c) Free-body diagrams with section taken to left of load **P.** (d) Free-body diagrams with section taken to right of load **P.** (e) Shear diagram. (f) Bending-moment diagram.

Determine the reactions at the supports from the free-body diagram of the entire beam (Fig. 5.7b). The magnitude of each reaction is equal to $P/2$.

Next cut the beam at a point *D* between *A* and *C* and draw the free-body diagrams of *AD* and *DB* (Fig. 5.7c). *Assuming that the shear and bending moment are positive,* we direct the internal forces **V** and **V′** and the internal couples **M** and **M′** as in Fig. 5.6*a*. Consider the free body *AD*. The sum of the vertical components and the sum of the moments about *D* of the forces acting on the free body are zero, so $V = +P/2$ and $M = +Px/2$. Both the shear and the bending moment are positive. This is checked by observing that the reaction at *A* tends to shear off and bend the beam at *D* as indicated in Figs. 5.6*b* and *c*. We plot *V* and *M* between *A* and *C* (Figs. 5.8*d* and *e*). The shear has a constant value $V = P/2$, while the bending moment increases linearly from $M = 0$ at $x = 0$ to $M = PL/4$ at $x = L/2$.

Cutting the beam at a point *E* between *C* and *B* and considering the free body *EB* (Fig. 5.7*d*), the sum of the vertical components and the sum of the moments about *E* of the forces acting on the free body are zero. Obtain $V = -P/2$ and $M = P(L - x)/2$. Therefore, the shear is negative, and the bending moment positive. This is checked by observing that the reaction at *B* bends the beam at *E* as in Fig. 5.6*c* but tends to shear it off in a manner opposite to that shown in Fig. 5.6*b*. The shear and bending-moment diagrams of Figs. 5.7*e* and *f* are completed by showing the shear with a constant value $V = -P/2$ between *C* and *B*, while the bending moment decreases linearly from $M = PL/4$ at $x = L/2$ to $M = 0$ at $x = L$.

Note from the previous Concept Application that when a beam is subjected only to concentrated loads, the shear is constant between loads and the bending moment varies linearly between loads. In such situations, the shear and bending-moment diagrams can be drawn easily once the values of V and M have been obtained at sections selected just to the left and just to the right of the points where the loads and reactions are applied (see Sample Prob. 5.1).

Concept Application 5.2

Draw the shear and bending-moment diagrams for a cantilever beam AB of span L supporting a uniformly distributed load w (Fig. 5.8a).

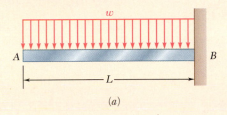

(a)

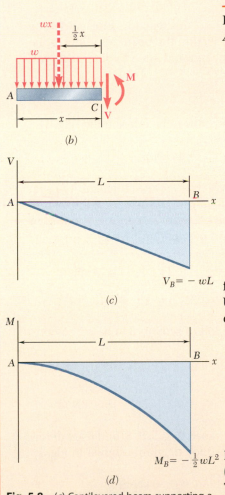

(b)

(c)

$$V_B = -wL$$

$$M_B = -\tfrac{1}{2}wL^2$$

(d)

Fig. 5.8 *(a)* Cantilevered beam supporting a uniformly distributed load. *(b)* Free-body diagram of section *AC*. *(c)* Shear diagram. *(d)* Bending-moment diagram.

Cut the beam at a point C, located between A and B, and draw the free-body diagram of AC (Fig. 5.8b), directing $\mathbf{V}$ and $\mathbf{M}$ as in Fig. 5.6a. Using the distance x from A to C and replacing the distributed load over AC by its resultant wx applied at the midpoint of AC, write

$$+\uparrow\Sigma F_y = 0: \qquad -wx - V = 0 \qquad V = -wx$$

$$+\curvearrowleft\Sigma M_C = 0: \qquad wx\left(\frac{x}{2}\right) + M = 0 \qquad M = -\frac{1}{2}wx^2$$

Note that the shear diagram is represented by an oblique straight line (Fig. 5.8c) and the bending-moment diagram by a parabola (Fig. 5.8d). The maximum values of V and M both occur at B, where

$$V_B = -wL \qquad M_B = -\tfrac{1}{2}wL^2$$

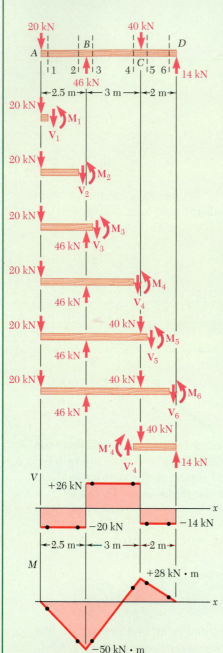

Fig. 1 Free-body diagram of beam, free-body diagrams of sections to left of cut, shear diagram, bending-moment diagram.

Sample Problem 5.1

For the timber beam and loading shown, draw the shear and bending-moment diagrams and determine the maximum normal stress due to bending.

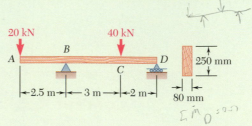

STRATEGY: After using statics to find the reaction forces, identify sections to be analyzed. You should section the beam at points to the immediate left and right of each concentrated force to determine values of V and M at these points.

MODELING and ANALYSIS:

Reactions. Considering the entire beam to be a free body (Fig. 1),

$$\mathbf{R}_B = 40 \text{ kN} \uparrow \qquad \mathbf{R}_D = 14 \text{ kN} \uparrow$$

Shear and Bending-Moment Diagrams. Determine the internal forces just to the right of the 20-kN load at A. Considering the stub of beam to the left of section *1* as a free body and assuming V and M to be positive (according to the standard convention), write

$$+\uparrow \Sigma F_y = 0: \qquad -20 \text{ kN} - V_1 = 0 \qquad V_1 = -20 \text{ kN}$$
$$+\circlearrowleft \Sigma M_1 = 0: \qquad (20 \text{ kN})(0 \text{ m}) + M_1 = 0 \qquad M_1 = 0$$

Next consider the portion to the left of section *2* to be a free body and write

$$+\uparrow \Sigma F_y = 0: \qquad -20 \text{ kN} - V_2 = 0 \qquad V_2 = -20 \text{ kN}$$
$$+\circlearrowleft \Sigma M_2 = 0: \qquad (20 \text{ kN})(2.5 \text{ m}) + M_2 = 0 \qquad M_2 = -50 \text{ kN·m}$$

The shear and bending moment at sections *3*, *4*, *5*, and *6* are determined in a similar way from the free-body diagrams shown in Fig. 1:

$$V_3 = +26 \text{ kN} \qquad M_3 = -50 \text{ kN·m}$$
$$V_4 = +26 \text{ kN} \qquad M_4 = +28 \text{ kN·m}$$
$$V_5 = -14 \text{ kN} \qquad M_5 = +28 \text{ kN·m}$$
$$V_6 = -14 \text{ kN} \qquad M_6 = 0$$

(continued)

For several of the latter sections, the results may be obtained more easily by considering the portion to the right of the section to be a free body. For example, for the portion of beam to the right of section *4*,

$$+\uparrow \Sigma F_y = 0: \qquad V_4 - 40\text{ kN} + 14\text{ kN} = 0 \qquad V_4 = +26\text{ kN}$$

$$+\curvearrowleft \Sigma M_4 = 0: \qquad -M_4 + (14\text{ kN})(2\text{ m}) = 0 \qquad M_4 = +28\text{ kN·m}$$

Now plot the six points shown on the shear and bending-moment diagrams. As indicated earlier, the shear is of constant value between concentrated loads, and the bending moment varies linearly.

Maximum Normal Stress.   This occurs at *B*, where |*M*| is largest. Use Eq. (5.4) to determine the section modulus of the beam:

$$S = \tfrac{1}{6}bh^2 = \tfrac{1}{6}(0.080\text{ m})(0.250\text{ m})^2 = 833.33 \times 10^{-6}\text{ m}^3$$

Substituting this value and $|M| = |M_B| = 50 \times 10^3$ N·m into Eq. (5.3) gives

$$\sigma_m = \frac{|M_B|}{S} = \frac{(50 \times 10^3\text{ N·m})}{833.33 \times 10^{-6}} = 60.00 \times 10^6\text{ Pa}$$

Maximum normal stress in the beam **= 60.0 MPa** ◀

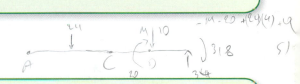

Sample Problem 5.2

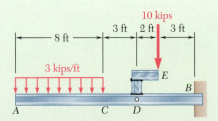

The structure shown consists of a W10 × 112 rolled-steel beam *AB* and two short members welded together and to the beam. (*a*) Draw the shear and bending-moment diagrams for the beam and the given loading. (*b*) Determine the maximum normal stress in sections just to the left and just to the right of point *D*.

STRATEGY: You should first replace the 10-kip load with an equivalent force-couple system at *D*. You can section the beam within each region of continuous load (including regions of no load) and find equations for the shear and bending moment.

MODELING and ANALYSIS:

Equivalent Loading of Beam.   The 10-kip load is replaced by an equivalent force-couple system at *D*. The reaction at *B* is determined by considering the beam to be free body (Fig. 1).

(continued)

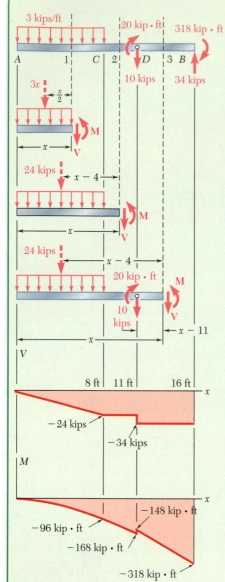

Fig. 1 Free-body diagram of beam, free-body diagrams of sections to left of cut, shear diagram, bending-moment diagram.

a. Shear and Bending-Moment Diagrams

From A to C. Determine the internal forces at a distance x from point A by considering the portion of beam to the left of section *1*. That part of the distributed load acting on the free body is replaced by its resultant, and

$$+\uparrow\Sigma F_y = 0: \qquad -3x - V = 0 \qquad V = -3x \text{ kips}$$
$$+\circlearrowleft\Sigma M_1 = 0: \qquad 3x(\tfrac{1}{2}x) + M = 0 \qquad M = -1.5x^2 \text{ kip·ft}$$

Since the free-body diagram shown in Fig. 1 can be used for all values of x smaller than 8 ft, the expressions obtained for V and M are valid in the region $0 < x < 8$ ft.

From C to D. Considering the portion of beam to the left of section *2* and again replacing the distributed load by its resultant,

$$+\uparrow\Sigma F_y = 0: \qquad -24 - V = 0 \qquad V = -24 \text{ kips}$$
$$+\circlearrowleft\Sigma M_2 = 0: \qquad 24(x-4) + M = 0 \qquad M = 96 - 24x \quad \text{kip·ft}$$

These expressions are valid in the region 8 ft $< x <$ 11 ft.

From D to B. Using the position of beam to the left of section *3*, the region 11 ft $< x <$ 16 ft is

$$V = -34 \text{ kips} \qquad M = 226 - 34x \quad \text{kip·ft}$$

The shear and bending-moment diagrams for the entire beam now can be plotted. Note that the couple of moment 20 kip·ft applied at point D introduces a discontinuity into the bending-moment diagram.

b. Maximum Normal Stress to the Left and Right of Point *D*.

From Appendix C for the W10 × 112 rolled-steel shape, $S = 126$ in^3 about the *X-X* axis.

To the left of D: $|M| = 168$ kip·ft $= 2016$ kip·in. Substituting for $|M|$ and S into Eq. (5.3), write

$$\sigma_m = \frac{|M|}{S} = \frac{2016 \text{ kip·in.}}{126 \text{ in}^3} = 16.00 \text{ ksi} \qquad \sigma_m = \mathbf{16.00 \text{ ksi}} \blacktriangleleft$$

To the right of D: $|M| = 148$ kip·ft $= 1776$ kip·in. Substituting for $|M|$ and S into Eq. (5.3), write

$$\sigma_m = \frac{|M|}{S} = \frac{1776 \text{ kip·in.}}{126 \text{ in}^3} = 14.10 \text{ ksi} \qquad \sigma_m = \mathbf{14.10 \text{ ksi}} \blacktriangleleft$$

REFLECT and THINK: It was not necessary to determine the reactions at the right end to draw the shear and bending-moment diagrams. However, having determined these at the start of the solution, they can be used as checks of the values at the right end of the shear and bending-moment diagrams.

Problems

5.1 through 5.6 For the beam and loading shown, (*a*) draw the shear and bending-moment diagrams, (*b*) determine the equations of the shear and bending-moment curves.

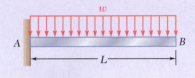

Fig. P5.1

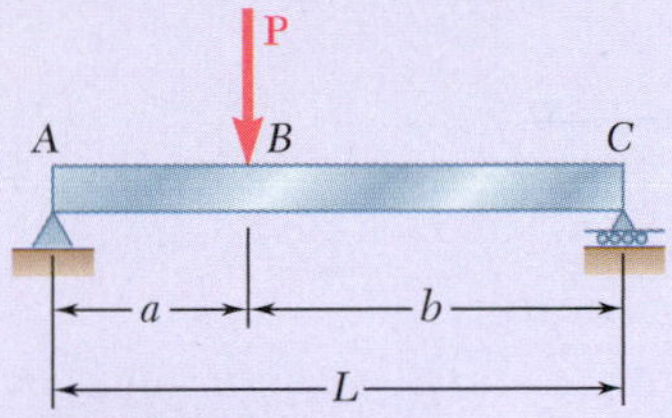

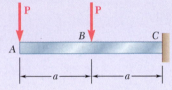

Fig. P5.2

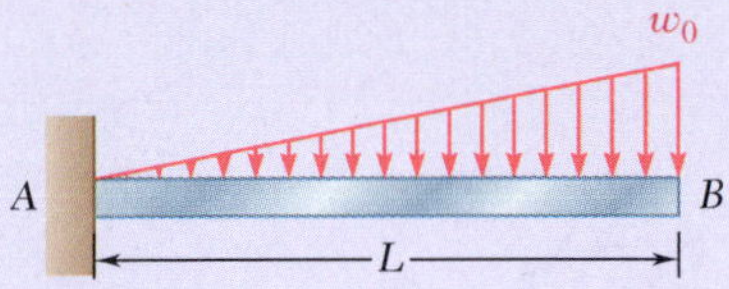

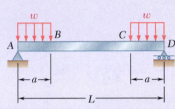

Fig. P5.3

Fig. P5.4

Fig. P5.5

Fig. P5.6

5.7 and 5.8 Draw the shear and bending-moment diagrams for the beam and loading shown, and determine the maximum absolute value (*a*) of the shear, (*b*) of the bending moment.

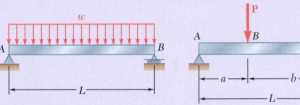

Fig. P5.7

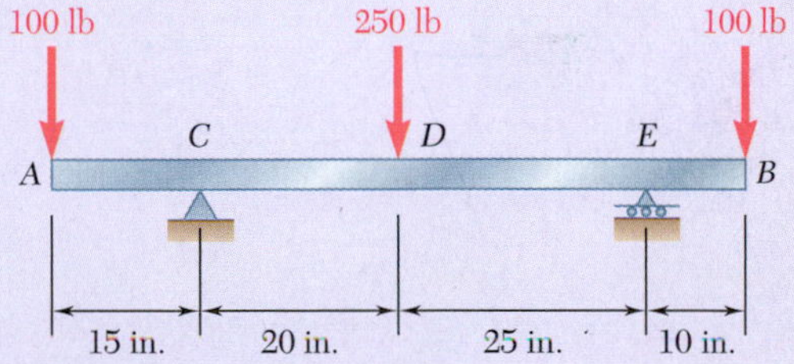

Fig. P5.8

5.9 and 5.10 Draw the shear and bending-moment diagrams for the beam and loading shown, and determine the maximum absolute value (*a*) of the shear, (*b*) of the bending moment.

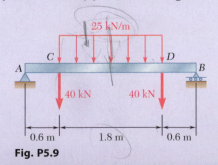

Fig. P5.9

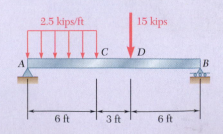

Fig. P5.10

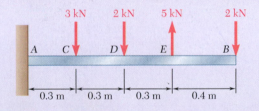

5.11 and 5.12 Draw the shear and bending-moment diagrams for the beam and loading shown, and determine the maximum absolute value (a) of the shear, (b) of the bending moment.

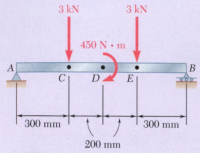

Fig. P5.11

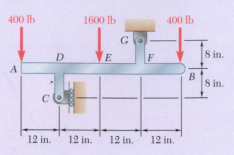

Fig. P5.12

5.13 and 5.14 Assuming that the reaction of the ground is uniformly distributed, draw the shear and bending-moment diagrams for the beam AB and determine the maximum absolute value (a) of the shear, (b) of the bending moment.

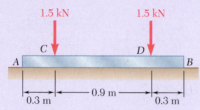

Fig. P5.13

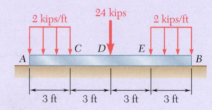

Fig. P5.14

5.15 and 5.16 For the beam and loading shown, determine the maximum normal stress due to bending on a transverse section at C.

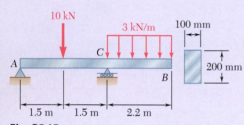

Fig. P5.15

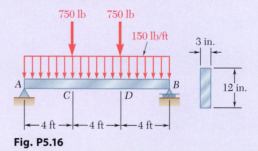

Fig. P5.16

5.17 For the beam and loading shown, determine the maximum normal stress due to bending on a transverse section at C.

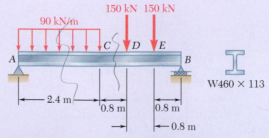

Fig. P5.17

5.18 For the beam and loading shown, determine the maximum normal stress due to bending on section *a-a*.

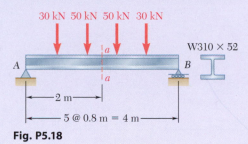

Fig. P5.18

5.19 and 5.20 For the beam and loading shown, determine the maximum normal stress due to bending on a transverse section at *C*.

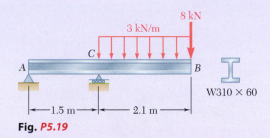

Fig. P5.19

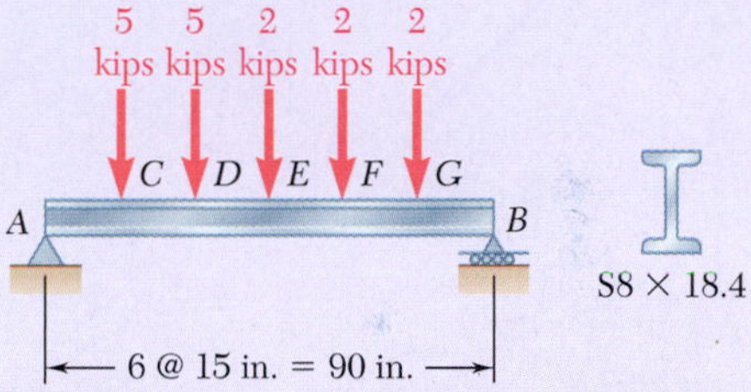

Fig. P5.20

5.21 Draw the shear and bending-moment diagrams for the beam and loading shown and determine the maximum normal stress due to bending.

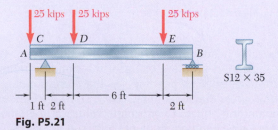

Fig. P5.21

5.22 and 5.23 Draw the shear and bending-moment diagrams for the beam and loading shown and determine the maximum normal stress due to bending.

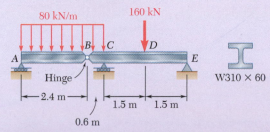

Fig. P5.22

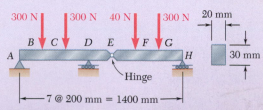

Fig. P5.23

5.24 and 5.25 Draw the shear and bending-moment diagrams for the beam and loading shown and determine the maximum normal stress due to bending.

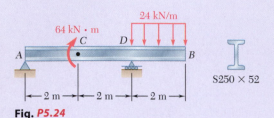

Fig. P5.24

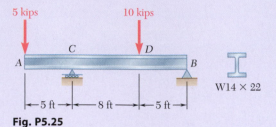

Fig. P5.25

5.26 Knowing that $W = 12$ kN, draw the shear and bending-moment diagrams for beam AB and determine the maximum normal stress due to bending.

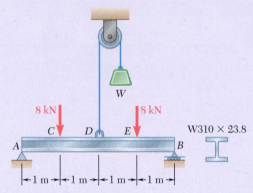

Figs. P5.26 and P5.27

5.27 Determine (*a*) the magnitude of the counterweight W for which the maximum absolute value of the bending moment in the beam is as small as possible, (*b*) the corresponding maximum normal stress due to bending. (*Hint:* Draw the bending-moment diagram and equate the absolute values of the largest positive and negative bending moments obtained.)

5.28 Determine (*a*) the distance a for which the absolute value of the bending moment in the beam is as small as possible, (*b*) the corresponding maximum normal stress due to bending. (See hint of Prob. 5.27.)

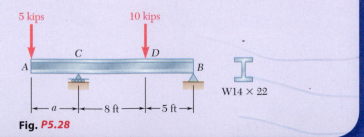

Fig. P5.28

5.29 Knowing that $P = Q = 480$ N, determine (*a*) the distance *a* for which the absolute value of the bending moment in the beam is as small as possible, (*b*) the corresponding maximum normal stress due to bending. (See hint of Prob. 5.27.)

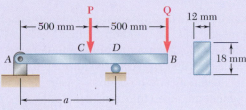

Fig. P5.29

5.30 Solve Prob. 5.29, assuming that P = 480 N and Q = 320 N.

5.31 Determine (*a*) the distance *a* for which the absolute value of the bending moment in the beam is as small as possible, (*b*) the corresponding maximum normal stress due to bending. (See hint of Prob. 5.27.)

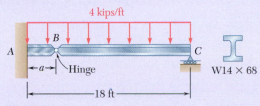

Fig. P5.31

5.32 A solid steel rod of diameter *d* is supported as shown. Knowing that for steel $\gamma = 490$ lb/ft^3, determine the smallest diameter *d* that can be used if the normal stress due to bending is not to exceed 4 ksi.

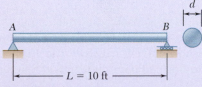

Fig. P5.32

5.33 A solid steel bar has a square cross section of side *b* and is supported as shown. Knowing that for steel $\rho = 7860$ kg/m^3, determine the dimension *b* for which the maximum normal stress due to bending is (*a*) 10 MPa, (*b*) 50 MPa.

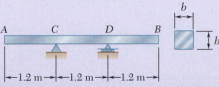

Fig. P5.33

5.2 RELATIONSHIPS BETWEEN LOAD, SHEAR, AND BENDING MOMENT

When a beam carries more than two or three concentrated loads, or when it carries distributed loads, the method outlined in Sec. 5.1 for plotting shear and bending moment can prove quite cumbersome. The construction of the shear diagram and, especially, of the bending-moment diagram will be greatly facilitated if certain relations existing between load, shear, and bending moment are taken into consideration.

For example, a simply supported beam AB is carrying a distributed load w per unit length (Fig. 5.9a), where C and C' are two points of the beam at a distance Δx from each other. The shear and bending moment at C is denoted by V and M, respectively, and is assumed to be positive. The shear and bending moment at C' is denoted by $V + \Delta V$ and $M + \Delta M$.

Detach the portion of beam CC' and draw its free-body diagram (Fig. 5.9b). The forces exerted on the free body include a load of magnitude $w\,\Delta x$ and internal forces and couples at C and C'. Since shear and bending moment are assumed to be positive, the forces and couples are directed as shown.

Relationships between Load and Shear. The sum of the vertical components of the forces acting on the free body CC' is zero, so

$$+\uparrow\Sigma F_y = 0: \qquad\qquad V - (V + \Delta V) - w\,\Delta x = 0$$
$$\Delta V = -w\,\Delta x$$

Dividing both members of the equation by Δx and then letting Δx approach zero,

$$\frac{dV}{dx} = -w \qquad\qquad (5.5)$$

Equation (5.5) indicates that, for a beam loaded as shown in Fig. 5.9a, the slope dV/dx of the shear curve is negative. The magnitude of the slope at any point is equal to the load per unit length at that point.

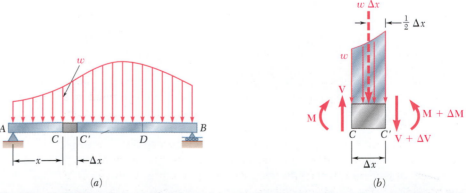

Fig. 5.9 (a) Simply supported beam subjected to a distributed load, with a small element between C and C', (b) free-body diagram of the element.

Integrating Eq. (5.5) between points C and D,

$$V_D - V_C = -\int_{x_C}^{x_D} w\, dx \qquad \textbf{(5.6a)}$$

$$V_D - V_C = -(\text{area under load curve between } C \text{ and } D) \qquad \textbf{(5.6b)}$$

This result is illustrated in Fig. 5.10b. Note that this result could be obtained by considering the equilibrium of the portion of beam CD, since the area under the load curve represents the total load applied between C and D.

Also, Eq. (5.5) is not valid at a point where a concentrated load is applied; the shear curve is discontinuous at such a point, as seen in Sec. 5.1. Similarly, Eqs. (5.6a) and (5.6b) are not valid when concentrated loads are applied between C and D, since they do not take into account the sudden change in shear caused by a concentrated load. Equations (5.6a) and (5.6b), should be applied only between successive concentrated loads.

Relationships between Shear and Bending Moment. Returning to the free-body diagram of Fig. 5.9b and writing that the sum of the moments about C' is zero, we have

$$+\curvearrowleft \Sigma M_{C'} = 0: \qquad (M + \Delta M) - M - V\,\Delta x + w\,\Delta x \frac{\Delta x}{2} = 0$$

$$\Delta M = V\,\Delta x - \frac{1}{2} w (\Delta x)^2$$

Dividing both members by Δx and then letting Δx approach zero,

$$\frac{dM}{dx} = V \qquad \textbf{(5.7)}$$

Equation (5.7) indicates that the slope dM/dx of the bending-moment curve is equal to the value of the shear. This is true at any point where the shear has a well-defined value (i.e., no concentrated load is applied). Equation (5.7) also shows that $V = 0$ at points where M is maximum. This property facilitates the determination of the points where the beam is likely to fail under bending.

Integrating Eq. (5.7) between points C and D,

$$M_D - M_C = \int_{x_C}^{x_D} V\, dx \qquad \textbf{(5.8a)}$$

$$M_D - M_C = \text{area under shear curve between } C \text{ and } D \qquad \textbf{(5.8b)}$$

This result is illustrated in Fig. 5.10c. Note that the area under the shear curve is positive where the shear is positive and negative where the shear is negative. Equations (5.8a) and (5.8b) are valid even when concentrated loads are applied between C and D, as long as the shear curve has been drawn correctly. The equations are not valid if a couple is applied at a point between C and D, since they do not take into account the sudden change in bending moment caused by a couple (see Sample Prob. 5.6).

In most engineering applications, one needs to know the value of the bending moment at only a few specific points. Once the shear diagram has been drawn and after M has been determined at one of the ends of the beam, the value of the bending moment can be obtained at any given point by computing the area under the shear curve and using Eq. (5.8b).

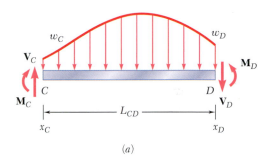

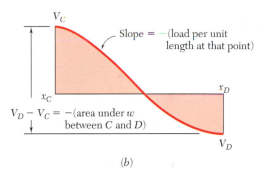

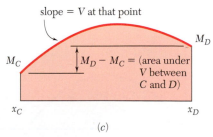

Fig. 5.10 Relationships between load, shear, and bending moment. (a) Section of loaded beam. (b) Shear curve for section. (c) Bending-moment curve for section.

For instance, since $M_A = 0$ for the beam of Concept Application 5.3, the maximum value of the bending moment for that beam is obtained simply by measuring the area of the shaded triangle of the positive portion of the shear diagram of Fig. 5.11c. So,

$$M_{\max} = \frac{1}{2}\frac{L}{2}\frac{wL}{2} = \frac{wL^2}{8}$$

Note that the load curve is a horizontal straight line, the shear curve an oblique straight line, and the bending-moment curve a parabola. If the load curve had been an oblique straight line (first degree), the shear curve would have been a parabola (second degree), and the bending-moment curve a cubic (third degree). The shear and bending-moment curves are always one and two degrees higher than the load curve, respectively. With this in mind, the shear and bending-moment diagrams can be drawn without actually determining the functions $V(x)$ and $M(x)$. The sketches will be more accurate if we make use of the fact that at any point where the curves are continuous, the slope of the shear curve is equal to $-w$ and the slope of the bending-moment curve is equal to V.

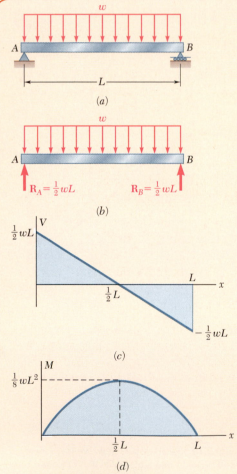

Fig. 5.11 (a) Simply supported beam with uniformly distributed load. (b) Free-body diagram. (c) Shear diagram. (d) Bending-moment diagram.

Concept Application 5.3

Draw the shear and bending-moment diagrams for the simply supported beam shown in Fig. 5.11a and determine the maximum value of the bending moment.

From the free-body diagram of the entire beam (Fig. 5.11b), we determine the magnitude of the reactions at the supports:

$$R_A = R_B = \tfrac{1}{2}wL$$

Next, draw the shear diagram. Close to the end A of the beam, the shear is equal to R_A, (that is, to $\tfrac{1}{2}wL$) which can be checked by considering as a free body a very small portion of the beam. Using Eq. (5.6a), the shear V at any distance x from A is

$$V - V_A = -\int_0^x w\,dx = -wx$$

$$V = V_A - wx = \tfrac{1}{2}wL - wx = w(\tfrac{1}{2}L - x)$$

Thus the shear curve is an oblique straight line that crosses the x axis at $x = L/2$ (Fig. 5.11c). Considering the bending moment, observe that $M_A = 0$. The value M of the bending moment at any distance x from A is obtained from Eq. (5.8a):

$$M - M_A = \int_0^x V\,dx$$

$$M = \int_0^x w(\tfrac{1}{2}L - x)dx = \tfrac{1}{2}w(Lx - x^2)$$

The bending-moment curve is a parabola. The maximum value of the bending moment occurs when $x = L/2$, since V (and thus dM/dx) is zero for this value of x. Substituting $x = L/2$ in the last equation, $M_{\max} = wL^2/8$ (Fig. 5.11d).

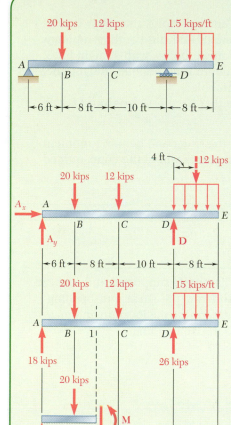

Sample Problem 5.3

Draw the shear and bending-moment diagrams for the beam and loading shown.

STRATEGY: The beam supports two concentrated loads and one distributed load. You can use the equations in this section between these loads and under the distributed load, but you should expect changes in the diagrams at the concentrated load points.

MODELING and ANALYSIS:

Reactions. Consider the entire beam as a free body as shown in Fig. 1.

$+\circlearrowleft \Sigma M_A = 0$:

$$D(24\,\text{ft}) - (20\,\text{kips})(6\,\text{ft}) - (12\,\text{kips})(14\,\text{ft}) - (12\,\text{kips})(28\,\text{ft}) = 0$$

$$D = +26\,\text{kips} \qquad\qquad \mathbf{D} = 26\,\text{kips} \uparrow$$

$+\uparrow \Sigma F_y = 0$: $\quad A_y - 20\,\text{kips} - 12\,\text{kips} + 26\,\text{kips} - 12\,\text{kips} = 0$

$$A_y = +18\,\text{kips} \qquad\qquad \mathbf{A}_y = 18\,\text{kips} \uparrow$$

$\xrightarrow{+} \Sigma F_x = 0$: $\quad A_x = 0 \qquad\qquad \mathbf{A}_x = 0$

Note that at both A and E the bending moment is zero. Thus, two points (indicated by dots) are obtained on the bending-moment diagram.

Shear Diagram. Since $dV/dx = -w$, between concentrated loads and reactions the slope of the shear diagram is zero (i.e., the shear is constant). The shear at any point is determined by dividing the beam into two parts and considering either part to be a free body. For example, using the portion of beam to the left of section 1, the shear between B and C is

$+\uparrow \Sigma F_y = 0$: $\qquad +18\,\text{kips} - 20\,\text{kips} - V = 0 \qquad V = -2\,\text{kips}$

Also, the shear is $+12$ kips just to the right of D and zero at end E. Since the slope $dV/dx = -w$ is constant between D and E, the shear diagram between these two points is a straight line.

Bending-Moment Diagram. Recall that the area under the shear curve between two points is equal to the change in bending moment between the same two points. For convenience, the area of each portion of the shear diagram is computed and indicated in parentheses on the diagram in Fig. 1. Since the bending moment M_A at the left end is known to be zero,

$M_B - M_A = +108$	$M_B = +108\,\text{kip}\cdot\text{ft}$
$M_C - M_B = -16$	$M_C = +92\,\text{kip}\cdot\text{ft}$
$M_D - M_C = -140$	$M_D = -48\,\text{kip}\cdot\text{ft}$
$M_E - M_D = +48$	$M_E = 0$

Since M_E is known to be zero, a check of the computations is obtained.

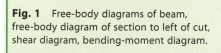

Fig. 1 Free-body diagrams of beam, free-body diagram of section to left of cut, shear diagram, bending-moment diagram.

(continued)

Between the concentrated loads and reactions, the shear is constant. Thus, the slope dM/dx is constant, and the bending-moment diagram is drawn by connecting the known points with straight lines. Between D and E where the shear diagram is an oblique straight line, the bending-moment diagram is a parabola.

From the V and M diagrams, note that $V_{max} = 18$ kips and $M_{max} = 108$ kip·ft.

REFLECT and THINK: As expected, the shear and bending-moment diagrams show abrupt changes at the points where the concentrated loads act.

Sample Problem 5.4

The W360 × 79 rolled-steel beam AC is simply supported and carries the uniformly distributed load shown. Draw the shear and bending-moment diagrams for the beam, and determine the location and magnitude of the maximum normal stress due to bending.

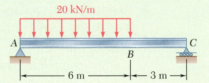

STRATEGY: A load is distributed over part of the beam. You can use the equations in this section in two parts: for the load and for the no-load regions. From the discussion in this section, you can expect the shear diagram will show an oblique line under the load, followed by a horizontal line. The bending-moment diagram should show a parabola under the load and an oblique line under the rest of the beam.

MODELING and ANALYSIS:

Reactions. Considering the entire beam as a free body (Fig. 1),

$$\mathbf{R}_A = 80 \text{ kN} \uparrow \qquad \mathbf{R}_C = 40 \text{ kN} \uparrow$$

Shear Diagram. The shear just to the right of A is $V_A = +80$ kN. Since the change in shear between two points is equal to *minus* the area under the load curve between the same two points, V_B is

$$V_B - V_A = -(20 \text{ kN/m})(6 \text{ m}) = -120 \text{ kN}$$
$$V_B = -120 + V_A = -120 + 80 = -40 \text{ kN}$$

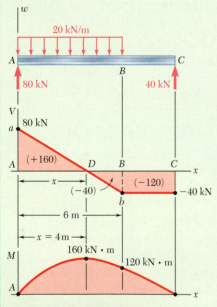

Fig. 1 Free-body diagram, shear diagram, bending-moment diagram.

(continued)

The slope $dV/dx = -w$ is constant between A and B, and the shear diagram between these two points is represented by a straight line. Between B and C, the area under the load curve is zero; therefore,

$$V_C - V_B = 0 \qquad V_C = V_B = -40\,\text{kN}$$

and the shear is constant between B and C.

Bending-Moment Diagram. Note that the bending moment at each end is zero. In order to determine the maximum bending moment, locate the section D of the beam where $V = 0$.

$$V_D - V_A = -wx$$

$$0 - 80\,\text{kN} = -(20\,\text{kN/m})x$$

Solving for x, $\qquad\qquad\qquad\qquad\qquad\qquad$ **$x = 4\,\text{m}$** ◄

The maximum bending moment occurs at point D, where $dM/dx = V = 0$. The areas of various portions of the shear diagram are computed and given (in parentheses). The area of the shear diagram between two points is equal to the change in bending moment between the same two points, giving

$$M_D - M_A = +\,160\,\text{kN·m} \qquad\qquad M_D = +160\,\text{kN·m}$$

$$M_B - M_D = -\;\;40\,\text{kN·m} \qquad\qquad M_B = +120\,\text{kN·m}$$

$$M_C - M_B = -\,120\,\text{kN·m} \qquad\qquad M_C = 0$$

The bending-moment diagram consists of an arc of parabola followed by a segment of straight line. The slope of the parabola at A is equal to the value of V at that point.

Maximum Normal Stress. This occurs at D, where $|M|$ is largest. From Appendix C, for a W360 × 79 rolled-steel shape, $S = 1270\,\text{mm}^3$ about a horizontal axis. Substituting this and $|M| = |M_D| = 160 \times 10^3\,\text{N·m}$ into Eq. (5.3),

$$\sigma_m = \frac{|M_D|}{S} = \frac{160 \times 10^3\,\text{N·m}}{1270 \times 10^{-6}\,\text{m}^3} = 126.0 \times 10^6\,\text{Pa}$$

Maximum normal stress in the beam = **126.0 MPa** ◄

Sample Problem 5.5

Sketch the shear and bending-moment diagrams for the cantilever beam shown in Fig. 1.

STRATEGY: Because there are no support reactions until the right end of the beam, you can rely solely on the equations from this section without needing to use free-body diagrams and equilibrium equations. Due to the non-uniform distributed load, you should expect the results to involve equations of higher degree, with a parabolic curve in the shear diagram and a cubic curve in the bending-moment diagram.

MODELING and ANALYSIS:

Shear Diagram. At the free end of the beam, $V_A = 0$. Between A and B, the area under the load curve is $\frac{1}{2} w_0 a$. Thus,

$$V_B - V_A = -\tfrac{1}{2} w_0 a \qquad V_B = -\tfrac{1}{2} w_0 a$$

Between B and C, the beam is not loaded, so $V_C = V_B$. At A, $w = w_0$. According to Eq. (5.5), the slope of the shear curve is $dV/dx = -w_0$, while at B the slope is $dV/dx = 0$. Between A and B, the loading decreases linearly, and the shear diagram is parabolic. Between B and C, $w = 0$, and the shear diagram is a horizontal line.

Bending-Moment Diagram. The bending moment M_A at the free end of the beam is zero. Compute the area under the shear curve to obtain.

$$M_B - M_A = -\tfrac{1}{3} w_0 a^2 \qquad M_B = -\tfrac{1}{3} w_0 a^2$$

$$M_C - M_B = -\tfrac{1}{2} w_0 a (L - a)$$

$$M_C = -\tfrac{1}{6} w_0 a (3L - a)$$

The sketch of the bending-moment diagram is completed by recalling that $dM/dx = V$. Between A and B, the diagram is represented by a cubic curve with zero slope at A and between B and C by a straight line.

REFLECT and THINK: Although not strictly required for the solution of this problem, determination of the support reactions would serve as an excellent check of the final values of the shear and bending-moment diagrams.

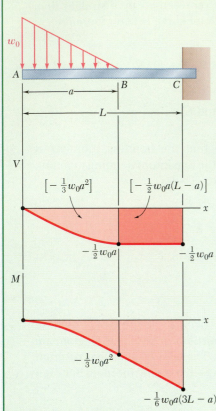

Fig. 1 Beam with load, shear diagram, bending-moment diagram.

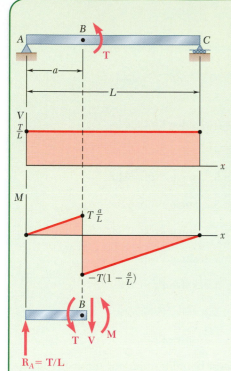

Fig. 1 Beam with load, shear diagram, bending-moment diagram, free-body diagram of section to left of *B*.

Sample Problem 5.6

The simple beam *AC* in Fig. 1 is loaded by a couple of moment *T* applied at point *B*. Draw the shear and bending-moment diagrams of the beam.

STRATEGY: The load supported by the beam is a concentrated couple. Since the only vertical forces are those associated with the support reactions, you should expect the shear diagram to be of constant value. However, the bending-moment diagram will have a discontinuity at *B* due to the couple.

MODELING and ANALYSIS:

The entire beam is taken as a free body.

$$\mathbf{R}_A = \frac{T}{L} \uparrow \qquad \mathbf{R}_C = \frac{T}{L} \downarrow$$

The shear at any section is constant and equal to T/L. Since a couple is applied at *B*, the bending-moment diagram is discontinuous at *B*. It is represented by two oblique straight lines and decreases suddenly at *B* by an amount equal to *T*. This discontinuity can be verified by equilibrium analysis. For example, considering the free body of the portion of the beam from *A* to just beyond the right of *B* as shown in Fig. 1, *M* is

$$+\curvearrowleft\Sigma M_B = 0: \quad -\frac{T}{L}a + T + M = 0 \quad M = -T\left(1 - \frac{a}{L}\right)$$

REFLECT and THINK: Notice that the applied couple results in a sudden change to the moment diagram at the point of application in the same way that a concentrated force results in a sudden change to the shear diagram.

Problems

5.34 Using the method of Sec. 5.2, solve Prob. 5.1*a*.

5.35 Using the method of Sec. 5.2, solve Prob. 5.2*a*.

5.36 Using the method of Sec. 5.2, solve Prob. 5.3*a*.

5.37 Using the method of Sec. 5.2, solve Prob. 5.4*a*.

5.38 Using the method of Sec. 5.2, solve Prob. 5.5*a*.

5.39 Using the method of Sec. 5.2, solve Prob. 5.6*a*.

5.40 Using the method of Sec. 5.2, solve Prob. 5.7.

5.41 Using the method of Sec. 5.2, solve Prob. 5.8.

5.42 Using the method of Sec. 5.2, solve Prob. 5.9.

5.43 Using the method of Sec. 5.2, solve Prob. 5.10.

5.44 and 5.45 Draw the shear and bending-moment diagrams for the beam and loading shown, and determine the maximum absolute value (*a*) of the shear, (*b*) of the bending moment.

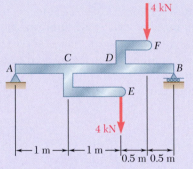

Fig. P5.44

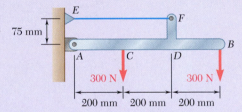

Fig. P5.45

5.46 Using the method of Sec. 5.2, solve Prob. 5.15.

5.47 Using the method of Sec. 5.2, solve Prob. 5.16.

5.48 Using the method of Sec. 5.2, solve Prob. 5.18.

5.49 Using the method of Sec. 5.2, solve Prob. 5.20.

5.50 and 5.51 Determine (*a*) the equations of the shear and bending-moment curves for the beam and loading shown, (*b*) the maximum absolute value of the bending moment in the beam.

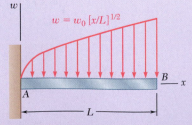

Fig. P5.50

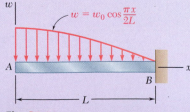

Fig. P5.51

5.52 and 5.53 Determine (*a*) the equations of the shear and bending-moment curves for the beam and loading shown, (*b*) the maximum absolute value of the bending moment in the beam.

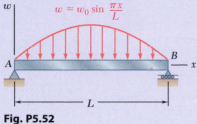

Fig. P5.52

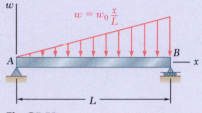

Fig. P5.53

5.54 and 5.55 Draw the shear and bending-moment diagrams for the beam and loading shown and determine the maximum normal stress due to bending.

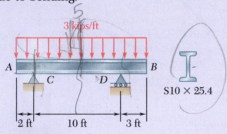

Fig. P5.54

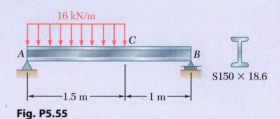

Fig. P5.55

5.56 and *5.57* Draw the shear and bending-moment diagrams for the beam and loading shown and determine the maximum normal stress due to bending.

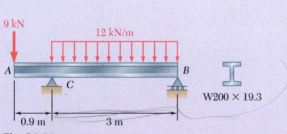

Fig. P5.56

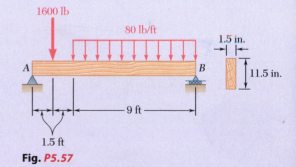

Fig. *P5.57*

5.58 and 5.59 Draw the shear and bending-moment diagrams for the beam and loading shown and determine the maximum normal stress due to bending.

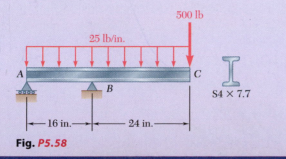

Fig. *P5.58*

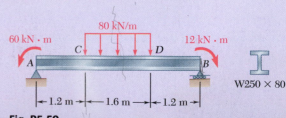

Fig. P5.59

5.60 Knowing that beam AB is in equilibrium under the loading shown, draw the shear and bending-moment diagrams and determine the maximum normal stress due to bending.

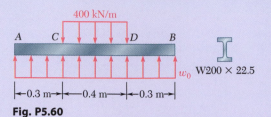

Fig. P5.60

5.61 Knowing that beam AB is in equilibrium under the loading shown, draw the shear and bending-moment diagrams and determine the maximum normal stress due to bending.

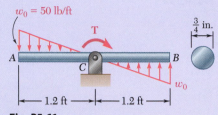

Fig. P5.61

*5.62 The beam AB supports two concentrated loads **P** and **Q**. The normal stress due to bending on the bottom edge of the beam is +55 MPa at D and +37.5 MPa at F. (*a*) Draw the shear and bending-moment diagrams for the beam. (*b*) Determine the maximum normal stress due to bending that occurs in the beam.

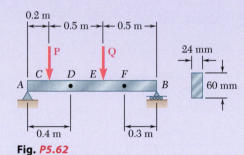

Fig. P5.62

*5.63 The beam AB supports a uniformly distributed load of 480 lb/ft and two concentrated loads **P** and **Q**. The normal stress due to bending on the bottom edge of the lower flange is +14.85 ksi at D and +10.65 ksi at E. (*a*) Draw the shear and bending-moment diagrams for the beam. (*b*) Determine the maximum normal stress due to bending that occurs in the beam.

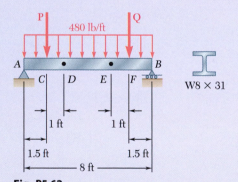

Fig. P5.63

*5.64 Beam AB supports a uniformly distributed load of 2 kN/m and two concentrated loads **P** and **Q**. It has been experimentally determined that the normal stress due to bending in the bottom edge of the beam is −56.9 MPa at A and −29.9 MPa at C. Draw the shear and bending-moment diagrams for the beam and determine the magnitudes of the loads **P** and **Q**.

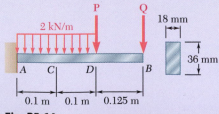

Fig. P5.64

5.3   DESIGN OF PRISMATIC BEAMS FOR BENDING

The design of a beam is usually controlled by the maximum absolute value $|M|_{max}$ of the bending moment that occurs in the beam. The largest normal stress σ_m in the beam is found at the surface of the beam in the critical section where $|M|_{max}$ occurs and is obtained by substituting $|M|_{max}$ for $|M|$ in Eq. (5.1) or Eq. (5.3).[†]

$$\sigma_m = \frac{|M|_{max}c}{I} \tag{5.1a}$$

$$\sigma_m = \frac{|M|_{max}}{S} \tag{5.3a}$$

A safe design requires that $\sigma_m \leq \sigma_{all}$, where σ_{all} is the allowable stress for the material used. Substituting σ_{all} for σ_m in (5.3a) and solving for S yields the minimum allowable value of the section modulus for the beam being designed:

$$S_{min} = \frac{|M|_{max}}{\sigma_{all}} \tag{5.9}$$

The design of common types of beams, such as timber beams of rectangular cross section and rolled-steel members of various cross-sectional shapes, is discussed in this section. A proper procedure should lead to the most economical design. This means that among beams of the same type and same material, and other things being equal, the beam with the smallest weight per unit length—and, thus, the smallest cross-sectional area—should be selected, since this beam will be the least expensive.

The design procedure generally includes the following steps:[‡]

Step 1. First determine the value of σ_{all} for the material selected from a table of properties of materials or from design specifications. You also can compute this value by dividing the ultimate strength σ_U of the material by an appropriate factor of safety (Sec. 1.5C). Assuming that the value of σ_{all} is the same in tension and in compression, proceed as follows.

Step 2. Draw the shear and bending-moment diagrams corresponding to the specified loading conditions, and determine the maximum absolute value $|M|_{max}$ of the bending moment in the beam.

Step 3. Determine from Eq. (5.9) the minimum allowable value S_{min} of the section modulus of the beam.

Step 4. For a timber beam, the depth h of the beam, its width b, or the ratio h/b characterizing the shape of its cross section probably will have been specified. The unknown dimensions can be selected by using Eq. (4.19), so b and h satisfy the relation $\frac{1}{6}bh^2 = S \geq S_{min}$.

[†]For beams that are not symmetrical with respect to their neutral surface, the largest of the distances from the neutral surface to the surfaces of the beam should be used for c in Eq. (5.1) and in the computation of the section modulus $S = I/c$.

[‡]It is assumed that all beams considered in this chapter are adequately braced to prevent lateral buckling and bearing plates are provided under concentrated loads applied to rolled-steel beams to prevent local buckling (crippling) of the web.

Step 5. For a rolled-steel beam, consult the appropriate table in Appendix C. Of the available beam sections, consider only those with a section modulus $S \geq S_{min}$ and select the section with the smallest weight per unit length. This is the most economical of the sections for which $S \geq S_{min}$. Note that this is not necessarily the section with the smallest value of S (see Concept Application 5.4). In some cases, the selection of a section may be limited by considerations such as the allowable depth of the cross section or the allowable deflection of the beam (see Chap. 9).

The previous discussion was limited to materials for which σ_{all} is the same in tension and compression. If σ_{all} is different, make sure to select the beam section where $\sigma_m \leq \sigma_{all}$ for both tensile and compressive stresses. If the cross section is not symmetric about its neutral axis, the largest tensile and the largest compressive stresses will not necessarily occur in the section where $|M|$ is maximum (one may occur where M is maximum and the other where M is minimum). Thus, step 2 should include the determination of both M_{max} and M_{min}, and step 3 should take into account both tensile and compressive stresses.

Finally, the design procedure described in this section takes into account only the normal stresses occurring on the surface of the beam. Short beams, especially those made of timber, may fail in shear under a transverse loading. The determination of shearing stresses in beams will be discussed in Chap. 6. Also, in rolled-steel beams normal stresses larger than those considered here may occur at the junction of the web with the flanges. This will be discussed in Chap. 8.

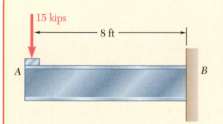

Fig. 5.12 Cantilevered wide-flange beam with end load.

Concept Application 5.4

Select a wide-flange beam to support the 15-kip load as shown in Fig. 5.12. The allowable normal stress for the steel used is 24 ksi.

1. The allowable normal stress is given: $\sigma_{all} = 24$ ksi.
2. The shear is constant and equal to 15 kips. The bending moment is maximum at B.

$$|M|_{max} = (15 \text{ kips})(8 \text{ ft}) = 120 \text{ kip·ft} = 1440 \text{ kip·in.}$$

3. The minimum allowable section modulus is

$$S_{min} = \frac{|M|_{max}}{\sigma_{all}} = \frac{1440 \text{ kip·in.}}{24 \text{ ksi}} = 60.0 \text{ in}^3$$

4. Referring to the table of *Properties of Rolled-Steel Shapes* in Appendix C, note that the shapes are arranged in groups of the same depth and are listed in order of decreasing weight. Choose the lightest beam in each group having a section modulus

(continued)

$S = I/c$ at least as large as S_{min} and record the results in the following table.

Shape	S, in^3
W21 × 44	81.6
W18 × 50	88.9
W16 × 40	64.7
W14 × 43	62.6
W12 × 50	64.2
W10 × 54	60.0

The most economical is the W16 × 40 shape since it weighs only 40 lb/ft, even though it has a larger section modulus than two of the other shapes. The total weight of the beam will be (8 ft) × (40 lb) = 320 lb. This weight is small compared to the 15,000-1b load and thus can be neglected in our analysis.

*Load and Resistance Factor Design.   This alternative method of design was applied to members under axial loading in Sec. 1.5D. It also can be applied to the design of beams in bending. Replace the loads P_D, P_L, and P_U in Eq. (1.27) by the bending moments M_D, M_L, and M_U:

$$\gamma_D M_D + \gamma_L M_L \leq \phi M_U \qquad \textbf{(5.10)}$$

The coefficients γ_D and γ_L are the *load factors,* and the coefficient ϕ is the *resistance factor*. The moments M_D and M_L are the bending moments due to the dead and the live loads respectively. M_U is equal to the product of the ultimate strength σ_U of the material and the section modulus S of the beam: $M_U = S\sigma_U$.

Sample Problem 5.7

A 12-ft-long overhanging timber beam AC with an 8-ft span AB is to be designed to support the distributed and concentrated loads shown. Knowing that timber of 4-in. nominal width (3.5-in. actual width) with a 1.75-ksi allowable stress is to be used, determine the minimum required depth h of the beam.

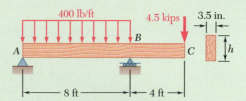

(continued)

STRATEGY: Draw the bending-moment diagram to find the absolute maximum bending-moment. Then, using this bending-moment, you can determine the required section properties that satisfy the given allowable stress.

MODELING and ANALYSIS:

Reactions. Consider the entire beam to be a free body (Fig. 1).

$$+\circlearrowleft\Sigma M_A = 0: B(8\text{ ft}) - (3.2\text{ kips})(4\text{ ft}) - (4.5\text{ kips})(12\text{ ft}) = 0$$

$$B = 8.35\text{ kips} \qquad \mathbf{B} = 8.35\text{ kips} \uparrow$$

$$\xrightarrow{+}\Sigma F_x = 0: \qquad A_x = 0$$

$$+\uparrow\Sigma F_y = 0: A_y + 8.35\text{ kips} - 3.2\text{ kips} - 4.5\text{ kips} = 0$$

$$A_y = -0.65\text{ kips} \qquad \mathbf{A} = 0.65\text{ kips} \downarrow$$

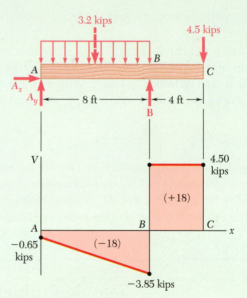

Fig. 1 Free-body diagram of beam and its shear diagram.

Shear Diagram. The shear just to the right of A is $V_A = A_y = -0.65$ kips. Since the change in shear between A and B is equal to *minus* the area under the load curve between these two points, V_B is obtained by

$$V_B - V_A = -(400\text{ lb/ft})(8\text{ ft}) = -3200\text{ lb} = -3.20\text{ kips}$$

$$V_B = V_A - 3.20\text{ kips} = -0.65\text{ kips} - 3.20\text{ kips} = -3.85\text{ kips}.$$

The reaction at B produces a sudden increase of 8.35 kips in V, resulting in a shear equal to 4.50 kips to the right of B. Since no load is applied between B and C, the shear remains constant between these two points.

(continued)

Determination of $|M|_{max}$**.**   Observe that the bending moment is equal to zero at both ends of the beam: $M_A = M_C = 0$. Between A and B, the bending moment decreases by an amount equal to the area under the shear curve, and between B and C it increases by a corresponding amount. Thus, the maximum absolute value of the bending moment is $|M|_{max} = 18.00$ kip·ft.

Minimum Allowable Section Modulus.   Substituting the values of σ_{all} and $|M|_{max}$ into Eq. (5.9) gives

$$S_{min} = \frac{|M|_{max}}{\sigma_{all}} = \frac{(18\text{ kip·ft})(12\text{ in./ft})}{1.75\text{ ksi}} = 123.43\text{ in}^3$$

Minimum Required Depth of Beam.   Recalling the formula developed in step 4 of the design procedure and substituting the values of b and S_{min}, we have

$$\tfrac{1}{6}bh^2 \geq S_{min} \qquad \tfrac{1}{6}(3.5\text{ in.})h^2 \geq 123.43\text{ in}^3 \qquad h \geq 14.546\text{ in.}$$

The minimum required depth of the beam is $h = 14.55$ in.   ◀

REFLECT and THINK: In practice, standard wood shapes are specified by nominal dimensions that are slightly larger than actual. In this case, specify a 4-in. × 16-in. member with the actual dimensions of 3.5 in. × 15.25 in.

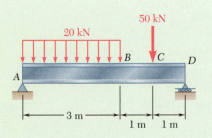

Sample Problem 5.8

A 5-m-long, simply supported steel beam AD is to carry the distributed and concentrated loads shown. Knowing that the allowable normal stress for the grade of steel is 160 MPa, select the wide-flange shape to be used.

STRATEGY: Draw the bending-moment diagram to find the absolute maximum bending moment. Then, using this moment, you can determine the required section modulus that satisfies the given allowable stress.

(continued)

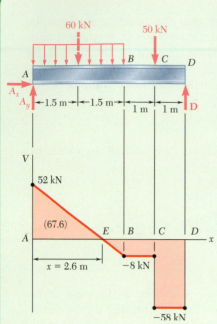

Fig. 1 Free-body diagram of beam and its shear diagram.

Shape	S, mm^3
W410 × 38.8	629
W360 × 32.9	475
W310 × 38.7	547
W250 × 44.8	531
W200 × 46.1	451

Fig. 2 Lightest shapes in each depth group that provide the required section modulus.

MODELING and ANALYSIS:

Reactions. Consider the entire beam to be a free body (Fig. 1).

$+\circlearrowleft\Sigma M_A = 0: D(5\,\text{m}) - (60\,\text{kN})(1.5\,\text{m}) - (50\,\text{kN})(4\,\text{m}) = 0$

$\qquad\qquad\qquad D = 58.0\,\text{kN} \qquad \mathbf{D} = 58.0\,\text{kN} \uparrow$

$+\rightarrow\Sigma F_x = 0: \qquad A_x = 0$

$+\uparrow\Sigma F_y = 0: A_y + 58.0\,\text{kN} - 60\,\text{kN} - 50\,\text{kN} = 0$

$\qquad\qquad\qquad A_y = 52.0\,\text{kN} \searrow \quad \mathbf{A} = 52.0\,\text{kN} \uparrow$

Shear Diagram. The shear just to the right of A is $V_A = A_y = +52.0\,\text{kN}$. Since the change in shear between A and B is equal to *minus* the area under the load curve between these two points,

$$V_B = 52.0\,\text{kN} - 60\,\text{kN} = -8\,\text{kN}$$

The shear remains constant between B and C, where it drops to $-58\,\text{kN}$, and keeps this value between C and D. Locate the section E of the beam where $V = 0$ by

$$V_E - V_A = -wx$$
$$0 - 52.0\,\text{kN} = -(20\,\text{kN/m})x$$

So, $x = 2.60\,\text{m}$.

Determination of $|M|_{max}$. The bending moment is maximum at E, where $V = 0$. Since M is zero at the support A, its maximum value at E is equal to the area under the shear curve between A and E. Therefore, $|M|_{max} = M_E = 67.6\,\text{kN·m}$.

Minimum Allowable Section Modulus. Substituting the values of σ_{all} and $|M|_{max}$ into Eq. (5.9) gives

$$S_{min} = \frac{|M|_{max}}{\sigma_{all}} = \frac{67.6\,\text{kN·m}}{160\,\text{MPa}} = 422.5 \times 10^{-6}\,\text{m}^3 = 422.5 \times 10^3\,\text{mm}^3$$

Selection of Wide-Flange Shape. From Appendix C, compile a list of shapes that have a section modulus larger than S_{min} and are also the lightest shape in a given depth group (Fig. 2).

The lightest shape available is   **W360 × 32.9** ◀

REFLECT and THINK: When a specific allowable normal stress is the sole design criterion for beams, the lightest acceptable shapes tend to be deeper sections. In practice, there will be other criteria to consider that may alter the final shape selection.

Problems

5.65 and *5.66* For the beam and loading shown, design the cross section of the beam, knowing that the grade of timber used has an allowable normal stress of 12 MPa.

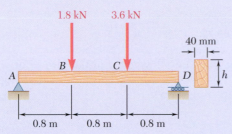

Fig. P5.65

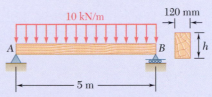

Fig. *P5.66*

_5.67_ and 5.68 For the beam and loading shown, design the cross section of the beam, knowing that the grade of timber used has an allowable normal stress of 1750 psi.

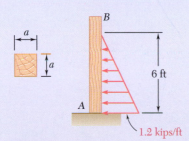

Fig. *P5.67*

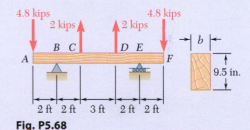

Fig. P5.68

5.69 and 5.70 For the beam and loading shown, design the cross section of the beam, knowing that the grade of timber used has an allowable normal stress of 12 MPa.

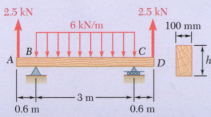

Fig. P5.69

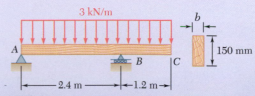

Fig. P5.70

5.71 and 5.72 Knowing that the allowable normal stress for the steel used is 24 ksi, select the most economical wide-flange beam to support the loading shown.

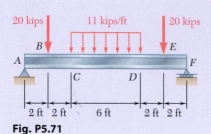

Fig. P5.71

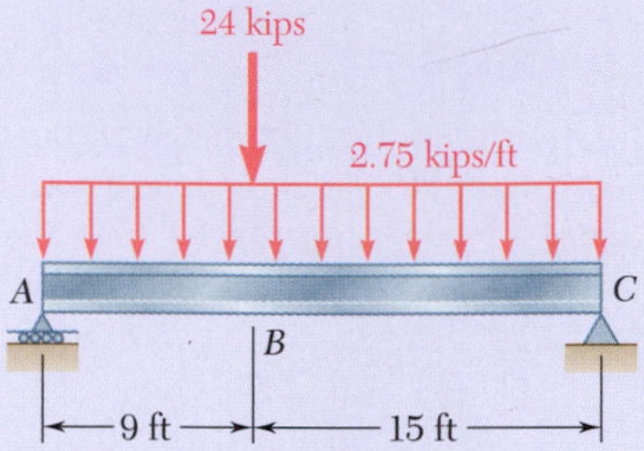

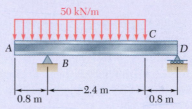

Fig. P5.72

5.73 and 5.74 Knowing that the allowable normal stress for the steel used is 160 MPa, select the most economical wide-flange beam to support the loading shown.

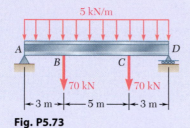

Fig. P5.73

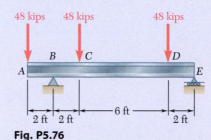

Fig. P5.74

5.75 and 5.76 Knowing that the allowable normal stress for the steel used is 24 ksi, select the most economical S-shape beam to support the loading shown.

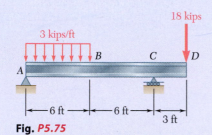

Fig. P5.75

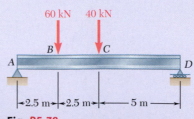

Fig. P5.76

5.77 and 5.78 Knowing that the allowable normal stress for the steel used is 160 MPa, select the most economical S-shape beam to support the loading shown.

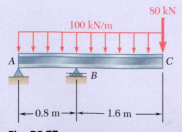

Fig. P5.77

Fig. P5.78

5.79 A steel pipe of 100-mm diameter is to support the loading shown. Knowing that the stock of pipes available has thicknesses varying from 6 mm to 24 mm in 3-mm increments, and that the allowable normal stress for the steel used is 150 MPa, determine the minimum wall thickness t that can be used.

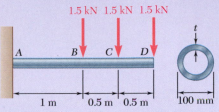

Fig. P5.79

5.80 Two metric rolled-steel channels are to be welded along their edges and used to support the loading shown. Knowing that the allowable normal stress for the steel used is 150 MPa, determine the most economical channels that can be used.

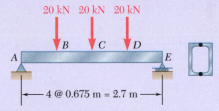

Fig. P5.80

5.81 Two rolled-steel channels are to be welded back to back and used to support the loading shown. Knowing that the allowable normal stress for the steel used is 30 ksi, determine the most economical channels that can be used.

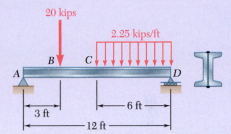

Fig. P5.81

5.82 Two L4 × 3 rolled-steel angles are bolted together and used to support the loading shown. Knowing that the allowable normal stress for the steel used is 24 ksi, determine the minimum angle thickness that can be used.

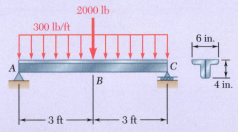

Fig. P5.82

5.83 Assuming the upward reaction of the ground to be uniformly distributed and knowing that the allowable normal stress for the steel used is 170 MPa, select the most economical wide-flange beam to support the loading shown.

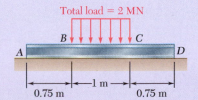

Fig. P5.83

5.84 Assuming the upward reaction of the ground to be uniformly distributed and knowing that the allowable normal stress for the steel used is 24 ksi, select the most economical wide-flange beam to support the loading shown.

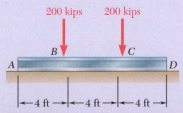

Fig. P5.84

5.85 Determine the largest permissible distributed load w for the beam shown, knowing that the allowable normal stress is $+80$ MPa in tension and -130 MPa in compression.

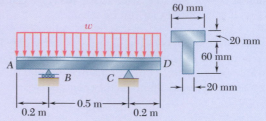

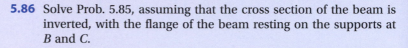

Fig. P5.85

5.86 Solve Prob. 5.85, assuming that the cross section of the beam is inverted, with the flange of the beam resting on the supports at B and C.

5.87 Determine the largest permissible value of **P** for the beam and loading shown, knowing that the allowable normal stress is $+8$ ksi in tension and -18 ksi in compression.

5.88 Solve Prob. 5.87, assuming that the T-shaped beam is inverted.

5.89 Beams AB, BC, and CD have the cross section shown and are pin-connected at B and C. Knowing that the allowable normal stress is $+110$ MPa in tension and -150 MPa in compression, determine (a) the largest permissible value of w if beam BC is not to be overstressed, (b) the corresponding maximum distance a for which the cantilever beams AB and CD are not overstressed.

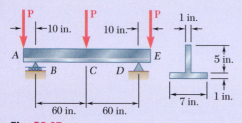

Fig. P5.87

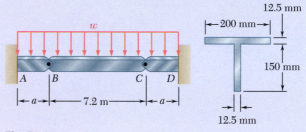

Fig. P5.89

5.90 Beams AB, BC, and CD have the cross section shown and are pin-connected at B and C. Knowing that the allowable normal stress is $+110$ MPa in tension and -150 MPa in compression, determine (a) the largest permissible value of **P** if beam BC is not to be overstressed, (b) the corresponding maximum distance a for which the cantilever beams AB and CD are not overstressed.

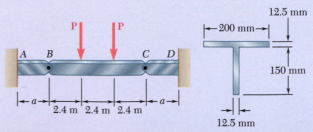

Fig. P5.90

5.91 Each of the three rolled-steel beams shown (numbered 1, 2, and 3) is to carry a 64-kip load uniformly distributed over the beam. Each of these beams has a 12-ft span and is to be supported by the two 24-ft rolled-steel girders *AC* and *BD*. Knowing that the allowable normal stress for the steel used is 24 ksi, select (*a*) the most economical S shape for the three beams, (*b*) the most economical W shape for the two girders.

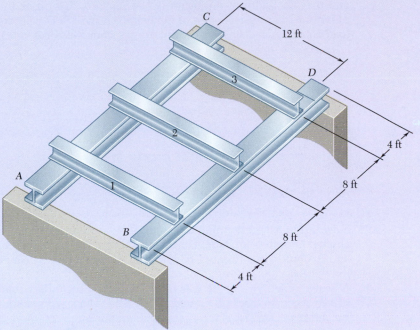

Fig. P5.91

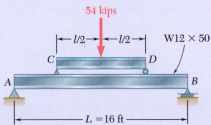

Fig. P5.92

5.92 A 54-kip load is to be supported at the center of the 16-ft span shown. Knowing that the allowable normal stress for the steel used is 24 ksi, determine (*a*) the smallest allowable length *l* of beam *CD* if the W12 × 50 beam *AB* is not to be overstressed, (*b*) the most economical W shape that can be used for beam *CD*. Neglect the weight of both beams.

5.93 A uniformly distributed load of 66 kN/m is to be supported over the 6-m span shown. Knowing that the allowable normal stress for the steel used is 140 MPa, determine (*a*) the smallest allowable length *l* of beam *CD* if the W460 × 74 beam *AB* is not to be overstressed, (*b*) the most economical W shape that can be used for beam *CD*. Neglect the weight of both beams.

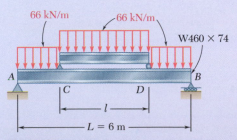

Fig. P5.93

*5.94 A roof structure consists of plywood and roofing material supported by several timber beams of length $L = 16$ m. The dead load carried by each beam, including the estimated weight of the beam, can be represented by a uniformly distributed load $w_D = 350$ N/m. The live load consists of a snow load, represented by a uniformly distributed load $w_L = 600$ N/m, and a 6-kN concentrated load **P** applied at the midpoint C of each beam. Knowing that the ultimate strength for the timber used is $\sigma_U = 50$ MPa and that the width of the beam is $b = 75$ mm, determine the minimum allowable depth h of the beams, using LRFD with the load factors $\gamma_D = 1.2$, $\gamma_L = 1.6$ and the resistance factor $\phi = 0.9$.

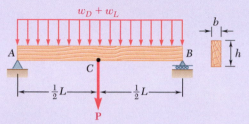

Fig. P5.94

*5.95 Solve Prob. 5.94, assuming that the 6-kN concentrated load **P** applied to each beam is replaced by 3-kN concentrated loads **P**$_1$ and **P**$_2$ applied at a distance of 4 m from each end of the beams.

*5.96 A bridge of length $L = 48$ ft is to be built on a secondary road whose access to trucks is limited to two-axle vehicles of medium weight. It will consist of a concrete slab and of simply supported steel beams with an ultimate strength $\sigma_U = 60$ ksi. The combined weight of the slab and beams can be approximated by a uniformly distributed load $w = 0.75$ kips/ft on each beam. For the purpose of the design, it is assumed that a truck with axles located at a distance $a = 14$ ft from each other will be driven across the bridge and that the resulting concentrated loads **P**$_1$ and **P**$_2$ exerted on each beam could be as large as 24 kips and 6 kips, respectively. Determine the most economical wide-flange shape for the beams, using LRFD with the load factors $\gamma_D = 1.25$, $\gamma_L = 1.75$ and the resistance factor $\phi = 0.9$. [*Hint:* It can be shown that the maximum value of $|M_L|$ occurs under the larger load when that load is located to the left of the center of the beam at a distance equal to $aP_2/2(P_1 + P_2)$.]

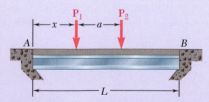

Fig. P5.96

*5.97 Assuming that the front and rear axle loads remain in the same ratio as for the truck of Prob. 5.96, determine how much heavier a truck could safely cross the bridge designed in that problem.

*5.4   SINGULARITY FUNCTIONS USED TO DETERMINE SHEAR AND BENDING MOMENT

Note that the shear and bending moment rarely can be described by single analytical functions. In the cantilever beam of Concept Application 5.2 (Fig. 5.8) that supported a uniformly distributed load w, the shear and bending moment *could* be represented by single analytical functions of $V = -wx$ and $M = -\frac{1}{2}wx^2$. This was due to the fact that *no discontinuity* existed in the loading of the beam. On the other hand, in the simply supported beam of Concept Application 5.1, which was loaded only at its midpoint C, the load **P** applied at C represented a *singularity* in the beam loading. This singularity resulted in discontinuities in the shear and bending moment and required the use of different analytical functions for V and M in the portions of beam to the left and right of point C. In Sample Prob. 5.2, the beam had to be divided into three portions, where different functions were used to represent the shear and the bending moment. This led to the graphical representation of the functions V and M provided by the shear and bending-moment diagrams and, later in Sec. 5.2, to a graphical method of integration to determine V and M from the distributed load w.

This section shows how the use of *singularity functions* makes it possible to represent the shear V and bending moment M with single mathematical expressions.

Consider the simply supported beam AB, with length of $2a$, that carries a uniformly distributed load w_0 extending from its midpoint C to its right-hand support B (Fig. 5.13). First, draw the free-body diagram of the entire beam (Fig. 5.14a). Replacing the distributed load with an equivalent concentrated load and summing moments about B,

$$+\circlearrowleft\Sigma M_B = 0: \qquad (w_0a)(\tfrac{1}{2}a) - R_A(2a) = 0 \qquad R_A = \tfrac{1}{4}w_0a$$

Next, cut the beam at a point D between A and C. From the free-body diagram of AD (Fig. 5.14b) and over the interval $0 < x < a$, the shear and bending moment are

$$V_1(x) = \tfrac{1}{4}w_0a \qquad \text{and} \qquad M_1(x) = \tfrac{1}{4}w_0ax$$

Cutting the beam at a point E between C and B, draw the free-body diagram of portion AE (Fig. 5.14c). Replacing the distributed load by an equivalent concentrated load,

$$+\uparrow\Sigma F_y = 0: \qquad \tfrac{1}{4}w_0a - w_0(x-a) - V_2 = 0$$

$$+\circlearrowleft\Sigma M_E = 0: \qquad -\tfrac{1}{4}w_0ax + w_0(x-a)[\tfrac{1}{2}(x-a)] + M_2 = 0$$

Over the interval $a < x < 2a$, the shear and bending moment are

$$V_2(x) = \tfrac{1}{4}w_0a - w_0(x-a) \qquad \text{and} \qquad M_2(x) = \tfrac{1}{4}w_0ax - \tfrac{1}{2}w_0(x-a)^2$$

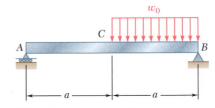

Fig. 5.13   Simply supported beam.

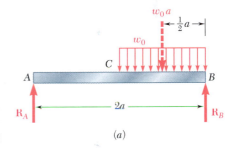

(a)

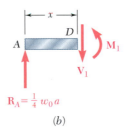

(b)

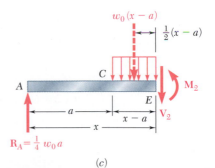

(c)

Fig. 5.14   Free-body diagrams at two sections required to draw shear and bending-moment diagrams.

The fact that the shear and bending moment are represented by different functions of x is due to the discontinuity in the loading of the beam. However, $V_1(x)$ and $V_2(x)$ can be represented by the single function

$$V(x) = \tfrac{1}{4}w_0 a - w_0\langle x - a \rangle \tag{5.11}$$

if the second term is included in the computations when $x \geq a$ and ignored when $x < a$. Therefore, *the brackets $\langle\,\rangle$ should be replaced by ordinary parentheses $(\,)$ when $x \geq a$ and by zero when $x < a$.* Using this convention, the bending moment can be represented at any point of the beam by

$$M(x) = \tfrac{1}{4}w_0 ax - \tfrac{1}{2}w_0\langle x - a \rangle^2 \tag{5.12}$$

The function within the brackets $\langle\,\rangle$ can be differentiated or integrated as if the brackets were replaced with ordinary parentheses. Instead of calculating the bending moment from free-body diagrams, the method indicated in Sec. 5.2 could be used, where the expression obtained for $V(x)$ is integrated to give

$$M(x) - M(0) = \int_0^x V(x)\, dx = \int_0^x \tfrac{1}{4}w_0 a\, dx - \int_0^x w_0\langle x - a \rangle\, dx$$

After integration and observing that $M(0) = 0$,

$$M(x) = \tfrac{1}{4}w_0 ax - \tfrac{1}{2}w_0\langle x - a \rangle^2$$

Furthermore, using the same convention, the distributed load at any point of the beam can be expressed as

$$w(x) = w_0\langle x - a \rangle^0 \tag{5.13}$$

Indeed, the brackets should be replaced by zero for $x < a$ and by parentheses for $x \geq a$. Thus, $w(x) = 0$ for $x < a$, and by defining the zero power of any number as unity, $\langle x - a \rangle^0 = (x - a)^0 = 1$ and $w(x) = w_0$ for $x \geq a$. Recall that the shear could have been obtained by integrating the function $-w(x)$. Observing that $V = \tfrac{1}{4}w_0 a$ for $x = 0$,

$$V(x) - V(0) = -\int_0^x w(x)\, dx = -\int_0^x w_0\langle x - a \rangle^0\, dx$$

$$V(x) - \tfrac{1}{4}w_0 a = -w_0\langle x - a \rangle^1$$

Solving for $V(x)$ and dropping the exponent 1,

$$V(x) = \tfrac{1}{4}w_0 a - w_0\langle x - a \rangle$$

The expressions $\langle x - a \rangle^0, \langle x - a \rangle, \langle x - a \rangle^2$ are called *singularity functions.* For $n \geq 0$,

$$\langle x - a \rangle^n = \begin{cases} (x - a)^n & \text{when } x \geq a \\ 0 & \text{when } x < a \end{cases} \tag{5.14}$$

Also note that whenever the quantity between brackets is positive or zero, the brackets should be replaced by ordinary parentheses. Whenever that quantity is negative, the bracket itself is equal to zero.

The three singularity functions corresponding to $n = 0$, $n = 1$, and $n = 2$ have been plotted in Fig. 5.15. Note that the function $\langle x - a \rangle^0$ is discontinuous at $x = a$ and is in the shape of a "step." For that reason, it is called the *step function*. According to Eq. (5.14) and using the zero power of any number as unity,[†]

$$\langle x - a \rangle^0 = \begin{cases} 1 & \text{when } x \geq a \\ 0 & \text{when } x < a \end{cases} \qquad \textbf{(5.15)}$$

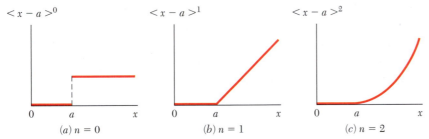

Fig. 5.15 Singularity functions.

It follows from the definition of singularity functions that

$$\int \langle x - a \rangle^n \, dx = \frac{1}{n + 1} \langle x - a \rangle^{n+1} \qquad \text{for } n \geq 0 \qquad \textbf{(5.16)}$$

and

$$\frac{d}{dx} \langle x - a \rangle^n = n \langle x - a \rangle^{n-1} \qquad \text{for } n \geq 1 \qquad \textbf{(5.17)}$$

Most of the beam loadings encountered in engineering practice can be broken down into the basic loadings shown in Fig. 5.16. When applicable, the corresponding functions $w(x)$, $V(x)$, and $M(x)$ are expressed in terms of singularity functions and plotted against a color background. A heavier color background is used to indicate the expression for each loading that is most easily obtained or remembered and from which the other functions can be obtained by integration.

After a given beam loading has been broken down into the basic loadings of Fig. 5.16, the functions $V(x)$ and $M(x)$ representing the shear and bending moment at any point of the beam can be obtained by adding the corresponding functions associated with each of the basic loadings

---

[†]Since $(x - a)^0$ is discontinuous at $x - a$, it can be argued that this function should be left undefined for $x = a$ or should be assigned both of the values 0 and 1 for $x = a$. However, defining $(x - a)^0$ as equal to 1 when $x = a$, as stated in (Eq. 5.15), has the advantage of being unambiguous. Thus it is easily applied to computer programming (see page 388).

and reactions. Since all of the distributed loadings shown in Fig. 5.16 are open-ended to the right, a distributed load that does not extend to the right end of the beam or is discontinuous should be replaced as shown in Fig. 5.17 by an equivalent combination of open-ended loadings. (See also Concept Application 5.5 and Sample Prob. 5.9.)

As you will see in Chap. 9, the use of singularity functions also simplifies the determination of beam deflections. It was in connection with

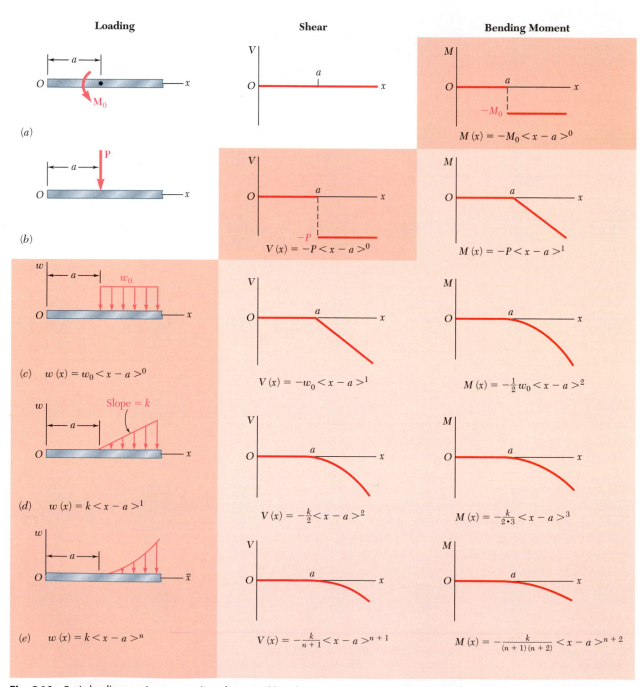

Fig. 5.16 Basic loadings and corresponding shears and bending moments expressed in terms of singularity functions.

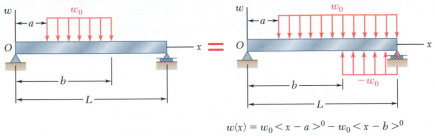

$$w(x) = w_0 <x - a>^0 - w_0 <x - b>^0$$

Fig. 5.17   Use of open-ended loadings to create a closed-ended loading.

that problem that the approach used in this section was first suggested in 1862 by the German mathematician A. Clebsch (1833–1872). However, the British mathematician and engineer W. H. Macaulay (1853–1936) is usually given credit for introducing the singularity functions in the form used here, and the brackets $\langle \ \rangle$ are called *Macaulay's brackets.*[†]

[†]W. H. Macaulay, "Note on the Deflection of Beams," *Messenger of Mathematics,* vol. 48, pp. 129–130, 1919.

Concept Application 5.5

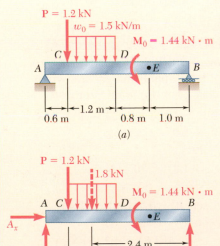

Fig. 5.18   (a) Simply supported beam with multiple loads. (b) Free-body diagram.

For the beam and loading shown (Fig. 5.18*a*) and using singularity functions, express the shear and bending moment as functions of the distance x from the support at A.

Determine the reaction at A by drawing the free-body diagram of the beam (Fig. 5.18*b*) and writing

$$\xrightarrow{+} \Sigma F_x = 0: \qquad A_x = 0$$

$$+\uparrow \Sigma M_B = 0: \qquad -A_y(3.6 \text{ m}) + (1.2 \text{ kN})(3 \text{ m})$$
$$+ (1.8 \text{ kN})(2.4 \text{ m}) + 1.44 \text{ kN·m} = 0$$
$$A_y = 2.60 \text{ kN}$$

Next, replace the given distributed load by two equivalent open-ended loads (Fig. 5.18*c*) and express the distributed load $w(x)$ as the sum of the corresponding step functions:

$$w(x) = +w_0\langle x - 0.6\rangle^0 - w_0\langle x - 1.8\rangle^0$$

The function $V(x)$ is obtained by integrating $w(x)$, reversing the + and − signs, and adding to the result the constants A_y and $-P\langle x - 0.6\rangle^0$, which represents the respective contributions to the shear of the reaction at A and of the concentrated load. (No other constant of integration is required.) Since the concentrated couple does not directly affect the shear, it should be ignored in this computation.

$$V(x) = -w_0\langle x - 0.6\rangle^1 + w_0\langle x - 1.8\rangle^1 + A_y - P\langle x - 0.6\rangle^0$$

(continued)

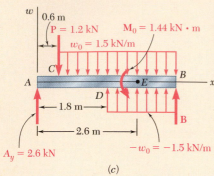

Fig. 5.18 *(cont.)* (c) Superposition of distributed loads.

In a similar way, the function $M(x)$ is obtained by integrating $V(x)$ and adding to the result the constant $-M_0\langle x - 2.6\rangle^0$, which represents the contribution of the concentrated couple to the bending moment. We have

$$M(x) = -\tfrac{1}{2}w_0\langle x - 0.6\rangle^2 + \tfrac{1}{2}w_0\langle x - 1.8\rangle^2$$
$$+ A_y x - P\langle x - 0.6\rangle^1 - M_0\langle x - 2.6\rangle^0$$

Substituting the numerical values of the reaction and loads into the expressions for $V(x)$ and $M(x)$ and being careful *not* to compute any product or expand any square involving a bracket, the expressions for the shear and bending moment at any point of the beam are

$$V(x) = -1.5\langle x - 0.6\rangle^1 + 1.5\langle x - 1.8\rangle^1 + 2.6 - 1.2\langle x - 0.6\rangle^0$$
$$M(x) = -0.75\langle x - 0.6\rangle^2 + 0.75\langle x - 1.8\rangle^2$$
$$+ 2.6x - 1.2\langle x - 0.6\rangle^1 - 1.44\langle x - 2.6\rangle^0$$

Concept Application 5.6

For the beam and loading of Concept Application 5.5, determine the numerical values of the shear and bending moment at the midpoint D.

Making $x = 1.8$ m in the equations found for $V(x)$ and $M(x)$ in Concept Application 5.5,

$$V(1.8) = -1.5\langle 1.2\rangle^1 + 1.5\langle 0\rangle^1 + 2.6 - 1.2\langle 1.2\rangle^0$$
$$M(1.8) = -0.75\langle 1.2\rangle^2 + 0.75\langle 0\rangle^2 + 2.6(1.8) - 1.2\langle 1.2\rangle^1 - 1.44\langle -0.8\rangle^0$$

Recall that whenever a quantity between brackets is positive or zero, the brackets should be replaced by ordinary parentheses, and whenever the quantity is negative, the bracket itself is equal to zero, so

$$V(1.8) = -1.5(1.2)^1 + 1.5(0)^1 + 2.6 - 1.2(1.2)^0$$
$$= -1.5(1.2) + 1.5(0) + 2.6 - 1.2(1)$$
$$= -1.8 + 0 + 2.6 - 1.2$$
$$V(1.8) = -0.4 \text{ kN}$$

and

$$M(1.8) = -0.75(1.2)^2 + 0.75(0)^2 + 2.6(1.8) - 1.2(1.2)^1 - 1.44(0)$$
$$= -1.08 + 0 + 4.68 - 1.44 - 0$$
$$M(1.8) = +2.16 \text{ kN·m}$$

Application to Computer Programming. Singularity functions are particularly well suited to computers. First note that the step function $\langle x - a\rangle^0$, which will be represented by the symbol STP, can be defined by an IF/THEN/ELSE statement as being equal to 1 for X ≥ A and to 0 otherwise. Any other singularity function $\langle x - a\rangle^n$, with $n \geq 1$, can be expressed as the product of the ordinary algebraic function $(x - a)^n$ and the step function $\langle x - a\rangle^0$.

When k different singularity functions are involved (such as $\langle x - a_i\rangle^n$ where $i = 1, 2, \ldots, k$) the corresponding step functions (STP(I), where $I = 1, 2, \ldots, K$) can be defined by a loop containing a single IF/THEN/ELSE statement.

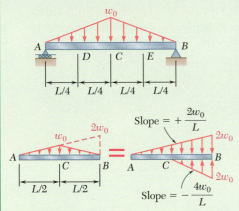

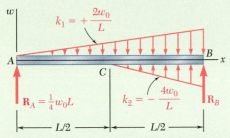

Fig. 1   Modeling the distributed load as the superposition of two distributed loads.

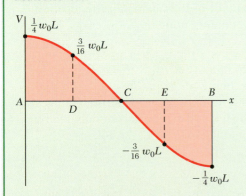

Fig. 2   Free body of beam with equivalent distributed load.

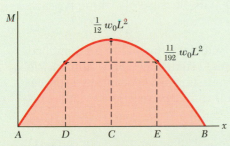

Fig. 3   Shear and bending-moment diagrams.

Sample Problem 5.9

For the beam and loading shown, determine (*a*) the equations defining the shear and bending moment at any point and (*b*) the shear and bending moment at points *C*, *D*, and *E*.

STRATEGY: After determining the support reactions, you can write equations for *w*, *V*, and *M*, beginning from the left end of the beam. Any abrupt changes in these parameters beyond the left end can be accommodated by adding appropriate singularity functions.

MODELING and ANALYSIS:

Reactions.   The total load is $\frac{1}{2}w_0L$. Due to symmetry, each reaction is equal to half that value as $\frac{1}{4}w_0L$.

Distributed Load.   The given distributed loading is replaced by two equivalent open-ended loadings as shown in Figs. 1 and 2. Using a singularity function to express the second loading,

$$w(x) = k_1 x + k_2\langle x - \tfrac{1}{2}L\rangle = \frac{2w_0}{L}x - \frac{4w_0}{L}\langle x - \tfrac{1}{2}L\rangle \qquad \textbf{(1)}$$

a. Equations for Shear and Bending Moment. $V(x)$ is obtained by integrating Eq. (1), changing the signs, and adding a constant equal to R_A:

$$V(x) = -\frac{w_0}{L}x^2 + \frac{2w_0}{L}\langle x - \tfrac{1}{2}L\rangle^2 + \tfrac{1}{4}w_0L \qquad \textbf{(2)} \blacktriangleleft$$

$M(x)$ is obtained by integrating Eq. (2). Since there is no concentrated couple, no constant of integration is needed, so

$$M(x) = -\frac{w_0}{3L}x^3 + \frac{2w_0}{3L}\langle x - \tfrac{1}{2}L\rangle^3 + \tfrac{1}{4}w_0Lx \qquad \textbf{(3)} \blacktriangleleft$$

b. Shear and Bending Moment at C, D, and E (Fig. 3)

At Point C: Making $x = \frac{1}{2}L$ in Eqs. (2) and (3) and recalling that whenever a quantity between brackets is positive or zero, the brackets can be replaced by parentheses:

$$V_C = -\frac{w_0}{L}(\tfrac{1}{2}L)^2 + \frac{2w_0}{L}\langle 0\rangle^2 + \tfrac{1}{4}w_0L \qquad V_C = 0 \blacktriangleleft$$

$$M_C = -\frac{w_0}{3L}(\tfrac{1}{2}L)^3 + \frac{2w_0}{3L}\langle 0\rangle^3 + \tfrac{1}{4}w_0L(\tfrac{1}{2}L) \qquad M_C = \frac{1}{12}w_0L^2 \blacktriangleleft$$

(continued)

At Point D: Making $x = \frac{1}{4}L$ in Eqs. (2) and (3) and recalling that a bracket containing a negative quantity is equal to zero gives

$$V_D = -\frac{w_0}{L}(\tfrac{1}{4}L)^2 + \frac{2w_0}{L}\langle -\tfrac{1}{4}L \rangle^2 + \tfrac{1}{4}w_0 L \qquad V_D = \frac{3}{16}w_0 L \qquad \blacktriangleleft$$

$$M_D = -\frac{w_0}{3L}(\tfrac{1}{4}L)^3 + \frac{2w_0}{3L}\langle -\tfrac{1}{4}L \rangle^3 + \tfrac{1}{4}w_0 L(\tfrac{1}{4}L) \qquad M_D = \frac{11}{192}w_0 L^2 \qquad \blacktriangleleft$$

At Point E: Making $x = \frac{3}{4}L$ in Eqs. (2) and (3) gives

$$V_E = -\frac{w_0}{L}(\tfrac{3}{4}L)^2 + \frac{2w_0}{L}\langle \tfrac{1}{4}L \rangle^2 + \tfrac{1}{4}w_0 L \qquad V_E = -\frac{3}{16}w_0 L \qquad \blacktriangleleft$$

$$M_E = -\frac{w_0}{3L}(\tfrac{3}{4}L)^3 + \frac{2w_0}{3L}\langle \tfrac{1}{4}L \rangle^3 + \tfrac{1}{4}w_0 L(\tfrac{3}{4}L) \qquad M_E = \frac{11}{192}w_0 L^2 \qquad \blacktriangleleft$$

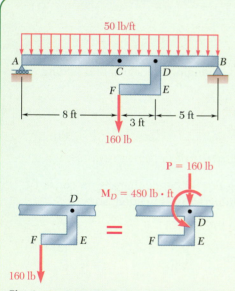

Fig. 1 Modeling the force at F as an equivalent force-couple at D.

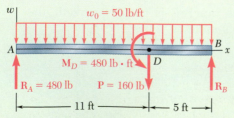

Fig. 2 Free-body diagram of beam, with equivalent force-couple at D.

Sample Problem 5.10

The rigid bar DEF is welded at point D to the steel beam AB. For the loading shown, determine (a) the equations defining the shear and bending moment at any point of the beam, (b) the location and magnitude of the largest bending moment.

STRATEGY: You can begin by first finding the support reactions and replacing the load on appendage DEF with an equivalent force-couple system. You can then write equations for w, V, and M, beginning from the left end of the beam. Any abrupt changes in these parameters beyond the left end can be accommodated by adding appropriate singularity functions.

MODELING and ANALYSIS:

Reactions. Consider the beam and bar as a free body and observe that the total load is 960 lb. Because of symmetry, each reaction is equal to 480 lb.

Modified Loading Diagram. Replace the 160-lb load applied at F by an equivalent force-couple system at D (Figs. 1 and 2.). We thus obtain a loading diagram consisting of a concentrated couple, three concentrated loads (including the two reactions), and a uniformly distributed load

$$w(x) = 50 \text{ lb/ft} \tag{1}$$

(continued)

a. Equations for Shear and Bending Moment.

$V(x)$ is obtained by integrating Eq. (1), changing the sign, and adding constants representing the respective contributions of $\mathbf{R}_A$ and $\mathbf{P}$ to the shear. Since $\mathbf{P}$ affects $V(x)$ when x is larger than 11 ft, use a step function to express its contribution.

$$V(x) = -50x + 480 - 160\langle x - 11 \rangle^0 \qquad \textbf{(2)} \quad \blacktriangleleft$$

Obtain $M(x)$ by integrating Eq. (2) and using a step function to represent the contribution of the concentrated couple $\mathbf{M}_D$:

$$M(x) = -25x^2 + 480x - 160\langle x - 11 \rangle^1 - 480\langle x - 11 \rangle^0 \ \textbf{(3)} \quad \blacktriangleleft$$

b. Largest Bending Moment.

Since M is maximum or minimum when $V = 0$, set $V = 0$ in Eq. (2) and solve that equation for x to find the location of the largest bending moment. Considering first values of x less than 11 ft, and noting that for such values the bracket is equal to zero:

$$-50x + 480 = 0 \qquad x = 9.60 \text{ ft}$$

Considering values of x larger than 11 ft, for which the bracket is equal to 1:

$$-50x + 480 - 160 = 0 \qquad x = 6.40 \text{ ft}$$

Since this value is *not* larger than 11 ft, it must be rejected. Thus, the value of x corresponding to the largest bending moment is

$$x_m = 9.60 \text{ ft} \quad \blacktriangleleft$$

Substituting this value for x into Eq. (3),

$$M_{\text{max}} = -25(9.60)^2 + 480(9.60) - 160\langle -1.40 \rangle^1 - 480\langle -1.40 \rangle^0$$

and recalling that brackets containing a negative quantity are equal to zero,

$$M_{\text{max}} = -25(9.60)^2 + 480(9.60) \qquad\qquad M_{\text{max}} = 2304 \text{ lb·ft} \quad \blacktriangleleft$$

The bending-moment diagram has been plotted (Fig. 3). Note the discontinuity at point D is due to the concentrated couple applied at that point. The values of M just to the left and just to the right of D are obtained by making $x = 11$ in Eq. (3) and replacing the step function $\langle x - 11 \rangle^0$ by 0 and 1, respectively.

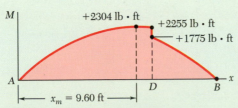

Fig. 3   Bending-moment diagram.

Problems

5.98 through 5.100 (*a*) Using singularity functions, write the equations defining the shear and bending moment for the beam and loading shown. (*b*) Use the equation obtained for *M* to determine the bending moment at point *C*, and check your answer by drawing the free-body diagram of the entire beam.

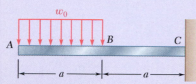

Fig. P5.98

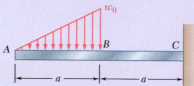

Fig. *P5.99*

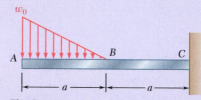

Fig. P5.100

5.101 through 5.103 (*a*) Using singularity functions, write the equations defining the shear and bending moment for the beam and loading shown. (*b*) Use the equation obtained for *M* to determine the bending moment at point *E*, and check your answer by drawing the free-body diagram of the portion of the beam to the right of *E*.

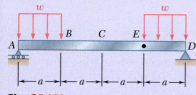

Fig. *P5.101*

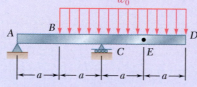

Fig. P5.102

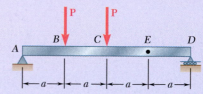

Fig. P5.103

5.104 and 5.105 (*a*) Using singularity functions, write the equations for the shear and bending moment for beam *ABC* under the loading shown. (*b*) Use the equation obtained for *M* to determine the bending moment just to the right of point *B*.

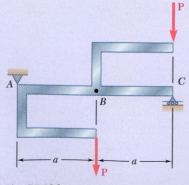

Fig. P5.104

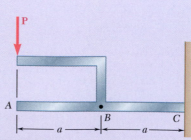

Fig. P5.105

5.106 through 5.109 (*a*) Using singularity functions, write the equations for the shear and bending moment for the beam and loading shown. (*b*) Determine the maximum value of the bending moment in the beam.

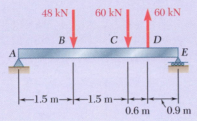

Fig. P5.106

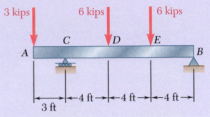

Fig. P5.107

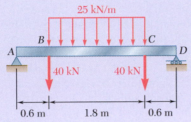

Fig. P5.108

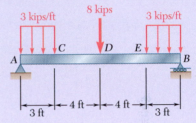

Fig. P5.109

5.110 and 5.111 (*a*) Using singularity functions, write the equations for the shear and bending moment for the beam and loading shown. (*b*) Determine the maximum normal stress due to bending.

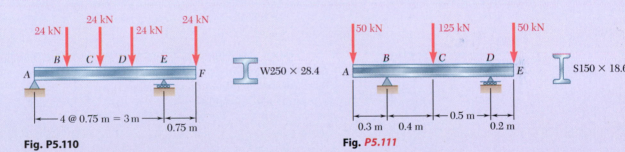

Fig. P5.110

Fig. P5.111

5.112 and 5.113 (*a*) Using singularity functions, find the magnitude and location of the maximum bending moment for the beam and loading shown. (*b*) Determine the maximum normal stress due to bending.

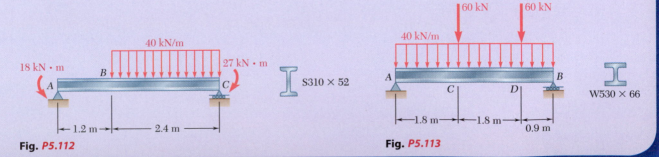

Fig. P5.112

Fig. P5.113

5.114 and 5.115 A beam is being designed to be supported and loaded as shown. (*a*) Using singularity functions, find the magnitude and location of the maximum bending moment in the beam. (*b*) Knowing that the allowable normal stress for the steel to be used is 24 ksi, find the most economical wide-flange shape that can be used.

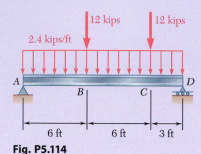

Fig. P5.114

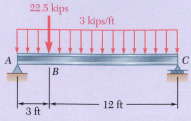

Fig. P5.115

5.116 and 5.117 A timber beam is being designed to be supported and loaded as shown. (*a*) Using singularity functions, find the magnitude and location of the maximum bending moment in the beam. (*b*) Knowing that the available stock consists of beams with an allowable normal stress of 12 MPa and a rectangular cross section of 30-mm width and depth *h* varying from 80 mm to 160 mm in 10-mm increments, determine the most economical cross section that can be used.

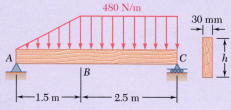

Fig. P5.116

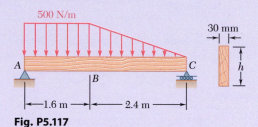

Fig. P5.117

5.118 through 5.121 Using a computer and step functions, calculate the shear and bending moment for the beam and loading shown. Use the specified increment Δ*L*, starting at point *A* and ending at the right-hand support.

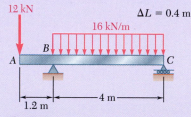

Fig. P5.118

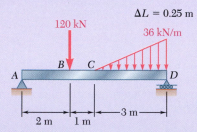

Fig. P5.119

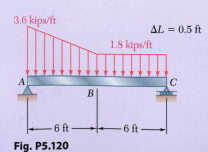

Fig. P5.120

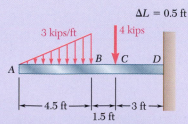

Fig. P5.121

5.122 and 5.123 For the beam and loading shown and using a computer and step functions, (*a*) tabulate the shear, bending moment, and maximum normal stress in sections of the beam from $x = 0$ to $x = L$, using the increments ΔL indicated, (*b*) using smaller increments if necessary, determine with a 2% accuracy the maximum normal stress in the beam. Place the origin of the x axis at end A of the beam.

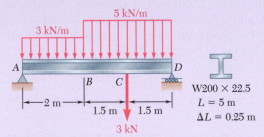

W200 × 22.5
$L = 5$ m
$\Delta L = 0.25$ m

Fig. P5.122

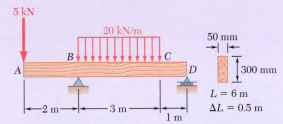

50 mm

300 mm

$L = 6$ m
$\Delta L = 0.5$ m

Fig. P5.123

5.124 and 5.125 For the beam and loading shown and using a computer and step functions, (*a*) tabulate the shear, bending moment, and maximum normal stress in sections of the beam from $x = 0$ to $x = L$, using the increments ΔL indicated, (*b*) using smaller increments if necessary, determine with a 2% accuracy the maximum normal stress in the beam. Place the origin of the x axis at end A of the beam.

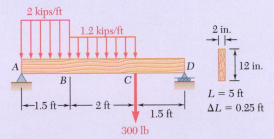

2 in.

12 in.

$L = 5$ ft
$\Delta L = 0.25$ ft

Fig. P5.124

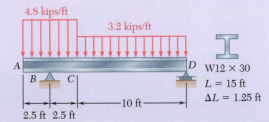

W12 × 30
$L = 15$ ft
$\Delta L = 1.25$ ft

Fig. P5.125

*5.5   NONPRISMATIC BEAMS

Prismatic beams, i.e., beams of uniform cross section, are designed so that the normal stresses in their critical sections are at most equal to the allowable value of the normal stress for the material being used. In all other sections, the normal stresses will be smaller (possibly much smaller) than their allowable value. Therefore, a prismatic beam is almost always overdesigned, and considerable savings can be made by using nonprismatic beams. The continuous spans shown in Photo 5.2 are examples of nonprismatic beams.

Since the maximum normal stresses σ_m usually control the design of a beam, the design of a nonprismatic beam is optimum if the section modulus $S = I/c$ of every cross section satisfies Eq. (5.3). Solving that equation for S,

$$S = \frac{|M|}{\sigma_{all}} \tag{5.18}$$

A beam designed in this manner is a *beam of constant strength*.

For a forged or cast structural or machine component, it is possible to vary the cross section of the component along its length and eliminate most of the unnecessary material (see Concept Application 5.7). For a timber or rolled-steel beam, it is not possible to vary the cross section of the beam. But considerable savings of material can be achieved by gluing wooden planks of appropriate lengths to a timber beam (see Sample Prob. 5.11) and using cover plates in portions of a rolled-steel beam where the bending moment is large (see Sample Prob. 5.12).

Photo 5.2   Bridge supported by nonprismatic beams.

Concept Application 5.7

A cast-aluminum plate of uniform thickness b is to support a uniformly distributed load w as shown in Fig. 5.19. (*a*) Determine the shape of the plate that will yield the most economical design. (*b*) Knowing that the allowable normal stress for the aluminum used is 72 MPa and that $b = 40$ mm, $L = 800$ mm, and $w = 135$ kN/m, determine the maximum depth h_0 of the plate.

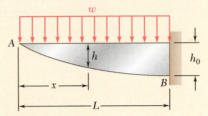

Fig. 5.19 Nonprismatic, cantilevered beam supporting a uniformly distributed load.

Bending Moment. Measuring the distance x from A and observing that $V_A = M_A = 0$, use Eqs. (5.6) and (5.8) for

$$V(x) = -\int_0^x w\,dx = -wx$$

$$M(x) = \int_0^x V(x)\,dx = -\int_0^x wx\,dx = -\tfrac{1}{2}wx^2$$

a. Shape of Plate. Recall that the modulus S of a rectangular cross section of width b and depth h is $S = \tfrac{1}{6}bh^2$. Carrying this value into Eq. (5.18) and solving for h^2,

$$h^2 = \frac{6|M|}{b\sigma_{all}} \tag{5.19}$$

and after substituting $|M| = \tfrac{1}{2}wx^2$,

$$h^2 = \frac{3wx^2}{b\sigma_{all}} \quad \text{or} \quad h = \left(\frac{3w}{b\sigma_{all}}\right)^{1/2} x \tag{5.20}$$

Since the relationship between h and x is linear, the lower edge of the plate is a straight line. Thus, the plate providing the most economical design is of *triangular shape.*

b. Maximum Depth h_0. Making $x = L$ in Eq. (5.20) and substituting the given data,

$$h_0 = \left[\frac{3(135 \text{ kN/m})}{(0.040 \text{ m})(72 \text{ MPa})}\right]^{1/2} (800 \text{ mm}) = 300 \text{ mm}$$

Sample Problem 5.11

A 12-ft-long beam made of a timber with an allowable normal stress of 2.40 ksi and an allowable shearing stress of 0.40 ksi is to carry two 4.8-kip loads located at its third points. As will be shown in Ch. 6, this beam of uniform rectangular cross section, 4 in. wide and 4.5 in. deep, would satisfy the allowable shearing stress requirement. Since such a beam would not satisfy the allowable normal stress requirement, it will be reinforced by gluing planks of the same timber, 4 in. wide and 1.25 in. thick, to the top and bottom of the beam in a symmetric manner. Determine (*a*) the required number of pairs of planks and (*b*) the length of the planks in each pair that will yield the most economical design.

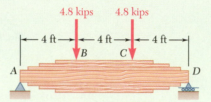

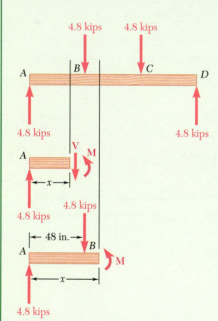

Fig. 1 Free-body diagrams of entire beam and sections.

STRATEGY: Since the moment is maximum and constant between the two concentrated loads (due to symmetry), you can analyze this region to determine the total number of reinforcing planks required. You can determine the cut-off points for each pair of planks by considering the range for which each reinforcing pair, combined with the rest of the section, meets the specified allowable normal stress.

MODELING and ANALYSIS:

Bending Moment. Draw the free-body diagram of the beam (Fig. 1) and find the expressions for the bending moment:

From *A* to *B* ($0 \le x \le 48$ in.): $M = (4.80 \text{ kips})x$
From *B* to *C* ($48 \text{ in.} \le x \le 96$ in.):

$$M = (4.80 \text{ kips})\, x - (4.80 \text{ kips})(x - 48 \text{ in.}) = 230.4 \text{ kip·in.}$$

a. Number of Pairs of Planks. Determine the required total depth of the reinforced beam between *B* and *C*. Recall from Sec. 5.3 that $S = \frac{1}{6}bh^2$ for a beam with a rectangular cross section of width *b* and depth *h*. Substituting this value into Eq. (5.19),

$$h^2 = \frac{6|M|}{b\sigma_{\text{all}}} \tag{1}$$

(continued)

Substituting the value obtained for M from B to C and the given values of b and σ_{all},

$$h^2 = \frac{6(230.4 \text{ kip·in.})}{(4 \text{ in.})(2.40 \text{ ksi})} = 144 \text{ in.}^2 \qquad h = 12.00 \text{ in.}$$

Since the original beam has a depth of 4.50 in., the planks must provide an additional depth of 7.50 in. Recalling that each pair of planks is 2.50 in. thick,

Required number of pairs of planks = 3 ◄

b. Length of Planks. The bending moment was found to be $M = (4.80 \text{ kips}) x$ in the portion AB of the beam. Substituting this expression and the given values of b and σ_{all} into Eq. (1) then solving for x, gives

$$x = \frac{(4 \text{ in.})(2.40 \text{ ksi})}{6 (4.80 \text{ kips})} h^2 \qquad x = \frac{h^2}{3 \text{ in.}} \tag{2}$$

Equation (2) defines the maximum distance x from end A at which a given depth h of the cross section is acceptable (Fig. 2). Making $h = 4.50$ in. you can find the distance x_1 from A at which the original prismatic beam is safe: $x_1 = 6.75$ in. From that point on, the original beam should be reinforced by the first pair of planks. Making $h = 4.50$ in. + 2.50 in. = 7.00 in. yields the distance $x_2 = 16.33$ in. from which the second pair of planks should be used, and making $h = 9.50$ in. yields the distance $x_3 = 30.08$ in. from which the third pair of planks should be used. The length l_i of the planks of the pair i, where $i = 1, 2, 3$, is obtained by subtracting $2x_i$ from the 144-in. length of the beam.

$$l_1 = 130.5 \text{ in.}, \; l_2 = 111.3 \text{ in.}, \; l_3 = 83.8 \text{ in.} \quad ◄$$

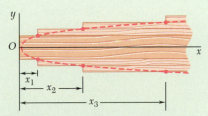

Fig. 2 Positions where planks must be added.

The corners of the various planks lie on the parabola defined by Eq. (2).

Sample Problem 5.12

Two steel plates, each 16 mm thick, are welded as shown to a W690 × 125 beam to reinforce it. Knowing that $\sigma_{all} = 160$ MPa for both the beam and the plates, determine the required value of (a) the length of the plates, (b) the width of the plates.

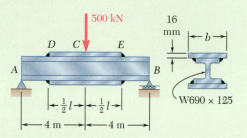

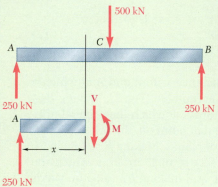

Fig. 1   Free-body diagrams of beam and section needed to find internal shear force and bending moment.

STRATEGY: To find the required length of the reinforcing plates, you can determine the extent of the beam that is not overstressed if left unreinforced. By considering the point of maximum moment, you can then size the reinforcing plates.

MODELING and ANALYSIS:

Bending Moment.   Find the reactions. From the free-body diagram in Fig. 1, using a portion of the beam of length $x \leq 4$ m, M is found between A and C as

$$M = (250\text{ kN})x \tag{1}$$

a. Required Length of Plates.   Determine the maximum allowable length x_m of the portion AD of the unreinforced beam. From Appendix C, the section modulus of a W690 × 125 beam is $S = 3490 \times 10^6$ mm^3 or $S = 3.49 \times 10^{-3}$ m^3. Substitute for S and σ_{all} into Eq. (5.17) and solve for M:

$$M = S\sigma_{all} = (3.49 \times 10^{-3}\text{ m}^3)(160 \times 10^3\text{ kN/m}^2) = 558.4\text{ kN·m}$$

Substituting for M in Eq. (1),

$$558.4\text{ kN·m} = (250\text{ kN})x_m \qquad x_m = 2.234\text{ m}$$

The required length l of the plates is obtained by subtracting $2x_m$ from the length of the beam:

$$l = 8\text{ m} - 2(2.234\text{ m}) = 3.532\text{ m} \qquad l = 3.53\text{ m} \blacktriangleleft$$

(continued)

b. Required Width of Plates. The maximum bending moment occurs in the midsection C of the beam. Making $x = 4$ m in Eq. (1), the bending moment in that section is

$$M = (250\,\text{kN})(4\,\text{m}) = 1000\,\text{kN·m}$$

In order to use Eq. (5.1), find the moment of inertia of the cross section of the reinforced beam with respect to a centroidal axis and the distance c from that axis to the outer surfaces of the plates (Fig. 2). From Appendix C, the moment of inertia of a W690 × 125 beam is $I_b = 1190 \times 10^6$ mm^4, and its depth is $d = 678$ mm. Using t as the thickness of one plate, b as its width, and $\bar{y}$ as the distance of its centroid from the neutral axis, the moment of inertia I_p of the two plates with respect to the neutral axis is

$$I_p = 2(\tfrac{1}{12}bt^3 + A\bar{y}^2) = (\tfrac{1}{6}t^3)b + 2\,bt(\tfrac{1}{2}d + \tfrac{1}{2}t)^2$$

Substituting $t = 16$ mm and $d = 678$ mm, we obtain $I_p = (3.854 \times 10^6$ mm$^3)\,b$. The moment of inertia I of the beam and plates is

$$I = I_b + I_p = 1190 \times 10^6\,\text{mm}^4 + (3.854 \times 10^6\,\text{mm}^3)b \qquad \textbf{(2)}$$

and the distance from the neutral axis to the surface is $c = \tfrac{1}{2}d + t = 355$ mm. Solving Eq. (5.1) for I and substituting the values of M, σ_{all}, and c,

$$I = \frac{|M|c}{\sigma_{\text{all}}} = \frac{(1000\,\text{kN·m})(355\,\text{mm})}{160\,\text{MPa}} = 2.219 \times 10^{-3}\,\text{m}^4 = 2219 \times 10^6\,\text{mm}^4$$

Replacing I by this value in Eq. (2) and solving for b,

$$2219 \times 10^6\,\text{mm}^4 = 1190 \times 10^6\,\text{mm}^4 + (3.854 \times 10^6\,\text{mm}^3)b$$

$$b = 267\,\text{mm} \qquad \blacktriangleleft$$

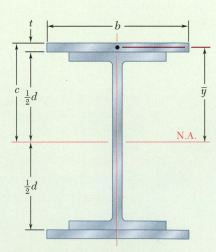

Fig. 2 Cross section of beam with plate reinforcement.

Problems

5.126 and 5.127 The beam AB, consisting of a cast-iron plate of uniform thickness b and length L, is to support the load shown. (*a*) Knowing that the beam is to be of constant strength, express h in terms of x, L, and h_0. (*b*) Determine the maximum allowable load if $L = 36$ in., $h_0 = 12$ in., $b = 1.25$ in., and $\sigma_{all} = 24$ ksi.

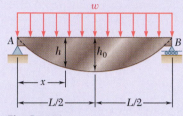

Fig. P5.126

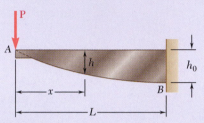

Fig. P5.127

5.128 and 5.129 The beam AB, consisting of a cast-iron plate of uniform thickness b and length L, is to support the distributed load $w(x)$ shown. (*a*) Knowing that the beam is to be of constant strength, express h in terms of x, L, and h_0. (*b*) Determine the smallest value of h_0 if $L = 750$ mm, $b = 30$ mm, $w_0 = 300$ kN/m, and $\sigma_{all} = 200$ MPa.

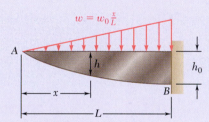

Fig. P5.128

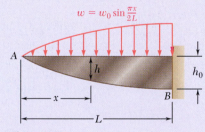

Fig. P5.129

5.130 and 5.131 The beam AB, consisting of an aluminum plate of uniform thickness b and length L, is to support the load shown. (*a*) Knowing that the beam is to be of constant strength, express h in terms of x, L, and h_0 for portion AC of the beam. (*b*) Determine the maximum allowable load if $L = 800$ mm, $h_0 = 200$ mm, $b = 25$ mm, and $\sigma_{all} = 72$ MPa.

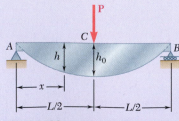

Fig. P5.130

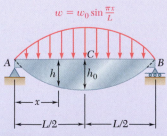

Fig. P5.131

5.132 and *5.133* A preliminary design on the use of a cantilever prismatic timber beam indicated that a beam with a rectangular cross section 2 in. wide and 10 in. deep would be required to safely support the load shown in part *a* of the figure. It was then decided to replace that beam with a built-up beam obtained by gluing together, as shown in part *b* of the figure, five pieces of the same timber as the original beam and of 2 × 2-in. cross section. Determine the respective lengths l_1 and l_2 of the two inner and outer pieces of timber that will yield the same factor of safety as the original design.

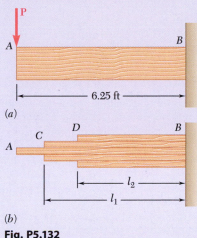

(a)

(b)

Fig. P5.132

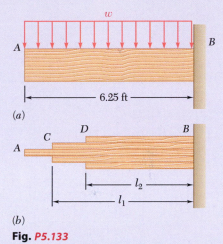

(a)

(b)

Fig. *P5.133*

5.134 and 5.135 A preliminary design on the use of a simply supported prismatic timber beam indicated that a beam with a rectangular cross section 50 mm wide and 200 mm deep would be required to safely support the load shown in part *a* of the figure. It was then decided to replace that beam with a built-up beam obtained by gluing together, as shown in part *b* of the figure, four pieces of the same timber as the original beam and of 50 × 50-mm cross section. Determine the length *l* of the two outer pieces of timber that will yield the same factor of safety as the original design.

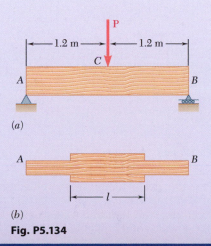

(a)

(b)

Fig. P5.134

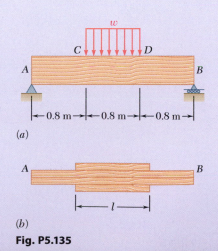

(a)

(b)

Fig. P5.135

5.136 and *5.137* A machine element of cast aluminum and in the shape of a solid of revolution of variable diameter d is being designed to support the load shown. Knowing that the machine element is to be of constant strength, express d in terms of x, L, and d_0.

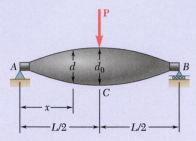

Fig. P5.136

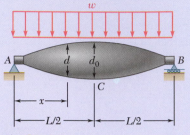

Fig. *P5.137*

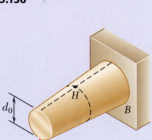

Fig. P5.138

5.138 A transverse force **P** is applied as shown at end A of the conical taper AB. Denoting by d_0 the diameter of the taper at A, show that the maximum normal stress occurs at point H, which is contained in a transverse section of diameter $d = 1.5\,d_0$.

5.139 A cantilever beam AB consisting of a steel plate of uniform depth h and variable width b is to support the distributed load w along its centerline AB. (a) Knowing that the beam is to be of constant strength, express b in terms of x, L, and b_0. (b) Determine the maximum allowable value of w if $L = 15$ in., $b_0 = 8$ in., $h = 0.75$ in., and $\sigma_{all} = 24$ ksi.

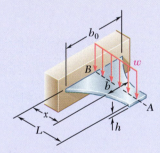

Fig. P5.139

5.140 Assuming that the length and width of the cover plates used with the beam of Sample Prob. 5.12 are, respectively, $l = 4$ m and $b = 285$ mm, and recalling that the thickness of each plate is 16 mm, determine the maximum normal stress on a transverse section (a) through the center of the beam, (b) just to the left of D.

5.141 Two cover plates, each $\frac{1}{2}$ in. thick, are welded to a W27 × 84 beam as shown. Knowing that $l = 10$ ft and $b = 10.5$ in., determine the maximum normal stress on a transverse section (a) through the center of the beam, (b) just to the left of D.

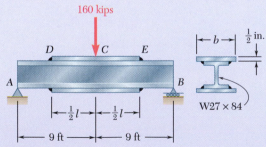

Fig. P5.141 and *P5.142*

5.142 Two cover plates, each $\frac{1}{2}$ in. thick, are welded to a W27 × 84 beam as shown. Knowing that $\sigma_{all} = 24$ ksi for both the beam and the plates, determine the required value of (a) the length of the plates, (b) the width of the plates.

5.143 Knowing that $\sigma_{all} = 150$ MPa, determine the largest concentrated load **P** that can be applied at end E of the beam shown.

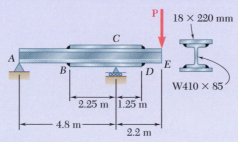

Fig. P5.143

5.144 Two cover plates, each 7.5 mm thick, are welded to a W460 × 74 beam as shown. Knowing that $l = 5$ m and $b = 200$ mm, determine the maximum normal stress on a transverse section (*a*) through the center of the beam, (*b*) just to the left of D.

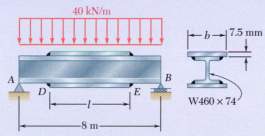

Fig. P5.144 and P5.145

5.145 Two cover plates, each 7.5 mm thick, are welded to a W460 × 74 beam as shown. Knowing that $\sigma_{all} = 150$ MPa for both the beam and the plates, determine the required value of (*a*) the length of the plates, (*b*) the width of the plates.

5.146 Two cover plates, each $\frac{5}{8}$ in. thick, are welded to a W30 × 99 beam as shown. Knowing that $l = 9$ ft and $b = 12$ in., determine the maximum normal stress on a transverse section (*a*) through the center of the beam, (*b*) just to the left of D.

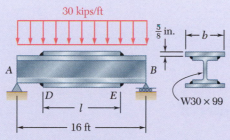

Fig. P5.146 and P5.147

5.147 Two cover plates, each $\frac{5}{8}$ in. thick, are welded to a W30 × 99 beam as shown. Knowing that $\sigma_{all} = 22$ ksi for both the beam and the plates, determine the required value of (*a*) the length of the plates, (*b*) the width of the plates.

5.148 For the tapered beam shown, determine (*a*) the transverse section in which the maximum normal stress occurs, (*b*) the largest distributed load *w* that can be applied, knowing that σ_{all} = 140 MPa.

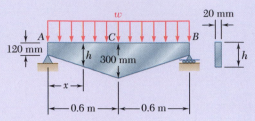

Fig. *P5.148* and P5.149

5.149 For the tapered beam shown, knowing that *w* = 160 kN/m, determine (*a*) the transverse section in which the maximum normal stress occurs, (*b*) the corresponding value of the normal stress.

5.150 For the tapered beam shown, determine (*a*) the transverse section in which the maximum normal stress occurs, (*b*) the largest distributed load *w* that can be applied, knowing that σ_{all} = 24 ksi.

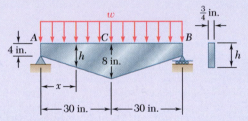

Fig. P5.150

5.151 For the tapered beam shown, determine (*a*) the transverse section in which the maximum normal stress occurs, (*b*) the largest concentrated load **P** that can be applied, knowing that σ_{all} = 24 ksi.

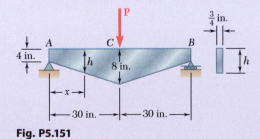

Fig. P5.151

Review and Summary

Design of Prismatic Beams

This chapter was devoted to the analysis and design of beams under transverse loadings consisting of concentrated or distributed loads. The beams are classified according to the way they are supported (Fig. 5.20). Only *statically determinate* beams were considered, where all support reactions can be determined by statics.

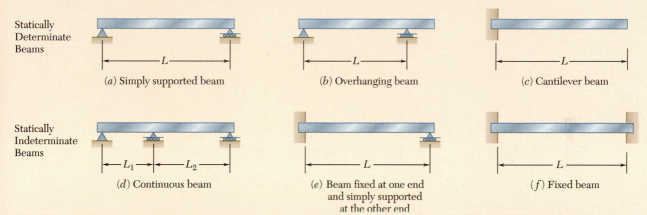

Statically Determinate Beams

(a) Simply supported beam

(b) Overhanging beam

(c) Cantilever beam

Statically Indeterminate Beams

(d) Continuous beam

(e) Beam fixed at one end and simply supported at the other end

(f) Fixed beam

Fig. 5.20 Common beam support configurations.

Normal Stresses Due to Bending

While transverse loadings cause both bending and shear in a beam, the normal stresses caused by bending are the dominant criterion in the design of a beam for strength [Sec. 5.1]. Therefore, this chapter dealt only with the determination of the normal stresses in a beam, the effect of shearing stresses being examined in the next one.

The flexure formula for the determination of the maximum value σ_m of the normal stress in a given section of the beam is

$$\sigma_m = \frac{|M|c}{I} \tag{5.1}$$

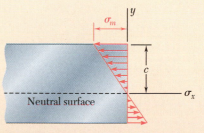

Fig. 5.21 Linear normal stress distribution for bending.

where I is the moment of inertia of the cross section with respect to a centroidal axis perpendicular to the plane of the bending couple $\mathbf{M}$ and c is the maximum distance from the neutral surface (Fig. 5.21). Introducing the elastic section modulus $S = I/c$ of the beam, the maximum value σ_m of the normal stress in the section can be expressed also as

$$\sigma_m = \frac{|M|}{S} \tag{5.3}$$

Shear and Bending-Moment Diagrams

From Eq. (5.1) it is seen that the maximum normal stress occurs in the section where $|M|$ is largest and at the point farthest from the neutral

(a) Internal forces
(positive shear and positive bending moment)

Fig. 5.22 Positive sign convention for internal shear and bending moment.

axis. The determination of the maximum value of $|M|$ and of the critical section of the beam in which it occurs is simplified if *shear diagrams* and *bending-moment diagrams are drawn*. These diagrams represent the variation of the shear and of the bending moment along the beam and are obtained by determining the values of V and M at selected points of the beam. These values are found by passing a section through the point and drawing the free-body diagram of either of the portions of beam. To avoid any confusion regarding the sense of the shearing force **V** and of the bending couple **M** (which act in opposite sense on the two portions of the beam), we follow the sign convention adopted earlier, as illustrated in Fig. 5.22.

Relationships Between Load, Shear, and Bending Moment

The construction of the shear and bending-moment diagrams is facilitated if the following relations are taken into account. Denoting by w the distributed load per unit length (assumed positive if directed downward)

$$\frac{dV}{dx} = -w \tag{5.5}$$

$$\frac{dM}{dx} = V \tag{5.7}$$

or in integrated form,

$$V_D - V_C = -(\text{area under load curve between } C \text{ and } D) \tag{5.6b}$$

$$M_D - M_C = \text{area under shear curve between } C \text{ and } D \tag{5.8b}$$

Equation (5.6b) makes it possible to draw the shear diagram of a beam from the curve representing the distributed load on that beam and V at one end of the beam. Similarly, Eq. (5.8b) makes it possible to draw the bending-moment diagram from the shear diagram and M at one end of the beam. However, concentrated loads introduce discontinuities in the shear diagram and concentrated couples in the bending-moment diagram, none of which is accounted for in these equations. The points of the beam where the bending moment is maximum or minimum are also the points where the shear is zero (Eq. 5.7).

Design of Prismatic Beams

Having determined σ_{all} for the material used and assuming that the design of the beam is controlled by the maximum normal stress in the beam, the minimum allowable value of the section modulus is

$$S_{\text{min}} = \frac{|M|_{\text{max}}}{\sigma_{\text{all}}} \tag{5.9}$$

For a timber beam of rectangular cross section, $S = \frac{1}{6}bh^2$, where b is the width of the beam and h its depth. The dimensions of the section, therefore, must be selected so that $\frac{1}{6}bh^2 \geq S_{\text{min}}$.

For a rolled-steel beam, consult the appropriate table in Appendix C. Of the available beam sections, consider only those with a section modulus $S \geq S_{\text{min}}$. From this group we normally select the section with the smallest weight per unit length.

Singularity Functions

An alternative method to determine the maximum values of the shear and bending moment is based on the *singularity functions* $\langle x - a \rangle^n$. For $n \geq 0$,

$$\langle x - a \rangle^n = \begin{cases} (x - a)^n & \text{when } x \geq a \\ 0 & \text{when } x < a \end{cases} \tag{5.14}$$

Step Function

Whenever the quantity between brackets is positive or zero, the brackets should be replaced by ordinary parentheses, and whenever that quantity is negative, the bracket itself is equal to zero. Also, singularity functions can be integrated and differentiated as ordinary binomials. The singularity function corresponding to $n = 0$ is discontinuous at $x = a$ (Fig. 5.23). This function is called the *step function*.

$$\langle x - a \rangle^0 = \begin{cases} 1 & \text{when } x \geq a \\ 0 & \text{when } x < a \end{cases} \tag{5.15}$$

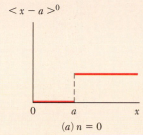

Fig. 5.23 Singular step function.

Using Singularity Functions to Express Shear and Bending Moment

The use of singularity functions makes it possible to represent the shear or the bending moment in a beam by a single expression. This is valid at any point of the beam. For example, the contribution to the shear of the concentrated load **P** applied at the midpoint C of a simply supported beam (Fig. 5.24) can be represented by $-P\langle x - \frac{1}{2}L \rangle^0$, since this expression is equal to zero to the left of C and to $-P$ to the right of C. Adding the reaction $R_A = \frac{1}{2}P$ at A, the shear at any point is

$$V(x) = \tfrac{1}{2}P - P\langle x - \tfrac{1}{2}L \rangle^0$$

The bending moment, obtained by integrating, is

$$M(x) = \tfrac{1}{2}Px - P\langle x - \tfrac{1}{2}L \rangle^1$$

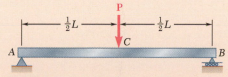

Fig. 5.24 Simply supported beam with a concentrated load at midpoint C.

Equivalent Open-Ended Loadings

The singularity functions representing the load, shear, and bending moment corresponding to various basic loadings were given in Fig. 5.16. A distributed load that does not extend to the right end of the beam or is discontinuous should be replaced by an equivalent combination of open-ended loadings. For instance, a uniformly distributed load extending from $x = a$ to $x = b$ (Fig. 5.25) is

$$w(x) = w_0\langle x - a \rangle^0 - w_0\langle x - b \rangle^0$$

The contribution of this load to the shear and bending moment is obtained through two successive integrations. Care should be used to include for $V(x)$ the contribution of concentrated loads and reactions, and for $M(x)$ the contribution of concentrated couples.

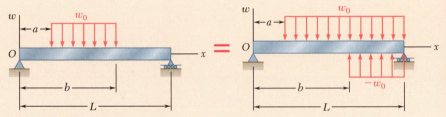

Fig. 5.25 Use of open-ended loadings to create a closed-ended loading.

Nonprismatic Beams

Nonprismatic beams are beams of variable cross section. By selecting the shape and size of the cross section so that its elastic section modulus $S = I/c$ varies along the beam in the same way as the bending moment M, beams can be designed where σ_m at each section is equal to σ_{all}. These are called *beams of constant strength,* and they provide a more effective use of the material than prismatic beams. Their section modulus at any section along the beam is

$$S = \frac{M}{\sigma_{\text{all}}} \tag{5.18}$$

Review Problems

5.152 Draw the shear and bending-moment diagrams for the beam and loading shown, and determine the maximum absolute value (*a*) of the shear, (*b*) of the bending moment.

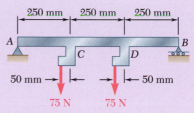

Fig. P5.152

5.153 Draw the shear and bending-moment diagrams for the beam and loading shown and determine the maximum normal stress due to bending.

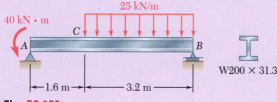

Fig. *P5.153*

5.154 Determine (*a*) the distance *a* for which the absolute value of the bending moment in the beam is as small as possible, (*b*) the corresponding maximum normal stress due to bending. (See hint of Prob. 5.27.)

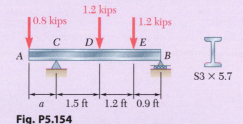

Fig. P5.154

5.155 For the beam and loading shown, determine the equations of the shear and bending-moment curves and the maximum absolute value of the bending moment in the beam, knowing that (*a*) $k = 1$, (*b*) $k = 0.5$.

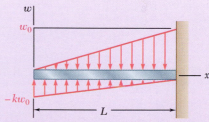

Fig. *P5.155*

5.156 Draw the shear and bending-moment diagrams for the beam and loading shown and determine the maximum normal stress due to bending.

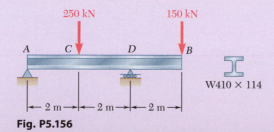

Fig. P5.156

5.157 Beam *AB*, of length *L* and square cross section of side *a*, is supported by a pivot at *C* and loaded as shown. (*a*) Check that the beam is in equilibrium. (*b*) Show that the maximum normal stress due to bending occurs at *C* and is equal to $w_0L^2/(1.5a)^3$.

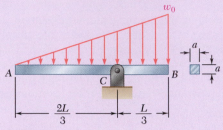

Fig. P5.157

5.158 For the beam and loading shown, design the cross section of the beam, knowing that the grade of timber used has an allowable normal stress of 1750 psi.

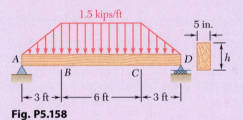

Fig. P5.158

5.159 Knowing that the allowable normal stress for the steel used is 24 ksi, select the most economical wide-flange beam to support the loading shown.

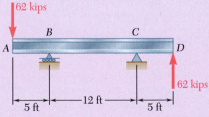

Fig. P5.159

5.160 Three steel plates are welded together to form the beam shown. Knowing that the allowable normal stress for the steel used is 22 ksi, determine the minimum flange width *b* that can be used.

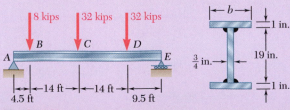

Fig. *P5.160*

5.161 (*a*) Using singularity functions, find the magnitude and location of the maximum bending moment for the beam and loading shown. (*b*) Determine the maximum normal stress due to bending.

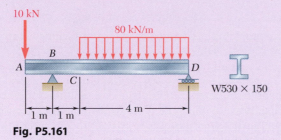

Fig. P5.161

5.162 The beam *AB*, consisting of an aluminum plate of uniform thickness *b* and length *L*, is to support the load shown. (*a*) Knowing that the beam is to be of constant strength, express *h* in terms of *x*, *L*, and h_0 for portion *AC* of the beam. (*b*) Determine the maximum allowable load if $L = 800$ mm, $h_0 = 200$ mm, $b = 25$ mm, and $\sigma_{all} = 72$ MPa.

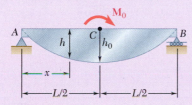

Fig. P5.162

5.163 A cantilever beam *AB* consisting of a steel plate of uniform depth *h* and variable width *b* is to support the concentrated load **P** at point *A*. (*a*) Knowing that the beam is to be of constant strength, express *b* in terms of *x*, *L*, and b_0. (*b*) Determine the smallest allowable value of *h* if $L = 300$ mm, $b_0 = 375$ mm, $P = 14.4$ kN, and $\sigma_{all} = 160$ MPa.

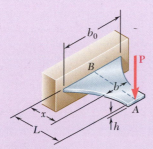

Fig. P5.163

Computer Problems

The following problems are designed to be solved with a computer.

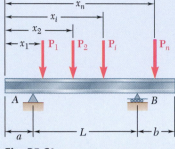

Fig. P5.C1

5.C1 Several concentrated loads P_i, $(i = 1, 2, \ldots, n)$ can be applied to a beam as shown. Write a computer program that can be used to calculate the shear, bending moment, and normal stress at any point of the beam for a given loading of the beam and a given value of its section modulus. Use this program to solve Probs. 5.18, 5.21, and 5.25. (*Hint:* Maximum values will occur at a support or under a load.)

5.C2 A timber beam is to be designed to support a distributed load and up to two concentrated loads as shown. One of the dimensions of its uniform rectangular cross section has been specified and the other is to be determined so that the maximum normal stress in the beam will not exceed a given allowable value σ_{all}. Write a computer program that can be used to calculate at given intervals ΔL the shear, the bending moment, and the smallest acceptable value of the unknown dimension. Apply this program to solve the following problems, using the intervals ΔL indicated: (*a*) Prob. 5.65 ($\Delta L = 0.1$ m), (*b*) Prob. 5.69 ($\Delta L = 0.3$ m), and (*c*) Prob. 5.70 ($\Delta L = 0.2$ m).

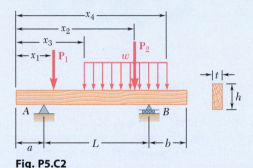

Fig. P5.C2

5.C3 Two cover plates, each of thickness t, are to be welded to a wide-flange beam of length L that is to support a uniformly distributed load w. Denoting by σ_{all} the allowable normal stress in the beam and in the plates, by d the depth of the beam, and by I_b and S_b, respectively, the moment of inertia and the section modulus of the cross section of the unreinforced beam about a horizontal centroidal axis, write a computer program that can be used to calculate the required value of (*a*) the length a of the plates, (*b*) the width b of the plates. Use this program to solve Prob. 5.145.

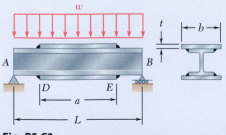

Fig. P5.C3

5.C4 Two 25-kip loads are maintained 6 ft apart as they are moved slowly across the 18-ft beam AB. Write a computer program and use it to calculate the bending moment under each load and at the midpoint C of the beam for values of x from 0 to 24 ft at intervals $\Delta x = 1.5$ ft.

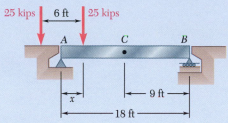

Fig. P5.C4

5.C5 Write a computer program that can be used to plot the shear and bending-moment diagrams for the beam and loading shown. Apply this program with a plotting interval $\Delta L = 0.2$ ft to the beam and loading of (*a*) Prob. 5.72, (*b*) Prob. 5.115.

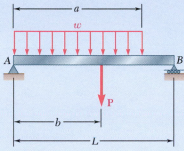

Fig. P5.C5

5.C6 Write a computer program that can be used to plot the shear and bending-moment diagrams for the beam and loading shown. Apply this program with a plotting interval $\Delta L = 0.025$ m to the beam and loading of Prob. 5.112.

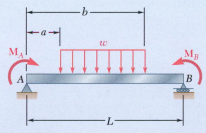

Fig. P5.C6

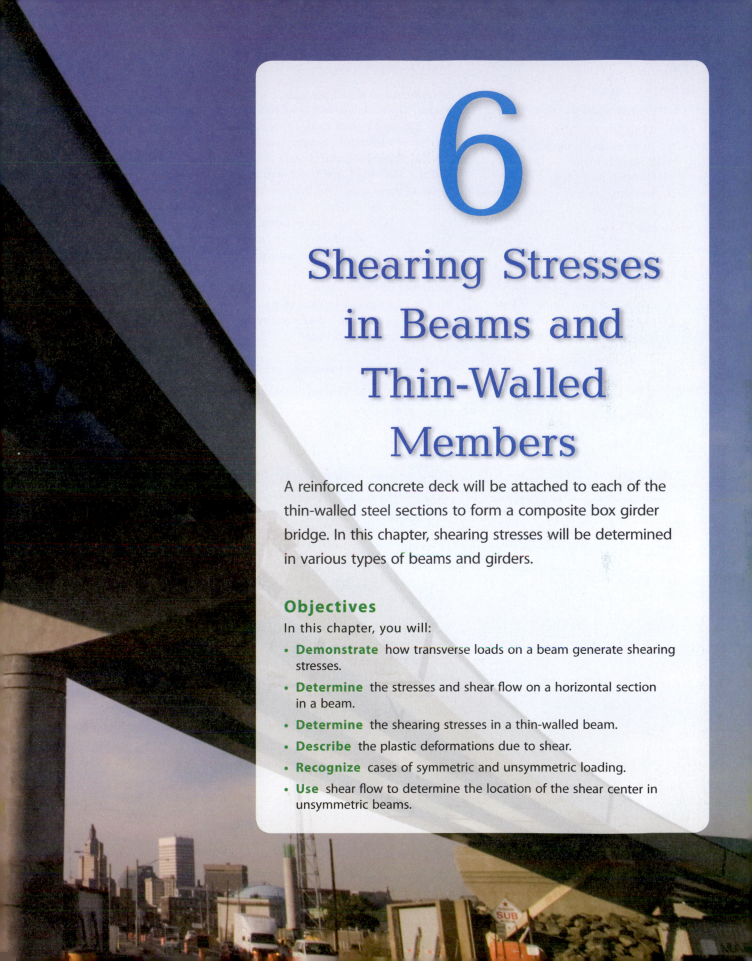

6

Shearing Stresses in Beams and Thin-Walled Members

A reinforced concrete deck will be attached to each of the thin-walled steel sections to form a composite box girder bridge. In this chapter, shearing stresses will be determined in various types of beams and girders.

Objectives

In this chapter, you will:

- **Demonstrate** how transverse loads on a beam generate shearing stresses.
- **Determine** the stresses and shear flow on a horizontal section in a beam.
- **Determine** the shearing stresses in a thin-walled beam.
- **Describe** the plastic deformations due to shear.
- **Recognize** cases of symmetric and unsymmetric loading.
- **Use** shear flow to determine the location of the shear center in unsymmetric beams.

Introduction

Shearing stresses are important, particularly in the design of short, stubby beams. Their analysis is the subject of the first part of this chapter.

Figure 6.1 graphically expresses the elementary normal and shearing forces exerted on a transverse section of a prismatic beam with a vertical plane of symmetry that are equivalent to the bending couple **M** and the shearing force **V**. Six equations can be written to express this. Three of these equations involve only the normal forces $\sigma_x\, dA$ and have been discussed in Sec. 4.2. These are Eqs. (4.1), (4.2), and (4.3), which express that the sum of the normal forces is zero and that the sums of their moments about the y and z axes are equal to zero and M, respectively. Three more equations involving the shearing forces $\tau_{xy}\, dA$ and $\tau_{xz}\, dA$ now can be written. One equation expresses that the sum of the moments of the shearing forces about the x axis is zero and can be dismissed as trivial in view of the symmetry of the beam with respect to the xy plane. The other two involve the y and z components of the elementary forces and are

$$y \text{ components:} \qquad \int \tau_{xy}\, dA = -V \qquad\qquad \textbf{(6.1)}$$

$$z \text{ components:} \qquad \int \tau_{xz}\, dA = 0 \qquad\qquad \textbf{(6.2)}$$

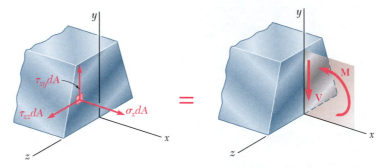

Fig. 6.1   All the stresses on elemental areas (left) sum to give the resultant shear V and bending moment M.

Equation (6.1) shows that vertical shearing stresses must exist in a transverse section of a beam under transverse loading. Equation (6.2) indicates that the average lateral shearing stress in any section is zero. However, this does not mean that the shearing stress τ_{xz} is zero everywhere.

Now consider a small cubic element located in the vertical plane of symmetry of the beam (where τ_{xz} must be zero) and examine the stresses exerted on its faces (Fig. 6.2). A normal stress σ_x and a shearing stress τ_{xy}

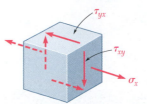

Fig. 6.2   Stress element from section of a transversely loaded beam.

are exerted on each of the two faces perpendicular to the x axis. But we know from Chapter 1 that when shearing stresses τ_{xy} are exerted on the vertical faces of an element, equal stresses must be exerted on the horizontal faces of the same element. Thus, the longitudinal shearing stresses must exist in any member subjected to a transverse loading. This is verified by considering a cantilever beam made of separate planks clamped together at the fixed end (Fig. 6.3a). When a transverse load **P** is applied to the free end of this composite beam, the planks slide with respect to each other (Fig. 6.3b). In contrast, if a couple **M** is applied to the free end of the same composite beam (Fig. 6.3c), the various planks bend into circular concentric arcs and do not slide with respect to each other. This verifies the fact that shear does not occur in a beam subjected to pure bending (see Sec. 4.3).

While sliding does not actually take place when a transverse load **P** is applied to a beam made of a homogeneous and cohesive material such as steel, the tendency to slide exists, showing that stresses occur on horizontal longitudinal planes as well as on vertical transverse planes. In timber beams, whose resistance to shear is weaker between fibers, failure due to shear occurs along a longitudinal plane rather than a transverse plane (Photo 6.1).

In Sec. 6.1A, a beam element of length Δx is considered that is bounded by one horizontal and two transverse planes. The shearing force **ΔH** exerted on its horizontal face will be determined, as well as the shear per unit length q, which is known as *shear flow*. An equation for the shearing stress in a beam with a vertical plane of symmetry is obtained in Sec. 6.1B and used in Sec. 6.1C to determine the shearing stresses in common types of beams. The distribution of stresses in a narrow rectangular beam is discussed further in Sec. 6.2.

The method in Sec. 6.1 is extended in Sec. 6.3 to cover the case of a beam element bounded by two transverse planes and a curved surface. This allows us to determine the shearing stresses at any point of a symmetric thin-walled member, such as the flanges of wide-flange beams and box beams in Sec. 6.4. The effect of plastic deformations on the magnitude and distribution of shearing stresses is discussed in Sec. 6.5.

In the Sec. 6.6, the unsymmetric loading of thin-walled members is considered and the concept of a *shear center* is introduced to determine the distribution of shearing stresses in such members.

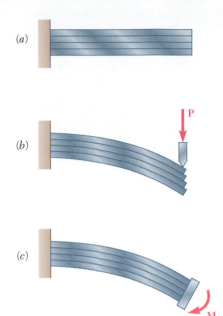

(a)

(b)

(c)

Fig. 6.3 (a) Beam made of planks to illustrate the role of shearing stresses. (b) Beam planks slide relative to each other when transversely loaded. (c) Bending moment causes deflection without sliding.

Photo 6.1 Longitudinal shear failure in timber beam loaded in the laboratory.

6.1 HORIZONTAL SHEARING STRESS IN BEAMS

6.1A Shear on the Horizontal Face of a Beam Element

Consider a prismatic beam AB with a vertical plane of symmetry that supports various concentrated and distributed loads (Fig. 6.4). At a distance x from end A, we detach from the beam an element $CDD'C'$ with length of Δx extending across the width of the beam from the upper surface to a

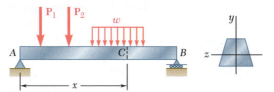

Fig. 6.4 Transversely loaded beam with vertical plane of symmetry.

horizontal plane located at a distance y_1 from the neutral axis (Fig. 6.5). The forces exerted on this element consist of vertical shearing forces $\mathbf{V}'_C$ and $\mathbf{V}'_D$, a horizontal shearing force $\Delta\mathbf{H}$ exerted on the lower face of the element, elementary horizontal normal forces $\sigma_C\,dA$ and $\sigma_D\,dA$, and possibly a load $w\,\Delta x$ (Fig. 6.6). The equilibrium equation for horizontal forces is

$$\overset{+}{\rightarrow}\Sigma F_x = 0: \qquad\qquad \Delta H + \int_{\mathfrak{a}}(\sigma_C - \sigma_D)\,dA = 0$$

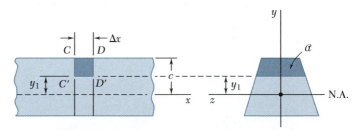

Fig. 6.5 Short segment of beam with stress element $CDD'C'$ defined.

where the integral extends over the shaded area $\mathfrak{a}$ of the section located above the line $y = y_1$. Solving this equation for ΔH and using Eq. (5.2), $\sigma = My/I$, to express the normal stresses in terms of the bending moments at C and D, provides

$$\Delta H = \frac{M_D - M_C}{I}\int_{\mathfrak{a}}y\,dA \qquad\qquad \textbf{(6.3)}$$

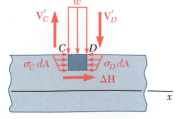

Fig. 6.6 Forces exerted on element $CCD'C'$.

The integral in Eq. (6.3) represents the *first moment* with respect to the neutral axis of the portion ⓐ of the cross section of the beam that is located above the line $y = y_1$ and will be denoted by Q. On the other hand, recalling Eq. (5.7), the increment $M_D - M_C$ of the bending moment is

$$M_D - M_C = \Delta M = (dM/dx)\, \Delta x = V\, \Delta x$$

Substituting into Eq. (6.3), the horizontal shear exerted on the beam element is

$$\Delta H = \frac{VQ}{I}\,\Delta x \qquad \text{(6.4)}$$

The same result is obtained if a free body the lower element $C'D'D''C''$ is used instead of the upper element $CDD'C'$ (Fig. 6.7), since the shearing forces $\Delta \mathbf{H}$ and $\Delta \mathbf{H}'$ exerted by the two elements on each other are equal and opposite. This leads us to observe that the first moment Q of the portion ⓐ' of the cross section located below the line $y = y_1$ (Fig. 6.7) is equal in magnitude and opposite in sign to the first moment of the portion ⓐ located above that line (Fig. 6.5). Indeed, the sum of these two moments is equal to the moment of the area of the entire cross section with respect to its centroidal axis and, thus must be zero. This property is sometimes used to simplify the computation of Q. Also note that Q is maximum for $y_1 = 0$, since the elements of the cross section located above the neutral axis contribute positively to the integral in Eq. (6.3) that defines Q, while the elements located below that axis contribute negatively.

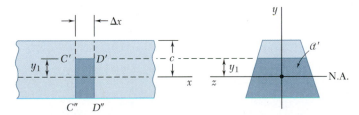

Fig. 6.7 Short segment of beam with stress element $C'D'D''C''$ defined.

The *horizontal shear per unit length*, which will be denoted by q, is obtained by dividing both members of Eq. (6.4) by Δx:

$$q = \frac{\Delta H}{\Delta x} = \frac{VQ}{I} \qquad \text{(6.5)}$$

Recall that Q is the first moment with respect to the neutral axis of the portion of the cross section located either above or below the point at which q is being computed and that I is the centroidal moment of inertia of the *entire* cross-sectional area. The horizontal shear per unit length q is also called the *shear flow* and will be discussed in Sec. 6.4.

Concept Application 6.1

A beam is made of three planks, 20 by 100 mm in cross section, and nailed together (Fig. 6.8a). Knowing that the spacing between nails is 25 mm and the vertical shear in the beam is $V = 500$ N, determine the shearing force in each nail.

Determine the horizontal force per unit length q exerted on the lower face of the upper plank. Use Eq. (6.5), where Q represents the first moment with respect to the neutral axis of the shaded area A shown in Fig. 6.8b, and I is the moment of inertia about the same axis of the entire cross-sectional area (Fig. 6.8c). Recalling that the first moment of an area with respect to a given axis is equal to the product of the area and of the distance from its centroid to the axis,[†]

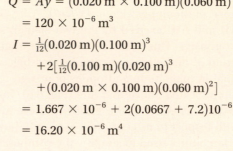

$$Q = A\bar{y} = (0.020 \text{ m} \times 0.100 \text{ m})(0.060 \text{ m})$$
$$= 120 \times 10^{-6} \text{ m}^3$$
$$I = \tfrac{1}{12}(0.020 \text{ m})(0.100 \text{ m})^3$$
$$+ 2[\tfrac{1}{12}(0.100 \text{ m})(0.020 \text{ m})^3$$
$$+ (0.020 \text{ m} \times 0.100 \text{ m})(0.060 \text{ m})^2]$$
$$= 1.667 \times 10^{-6} + 2(0.0667 + 7.2)10^{-6}$$
$$= 16.20 \times 10^{-6} \text{ m}^4$$

Fig. 6.8 (a) Composite beam made of three boards nailed together. (b) Cross section for computing Q. (c) Cross section for computing moment of inertia.

Substituting into Eq. (6.5),

$$q = \frac{VQ}{I} = \frac{(500 \text{ N})(120 \times 10^{-6} \text{ m}^3)}{16.20 \times 10^{-6} \text{ m}^4} = 3704 \text{ N/m}$$

Since the spacing between the nails is 25 mm, the shearing force in each nail is

$$F = (0.025 \text{ m})q = (0.025 \text{ m})(3704 \text{ N/m}) = 92.6 \text{ N}$$

6.1B Shearing Stresses in a Beam

Consider again a beam with a vertical plane of symmetry that is subjected to various concentrated or distributed loads applied in that plane. If, through two vertical cuts and one horizontal cut, an element of length Δx is detached from the beam (Fig. 6.9), the magnitude ΔH of the shearing force exerted on the horizontal face of the element can be obtained from Eq. (6.4). The *average shearing stress* τ_{ave} on that face of the element is obtained by dividing ΔH by the area ΔA of the face. Observing that $\Delta A = t \Delta x$, where t is the width of the element at the cut, we write

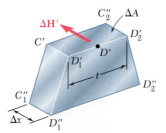

Fig. 6.9 Stress element $C'D'D''C''$ showing the shear force on a horizontal plane.

$$\tau_{ave} = \frac{\Delta H}{\Delta A} = \frac{VQ}{I} \frac{\Delta x}{t \Delta x}$$

[†]See Appendix A.

or

$$\tau_{\text{ave}} = \frac{VQ}{It} \qquad (6.6)$$

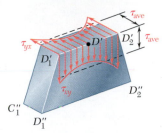

Fig. 6.10 Stress element $C'D'D''C''$ showing the shearing stress distribution along $D_1' D_2'$.

Note that since the shearing stresses τ_{xy} and τ_{yx} exerted on a transverse and a horizontal plane through D' are equal, the expression also represents the average value of τ_{xy} along the line $D_1' D_2'$ (Fig. 6.10).

Observe that $\tau_{yx} = 0$ on the upper and lower faces of the beam, since no forces are exerted on these faces. It follows that $\tau_{xy} = 0$ along the upper and lower edges of the transverse section (Fig. 6.11). Also note that while Q is maximum for $y = 0$ (see Sec. 6.1A), τ_{ave} may not be maximum along the neutral axis, since τ_{ave} depends upon the width t of the section as well as upon Q.

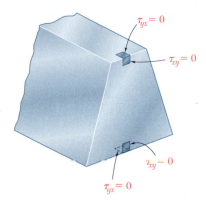

Fig. 6.11 Beam cross section showing that the shearing stress is zero at the top and bottom of the beam.

As long as the width of the beam cross section remains small compared to its depth, the shearing stress varies only slightly along the line $D_1' D_2'$ (Fig. 6.10), and Eq. (6.6) can be used to compute τ_{xy} at any point along $D_1' D_2'$. Actually, τ_{xy} is larger at points D_1' and D_2' than at D', but the theory of elasticity shows[†] that, for a beam of rectangular section of width b and depth h, and as long as $b \le h/4$, the value of the shearing stress at points C_1 and C_2 (Fig. 6.12) does not exceed by more than 0.8% the average value of the stress computed along the neutral axis.

On the other hand, for large values of b/h, τ_{max} of the stress at C_1 and C_2 may be many times larger then the average value τ_{ave} computed along the neutral axis, as shown in the following table.

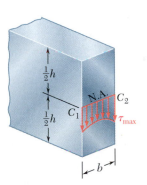

Fig. 6.12 Shearing stress distribution along neutral axis of rectangular beam cross section.

b/h	0.25	0.5	1	2	4	6	10	20	50
$\tau_{\text{max}}/\tau_{\text{ave}}$	1.008	1.033	1.126	1.396	1.988	2.582	3.770	6.740	15.65
$\tau_{\text{min}}/\tau_{\text{ave}}$	0.996	0.983	0.940	0.856	0.805	0.800	0.800	0.800	0.800

[†]See S. P. Timoshenko and J. N. Goodier, *Theory of Elasticity*, McGraw-Hill, New York, 3d ed., 1970, sec. 124.

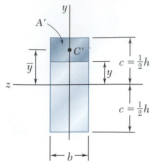

Fig. 6.13 Geometric terms for rectangular section used to calculate shearing stress.

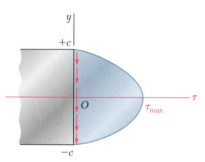

Fig. 6.14 Shearing stress distribution on transverse section of rectangular beam.

6.1C Shearing Stresses τ_{xy} In Common Beam Types

In the preceding section for a *narrow rectangular beam* (i.e., a beam of rectangular section of width b and depth h with $b \le \frac{1}{4}h$), the variation of the shearing stress τ_{xy} across the width of the beam is less than 0.8% of τ_{ave}. Therefore, Eq. (6.6) is used in practical applications to determine the shearing stress at any point of the cross section of a narrow rectangular beam, and

$$\tau_{xy} = \frac{VQ}{It} \qquad (6.7)$$

where t is equal to the width b of the beam and Q is the first moment with respect to the neutral axis of the shaded area A (Fig. 6.13).

Observing that the distance from the neutral axis to the centroid C' of A is $\bar{y} = \frac{1}{2}(c + y)$ and recalling that $Q = A\bar{y}$,

$$Q = A\bar{y} = b(c - y)\tfrac{1}{2}(c + y) = \tfrac{1}{2}b(c^2 - y^2) \qquad (6.8)$$

Recalling that $I = bh^3/12 = \frac{2}{3}bc^3$,

$$\tau_{xy} = \frac{VQ}{Ib} = \frac{3}{4}\frac{c^2 - y^2}{bc^3}V$$

or noting that the cross-sectional area of the beam is $A = 2bc$,

$$\tau_{xy} = \frac{3}{2}\frac{V}{A}\left(1 - \frac{y^2}{c^2}\right) \qquad (6.9)$$

Equation (6.9) shows that the distribution of shearing stresses in a transverse section of a rectangular beam is *parabolic* (Fig. 6.14). As observed in the preceding section, the shearing stresses are zero at the top and bottom of the cross section ($y = \pm c$). Making $y = 0$ in Eq. (6.9), the value of the maximum shearing stress in a given section of a *narrow rectangular beam* is

$$\tau_{\text{max}} = \frac{3}{2}\frac{V}{A} \qquad (6.10)$$

This relationship shows that the maximum value of the shearing stress in a beam of rectangular cross section is 50% larger than the value V/A obtained by wrongly assuming a uniform stress distribution across the entire cross section.

In an *American standard beam* (S-beam) or a *wide-flange beam* (W-beam), Eq. (6.6) can be used to determine the average value of the shearing stress τ_{xy} over a section aa' or bb' of the transverse cross section of the beam (Figs. 6.15a and b). So

$$\tau_{\text{ave}} = \frac{VQ}{It} \qquad (6.6)$$

where V is the vertical shear, t is the width of the section at the elevation considered, Q is the first moment of the shaded area with respect to the neutral axis cc', and I is the moment of inertia of the entire cross-sectional area about cc'. Plotting τ_{ave} against the vertical distance y provides the curve shown in Fig. 6.15c. Note the discontinuities existing in this curve, which reflect the difference between the values of t corresponding respectively to the flanges $ABGD$ and $A'B'G'D'$ and to the web $EFF'E'$.

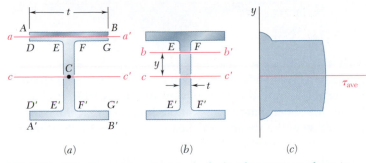

Fig. 6.15 Wide-flange beam. (a) Area for finding first moment of area in flange. (b) Area for finding first moment of area in web. (c) Shearing stress distribution.

In the web, the shearing stress τ_{xy} varies only very slightly across the section bb' and is assumed to be equal to its average value τ_{ave}. This is not true, however, for the flanges. For example, considering the horizontal line $DEFG$, note that τ_{xy} is zero between D and E and between F and G, since these two segments are part of the free surface of the beam. However, the value of τ_{xy} between E and F is non-zero and can be obtained by making $t = EF$ in Eq. (6.6). In practice, one usually assumes that the entire shear load is carried by the web and that a good approximation of the maximum value of the shearing stress in the cross section can be obtained by dividing V by the cross-sectional area of the web.

$$\tau_{max} = \frac{V}{A_{web}} \qquad \textbf{(6.11)}$$

However, while the vertical component τ_{xy} of the shearing stress in the flanges can be neglected, its horizontal component τ_{xz} has a significant value that will be determined in Sec. 6.4.

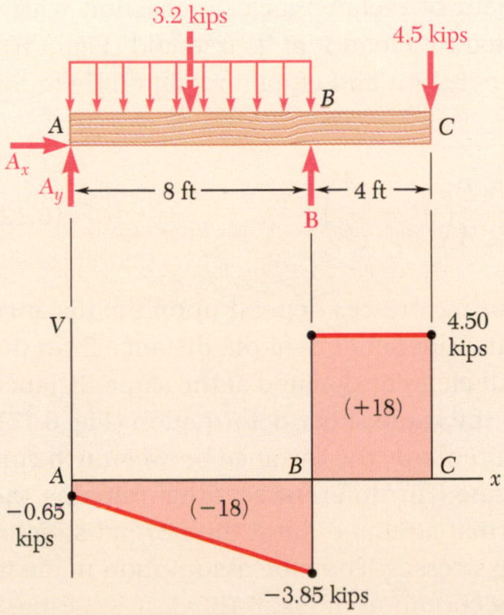

Concept Application 6.2

Knowing that the allowable shearing stress for the timber beam of Sample Prob. 5.7 is $\tau_{all} = 0.250$ ksi, check that the design is acceptable from the point of view of the shearing stresses.

Recall from the shear diagram of Sample Prob. 5.7 that $V_{max} = 4.50$ kips. The actual width of the beam was given as $b = 3.5$ in., and the value obtained for its depth was $h = 14.55$ in. Using Eq. (6.10) for the maximum shearing stress in a narrow rectangular beam,

$$\tau_{max} = \frac{3}{2}\frac{V}{A} = \frac{3}{2}\frac{V}{bh} = \frac{3(4.50 \text{ kips})}{2(3.5 \text{ in.})(14.55 \text{ in.})} = 0.1325 \text{ ksi}$$

Since $\tau_{max} < \tau_{all}$, the design obtained in Sample Prob. 5.7 is acceptable.

Fig. 5.19 *(repeated)*

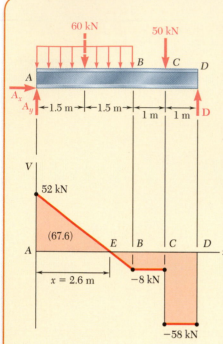

Fig. 5.20 (repeated)

Concept Applications 6.3

Knowing that the allowable shearing stress for the steel beam of Sample Prob. 5.8 is $\tau_{all} = 90$ MPa, check that the W360 × 32.9 shape obtained is acceptable from the point of view of the shearing stresses.

Recall from the shear diagram of Sample Prob. 5.8 that the maximum absolute value of the shear in the beam is $|V|_{max} = 58$ kN. It may be assumed that the entire shear load is carried by the web and that the maximum value of the shearing stress in the beam can be obtained from Eq. (6.11). From Appendix C, for a W360 × 32.9 shape, the depth of the beam and the thickness of its web are $d = 348$ mm and $t_w = 5.84$ mm. Thus,

$$A_{web} = d\,t_w = (348 \text{ mm})(5.84 \text{ mm}) = 2032 \text{ mm}^2$$

Substituting $|V|_{max}$ and A_{web} into Eq. (6.11),

$$\tau_{max} = \frac{|V|_{max}}{A_{web}} = \frac{58 \text{ kN}}{2032 \text{ mm}^2} = 28.5 \text{ MPa}$$

Since $\tau_{max} < \tau_{all}$, the design obtained in Sample Prob. 5.8 is acceptable.

*6.2 DISTRIBUTION OF STRESSES IN A NARROW RECTANGULAR BEAM

Consider a narrow cantilever beam of rectangular cross section with a width of b and depth of h subjected to a load **P** at its free end (Fig. 6.16). Since the shear V in the beam is constant and equal in magnitude to the load **P**, Eq. (6.9) yields

$$\tau_{xy} = \frac{3}{2}\frac{P}{A}\left(1 - \frac{y^2}{c^2}\right) \tag{6.12}$$

Note from Eq. (6.12) that the shearing stresses depend upon the distance y from the neutral surface. They are independent of the distance from the point of application of the load. All elements located at the same distance from the neutral surface undergo the same shear deformation (Fig. 6.17). While plane sections do *not* remain plane, the distance between two corresponding points D and D' located in different sections remains the same. This indicates that the normal strains ϵ_x, and the normal stresses σ_x, are unaffected by the shearing stresses. Thus the assumption made in Chap. 5 is justified for the loading condition of Fig. 6.16.

We therefore conclude that this analysis of the stresses in a cantilever beam of rectangular cross section subjected to a concentrated load **P**

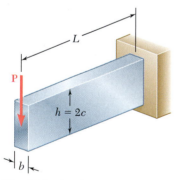

Fig. 6.16 Cantilever beam with rectangular cross section.

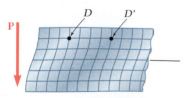

Fig. 6.17 Deformation of segment of cantilever beam.

at its free end is valid. The correct values of the shearing stresses in the beam are given by Eq. (6.12), and the normal stresses at a distance x from the free end are obtained by making $M = -Px$ in Eq. (5.2). So

$$\sigma_x = +\frac{Pxy}{I} \tag{6.13}$$

The validity of the this statement depends upon the end conditions. If Eq. (6.12) is to apply everywhere, the load P must be distributed parabolically over the free-end section. Also, the fixed-end support must allow the type of shear deformation indicated in Fig. 6.17. The resulting model (Fig. 6.18) is highly unlikely to be encountered in practice. However, it follows from

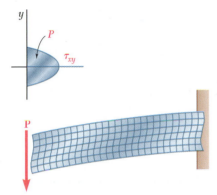

Fig. 6.18   Deformation of cantilever beam with concentrated load, with a parabolic shearing stress distribution.

Saint-Venant's principle that for other modes of application of the load and for other types of fixed-end supports, Eqs. (6.12) and (6.13) provide the correct distribution of stresses, except close to either end of the beam.

When a beam of rectangular cross section is subjected to several concentrated loads (Fig. 6.19), the principle of superposition can be used to determine the normal and shearing stresses in sections located between the points of application. However, since the loads $\mathbf{P}_2$, $\mathbf{P}_3$, etc. are applied on the surface of the beam and are not assumed to be distributed parabolically throughout the cross section, the results cease to be valid in the immediate vicinity of the points of application of the loads.

When the beam is subjected to a distributed load (Fig. 6.20), both the shear and shearing stress at a given elevation y vary with the distance from the end of the beam. The shear deformation results show that the distance between two corresponding points of different cross sections, such as D_1 and D'_1, or D_2 and D'_2, depends upon their elevation. As a result, the assumption that plane sections remain plane, as in Eqs. (6.12) and (6.13), must be rejected for the loading condition of Fig. 6.20. However, the error involved is small for the values of the span-depth ratio encountered in practice.

In portions of the beam located under a concentrated or distributed load, normal stresses σ_y are exerted on the horizontal faces of a cubic element of material in addition to the stresses τ_{xy} shown in Fig. 6.2.

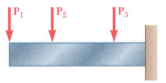

Fig. 6.19   Cantilever beam with multiple loads.

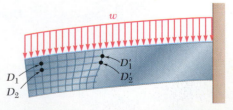

Fig. 6.20   Deformation of cantilever beam with distributed load.

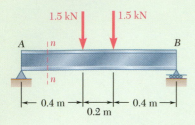

Sample Problem 6.1

Beam *AB* is made of three plates glued together and is subjected, in its plane of symmetry, to the loading shown. Knowing that the width of each glued joint is 20 mm, determine the average shearing stress in each joint at section *n–n* of the beam. The location of the centroid of the section is given in Fig. 1 and the centroidal moment of inertia is known to be $I = 8.63 \times 10^{-6} \text{ m}^4$.

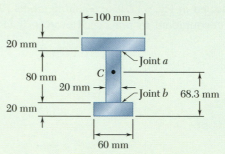

Fig. 1 Cross section dimensions with location of centroid.

STRATEGY: A free-body diagram is first used to determine the shear at the required section. Eq. (6.7) is then used to determine the average shearing stress in each joint.

MODELING:

Vertical Shear at Section *n–n*. As shown in the free-body diagram in Fig. 2, the beam and loading are both symmetric with respect to the center of the beam. Thus, we have $\mathbf{A} = \mathbf{B} = 1.5 \text{ kN} \uparrow$.

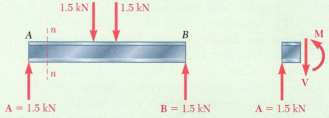

Fig. 2 Free-body diagram of beam and segment of beam to left of section *n–n*.

Drawing the free-body diagram of the portion of the beam to the left of section *n–n* (Fig. 2), we write

$$+\uparrow \Sigma F_y = 0: \qquad 1.5 \text{ kN} - V = 0 \qquad V = 1.5 \text{ kN}$$

(continued)

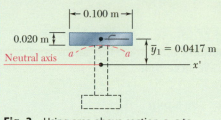

Fig. 3 Using area above section *a–a* to find *Q*.

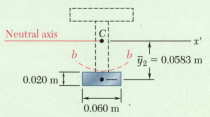

Fig. 4 Using area below section *b–b* to find *Q*.

ANALYSIS:

Shearing Stress in Joint *a*. Using Fig. 3, pass the section *a–a* through the glued joint and separate the cross-sectional area into two parts. We choose to determine Q by computing the first moment with respect to the neutral axis of the area above section *a–a*.

$$Q = A\bar{y}_1 = [(0.100 \text{ m})(0.020 \text{ m})](0.0417 \text{ m}) = 83.4 \times 10^{-6} \text{ m}^3$$

Recalling that the width of the glued joint is $t = 0.020$ m, we use Eq. (6.7) to determine the average shearing stress in the joint.

$$\tau_{\text{ave}} = \frac{VQ}{It} = \frac{(1500 \text{ N})(83.4 \times 10^{-6} \text{ m}^3)}{(8.63 \times 10^{-6} \text{ m}^4)(0.020 \text{ m})} \qquad \tau_{\text{ave}} = 725 \text{ kPa} \blacktriangleleft$$

Shearing Stress in Joint *b*. Using Fig. 4, now pass section *b–b* and compute Q by using the area below the section.

$$Q = A\bar{y}_2 = [(0.060 \text{ m})(0.020 \text{ m})](0.0583 \text{ m}) = 70.0 \times 10^{-6} \text{ m}^3$$

$$\tau_{\text{ave}} = \frac{VQ}{It} = \frac{(1500 \text{ N})(70.0 \times 10^{-6} \text{ m}^3)}{(8.63 \times 10^{-6} \text{ m}^4)(0.020 \text{ m})} \qquad \tau_{\text{ave}} = 608 \text{ kPa} \blacktriangleleft$$

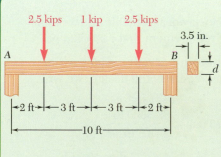

Sample Problem 6.2

A timber beam *AB* of span 10 ft and nominal width 4 in. (actual width = 3.5 in.) is to support the three concentrated loads shown. Knowing that for the grade of timber used $\sigma_{\text{all}} = 1800$ psi and $\tau_{\text{all}} = 120$ psi, determine the minimum required depth *d* of the beam.

STRATEGY: A free-body diagram with the shear and bending-moment diagrams is used to determine the maximum shear and bending moment. The resulting design must satisfy both allowable stresses. Start by assuming that one allowable stress criterion governs, and solve for the required depth *d*. Then use this depth with the other criterion to determine if it is also satisfied. If this stress is greater than the allowable, revise the design using the second criterion.

(continued)

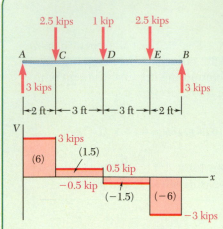

Fig. 1 Free-body diagram of beam with shear and bending-moment diagrams.

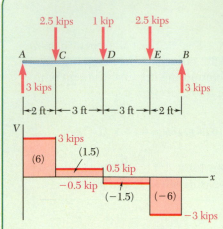

Fig. 2 Section of beam having depth d.

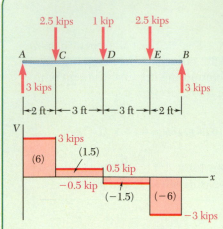

4 in. × 12 in. nominal size

Fig. 3 Design cross section.

MODELING:

Maximum Shear and Bending Moment. The free-body diagram is used to determine the reactions and draw the shear and bending-moment diagrams in Fig. 1. We note that

$$M_{max} = 7.5 \text{ kip·ft} = 90 \text{ kip·in.}$$

$$V_{max} = 3 \text{ kips}$$

ANALYSIS:

Design Based on Allowable Normal Stress. We first express the elastic section modulus S in terms of the depth d (Fig. 2). We have

$$I = \frac{1}{12}bd^3 \qquad S = \frac{I}{c} = \frac{1}{6}bd^2 = \frac{1}{6}(3.5)d^2 = 0.5833d^2$$

For $M_{max} = 90$ kip·in. and $\sigma_{all} = 1800$ psi, we write

$$S = \frac{M_{max}}{\sigma_{all}} \qquad 0.5833d^2 = \frac{90 \times 10^3 \text{ lb·in.}}{1800 \text{ psi}}$$

$$d^2 = 85.7 \qquad d = 9.26 \text{ in.}$$

We have satisfied the requirement that $\sigma_m \leq 1800$ psi.

Check Shearing Stress. For $V_{max} = 3$ kips and $d = 9.26$ in., we find

$$\tau_m = \frac{3}{2}\frac{V_{max}}{A} = \frac{3}{2}\frac{3000 \text{ lb}}{(3.5 \text{ in.})(9.26 \text{ in.})} \qquad \tau_m = 138.8 \text{ psi}$$

Since $\tau_{all} = 120$ psi, the depth $d = 9.26$ in. is *not* acceptable and we must redesign the beam on the basis of the requirement that $\tau_m \leq 120$ psi.

Design Based on Allowable Shearing Stress. Since we now know that the allowable shearing stress controls the design, we write

$$\tau_m = \tau_{all} = \frac{3}{2}\frac{V_{max}}{A} \qquad 120 \text{ psi} = \frac{3}{2}\frac{3000 \text{ lb}}{(3.5 \text{ in.})d}$$

$$d = 10.71 \text{ in.} \blacktriangleleft$$

The normal stress is, of course, less than $\sigma_{all} = 1800$ psi, and the depth of 10.71 in. is fully acceptable.

REFLECT and THINK: Since timber is normally available in nominal depth increments of 2 in., a 4 × 12-in. standard size timber should be used. The actual cross section would then be 3.5 × 11.25 in. (Fig. 3).

Problems

6.1 Three full-size 50 × 100-mm boards are nailed together to form a beam that is subjected to a vertical shear of 1500 N. Knowing that the allowable shearing force in each nail is 400 N, determine the largest longitudinal spacing s that can be used between each pair of nails.

6.2 For the built-up beam of Prob. 6.1, determine the allowable shear if the spacing between each pair of nails is $s = 45$ mm.

6.3 Three boards, each 2 in. thick, are nailed together to form a beam that is subjected to a vertical shear. Knowing that the allowable shearing force in each nail is 150 lb, determine the allowable shear if the spacing s between the nails is 3 in.

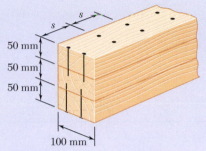

Fig. P6.1

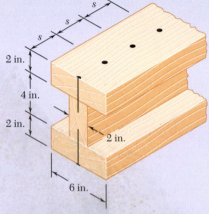

Fig. P6.3

6.4 A square box beam is made of two 20 × 80-mm planks and two 20 × 120-mm planks nailed together as shown. Knowing that the spacing between the nails is $s = 30$ mm and that the vertical shear in the beam is $V = 1200$ N, determine (a) the shearing force in each nail, (b) the maximum shearing stress in the beam.

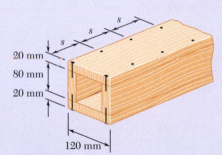

Fig. P6.4

6.5 The American Standard rolled-steel beam shown has been reinforced by attaching to it two 16 × 200-mm plates, using 18-mm-diameter bolts spaced longitudinally every 120 mm. Knowing that the average allowable shearing stress in the bolts is 90 MPa, determine the largest permissible vertical shearing force.

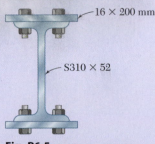

16 × 200 mm

S310 × 52

Fig. P6.5

6.6 The beam shown is fabricated by connecting two channel shapes and two plates, using bolts of $\frac{3}{4}$-in. diameter spaced longitudinally every 7.5 in. Determine the average shearing stress in the bolts caused by a shearing force of 25 kips parallel to the y axis.

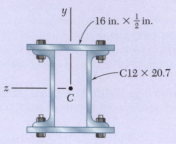

16 in. $\times \frac{1}{2}$ in.

C12 $\times$ 20.7

Fig. P6.6

6.7 A column is fabricated by connecting the rolled-steel members shown by bolts of $\frac{3}{4}$-in. diameter spaced longitudinally every 5 in. Determine the average shearing stress in the bolts caused by a shearing force of 30 kips parallel to the y axis.

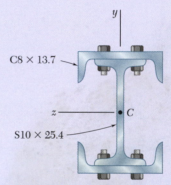

C8 $\times$ 13.7

S10 $\times$ 25.4

Fig. P6.7

6.8 The composite beam shown is fabricated by connecting two W6 $\times$ 20 rolled-steel members, using bolts of $\frac{5}{8}$-in. diameter spaced longitudinally every 6 in. Knowing that the average allowable shearing stress in the bolts is 10.5 ksi, determine the largest allowable vertical shear in the beam.

Fig. P6.8

6.9 through 6.12 For beam and loading shown, consider section *n–n* and determine (*a*) the largest shearing stress in that section, (*b*) the shearing stress at point *a*.

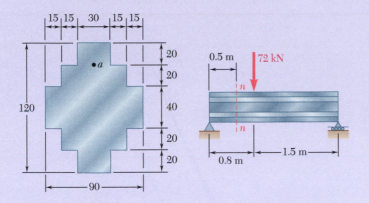

Dimensions in mm

Fig. P6.9

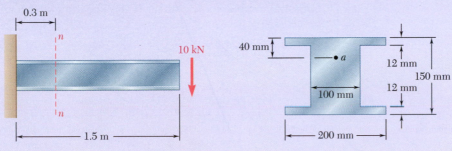

Fig. *P6.10*

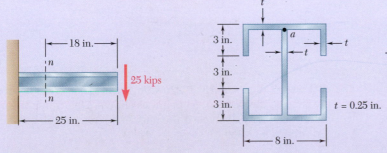

Fig. P6.11

$t = 0.25$ in.

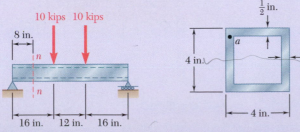

Fig. P6.12

6.13 and 6.14 For a beam having the cross section shown, determine the largest allowable vertical shear if the shearing stress is not to exceed 60 MPa.

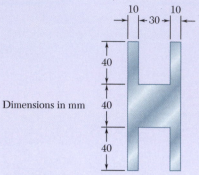

Dimensions in mm

Fig. P6.13

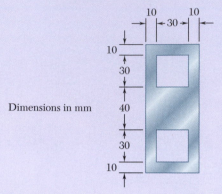

Dimensions in mm

Fig. P6.14

6.15 For a timber beam having the cross section shown, determine the largest allowable vertical shear if the shearing stress is not to exceed 150 psi.

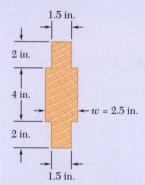

Fig. P6.15

6.16 Two steel plates of 12 × 220-mm rectangular cross section are welded to the W250 × 58 beam as shown. Determine the largest allowable vertical shear if the shearing stress in the beam is not to exceed 90 MPa.

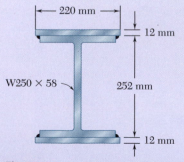

Fig. P6.16

6.17 Two W8 × 31 rolled sections may be welded at A and B in either of the two ways shown in order to form a composite beam. Knowing that for each weld the allowable shearing force is 3000 lb per inch of weld, determine for each arrangement the maximum allowable vertical shear in the composite beam.

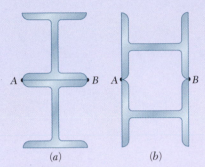

(a)   (b)

Fig. P6.17

6.18 For the beam and loading shown, determine the minimum required width b, knowing that for the grade of timber used, $\sigma_{all} = 12$ MPa and $\tau_{all} = 825$ kPa.

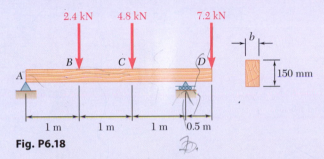

Fig. P6.18

6.19 A timber beam AB of length L and rectangular cross section carries a single concentrated load $\mathbf{P}$ at its midpoint C. (a) Show that the ratio τ_m/σ_m of the maximum values of the shearing and normal stresses in the beam is equal to $h/2L$, where h and L are, respectively, the depth and the length of the beam. (b) Determine the depth h and the width b of the beam, knowing that $L = 2$ m, $P = 40$ kN, $\tau_m = 960$ kPa, and $\sigma_m = 12$ MPa.

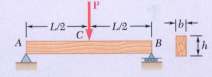

Fig. P6.19

6.20 A timber beam AB of length L and rectangular cross section carries a uniformly distributed load w and is supported as shown. (a) Show that the ratio τ_m/σ_m of the maximum values of the shearing and normal stresses in the beam is equal to $2h/L$, where h and L are, respectively, the depth and the length of the beam. (b) Determine the depth h and the width b of the beam, knowing that $L = 5$ m, $w = 8$ kN/m, $\tau_m = 1.08$ MPa, and $\sigma_m = 12$ MPa.

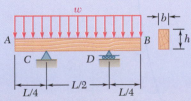

Fig. P6.20

6.21 and 6.22 For the beam and loading shown, consider section n–n and determine the shearing stress at (a) point a, (b) point b.

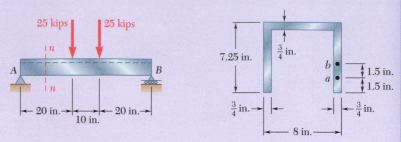

Fig. P6.21 and P6.23

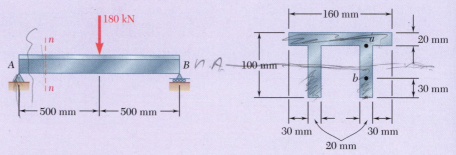

Fig. P6.22 and P6.24

6.23 and 6.24 For the beam and loading shown, determine the largest shearing stress in section n–n. $\rightarrow$ max at N.A

6.25 through 6.28 A beam having the cross section shown is subjected to a vertical shear **V**. Determine (a) the horizontal line along which the shearing stress is maximum, (b) the constant k in the following expression for the maximum shearing stress

$$\tau_{max} = k\frac{V}{A}$$

where A is the cross-sectional area of the beam.

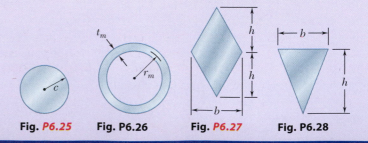

Fig. *P6.25*   Fig. P6.26   Fig. *P6.27*   Fig. P6.28

6.3   LONGITUDINAL SHEAR ON A BEAM ELEMENT OF ARBITRARY SHAPE

Consider a box beam obtained by nailing together four planks, as shown in Fig. 6.21a. Sec. 6.1A showed how to determine the shear per unit length q on the horizontal surfaces along which the planks are joined. But could q be determined if the planks are joined along *vertical* surfaces, as shown in Fig. 6.21b? Section 6.2 showed the distribution of the vertical components τ_{xy} of the stresses on a transverse section of a W- or S-beam. These stresses had a fairly constant value in the web of the beam and were negligible in its flanges. But what about the *horizontal* components τ_{xz} of the stresses in the flanges? The procedure developed in Sec. 6.1A to determine the shear per unit length q applies to the cases just described.

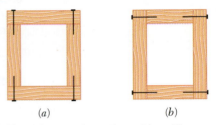

Fig. 6.21   Box beam formed by nailing planks together.

Consider the prismatic beam AB of Fig. 6.4, which has a vertical plane of symmetry and supports the loads shown. At a distance x from end A, detach an element $CDD'C'$ with a length of Δx. However, this element now extends from two sides of the beam to an arbitrary curved surface (Fig. 6.22). The forces exerted on the element include vertical shearing

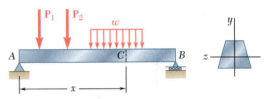

Fig. 6.4   (*repeated*) Beam example.

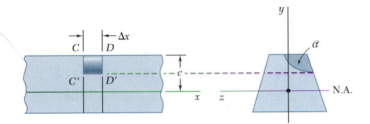

Fig. 6.22   Short segment of beam with element $CDD'C'$ of length Δx.

forces $\mathbf{V'}_C$ and $\mathbf{V'}_D$, elementary horizontal normal forces $\sigma_C\,dA$ and $\sigma_D\,dA$, possibly a load $w\,\Delta x$, and a longitudinal shearing force $\Delta \mathbf{H}$, which represent the resultant of the elementary longitudinal shearing forces exerted on the curved surface (Fig. 6.23). The equilibrium equation is

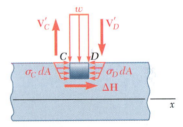

Fig. 6.23   Forces exerted on element $CDD'C'$.

$$\overset{+}{\rightarrow}\Sigma F_x = 0: \qquad \Delta H + \int_{\mathcal{C}} (\sigma_C - \sigma_D)\,dA = 0$$

where the integral is to be computed over the shaded area $\mathcal{C}$ of the section in Fig. 6.22. This equation is the same as the one in Sec. 6.1A, but the shaded area $\mathcal{C}$ now extends to the curved surface.

The longitudinal shear exerted on the beam element is

$$\Delta H = \frac{VQ}{I}\Delta x \qquad\qquad \textbf{(6.4)}$$

where I is the centroidal moment of inertia of the entire section, Q is the first moment of the shaded area $\mathcal{C}$ with respect to the neutral axis, and V is the vertical shear in the section. Dividing both members of Eq. (6.4) by Δx, the horizontal shear per unit length or shear flow is

$$q = \frac{\Delta H}{\Delta x} = \frac{VQ}{I} \qquad\qquad \textbf{(6.5)}$$

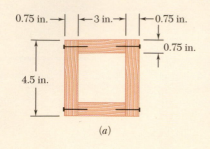

(a)

Concept Application 6.4

A square box beam is made of two 0.75 × 3-in. planks and two 0.75 × 4.5-in. planks nailed together, as shown (Fig. 6.24a). Knowing that the spacing between nails is 1.75 in. and that the beam is subjected to a vertical shear with a magnitude of $V = 600$ lb, determine the shearing force in each nail.

Isolate the upper plank and consider the total force per unit length q exerted on its two edges. Use Eq. (6.5), where Q represents the first moment with respect to the neutral axis of the shaded area A' shown in Fig. 6.24b and I is the moment of inertia about the same axis of the entire cross-sectional area of the box beam (Fig. 6.24c).

$$Q = A'\bar{y} = (0.75 \text{ in.})(3 \text{ in.})(1.875 \text{ in.}) = 4.22 \text{ in}^3$$

Recalling that the moment of inertia of a square of side a about a centroidal axis is $I = \frac{1}{12}a^4$,

$$I = \frac{1}{12}(4.5 \text{ in.})^4 - \frac{1}{12}(3 \text{ in.})^4 = 27.42 \text{ in}^4$$

Substituting into Eq. (6.5),

$$q = \frac{VQ}{I} = \frac{(600 \text{ lb})(4.22 \text{ in}^3)}{27.42 \text{ in}^4} = 92.3 \text{ lb/in.}$$

Because both the beam and the upper plank are symmetric with respect to the vertical plane of loading, equal forces are exerted on both edges of the plank. The force per unit length on each of these edges is thus $\frac{1}{2}q = \frac{1}{2}(92.3) = 46.15$ lb/in. Since the spacing between nails is 1.75 in., the shearing force in each nail is

$$F = (1.75 \text{ in.})(46.15 \text{ lb/in.}) = 80.8 \text{ lb}$$

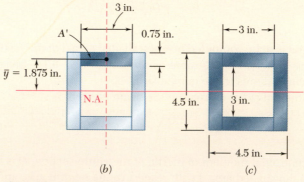

Fig. 6.24 (a) Box beam made from planks nailed together. (b) Geometry for finding first moment of area of top plank. (c) Geometry for finding the moment of inertia of entire cross section.

6.4   SHEARING STRESSES IN THIN-WALLED MEMBERS

We saw in the preceding section that Eq. (6.4) may be used to determine the longitudinal shear $\Delta \mathbf{H}$ exerted on the walls of a beam element of arbitrary shape and Eq. (6.5) to determine the corresponding shear flow q. Equations (6.4) and (6.5) are used in this section to calculate both the shear flow and the average shearing stress in thin-walled members such as the flanges of wide-flange beams (Photo 6.2), box beams, or the walls of structural tubes (Photo 6.3).

Photo 6.2   Wide-flange beams.

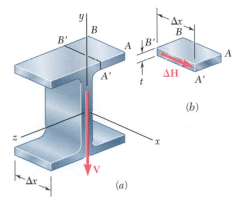

Photo 6.3   Structural tubes.

Consider a segment of length Δx of a wide-flange beam (Fig. 6.25a) where $\mathbf{V}$ is the vertical shear in the transverse section shown. Detach an element $ABB'A'$ of the upper flange (Fig. 6.25b). The longitudinal shear $\Delta \mathbf{H}$ exerted on that element can be obtained from Eq. (6.4):

$$\Delta H = \frac{VQ}{I} \Delta x \tag{6.4}$$

Dividing ΔH by the area $\Delta A = t \, \Delta x$ of the cut, the average shearing stress exerted on the element is the same expression obtained in Sec. 6.1B for a horizontal cut:

$$\tau_{\text{ave}} = \frac{VQ}{It} \tag{6.6}$$

Note that τ_{ave} now represents the average value of the shearing stress τ_{zx} over a vertical cut, but since the thickness t of the flange is small, there is very little variation of τ_{zx} across the cut. Recalling that $\tau_{xz} = \tau_{zx}$,

Fig. 6.25   (a) Wide-flange beam section with vertical shear V. (b) Segment of flange with longitudinal shear ΔH.

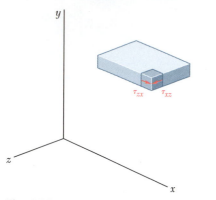

Fig. 6.26   Stress element within flange segment.

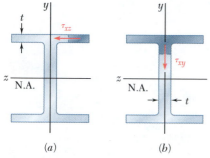

(a)   (b)

Fig. 6.27   Wide-flange beam sections showing shearing stress (a) in flange and (b) in web. The shaded area is that used for calculating the first moment of area.

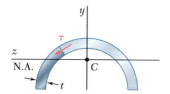

Fig. 6.29   Half pipe section showing shearing stress, and shaded area for calculating first moment of area.

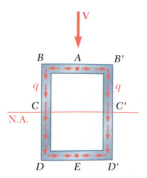

Fig. 6.30   Shear flow, q, in a box beam section.

(Fig. 6.26), the horizontal component τ_{xz} of the shearing stress at any point of a transverse section of the flange can be obtained from Eq. (6.6), where Q is the first moment of the shaded area about the neutral axis (Fig. 6.27a). A similar result was obtained for the vertical component τ_{xy} of the shearing stress in the web (Fig. 6.27b). Equation (6.6) can be used to determine shearing stresses in box beams (Fig. 6.28), half pipes (Fig. 6.29), and other thin-walled members, as long as the loads are applied in a plane of symmetry. In each case, the cut must be perpendicular to the surface of the member, and Eq. (6.6) will yield the component of the shearing stress in the direction tangent to that surface. (The other component is assumed to be equal to zero, because of the proximity of the two free surfaces.)

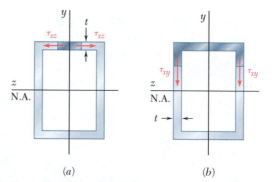

(a)   (b)

Fig. 6.28   Box beam showing shearing stress (a) in flange, (b) in web. Shaded area is that used for calculating the first moment of area.

Comparing Eqs. (6.5) and (6.6), the product of the shearing stress τ at a given point of the section and the thickness t at that point is equal to q. Since V and I are constant, q depends only upon the first moment Q and easily can be sketched on the section. For a box beam (Fig. 6.30), q grows smoothly from zero at A to a maximum value at C and C' on the neutral axis and decreases back to zero as E is reached. There is no sudden variation in the magnitude of q as it passes a corner at B, D, B', or D', and the sense of q in the horizontal portions of the section is easily obtained from its sense in the vertical portions (the sense of the shear **V**). In a wide-flange section (Fig. 6.31), the values of q in portions AB and A'B of the upper flange are distributed symmetrically. At B in the web, q corresponds to the two halves of the flange, which must be combined to obtain the value of q at the top of the web. After reaching a maximum value at C on the neutral axis, q decreases and splits into two equal parts at D, which corresponds at D to the two halves of the lower flange. The shear per unit length q is commonly called the *shear flow* and reflects the similarity between the properties of q just described and some of the characteristics of a fluid flow through an open channel or pipe.[†]

So far, all of the loads were applied in a plane of symmetry of the member. In the case of members possessing two planes of symmetry (Fig. 6.27 or 6.30), any load applied through the centroid of a given cross

---

[†]Recall that the concept of shear flow was used to analyze the distribution of shearing stresses in thin-walled hollow shafts (Sec. 3.10). However, while the shear flow in a hollow shaft is constant, the shear flow in a member under a transverse loading is not.

section can be resolved into components along the two axes of symmetry. Each component will cause the member to bend in a plane of symmetry, and the corresponding shearing stresses can be obtained from Eq. (6.6). The principle of superposition can then be used to determine the resulting stresses.

However, if the member possesses no plane of symmetry or a single plane of symmetry and is subjected to a load that is not contained in that plane, that member is observed to *bend and twist* at the same time—except when the load is applied at a specific point called the *shear center*. The shear center normally does *not* coincide with the centroid of the cross section. The shear center of various thin-walled shapes is discussed in Sec. 6.6.

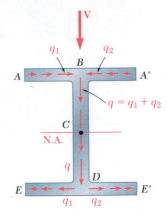

Fig. 6.31 Shear flow, q, in a wide-flange beam section.

*6.5 PLASTIC DEFORMATIONS

Consider a cantilever beam AB with a length of L and a rectangular cross section subjected to a concentrated load **P** at its free end A (Fig. 6.32). The largest bending moment occurs at the fixed end B and is equal to $M = PL$. As long as this value does not exceed the maximum elastic moment M_Y (i.e., $PL \leq M_Y$), the normal stress σ_x will not exceed the yield strength σ_Y anywhere in the beam. However, as P is increased beyond M_Y/L, yield is initiated at points B and B' and spreads toward the free end of the beam. Assuming the material is elastoplastic and considering a cross section CC' located a distance x from the free end A of the beam (Fig. 6.33), the half-thickness y_Y of the elastic core in that section is obtained by making $M = Px$ in Eq. (4.38). Thus,

$$Px = \frac{3}{2}M_Y\left(1 - \frac{1}{3}\frac{y_Y^2}{c^2}\right) \tag{6.14}$$

where c is the half-depth of the beam. Plotting y_Y against x gives the boundary between the elastic and plastic zones.

As long as $PL < \frac{3}{2}M_Y$, the parabola from Eq. (6.14) intersects the line BB', as shown in Fig. 6.33. However, when PL reaches the value $\frac{3}{2}M_Y$ ($PL = M_p$) where M_p is the plastic moment, Eq. (6.14) yields $y_Y = 0$ for $x = L$, which shows that the vertex of the parabola is now located in section BB' and that this section has become fully plastic (Fig. 6.34). Recalling Eq. (4.40), the radius of curvature ρ of the neutral surface at that point is equal to zero, indicating the presence of a sharp bend in the beam at its fixed end. Thus, a *plastic hinge* has developed at that point. The load $P = M_p/L$ is the largest load that can be supported by the beam.

This discussion is based only on the analysis of the normal stresses in the beam. Now examine the distribution of the shearing stresses in a section

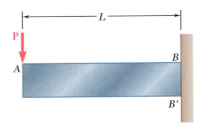

Fig. 6.32 Cantilever beam having maximum moment PL at section B-B'. As long as $PL \leq M_Y$, the beam remains elastic.

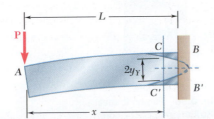

Fig. 6.33 Cantilever beam exhibiting partial yielding, showing the elastic core at section C–C'.

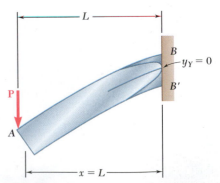

Fig. 6.34 Fully plastic cantilever beam having $PL = M_p = 1.5\, M_Y$.

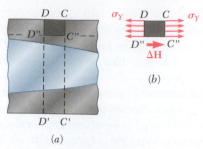

Fig. 6.35 (a) Beam segment in partially plastic area. (b) Element $DCC''D''$ is fully plastic.

that has become partly plastic. Consider the portion of beam $CC''D''D$ located between the transverse sections CC' and DD' and above the horizontal plane $D''C''$ (Fig. 6.35a). If this portion is located entirely in the plastic zone, the normal stresses exerted on the faces CC'' and DD'' will be uniformly distributed and equal to the yield strength σ_Y (Fig. 6.35b). The equilibrium of the free body $CC''D''D$ requires that the horizontal shearing force ΔH exerted on its lower face is equal to zero. The average value of the horizontal shearing stress τ_{yx} across the beam at C'' is also zero, as well as the average value of the vertical shearing stress τ_{xy}. Thus, the vertical shear $V = P$ in section CC' must be distributed entirely over the portion EE' of the section located within the elastic zone (Fig. 6.36). The distribution of the shearing stresses over EE' is the same as that in an elastic rectangular beam with the same width b as beam AB and depth equal to the thickness $2y_Y$ of the elastic zone.[†] The area $2by_Y$ of the elastic portion of the cross section A' gives

$$\tau_{xy} = \frac{3}{2}\frac{P}{A'}\left(1 - \frac{y^2}{y_Y^2}\right) \tag{6.15}$$

The maximum value of the shearing stress occurs for $y = 0$ and is

$$\tau_{\max} = \frac{3}{2}\frac{P}{A'} \tag{6.16}$$

As the area A' of the elastic portion of the section decreases, $\tau_{\max}$ increases and eventually reaches the yield strength in shear τ_Y. Thus, shear contributes to the ultimate failure of the beam. A more exact analysis of this mode of failure should take into account the combined effect of the normal and shearing stresses.

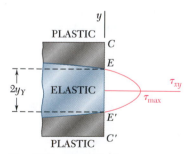

Fig. 6.36 Parabolic shear distribution in elastic core.

[†]See Prob. 6.60.

Sample Problem 6.3

Knowing that the vertical shear is 50 kips in a W10 × 68 rolled-steel beam, determine the horizontal shearing stress in the top flange at a point *a* located 4.31 in. from the edge of the beam. The dimensions and other geometric data of the rolled-steel section are given in Appendix C.

STRATEGY: Determine the horizontal shearing stress at the required section.

MODELING and ANALYSIS:

As shown in Fig. 1, we isolate the shaded portion of the flange by cutting along the dashed line that passes through point *a*.

$$Q = (4.31 \text{ in.})(0.770 \text{ in.})(4.815 \text{ in.}) = 15.98 \text{ in}^3$$

$$\tau = \frac{VQ}{It} = \frac{(50 \text{ kips})(15.98 \text{ in}^3)}{(394 \text{ in}^4)(0.770 \text{ in.})} \qquad \tau = 2.63 \text{ ksi} \quad \blacktriangleleft$$

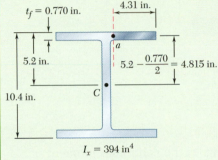

$t_f = 0.770$ in.   4.31 in.

5.2 in.

$5.2 - \dfrac{0.770}{2} = 4.815$ in.

10.4 in.

C

a

$I_x = 394$ in^4

Fig. 1 Cross section dimensions for W10 × 68 steel beam.

Sample Problem 6.4

Solve Sample Prob. 6.3, assuming that 0.75 × 12-in. plates have been attached to the flanges of the W10 × 68 beam by continuous fillet welds as shown.

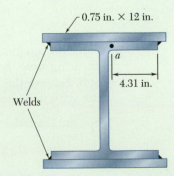

0.75 in. × 12 in.

a

4.31 in.

Welds

STRATEGY: Calculate the properties for the composite beam and then determine the shearing stress at the required section.

(continued)

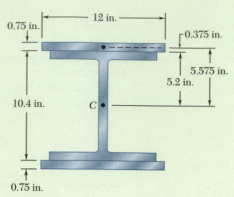

Fig. 1   Cross section dimensions for calculating moment of inertia.

MODELING and ANALYSIS:

For the composite beam shown in Fig. 1, the centroidal moment of inertia is

$$I = 394 \text{ in}^4 + 2\left[\tfrac{1}{12}(12 \text{ in.})(0.75 \text{ in.})^3 + (12 \text{ in.})(0.75 \text{ in.})(5.575 \text{ in.})^2\right]$$
$$I = 954 \text{ in}^4$$

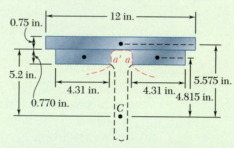

Fig. 2   Dimensions used to find first moment of area and shearing stress at flange-web junction.

Since the top plate and the flange are connected only at the welds, the shearing stress is found at a by passing a section through the flange at a, *between* the plate and the flange, and again through the flange at the symmetric point a' (Fig. 2).

For the shaded area,

$$t = 2t_f = 2(0.770 \text{ in.}) = 1.540 \text{ in.}$$

$$Q = 2[(4.31 \text{ in.})(0.770 \text{ in.})(4.815 \text{ in.})] + (12 \text{ in.})(0.75 \text{ in.})(5.575 \text{ in.})$$

$$Q = 82.1 \text{ in}^3$$

$$\tau = \frac{VQ}{It} = \frac{(50 \text{ kips})(82.1 \text{ in}^3)}{(954 \text{ in}^4)(1.540 \text{ in.})} \qquad\qquad \tau = 2.79 \text{ ksi} \ \blacktriangleleft$$

Sample Problem 6.5

The thin-walled extruded beam shown is made of aluminum and has a uniform 3-mm wall thickness. Knowing that the shear in the beam is 5 kN, determine (*a*) the shearing stress at point *A*, (*b*) the maximum shearing stress in the beam. *Note:* The dimensions given are to lines midway between the outer and inner surfaces of the beam.

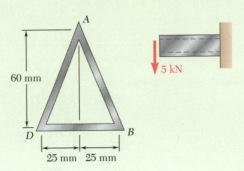

STRATEGY: Determine the location of the centroid and then calculate the moment of inertia. Calculate the two required stresses.

MODELING and ANALYSIS:

Centroid.   Using Fig. 1, we note that $AB = AD = 65$ mm.

$$\overline{Y} = \frac{\Sigma \, \overline{y} A}{\Sigma \, A} = \frac{2[(65 \text{ mm})(3 \text{ mm})(30 \text{ mm})]}{2[(65 \text{ mm})(3 \text{ mm})] + (50 \text{ mm})(3 \text{ mm})}$$

$$\overline{Y} = 21.67 \text{ mm}$$

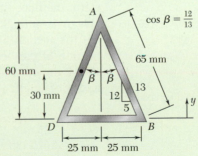

Fig. 1   Section dimensions for finding centroid.

(continued)

Centroidal Moment of Inertia. Each side of the thin-walled beam can be considered as a parallelogram (Fig. 2), and we recall that for the case shown $I_{nn} = bh^3/12$, where b is measured parallel to the axis nn. Using Fig. 3 we write

$$b = (3\text{ mm})/\cos\beta = (3\text{ mm})/(12/13) = 3.25\text{ mm}$$

$$I = \Sigma(\bar{I} + Ad^2) = 2[\tfrac{1}{12}(3.25\text{ mm})(60\text{ mm})^3$$
$$+ (3.25\text{ mm})(60\text{ mm})(8.33\text{ mm})^2] + [\tfrac{1}{12}(50\text{ mm})(3\text{ mm})^3$$
$$+ (50\text{ mm})(3\text{ mm})(21.67\text{ mm})^2]$$
$$I = 214.6 \times 10^3\text{ mm}^4 \qquad I = 0.2146 \times 10^{-6}\text{ m}^4$$

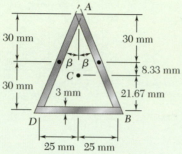

Fig. 2 Dimensions locating centroid.

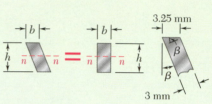

Fig. 3 Determination of horizontal width for side elements.

a. Shearing Stress at A. If a shearing stress τ_A occurs at A, the shear flow will be $q_A = \tau_A t$ and must be directed in one of the two ways shown in Fig 4. But the cross section and the loading are symmetric about a vertical line through A, and thus the shear flow must also be symmetric. Since neither of the possible shear flows is symmetric, we conclude that

$$\tau_A = 0. \qquad \blacktriangleleft$$

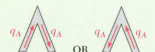

Fig. 4 Possible directions for shear flow at A.

b. Maximum Shearing Stress. Since the wall thickness is constant, the maximum shearing stress occurs at the neutral axis, where Q is maximum. Since we know that the shearing stress at A is zero, we cut the section along the dashed line shown and isolate the shaded portion of the beam (Fig. 5). In order to obtain the largest shearing stress, the cut at the neutral axis is made perpendicular to the sides and is of length $t = 3$ mm.

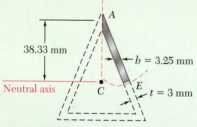

Fig. 5 Section for finding the maximum shearing stress.

$$Q = [(3.25\text{ mm})(38.33\text{ mm})]\left(\frac{38.33\text{ mm}}{2}\right) = 2387\text{ mm}^3$$

$$Q = 2.387 \times 10^{-6}\text{ m}^3$$

$$\tau_E = \frac{VQ}{It} = \frac{(5\text{ kN})(2.387 \times 10^{-6}\text{ m}^3)}{(0.2146 \times 10^{-6}\text{ m}^4)(0.003\text{ m})} \qquad \tau_{\max} = \tau_E = 18.54\text{ MPa} \qquad \blacktriangleleft$$

Problems

6.29 The built-up timber beam shown is subjected to a vertical shear of 1200 lb. Knowing that the allowable shearing force in the nails is 75 lb, determine the largest permissible spacing s of the nails.

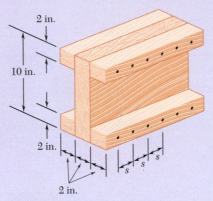

Fig. P6.29

6.30 The built-up beam shown is made by gluing together two 20×250-mm plywood strips and two 50×100-mm planks. Knowing that the allowable average shearing stress in the glued joints is 350 kPa, determine the largest permissible vertical shear in the beam.

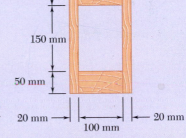

Fig. P6.30

6.31 The built-up beam was made by gluing together several wooden planks. Knowing that the beam is subjected to a 1200-lb vertical shear, determine the average shearing stress in the glued joint (*a*) at *A*, (*b*) at *B*.

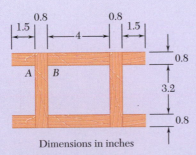

Dimensions in inches

Fig. P6.31

6.32 Several wooden planks are glued together to form the box beam shown. Knowing that the beam is subjected to a vertical shear of 3 kN, determine the average shearing stress in the glued joint (*a*) at *A*, (*b*) at *B*.

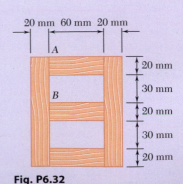

Fig. P6.32

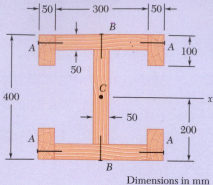

Fig. P6.33

Dimensions in mm

6.33 The built-up wooden beam shown is subjected to a vertical shear of 8 kN. Knowing that the nails are spaced longitudinally every 60 mm at A and every 25 mm at B, determine the shearing force in the nails (a) at A, (b) at B. (*Given*: $I_x = 1.504 \times 10^9$ mm^4.)

6.34 Knowing that a W360 × 122 rolled-steel beam is subjected to a 250-kN vertical shear, determine the shearing stress (a) at point A, (b) at the centroid C of the section.

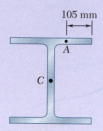

Fig. P6.34

6.35 and 6.36 An extruded aluminum beam has the cross section shown. Knowing that the vertical shear in the beam is 150 kN, determine the shearing stress at (a) point a, (b) point b.

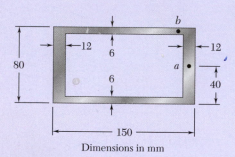

Dimensions in mm

Fig. P6.35

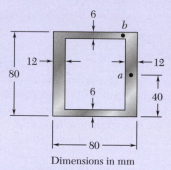

Dimensions in mm

Fig. P6.36

6.37 Knowing that a given vertical shear **V** causes a maximum shearing stress of 75 MPa in an extruded beam having the cross section shown, determine the shearing stress at the three points indicated.

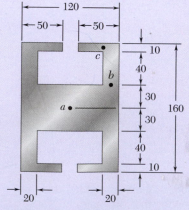

Dimensions in mm

Fig. P6.37

448

6.38 The vertical shear is 1200 lb in a beam having the cross section shown. Knowing that $d = 4$ in., determine the shearing stress at (a) point a, (b) point b.

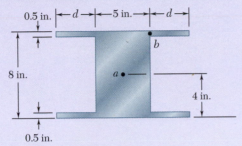

Fig. P6.38 and P6.39

6.39 The vertical shear is 1200 lb in a beam having the cross section shown. Determine (a) the distance d for which $\tau_a = \tau_b$, (b) the corresponding shearing stress at points a and b.

6.40 and 6.41 The extruded aluminum beam has a uniform wall thickness of $\frac{1}{8}$ in. Knowing that the vertical shear in the beam is 2 kips, determine the corresponding shearing stress at each of the five points indicated.

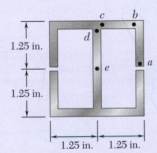

Fig. P6.40

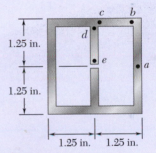

Fig. P6.41

6.42 Knowing that a given vertical shear **V** causes a maximum shearing stress of 50 MPa in a thin-walled member having the cross section shown, determine the corresponding shearing stress at (a) point a, (b) point b, (c) point c.

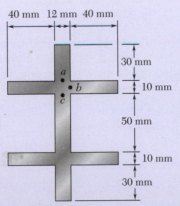

Fig. P6.42

6.43 Three planks are connected as shown by bolts of $\frac{3}{8}$-in. diameter spaced every 6 in. along the longitudinal axis of the beam. For a vertical shear of 2.5 kips, determine the average shearing stress in the bolts.

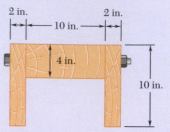

Fig. P6.43

6.44 A beam consists of three planks connected as shown by steel bolts with a longitudinal spacing of 225 mm. Knowing that the shear in the beam is vertical and equal to 6 kN and that the allowable average shearing stress in each bolt is 60 MPa, determine the smallest permissible bolt diameter that can be used.

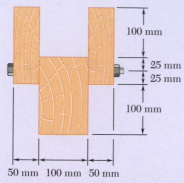

Fig. P6.44

6.45 A beam consists of five planks of 1.5 × 6-in. cross section connected by steel bolts with a longitudinal spacing of 9 in. Knowing that the shear in the beam is vertical and equal to 2000 lb and that the allowable average shearing stress in each bolt is 7500 psi, determine the smallest permissible bolt diameter that can be used.

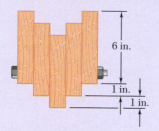

Fig. P6.45

6.46 Four L102 × 102 × 9.5 steel angle shapes and a 12 × 400-mm plate are bolted together to form a beam with the cross section shown. The bolts are of 22-mm diameter and are spaced longitudinally every 120 mm. Knowing that the beam is subjected to a vertical shear of 240 kN, determine the average shearing stress in each bolt.

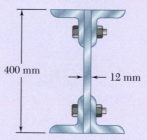

Fig. P6.46

6.47 A plate of $\frac{1}{4}$-in. thickness is corrugated as shown and then used as a beam. For a vertical shear of 1.2 kips, determine (*a*) the maximum shearing stress in the section, (*b*) the shearing stress at point *B*. Also sketch the shear flow in the cross section.

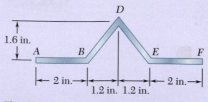

Fig. *P6.47*

6.48 A plate of 2-mm thickness is bent as shown and then used as a beam. For a vertical shear of 5 kN, determine the shearing stress at the five points indicated and sketch the shear flow in the cross section.

6.49 An extruded beam has the cross section shown and a uniform wall thickness of 3 mm. For a vertical shear of 10 kN, determine (*a*) the shearing stress at point *A*, (*b*) the maximum shearing stress in the beam. Also sketch the shear flow in the cross section.

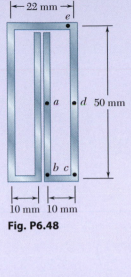

Fig. P6.48

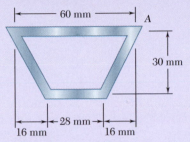

Fig. P6.49

6.50 A plate of thickness *t* is bent as shown and then used as a beam. For a vertical shear of 600 lb, determine (*a*) the thickness *t* for which the maximum shearing stress is 300 psi, (*b*) the corresponding shearing stress at point *E*. Also sketch the shear flow in the cross section.

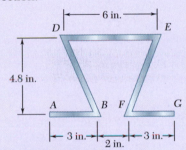

Fig. *P6.50*

6.51 The design of a beam calls for connecting two vertical rectangular $\frac{3}{8} \times 4$-in. plates by welding them to two horizontal $\frac{1}{2} \times 2$-in. plates as shown. For a vertical shear **V**, determine the dimension *a* for which the shear flow through the welded surfaces is maximum.

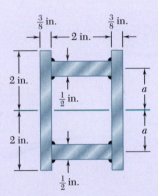

Fig. P6.51

6.52 The cross section of an extruded beam is a hollow square of side *a* = 3 in. and thickness *t* = 0.25 in. For a vertical shear of 15 kips, determine the maximum shearing stress in the beam and sketch the shear flow in the cross section.

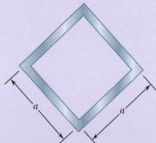

Fig. P6.52

6.53 An extruded beam has a uniform wall thickness t. Denoting by **V** the vertical shear and by A the cross-sectional area of the beam, express the maximum shearing stress as $\tau_{max} = k(V/A)$ and determine the constant k for each of the two orientations shown.

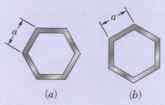

Fig. P6.53

6.54 (*a*) Determine the shearing stress at point P of a thin-walled pipe of the cross section shown caused by a vertical shear **V**. (*b*) Show that the maximum shearing stress occurs for $\theta = 90°$ and is equal to $2V/A$, where A is the cross-sectional area of the pipe.

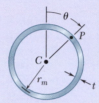

Fig. P6.54

6.55 For a beam made of two or more materials with different moduli of elasticity, show that Eq. (6.6)

$$\tau_{ave} = \frac{VQ}{It}$$

remains valid provided that both Q and I are computed by using the transformed section of the beam (see Sec. 4.4) and provided further that t is the actual width of the beam where τ_{ave} is computed.

6.56 and 6.57 A composite beam is made by attaching the timber and steel portions shown with bolts of 12-mm diameter spaced longitudinally every 200 mm. The modulus of elasticity is 10 GPa for the wood and 200 GPa for the steel. For a vertical shear of 4 kN, determine (*a*) the average shearing stress in the bolts, (*b*) the shearing stress at the center of the cross section. (*Hint*: Use the method indicated in Prob. 6.55.)

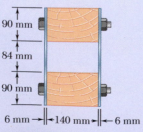

Fig. P6.56

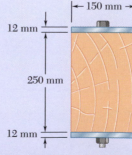

Fig. P6.57

6.58 and 6.59 A steel bar and an aluminum bar are bonded together as shown to form a composite beam. Knowing that the vertical shear in the beam is 4 kips and that the modulus of elasticity is 29×10^6 psi for the steel and 10.6×10^6 psi for the aluminum, determine (*a*) the average shearing stress at the bonded surface, (*b*) the maximum shearing stress in the beam. (*Hint*: Use the method indicated in Prob. 6.55.)

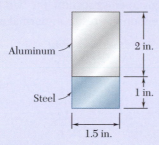

Aluminum

Steel

2 in.

1 in.

1.5 in.

Fig. P6.58

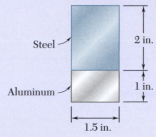

Steel

Aluminum

2 in.

1 in.

1.5 in.

Fig. P6.59

6.60 Consider the cantilever beam *AB* discussed in Sec. 6.5 and the portion *ACKJ* of the beam that is located to the left of the transverse section *CC′* and above the horizontal plane *JK*, where *K* is a point at a distance $y < y_Y$ above the neutral axis (Fig. P6.60). (*a*) Recalling that $\sigma_x = \sigma_Y$ between *C* and *E* and $\sigma_x = (\sigma_Y / y_Y)y$ between *E* and *K*, show that the magnitude of the horizontal shearing force **H** exerted on the lower face of the portion of beam *ACKJ* is

$$H = \frac{1}{2} b \sigma_Y \left(2c - y_Y - \frac{y^2}{y_Y} \right)$$

(*b*) Observing that the shearing stress at *K* is

$$\tau_{xy} = \lim_{\Delta A \to 0} \frac{\Delta H}{\Delta A} = \lim_{\Delta x \to 0} \frac{1}{b} \frac{\Delta H}{\Delta x} = \frac{1}{b} \frac{\partial H}{\partial x}$$

and recalling that y_Y is a function of x defined by Eq. (6.14), derive Eq. (6.15).

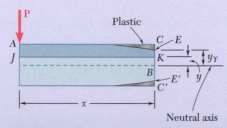

Fig. P6.60

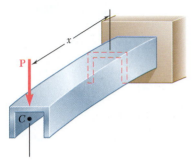

Fig. 6.37 Cantilevered channel beam with vertical plane of symmetry.

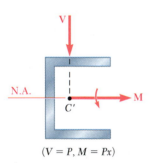

$(V = P, M = Px)$

Fig. 6.39 Load perpendicular to plane of symmetry.

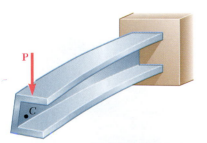

Fig. 6.40 Deformation of channel when not loaded in plane of symmetry.

*6.6   UNSYMMETRIC LOADING OF THIN-WALLED MEMBERS AND SHEAR CENTER

Our analysis of the effects of transverse loadings has been limited to members possessing a vertical plane of symmetry and to loads applied in that plane. The members were observed to bend in the plane of loading (Fig. 6.37), and in any given cross section, the bending couple **M** and the shear **V** (Fig. 6.38) were found to result in normal and shearing stresses:

$$\sigma_x = -\frac{My}{I} \tag{4.16}$$

and

$$\tau_{\text{ave}} = \frac{VQ}{It} \tag{6.6}$$

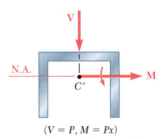

$(V = P, M = Px)$

Fig. 6.38   Load applied in vertical plane of symmetry.

In this section, the effects of transverse loads on *thin-walled members that do not possess a vertical plane of symmetry* are examined. Assume that the channel member of Fig. 6.37 has been rotated through 90° and that the line of action of **P** still passes through the centroid of the end section. The couple vector **M** representing the bending moment in a given cross section is still directed along a principal axis of the section (Fig. 6.39), and the neutral axis will coincide with that axis (see Sec. 4.8). Equation (4.16) can be used to compute the normal stresses in the section. However, Eq. (6.6) cannot be used to determine the shearing stresses, since this equation was derived for a member possessing a vertical plane of symmetry (see Sec. 6.4). Actually, the member will be observed to *bend and twist* under the applied load (Fig. 6.40), and the resulting distribution of shearing stresses will be quite different from that given by Eq. (6.6).

Is it possible to apply the vertical load **P** so that the channel member of Fig. 6.40 will *bend without twisting*? If so, where should the load **P** be applied? If the member bends without twisting, the shearing stress at any point of a given cross section can be obtained from Eq. (6.6), where Q is the first moment of the shaded area with respect to the neutral axis (Fig. 6.41a) and the distribution of stresses is as shown in Fig. 6.41b with $\tau = 0$ at both A and E. The shearing force exerted on a small element of cross-sectional area $dA = t\,ds$ is $dF = \tau\,dA = \tau t\,ds$ or $dF = q\,ds$ (Fig. 6.42a), where q is the shear flow $q = \tau t = VQ/I$. The resultant of the shearing

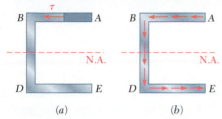

Fig. 6.41 Shearing stress and shear flow as a result of unsymmetric loading. (*a*) Shearing stress. (*b*) Shear flow *q*.

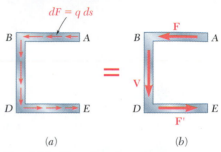

Fig. 6.42 Shear flow in each element results in a vertical shear and couple. (*a*) Shear flow *q*. (*b*) Resultant forces on elements.

forces exerted on the elements of the upper flange *AB* of the channel is a horizontal force **F** (Fig. 6.42*b*) of magnitude

$$F = \int_A^B q \, ds \tag{6.17}$$

Because of the symmetry of the channel section about its neutral axis, the resultant of the shearing forces exerted on the lower flange *DE* is a force **F′** of the same magnitude as **F** but of opposite sense. The resultant of the shearing forces exerted on the web *BD* must be equal to the vertical shear **V** in the section:

$$V = \int_B^D q \, ds \tag{6.18}$$

The forces **F** and **F′** form a couple of moment *Fh*, where *h* is the distance between the center lines of the flanges *AB* and *DE* (Fig. 6.43*a*). This couple can be eliminated if the vertical shear **V** is moved to the left through a distance *e* so the moment of **V** about *B* is equal to *Fh* (Fig. 6.43*b*). Thus, *Ve* = *Fh* or

$$e = \frac{Fh}{V} \tag{6.19}$$

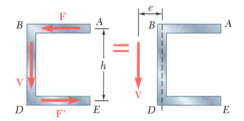

(*a*) Resultant forces on elements

(*b*) Placement of **V** to eliminate twisting

Fig. 6.43 Resultant force-couple for bending without twisting, and relocation of **V** to create same effect.

When the force **P** is applied at a distance *e* to the left of the center line of the web *BD*, the member bends in a vertical plane without twisting (Fig. 6.44).

The point *O* where the line of action of **P** intersects the axis of symmetry of the end section is the *shear center* of that section. In the case of an oblique load **P** (Fig. 6.45*a*), the member will also be free of twist if the load **P** is applied at the shear center of the section. The load **P** then can be resolved into two components **P**$_z$ and **P**$_y$ (Fig. 6.45*b*) corresponding to the load conditions of Figs. 6.37 and 6.44, neither of which causes the member to twist.

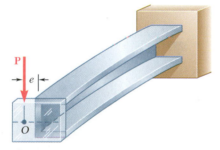

Fig. 6.44 Placement of load to eliminate twisting through the use of an attached bracket.

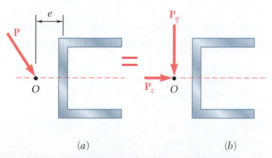

Fig. 6.45 (*a*) Oblique load applied at shear center will not cause twist, since (*b*) it can be resolved into components that do not cause twist.

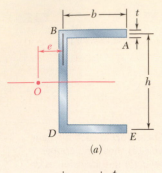

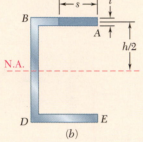

Fig. 6.46 (a) Channel section. (b) Flange segment used for calculation of shear flow.

Concept Application 6.5

Determine the shear center O of a channel section of uniform thickness (Fig. 6.46a), knowing that $b = 4$ in., $h = 6$ in., and $t = 0.15$ in.

Assuming that the member does not twist, determine the shear flow q in flange AB at a distance s from A (Fig. 6.46b). Recalling Eq. (6.5) and observing that the first moment Q of the shaded area with respect to the neutral axis is $Q = (st)(h/2)$,

$$q = \frac{VQ}{I} = \frac{Vsth}{2I} \tag{6.20}$$

where V is the vertical shear and I is the moment of inertia of the section with respect to the neutral axis.

Recalling Eq. (6.17), the magnitude of the shearing force **F** exerted on flange AB is found by integrating the shear flow q from A to B

$$F = \int_0^b q\,ds = \int_0^b \frac{Vsth}{2I}\,ds = \frac{Vth}{2I}\int_0^b s\,ds$$

$$F = \frac{Vthb^2}{4I} \tag{6.21}$$

The distance e from the center line of the web BD to the shear center O can be obtained from Eq. (6.19):

$$e = \frac{Fh}{V} = \frac{Vthb^2}{4I}\frac{h}{V} = \frac{th^2b^2}{4I} \tag{6.22}$$

The moment of inertia I of the channel section can be expressed as

$$I = I_{\text{web}} + 2I_{\text{flange}}$$

$$= \frac{1}{12}th^3 + 2\left[\frac{1}{12}bt^3 + bt\left(\frac{h}{2}\right)^2\right]$$

Neglecting the term containing t^3, which is very small, gives

$$I = \tfrac{1}{12}th^3 + \tfrac{1}{2}tbh^2 = \tfrac{1}{12}th^2(6b + h) \tag{6.23}$$

Substituting this expression into Eq. (6.22) gives

$$e = \frac{3b^2}{6b + h} = \frac{b}{2 + \dfrac{h}{3b}} \tag{6.24}$$

Note that the distance e does not depend upon t and can vary from 0 to $b/2$, depending upon the value of the ratio $h/3b$. For the given channel section,

$$\frac{h}{3b} = \frac{6\text{ in.}}{3(4\text{ in.})} = 0.5$$

and

$$e = \frac{4\text{ in.}}{2 + 0.5} = 1.6\text{ in.}$$

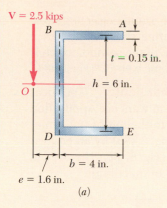

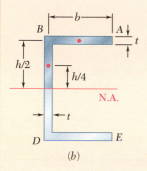

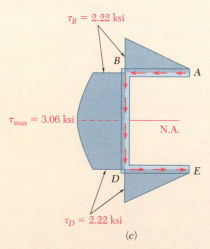

Fig. 6.47 (a) Channel section loaded at shear center. (b) Section used to find the maximum shearing stress. (c) Shearing stress distribution.

Concept Application 6.6

For the channel section of Concept Application 6.5, determine the distribution of the shearing stresses caused by a 2.5-kip vertical shear **V** applied at the shear center O (Fig. 6.47a).

Shearing Stresses in Flanges. Since **V** is applied at the shear center, there is no torsion, and the stresses in flange AB are obtained from Eq. (6.20), so

$$\tau = \frac{q}{t} = \frac{VQ}{It} = \frac{Vh}{2I}s \qquad \textbf{(6.25)}$$

which shows that the stress distribution in flange AB is linear. Letting $s = b$ and substituting for I from Eq. (6.23), we obtain the value of the shearing stress at B:

$$\tau_B = \frac{Vhb}{2(\frac{1}{12}th^2)(6b + h)} = \frac{6Vb}{th(6b + h)} \qquad \textbf{(6.26)}$$

Letting $V = 2.5$ kips and using the given dimensions,

$$\tau_B = \frac{6(2.5\text{ kips})(4\text{ in.})}{(0.15\text{ in.})(6\text{ in.})(6 \times 4\text{ in.} + 6\text{ in.})}$$
$$= 2.22\text{ ksi}$$

Shearing Stresses in Web. The distribution of the shearing stresses in the web BD is parabolic, as in the case of a W-beam, and the maximum stress occurs at the neutral axis. Computing the first moment of the upper half of the cross section with respect to the neutral axis (Fig. 6.47b),

$$Q = bt(\tfrac{1}{2}h) + \tfrac{1}{2}ht(\tfrac{1}{4}h) = \tfrac{1}{8}ht(4b + h) \qquad \textbf{(6.27)}$$

Substituting for I and Q from Eqs. (6.23) and (6.27), respectively, into the expression for the shearing stress,

$$\tau_{max} = \frac{VQ}{It} = \frac{V(\frac{1}{8}ht)(4b + h)}{\frac{1}{12}th^2(6b + h)t} = \frac{3V(4b + h)}{2th(6b + h)}$$

or with the given data,

$$\tau_{max} = \frac{3(2.5\text{ kips})(4 \times 4\text{ in.} + 6\text{ in.})}{2(0.15\text{ in.})(6\text{ in.})(6 \times 4\text{ in.} + 6\text{ in.})}$$
$$= 3.06\text{ ksi}$$

Distribution of Stresses Over the Section. The distribution of the shearing stresses over the entire channel section has been plotted in Fig. 6.47c.

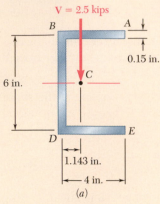

Fig. 6.48 (*a*) Channel section loaded at centroid (not shear center).

Concept Application 6.7

For the channel section of Concept Application 6.5, and neglecting stress concentrations, determine the maximum shearing stress caused by a 2.5-kip vertical shear **V** applied at the centroid *C* of the section, which is located 1.143 in. to the right of the center line of the web *BD* (Fig. 6.48*a*).

Equivalent Force-Couple System at Shear Center. The shear center *O* of the cross section was determined in Concept Application 6.5 and found to be at a distance $e = 1.6$ in. to the left of the center line of the web *BD*. We replace the shear **V** (Fig. 6.48*b*) by an equivalent force-couple system at the shear center *O* (Fig. 6.48*c*). This system consists of a 2.5-kip force **V** and of a torque **T** of magnitude

$$T = V(OC) = (2.5 \text{ kips})(1.6 \text{ in.} + 1.143 \text{ in.})$$
$$= 6.86 \text{ kip·in.}$$

Stresses Due to Bending. The 2.5-kip force **V** causes the member to bend, and the corresponding distribution of shearing stresses in the section (Fig. 6.48*d*) was determined in Concept Application 6.6. Recall that the maximum value of the stress due to this force was found to be

$$(\tau_{max})_{bending} = 3.06 \text{ ksi}$$

Stresses Due to Twisting. The torque **T** causes the member to twist, and the corresponding distribution of stresses is shown in Fig. 6.48*e*. Recall from Chap. 3 that the membrane analogy shows that in a thin-walled member of uniform thickness, the stress caused by a torque **T** is maximum along the edge of the section. Using Eqs. (3.42) and (3.40) with

$$a = 4 \text{ in.} + 6 \text{ in.} + 4 \text{ in.} = 14 \text{ in.}$$
$$b = t = 0.15 \text{ in.} \qquad b/a = 0.0107$$

So,

$$c_1 = \tfrac{1}{3}(1 - 0.630b/a) = \tfrac{1}{3}(1 - 0.630 \times 0.0107) = 0.331$$

$$(\tau_{max})_{twisting} = \frac{T}{c_1 ab^2} = \frac{6.86 \text{ kip·in.}}{(0.331)(14 \text{ in.})(0.15 \text{ in.})^2} = 65.8 \text{ ksi}$$

Combined Stresses. The maximum shearing stress due to the combined bending and twisting occurs at the neutral axis on the inside surface of the web and is

$$\tau_{max} = 3.06 \text{ ksi} + 65.8 \text{ ksi} = 68.9 \text{ ksi}$$

As a practical observation, this exceeds the shearing stress at yield for commonly available steels. This analysis demonstrates the potentially large effect that torsion can have on the shearing stresses in channels and similar structural shapes.

(continued)

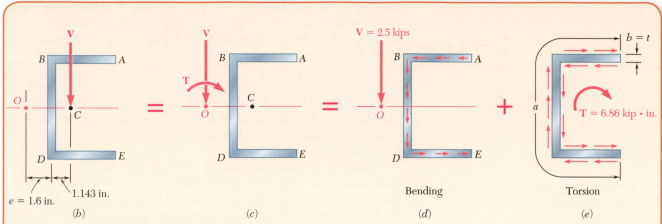

Fig. 6.48 (*cont.*) (*b*) Load at the centroid (*c*) is equivalent to a force-torque at the shear center, which is the superposition of shearing stress due to (*d*) bending and (*e*) torsion.

We now consider thin-walled members possessing no plane of symmetry. Consider an angle shape subjected to a vertical load **P.** If the member is oriented in such a way that the load **P** is perpendicular to one of the principal centroidal axes Cz of the cross section, the couple vector **M** representing the bending moment in a given section will be directed along Cz (Fig. 6.49), and the neutral axis will coincide with that axis (see Sec. 4.8). Equation (4.16) is applicable and can be used to compute the normal stresses in the section. We will now determine where the load **P** should be applied so that Eq. (6.6) can be used to determine the shearing stresses in the section, i.e., so that the member is to *bend without twisting.*

Assume that the shearing stresses in the section are defined by Eq. (6.6). As in the channel member, the elementary shearing forces exerted on the section can be expressed as $dF = q\,ds$, with $q = VQ/I$, where Q represents a first moment with respect to the neutral axis (Fig. 6.50*a*). The resultant of the shearing forces exerted on portion OA of the cross section is force $\mathbf{F}_1$ directed along OA, and the resultant of the shearing forces exerted on portion OB is a force $\mathbf{F}_2$ along OB (Fig. 6.50*b*). Since both $\mathbf{F}_1$ and $\mathbf{F}_2$ pass through point O at the corner of

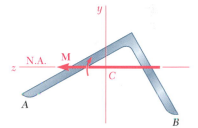

Fig. 6.49 Beam without plane of symmetry subject to bending moment.

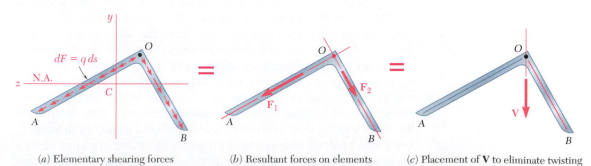

(*a*) Elementary shearing forces   (*b*) Resultant forces on elements   (*c*) Placement of **V** to eliminate twisting

Fig. 6.50 Determination of shear center, O, in an angle shape.

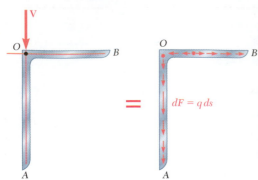

Fig. 6.51 Vertically loaded angle section and resulting shear flow.

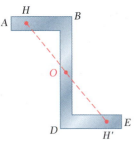

Fig. 6.52 Z section has centroid and shear center coinciding.

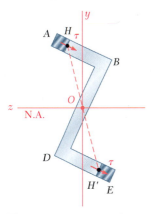

Fig. 6.53 Neutral axis location for load applied in a plane perpendicular to principal axis z.

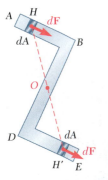

Fig. 6.54 For member bending without twisting, equal and opposite moments about O occur for any pair of symmetric elements.

the angle, their own resultant, which is the shear **V** in the section, must also pass through O (Fig. 6.50c). The member will not be twisted if the line of action of the load **P** passes through the corner O of the section in which it is applied.

The same reasoning can be applied when load **P** is perpendicular to the other principal centroidal axis Cy of the angle section. Since any load **P** applied at the corner O of a cross section also can be resolved into components perpendicular to the principal axes, the member will not be twisted if each load is applied at the corner O of a cross section. Thus, O is the shear center of the section.

Angle shapes with one vertical and one horizontal leg are encountered in many structures. Such members will not be twisted if vertical loads are applied along the center line of their vertical leg. Note from Fig. 6.51 that the resultant of the elementary shearing forces exerted on the vertical portion OA of a given section will be equal to the shear **V,** while the resultant of the shearing forces on the horizontal portion OB will be zero:

$$\int_O^A q\,ds = V \qquad \int_O^B q\,ds = 0$$

This does *not* mean that there will be no shearing stress in the horizontal leg of the member. By resolving the shear **V** into components perpendicular to the principal centroidal axes of the section and computing the shearing stress at every point, τ is zero at only one point between O and B (see Sample Prob. 6.6).

Another type of thin-walled member frequently encountered in practice is the Z shape. While the cross section of a Z shape does not possess any axis of symmetry, it does possess a *center of symmetry* O (Fig. 6.52). This means that any point H of the cross section corresponds another point H', so that the segment of straight line HH' is bisected by O. Clearly, the center of symmetry O coincides with the centroid of the cross section. As we will now demonstrate, point O is also the shear center of the cross section.

As for an angle shape, we assume that the loads are applied in a plane perpendicular to one of the principal axes of the section, so that this axis is also the neutral axis of the section (Fig. 6.53). We further assume that the shearing stresses in the section are defined by Eq. (6.6), where the member is bent without being twisted. Denoting by Q the first moment about the neutral axis of portion AH of the cross section and by Q' the first moment of portion EH', we note that $Q' = -Q$. Thus, the shearing stresses at H and H' have the same magnitude and the same direction, and the shearing forces exerted on small elements of area dA located respectively at H and H' are equal forces that have equal and opposite moments about O (Fig. 6.54). Since this is true for any pair of symmetric elements, the resultant of the shearing forces exerted on the section has a zero moment about O. This means that the shear **V** in the section is directed along a line that passes through O. Since this analysis can be repeated when the loads are applied in a plane perpendicular to the other principal axis, point O is the shear center of the section.

Sample Problem 6.6

Determine the distribution of shearing stresses in the thin-walled angle shape DE of uniform thickness t for the loading shown.

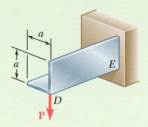

STRATEGY: Locate the centroid of the cross section and determine the two principal moments of inertia. Resolve the load **P** into components parallel to the principal axes, equal to the shear forces. The two sets of shearing stresses are then calculated at locations along the two angle legs. These are then superposed to obtain the shearing stress distribution.

MODELING and ANALYSIS:

Shear Center.   We recall from Sec. 6.6 that the shear center of the cross section of a thin-walled angle shape is located at its corner. Since the load **P** is applied at D, it causes bending but no twisting of the shape.

Principal Axes.   We locate the centroid C of a given cross section AOB (Fig. 1). Since the y' axis is an axis of symmetry, the y' and z' axes are the principal centroidal axes of the section. We recall that for the parallelogram shown (Fig. 2), $I_{nn} = \frac{1}{12} bh^3$ and $I_{mm} = \frac{1}{3} bh^3$. Considering each leg of the section as a parallelogram, we now determine the centroidal moments of inertia $I_{y'}$ and $I_{z'}$:

$$I_{y'} = 2\left[\frac{1}{3}\left(\frac{t}{\cos 45°}\right)(a\cos 45°)^3\right] = \frac{1}{3}ta^3$$

$$I_{z'} = 2\left[\frac{1}{12}\left(\frac{t}{\cos 45°}\right)(a\cos 45°)^3\right] = \frac{1}{12}ta^3$$

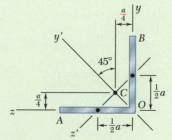

Fig. 1 Angle section with principal axes y' and z'.

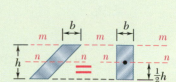

Fig. 2 Parallelogram and equivalent rectangle for determining moments of inertia.

(continued)

Superposition.   The shear **V** in the section is equal to the load **P**. As shown in Fig. 3, we resolve it into components parallel to the principal axes.

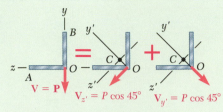

Fig. 3   Resolution of the load into components parallel to principal axes.

Shearing Stresses Due to $V_{y'}$.
Using Fig. 4, we determine the shearing stress at point e of coordinate y:

$$\bar{y}' = \tfrac{1}{2}(a + y)\cos 45° - \tfrac{1}{2}a\cos 45° = \tfrac{1}{2}y\cos 45°$$

$$Q = t(a - y)\bar{y}' = \tfrac{1}{2}t(a - y)y\cos 45°$$

$$\tau_1 = \frac{V_{y'}Q}{I_{z'}t} = \frac{(P\cos 45°)[\tfrac{1}{2}t(a - y)y\cos 45°]}{(\tfrac{1}{12}ta^3)t} = \frac{3P(a - y)y}{ta^3}$$

The shearing stress at point f is represented by a similar function of z.

Shearing Stresses Due to $V_{z'}$.
Using Fig. 5, reconsider point e:

$$\bar{z}' = \tfrac{1}{2}(a + y)\cos 45°$$

$$Q = (a - y)t\bar{z}' = \tfrac{1}{2}(a^2 - y^2)t\cos 45°$$

$$\tau_2 = \frac{V_{z'}Q}{I_y t} = \frac{(P\cos 45°)[\tfrac{1}{2}(a^2 - y^2)t\cos 45°]}{(\tfrac{1}{3}ta^3)t} = \frac{3P(a^2 - y^2)}{4ta^3}$$

The shearing stress at point f is represented by a similar function of z.

Combined Stresses.   *Along the Vertical Leg.*
The shearing stress at point e is

$$\tau_e = \tau_2 + \tau_1 = \frac{3P(a^2 - y^2)}{4ta^3} + \frac{3P(a - y)y}{ta^3} = \frac{3P(a - y)}{4ta^3}[(a + y) + 4y]$$

$$\tau_e = \frac{3P(a - y)(a + 5y)}{4ta^3} \quad \blacktriangleleft$$

Along the Horizontal Leg.
The shearing stress at point f is

$$\tau_f = \tau_2 - \tau_1 = \frac{3P(a^2 - z^2)}{4ta^3} - \frac{3P(a - z)z}{ta^3} = \frac{3P(a - z)}{4ta^3}[(a + z) - 4z]$$

$$\tau_f = \frac{3P(a - z)(a - 3z)}{4ta^3} \quad \blacktriangleleft$$

REFLECT and THINK: The combined stresses are plotted in Fig. 6.

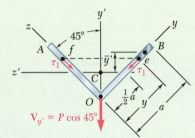

Fig. 4   Load component in plane of symmetry.

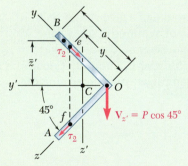

Fig. 5   Load component perpendicular to plane of symmetry.

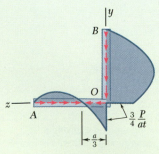

Fig. 6   Shearing stress distribution.

Problems

6.61 through 6.64 Determine the location of the shear center O of a thin-walled beam of uniform thickness having the cross section shown.

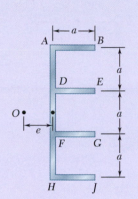

Fig. P6.61

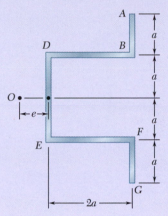

Fig. P6.62

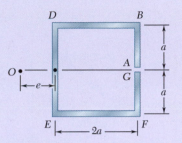

Fig. P6.63

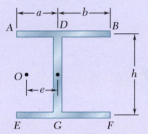

Fig. P6.64

6.65 through 6.68 An extruded beam has the cross section shown. Determine (*a*) the location of the shear center O, (*b*) the distribution of the shearing stresses caused by the vertical shearing force **V** shown applied at O.

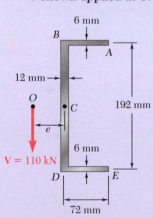

Fig. P6.65

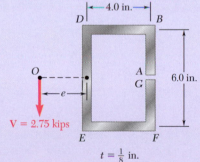

Fig. *P6.66*

$$t = \tfrac{1}{8} \text{ in.}$$

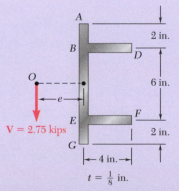

$$t = \tfrac{1}{8} \text{ in.}$$

Fig. *P6.67*

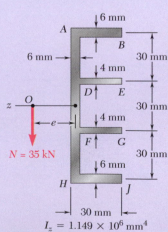

$$I_z = 1.149 \times 10^6 \text{ mm}^4$$

Fig. P6.68

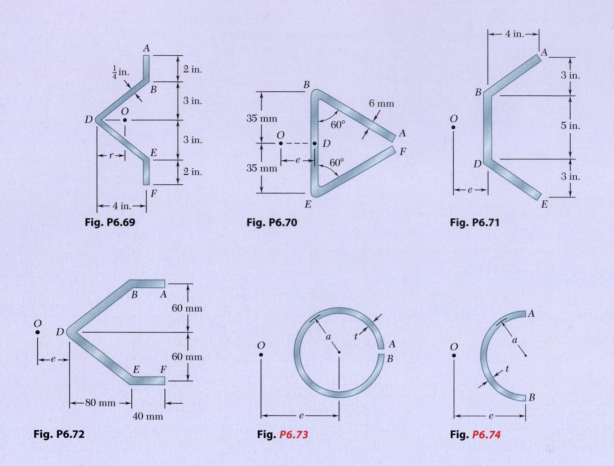

Fig. P6.69

Fig. P6.70

Fig. P6.71

Fig. P6.72

Fig. P6.73

Fig. P6.74

6.75 and 6.76 A thin-walled beam has the cross section shown. Determine the location of the shear center O of the cross section.

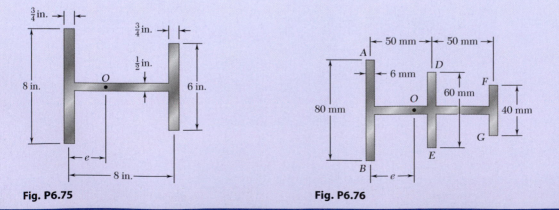

Fig. P6.75

Fig. P6.76

6.77 and 6.78 A thin-walled beam of uniform thickness has the cross section shown. Determine the dimension b for which the shear center O of the cross section is located at the point indicated.

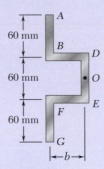

Fig. P6.77

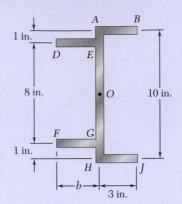

Fig. P6.78

6.79 For the angle shape and loading of Sample Prob. 6.6, check that $\int q \, dz = 0$ along the horizontal leg of the angle and $\int q \, dy = P$ along its vertical leg.

6.80 For the angle shape and loading of Sample Prob. 6.6, (a) determine the points where the shearing stress is maximum and the corresponding values of the stress, (b) verify that the points obtained are located on the neutral axis corresponding to the given loading.

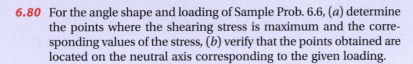

*6.81 Determine the distribution of the shearing stresses along line $D'B'$ in the horizontal leg of the angle shape for the loading shown. The x' and y' axes are the principal centroidal axes of the cross section.

*6.82 For the angle shape and loading of Prob. 6.81, determine the distribution of the shearing stresses along line $D'A'$ in the vertical leg.

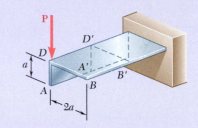

*6.83 A steel plate, 160 mm wide and 8 mm thick, is bent to form the channel shown. Knowing that the vertical load $\mathbf{P}$ acts at a point in the midplane of the web of the channel, determine (a) the torque $\mathbf{T}$ that would cause the channel to twist in the same way that it does under the load $\mathbf{P}$, (b) the maximum shearing stress in the channel caused by the load $\mathbf{P}$.

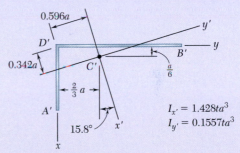

$I_{x'} = 1.428ta^3$
$I_{y'} = 0.1557ta^3$

Fig. P6.81

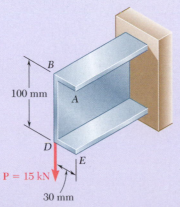

Fig. P6.83

*6.84 Solve Prob. 6.83, assuming that a 6-mm-thick plate is bent to form the channel shown.

*6.85 The cantilever beam *AB*, consisting of half of a thin-walled pipe of 1.25-in. mean radius and $\frac{3}{8}$-in. wall thickness, is subjected to a 500-lb vertical load. Knowing that the line of action of the load passes through the centroid *C* of the cross section of the beam, determine (*a*) the equivalent force-couple system at the shear center of the cross section, (*b*) the maximum shearing stress in the beam. (*Hint*: The shear center *O* of this cross section was shown in Prob. 6.74 to be located twice as far from its vertical diameter as its centroid *C*.)

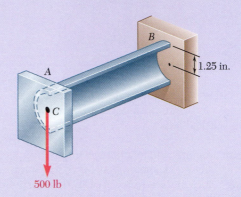

Fig. P6.85

*6.86 Solve Prob. 6.85, assuming that the thickness of the beam is reduced to $\frac{1}{4}$ in.

*6.87 The cantilever beam shown consists of a Z shape of $\frac{1}{4}$-in. thickness. For the given loading, determine the distribution of the shearing stresses along line *A′B′* in the upper horizontal leg of the Z shape. The *x′* and *y′* axes are the principal centroidal axes of the cross section, and the corresponding moments of inertia are $I_{x'} = 166.3$ in⁴ and $I_{y'} = 13.61$ in⁴.

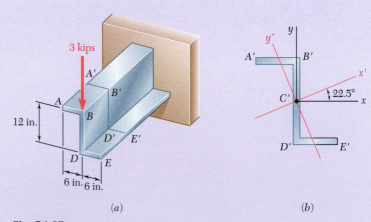

Fig. P6.87

*6.88 For the cantilever beam and loading of Prob. 6.87, determine the distribution of the shearing stresses along line *B′D′* in the vertical web of the Z shape.

Review and Summary

Stresses on a Beam Element

A small element located in the vertical plane of symmetry of a beam under a transverse loading was considered (Fig. 6.55), and it was found that normal stresses σ_x and shearing stresses τ_{xy} are exerted on the transverse faces of that element, while shearing stresses τ_{yx}, equal in magnitude to τ_{xy}, are exerted on its horizontal faces.

Fig. 6.55 Stress element from section of transversely loaded beam.

Horizontal Shear

For a prismatic beam AB with a vertical plane of symmetry supporting various concentrated and distributed loads (Fig. 6.56), at a distance x from end A we can detach an element $CDD'C'$ of length Δx that extends

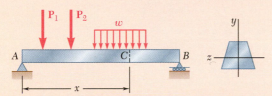

Fig. 6.56 Transversely loaded beam with vertical plane of symmetry.

across the width of the beam from the upper surface of the beam to a horizontal plane located at a distance y_1 from the neutral axis (Fig. 6.57). The magnitude of the shearing force $\Delta \mathbf{H}$ exerted on the lower face of the beam element is

$$\Delta H = \frac{VQ}{I} \Delta x \qquad (6.4)$$

where V = vertical shear in the given transverse section
Q = first moment with respect to the neutral axis of the shaded portion ⓐ of the section
I = centroidal moment of inertia of the entire cross-sectional area

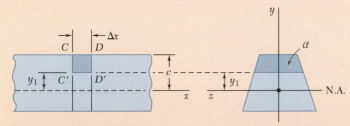

Fig. 6.57 Short segment of beam with stress element $CDD'C'$.

Shear Flow

The *horizontal shear per unit length* or *shear flow*, denoted by the letter q, is obtained by dividing both members of Eq. (6.4) by Δx:

$$q = \frac{\Delta H}{\Delta x} = \frac{VQ}{I} \tag{6.5}$$

Shearing Stresses in a Beam

Dividing both members of Eq. (6.4) by the area ΔA of the horizontal face of the element and observing that $\Delta A = t\,\Delta x$, where t is the width of the element at the cut, the *average shearing stress* on the horizontal face of the element is

$$\tau_{\text{ave}} = \frac{VQ}{It} \tag{6.6}$$

Since the shearing stresses τ_{xy} and τ_{yx} are exerted on a transverse and a horizontal plane through D' and are equal, Eq. (6.6) also represents the average value of τ_{xy} along the line $D'_1 D'_2$ (Fig. 6.58).

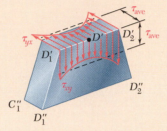

Fig. 6.58 Shearing stress distribution across horizontal and transverse planes.

Shearing Stresses in a Beam of Rectangular Cross Section

The distribution of shearing stresses in a beam of rectangular cross section was found to be parabolic, and the maximum stress, which occurs at the center of the section, is

$$\tau_{\max} = \frac{3}{2}\frac{V}{A} \tag{6.10}$$

where A is the area of the rectangular section. For wide-flange beams, a good approximation of the maximum shearing stress is obtained by dividing the shear V by the cross-sectional area of the web.

Longitudinal Shear on Curved Surface

Equations (6.4) and (6.5) can be used to determine the longitudinal shearing force $\Delta \mathbf{H}$ and the shear flow q exerted on a beam element if the element is bounded by an arbitrary curved surface instead of a horizontal plane (Fig. 6.59).

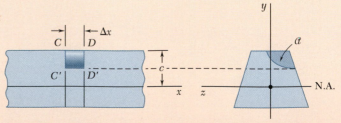

Fig. 6.59 Segment of beam showing element CDD'C' of length Δx.

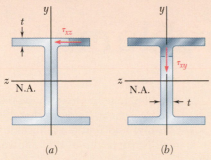

Fig. 6.60 Wide-flange beam sections showing shearing stress (*a*) in flange, (*b*) in web. The shaded area is that used for calculating the first moment of area.

Shearing Stresses in Thin-Walled Members

We found that we could extend the use of Eq. (6.6) to determine the average shearing stress in both the webs and flanges of thin-walled members, such as wide-flange beams and box beams (Fig. 6.60).

Plastic Deformations

Once plastic deformation has been initiated, additional loading causes plastic zones to penetrate into the elastic core of a beam. Because shearing stresses can occur only in the elastic core of a beam, both an increase in loading and the resulting decrease in the size of the elastic core contribute to an increase in shearing stresses.

Unsymmetric Loading and Shear Center

Prismatic members that are *not* loaded in their plane of symmetry will have both bending and twisting. Twisting is prevented if the load is applied at the point O of the cross section. This point is known as the *shear center*, where the loads may be applied so the member only bends (Fig. 6.61). If the loads are applied at that point,

$$\sigma_x = -\frac{My}{I} \qquad \tau_{\text{ave}} = \frac{VQ}{It} \qquad \textbf{(4.16, 6.6)}$$

The principle of superposition can be used to find the stresses in unsymmetric thin-walled members such as channels, angles, and extruded beams.

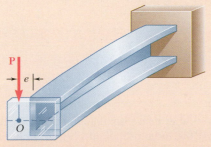

Fig. 6.61 Placement of load to eliminate twisting through the use of an attached bracket.

Review Problems

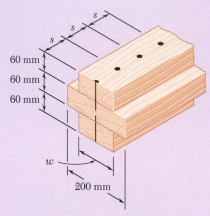

Fig. P6.89

6.89 Three boards are nailed together to form the beam shown, which is subjected to a vertical shear. Knowing that the spacing between the nails is $s = 75$ mm and that the allowable shearing force in each nail is 400 N, determine the allowable shear when $w = 120$ mm.

6.90 For the beam and loading shown, consider section n–n and determine (a) the largest shearing stress in that section, (b) the shearing stress at point a.

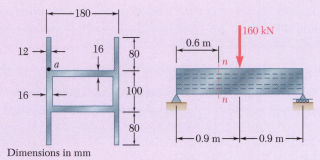

Dimensions in mm

Fig. *P6.90*

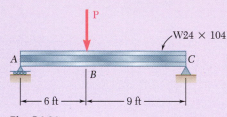

Fig. P6.91

6.91 For the wide-flange beam with the loading shown, determine the largest **P** that can be applied, knowing that the maximum normal stress is 24 ksi and the largest shearing stress, using the approximation $\tau_m = V/A_{\text{web}}$, is 14.5 ksi.

6.92 For the beam and loading shown, consider section n–n and determine the shearing stress at (a) point a, (b) point b.

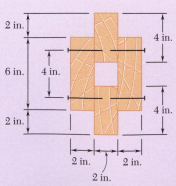

Fig. P6.93

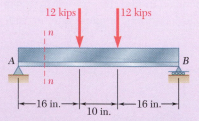

Fig. *P6.92*

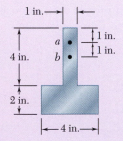

6.93 The built-up timber beam is subjected to a 1500-lb vertical shear. Knowing that the longitudinal spacing of the nails is $s = 2.5$ in. and that each nail is 3.5 in. long, determine the shearing force in each nail.

6.94 Knowing that a given vertical shear **V** causes a maximum shearing stress of 75 MPa in the hat-shaped extrusion shown, determine the corresponding shearing stress at (*a*) point *a*, (*b*) point *b*.

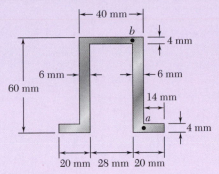

Fig. P6.94

6.95 Three planks are connected as shown by bolts of 14-mm diameter spaced every 150 mm along the longitudinal axis of the beam. For a vertical shear of 10 kN, determine the average shearing stress in the bolts.

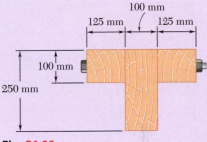

Fig. P6.95

6.96 Three 1 × 18-in. steel plates are bolted to four L6 × 6 × 1 angles to form a beam with the cross section shown. The bolts have a $\frac{7}{8}$-in. diameter and are spaced longitudinally every 5 in. Knowing that the allowable average shearing stress in the bolts is 12 ksi, determine the largest permissible vertical shear in the beam. (*Given: I_x = 6123 in.^{4}.*)

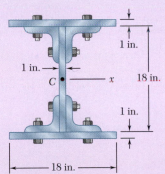

Fig. P6.96

6.97 The composite beam shown is made by welding C200 × 17.1 rolled-steel channels to the flanges of a W250 × 80 wide-flange rolled-steel shape. Knowing that the beam is subjected to a vertical shear of 200 kN, determine (*a*) the horizontal shearing force per meter at each weld, (*b*) the shearing stress at point *a* of the flange of the wide-flange shape.

Fig. P6.97

6.98 The design of a beam requires welding four horizontal plates to a vertical 0.5 × 5-in. plate as shown. For a vertical shear **V**, determine the dimension h for which the shear flow through the welded surfaces is maximum.

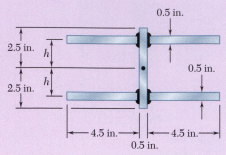

Fig. P6.98

6.99 A thin-walled beam of uniform thickness has the cross section shown. Determine the dimension b for which the shear center O of the cross section is located at the point indicated.

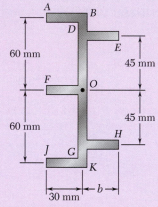

Fig. P6.99

6.100 Determine the location of the shear center O of a thin-walled beam of uniform thickness having the cross section shown.

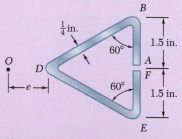

Fig. P6.100

Computer Problems

The following problems are designed to be solved with a computer.

6.C1 A timber beam is to be designed to support a distributed load and up to two concentrated loads as shown. One of the dimensions of its uniform rectangular cross section has been specified, and the other is to be determined so that the maximum normal stress and the maximum shearing stress in the beam will not exceed given allowable values σ_{all} and τ_{all}. Measuring x from end A and using either SI or U.S. customary units, write a computer program to calculate for successive cross sections, from $x = 0$ to $x = L$ and using given increments Δx, the shear, the bending moment, and the smallest value of the unknown dimension that satisfies in that section (1) the allowable normal stress requirement and (2) the allowable shearing stress requirement. Use this program to solve Prob. 5.65, assuming $\sigma_{all} = 12$ MPa and $\tau_{all} = 825$ kPa and using $\Delta x = 0.1$ m.

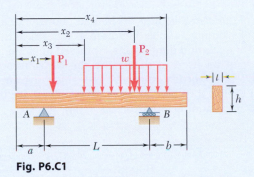

Fig. P6.C1

6.C2 A cantilever timber beam AB of length L and of uniform rectangular section shown supports a concentrated load **P** at its free end and a uniformly distributed load w along its entire length. Write a computer program to determine the length L and the width b of the beam for which both the maximum normal stress and the maximum shearing stress in the beam reach their largest allowable values. Assuming $\sigma_{all} = 1.8$ ksi and $\tau_{all} = 120$ psi, use this program to determine the dimensions L and b when (a) $P = 1000$ lb and $w = 0$, (b) $P = 0$ and $w = 12.5$ lb/in., and (c) $P = 500$ lb and $w = 12.5$ lb/in.

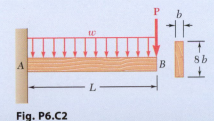

Fig. P6.C2

6.C3 A beam having the cross section shown is subjected to a vertical shear **V.** Write a computer program that, for loads and dimensions expressed in either SI or U.S. customary units, can be used to calculate the shearing stress along the line between any two adjacent rectangular areas forming the cross section. Use this program to solve (*a*) Prob. 6.10, (*b*) Prob. 6.12, (*c*) Prob. 6.22.

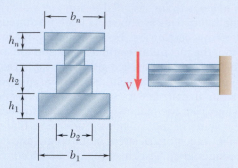

Fig. P6.C3

6.C4 A plate of uniform thickness *t* is bent as shown into a shape with a vertical plane of symmetry and is then used as a beam. Write a computer program that, for loads and dimensions expressed in either SI or U.S. customary units, can be used to determine the distribution of shearing stresses caused by a vertical shear **V.** Use this program (*a*) to solve Prob. 6.47, (*b*) to find the shearing stress at a point *E* for the shape and load of Prob. 6.50, assuming a thickness $t = \frac{1}{4}$ in.

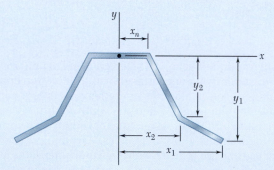

Fig. P6.C4

6.C5   The cross section of an extruded beam is symmetric with respect to the x axis and consists of several straight segments as shown. Write a computer program that, for loads and dimensions expressed in either SI or U.S. customary units, can be used to determine (*a*) the location of the shear center O, (*b*) the distribution of shearing stresses caused by a vertical force applied at O. Use this program to solve Prob. 6.70.

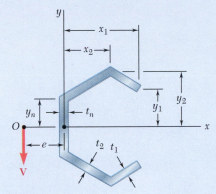

Fig. P6.C5

6.C6   A thin-walled beam has the cross section shown. Write a computer program that, for loads and dimensions expressed in either SI or U.S. customary units, can be used to determine the location of the shear center O of the cross section. Use the program to solve Prob. 6.75.

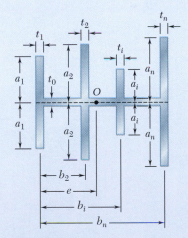

Fig. P6.C6

7

Transformations of Stress and Strain

The aircraft wing shown is being tested to determine how forces due to lift are distributed through the wing. This chapter will examine methods for determining maximum stresses and strains at any point in a structure such as this, as well as study the stress conditions necessary to cause failure.

Objectives

In this chapter, you will:

- **Apply** stress transformation equations to plane stress situations to determine any stress component at a point.

- **Apply** the alternative Mohr's circle approach to perform plane stress transformations.

- **Use** transformation techniques to identify key components of stress, such as principal stresses.

- **Extend** Mohr's circle analysis to examine three-dimensional states of stress.

- **Examine** theories of failure for ductile and brittle materials.

- **Analyze** plane stress states in thin-walled pressure vessels.

- **Extend** Mohr's circle analysis to examine the transformation of strain.

Introduction

The most general state of stress at a given point Q is represented by six components (Sec. 1.4). Three of these components, σ_x, σ_y, and σ_z, are the normal stresses exerted on the faces of a small cubic element centered at Q with the same orientation as the coordinate axes (Fig. 7.1a). The other three, τ_{xy}, τ_{yz}, and τ_{zx},[†] are the components of the shearing stresses on the same element. The same state of stress will be represented by a different set of components if the coordinate axes are rotated (Fig. 7.1b). The first part of this chapter determines how the components of stress are transformed under a rotation of the coordinate axes. The second part of the chapter is devoted to a similar analysis of the transformation of strain components.

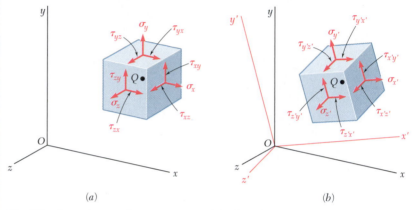

Fig. 7.1 General state of stress at a point: (a) referred to {xyz}, (b) referred to {$x'y'z'$}.

Our discussion of the transformation of stress will deal mainly with *plane stress*, i.e., with a situation in which two of the faces of the cubic element are free of any stress. If the z axis is chosen perpendicular to these faces, $\sigma_z = \tau_{zx} = \tau_{zy} = 0$, and the only remaining stress components are σ_x, σ_y, and τ_{xy} (Fig. 7.2). This situation occurs in a thin plate subjected to forces acting in the midplane of the plate (Fig. 7.3). It also occurs on the

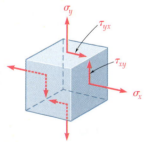

Fig. 7.2 Non-zero stress components for state of plane stress.

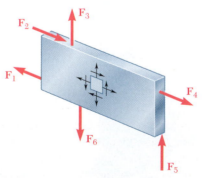

Fig. 7.3 Example of plane stress: thin plate subjected to only in-plane loads.

[†]Recall that $\tau_{yx} = \tau_{xy}$, $\tau_{zy} = \tau_{yz}$, and $\tau_{xz} = \tau_{zx}$ (Sec. 1.4).

free surface of a structural element or machine component where any point of the surface of that element or component is not subjected to an external force (Fig. 7.4).

In Sec. 7.1A, a state of plane stress at a given point Q is characterized by the stress components σ_x, σ_y, and τ_{xy} associated with the element shown in Fig. 7.5a. Components $\sigma_{x'}$, $\sigma_{y'}$, and $\tau_{x'y'}$ associated with that element after it has been rotated through an angle θ about the z axis (Fig. 7.5b) will then be determined. In Sec. 7.1B, the value θ_p of θ will be found, where the stresses $\sigma_{x'}$ and $\sigma_{y'}$ are the maximum and minimum stresses. These values of the normal stress are the *principal stresses* at point Q, and the faces of the corresponding element define the *principal planes of stress* at that point. The angle of rotation θ_s for which the shearing stress is maximum also is discussed.

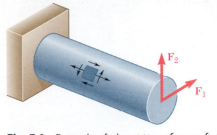

Fig. 7.4 Example of plane stress: free surface of a structural component.

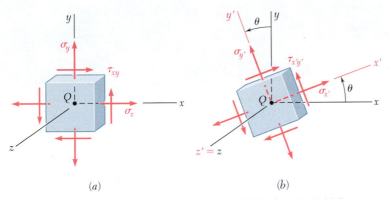

Fig. 7.5 State of plane stress: (a) referred to {xyz}, (b) referred to {x'y'z'}.

In Sec. 7.2, an alternative method to solve problems involving the transformation of plane stress, based on the use of *Mohr's circle*, is presented.

In Sec. 7.3, the *three-dimensional state of stress* at a given point is discussed, and the normal stress on a plane of arbitrary orientation at that point is determined. In Sec. 7.4, the rotations of a cubic element about each of the principal axes of stress and the corresponding transformations of stress are described by three different Mohr's circles. For a state of *plane stress* at a given point, the maximum value of the shearing stress obtained using rotations in the plane of stress does not necessarily represent the maximum shearing stress at that point. This make it necessary to distinguish *in-plane* and *out-of-plane* maximum shearing stresses.

Yield criteria for ductile materials under plane stress are discussed in Sec. 7.5A. To predict whether a material yields at some critical point under given load conditions, the principal stresses σ_a and σ_b will be determined at that point, and then used with the yield strength σ_Y of the material to evaluate a certain criterion. Two criteria in common use are the *maximum-shearing-strength criterion* and the *maximum-distortion-energy criterion*. In Sec. 7.5B, *fracture criteria* for brittle materials under plane stress are developed using the principal stresses σ_a and σ_b at some critical point and the ultimate strength σ_U of the material. Two criteria discussed here are the *maximum-normal-stress* criterion and *Mohr's criterion*.

Thin-walled pressure vessels are an important application of the analysis of plane stress. Stresses in both cylindrical and spherical pressure vessels (Photos 7.1 and 7.2) are discussed in Sec. 7.6.

Photo 7.1 Cylindrical pressure vessels.

Photo 7.2 Spherical pressure vessel.

Section 7.7 is devoted to the *transformation of plane strain* and *Mohr's circle for plane strain*. In Sec. 7.8, the three-dimensional analysis of strain shows how Mohr's circles can be used to determine the maximum shearing strain at a given point. These two particular cases are of special interest and should not be confused: the case of *plane strain* and the case of *plane stress*.

The application of *strain gages* to measure the normal strain on the surface of a structural element or machine component is considered in Sec. 7.9. The components ϵ_x, ϵ_y, and γ_{xy} characterizing the state of strain at a given point are computed from the measurements made with three strain gages forming a *strain rosette*.

7.1 TRANSFORMATION OF PLANE STRESS

7.1A Transformation Equations

Assume that a state of plane stress exists at point Q (with $\sigma_z = \tau_{zx} = \tau_{zy} = 0$) and is defined by the stress components σ_x, σ_y, and τ_{xy} associated with the element shown in Fig. 7.5a. The stress components $\sigma_{x'}$, $\sigma_{y'}$, and $\tau_{x'y'}$ associated with the element are determined after it has been rotated through an angle θ about the z axis (Fig. 7.5b). These components are given in terms of σ_x, σ_y, τ_{xy}, and θ.

In order to determine the normal stress $\sigma_{x'}$ and shearing stress $\tau_{x'y'}$ exerted on the face perpendicular to the x' axis, consider a prismatic element with faces perpendicular to the x, y, and x' axes (Fig. 7.6a). If the area of the oblique face is ΔA, the areas of the vertical and horizontal faces are equal to $\Delta A \cos \theta$ and $\Delta A \sin \theta$, respectively. The *forces* exerted on the three faces are as shown in Fig. 7.6b. (No forces are exerted on the

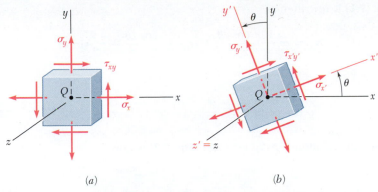

Fig. 7.5 (repeated) State of plane stress: (a) referred to {xyz}, (b) referred to {x'y'z'}.

triangular faces of the element, since the corresponding normal and shearing stresses are assumed equal to zero.) Using components along the x' and y' axes, the equilibrium equations are

$$\sum F_{x'} = 0: \quad \sigma_{x'} \Delta A - \sigma_x(\Delta A \cos \theta) \cos \theta - \tau_{xy}(\Delta A \cos \theta) \sin \theta$$
$$- \sigma_y(\Delta A \sin \theta) \sin \theta - \tau_{xy}(\Delta A \sin \theta) \cos \theta = 0$$

$$\sum F_{y'} = 0: \quad \tau_{x'y'} \Delta A + \sigma_x(\Delta A \cos \theta) \sin \theta - \tau_{xy}(\Delta A \cos \theta) \cos \theta$$
$$- \sigma_y(\Delta A \sin \theta) \cos \theta + \tau_{xy}(\Delta A \sin \theta) \sin \theta = 0$$

Solving the first equation for $\sigma_{x'}$ and the second for $\tau_{x'y'}$,

$$\sigma_{x'} = \sigma_x \cos^2 \theta + \sigma_y \sin^2 \theta + 2\tau_{xy} \sin \theta \cos \theta \tag{7.1}$$

$$\tau_{x'y'} = -(\sigma_x - \sigma_y) \sin \theta \cos \theta + \tau_{xy}(\cos^2 \theta - \sin^2 \theta) \tag{7.2}$$

Recalling the trigonometric relations

$$\sin 2\theta = 2 \sin \theta \cos \theta \quad \cos 2\theta = \cos^2 \theta - \sin^2 \theta \tag{7.3}$$

and

$$\cos^2 \theta = \frac{1 + \cos 2\theta}{2} \quad \sin^2 \theta = \frac{1 - \cos 2\theta}{2} \tag{7.4}$$

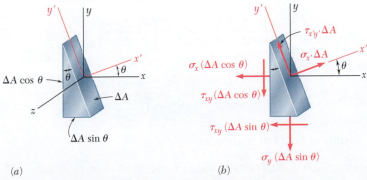

Fig. 7.6 Stress transformation equations are determined by considering an arbitrary prismatic wedge element. (a) Geometry of the element. (b) Free-body diagram.

Eq. (7.1) is rewritten as

$$\sigma_{x'} = \sigma_x \frac{1 + \cos 2\theta}{2} + \sigma_y \frac{1 - \cos 2\theta}{2} + \tau_{xy} \sin 2\theta$$

or

$$\sigma_{x'} = \frac{\sigma_x + \sigma_y}{2} + \frac{\sigma_x - \sigma_y}{2} \cos 2\theta + \tau_{xy} \sin 2\theta \tag{7.5}$$

Using the relationships of Eq. (7.3), Eq. (7.2) is now

$$\tau_{x'y'} = -\frac{\sigma_x - \sigma_y}{2} \sin 2\theta + \tau_{xy} \cos 2\theta \tag{7.6}$$

The normal stress $\sigma_{y'}$ is obtained by replacing θ in Eq. (7.5) by the angle $\theta + 90°$ that the y' axis forms with the x axis. Since $\cos(2\theta + 180°) = -\cos 2\theta$ and $\sin(2\theta + 180°) = -\sin 2\theta$,

$$\sigma_{y'} = \frac{\sigma_x + \sigma_y}{2} - \frac{\sigma_x - \sigma_y}{2} \cos 2\theta - \tau_{xy} \sin 2\theta \tag{7.7}$$

Adding Eqs. (7.5) and (7.7) member to member,

$$\sigma_{x'} + \sigma_{y'} = \sigma_x + \sigma_y \tag{7.8}$$

Since $\sigma_z = \sigma_{z'} = 0$, we thus verify for plane stress that the sum of the normal stresses exerted on a cubic element of material is independent of the orientation of that element.[†]

7.1B Principal Stresses and Maximum Shearing Stress

Equations (7.5) and (7.6) are the parametric equations of a circle. This means that, if a set of rectangular axes is used to plot a point M of abscissa $\sigma_{x'}$ and ordinate $\tau_{x'y'}$ for any given parameter θ, all of the points obtained will lie on a circle. To establish this property, we eliminate θ from Eqs. (7.5) and (7.6) by first transposing $(\sigma_x + \sigma_y)/2$ in Eq. (7.5) and squaring both members of the equation, then squaring both members of Eq. (7.6), and finally adding member to member the two equations obtained:

$$\left(\sigma_{x'} - \frac{\sigma_x + \sigma_y}{2}\right)^2 + \tau_{x'y'}^2 = \left(\frac{\sigma_x - \sigma_y}{2}\right)^2 + \tau_{xy}^2 \tag{7.9}$$

Setting

$$\sigma_{ave} = \frac{\sigma_x + \sigma_y}{2} \qquad \text{and} \qquad R = \sqrt{\left(\frac{\sigma_x - \sigma_y}{2}\right)^2 + \tau_{xy}^2} \tag{7.10}$$

the identity of Eq. (7.9) is given as

$$(\sigma_{x'} - \sigma_{ave})^2 + \tau_{x'y'}^2 = R^2 \tag{7.11}$$

[†]This verifies the property of dilatation as discussed in the first footnote of Sec. 2.6.

which is the equation of a circle of radius R centered at the point C of abscissa σ_{ave} and ordinate 0 (Fig. 7.7). Due to the symmetry of the circle about the horizontal axis, the same result is obtained if a point N of abscissa $\sigma_{x'}$ and ordinate $-\tau_{x'y'}$ is plotted instead of M. (Fig. 7.8). This property will be used in Sec. 7.2.

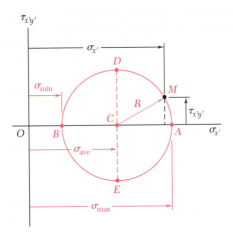

Fig. 7.7 Circular relationship of transformed stresses.

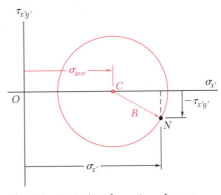

Fig. 7.8 Equivalent formation of stress transformation circle.

The points A and B where the circle of Fig. 7.7 intersects the horizontal axis are of special interest: point A corresponds to the maximum value of the normal stress $\sigma_{x'}$, while point B corresponds to its minimum value. Both points also correspond to a zero value of the shearing stress $\tau_{x'y'}$. Thus, the values θ_p of the parameter θ which correspond to points A and B can be obtained by setting $\tau_{x'y'} = 0$ in Eq. (7.6).[†]

$$\tan 2\theta_p = \frac{2\tau_{xy}}{\sigma_x - \sigma_y} \tag{7.12}$$

This equation defines two values $2\theta_p$ that are 180° apart and thus two values θ_p that are 90° apart. Either value can be used to determine the orientation of the corresponding element (Fig. 7.9). The planes containing the faces of the element obtained in this way are the *principal planes of stress* at point Q, and the corresponding values σ_{max} and σ_{min} exerted on these planes are the *principal stresses* at Q. Since both values θ_p defined by Eq. (7.12) are obtained by setting $\tau_{x'y'} = 0$ in Eq. (7.6), it is clear that no shearing stress is exerted on the principal planes.

From Fig. 7.7,

$$\sigma_{max} = \sigma_{ave} + R \quad \text{and} \quad \sigma_{min} = \sigma_{ave} - R \tag{7.13}$$

Substituting for σ_{ave} and R from Eq. (7.10),

$$\sigma_{max, min} = \frac{\sigma_x + \sigma_y}{2} \pm \sqrt{\left(\frac{\sigma_x - \sigma_y}{2}\right)^2 + \tau_{xy}^2} \tag{7.14}$$

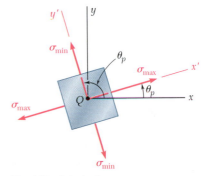

Fig. 7.9 Principal stresses.

[†]This relationship also can be obtained by differentiating $\sigma_{x'}$ in Eq. (7.5) and setting the derivative equal to zero: $d\sigma_{x'}/d\theta = 0$.

Unless it is possible to tell by inspection which of these principal planes is subjected to σ_{max} and which is subjected to σ_{min}, it is necessary to substitute one of the values θ_p into Eq. (7.5) in order to determine which corresponds to the maximum value of the normal stress.

Referring again to Fig. 7.7, points D and E located on the vertical diameter of the circle correspond to the largest value of the shearing stress $\tau_{x'y'}$. Since the abscissa of points D and E is $\sigma_{ave} = (\sigma_x + \sigma_y)/2$, the values θ_s of the parameter θ corresponding to these points are obtained by setting $\sigma_{x'} = (\sigma_x + \sigma_y)/2$ in Eq. (7.5). The sum of the last two terms in that equation must be zero. Thus, for $\theta = \theta_s$,[†]

$$\frac{\sigma_x - \sigma_y}{2} \cos 2\theta_s + \tau_{xy} \sin 2\theta_s = 0$$

or

$$\tan 2\theta_s = -\frac{\sigma_x - \sigma_y}{2\tau_{xy}} \tag{7.15}$$

This equation defines two values $2\theta_s$ that are 180° apart, and thus two values θ_s that are 90° apart. Either of these values can be used to determine the orientation of the element corresponding to the maximum shearing stress (Fig. 7.10). Fig. 7.7 shows that the maximum value of the shearing stress is equal to the radius R of the circle. Recalling the second of Eqs. (7.10),

$$\tau_{max} = \sqrt{\left(\frac{\sigma_x - \sigma_y}{2}\right)^2 + \tau_{xy}^2} \tag{7.16}$$

As observed earlier, the normal stress corresponding to the condition of maximum shearing stress is

$$\sigma' = \sigma_{ave} = \frac{\sigma_x + \sigma_y}{2} \tag{7.17}$$

Comparing Eqs. (7.12) and (7.15), $\tan 2\theta_s$ is the negative reciprocal of $\tan 2\theta_p$. Thus, angles $2\theta_s$ and $2\theta_p$ are 90° apart, and therefore angles θ_s and θ_p are 45° apart. Thus, *the planes of maximum shearing stress are at 45° to the principal planes*. This confirms the results found in Sec. 1.4 for a centric axial load (Fig. 1.38) and in Sec. 3.1C for a torsional load (Fig. 3.17).

Be aware that the analysis of the transformation of plane stress has been limited to rotations *in the plane of stress*. If the cubic element of Fig. 7.5 is rotated about an axis other than the z axis, its faces may be subjected to shearing stresses larger than defined by Eq. (7.16). In Sec. 7.3, this occurs when the principal stresses in Eq. (7.14) have the same sign (i.e., either both tensile or both compressive). In these cases, the value given by Eq. (7.16) is referred to as the maximum *in-plane* shearing stress.

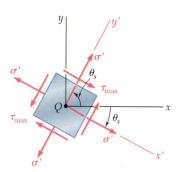

Fig. 7.10 Maximum shearing stress.

[†]This relationship also can be obtained by differentiating $\tau_{x'y'}$ in Eq. (7.6) and setting the derivative equal to zero: $d\tau_{x'y'}/d\theta = 0$.

Concept Application 7.1

For the state of plane stress shown in Fig. 7.11a, determine (a) the principal planes, (b) the principal stresses, (c) the maximum shearing stress and the corresponding normal stress.

a. Principal Planes. Following the usual sign convention, the stress components are

$$\sigma_x = +50 \text{ MPa} \qquad \sigma_y = -10 \text{ MPa} \qquad \tau_{xy} = +40 \text{ MPa}$$

Substituting into Eq. (7.12),

$$\tan 2\theta_p = \frac{2\tau_{xy}}{\sigma_x - \sigma_y} = \frac{2(+40)}{50 - (-10)} = \frac{80}{60}$$

$$2\theta_p = 53.1° \quad \text{and} \quad 180° + 53.1° = 233.1°$$

$$\theta_p = 26.6° \quad \text{and} \quad 116.6°$$

b. Principal Stresses. Equation (7.14) yields

$$\sigma_{\text{max, min}} = \frac{\sigma_x + \sigma_y}{2} \pm \sqrt{\left(\frac{\sigma_x - \sigma_y}{2}\right)^2 + \tau_{xy}^2}$$

$$= 20 \pm \sqrt{(30)^2 + (40)^2}$$

$$\sigma_{\text{max}} = 20 + 50 = 70 \text{ MPa}$$

$$\sigma_{\text{min}} - 20 - 50 = -30 \text{ MPa}$$

The principal planes and principal stresses are shown in Fig. 7.11b. Making $2\theta = 53.1°$ in Eq. (7.5), it is confirmed that the normal stress exerted on face BC of the element is the maximum stress:

$$\sigma_{x'} = \frac{50 - 10}{2} + \frac{50 + 10}{2} \cos 53.1° + 40 \sin 53.1°$$

$$= 20 + 30 \cos 53.1° + 40 \sin 53.1° = 70 \text{ MPa} = \sigma_{\text{max}}$$

c. Maximum Shearing Stress. Equation (7.16) yields

$$\tau_{\text{max}} = \sqrt{\left(\frac{\sigma_x - \sigma_y}{2}\right)^2 + \tau_{xy}^2} = \sqrt{(30)^2 + (40)^2} = 50 \text{ MPa}$$

Since σ_{max} and σ_{min} have opposite signs, τ_{max} actually represents the maximum value of the shearing stress at the point. The orientation of the planes of maximum shearing stress and the sense of the shearing stresses are determined by passing a section along the diagonal plane AC of the element of Fig. 7.11b. Since the faces AB and BC of the element are in the principal planes, the diagonal plane AC must be one of the planes of maximum shearing stress (Fig. 7.11c). Furthermore, the equilibrium conditions for the prismatic element ABC require that the shearing stress exerted on AC be directed as shown. The cubic element corresponding to the maximum shearing stress is shown in Fig. 7.11d. The normal stress on each of the four faces of the element is given by Eq. (7.17):

$$\sigma' = \sigma_{\text{ave}} = \frac{\sigma_x + \sigma_y}{2} = \frac{50 - 10}{2} = 20 \text{ MPa}$$

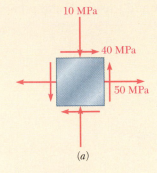

(a)

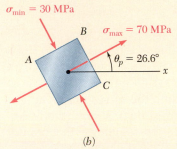

(b)

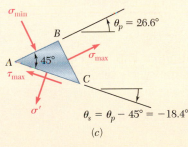

(c)

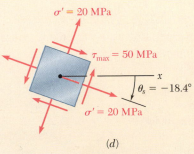

(d)

Fig. 7.11 (a) Plane stress element. (b) Plane stress element oriented in principal directions. (c) Plane stress element showing principal and maximum shear planes. (d) Plane stress element showing maximum shear orientation.

Sample Problem 7.1

A single horizontal force **P** with a magnitude of 150 lb is applied to end D of lever ABD. Knowing that portion AB of the lever has a diameter of 1.2 in., determine (a) the normal and shearing stresses located at point H and having sides parallel to the x and y axes, (b) the principal planes and principal stresses at point H.

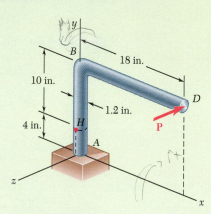

STRATEGY: You can begin by determining the forces and couples acting on the section containing the point of interest, and then use them to calculate the normal and shearing stresses acting at that point. These stresses can then be transformed to obtain the principal stresses and their orientation.

MODELING and ANALYSIS:

Force-Couple System. We replace the force **P** by an equivalent force-couple system at the center C of the transverse section containing point H (Fig.1):

$$P = 150\,\text{lb} \qquad T = (150\,\text{lb})(18\,\text{in.}) = 2.7\,\text{kip·in.}$$

$$M_x = (150\,\text{lb})(10\,\text{in.}) = 1.5\,\text{kip·in.}$$

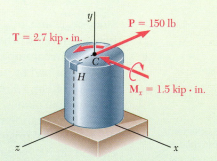

Fig. 1 Equivalent force-couple system acting on transverse section containing point H.

(continued)

a. Stresses σ_x, σ_y, τ_{xy} at Point H. Using the sign convention shown in Fig. 7.2, the sense and the sign of each stress component are found by carefully examining the force-couple system at point C (Fig. 1):

$$\sigma_x = 0 \quad \sigma_y = +\frac{Mc}{I} = +\frac{(1.5 \text{ kip·in.})(0.6 \text{ in.})}{\frac{1}{4}\pi \,(0.6 \text{ in.})^4} \qquad \sigma_y = +8.84 \text{ ksi} \blacktriangleleft$$

$$\tau_{xy} = +\frac{Tc}{J} = +\frac{(2.7 \text{ kip·in.})(0.6 \text{ in.})}{\frac{1}{2}\pi \,(0.6 \text{ in.})^4} \qquad \tau_{xy} = +7.96 \text{ ksi} \blacktriangleleft$$

We note that the shearing force **P** does not cause any shearing stress at point H. The general plane stress element (Fig. 2) is completed to reflect these stress results (Fig. 3).

b. Principal Planes and Principal Stresses. Substituting the values of the stress components into Eq. (7.12), the orientation of the principal planes is

$$\tan 2\theta_p = \frac{2\tau_{xy}}{\sigma_x - \sigma_y} = \frac{2(7.96)}{0 - 8.84} = -1.80$$

$$2\theta_p = -61.0° \quad \text{and} \quad 180° - 61.0° = +119°$$

$$\theta_p = -30.5° \quad \text{and} \quad +59.5° \blacktriangleleft$$

Substituting into Eq. (7.14), the magnitudes of the principal stresses are

$$\sigma_{\text{max, min}} = \frac{\sigma_x + \sigma_y}{2} \pm \sqrt{\left(\frac{\sigma_x - \sigma_y}{2}\right)^2 + \tau_{xy}^2}$$

$$= \frac{0 + 8.84}{2} \pm \sqrt{\left(\frac{0 - 8.84}{2}\right)^2 + (7.96)^2} = +4.42 \pm 9.10$$

$$\sigma_{\text{max}} = +13.52 \text{ ksi} \blacktriangleleft$$

$$\sigma_{\text{min}} = -4.68 \text{ ksi} \blacktriangleleft$$

Considering face ab of the element shown, $\theta_p = -30.5°$ in Eq. (7.5) and $\sigma_{x'} = -4.68$ ksi. The principal stresses are as shown in Fig. 4.

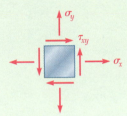

Fig. 2 General plane stress element (showing positive directions).

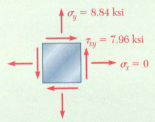

Fig. 3 Stress element at point H.

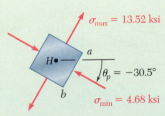

Fig. 4 Stress element at point H oriented in principal directions.

Problems

7.1 through 7.4 For the given state of stress, determine the normal and shearing stresses exerted on the oblique face of the shaded triangular element shown. Use a method of analysis based on the equilibrium of that element, as was done in the derivations of Sec. 7.1A.

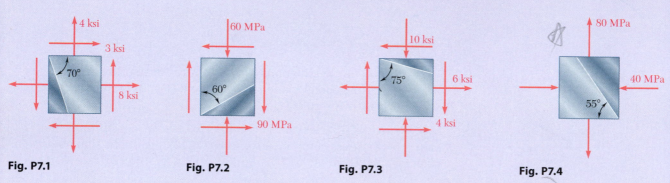

Fig. P7.1   **Fig. P7.2**   **Fig. P7.3**   **Fig. P7.4**

7.5 through 7.8 For the given state of stress, determine (*a*) the principal planes, (*b*) the principal stresses.

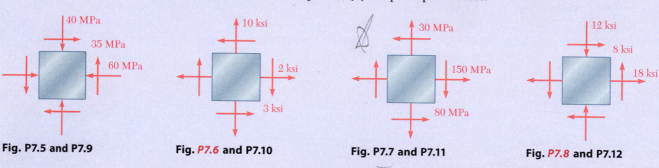

Fig. P7.5 and P7.9   **Fig. P7.6 and P7.10**   **Fig. P7.7 and P7.11**   **Fig. P7.8 and P7.12**

7.9 through 7.12 For the given state of stress, determine (*a*) the orientation of the planes of maximum in-plane shearing stress, (*b*) the maximum in-plane shearing stress, (*c*) the corresponding normal stress.

7.13 through 7.16 For the given state of stress, determine the normal and shearing stresses after the element shown has been rotated through (*a*) 25° clockwise, (*b*) 10° counterclockwise.

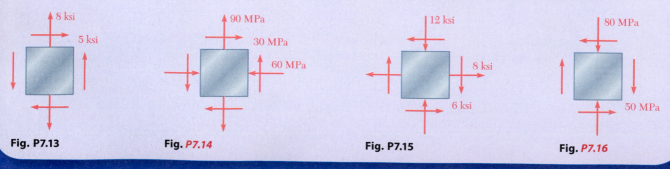

Fig. P7.13   **Fig. P7.14**   **Fig. P7.15**   **Fig. P7.16**

7.17 and 7.18 The grain of a wooden member forms an angle of 15° with the vertical. For the state of stress shown, determine (*a*) the in-plane shearing stress parallel to the grain, (*b*) the normal stress perpendicular to the grain.

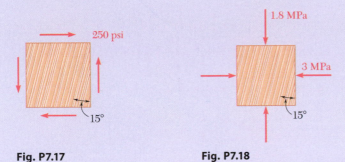

Fig. P7.17   Fig. P7.18

7.19 Two wooden members of 80 × 120-mm uniform rectangular cross section are joined by the simple glued scarf splice shown. Knowing that $\beta = 22°$ and that the maximum allowable stresses in the joint are, respectively, 400 kPa in tension (perpendicular to the splice) and 600 kPa in shear (parallel to the splice), determine the largest centric load **P** that can be applied.

7.20 Two wooden members of 80 × 120-mm uniform rectangular cross section are joined by the simple glued scarf splice shown. Knowing that $\beta = 25°$ and that centric loads of magnitude $P = 10$ kN are applied to the members as shown, determine (*a*) the in-plane shearing stress parallel to the splice, (*b*) the normal stress perpendicular to the splice.

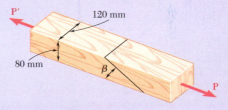

Fig. P7.19 and *P7.20*

7.21 The centric force **P** is applied to a short post as shown. Knowing that the stresses on plane *a-a* are $\sigma = -15$ ksi and $\tau = 5$ ksi, determine (*a*) the angle β that plane *a-a* forms with the horizontal, (*b*) the maximum compressive stress in the post.

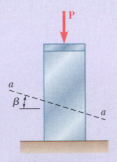

Fig. P7.21

7.22 Two members of uniform cross section 50 × 80 mm are glued together along plane *a-a* that forms an angle of 25° with the horizontal. Knowing that the allowable stresses for the glued joint are $\sigma = 800$ kPa and $\tau = 600$ kPa, determine the largest centric load **P** that can be applied.

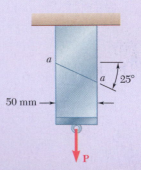

Fig. *P7.22*

7.23 The axle of an automobile is acted upon by the forces and couple shown. Knowing that the diameter of the solid axle is 32 mm, determine (*a*) the principal planes and principal stresses at point *H* located on top of the axle, (*b*) the maximum shearing stress at the same point.

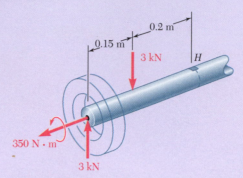

Fig. P7.23

7.24 A 400-lb vertical force is applied at *D* to a gear attached to the solid 1-in. diameter shaft *AB*. Determine the principal stresses and the maximum shearing stress at point *H* located as shown on top of the shaft.

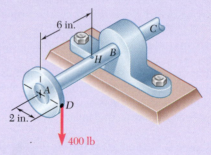

Fig. P7.24

7.25 A mechanic uses a crowfoot wrench to loosen a bolt at *E*. Knowing that the mechanic applies a vertical 24-lb force at *A*, determine the principal stresses and the maximum shearing stress at point *H* located as shown on top of the $\frac{3}{4}$-in. diameter shaft.

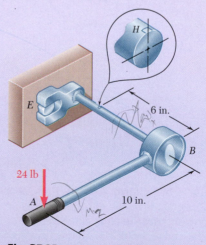

Fig. P7.25

7.26 The steel pipe *AB* has a 102-mm outer diameter and a 6-mm wall thickness. Knowing that arm *CD* is rigidly attached to the pipe, determine the principal stresses and the maximum shearing stress at point *K*.

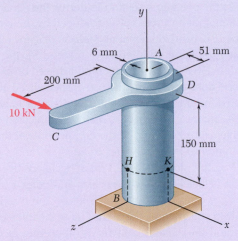

Fig. P7.26

7.27 For the state of plane stress shown, determine the largest value of σ_y for which the maximum in-plane shearing stress is equal to or less than 75 MPa.

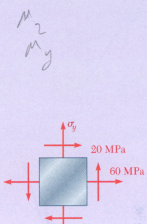

Fig. P7.27

7.28 For the state of plane stress shown, determine (*a*) the largest value of τ_{xy} for which the maximum in-plane shearing stress is equal to or less than 12 ksi, (*b*) the corresponding principal stresses.

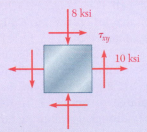

Fig. *P7.28*

7.29 For the state of plane stress shown, determine (*a*) the value of τ_{xy} for which the in-plane shearing stress parallel to the weld is zero, (*b*) the corresponding principal stresses.

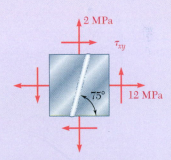

Fig. P7.29

7.30 Determine the range of values of σ_x for which the maximum in-plane shearing stress is equal to or less than 10 ksi.

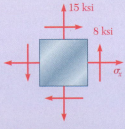

Fig. *P7.30*

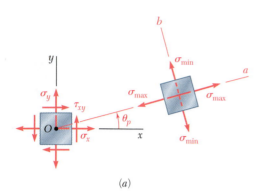

(a)

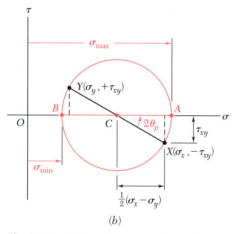

(b)

Fig. 7.12 (a) Plane stress element and the orientation of principal planes. (b) Corresponding Mohr's circle.

7.2 MOHR'S CIRCLE FOR PLANE STRESS

The circle used in the preceding section to derive the equations relating to the transformation of plane stress was introduced by the German engineer Otto Mohr (1835–1918) and is known as *Mohr's circle* for plane stress. This circle can be used to obtain an alternative method for the solution of the problems considered in Sec. 7.1. This method is based on simple geometric considerations and does not require the use of specialized equations. While originally designed for graphical solutions, a calculator may also be used.

Consider a square element of a material subjected to plane stress (Fig. 7.12a), and let σ_x, σ_y, and τ_{xy} be the components of the stress exerted on the element. A point X of coordinates σ_x and $-\tau_{xy}$ and a point Y of coordinates σ_y and $+\tau_{xy}$ are plotted (Fig. 7.12b). If τ_{xy} is positive, as assumed in Fig. 7.12a, point X is located below the σ axis and point Y above, as shown in Fig. 7.12b. If τ_{xy} is negative, X is located above the σ axis and Y below. Joining X and Y by a straight line, the point C is at the intersection of line XY with the σ axis, and the circle is drawn with its center at C and having a diameter XY. The abscissa of C and the radius of the circle are respectively equal to σ_{ave} and R in Eqs. (7.10). The circle obtained is Mohr's circle for plane stress. Thus, the abscissas of points A and B where the circle intersects the σ axis represent the principal stresses σ_{max} and σ_{min} at the point considered.

Since tan $(XCA) = 2\tau_{xy}/(\sigma_x - \sigma_y)$, the angle XCA is equal in magnitude to one of the angles $2\theta_p$ that satisfy Eq. (7.12). Thus, the angle θ_p in Fig. 7.12a defines the orientation of the principal plane corresponding to point A in Fig. 7.12b and can be obtained by dividing the angle XCA measured on Mohr's circle in half. If $\sigma_x > \sigma_y$ and $\tau_{xy} > 0$, as in the case considered here, the rotation that brings CX into CA is counterclockwise. But, in that case, the angle θ_p obtained from Eq. (7.12) and defining the direction of the normal Oa to the principal plane is positive; thus, the rotation bringing Ox into Oa is also counterclockwise. Therefore, the senses of rotation in both parts of Fig. 7.12 are the same. So, if a counterclockwise rotation through $2\theta_p$ is required to bring CX into CA on Mohr's circle, a counterclockwise rotation through θ_p will bring Ox into Oa in Fig. 7.12a.[†]

Since Mohr's circle is uniquely defined, the same circle can be obtained from the stress components $\sigma_{x'}$, $\sigma_{y'}$, and $\tau_{x'y'}$, which correspond to the x' and y' axes shown in Fig. 7.13a. Point X' of coordinates $\sigma_{x'}$ and $-\tau_{x'y'}$ and point Y' of coordinates $\sigma_{y'}$ and $+\tau_{x'y'}$ are located on Mohr's circle, and the angle $X'CA$ in Fig. 7.13b must be equal to twice the angle $x'Oa$ in Fig. 7.13a. Since the angle XCA is twice the angle xOa, the angle XCX' in Fig. 7.13b is twice the angle xOx' in Fig. 7.13a. Thus the diameter $X'Y'$ defining the normal and shearing stresses $\sigma_{x'}$, $\sigma_{y'}$, and $\tau_{x'y'}$ is obtained by rotating the diameter XY through an angle equal to twice the angle θ formed by the x' and x axes in Fig. 7.13a. The rotation that brings XY into the diameter $X'Y'$ in Fig. 7.13b has the same sense as the rotation that brings the xy axes into the $x'y'$ axes in Fig. 7.13a.

This property can be used to verify that planes of maximum shearing stress are at 45° to the principal planes. Indeed, points D and E on

[†]This is due to the fact that we are using the circle of Fig 7.8 rather than the circle of Fig. 7.7 as Mohr's circle.

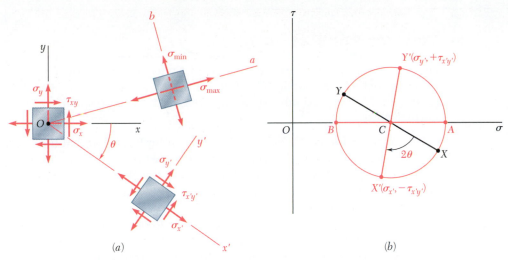

Fig. 7.13 (a) Stress element referenced to xy axes, transformed to obtain components referenced to $x'y'$ axes. (b) Corresponding Mohr's circle.

Mohr's circle correspond to the planes of maximum shearing stress, while A and B correspond to the principal planes (Fig. 7.14b). Since the diameters AB and DE of Mohr's circle are at 90° to each other, the faces of the corresponding elements are at 45° to each other (Fig. 7.14a).

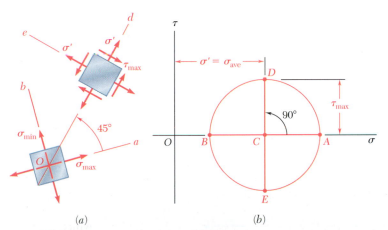

Fig. 7.14 (a) Stress elements showing orientation of planes of maximum shearing stress relative to principal planes. (b) Corresponding Mohr's circle.

The construction of Mohr's circle for plane stress is simplified if each face of the element used to define the stress components is considered separately. From Figs. 7.12 and 7.13, when the shearing stress exerted *on a given face* tends to rotate the element *clockwise*, the point on Mohr's circle corresponding to that face is located *above* the σ axis. When the shearing stress on a given face tends to rotate the element *counterclockwise*, the point corresponding to that face is located *below* the σ axis (Fig. 7.15).[†] As far as the normal stresses are concerned, the usual convention holds, so that a tensile stress is positive and is plotted to the right, while a compressive stress is considered negative and is plotted to the left.

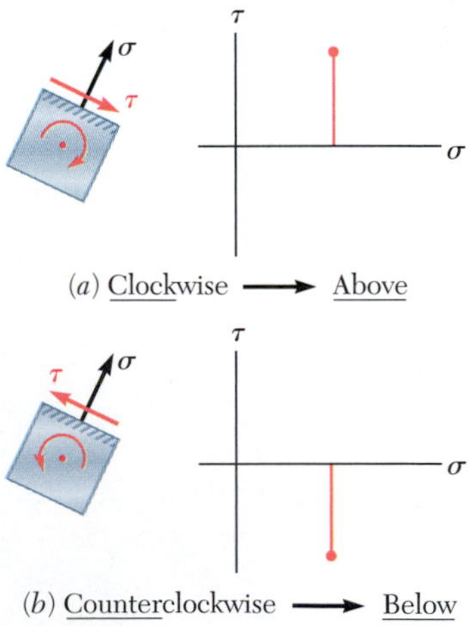

(a) Clockwise ⟶ Above

(b) Counterclockwise ⟶ Below

Fig. 7.15 Convention for plotting shearing stress on Mohr's circle.

[†]To remember this convention, think "In the kitchen, the *clock* is above, and the *counter* is below."

Concept Application 7.2

For the state of plane stress considered in Concept Application 7.1, (*a*) construct Mohr's circle, (*b*) determine the principal stresses, (*c*) determine the maximum shearing stress and the corresponding normal stress.

a. Construction of Mohr's Circle. Note from Fig. 7.16*a* that the normal stress exerted on the face oriented toward the *x* axis is tensile (positive) and the shearing stress tends to rotate the element counterclockwise. Therefore, point *X* of Mohr's circle is plotted to the right of the vertical axis and below the horizontal axis (Fig. 7.16*b*). A similar inspection of the normal and shearing stresses exerted on the upper face of the element shows that point *Y* should be plotted to the left of the vertical axis and above the horizontal axis. Drawing the line *XY*, the center *C* of Mohr's circle is found. Its abscissa is

$$\sigma_{ave} = \frac{\sigma_x + \sigma_y}{2} = \frac{50 + (-10)}{2} = 20 \text{ MPa}$$

Since the sides of the shaded triangle are

$$CF = 50 - 20 = 30 \text{ MPa} \quad \text{and} \quad FX = 40 \text{ MPa}$$

the radius of the circle is

$$R = CX = \sqrt{(30)^2 + (40)^2} = 50 \text{ MPa}$$

b. Principal Planes and Principal Stresses. The principal stresses are

$$\sigma_{max} = OA = OC + CA = 20 + 50 = 70 \text{ MPa}$$

$$\sigma_{min} = OB = OC - BC = 20 - 50 = -30 \text{ MPa}$$

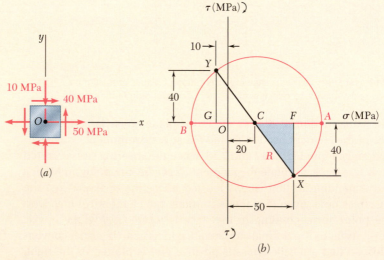

Fig. 7.16 (*a*) Plane stress element. (*b*) Corresponding Mohr's circle.

(continued)

Recalling that the angle *ACX* represents $2\theta_p$ (Fig. 7.16*b*),

$$\tan 2\theta_p = \frac{FX}{CF} = \frac{40}{30}$$

$$2\theta_p = 53.1° \qquad \theta_p = 26.6°$$

Since the rotation that brings *CX* into *CA* in Fig. 7.16*d* is counterclockwise, the rotation that brings *Ox* into the axis *Oa* corresponding to σ_{max} in Fig. 7.16*c* is also counterclockwise.

c. Maximum Shearing Stress. Since a further rotation of 90° counterclockwise brings *CA* into *CD* in Fig. 7.16*d*, a further rotation of 45° counterclockwise will bring the axis *Oa* into the axis *Od* corresponding to the maximum shearing stress in Fig. 7.16*d*. Note from Fig. 7.16*d* that $\tau_{max} = R = 50$ MPa and the corresponding normal stress is $\sigma' = \sigma_{ave} = 20$ MPa. Since point *D* is located above the σ axis in Fig. 7.16*c*, the shearing stresses exerted on the faces perpendicular to *Od* in Fig. 7.16*d* must be directed so that they will tend to rotate the element clockwise.

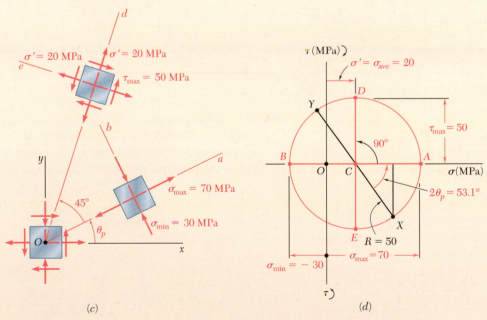

(*c*) (*d*)

Fig. 7.16 (*cont.*) (*c*) Stress element orientations for principal and maximum shearing stresses. (*d*) Mohr's circle used to determine principal and maximum shearing stresses.

Mohr's circle provides a convenient way of checking the results obtained earlier for stresses under a centric axial load (Sec. 1.4) and under a torsional load (Sec. 3.1C). In the first case (Fig. 7.17a), $\sigma_x = P/A$, $\sigma_y = 0$, and $\tau_{xy} = 0$. The corresponding points X and Y define a circle of radius $R = P/2A$ that passes through the origin of coordinates (Fig. 7.17b). Points D and E yield the orientation of the planes of maximum shearing stress (Fig. 7.17c), as well as τ_{max} and the corresponding normal stresses σ':

$$\tau_{max} = \sigma' = R = \frac{P}{2A} \tag{7.18}$$

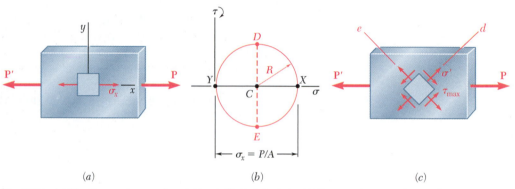

(a)　　　　　　(b)　　　　　　(c)

Fig. 7.17 (a) Member under centric axial load. (b) Mohr's circle. (c) Element showing planes of maximum shearing stress.

In the case of torsion (Fig. 7.18a), $\sigma_x = \sigma_y = 0$ and $\tau_{xy} = \tau_{max} = Tc/J$. Therefore, points X and Y are located on the τ axis, and Mohr's circle has a radius of $R = Tc/J$ centered at the origin (Fig. 7.18b). Points A and B define the principal planes (Fig. 7.18c) and the principal stresses:

$$\sigma_{max, min} = \pm R = \pm \frac{Tc}{J} \tag{7.19}$$

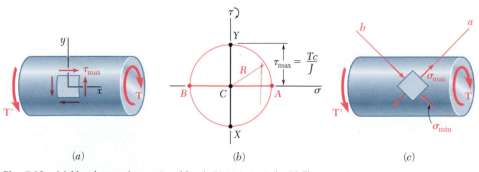

(a)　　　　　　(b)　　　　　　(c)

Fig. 7.18 (a) Member under torsional load. (b) Mohr's circle. (c) Element showing orientation of principal stresses.

Sample Problem 7.2

For the state of plane stress shown determine (*a*) the principal planes and the principal stresses, (*b*) the stress components exerted on the element obtained by rotating the given element counterclockwise through 30°.

STRATEGY: Since the given state of stress represents two points on Mohr's circle, you can use these points to generate the circle. The state of stress on any other plane, including the principal planes, can then be readily determined through the geometry of the circle.

MODELING and ANALYSIS:

Construction of Mohr's Circle (Fig 1). On a face perpendicular to the *x* axis, the normal stress is tensile, and the shearing stress tends to rotate the element clockwise. Thus, *X* is plotted at a point 100 units to the right of the vertical axis and 48 units above the horizontal axis. By examining the stress components on the upper face, point $Y(60, -48)$ is plotted. Join points *X* and *Y* by a straight line to define the center *C* of Mohr's circle. The abscissa of *C*, which represents σ_{ave}, and the radius *R* of the circle, can be measured directly or calculated as

$$\sigma_{ave} = OC = \tfrac{1}{2}(\sigma_x + \sigma_y) = \tfrac{1}{2}(100 + 60) = 80\ \text{MPa}$$

$$R = \sqrt{(CF)^2 + (FX)^2} = \sqrt{(20)^2 + (48)^2} = 52\ \text{MPa}$$

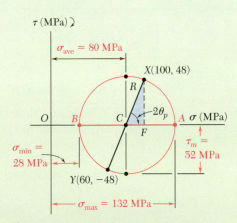

Fig. 1 Mohr's circle for given stress state.

a. Principal Planes and Principal Stresses. We rotate the diameter *XY* clockwise through $2\theta_p$ until it coincides with the diameter *AB*. Thus,

$$\tan 2\theta_p = \frac{XF}{CF} = \frac{48}{20} = 2.4 \quad 2\theta_p = 67.4° \downarrow \quad \theta_p = 33.7° \downarrow \quad \blacktriangleleft$$

(continued)

The principal stresses are represented by the abscissas of points A and B:

$$\sigma_{max} = OA = OC + CA = 80 + 52 \qquad \sigma_{max} = +132 \text{ MPa} \blacktriangleleft$$

$$\sigma_{min} = OB = OC - BC = 80 - 52 \qquad \sigma_{min} = +28 \text{ MPa} \blacktriangleleft$$

Since the rotation that brings XY into AB is clockwise, the rotation that brings Ox into the axis Oa corresponding to σ_{max} is also clockwise; we obtain the orientation shown in Fig. 2 for the principal planes.

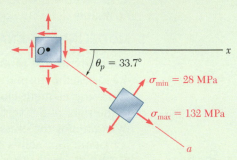

Fig. 2   Orientation of principal stress element.

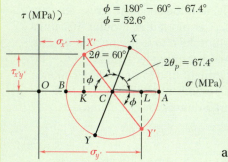

Fig. 3   Mohr's circle analysis for element rotation of 30° counterclockwise.

b. Stress Components on Element Rotated 30°↶.
Points X' and Y' on Mohr's circle that correspond to the stress components on the rotated element are obtained by rotating XY counterclockwise through $2\theta = 60°$ (Fig. 3). We find

$$\phi = 180° - 60° - 67.4° \qquad\qquad \phi = 52.6° \blacktriangleleft$$

$$\sigma_{x'} = OK = OC - KC = 80 - 52\cos 52.6° \qquad \sigma_{x'} = +48.4 \text{ MPa} \blacktriangleleft$$

$$\sigma_{y} = OL = OC + CL = 80 + 52\cos 52.6° \qquad \sigma_{y} = +111.6 \text{ MPa} \blacktriangleleft$$

$$\tau_{x'y'} = KX' = 52\sin 52.6° \qquad\qquad \tau_{x'y'} = 41.3 \text{ MPa} \blacktriangleleft$$

Since X' is located above the horizontal axis, the shearing stress on the face perpendicular to Ox' tends to rotate the element clockwise. The stresses, along with their orientation, are shown in Fig. 4.

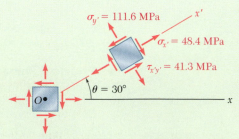

Fig. 4   Stress components obtained by rotating original element 30° counterclockwise.

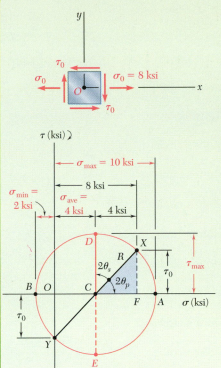

Fig. 1   Mohr's circle for given state of stress.

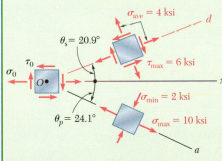

Fig. 2   Orientation of principal and maximum shearing stress planes for assumed sense of τ_0.

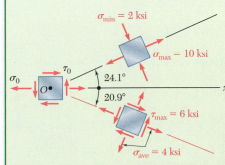

Fig. 3   Orientation of principal and maximum shearing stress planes for opposite sense of τ_0.

Sample Problem 7.3

A state of plane stress consists of a tensile stress $\sigma_0 = 8$ ksi exerted on vertical surfaces and of unknown shearing stresses. Determine (*a*) the magnitude of the shearing stress τ_0 for which the largest normal stress is 10 ksi, (*b*) the corresponding maximum shearing stress.

STRATEGY: You can use the normal stresses on the given element to determine the average normal stress, thereby establishing the center of Mohr's circle. Knowing that the given maximum normal stress is also a principal stress, you can use this to complete the construction of the circle.

MODELING and ANALYSIS:

Construction of Mohr's Circle (Fig.1).   Assume that the shearing stresses act in the senses shown. Thus, the shearing stress τ_0 on a face perpendicular to the x axis tends to rotate the element clockwise, and point X of coordinates 8 ksi and τ_0 is plotted above the horizontal axis. Considering a horizontal face of the element, $\sigma_y = 0$ and τ_0 tends to rotate the element counterclockwise. Thus, Y is plotted at a distance τ_0 below O.

The abscissa of the center C of Mohr's circle is

$$\sigma_{ave} = \tfrac{1}{2}(\sigma_x + \sigma_y) = \tfrac{1}{2}(8 + 0) = 4 \text{ ksi}$$

The radius R of the circle is found by observing that $\sigma_{max} = 10$ ksi and is represented by the abscissa of point A:

$$\sigma_{max} = \sigma_{ave} + R$$
$$10 \text{ ksi} = 4 \text{ ksi} + R \qquad R = 6 \text{ ksi}$$

a. Shearing Stress τ_0.   Considering the right triangle CFX,

$$\cos 2\theta_p = \frac{CF}{CX} = \frac{CF}{R} = \frac{4 \text{ ksi}}{6 \text{ ksi}} \qquad 2\theta_p = 48.2° \downarrow \qquad \theta_p = 24.1° \downarrow$$

$$\tau_0 = FX = R \sin 2\theta_p = (6 \text{ ksi}) \sin 48.2° \qquad \tau_0 = 4.47 \text{ ksi} \blacktriangleleft$$

b. Maximum Shearing Stress.   The coordinates of point D of Mohr's circle represent the maximum shearing stress and the corresponding normal stress.

$$\tau_{max} = R = 6 \text{ ksi} \qquad \tau_{max} = 6 \text{ ksi} \blacktriangleleft$$
$$2\theta_s = 90° - 2\theta_p = 90° - 48.2° = 41.8° \nwarrow \qquad \theta_x = 20.9° \nwarrow$$

The maximum shearing stress is exerted on an element that is oriented as shown in Fig. 2. (The element upon which the principal stresses are exerted is also shown.)

REFLECT and THINK.   If our original assumption regarding the sense of τ_0 was reversed, we would obtain the same circle and the same answers, but the orientation of the elements would be as shown in Fig. 3.

Problems

7.31 Solve Probs. 7.5 and 7.9, using Mohr's circle.

7.32 Solve Probs. 7.7 and 7.11, using Mohr's circle.

7.33 Solve Prob. 7.10, using Mohr's circle.

7.34 Solve Prob. 7.12, using Mohr's circle.

7.35 Solve Prob. 7.13, using Mohr's circle.

7.36 Solve Prob. 7.14, using Mohr's circle.

7.37 Solve Prob. 7.15, using Mohr's circle.

7.38 Solve Prob. 7.16, using Mohr's circle.

7.39 Solve Prob. 7.17, using Mohr's circle.

7.40 Solve Prob. 7.18, using Mohr's circle.

7.41 Solve Prob. 7.19, using Mohr's circle.

7.42 Solve Prob. 7.20, using Mohr's circle.

7.43 Solve Prob. 7.21, using Mohr's circle.

7.44 Solve Prob. 7.22, using Mohr's circle.

7.45 Solve Prob. 7.23, using Mohr's circle.

7.46 Solve Prob. 7.24, using Mohr's circle.

7.47 Solve Prob. 7.25, using Mohr's circle.

7.48 Solve Prob. 7.26, using Mohr's circle.

7.49 Solve Prob. 7.27, using Mohr's circle.

7.50 Solve Prob. 7.28, using Mohr's circle.

7.51 Solve Prob. 7.29, using Mohr's circle.

7.52 Solve Prob. 7.30, using Mohr's circle.

7.53 Solve Prob. 7.29, using Mohr's circle and assuming that the weld forms an angle of 60° with the horizontal.

7.54 and 7.55 Determine the principal planes and the principal stresses for the state of plane stress resulting from the superposition of the two states of stress shown.

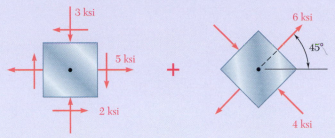

Fig. P7.54

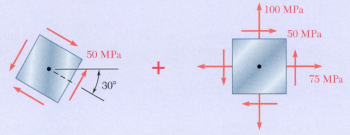

Fig. P7.55

7.56 and 7.57 Determine the principal planes and the principal stresses for the state of plane stress resulting from the superposition of the two states of stress shown.

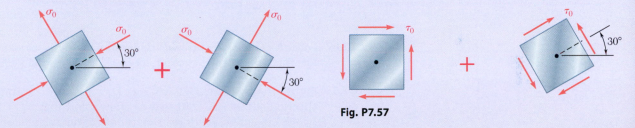

Fig. P7.56

Fig. P7.57

7.58 For the element shown, determine the range of values of τ_{xy} for which the maximum tensile stress is equal to or less than 60 MPa.

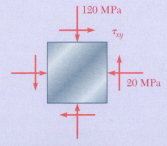

Fig. P7.58 and P7.59

7.59 For the element shown, determine the range of values of τ_{xy} for which the maximum in-plane shearing stress is equal to or less than 150 MPa.

7.60 For the state of stress shown, determine the range of values of θ for which the magnitude of the shearing stress $\tau_{x'y'}$ is equal to or less than 8 ksi.

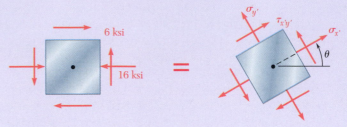

Fig. P7.60

7.61 For the state of stress shown, determine the range of values of θ for which the normal stress $\sigma_{x'}$ is equal to or less than 50 MPa.

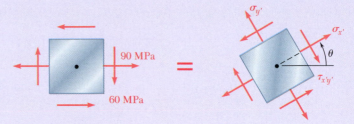

Fig. P7.61 and P7.62

7.62 For the state of stress shown, determine the range of values of θ for which the normal stress $\sigma_{x'}$ is equal to or less than 100 MPa.

7.63 For the state of stress shown, it is known that the normal and shearing stresses are directed as shown and that $\sigma_x = 14$ ksi, $\sigma_y = 9$ ksi, and $\sigma_{min} = 5$ ksi. Determine (*a*) the orientation of the principal planes, (*b*) the principal stress σ_{max}, (*c*) the maximum in-plane shearing stress.

7.64 The Mohr's circle shown corresponds to the state of stress given in Fig. 7.5*a* and *b*. Noting that $\sigma_{x'} = OC + (CX') \cos(2\theta_p - 2\theta)$ and that $\tau_{x'y'} = (CX') \sin(2\theta_p - 2\theta)$, derive the expressions for $\sigma_{x'}$ and $\tau_{x'y'}$ given in Eqs. (7.5) and (7.6), respectively. [*Hint:* Use $\sin(A + B) = \sin A \cos B + \cos A \sin B$ and $\cos(A + B) = \cos A \cos B - \sin A \sin B$.]

7.65 (*a*) Prove that the expression $\sigma_{x'}\sigma_{y'} - \tau_{x'y'}^2$, where $\sigma_{x'}$, $\sigma_{y'}$, and $\tau_{x'y'}$ are components of the stress along the rectangular axes x' and y', is independent of the orientation of these axes. Also, show that the given expression represents the square of the tangent drawn from the origin of the coordinates to Mohr's circle. (*b*) Using the invariance property established in part *a*, express the shearing stress τ_{xy} in terms of σ_x, σ_y, and the principal stresses σ_{max} and σ_{min}.

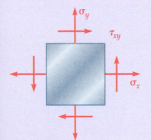

Fig. P7.63

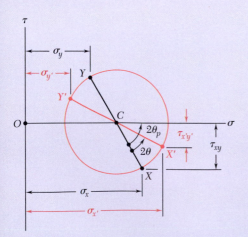

Fig. P7.64

7.3 GENERAL STATE OF STRESS

In the preceding sections, we have assumed a state of plane stress with $\sigma_z = \tau_{zx} = \tau_{zy} = 0$, and have considered only transformations of stress associated with a rotation about the z axis. We will now consider the general state of stress represented in Fig. 7.1a and the transformation of stress associated with the rotation of axes shown in Fig. 7.1b. However, our analysis will be limited to the determination of the *normal stress* σ_n on a plane of arbitrary orientation.

Three of the faces in the tetrahedron shown in Fig. 7.19 are parallel to the coordinate planes, while the fourth face, ABC, is perpendicular to the line QN. Denoting the area of face ABC as ΔA and the direction cosines of line QN as λ_x, λ_y, λ_z, the areas of the faces perpendicular to the x, y, and z axes are $(\Delta A)\lambda_x$, $(\Delta A)\lambda_y$, and $(\Delta A)\lambda_z$. If the state of stress at point Q is defined by the stress components σ_x, σ_y, σ_z, τ_{xy}, τ_{yz}, and τ_{zx}, the *forces* exerted on the faces parallel to the coordinate planes are obtained by multiplying the appropriate stress components by the area of each face (Fig. 7.20). On the other hand, the forces exerted on face ABC consist of a normal force of magnitude $\sigma_n \Delta A$ directed along QN and a shearing force with a magnitude $\tau \Delta A$ perpendicular to QN but of unknown direction. Since QBC, QCA, and QAB face the negative x, y, and z axes respectively, the forces exerted must be shown with negative senses.

The sum of the components along QN of all the forces acting on the tetrahedron is zero. The component along QN of a force parallel to the x axis is obtained by multiplying the magnitude of that force by the direction cosine λ_x. The components of forces parallel to the y and z axes are obtained in a similar way. Thus,

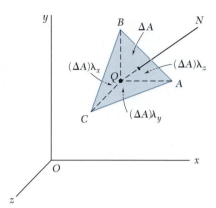

Fig. 7.19 Stress tetrahedron at point Q with three faces parallel to the coordinate planes.

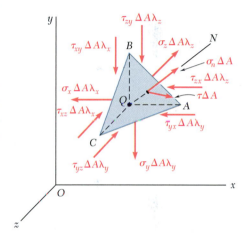

Fig. 7.20 Free-body diagram of stress tetrahedron at point Q.

$$\Sigma F_n = 0: \quad \sigma_n \Delta A - (\sigma_x \Delta A\, \lambda_x)\lambda_x - (\tau_{xy} \Delta A\, \lambda_x)\lambda_y - (\tau_{xz} \Delta A\, \lambda_x)\lambda_z$$
$$- (\tau_{yx} \Delta A\, \lambda_y)\lambda_x - (\sigma_y \Delta A\, \lambda_y)\lambda_y - (\tau_{yz} \Delta A\, \lambda_y)\lambda_z$$
$$- (\tau_{zx} \Delta A\, \lambda_z)\lambda_x - (\tau_{zy} \Delta A\, \lambda_z)\lambda_y - (\sigma_z \Delta A\, \lambda_z)\lambda_z = 0$$

Dividing through by ΔA and solving for σ_n gives

$$\sigma_n = \sigma_x \lambda_x^2 + \sigma_y \lambda_y^2 + \sigma_z \lambda_z^2 + 2\tau_{xy}\lambda_x\lambda_y + 2\tau_{yz}\lambda_y\lambda_z + 2\tau_{zx}\lambda_z\lambda_x \quad \text{(7.20)}$$

Note that the equation for the normal stress σ_n is a *quadratic form* in λ_x, λ_y, and λ_z. The coordinate axes are found when the right-hand member of Eq. (7.20) reduces to the three terms containing the squares of the direction cosines.[†] Calling these axes a, b, and c, the corresponding normal stresses σ_a, σ_b, and σ_c, and the direction cosines of QN with respect to these axes λ_a, λ_b, and λ_c. gives

$$\sigma_n = \sigma_a \lambda_a^2 + \sigma_b \lambda_b^2 + \sigma_c \lambda_c^2 \quad \text{(7.21)}$$

[†]In Sec. 9.16 of F. P. Beer and E. R. Johnston, *Vector Mechanics for Engineers*, 10th ed., McGraw-Hill Book Company, 2013, a similar quadratic form is found to represent the moment of inertia of a rigid body with respect to an arbitrary axis. It is shown in Sec. 9.17 that this form is associated with a *quadric surface* and reducing the quadratic form to terms containing only the squares of the direction cosines is equivalent to determining the principal axes of that surface.

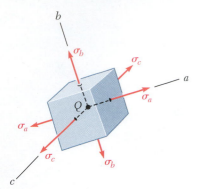

Fig. 7.21 General stress element oriented to principal axes.

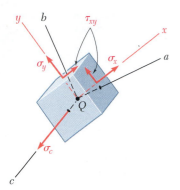

Fig. 7.22 Stress element rotated about c axis.

The coordinate axes a, b, c are the *principal axes of stress*. Since their orientation depends upon the state of stress at Q and thus upon the position of Q, these axes are represented in Fig. 7.21 as attached to Q. The corresponding coordinate planes are known as the *principal planes of stress*, and the corresponding normal stresses σ_a, σ_b, and σ_c are the *principal stresses* at Q.[†]

7.4 THREE-DIMENSIONAL ANALYSIS OF STRESS

If the element in Fig. 7.21 is rotated about one of the principal axes at Q, say the c axis (Fig. 7.22), the corresponding transformation of stress can be analyzed using Mohr's circle as a transformation of plane stress. The shearing stresses exerted on the faces perpendicular to the c axis remain equal to zero. The normal stress σ_c is perpendicular to the plane ab where the transformation takes place and does not affect this transformation. Therefore, the circle of diameter AB is used to determine the normal and shearing stresses exerted on the faces of the element as it is rotated about the c axis (Fig. 7.23). Similarly, circles of diameter BC and CA can be used to determine the stresses on the element as it is rotated about the a and b axes, respectively. While this analysis is limited to rotations about the principal axes, it could be shown that any other transformation of axes would lead to stresses represented in Fig. 7.23 by a point located within

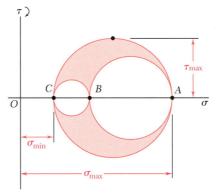

Fig. 7.23 Mohr's circles for general state of stress.

the shaded area. Thus, the radius of the largest circle yields the maximum value of the shearing stress at point Q. Noting that the diameter of that circle is equal to the difference between $\sigma_{\max}$ and $\sigma_{\min}$,

$$\tau_{\max} = \tfrac{1}{2}|\sigma_{\max} - \sigma_{\min}| \tag{7.22}$$

where $\sigma_{\max}$ and $\sigma_{\min}$ represent the *algebraic* values of the maximum and minimum stresses at point Q.

[†]For a discussion of the determination of the principal planes of stress and of the principal stresses, see S. P. Timoshenko and J. N. Goodier, *Theory of Elasticity*, 3d ed., McGraw-Hill Book Company, 1970, Sec. 77.

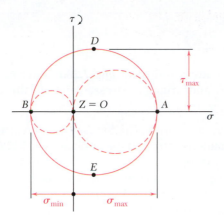

Fig. 7.24 Three-dimensional Mohr's circles for state of plane stress where $\sigma_a > 0 > \sigma_b$.

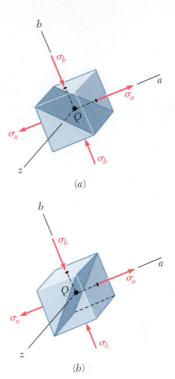

Fig. 7.25 In-plane maximum shearing stress for an element having a principal axis aligned with the z-axis. (*a*) 45° clockwise from principal axis *a*. (*b*) 45° counterclockwise from principal axis *a*.

Recall that in *plane stress*, if the *x* and *y* axes are selected, we have $\sigma_z = \tau_{zx} = \tau_{zy} = 0$. This means that the *z* axis (i.e., the axis perpendicular to the plane of stress) is one of the three principal axes of stress. In a Mohr-circle diagram, this axis corresponds to the origin *O*, where $\sigma = \tau = 0$. The other two principal axes correspond to points *A* and *B* where Mohr's circle for the *xy* plane intersects the σ axis. If *A* and *B* are located on opposite sides of the origin *O* (Fig. 7.24), the corresponding principal stresses represent the maximum and minimum normal stresses at point *Q*, and the maximum shearing stress is equal to the maximum "in-plane" shearing stress. Recall that in Sec. 7.1B the planes of maximum shearing stress correspond to points *D* and *E* of Mohr's circle and are at 45° to the principal planes corresponding to points *A* and *B*. These are shown in the shaded diagonal planes of Figs. 7.25*a* and *b*.

However, if *A* and *B* are on the same side of *O*, where σ_a and σ_b have the same sign, the circle defining σ_{max}, σ_{min}, and τ_{max} is *not* the circle corresponding to a transformation of stress within the *xy* plane. If $\sigma_a > \sigma_b > 0$, as assumed in Fig. 7.26, $\sigma_{max} = \sigma_a$, $\sigma_{min} = 0$, and τ_{max} is equal to the radius of the circle defined by points *O* and *A*. Thus, $\tau_{max} = \frac{1}{2}\sigma_{max}$. The normals *Qd'* and *Qe'* to the planes of maximum shearing stress are obtained by rotating the axis *Qa* through 45° within the *za* plane. These planes of maximum shearing stress are shown in the shaded diagonal planes of Figs. 7.27*a* and *b*.

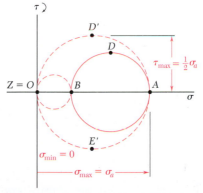

Fig. 7.26 Three-dimensional Mohr's circles for state of plane stress where $\sigma_a > \sigma_b > 0$.

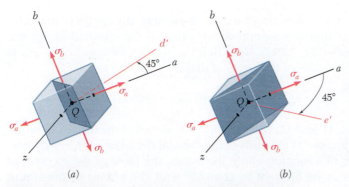

Fig. 7.27 Out-of-plane of maximum shearing stress for plane stress element. (*a*) 45° counterclockwise from principal axis *a*. (*b*) 45° clockwise from principal axis *a*.

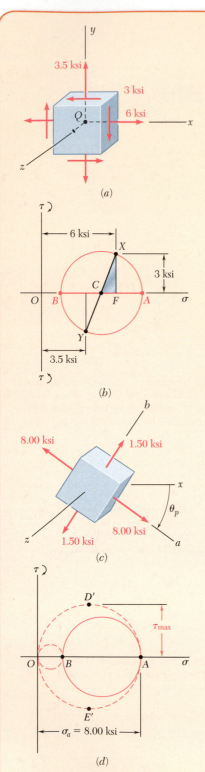

Fig. 7.28 (a) Plane stress element.
(b) Mohr's circle for stress transformation in
xy plane. (c) Orientation of principal stresses.
(d) Three-dimensional Mohr's circles.

Concept Application 7.3

For the state of plane stress shown in Fig. 7.28a, determine (a) the three principal planes and principal stresses and (b) the maximum shearing stress.

a. Principal Planes and Principal Stresses.   Construct Mohr's circle for the transformation of stress in the xy plane (Fig. 7.28b). Point X is plotted 6 units to the right of the τ axis and 3 units above the σ axis (since the corresponding shearing stress tends to rotate the element clockwise). Point Y is plotted 3.5 units to the right of the τ axis and 3 units below the σ axis. Drawing the line XY, the center C of Mohr's circle is found for the xy plane. Its abscissa is

$$\sigma_{ave} = \frac{\sigma_x + \sigma_y}{2} = \frac{6 + 3.5}{2} = 4.75 \text{ ksi}$$

Since the sides of the right triangle CFX are CF = 6 − 4.75 = 1.25 ksi and FX = 3 ksi, the radius of the circle is

$$R = CX = \sqrt{(1.25)^2 + (3)^2} = 3.25 \text{ ksi}$$

The principal stresses in the plane of stress are

$$\sigma_a = OA = OC + CA = 4.75 + 3.25 = 8.00 \text{ ksi}$$
$$\sigma_b = OB = OC - BC = 4.75 - 3.25 = 1.50 \text{ ksi}$$

Since the faces of the element perpendicular to the z axis are free of stress, they define one of the principal planes, and the corresponding principal stress is $\sigma_z = 0$. The other two principal planes are defined by points A and B on Mohr's circle. The angle θ_p through which the element should be rotated about the z axis to bring its faces to coincide with these planes (Fig. 7.28c) is half the angle ACX.

$$\tan 2\theta_p = \frac{FX}{CF} = \frac{3}{1.25}$$
$$2\theta_p = 67.4° \downarrow \qquad \theta_p = 33.7° \downarrow$$

b. Maximum Shearing Stress.   Now draw the circles of diameter OB and OA that correspond to rotations of the element about the a and b axes (Fig. 7.28d). Note that the maximum shearing stress is equal to the radius of the circle of diameter OA. Thus,

$$\tau_{max} = \tfrac{1}{2}\sigma_a = \tfrac{1}{2}(8.00 \text{ ksi}) = 4.00 \text{ ksi}$$

Since points D′ and E′, which define the planes of maximum shearing stress, are located at the ends of the vertical diameter of the circle corresponding to a rotation about the b axis, the faces of the element of Fig. 7.28c can be brought to coincide with the planes of maximum shearing stress through a rotation of 45° about the b axis.

*7.5 THEORIES OF FAILURE

7.5A Yield Criteria for Ductile Materials

Structural elements and machine components made of a ductile material are usually designed so that the material will not yield under the expected loading conditions. When the element or component is under uniaxial stress (Fig. 7.29), the value of the normal stress σ_x that causes the material to yield is obtained from a tensile test of the same material, since the test specimen and the structural element or machine component are in the same state of stress. Thus, regardless of the actual mechanism that causes the material to yield, the element or component will be safe as long as $\sigma_x < \sigma_Y$, where σ_Y is the yield strength of the test specimen.

On the other hand, when a structural element or machine component is in a state of plane stress (Fig. 7.30a), it is convenient to use one of the methods developed earlier to determine the principal stresses σ_a and σ_b at any given point (Fig. 7.30b). The material can then be considered to be in a state of biaxial stress at that point. Since this state is different from the state of uniaxial stress, it is not possible to predict from such a test whether or not the structural element or machine component under investigation will fail. Some criterion regarding the actual mechanism of failure of the material must be established that will make it possible to compare the effects of both states of stress. The purpose of this section is to present the two yield criteria most frequently used for ductile materials.

Maximum-Shearing-Stress Criterion. This criterion is based on the observation that yield in ductile materials is caused by slippage of the material along oblique surfaces and is due primarily to shearing stresses (see. Sec. 2.1B). According to this criterion, a structural component is safe as long as the maximum value τ_{max} of the shearing stress in that component remains smaller than the corresponding shearing stress in a tensile-test specimen of the same material as the specimen starts to yield.

Recalling from Sec. 1.3 that the maximum value of the shearing stress under a centric axial load is equal to half the value of the corresponding normal stress, we conclude that the maximum shearing stress in a tensile-test specimen is $\frac{1}{2}\sigma_Y$ as the specimen starts to yield. On the other hand, Sec. 7.4 showed, for plane stress, that τ_{max} of the shearing stress is equal to $\frac{1}{2}|\sigma_{max}|$ if the principal stresses are either both positive or both negative and to $\frac{1}{2}|\sigma_{max} - \sigma_{min}|$ if the maximum stress is positive and the minimum stress is negative. Thus, if the principal stresses σ_a and σ_b have the same sign, the maximum-shearing-stress criterion gives

$$|\sigma_a| < \sigma_Y \qquad |\sigma_b| < \sigma_Y \qquad \text{(7.23)}$$

If the principal stresses σ_a and σ_b have opposite signs, the maximum-shearing-stress criterion yields

$$|\sigma_a - \sigma_b| < \sigma_Y \qquad \text{(7.24)}$$

These relationships have been represented graphically in Fig. 7.31. Any given state of stress is represented by a point of coordinates σ_a and σ_b, where σ_a and σ_b are the two principal stresses. If this point falls within the area shown, the structural component is safe. If it falls outside this area,

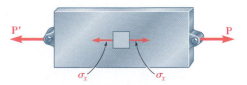

Fig. 7.29 Structural element under uniaxial stress.

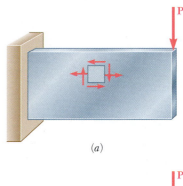

(a)

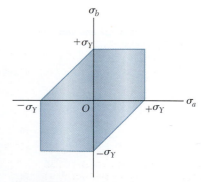

(b)

Fig. 7.30 Structural element in a state of plane stress. (a) Stress element referred to coordinate axes. (b) Stress element referred to principal axes.

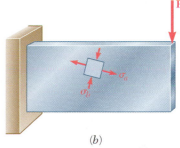

Fig. 7.31 Tresca's hexagon for maximum-shearing-stress criterion.

the component fails as a result of yield in the material. The hexagon associated with the initiation of yield is known as *Tresca's hexagon* after the French engineer Henri Edouard Tresca (1814–1885).

Maximum-Distortion-Energy Criterion.

This criterion is based on the determination of the distortion energy in a given material. This is the energy associated with changes in shape in that material (as opposed to the energy associated with changes in volume in the same material). This criterion is also known as the *von Mises criterion* after the German-American applied mathematician Richard von Mises (1883–1953). Here, a given structural component is safe as long as the maximum value of the distortion energy per unit volume in that material remains smaller than the distortion energy per unit volume required to cause yield in a tensile-test specimen of the same material. The distortion energy per unit volume in an isotropic material under plane stress is

$$u_d = \frac{1}{6G}(\sigma_a^2 - \sigma_a\sigma_b + \sigma_b^2) \qquad (7.25)$$

where σ_a and σ_b are the principal stresses and G is the modulus of rigidity. In a tensile-test specimen that is starting to yield, $\sigma_a = \sigma_Y$, $\sigma_b = 0$, and $(u_d)_Y = \sigma_Y^2/6G$. Thus, the maximum-distortion-energy criterion indicates that the structural component is safe as long as $u_d < (u_d)_Y$, or

$$\sigma_a^2 - \sigma_a\sigma_b + \sigma_b^2 < \sigma_Y^2 \qquad (7.26)$$

where the point of coordinates σ_a and σ_b falls within the area shown in Fig. 7.32. This area is bounded by the ellipse

$$\sigma_a^2 - \sigma_a\sigma_b + \sigma_b^2 = \sigma_Y^2 \qquad (7.27)$$

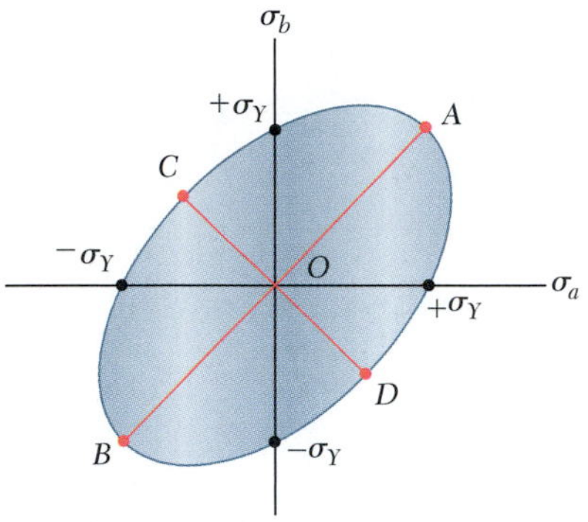

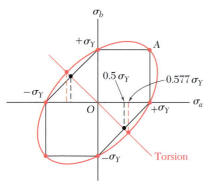

Fig. 7.32 Von Mises surface based on maximum-distortion-energy criterion.

which intersects the coordinate axes at $\sigma_a = \pm\sigma_Y$ and $\sigma_b = \pm\sigma_Y$. The major axis of the ellipse bisects the first and third quadrants and extends from A ($\sigma_a = \sigma_b = \sigma_Y$) to B ($\sigma_a = \sigma_b = -\sigma_Y$), while its minor axis extends from C ($\sigma_a = -\sigma_b = -0.577\sigma_Y$) to D ($\sigma_a = -\sigma_b = 0.577\sigma_Y$).

The maximum-shearing-stress criterion and the maximum-distortion-energy criterion are compared in Fig. 7.33. The ellipse passes through the vertices of the hexagon. Thus, for the states of stress represented by these six points, the two criteria give the same results. For any other state of stress, the maximum-shearing-stress criterion is more conservative than the maximum-distortion-energy criterion, since the hexagon is located within the ellipse.

A state of stress of particular interest is associated with yield in a torsion test. Recall from Fig. 7.18 that, for torsion, $\sigma_{min} = -\sigma_{max}$. Thus, the corresponding points in Fig. 7.33 are located on the bisector of the second and fourth quadrants. It follows that yield occurs in a torsion test when $\sigma_a = -\sigma_b = \pm 0.5\sigma_Y$ according to the maximum-shearing-stress criterion and $\sigma_a = -\sigma_b = \pm 0.577\sigma_Y$ according to the maximum-distortion-energy criterion. But again recalling Fig. 7.18, σ_a and σ_b must be equal in magnitude to τ_{max}, which is obtained from a torsion test for the yield strength τ_Y of the material. Since the yield strength σ_Y in tension and τ_Y in shear are given for various ductile materials in Appendix B, the ratio τ_Y/σ_Y can be determined for these materials where the range is from 0.55 to 0.60. Thus, the maximum-distortion-energy criterion appears somewhat more accurate than the maximum-shearing-stress criterion for predicting yield in torsion.

Fig. 7.33 Comparison of Tresca and von Mises criteria.

7.5B Fracture Criteria for Brittle Materials Under Plane Stress

When brittle materials are subjected to a tensile test, they fail suddenly through rupture—or fracture—without any prior yielding. When a structural element or machine component made of a brittle material is under uniaxial tensile stress, the normal stress that causes it to fail is equal to the ultimate strength σ_U as determined from a tensile test, since both the specimen and the element or component are in the same state of stress. However, when a structural element or machine component is in a state of plane stress, it is found convenient to determine the principal stresses σ_a and σ_b at any given point and to use one of the criteria presented in this section to predict whether or not the structural element or machine component will fail.

Maximum-Normal-Stress Criterion. According to this criterion, a given structural component fails when the maximum normal stress reaches the ultimate strength σ_U obtained from the tensile test of a specimen of the same material. Thus, the structural component will be safe as long as the absolute values of the principal stresses σ_a and σ_b are both less than σ_U:

$$|\sigma_a| < \sigma_U \qquad |\sigma_b| < \sigma_U \tag{7.28}$$

The maximum-normal-stress criterion is shown graphically in Fig. 7.34. If the point obtained by plotting the values σ_a and σ_b of the principal stresses falls within the square area shown, the structural component is safe. If it falls outside that area, the component will fail.

   The maximum-normal-stress criterion is known as *Coulomb's criterion* after the French physicist Charles Augustin de Coulomb (1736–1806). This criterion suffers from an important shortcoming: it is based on the assumption that the ultimate strength of the material is the same in tension and in compression. As noted in Sec. 2.1B, this is seldom the case because the presence of flaws in the material, such as microscopic cracks or cavities, tends to weaken the material in tension, while not appreciably affecting its resistance to compressive failure. This criterion also makes no allowance for effects other than those of the normal stresses on the failure mechanism of the material.[†]

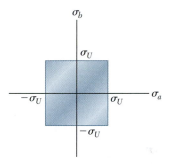

Fig. 7.34 Coulomb's surface for maximum-normal-stress criterion.

[†]Another failure criterion known as the *maximum-normal-strain criterion,* or Saint-Venant's criterion, was widely used during the nineteenth century. According to this criterion, a given structural component is safe as long as the maximum value of the normal strain in that component remains smaller than the value ϵ_U of the strain at which a tensile-test specimen of the same material will fail. But, as will be shown in Sec. 7.8, the strain is maximum along one of the principal axes of stress, if the deformation is elastic and the material homogeneous and isotropic. Thus, denoting by ϵ_a and ϵ_b the values of the normal strain along the principal axes in the plane of stress, we write

$$|\epsilon_a| < \epsilon_U \qquad |\epsilon_b| < \epsilon_U \tag{7.29}$$

Making use of the generalized Hooke's law (Sec. 2.5), we could express these relations in terms of the principal stresses σ_a and σ_b and the ultimate strength σ_U of the material. We would find that, according to the maximum-normal-strain criterion, the structural component is safe as long as the point obtained by plotting σ_a and σ_b falls within the area shown in Fig. 7.35, where ν is Poisson's ratio for the given material.

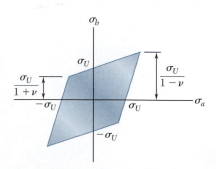

Fig. 7.35 Saint-Venant's surface for maximum-normal-strain criterion.

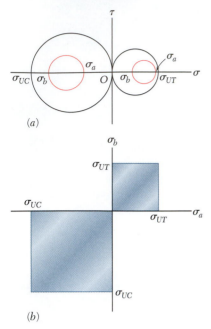

Fig. 7.36 Mohr's criterion for brittle
materials having different ultimate
strengths in tension and compression.
(a) Mohr's circles for uniaxial compression
(left) and tension (right) tests at rupture.
(b) Safe stress states when σ_a and σ_b have
the same sign.

Mohr's Criterion. Suggested by the German engineer Otto Mohr, this criterion is used to predict the effect of a given state of plane stress on a brittle material when the results of various types of tests are available.

Assume that tensile compressive tests have been conducted on a given material and that σ_{UT} and σ_{UC} of the ultimate strength in tension and compression have been determined. The state of stress corresponding to the rupture of the tensile-test specimen is represented on a Mohr-circle diagram where the circle intersects the horizontal axis at O and σ_{UT} (Fig. 7.36a). Similarly, the state of stress corresponding to the failure of the compressive-test specimen is represented by the circle intersecting the horizontal axis at O and σ_{UC}. Clearly, a state of stress represented by a circle entirely contained in either of these circles will be safe. Thus, if both principal stresses are positive, the state of stress is safe as long as $\sigma_a < \sigma_{UT}$ and $\sigma_b < \sigma_{UT}$. If both principal stresses are negative, the state of stress is safe as long as $|\sigma_a| < |\sigma_{UC}|$ and $|\sigma_b| < |\sigma_{UC}|$. Plotting the point of coordinates σ_a and σ_b (Fig. 7.36b), the state of stress is safe as long as that point falls within one of the square areas shown in that figure.

In order to analyze σ_a and σ_b when they have opposite signs, assume that a torsion test has been conducted on the material and that its ultimate strength in shear, τ_U, has been determined. Drawing the circle centered at O representing the state of stress corresponding to the failure of the torsion-test specimen (Fig. 7.37a), observe that any state of stress represented by a circle entirely contained in that circle is also safe. According to Mohr's criterion, a state of stress is safe if it is represented by a circle located entirely within the area bounded by the envelope of the circles corresponding to the available data. The remaining portions of the principal-stress diagram are obtained by drawing various circles tangent to this envelope, determining the corresponding values of σ_a and σ_b, and plotting the points of coordinates σ_a and σ_b (Fig. 7.37b).

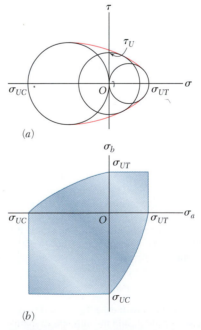

Fig. 7.37 Mohr's criterion for brittle materials. (a) Mohr's circles for uniaxial compression (left), torsion (middle), and uniaxial tension (right) tests at rupture. (b) Envelope of safe stress states.

More accurate diagrams can be drawn when test results correspond-ing to various states of stress are available. If the only available data con-sists of the ultimate strengths σ_{UT} and σ_{UC}, the envelope in Fig. 7.37a is replaced by the tangents AB and $A'B'$ to the circles corresponding to fail-ure in tension and compression (Fig. 7.38a). From the similar triangles in Fig. 7.38, the abscissa of the center C of a circle tangent to AB and $A'B'$ is linearly related to its radius R. Since $\sigma_a = OC + R$ and $\sigma_b = OC - R$, σ_a and σ_b are also related linearly. Thus, the shaded area corresponding to this simplified Mohr's criterion is bounded by straight lines in the second and fourth quadrants (Fig. 7.38b).

In order to determine whether a structural component is safe under a given load, the state of stress should be calculated at all critical points of the component (i.e., where stress concentrations are likely to occur). This can be done by using the stress-concentration factors given in Figs. 2.52, 3.28, 4.24, and 4.25. However, there are many instances when the theory of elasticity must be used to determine the state of stress at a critical point.

Special care should be taken when *macroscopic cracks* are detected in a structural component. While it can be assumed that the test specimen used to determine the ultimate tensile strength of the material contained the same type of flaws (i.e., *microscopic* cracks or cavities) as the structural component, the specimen was certainly free of any noticeable macro-scopic cracks. When a crack is detected in a structural component, it is necessary to determine whether that crack will propagate under the expected load and cause the component to fail or will remain stable. This requires an analysis involving the energy associated with the growth of the crack. Such an analysis is beyond the scope of this text and should be carried out using by the methods of fracture mechanics.

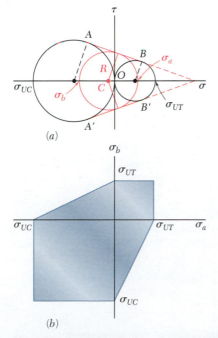

Fig. 7.38 Simplified Mohr's criterion for brittle materials. (*a*) Mohr's circles for uniaxial compression (left), torsion (middle), and uniaxial tension (right) tests at rupture. (*b*) Envelope of safe stress states.

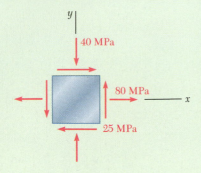

Sample Problem 7.4

The state of plane stress shown occurs at a critical point of a steel machine component. As a result of several tensile tests, the tensile yield strength is $\sigma_Y = 250$ MPa for the grade of steel used. Determine the factor of safety with respect to yield using (*a*) the maximum-shearing-stress criterion, (*b*) the maximum-distortion-energy criterion.

STRATEGY: Draw Mohr's circle from the given state of plane stress. Analyzing this circle to obtain the principal stresses and the maximum shearing stress, you can then apply the maximum-shearing-stress and maximum-distortion-energy criteria.

MODELING and ANALYSIS:

Mohr's Circle. We construct Mohr's circle (Fig. 1) for the given state of stress and find

$$\sigma_{ave} = OC = \tfrac{1}{2}(\sigma_x + \sigma_y) = \tfrac{1}{2}(80 - 40) = 20 \text{ MPa}$$

$$\tau_m = R = \sqrt{(CF)^2 + (FX)^2} = \sqrt{(60)^2 + (25)^2} = 65 \text{ MPa}$$

Principal Stresses

$$\sigma_a = OC + CA = 20 + 65 = +85 \text{ MPa}$$
$$\sigma_b = OC - BC = 20 - 65 = -45 \text{ MPa}$$

a. Maximum-Shearing-Stress Criterion. Since the tensile strength is $\sigma_Y = 250$ MPa, the corresponding shearing stress at yield is

$$\tau_Y = \tfrac{1}{2}\sigma_Y = \tfrac{1}{2}(250 \text{ MPa}) = 125 \text{ MPa}$$

For $\tau_m = 65$ MPa, $F.S. = \dfrac{\tau_Y}{\tau_m} = \dfrac{125 \text{ MPa}}{65 \text{ MPa}}$ *F.S.* = 1.92 ◄

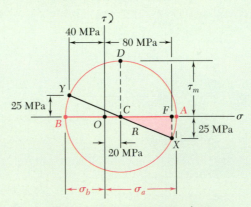

Fig. 1 Mohr's circle for given stress element.

(continued)

b. Maximum-Distortion-Energy Criterion. Introducing a factor of safety into Eq. (7.26) gives

$$\sigma_a^2 - \sigma_a \sigma_b + \sigma_b^2 = \left(\frac{\sigma_Y}{F.S.}\right)^2$$

For $\sigma_a = +85$ MPa, $\sigma_b = -45$ MPa, and $\sigma_Y = 250$ MPa, we have

$$(85)^2 - (85)(-45) + (45)^2 = \left(\frac{250}{F.S.}\right)^2$$

$$114.3 = \frac{250}{F.S.} \qquad F.S. = 2.19 \quad \blacktriangleleft$$

REFLECT and THINK. For a ductile material with $\sigma_Y = 250$ MPa, we have drawn the hexagon associated with the maximum-shearing-stress criterion and the ellipse associated with the maximum-distortion-energy criterion (Fig. 2). The given state of plane stress is represented by point H with coordinates $\sigma_a = 85$ MPa and $\sigma_b = -45$ MPa. The straight line drawn through points O and H intersects the hexagon at point T and the ellipse at point M. For each criterion, $F.S.$ is verified by measuring the line segments indicated and computing their ratios:

$$(a)\ F.S. = \frac{OT}{OH} = 1.92 \qquad (b)\ F.S. = \frac{OM}{OH} = 2.19$$

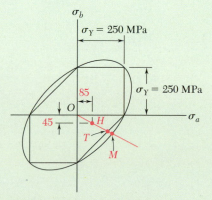

Fig. 2 Tresca and von Mises envelopes and given stress state (point H).

Problems

7.66 For the state of plane stress shown, determine the maximum shearing stress when (*a*) $\sigma_x = 14$ ksi and $\sigma_y = 4$ ksi, (*b*) $\sigma_x = 21$ ksi and $\sigma_y = 14$ ksi. (*Hint:* Consider both in-plane and out-of-plane shearing stresses.)

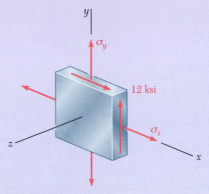

Fig. P7.66 and *P7.67*

7.67 For the state of plane stress shown, determine the maximum shearing stress when (*a*) $\sigma_x = 20$ ksi and $\sigma_y = 10$ ksi, (*b*) $\sigma_x = 12$ ksi and $\sigma_y = 5$ ksi. (*Hint:* Consider both in-plane and out-of-plane shearing stresses.)

7.68 For the state of stress shown, determine the maximum shearing stress when (*a*) $\sigma_y = 40$ MPa, (*b*) $\sigma_y = 120$ MPa. (*Hint:* Consider both in-plane and out-of-plane shearing stresses.)

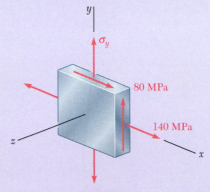

Fig. P7.68 and P7.69

7.69 For the state of stress shown, determine the maximum shearing stress when (*a*) $\sigma_y = 20$ MPa, (*b*) $\sigma_y = 140$ MPa. (*Hint:* Consider both in-plane and out-of-plane shearing stresses.)

7.70 and 7.71 For the state of stress shown, determine the maximum shearing stress when (a) $\sigma_z = 0$, (b) $\sigma_z = +60$ MPa, (c) $\sigma_z = -60$ MPa.

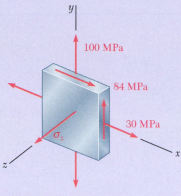

Fig. P7.70

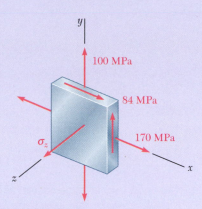

Fig. P7.71

7.72 and 7.73 For the state of stress shown, determine the maximum shearing stress when (a) $\tau_{yz} = 17.5$ ksi, (b) $\tau_{yz} = 8$ ksi, (c) $\tau_{yz} = 0$.

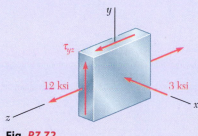

Fig. P7.72

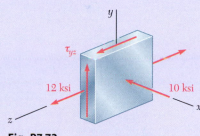

Fig. P7.73

7.74 For the state of stress shown, determine the value of τ_{xy} for which the maximum shearing stress is (a) 9 ksi, (b) 12 ksi.

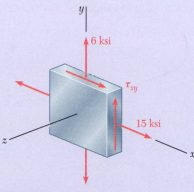

Fig. P7.74

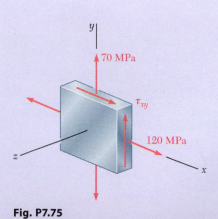

Fig. P7.75

7.75 For the state of stress shown, determine the value of τ_{xy} for which the maximum shearing stress is 80 MPa.

7.76 For the state of stress shown, determine two values of σ_y for which the maximum shearing stress is 73 MPa.

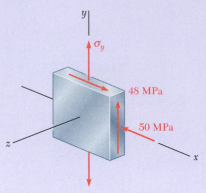

Fig. P7.76

7.77 For the state of stress shown, determine two values of σ_y for which the maximum shearing stress is 10 ksi.

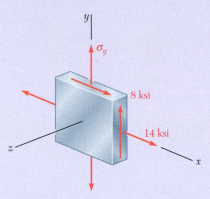

Fig. P7.77

7.78 For the state of stress shown, determine the range of values of τ_{xz} for which the maximum shearing stress is equal to or less than 60 MPa.

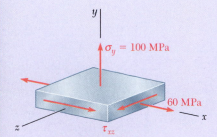

Fig. P7.78

Fig. P7.79

7.79 For the state of stress shown, determine two values of σ_y for which the maximum shearing stress is 80 MPa.

*7.80 For the state of stress of Prob. 7.69, determine (*a*) the value of σ_y for which the maximum shearing stress is as small as possible, (*b*) the corresponding value of the shearing stress.

7.81 The state of plane stress shown occurs in a machine component made of a steel with $\sigma_Y = 325$ MPa. Using the maximum-distortion-energy criterion, determine whether yield will occur when (*a*) $\sigma_0 = 200$ MPa, (*b*) $\sigma_0 = 240$ MPa, (*c*) $\sigma_0 = 280$ MPa. If yield does not occur, determine the corresponding factor of safety.

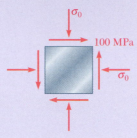

Fig. P7.81

7.82 Solve Prob. 7.81, using the maximum-shearing-stress criterion.

7.83 The state of plane stress shown occurs in a machine component made of a steel with $\sigma_Y = 45$ ksi. Using the maximum-distortion-energy criterion, determine whether yield will occur when (*a*) $\tau_{xy} = 9$ ksi, (*b*) $\tau_{xy} = 18$ ksi, (*c*) $\tau_{xy} = 20$ ksi. If yield does not occur, determine the corresponding factor of safety.

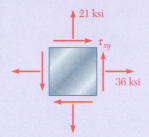

Fig. P7.83

7.84 Solve Prob. 7.83, using the maximum-shearing-stress criterion.

7.85 The 38-mm-diameter shaft *AB* is made of a grade of steel for which the yield strength is $\sigma_Y = 250$ MPa. Using the maximum-shearing-stress criterion, determine the magnitude of the torque **T** for which yield occurs when *P* = 240 kN.

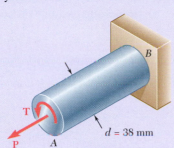

Fig. P7.85

7.86 Solve Prob. 7.85, using the maximum-distortion-energy criterion.

7.87 The 1.5-in.-diameter shaft *AB* is made of a grade of steel with a 42-ksi tensile yield stress. Using the maximum-shearing-stress criterion, determine the magnitude of the torque **T** for which yield occurs when *P* = 60 kips.

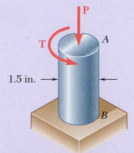

Fig. P7.87

7.88 Solve Prob. 7.87, using the maximum-distortion-energy criterion.

7.89 and 7.90 The state of plane stress shown is expected to occur in an aluminum casting. Knowing that for the aluminum alloy used $\sigma_{UT} = 80$ MPa and $\sigma_{UC} = 200$ MPa and using Mohr's criterion, determine whether rupture of the casting will occur.

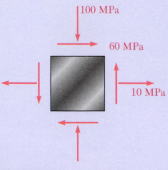

Fig. P7.89

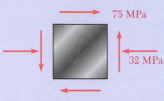

Fig. P7.90

7.91 and 7.92 The state of plane stress shown is expected to occur in an aluminum casting. Knowing that for the aluminum alloy used $\sigma_{UT} = 10$ ksi and $\sigma_{UC} = 30$ ksi and using Mohr's criterion, determine whether rupture of the casting will occur.

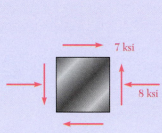

Fig. P7.91

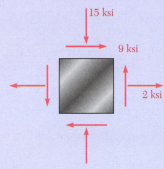

Fig. P7.92

7.93 The state of plane stress shown will occur at a critical point in an aluminum casting that is made of an alloy for which $\sigma_{UT} = 10$ ksi and $\sigma_{UC} = 25$ ksi. Using Mohr's criterion, determine the shearing stress τ_0 for which failure should be expected.

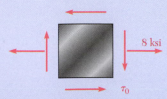

Fig. *P7.93*

7.94 The state of plane stress shown will occur at a critical point in a pipe made of an aluminum alloy for which $\sigma_{UT} = 75$ MPa and $\sigma_{UC} = 150$ MPa. Using Mohr's criterion, determine the shearing stress τ_0 for which failure should be expected.

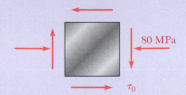

Fig. P7.94

7.95 The cast-aluminum rod shown is made of an alloy for which $\sigma_{UT} = 70$ MPa and $\sigma_{UC} = 175$ MPa. Knowing that the magnitude T of the applied torques is slowly increased and using Mohr's criterion, determine the shearing stress τ_0 that should be expected at rupture.

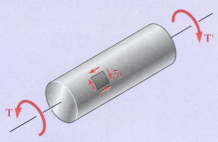

Fig. P7.95

7.96 The cast-aluminum rod shown is made of an alloy for which $\sigma_{UT} = 60$ MPa and $\sigma_{UC} = 120$ MPa. Using Mohr's criterion, determine the magnitude of the torque **T** for which failure should be expected.

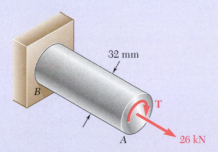

Fig. P7.96

7.97 A machine component is made of a grade of cast iron for which $\sigma_{UT} = 8$ ksi and $\sigma_{UC} = 20$ ksi. For each of the states of stress shown and using Mohr's criterion, determine the normal stress σ_0 at which rupture of the component should be expected.

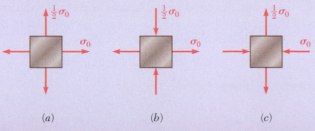

(a) (b) (c)

Fig. P7.97

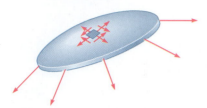

Fig. 7.39 Assumed stress distribution in thin-walled pressure vessels.

7.6 STRESSES IN THIN-WALLED PRESSURE VESSELS

Thin-walled pressure vessels provide an important application of the analysis of plane stress. Since their walls offer little resistance to bending, it can be assumed that the internal forces exerted on a given portion of wall are tangent to the surface of the vessel (Fig. 7.39). The resulting stresses on an element of wall will be contained in a plane tangent to the surface of the vessel.

This analysis of stresses in thin-walled pressure vessels is limited to two types of vessels: cylindrical and spherical (Photos 7.3 and 7.4).

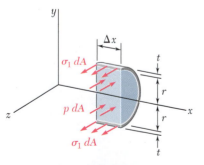

Photo 7.3 Cylindrical pressure vessels for liquid propane.

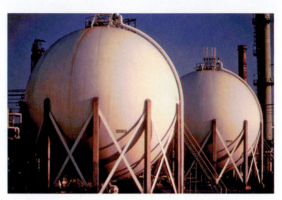

Photo 7.4 Spherical pressure vessels at a chemical plant.

Fig. 7.40 Pressurized cylindrical vessel.

Cylindrical Pressure Vessels. Consider a cylindrical vessel with an inner radius r and a wall thickness t containing a fluid under pressure (Fig. 7.40). The stresses exerted on a small element of wall with sides respectively parallel and perpendicular to the axis of the cylinder will be determined. Because of the axisymmetry of the vessel and its contents, no shearing stress is exerted on the element. The normal stresses σ_1 and σ_2 shown in Fig. 7.40 are therefore principal stresses. The stress σ_1 is called the *hoop stress*, because it is the type of stress found in hoops used to hold together the various slats of a wooden barrel. Stress σ_2 is called the *longitudinal stress*.

To determine the hoop stress σ_1, detach a portion of the vessel and its contents bounded by the xy plane and by two planes parallel to the yz plane at a distance Δx from each other (Fig. 7.41). The forces parallel to the z axis acting on the free body consist of the elementary internal forces $\sigma_1 \, dA$ on the wall sections and the elementary pressure forces $p \, dA$ exerted on the portion of fluid included in the free body. Note that the *gage pressure* of the fluid p is the excess of the inside pressure over the outside atmospheric pressure. The resultant of the internal forces $\sigma_1 \, dA$ is equal to the product of σ_1 and the cross-sectional area $2t \, \Delta x$ of the wall, while the resultant of the pressure forces $p \, dA$ is equal to the product of p and the area $2r \, \Delta x$. The equilibrium equation $\Sigma F_z = 0$ gives

$$\Sigma F_z = 0: \qquad \sigma_1(2t \, \Delta x) - p(2r \, \Delta x) = 0$$

and solving for the hoop stress σ_1,

$$\sigma_1 = \frac{pr}{t} \tag{7.30}$$

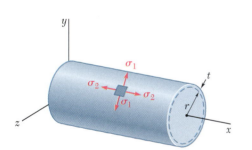

Fig. 7.41 Free-body diagram to determine hoop stress in a cylindrical pressure vessel.

To determine the longitudinal stress σ_2, pass a section perpendicular to the x axis and consider the free body consisting of the portion of the vessel and its contents located to the left of the section (Fig. 7.42). The forces acting on this free body are the elementary internal forces $\sigma_2\,dA$ on the wall section and the elementary pressure forces $p\,dA$ exerted on the portion of fluid included in the free body. Noting that the area of the fluid section is πr^2 and that the area of the wall section can be obtained by multiplying the circumference $2\pi r$ of the cylinder by its wall thickness t, the equilibrium equation is:[†]

$$\Sigma F_x = 0: \qquad\qquad \sigma_2(2\pi rt) - p(\pi r^2) = 0$$

and solving for the longitudinal stress σ_2,

$$\sigma_2 = \frac{pr}{2t} \tag{7.31}$$

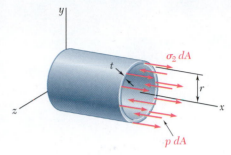

Fig. 7.42 Free-body diagram to determine longitudinal stress.

Note from Eqs. (7.30) and (7.31) that the hoop stress σ_1 is twice as large as the longitudinal stress σ_2:

$$\sigma_1 = 2\sigma_2 \tag{7.32}$$

Drawing Mohr's circle through the points A and B that correspond to the principal stresses σ_1 and σ_2 (Fig. 7.43), and recalling that the maximum in-plane shearing stress is equal to the radius of this circle, we obtain

$$\tau_{\text{max(in plane)}} = \tfrac{1}{2}\sigma_2 = \frac{pr}{4t} \tag{7.33}$$

This stress corresponds to points D and E and is exerted on an element obtained by rotating the original element of Fig. 7.40 through 45° *within*

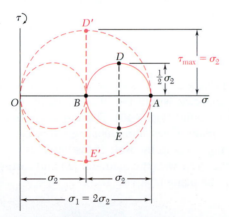

Fig. 7.43 Mohr's circle for element of cylindrical pressure vessel.

[†]Using the mean radius of the wall section, $r_m = r + \tfrac{1}{2}t$, to compute the resultant of the forces, a more accurate value of the longitudinal stress is

$$\sigma_2 = \frac{pr}{2t}\,\frac{1}{1 + \dfrac{t}{2r}}$$

However, for a thin-walled pressure vessel, the term $t/2r$ is sufficiently small to allow the use of Eq. (7.31) for engineering design and analysis. If a pressure vessel is not thin-walled (i.e., if $t/2r$ is not small), the stresses σ_1 and σ_2 vary across the wall and must be determined by the methods of the theory of elasticity.

the plane tangent to the surface of the vessel. However, the maximum shearing stress in the wall of the vessel is larger. It is equal to the radius of the circle of diameter OA and corresponds to a rotation of 45° about a longitudinal axis and *out of the plane* of stress.[†]

$$\tau_{max} = \sigma_2 = \frac{pr}{2t} \tag{7.34}$$

Spherical Pressure Vessels.   Now consider a spherical vessel of inner radius r and wall thickness t, containing a fluid under a gage pressure p. For reasons of symmetry, the stresses exerted on the four faces of a small element of wall must be equal (Fig. 7.44).

$$\sigma_1 = \sigma_2 \tag{7.35}$$

Fig. 7.44   Pressurized spherical vessel.

To determine the stress, pass a section through the center C of the vessel and consider the free body consisting of the portion of the vessel and its contents located to the left of the section (Fig. 7.45). The equation of equilibrium for this free body is the same as for the free body of Fig. 7.42. So for a spherical vessel,

$$\sigma_1 = \sigma_2 = \frac{pr}{2t} \tag{7.36}$$

Since the principal stresses σ_1 and σ_2 are equal, Mohr's circle for transformations of stress within the plane tangent to the surface of the vessel reduces to a point (Fig. 7.46). The in-plane normal stress is constant, and the in-plane maximum shearing stress is zero. However, the maximum shearing stress in the wall of the vessel is not zero; it is equal to the radius of the circle with the diameter OA and corresponds to a rotation of 45° out of the plane of stress. Thus,

$$\tau_{max} = \tfrac{1}{2}\sigma_1 = \frac{pr}{4t} \tag{7.37}$$

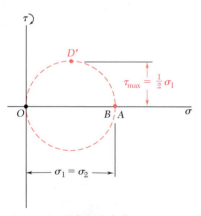

Fig. 7.45   Free-body diagram to determine spherical pressure vessel stress.

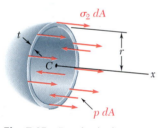

Fig. 7.46   Mohr's circle for element of spherical pressure vessel.

[†]While the third principal stress is zero on the outer surface of the vessel, it is equal to $-p$ on the inner surface and is represented by a point $C(-p, 0)$ on a Mohr-circle diagram. Thus, close to the inside surface of the vessel, the maximum shearing stress is equal to the radius of a circle of diameter CA, or

$$\tau_{max} = \frac{1}{2}(\sigma_1 + p) = \frac{pr}{2t}\left(1 + \frac{t}{r}\right)$$

However, for a thin-walled vessel, t/r is small, and the variation of τ_{max} across the wall section can be neglected. This also applies to spherical pressure vessels.

Sample Problem 7.5

A compressed-air tank is supported by two cradles as shown. One of the cradles is designed so that it does not exert any longitudinal force on the tank. The cylindrical body of the tank has a 30-in. outer diameter and is made of a $\frac{3}{8}$-in. steel plate by butt welding along a helix that forms an angle of 25° with a transverse plane. The end caps are spherical and have a uniform wall thickness of $\frac{5}{16}$ in. For an internal gage pressure of 180 psi, determine (*a*) the normal stress and the maximum shearing stress in the spherical caps, (*b*) the stresses in directions perpendicular and parallel to the helical weld.

STRATEGY: Using the equations for thin-walled pressure vessels, you can determine the state of plane stress at any point within the spherical end cap and within the cylindrical body. You can then plot the corresponding Mohr's circles and use them to determine the stress components of interest.

MODELING and ANALYSIS:

a. Spherical Cap. The state of stress within any point in the spherical cap is shown in Fig. 1. Using Eq. (7.36), we write

$$p = 180\,\text{psi}, \; t = \tfrac{5}{16}\,\text{in.} = 0.3125\,\text{in.}, \; r = 15 - 0.3125 = 14.688\,\text{in.}$$

$$\sigma_1 = \sigma_2 = \frac{pr}{2t} = \frac{(180\,\text{psi})(14.688\,\text{in.})}{2(0.3125\,\text{in.})} \qquad \sigma = 4230\,\text{psi} \; \blacktriangleleft$$

We note that for stresses in a plane tangent to the cap, Mohr's circle reduces to a point (*A*, *B*) on the horizontal axis, and that all in-plane shearing stresses are zero (Fig. 2). On the surface of the cap, the third principal stress is zero and corresponds to point *O*. On a Mohr's circle with a diameter of *AO*, point *D'* represents the maximum shearing stress that occurs on planes at 45° to the plane tangent to the cap.

$$\tau_{\text{max}} = \tfrac{1}{2}(4230\,\text{psi}) \qquad \tau_{\text{max}} = 2115\,\text{psi} \; \blacktriangleleft$$

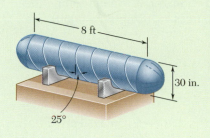

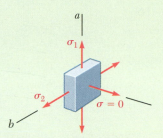

Fig. 1 State of stress at any point in spherical cap.

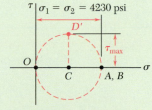

Fig. 2 Mohr's circle for stress element in spherical cap.

(continued)

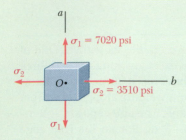

Fig. 3 State of stress at any point in cylindrical body.

b. Cylindrical Body of the Tank.

The state of stress within any point in the cylindrical body is as shown in Fig. 3. We determine the hoop stress σ_1 and the longitudinal stress σ_2 using Eqs. (7.30) and (7.32). We write

$$p = 180 \text{ psi}, \quad t = \tfrac{3}{8} \text{ in.} = 0.375 \text{ in.}, \quad r = 15 - 0.375 = 14.625 \text{ in.}$$

$$\sigma_1 = \frac{pr}{t} = \frac{(180 \text{ psi})(14.625 \text{ in.})}{0.375 \text{ in.}} = 7020 \text{ psi} \qquad \sigma_2 = \tfrac{1}{2}\sigma_1 = 3510 \text{ psi}$$

$$\sigma_{\text{ave}} = \tfrac{1}{2}(\sigma_1 + \sigma_2) = 5265 \text{ psi} \qquad R = \tfrac{1}{2}(\sigma_1 - \sigma_2) = 1755 \text{ psi}$$

Stresses at the Weld. Noting that both the hoop stress and the longitudinal stress are principal stresses, we draw Mohr's circle as shown in Fig 4.

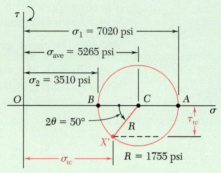

Fig. 4 Mohr's circle for stress element in cylindrical body.

An element having a face parallel to the weld is obtained by rotating the face perpendicular to the axis Ob (Fig. 3) counterclockwise through 25°. Therefore, on Mohr's circle (Fig. 4), point X' corresponds to the stress components on the weld by rotating radius CB counterclockwise through $2\theta = 50°$.

$$\sigma_w = \sigma_{\text{ave}} - R\cos 50° = 5265 - 1755 \cos 50° \qquad \sigma_w = +4140 \text{ psi} \; \blacktriangleleft$$

$$\tau_w = R\sin 50° = 1755 \sin 50° \qquad \tau_w = 1344 \text{ psi} \; \blacktriangleleft$$

Since X' is below the horizontal axis, τ_w tends to rotate the element counterclockwise. The stress components on the weld are shown in Fig. 5.

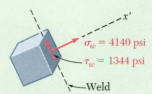

Fig. 5 Stress components on the weld.

Problems

7.98 A spherical pressure vessel has an outer diameter of 3 m and a wall thickness of 12 mm. Knowing that for the steel used σ_{all} = 80 MPa, E = 200 GPa, and ν = 0.29, determine (a) the allowable gage pressure, (b) the corresponding increase in the diameter of the vessel.

7.99 A spherical gas container having an inner diameter of 5 m and a wall thickness of 24 mm is made of steel for which E = 200 GPa and ν = 0.29. Knowing that the gage pressure in the container is increased from zero to 1.8 MPa, determine (a) the maximum normal stress in the container, (b) the corresponding increase in the diameter of the container.

7.100 The maximum gage pressure is known to be 1150 psi in a spherical steel pressure vessel having a 10-in. outer diameter and a 0.25-in. wall thickness. Knowing that the ultimate stress in the steel used is σ_U = 60 ksi, determine the factor of safety with respect to tensile failure.

7.101 A spherical pressure vessel of 750-mm outer diameter is to be fabricated from a steel having an ultimate stress σ_U = 400 MPa. Knowing that a factor of safety of 4.0 is desired and that the gage pressure can reach 4.2 MPa, determine the smallest wall thickness that should be used.

7.102 A spherical gas container made of steel has a 20-ft outer diameter and a wall thickness of $\frac{7}{16}$ in. Knowing that the internal pressure is 75 psi, determine the maximum normal stress and the maximum shearing stress in the container.

7.103 A basketball has a 300-mm outer diameter and a 3-mm wall thickness. Determine the normal stress in the wall when the basketball is inflated to a 120-kPa gage pressure.

7.104 The unpressurized cylindrical storage tank shown has a 5-mm wall thickness and is made of steel having a 400-MPa ultimate strength in tension. Determine the maximum height h to which it can be filled with water if a factor of safety of 4.0 is desired. (Density of water = 1000 kg/m^3.)

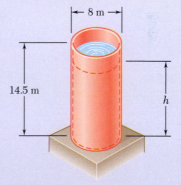

Fig. P7.104

7.105 For the storage tank of Prob. 7.104, determine the maximum normal stress and the maximum shearing stress in the cylindrical wall when the tank is filled to capacity (h = 14.5 m).

7.106 The bulk storage tank shown in Photo 7.3 has an outer diameter of 3.3 m and a wall thickness of 18 mm. At a time when the internal pressure of the tank is 1.5 MPa, determine the maximum normal stress and the maximum shearing stress in the tank.

7.107 A standard-weight steel pipe of 12-in. nominal diameter carries water under a pressure of 400 psi. (*a*) Knowing that the outside diameter is 12.75 in. and the wall thickness is 0.375 in., determine the maximum tensile stress in the pipe. (*b*) Solve part *a*, assuming an extra-strong pipe is used, of 12.75-in. outside diameter and 0.5-in. wall thickness.

7.108 A cylindrical storage tank contains liquefied propane under a pressure of 1.5 MPa at a temperature of 38°C. Knowing that the tank has an outer diameter of 320 mm and a wall thickness of 3 mm, determine the maximum normal stress and the maximum shearing stress in the tank.

7.109 Determine the largest internal pressure that can be applied to a cylindrical tank of 5.5-ft outer diameter and $\frac{5}{8}$-in. wall thickness if the ultimate normal stress of the steel used is 65 ksi and a factor of safety of 5.0 is desired.

7.110 A steel penstock has a 36-in. outer diameter, a 0.5-in. wall thickness, and connects a reservoir at *A* with a generating station at *B*. Knowing that the specific weight of water is 62.4 lb/ft^3, determine the maximum normal stress and the maximum shearing stress in the penstock under static conditions.

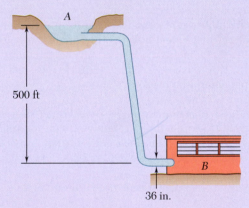

Fig. *P7.110* and P7.111

7.111 A steel penstock has a 36-in. outer diameter and connects a reservoir at *A* with a generating station at *B*. Knowing that the specific weight of water is 62.4 lb/ft^3 and that the allowable normal stress in the steel is 12.5 ksi, determine the smallest thickness that can be used for the penstock.

7.112 The cylindrical portion of the compressed-air tank shown is fabricated of 8-mm-thick plate welded along a helix forming an angle $\beta = 30°$ with the horizontal. Knowing that the allowable stress normal to the weld is 75 MPa, determine the largest gage pressure that can be used in the tank.

7.113 For the compressed-air tank of Prob. 7.112, determine the gage pressure that will cause a shearing stress parallel to the weld of 30 MPa.

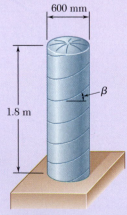

Fig. P7.112

7.114 The steel pressure tank shown has a 750-mm inner diameter and a 9-mm wall thickness. Knowing that the butt-welded seams form an angle $\beta = 50°$ with the longitudinal axis of the tank and that the gage pressure in the tank is 1.5 MPa, determine, (*a*) the normal stress perpendicular to the weld, (*b*) the shearing stress parallel to the weld.

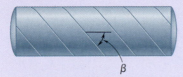

Fig. P7.114 and P7.115

7.115 The pressurized tank shown was fabricated by welding strips of plate along a helix forming an angle β with a transverse plane. Determine the largest value of β that can be used if the normal stress perpendicular to the weld is not to be larger than 85 percent of the maximum stress in the tank.

7.116 Square plates, each of 0.5-in. thickness, can be bent and welded together in either of the two ways shown to form the cylindrical portion of a compressed-air tank. Knowing that the allowable normal stress perpendicular to the weld is 12 ksi, determine the largest allowable gage pressure in each case.

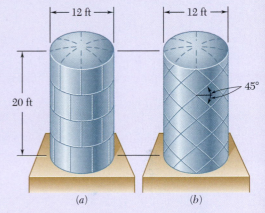

Fig. *P7.116*

7.117 The pressure tank shown has a 0.375-in. wall thickness and butt-welded seams forming an angle $\beta = 20°$ with a transverse plane. For a gage pressure of 85 psi, determine, (*a*) the normal stress perpendicular to the weld, (*b*) the shearing stress parallel to the weld.

7.118 For the tank of Prob. 7.117, determine the largest allowable gage pressure, knowing that the allowable normal stress perpendicular to the weld is 18 ksi and the allowable shearing stress parallel to the weld is 10 ksi.

7.119 For the tank of Prob. 7.117, determine the range of values of β that can be used if the shearing stress parallel to the weld is not to exceed 1350 psi when the gage pressure is 85 psi.

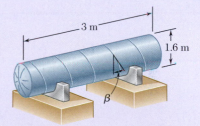

Fig. P7.117

7.120 A pressure vessel of 10-in. inner diameter and 0.25-in. wall thickness is fabricated from a 4-ft section of spirally-welded pipe *AB* and is equipped with two rigid end plates. The gage pressure inside the vessel is 300 psi and 10-kip centric axial forces **P** and **P′** are applied to the end plates. Determine (*a*) the normal stress perpendicular to the weld, (*b*) the shearing stress parallel to the weld.

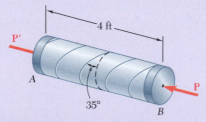

Fig. P7.120

7.121 Solve Prob. 7.120, assuming that the magnitude P of the two forces is increased to 30 kips.

7.122 A torque of magnitude $T = 12$ kN·m is applied to the end of a tank containing compressed air under a pressure of 8 MPa. Knowing that the tank has a 180-mm inner diameter and a 12-mm wall thickness, determine the maximum normal stress and the maximum shearing stress in the tank.

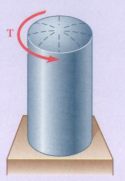

Fig. P7.122 and *P7.123*

7.123 The tank shown has a 180-mm inner diameter and a 12-mm wall thickness. Knowing that the tank contains compressed air under a pressure of 8 MPa, determine the magnitude T of the applied torque for which the maximum normal stress is 75 MPa.

7.124 The compressed-air tank AB has a 250-mm outside diameter and an 8-mm wall thickness. It is fitted with a collar by which a 40-kN force **P** is applied at B in the horizontal direction. Knowing that the gage pressure inside the tank is 5 MPa, determine the maximum normal stress and the maximum shearing stress at point K.

7.125 In Prob. 7.124, determine the maximum normal stress and the maximum shearing stress at point L.

7.126 A brass ring of 5-in. outer diameter and 0.25-in. thickness fits exactly inside a steel ring of 5-in. inner diameter and 0.125-in. thickness when the temperature of both rings is 50°F. Knowing that the temperature of both rings is then raised to 125°F, determine (*a*) the tensile stress in the steel ring, (*b*) the corresponding pressure exerted by the brass ring on the steel ring.

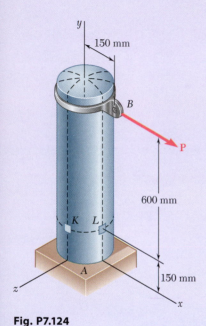

Fig. P7.124

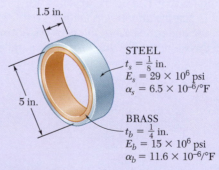

Fig. P7.126

7.127 Solve Prob. 7.126, assuming that the brass ring is 0.125 in. thick and the steel ring is 0.25 in. thick.

*7.7 TRANSFORMATION OF PLANE STRAIN

7.7A Transformation Equations

Transformations of *strain* under a rotation of the coordinate axes will now be considered. Our analysis will first be limited to states of *plane strain*. These are situations where the deformations of the material take place within parallel planes and are the same in each of these planes. If the z axis is chosen perpendicular to the planes in which the deformations take place, $\epsilon_z = \gamma_{zx} = \gamma_{zy} = 0$, and the only remaining strain components are ϵ_x, ϵ_y, and γ_{xy}. This occurs in a plate subjected to uniformly distributed loads along its edges and restrained from expanding or contracting laterally by smooth, rigid, and fixed supports (Fig. 7.47). It is also found in a bar of infinite length subjected to uniformly distributed loads on its sides, because by reason of symmetry, the elements located in a transverse plane cannot move out of that plane. This idealized model shows that a long bar subjected to uniformly distributed transverse loads (Fig. 7.48) is in a state of plane strain in any given transverse section that is not located too close to either end of the bar.[†]

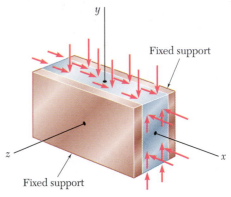

Fig. 7.47 Plane strain example: laterally restrained by fixed supports.

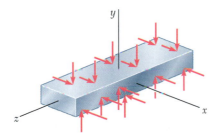

Fig. 7.48 Plane strain example: bar of infinite length in z direction.

Assume that a state of plane strain exists at point Q (with $\epsilon_z = \gamma_{zx} = \gamma_{zy} = 0$) and that it is defined by the strain components ϵ_z, ϵ_y, and γ_{xy} associated with the x and y axes. Recalling Secs. 2.5 and 2.7, a square element of center Q with sides of a length Δs and parallel to the x and y axes is deformed into a parallelogram where the sides are now equal to $\Delta s (1 + \epsilon_x)$ and $\Delta s (1 + \epsilon_y)$, forming angles of $\frac{\pi}{2} - \gamma_{xy}$ and $\frac{\pi}{2} + \gamma_{xy}$ with each other (Fig. 7.49). As a result of the deformations of the other elements located in the xy plane, the element can also undergo a rigid-body motion, but such a motion is irrelevant to the strains at point Q and will be ignored in this analysis. Our purpose is to determine in terms of ϵ_x, ϵ_y, γ_{xy}, and θ the strain components $\epsilon_{x'}$, $\epsilon_{y'}$, and $\gamma_{x'y'}$ associated with the frame of reference $x'y'$ obtained by rotating the x and y axes through angle θ. As shown in Fig. 7.50, these new strain components define the parallelogram into which a square with sides parallel to the x' and y' axes is deformed.

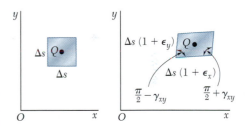

Fig. 7.49 Plane strain element: undeformed and deformed.

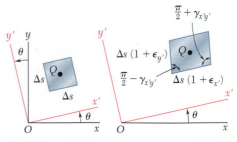

Fig. 7.50 Transformation of plane strain element in undeformed and deformed orientations.

[†]A state of *plane strain* and a state of *plane stress* do not occur simultaneously, except for ideal materials with a Poisson ratio equal to zero. The constraints placed on the elements of the plate of Fig. 7.47 and of the bar of Fig. 7.48 result in a stress σ_z different from zero. On the other hand, in the case of the plate of Fig. 7.3, the absence of any lateral restraint results in $\sigma_z = 0$ and $\epsilon_z \neq 0$.

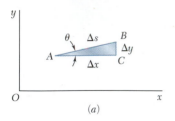

(a)

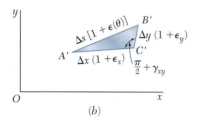

(b)

Fig. 7.51 Evaluating strain along line *AB*. (*a*) Undeformed; (*b*) deformed.

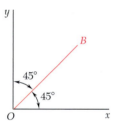

Fig. 7.52 Bisector *OB*.

The normal strain $\epsilon(\theta)$ along a line *AB* forms an arbitrary angle θ with the *x* axis. This strain is determined by using the right triangle *ABC*, which has *AB* for hypothenuse (Fig. 7.51*a*), and the oblique triangle *A'B'C'* into which triangle *ABC* is deformed (Fig. 7.51*b*). With the length of *AB* denoted as Δs, the length of *A'B'* is $\Delta s\,[1 + \epsilon(\theta)]$. Similarly, using Δx and Δy as the lengths of sides *AC* and *CB*, the lengths of *A'C'* and *C'B'* are $\Delta x\,(1 + \epsilon_x)$ and $\Delta y\,(1 + \epsilon_y)$, respectively. Recall from Fig. 7.49 that the right angle at *C* in Fig. 7.51*a* deforms into an angle equal to $\frac{\pi}{2} + \gamma_{xy}$ in Fig. 7.51*b*, and apply the law of cosines to triangle *A'B'C'* to obtain

$$(A'B')^2 = (A'C')^2 + (C'B')^2 - 2(A'C')(C'B') \cos\left(\frac{\pi}{2} + \gamma_{xy}\right)$$

$$(\Delta s)^2[1 + \epsilon(\theta)]^2 = (\Delta x)^2(1 + \epsilon_x)^2 + (\Delta y)^2(1 + \epsilon_y)^2$$

$$-2(\Delta x)(1 + \epsilon_x)(\Delta y)(1 + \epsilon_y) \cos\left(\frac{\pi}{2} + \gamma_{xy}\right) \qquad \textbf{(7.38)}$$

But from Fig. 7.51*a*,

$$\Delta x = (\Delta s) \cos\theta \qquad \Delta y = (\Delta s) \sin\theta \qquad \textbf{(7.39)}$$

and since γ_{xy} is very small,

$$\cos\left(\frac{\pi}{2} + \gamma_{xy}\right) = -\sin\gamma_{xy} \approx -\gamma_{xy} \qquad \textbf{(7.40)}$$

Substituting from Eqs. (7.39) and (7.40) into Eq. (7.38), recalling that $\cos^2\theta + \sin^2\theta = 1$, and neglecting second-order terms in $\epsilon(\theta)$, ϵ_x, ϵ_y, and γ_{xy} gives

$$\epsilon(\theta) = \epsilon_x \cos^2\theta + \epsilon_y \sin^2\theta + \gamma_{xy} \sin\theta \cos\theta \qquad \textbf{(7.41)}$$

Equation (7.41) enables us to determine the normal strain $\epsilon(\theta)$ in any direction *AB* in terms of the strain components ϵ_x, ϵ_y, γ_{xy}, and the angle θ that *AB* forms with the *x* axis. We check that for $\theta = 0$, Eq. (7.41) yields $\epsilon(0) = \epsilon_x$ and for $\theta = 90°$, it yields $\epsilon(90°) = \epsilon_y$. On the other hand, making $\theta = 45°$ in Eq. (7.41), we obtain the normal strain in the direction of the bisector *OB* of the angle formed by the *x* and *y* axes (Fig. 7.52). Denoting this strain by ϵ_{OB}, we write

$$\epsilon_{OB} = \epsilon(45°) = \tfrac{1}{2}(\epsilon_x + \epsilon_y + \gamma_{xy}) \qquad \textbf{(7.42)}$$

Solving Eq. (7.42) for γ_{xy},

$$\boxed{\gamma_{xy} = 2\epsilon_{OB} - (\epsilon_x + \epsilon_y)} \qquad \textbf{(7.43)}$$

This relationship makes it possible to express the *shearing strain* associated with a given pair of rectangular axes in terms of the *normal strains* measured along these axes and their bisector. It plays a fundamental role in the present derivation and will also be used in Sec. 7.9 for the experimental determination of shearing strains.

The main purpose of this section is to express the strain components associated with the frame of reference *x'y'* of Fig. 7.50 in terms of the angle θ and the strain components ϵ_x, ϵ_y, and γ_{xy} associated with the *x* and *y* axes. Thus, we note that the normal strain $\epsilon_{x'}$ along the *x'* axis is given by

Eq. (7.41). Using the trigonometric relationships in Eqs. (7.3) and (7.4), the alternative form of Eq. (7.41) is

$$\epsilon_{x'} = \frac{\epsilon_x + \epsilon_y}{2} + \frac{\epsilon_x - \epsilon_y}{2} \cos 2\theta + \frac{\gamma_{xy}}{2} \sin 2\theta \qquad \textbf{(7.44)}$$

The normal strain along the y' axis is obtained by replacing θ with $\theta + 90°$. Since $\cos(2\theta + 180°) = -\cos 2\theta$ and $\sin(2\theta + 180°) = -\sin 2\theta$,

$$\epsilon_{y'} = \frac{\epsilon_x + \epsilon_y}{2} - \frac{\epsilon_x - \epsilon_y}{2} \cos 2\theta - \frac{\gamma_{xy}}{2} \sin 2\theta \qquad \textbf{(7.45)}$$

Adding Eqs. (7.44) and (7.45) member to member gives

$$\epsilon_{x'} + \epsilon_{y'} = \epsilon_x + \epsilon_y \qquad \textbf{(7.46)}$$

Since $\epsilon_z = \epsilon_{z'} = 0$, the sum of the normal strains associated with a cubic element of material is independent of the orientation of that element in plane strain.[†]

Replacing θ by $\theta + 45°$ in Eq. (7.44), an expression is obtained for the normal strain along the bisector OB' of the angle formed by the x' and y' axes. Since $\cos(2\theta + 90°) = -\sin 2\theta$ and $\sin(2\theta + 90°) = \cos 2\theta$,

$$\epsilon_{OB'} = \frac{\epsilon_x + \epsilon_y}{2} - \frac{\epsilon_x - \epsilon_y}{2} \sin 2\theta + \frac{\gamma_{xy}}{2} \cos 2\theta \qquad \textbf{(7.47)}$$

Writing Eq. (7.43) with respect to the x' and y' axes, the shearing strain $\gamma_{x'y'}$ is expressed in terms of the normal strains measured along the x' and y' axes and the bisector OB':

$$\gamma_{x'y'} = 2\epsilon_{OB'} - (\epsilon_{x'} + \epsilon_{y'}) \qquad \textbf{(7.48)}$$

Substituting from Eqs. (7.46) and (7.47) into Eq. (7.48) gives

$$\gamma_{x'y'} = -(\epsilon_x - \epsilon_y) \sin 2\theta + \gamma_{xy} \cos 2\theta \qquad \textbf{(7.49a)}$$

Equations (7.44), (7.45), and (7.49a) are the desired equations defining the transformation of plane strain under a rotation of axes in the plane of strain. Dividing all terms in Eq. (7.49a) by 2, the alternative form is

$$\frac{\gamma_{x'y'}}{2} = -\frac{\epsilon_x - \epsilon_y}{2} \sin 2\theta + \frac{\gamma_{xy}}{2} \cos 2\theta \qquad \textbf{(7.49b)}$$

Observe that Eqs. (7.44), (7.45), and (7.49b) for the transformation of plane strain closely resemble those for the transformation of plane stress (Sec 7.1). While the former can be obtained from the latter by replacing the normal stresses by the corresponding normal strains, it should be noted that the shearing stresses τ_{xy} and $\tau_{x'y'}$ should be replaced by *half* of the corresponding shearing strains (i.e., by $\frac{1}{2}\gamma_{xy}$ and $\frac{1}{2}\gamma_{x'y'}$).

[†]Cf. first footnote on page 98.

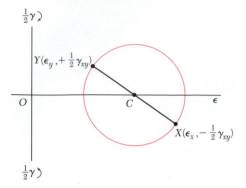

Fig. 7.53 Mohr's circle for plane strain.

7.7B Mohr's Circle for Plane Strain

Since the equations for the transformation of plane strain are of the same form as those for plane stress, Mohr's circle can be used for analysis of plane strain. Given the strain components ϵ_x, ϵ_y, and γ_{xy} defining the deformation in Fig. 7.49, point $X(\epsilon_x, -\frac{1}{2}\gamma_{xy})$ of abscissa equal to the normal strain ϵ_x and of ordinate equal to minus half the shearing strain γ_{xy}, and point $Y(\epsilon_y, +\frac{1}{2}\gamma_{xy})$ are plotted (Fig. 7.53). Drawing the diameter XY, the center C of Mohr's circle for plane strain is defined. The abscissa of C and the radius R of the circle are

$$\epsilon_{ave} = \frac{\epsilon_x + \epsilon_y}{2} \quad \text{and} \quad R = \sqrt{\left(\frac{\epsilon_x - \epsilon_y}{2}\right)^2 + \left(\frac{\gamma_{xy}}{2}\right)^2} \tag{7.50}$$

If γ_{xy} is positive, as assumed in Fig. 7.49, points X and Y are plotted below and above the horizontal axis in Fig. 7.53. But in the absence of any overall rigid-body rotation, the side of the element in Fig. 7.49 that is associated with ϵ_x rotates counterclockwise, while the side associated with ϵ_y rotates clockwise. Thus, if the shear deformation causes a given side to rotate *clockwise*, the corresponding point on Mohr's circle for plane strain is plotted *above* the horizontal axis, and if the deformation causes the side to rotate *counterclockwise*, the corresponding point is plotted *below* the horizontal axis. This convention matches the convention used to draw Mohr's circle for plane stress.

Points A and B where Mohr's circle intersects the horizontal axis correspond to the *principal strains* ϵ_{max} and ϵ_{min} (Fig. 7.54a). Thus,

$$\epsilon_{max} = \epsilon_{ave} + R \quad \text{and} \quad \epsilon_{min} = \epsilon_{ave} - R \tag{7.51}$$

where ϵ_{ave} and R are defined by Eqs. (7.50). The corresponding value θ_p of angle θ is obtained by observing that the shearing strain is zero for A and B. Setting $\gamma_{x'y'} = 0$ in Eq. (7.49a),

$$\tan 2\theta_p = \frac{\gamma_{xy}}{\epsilon_x - \epsilon_y} \tag{7.52}$$

The corresponding axes a and b in Fig. 7.54b are the *principal axes of strain*. Angle θ_p, which defines the direction of the principal axis Oa in Fig. 7.54b corresponding to point A in Fig. 7.54a, is equal to half of the

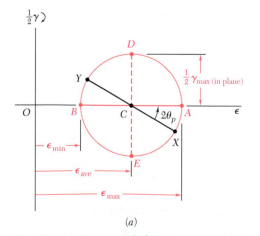

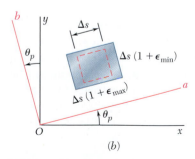

(a) $\qquad\qquad$ (b)

Fig. 7.54 (a) Mohr's circle for plane strain, showing principal strains and maximum in-plane shearing strain. (b) Strain element oriented to principal directions.

angle *XCA* measured on Mohr's circle, and the rotation that brings *Ox* into *Oa* has the same sense as the rotation that brings the diameter *XY* of Mohr's circle into the diameter *AB*.

Recall from Sec. 2.7 that in the elastic deformation of a homogeneous, isotropic material, Hooke's law for shearing stress and strain applies and yields $\tau_{xy} = G\gamma_{xy}$ for any pair of rectangular x and y axes. Thus, $\gamma_{xy} = 0$ when $\tau_{xy} = 0$, which indicates that the principal axes of strain coincide with the principal axes of stress.

The maximum in-plane shearing strain is defined by points *D* and *E* in Fig. 7.54a. This is equal to the diameter of Mohr's circle. From the second of Eqs. (7.50),

$$\gamma_{\max\,(\text{in plane})} = 2R = \sqrt{(\epsilon_x - \epsilon_y)^2 + \gamma_{xy}^2} \qquad (7.53)$$

Finally, points X' and Y', which define the components of strain corresponding to a rotation of the coordinate axes through an angle θ (Fig. 7.50), are obtained by rotating the diameter *XY* of Mohr's circle in the same sense through an angle 2θ (Fig. 7.55).

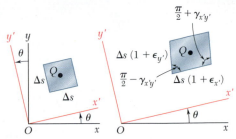

Fig. 7.50 *(repeated)* Transformation of plane strain element in undeformed and deformed orientations.

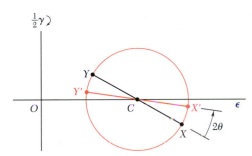

Fig. 7.55 Strains on arbitrary planes X' and Y' referenced to original planes X and Y on Mohr's circle.

Concept Application 7.4

For a material in a state of plane strain, it is found that the horizontal side of a 10 × 10-mm square elongates by 4 μm, its vertical side remains unchanged, and the angle at the lower-left corner increases by 0.4×10^{-3} rad (Fig. 7.56a). Determine (a) the principal axes and principal strains and (b) the maximum shearing strain and the corresponding normal strain.

a. Principal Axes and Principal Strains. Determine the coordinates of points X and Y on Mohr's circle for strain.

$$\epsilon_x = \frac{+4 \times 10^{-6}\,\text{m}}{10 \times 10^{3}\,\text{m}} = +400\,\mu \qquad \epsilon_y = 0 \qquad \left|\frac{\gamma_{xy}}{2}\right| = 200\,\mu$$

Since the side of the square associated with ϵ_x rotates *clockwise*, point X of coordinates ϵ_x and $|\gamma_{xy}/2|$ is plotted *above* the horizontal axis. Since $\epsilon_y = 0$ and the corresponding side rotates *counterclockwise*, point Y is

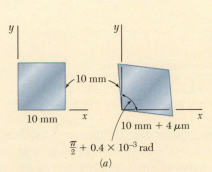

Fig. 7.56 Analysis of plane strain state. (a) Strain element: undeformed and deformed.

(continued)

plotted directly *below* the origin (Fig. 7.56*b*). Drawing the diameter *XY*, determine the center *C* of Mohr's circle and its radius *R*.

$$OC = \frac{\epsilon_x + \epsilon_y}{2} = 200\,\mu \qquad OY = 200\,\mu$$

$$R = \sqrt{(OC)^2 + (OY)^2} = \sqrt{(200\,\mu)^2 + (200\,\mu)^2} = 283\,\mu$$

The principal strains are defined by the abscissas of points *A* and *B*.

$$\epsilon_a = OA = OC + R = 200\,\mu + 283\,\mu = 483\,\mu$$
$$\epsilon_b = OB = OC - R = 200\,\mu - 283\,\mu = -83\,\mu$$

The principal axes *Oa* and *Ob* are shown in Fig. 7.56*c*. Since *OC = OY*, the angle at *C* in triangle *OCY* is 45°. Thus, the angle $2\theta_p$ that brings *XY* into *AB* is 45°↓ and angle θ_p bringing *Ox* into *Oa* is 22.5°↓.

b. Maximum Shearing Strain. Points *D* and *E* define the maximum in-plane shearing strain which, since the principal strains have opposite signs, is also the actual maximum shearing strain (see Sec. 7.8).

$$\frac{\gamma_{max}}{2} = R = 283\,\mu \qquad \gamma_{max} = 566\,\mu$$

The corresponding normal strains are both equal to

$$\epsilon' = OC = 200\,\mu$$

The axes of maximum shearing strain are shown in Fig. 7.56*d*.

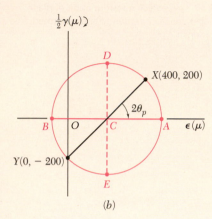

(*b*)

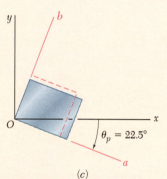

(*c*)

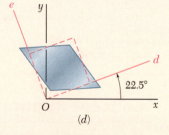

(*d*)

Fig. 7.56 (*cont.*) (*b*) Mohr's circle for given plane strain element. (*c*) Undeformed and deformed principal strain elements. (*d*) Undeformed and deformed maximum shearing strain elements.

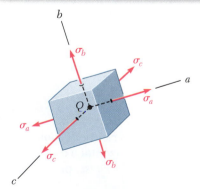

Fig. 7.21 (*repeated*) General stress element oriented to principal axes.

*7.8 THREE-DIMENSIONAL ANALYSIS OF STRAIN

We saw in Sec. 7.3 that, in the most general case of stress, we can determine three coordinate axes *a*, *b*, and *c*, called the principal axes of stress. A small cubic element with faces perpendicular to these axes is free of shearing stresses (Fig. 7.21), as $\tau_{ab} = \tau_{bc} = \tau_{ca} = 0$. Hooke's law for shearing stress and strain applies when the deformation is elastic and the material homogeneous and isotropic. Thus, $\gamma_{ab} = \gamma_{bc} = \gamma_{ca} = 0$, so the axes *a*, *b*, and *c* are also *principal axes of strain*. A small cube with sides equal to unity, centered at *Q*, and with faces perpendicular to the principal axes is deformed into a rectangular parallelepiped with sides $1 + \epsilon_a$, $1 + \epsilon_b$, and $1 + \epsilon_c$ (Fig. 7.57).

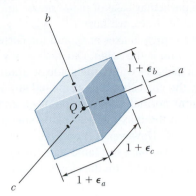

Fig. 7.57 Strain element oriented to directions of principal axes.

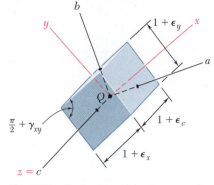

Fig. 7.58 Strain element having one axis coincident with a principal strain axis.

If the element of Fig. 7.57 is rotated about one of the principal axes at Q, say the c axis (Fig. 7.58), the method of analysis for the transformation of plane strain also can be used to determine the strain components ϵ_x, ϵ_y, and γ_{xy} associated with the faces perpendicular to the c axis, since this method did not involve any of the other strain components.[†] Therefore, Mohr's circle is drawn through the points A and B corresponding to the principal axes a and b (Fig. 7.59). Similarly, circles of diameters BC and CA are used to analyze the transformation of strain as the element is rotated about the a and b axes, respectively.

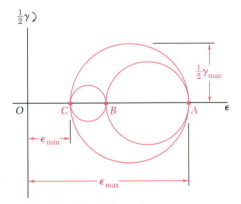

Fig. 7.59 Mohr's circle for three-dimensional analysis of strain.

The three-dimensional analysis of strain using Mohr's circle is limited here to rotations about principal axes (as for the analysis of stress) and is used to determine the maximum shearing strain γ_{max} at point Q. Since γ_{max} is equal to the diameter of the largest of the three circles shown in Fig. 7.59,

$$\gamma_{max} = |\epsilon_{max} - \epsilon_{min}| \tag{7.54}$$

[†]The other four faces of the element remain rectangular, and the edges parallel to the c axis remain unchanged.

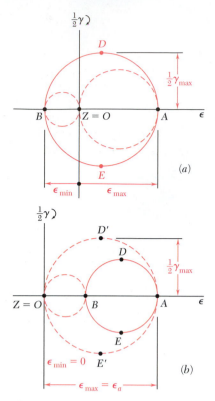

Fig. 7.60 Possible configurations of Mohr's circle for plane strain. (*a*) Principal strains having mixed signs. (*b*) Principal strains having positive signs.

where ϵ_{max} and ϵ_{min} represent the *algebraic* values of the maximum and minimum strains at point Q.

Returning to the particular case of *plane strain*, and selecting the x and y axes in the plane of strain, we have $\epsilon_z = \gamma_{zx} = \gamma_{zy} = 0$. Thus, the z axis is one of the three principal axes at Q, and the corresponding point in the Mohr's circle diagram is the origin O, where $\epsilon = \gamma = 0$. If points A and B defining the principal axes within the plane of strain fall on opposite sides of O (Fig. 7.60*a*), the corresponding principal strains represent the maximum and minimum normal strains at point Q, and the maximum shearing strain is equal to the maximum in-plane shearing strain corresponding to points D and E. However, if A and B are on the same side of O (Fig. 7.60*b*), so that ϵ_a and ϵ_b have the same sign, the maximum shearing strain is defined by points D' and E' on the circle of diameter OA, and $\gamma_{max} = \epsilon_{max}$.

Now consider the particular case of *plane stress* encountered in a thin plate or on the free surface of a structural element or machine component. Selecting the x and y axes in the plane of stress, $\sigma_z = \tau_{zx} = \tau_{zy} = 0$, and the z axis is a principal axis of stress. If the deformation is elastic and the material is homogeneous and isotropic, Hooke's law shows that $\gamma_{zx} = \gamma_{zy} = 0$. Thus, the z axis is also a principal axis of strain, and Mohr's circle can be used to analyze the transformation of strain in the xy plane. However, as we shall see presently, Hooke's law does *not* show that $\epsilon_z = 0$; indeed, a state of plane stress does not, in general, result in a state of plane strain.

Using a and b as the principal axes within the plane of stress and c as the principal axis perpendicular to that plane, we let $\sigma_x = \sigma_a$, $\sigma_y = \sigma_b$, and $\sigma_z = 0$ in Eqs. (2.20) for the generalized Hooke's law (Sec. 2.5), and obtain

$$\epsilon_a = \frac{\sigma_a}{E} - \frac{\nu \sigma_b}{E} \qquad (7.55)$$

$$\epsilon_b = -\frac{\nu \sigma_a}{E} + \frac{\sigma_b}{E} \qquad (7.56)$$

$$\epsilon_c = -\frac{\nu}{E}(\sigma_a + \sigma_b) \qquad (7.57)$$

Adding Eqs. (7.55) and (7.56) member to member gives

$$\epsilon_a + \epsilon_b = \frac{1 - \nu}{E}(\sigma_a + \sigma_b) \qquad (7.58)$$

Solving Eq. (7.58) for $\sigma_a + \sigma_b$ and substituting into Eq. (7.57), we write

$$\epsilon_c = -\frac{\nu}{1 - \nu}(\epsilon_a + \epsilon_b) \qquad (7.59)$$

The relationship obtained defines the third principal strain in terms of the in-plane principal strains. If B is located between A and C on the Mohr's circle diagram (Fig. 7.61), the maximum shearing strain is equal to the diameter CA of the circle corresponding to a rotation about the b axis, out of the plane of stress.

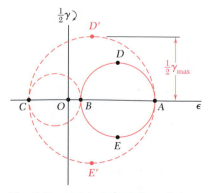

Fig. 7.61 Mohr's circle strain analysis for plane stress.

Concept Application 7.5

As a result of measurements made on the surface of a machine component with strain gages oriented in various ways, it has been established that the principal strains on the free surface are $\epsilon_a = +400 \times 10^{-6}$ in./in. and $\epsilon_b = -50 \times 10^{-6}$ in./in. Knowing that Poisson's ratio for the given material is $\nu = 0.30$, determine (a) the maximum in-plane shearing strain, (b) the true value of the maximum shearing strain near the surface of the component.

a. Maximum In-Plane Shearing Strain. Draw Mohr's circle through points A and B corresponding to the given principal strains (Fig. 7.62a). The maximum in-plane shearing strain is defined by points D and E and is equal to the diameter of Mohr's circle:

$$\gamma_{max\,(in\,plane)} = 400 \times 10^{-6} + 50 \times 10^{-6} = 450 \times 10^{-6}\ \text{rad}$$

b. Maximum Shearing Strain. Determine the third principal strain ϵ_c. Since a state of plane *stress* is on the surface of the machine component, Eq. (7.59) gives

$$\epsilon_c = -\frac{\nu}{1-\nu}(\epsilon_a + \epsilon_b)$$

$$= -\frac{0.30}{0.70}(400 \times 10^{-6} - 50 \times 10^{-6}) = -150 \times 10^{-6}\ \text{in./in.}$$

Draw Mohr's circles through A and C and through B and C (Fig. 7.62b), and find that the maximum shearing strain is equal to the diameter of the circle CA:

$$\gamma_{max} = 400 \times 10^{-6} + 150 \times 10^{-6} = 550 \times 10^{-6}\ \text{rad}$$

Note that even though ϵ_a and ϵ_b have opposite signs, the maximum in-plane shearing strain does not represent the true maximum shearing strain.

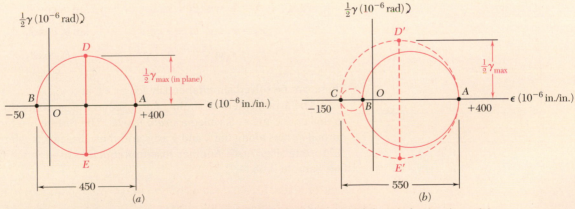

Fig. 7.62 Using Mohr's circle to determine maximum shearing strain. (a) Mohr's circle for the plane of the given strains. (b) Three-dimensional Mohr's circle for strain.

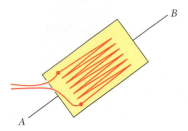

Fig. 7.63 Electrical strain gage.

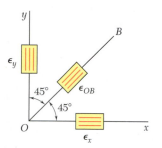

Fig. 7.64 Strain rosette that measures normal strains in direction of x, y, and bisector OB.

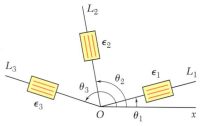

Fig. 7.65 Generalized strain gage rosette arrangement.

*7.9   MEASUREMENTS OF STRAIN; STRAIN ROSETTE

The normal strain can be determined in any given direction on the surface of a structural element or machine component by scribing two gage marks A and B across a line drawn in the desired direction and measuring the length of the segment AB before and after the load has been applied. If L is the undeformed length of AB and δ its deformation, the normal strain along AB is $\epsilon_{AB} = \delta/L$.

A more convenient and accurate method for measuring normal strains is provided by electrical strain gages. A typical electrical strain gage consists of a length of thin wire arranged as shown in Fig. 7.63 and cemented to two pieces of paper. In order to measure the strain ϵ_{AB} of a given material in the direction AB, the gage is cemented to the surface of the material with the wire folds running parallel to AB. As the material elongates, the wire increases in length and decreases in diameter, causing the electrical resistance of the gage to increase. By measuring the current passing through a properly calibrated gage, the strain ϵ_{AB} can be determined accurately and continuously as the load is increased.

The strain components ϵ_x and ϵ_y can be determined at a given point of the free surface of a material by simply measuring the normal strain along the x and y axes drawn through that point. Recalling Eq. (7.43), we note that a third measurement of normal strain, made along the bisector OB of the angle formed by the x and y axes, enables us to determine the shearing strain γ_{xy} as well (Fig. 7.64):

$$\gamma_{xy} = 2\epsilon_{OB} - (\epsilon_x + \epsilon_y) \tag{7.43}$$

The strain components ϵ_x, ϵ_y, and γ_{xy} at a given point also can be obtained from normal strain measurements made along *any three lines* drawn through that point (Fig. 7.65). Denoting respectively by θ_1, θ_2, and θ_3 the angle each of the three lines forms with the x axis, by ϵ_1, ϵ_2, and ϵ_3 the corresponding strain measurements, and substituting into Eq. (7.41), we write the three equations

$$\epsilon_1 = \epsilon_x \cos^2 \theta_1 + \epsilon_y \sin^2 \theta_1 + \gamma_{xy} \sin \theta_1 \cos \theta_1$$
$$\epsilon_2 = \epsilon_x \cos^2 \theta_2 + \epsilon_y \sin^2 \theta_2 + \gamma_{xy} \sin \theta_2 \cos \theta_2 \tag{7.60}$$
$$\epsilon_3 = \epsilon_x \cos^2 \theta_3 + \epsilon_y \sin^2 \theta_3 + \gamma_{xy} \sin \theta_3 \cos \theta_3$$

These can be solved simultaneously for ϵ_x, ϵ_y, and γ_{xy}.[†]

The arrangement of strain gages used to measure the three normal strains ϵ_1, ϵ_2, and ϵ_3 is called a *strain rosette*. The rosette used to measure normal strains along the x and y axes and their bisector is referred to as a 45° rosette (Fig. 7.64). Another rosette frequently used is the 60° rosette (see Sample Prob. 7.7).

[†]It should be noted that the free surface on which the strain measurements are made is in a state of *plane stress*, while Eqs. (7.41) and (7.43) were derived for a state of *plane strain*. However, as observed earlier the normal to the free surface is a principal axis of strain, and the derivations given in Sec. 7.7A remain valid.

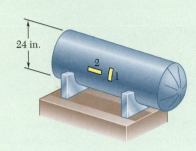

24 in.

Sample Problem 7.6

A cylindrical storage tank used to transport gas under pressure has an inner diameter of 24 in. and a wall thickness of $\frac{3}{4}$ in. Strain gages attached to the surface of the tank in transverse and longitudinal directions indicate strains of 255×10^{-6} and 60×10^{-6} in./in., respectively. Knowing that a torsion test has shown that the modulus of rigidity of the material used in the tank is $G = 11.2 \times 10^{6}$ psi, determine (*a*) the gage pressure inside the tank, (*b*) the principal stresses and the maximum shearing stress in the wall of the tank.

STRATEGY: You can use the given measured strains to plot Mohr's circle for strain, and use this circle to determine the maximum in-plane shearing strain. Applying Hooke's law to obtain the corresponding maximum in-plane shearing stress, you can then determine the gage pressure in the tank through the appropriate thin-walled pressure vessel equation, as well as develop Mohr's circle for stress to determine the principal stresses and the maximum shearing stress.

MODELING and ANALYSIS:

a. Gage Pressure Inside Tank. The given strains are the principal strains at the surface of the tank. Plotting the corresponding points *A* and *B*, draw Mohr's circle for strain (Fig. 1). The maximum in-plane shearing strain is equal to the diameter of the circle.

$$\gamma_{\text{max (in plane)}} = \epsilon_1 - \epsilon_2 = 255 \times 10^{-6} - 60 \times 10^{-6} = 195 \times 10^{-6}\ \text{rad}$$

From Hooke's law for shearing stress and strain,

$$\begin{aligned}
\tau_{\text{max (in plane)}} &= G\gamma_{\text{max (in plane)}} \\
&= (11.2 \times 10^{6}\ \text{psi})(195 \times 10^{-6}\ \text{rad}) \\
&= 2184\ \text{psi} = 2.184\ \text{ksi}
\end{aligned}$$

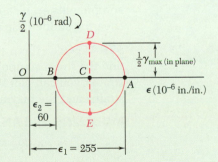

Fig. 1 Mohr's circle for measured strains.

(continued)

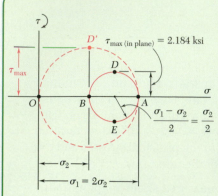

Fig. 2 Three-dimensional Mohr's circles for vessel stress components.

Substituting this and the given data in Eq. (7.33),

$$\tau_{\max(\text{in plane})} = \frac{pr}{4t} \qquad 2184 \text{ psi} = \frac{p(12 \text{ in.})}{4(0.75 \text{ in.})}$$

Solving for the gage pressure p,

$$p = 546 \text{ psi} \blacktriangleleft$$

b. Principal Stresses and Maximum Shearing Stress. Recalling that for a thin-walled cylindrical pressure vessel $\sigma_1 = 2\sigma_2$, we draw Mohr's circle for stress (Fig. 2) and obtain

$$\sigma_2 = 2\tau_{\max(\text{in plane})} = 2(2.184 \text{ ksi}) = 4.368 \text{ ksi} \qquad \sigma_2 = 4.37 \text{ ksi} \blacktriangleleft$$
$$\sigma_1 = 2\sigma_2 = 2(4.368 \text{ ksi}) \qquad\qquad\qquad \sigma_1 = 8.74 \text{ ksi} \blacktriangleleft$$

The maximum shearing stress is equal to the radius of the circle of diameter OA and corresponds to a rotation of 45° about a longitudinal axis.

$$\tau_{\max} = \tfrac{1}{2}\sigma_1 = \sigma_2 = 4.368 \text{ ksi} \qquad \tau_{\max} = 4.37 \text{ ksi} \blacktriangleleft$$

Sample Problem 7.7

Using a 60° rosette, the following strains have been measured at point Q on the surface of a steel machine base:

$$\epsilon_1 = 40\,\mu \qquad \epsilon_2 = 980\,\mu \qquad \epsilon_3 = 330\,\mu$$

Using the coordinate axes shown, determine at point Q (a) the strain components ϵ_x, ϵ_y, and γ_{xy}, (b) the principal strains, (c) the maximum shearing strain. (Use $\nu = 0.29$.)

STRATEGY: From the given strain rosette measurements, you can find the strain components ϵ_x, ϵ_y, and γ_{xy} using Eq. (7.60). Using these strains, you can plot Mohr's circle for strain to determine the principal strains and the maximum shearing strain.

MODELING and ANALYSIS:

a. Strain Components ϵ_x, ϵ_y, γ_{xy}. For the coordinate axes shown

$$\theta_1 = 0 \qquad \theta_2 = 60° \qquad \theta_3 = 120°$$

(continued)

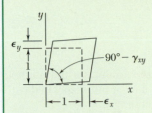

Fig. 1 Undeformed and deformed strain elements at Q.

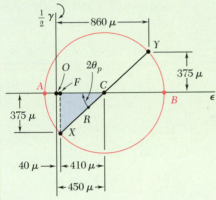

Fig. 2. Mohr's circle used to determine principal strains.

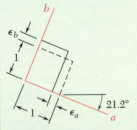

Fig. 3. Undeformed and deformed principal strain element at Q.

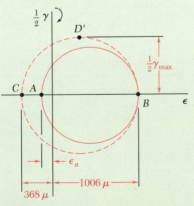

Fig. 4 Three-dimensional Mohr's circles used to determine maximum shearing strain.

Substituting these into Eqs. (7.60), gives

$$\epsilon_1 = \epsilon_x(1) \qquad + \epsilon_y(0) \qquad + \gamma_{xy}(0)(1)$$

$$\epsilon_2 = \epsilon_x(0.500)^2 + \epsilon_y(0.866)^2 + \gamma_{xy}(0.866)(0.500)$$

$$\epsilon_3 = \epsilon_x(-0.500)^2 + \epsilon_y(0.866)^2 + \gamma_{xy}(0.866)(-0.500)$$

Solving these equations for ϵ_x, ϵ_y, and γ_{xy},

$$\epsilon_x = \epsilon_1 \qquad \epsilon_y = \tfrac{1}{3}(2\epsilon_2 + 2\epsilon_3 - \epsilon_1) \qquad \gamma_{xy} = \frac{\epsilon_2 - \epsilon_3}{0.866}$$

Substituting for ϵ_1, ϵ_2, and ϵ_3,

$$\epsilon_x = 40\,\mu \qquad \epsilon_y = \tfrac{1}{3}[2(980) + 2(330) - 40] \qquad \epsilon_y = +860\,\mu \;\blacktriangleleft$$

$$\gamma_{xy} = (980 - 330)/0.866 \qquad \gamma_{xy} = 750\,\mu \;\blacktriangleleft$$

These strains are indicated on the element shown in Fig. 1.

b. Principal Strains. The side of the element associated with ϵ_x rotates counterclockwise; thus, point X is plotted below the horizontal axis, as $X(40, -375)$. Then $Y(860, +375)$ is plotted and Mohr's circle is drawn (Fig. 2).

$$\epsilon_{\text{ave}} = \tfrac{1}{2}(860\,\mu + 40\,\mu) = 450\,\mu$$

$$R = \sqrt{(375\,\mu)^2 + (410\,\mu)^2} = 556\,\mu$$

$$\tan 2\theta_p = \frac{375\,\mu}{410\,\mu} \qquad 2\theta_p = 42.4° \qquad \theta_p = 21.2°$$

Points A and B correspond to the principal strains,

$$\epsilon_a = \epsilon_{\text{ave}} - R = 450\,\mu - 556\,\mu \qquad \epsilon_a = -106\,\mu \;\blacktriangleleft$$

$$\epsilon_b = \epsilon_{\text{ave}} + R = 450\,\mu + 556\,\mu \qquad \epsilon_b = +1006\,\mu \;\blacktriangleleft$$

These strains are indicated on the element shown in Fig. 3. Since $\sigma_z = 0$ on the surface, Eq. (7.59) is used to find the principal strain ϵ_c:

$$\epsilon_c = -\frac{\nu}{1-\nu}(\epsilon_a + \epsilon_b) = -\frac{0.29}{1 - 0.29}(-106\,\mu + 1006\,\mu) \quad \epsilon_c = -368\,\mu \;\blacktriangleleft$$

c. Maximum Shearing Strain. Plotting point C and drawing Mohr's circle through points B and C (Fig. 4), we obtain point D' and write

$$\tfrac{1}{2}\gamma_{\text{max}} = \tfrac{1}{2}(1006\,\mu + 368\,\mu) \qquad \gamma_{\text{max}} = 1374\,\mu \;\blacktriangleleft$$

Problems

7.128 through 7.131 For the given state of plane strain, use the method of Sec. 7.7A to determine the state of plane strain associated with axes x' and y' rotated through the given angle θ.

	ϵ_x	ϵ_y	γ_{xy}	θ
7.128 and 7.132	-800μ	$+450\mu$	$+200\mu$	25°↙
7.129 and 7.133	$+240\mu$	$+160\mu$	$+150\mu$	60°↙
7.130 and 7.134	-500μ	$+250\mu$	0	15°↖
7.131 and 7.135	0	$+320\mu$	-100μ	30°↖

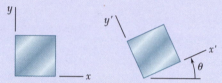

Fig. P7.128 through P7.135

7.132 through 7.135 For the given state of plane strain, use Mohr's circle to determine the state of plane strain associated with axes x' and y' rotated through the given angle θ.

7.136 through 7.139 The following state of strain has been measured on the surface of a thin plate. Knowing that the surface of the plate is unstressed, determine (*a*) the direction and magnitude of the principal strains, (*b*) the maximum in-plane shearing strain, (*c*) the maximum shearing strain. (Use $\nu = \frac{1}{3}$)

	ϵ_x	ϵ_y	γ_{xy}
7.136	-260μ	-60μ	$+480\mu$
7.137	-600μ	-400μ	$+350\mu$
7.138	$+160\mu$	-480μ	-600μ
7.139	$+30\mu$	$+570\mu$	$+720\mu$

7.140 through 7.143 For the given state of plane strain, use Mohr's circle to determine (*a*) the orientation and magnitude of the principal strains, (*b*) the maximum in-plane strain, (*c*) the maximum shearing strain.

	ϵ_x	ϵ_y	γ_{xy}
7.140	$+60\mu$	$+240\mu$	-50μ
7.141	$+400\mu$	$+200\mu$	$+375\mu$
7.142	$+300\mu$	$+60\mu$	$+100\mu$
7.143	-180μ	-260μ	$+315\mu$

7.144 Determine the strain ϵ_x, knowing that the following strains have been determined by use of the rosette shown:

$$\epsilon_1 = +480\mu \qquad \epsilon_2 = -120\mu \qquad \epsilon_3 = +80\mu$$

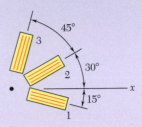

7.145 The strains determined by the use of the rosette shown during the test of a machine element are

$$\epsilon_1 = +600\mu \qquad \epsilon_2 = +450\mu \qquad \epsilon_3 = -75\mu$$

Fig. P7.144

Determine (*a*) the in-plane principal strains, (*b*) the in-plane maximum shearing strain.

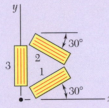

Fig. P7.145

7.146 The rosette shown has been used to determine the following strains at a point on the surface of a crane hook:

$$\epsilon_1 = +420 \times 10^{-6} \text{ in./in.} \qquad \epsilon_2 = -45 \times 10^{-6} \text{ in./in.}$$
$$\epsilon_4 = +165 \times 10^{-6} \text{ in./in.}$$

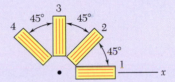

Fig. P7.146

(*a*) What should be the reading of gage 3? (*b*) Determine the principal strains and the maximum in-plane shearing strain.

7.147 Using a 45° rosette, the strains ϵ_1, ϵ_2, and ϵ_3 have been determined at a given point. Using Mohr's circle, show that the principal strains are:

$$\epsilon_{\text{max, min}} = \frac{1}{2}(\epsilon_1 + \epsilon_3) \pm \frac{1}{\sqrt{2}}[(\epsilon_1 - \epsilon_2)^2 + (\epsilon_2 - \epsilon_3)^2]^{\frac{1}{2}}$$

(*Hint:* The shaded triangles are congruent.)

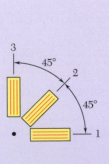

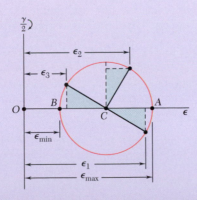

Fig. P7.147

7.148 Show that the sum of the three strain measurements made with a 60° rosette is independent of the orientation of the rosette and equal to

$$\epsilon_1 + \epsilon_2 + \epsilon_3 = 3\epsilon_{avg}$$

where ϵ_{avg} is the abscissa of the center of the corresponding Mohr's circle.

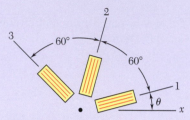

Fig. P7.148

7.149 The strains determined by the use of the rosette attached as shown during the test of a machine element are

$$\epsilon_1 = -93.1 \times 10^{-6} \text{ in./in.} \qquad \epsilon_2 = +385 \times 10^{-6} \text{ in./in.}$$
$$\epsilon_3 = +210 \times 10^{-6} \text{ in./in.}$$

Determine (*a*) the orientation and magnitude of the principal strains in the plane of the rosette, (*b*) the maximum in-plane shearing strain.

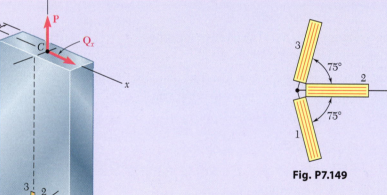

Fig. P7.149

7.150 A centric axial force **P** and a horizontal force **Q**$_x$ are both applied at point C of the rectangular bar shown. A 45° strain rosette on the surface of the bar at point A indicates the following strains:

$$\epsilon_1 = -60 \times 10^{-6} \text{ in./in.} \qquad \epsilon_2 = +240 \times 10^{-6} \text{ in./in.}$$
$$\epsilon_3 = +200 \times 10^{-6} \text{ in./in.}$$

Knowing that $E = 29 \times 10^6$ psi and $\nu = 0.30$, determine the magnitudes of **P** and **Q**$_x$.

7.151 Solve Prob. 7.150, assuming that the rosette at point A indicates the following strains:

$$\epsilon_1 = -30 \times 10^{-6} \text{ in./in.} \qquad \epsilon_2 = +250 \times 10^{-6} \text{ in./in.}$$
$$\epsilon_3 = +100 \times 10^{-6} \text{ in./in.}$$

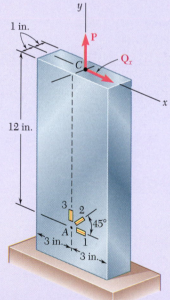

Fig. P7.150

7.152 A single strain gage is cemented to a solid 4-in.-diameter steel shaft at an angle $\beta = 25°$ with a line parallel to the axis of the shaft. Knowing that $G = 11.5 \times 10^6$ psi, determine the torque **T** indicated by a gage reading of 300×10^{-6} in./in.

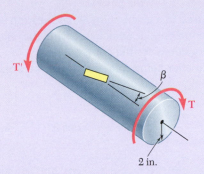

Fig. P7.152

7.153 Solve Prob. 7.152, assuming that the gage forms an angle $\beta = 35°$ with a line parallel to the axis of the shaft.

7.154 A single strain gage forming an angle $\beta = 18°$ with a horizontal plane is used to determine the gage pressure in the cylindrical steel tank shown. The cylindrical wall of the tank is 6 mm thick, has a 600-mm inside diameter, and is made of a steel with $E = 200$ GPa and $\nu = 0.30$. Determine the pressure in the tank indicated by a strain gage reading of 280μ.

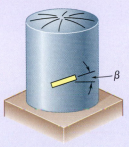

Fig. P7.154

7.155 Solve Prob. 7.154, assuming that the gage forms an angle $\beta = 35°$ with a horizontal plane.

7.156 The given state of plane stress is known to exist on the surface of a machine component. Knowing that $E = 200$ GPa and $G = 77.2$ GPa, determine the direction and magnitude of the three principal strains (*a*) by determining the corresponding state of strain [use Eq. (2.43) and Eq. (2.38)] and then using Mohr's circle for strain, (*b*) by using Mohr's circle for stress to determine the principal planes and principal stresses and then determining the corresponding strains.

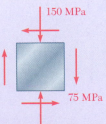

Fig. P7.156

7.157 The following state of strain has been determined on the surface of a cast-iron machine part:

$$\epsilon_x = -720\mu \qquad \epsilon_y = -400\mu \qquad \gamma_{xy} = +660\mu$$

Knowing that $E = 69$ GPa and $G = 28$ GPa, determine the principal planes and principal stresses (*a*) by determining the corresponding state of plane stress [use Eq. (2.36), Eq. (2.43), and the first two equations of Prob. 2.73] and then using Mohr's circle for stress, (*b*) by using Mohr's circle for strain to determine the orientation and magnitude of the principal strains and then determining the corresponding stresses.

Review and Summary

Transformation of Plane Stress

A state of *plane stress* at a given point Q has nonzero values for σ_x, σ_y, and τ_{xy}. The stress components associated with the element are shown in Fig. 7.66a. The equations for the components $\sigma_{x'}$, $\sigma_{y'}$, and $\tau_{x'y'}$ associated with that element after being rotated through an angle θ about the z axis (Fig. 7.66b) are

$$\sigma_{x'} = \frac{\sigma_x + \sigma_y}{2} + \frac{\sigma_x - \sigma_y}{2}\cos 2\theta + \tau_{xy}\sin 2\theta \tag{7.5}$$

$$\sigma_{y'} = \frac{\sigma_x + \sigma_y}{2} - \frac{\sigma_x - \sigma_y}{2}\cos 2\theta - \tau_{xy}\sin 2\theta \tag{7.7}$$

$$\tau_{x'y'} = -\frac{\sigma_x - \sigma_y}{2}\sin 2\theta + \tau_{xy}\cos 2\theta \tag{7.6}$$

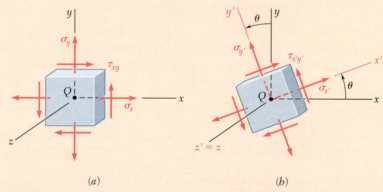

(a)   (b)

Fig. 7.66 State of plane stress. (a) Referred to {x y z}. (b) Referred to {x'y'z'}.

The values θ_p of the angle of rotation that correspond to the maximum and minimum values of the normal stress at point Q are

$$\tan 2\theta_p = \frac{2\tau_{xy}}{\sigma_x - \sigma_y} \tag{7.12}$$

Principal Planes and Principal Stresses

The two values obtained for θ_p are 90° apart (Fig. 7.67) and define the *principal planes of stress* at point Q. The corresponding values of the normal stress are called the *principal stresses* at Q:

$$\sigma_{\text{max, min}} = \frac{\sigma_x + \sigma_y}{2} \pm \sqrt{\left(\frac{\sigma_x - \sigma_y}{2}\right)^2 + \tau_{xy}^2} \tag{7.14}$$

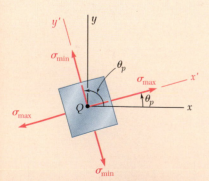

Fig. 7.67 Principal stresses.

The corresponding shearing stress is zero.

Maximum In-Plane Shearing Stress

The angle θ for the largest value of the shearing stress θ_s is found using

$$\tan 2\theta_s = -\frac{\sigma_x - \sigma_y}{2\tau_{xy}} \tag{7.15}$$

The two values obtained for θ_s are 90° apart (Fig. 7.68). However, the planes of maximum shearing stress are at 45° to the principal planes. The maximum value of the shearing stress *in the plane of stress is*

$$\tau_{max} = \sqrt{\left(\frac{\sigma_x - \sigma_y}{2}\right)^2 + \tau_{xy}^2} \tag{7.16}$$

and the corresponding value of the normal stresses is

$$\sigma' = \sigma_{ave} = \frac{\sigma_x + \sigma_y}{2} \tag{7.17}$$

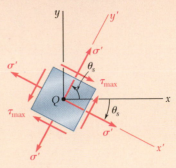

Fig. 7.68 Maximum shearing stress.

Mohr's Circle for Stress

Mohr's circle provides an alternative method for the analysis of the transformation of plane stress based on simple geometric considerations. Given the state of stress shown in the left element in Fig. 7.69a, point X of

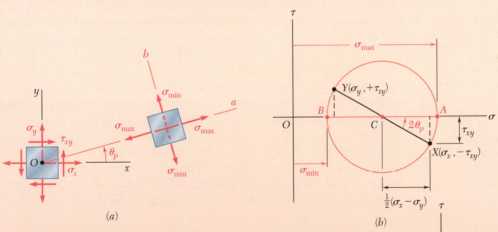

Fig. 7.69 (a) Plane stress element, and the orientation of principal planes. (b) Corresponding Mohr's circle.

coordinates σ_x, $-\tau_{xy}$ and point Y of coordinates σ_y, $+\tau_{xy}$ are plotted in Fig. 7.69b. Drawing the circle of diameter XY provides Mohr's circle. The abscissas of the points of intersection A and B of the circle with the horizontal axis represent the principal stresses, and the angle of rotation bringing the diameter XY into AB is twice the angle θ_p defining the principal planes, as shown in the right element of Fig. 7.69a. The diameter DE defines the maximum shearing stress and the orientation of the corresponding plane (Fig. 7.70).

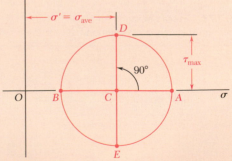

Fig. 7.70 Maximum shearing stress is oriented $\pm45°$ from principal directions.

General State of Stress

A *general state of stress* is characterized by six stress components, where the normal stress on a plane of arbitrary orientation can be expressed as a quadratic form of the direction cosines of the normal to that plane. This proves the existence of three *principal axes of stress* and three *principal stresses* at any given point. Rotating a small cubic element about each of the three principal axes was used to draw the corresponding Mohr's circles that yield the values of σ_{max}, σ_{min}, and τ_{max} (Fig. 7.71). In the case of

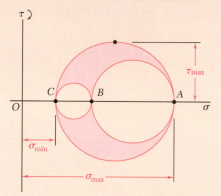

Fig. 7.71 Three-dimensional Mohr's circles for general state of stress.

plane stress when the x and y axes are selected in the plane of stress, point C coincides with the origin O. If A and B are located on opposite sides of O, the maximum shearing stress is equal to the maximum in-plane shearing stress. If A and B are located on the same side of O, this is not the case. For instance if $\sigma_a > \sigma_b > 0$, the maximum shearing stress is equal to $\frac{1}{2}\sigma_a$ and corresponds to a rotation out of the plane of stress (Fig. 7.72).

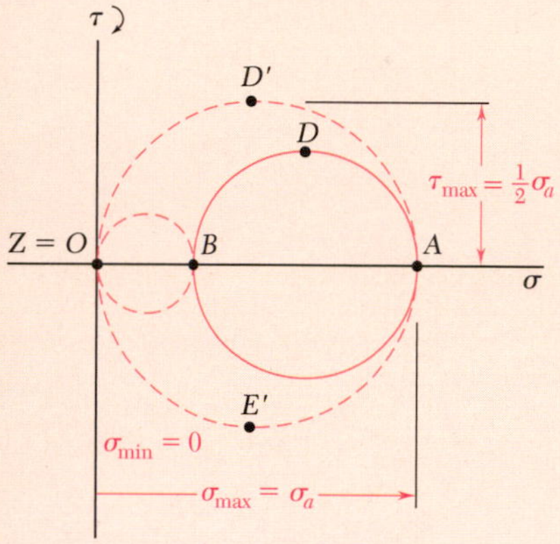

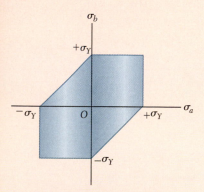

Fig. 7.72 Three-dimensional Mohr's circles for plane stress having two positive principal stresses.

Yield Criteria for Ductile Materials

To predict whether a structural or machine component will fail at some critical point due to yield in the material, the principal stresses σ_a and σ_b at that point for the given loading condition are determined. The point of coordinates σ_a and σ_b is plotted, and if this point falls within a certain area, the component is safe. If it falls outside, the component will fail. The area used with the maximum-shearing-stress criterion is shown in Fig. 7.73, and the area used with the maximum-distortion-energy criterion in Fig. 7.74. Both areas depend upon the value of the yield strength σ_Y of the material.

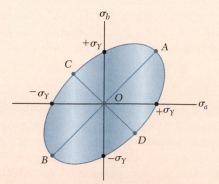

Fig. 7.73 Tresca's hexagon for maximum shearing-stress criterion.

Fig. 7.74 Von Mises surface based on maximum-distortion-energy criterion.

Fracture Criteria for Brittle Materials

The most commonly used method to predict failure of brittle materials is the fracture-based *Mohr's criterion*, which uses the results of various tests for a given material. The shaded area shown in Fig. 7.75 is used when the ultimate strengths σ_{UT} and σ_{UC} have been determined, respectively, from a tension and a compression test. The principal stresses σ_a and σ_b are determined at a given point, and if the corresponding point falls within the shaded area, the component is safe, and if it falls outside, the component will rupture.

Cylindrical Pressure Vessels

The stresses in *thin-walled pressure vessels* and equations relating to the stresses in the walls and the *gage pressure p* in the fluid were discussed. For a *cylindrical vessel* of inside radius r and thickness t (Fig. 7.76), the *hoop stress* σ_1 and the *longitudinal stress* σ_2 are

$$\sigma_1 = \frac{pr}{t} \qquad \sigma_2 = \frac{pr}{2t} \qquad \textbf{(7.30, 7.31)}$$

The *maximum shearing* stress occurs out of the plane of stress and is

$$\tau_{\max} = \sigma_2 = \frac{pr}{2t} \qquad \textbf{(7.34)}$$

Spherical Pressure Vessels

For a *spherical vessel* of inside radius r and thickness t (Fig. 7.77), the two principal stresses are equal:

$$\sigma_1 = \sigma_2 = \frac{pr}{2t} \qquad \textbf{(7.36)}$$

Again, the *maximum shearing* stress occurs out of the plane of stress and is

$$\tau_{\max} = \tfrac{1}{2}\sigma_1 = \frac{pr}{4t} \qquad \textbf{(7.37)}$$

Transformation of Plane Strain

The last part of the chapter was devoted to the *transformation of strain*. We discussed the transformation of *plane strain* and introduced *Mohr's circle for plane strain*. The discussion was similar to the corresponding discussion of the transformation of stress, except that, where the shearing stress τ was used, we now used $\frac{1}{2}\gamma$, that is, *half the shearing strain*. The formulas obtained for the transformation of strain under a rotation of axes through an angle θ were

$$\epsilon_{x'} = \frac{\epsilon_x + \epsilon_y}{2} + \frac{\epsilon_x - \epsilon_y}{2}\cos 2\theta + \frac{\gamma_{xy}}{2}\sin 2\theta \qquad \textbf{(7.44)}$$

$$\epsilon_{y'} = \frac{\epsilon_x + \epsilon_y}{2} - \frac{\epsilon_x - \epsilon_y}{2}\cos 2\theta - \frac{\gamma_{xy}}{2}\sin 2\theta \qquad \textbf{(7.45)}$$

$$\gamma_{x'y'} = -(\epsilon_x - \epsilon_y)\sin 2\theta + \gamma_{xy}\cos 2\theta \qquad \textbf{(7.49)}$$

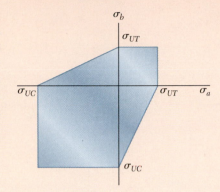

Fig. 7.75 Simplified Mohr's criterion for brittle materials.

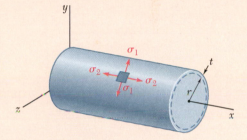

Fig. 7.76 Pressurized cylindrical vessel.

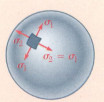

Fig. 7.77 Pressurized spherical vessel.

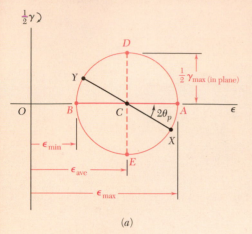

(a)

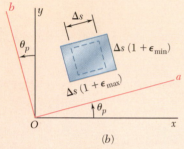

(b)

Fig. 7.78 (a) Mohr's circle for plane strain, showing principal strains and maximum in-plane shearing strain. (b) Strain element oriented to principal directions.

Mohr's Circle for Strain

Using Mohr's circle for strain (Fig. 7.78), the relationships defining the angle of rotation θ_p corresponding to the *principal axes of strain* and the values of the *principal strains* ϵ_{max} and ϵ_{min} are

$$\tan 2\theta_p = \frac{\gamma_{xy}}{\epsilon_x - \epsilon_y} \tag{7.52}$$

$$\epsilon_{max} = \epsilon_{ave} + R \quad \text{and} \quad \epsilon_{min} = \epsilon_{ave} - R \tag{7.51}$$

where

$$\epsilon_{ave} = \frac{\epsilon_x + \epsilon_y}{2} \quad \text{and} \quad R = \sqrt{\left(\frac{\epsilon_x - \epsilon_y}{2}\right)^2 + \left(\frac{\gamma_{xy}}{2}\right)^2} \tag{7.50}$$

The *maximum shearing strain* for a rotation in the plane of strain is

$$\gamma_{max(\text{in plane})} = 2R = \sqrt{(\epsilon_x - \epsilon_y)^2 + \gamma_{xy}^2} \tag{7.53}$$

In *plane stress*, the principal strain ϵ_c in a direction perpendicular to the plane of stress is expressed in terms of the in-plane principal strains ϵ_a and ϵ_b:

$$\epsilon_c = -\frac{\nu}{1 - \nu}(\epsilon_a + \epsilon_b) \tag{7.59}$$

Strain Gages and Strain Rosette

Strain gages are used to measure the normal strain on the surface of a structural element or machine component. A *strain rosette* consists of three gages aligned along lines forming angles θ_1, θ_2, and θ_3 with the x axis (Fig. 7.79). The relationships among the measurements ϵ_1, ϵ_2, ϵ_3 of the gages and the components ϵ_x, ϵ_y, γ_{xy} characterizing the state of strain at that point are

$$\epsilon_1 = \epsilon_x \cos^2 \theta_1 + \epsilon_y \sin^2 \theta_1 + \gamma_{xy} \sin \theta_1 \cos \theta_1$$

$$\epsilon_2 = \epsilon_x \cos^2 \theta_2 + \epsilon_y \sin^2 \theta_2 + \gamma_{xy} \sin \theta_2 \cos \theta_2 \tag{7.60}$$

$$\epsilon_3 = \epsilon_x \cos^2 \theta_3 + \epsilon_y \sin^2 \theta_3 + \gamma_{xy} \sin \theta_3 \cos \theta_3$$

These equations can be solved for ϵ_x, ϵ_y, and γ_{xy} once ϵ_1, ϵ_2, and ϵ_3 have been determined.

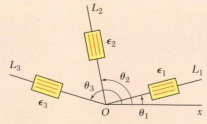

Fig. 7.79 Generalized strain gage rosette arrangement.

Review Problems

7.158 A steel pipe of 12-in. outer diameter is fabricated from $\frac{1}{4}$-in.-thick plate by welding along a helix that forms an angle of 22.5° with a plane perpendicular to the axis of the pipe. Knowing that a 40-kip axial force **P** and an 80-kip·in. torque **T,** each directed as shown, are applied to the pipe, determine the normal and in-plane shearing stresses in directions, respectively, normal and tangential to the weld.

7.159 Two steel plates of uniform cross section 10 × 80 mm are welded together as shown. Knowing that centric 100-kN forces are applied to the welded plates and that $\beta = 25°$, determine (a) the in-plane shearing stress parallel to the weld, (b) the normal stress perpendicular to the weld.

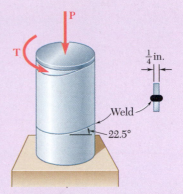

Fig. P7.158

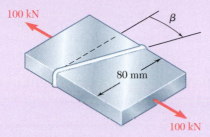

Fig. P7.159 and *P7.160*

7.160 Two steel plates of uniform cross section 10 × 80 mm are welded together as shown. Knowing that centric 100-kN forces are applied to the welded plates and that the in-plane shearing stress parallel to the weld is 30 MPa, determine (a) the angle β, (b) the corresponding normal stress perpendicular to the weld.

7.161 Determine the principal planes and the principal stresses for the state of plane stress resulting from the superposition of the two states of stress shown.

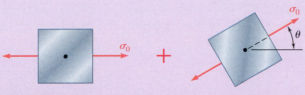

Fig. P7.161

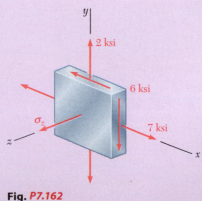

Fig. *P7.162*

7.162 For the state of stress shown, determine the maximum shearing stress when (a) $\sigma_z = +4$ ksi, (b) $\sigma_z = -4$ ksi, (c) $\sigma_z = 0$.

7.163 For the state of stress shown, determine the value of τ_{xy} for which the maximum shearing stress is (a) 60 MPa, (b) 78 MPa.

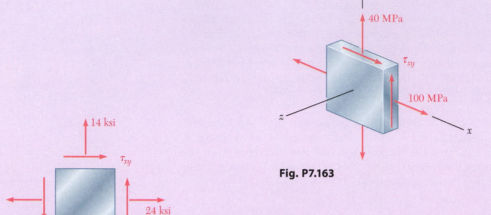

Fig. P7.163

Fig. P7.164

7.164 The state of plane stress shown occurs in a machine component made of a steel with $\sigma_Y = 30$ ksi. Using the maximum-distortion-energy criterion, determine whether yield will occur when (a) $\tau_{xy} = 6$ ksi, (b) $\tau_{xy} = 12$ ksi, (c) $\tau_{xy} = 14$ ksi. If yield does not occur, determine the corresponding factor of safety.

7.165 The compressed-air tank AB has an inner diameter of 450 mm and a uniform wall thickness of 6 mm. Knowing that the gage pressure inside the tank is 1.2 MPa, determine the maximum normal stress and the maximum in-plane shearing stress at point a on the top of the tank.

Fig. P7.165

7.166 For the compressed-air tank and loading of Prob. 7.165, determine the maximum normal stress and the maximum in-plane shearing stress at point b on the top of the tank.

7.167 The brass pipe *AD* is fitted with a jacket used to apply a hydro-static pressure of 500 psi to portion *BC* of the pipe. Knowing that the pressure inside the pipe is 100 psi, determine the maximum normal stress in the pipe.

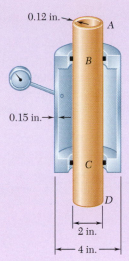

Fig. P7.167

7.168 For the assembly of Prob. 7.167, determine the normal stress in the jacket (*a*) in a direction perpendicular to the longitudinal axis of the jacket, (*b*) in a direction parallel to that axis.

7.169 Determine the largest in-plane normal strain, knowing that the following strains have been obtained by the use of the rosette shown:

$$\epsilon_1 = -50 \times 10^{-6} \text{ in./in.} \qquad \epsilon_2 = +360 \times 10^{-6} \text{ in./in.}$$
$$\epsilon_3 = +315 \times 10^{-6} \text{ in./in.}$$

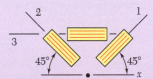

Fig. P7.169

Computer Problems

The following problems are to be solved with a computer.

7.C1 A state of plane stress is defined by the stress components σ_x, σ_y, and τ_{xy} associated with the element shown in Fig. P7.C1a. (*a*) Write a computer program that can be used to calculate the stress components $\sigma_{x'}$, $\sigma_{y'}$, and $\tau_{x'y'}$ associated with the element after it has rotated through an angle θ about the z axis (Fig. P7.C1b). (*b*) Use this program to solve Probs. 7.13 through 7.16.

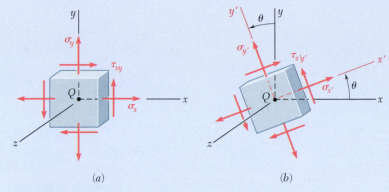

(*a*)   (*b*)

Fig. P7.C1

7.C2 A state of plane stress is defined by the stress components σ_x, σ_y, and τ_{xy} associated with the element shown in Fig. P7.C1a. (*a*) Write a computer program that can be used to calculate the principal axes, the principal stresses, the maximum in-plane shearing stress, and the maximum shearing stress. (*b*) Use this program to solve Probs. 7.5, 7.9, 7.68, and 7.69.

7.C3 (*a*) Write a computer program that, for a given state of plane stress and a given yield strength of a ductile material, can be used to determine whether the material will yield. The program should use both the maximum-shearing-stress criterion and the maximum-distortion-energy criterion. It should also print the values of the principal stresses and, if the material does not yield, calculate the factor of safety. (*b*) Use this program to solve Probs. 7.81, 7.82, and 7.164.

7.C4 (*a*) Write a computer program based on Mohr's fracture criterion for brittle materials that, for a given state of plane stress and given values of the ultimate stress of the material in tension and compression, can be used to determine whether rupture will occur. The program should also print the values of the principal stresses. (*b*) Use this program to solve Probs. 7.91 and 7.92 and to check the answers to Probs. 7.93 and 7.94.

7.C5 A state of plane strain is defined by the strain components ϵ_x, ϵ_y, and γ_{xy} associated with the x and y axes. (*a*) Write a computer program that can be used to calculate the strain components $\epsilon_{x'}$, $\epsilon_{y'}$, and $\gamma_{x'y'}$ associated with the frame of reference $x'y'$ obtained by rotating the x and y axes through an angle θ. (*b*) Use this program to solve Probs. 7.129 and 7.131.

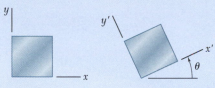

Fig. P7.C5

7.C6 A state of strain is defined by the strain components ϵ_x, ϵ_y, and γ_{xy} associated with the x and y axes. (*a*) Write a computer program that can be used to determine the orientation and magnitude of the principal strains, the maximum in-plane shearing strain, and the maximum shearing strain. (*b*) Use this program to solve Probs. 7.136 through 7.139.

7.C7 A state of plane strain is defined by the strain components ϵ_x, ϵ_y, and γ_{xy} measured at a point. (*a*) Write a computer program that can be used to determine the orientation and magnitude of the principal strains, the maximum in-plane shearing strain, and the magnitude of the shearing strain. (*b*) Use this program to solve Probs. 7.140 through 7.143.

7.C8 A rosette consisting of three gages forming angles of θ_1, θ_2, and θ_3 with the x axis is attached to the free surface of a machine component made of a material with a given Poisson's ratio ν. (*a*) Write a computer program that, for given readings ϵ_1, ϵ_2, and ϵ_3 of the gages, can be used to calculate the strain components associated with the x and y axes and to determine the orientation and magnitude of the three principal strains, the maximum in-plane shearing strain, and the maximum shearing strain. (*b*) Use this program to solve Probs. 7.144, 7.145, 7.146, and 7.169.

8

Principal Stresses under a Given Loading

Due to gravity and wind load, the signpost support column is subjected simultaneously to compression, bending, and torsion. This chapter will examine the stresses resulting from such combined loadings.

Objectives

In this chapter, you will:

- **Describe** how stress components vary throughout a beam.
- **Identify** key stress analysis locations in an I-shaped beam.
- **Design** transmission shafts subject to transverse loads and torques.
- **Describe** the stresses throughout a member arising from combined loads.

Introduction

Introduction

In the first part of this chapter, you will apply to the design of beams and shafts the knowledge that you acquired in Chap. 7 on the transformation of stresses. In the second part of the chapter, you will learn how to determine the principal stresses in structural members and machine elements under given loading conditions.

The maximum normal stress σ_m that occurs in a beam under a transverse load (Fig. 8.1a) and whether this value exceeds the allowable stress σ_{all} for the given material has been studied in Chap. 5. If the allowable stress is exceeded, the design of the beam is not acceptable. While the danger for a brittle material is actually to fail in tension, the danger for a ductile material is to fail in shear (Fig. 8.1b). Thus, a situation where $\sigma_m > \sigma_{all}$ indicates that $|M|_{max}$ is too large for the cross section selected,

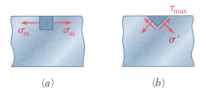

Fig. 8.1   Stress elements where normal stress is maximum in a transversely-loaded beam. (a) Element showing maximum normal stress. (b) Element showing corresponding maximum shearing stress.

Fig. 8.2   Stress elements where shearing stress is maximum in a transversely-loaded beam. (a) Element showing maximum shearing stress. (b) Element showing corresponding maximum normal stress.

but it does not provide any information on the actual mechanism of failure. Similarly, $\tau_m > \tau_{all}$ indicates that $|V|_{max}$ is too large for the cross section selected. While the danger for a ductile material is actually to fail in shear (Fig. 8.2a), the danger for a brittle material is to fail in tension under the principal stresses (Fig. 8.2b). The distribution of the principal stresses in a beam is discussed in Sec. 8.1.

Depending on the shape of the beam's cross section and the value of the shear V in the critical section where $|M| = |M|_{max}$, the largest value of the normal stress may not necessarily occur at the top or bottom, but at some other point within the section. In Sec. 8.1, a combination of large values of σ_x and τ_{xy} near the junction of the web and the flanges of a W- or S-beam can result in a value of the principal stress σ_{max} (Fig. 8.3) that is larger than the value of σ_m on the surface of the beam.

Section 8.2 covers the design of transmission shafts subjected to transverse loads and torques. The effects of both normal stresses due to bending and shearing stresses due to torsion are discussed.

In Sec. 8.3, the stresses are determined at a given point K of a body of arbitrary shape subjected to combined loading. First, the given load is

Fig. 8.3   Principal stress element at the junction of a flange and web in an I-shaped beam.

reduced to forces and couples in the section containing K. Next, the normal and shearing stresses at K are calculated. Finally, the principal planes, principal stresses, and maximum shearing stress are found, using one of the methods for transformation of stresses (Chap. 7).

8.1   PRINCIPAL STRESSES IN A BEAM

Consider a prismatic beam AB subjected to some arbitrary transverse loads (Fig. 8.4). The shear and bending moment in a section through a

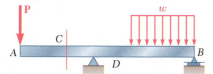

Fig. 8.4   Transversely loaded prismatic beam.

given point C are denoted by V and M, respectively. Recall from Chaps. 5 and 6 that, within the elastic limit, the stresses on a small element with faces perpendicular to the x and y axes reduce to the normal stresses $\sigma_m = Mc/I$ if the element is at the free surface of the beam and to the shearing stresses $\tau_m = VQ/It$ if the element is at the neutral surface (Fig. 8.5).

At any other point of the cross section, an element is subjected simultaneously to the normal stresses

$$\sigma_x = -\frac{My}{I} \tag{8.1}$$

where y is the distance from the neutral surface and I is the centroidal moment of inertia of the section, and to the shearing stresses

$$\tau_{xy} = -\frac{VQ}{It} \tag{8.2}$$

where Q is the first moment about the neutral axis of the portion of the cross-sectional area located above the point where the stresses are computed, and t is the width of the cross section at that point. Either of the methods of analysis presented in Chap. 7 can be used to obtain the principal stresses at any point of the cross section (Fig. 8.6).

The following question now arises: can the maximum normal stress σ_{max} at some point within the cross section be larger than $\sigma_m = Mc/I$ at the surface of the beam? If it can, then determining the largest normal stress in the beam involves more than the computation of $|M|_{max}$ and the use of Eq. (8.1). An answer to this question is obtained by investigating the distribution of the principal stresses in a narrow rectangular cantilever beam subjected to a concentrated load $\mathbf{P}$ at its free end (Fig. 8.7). Recall from Sec. 6.2 that the normal and shearing stresses at a distance x from the load $\mathbf{P}$ and at a distance y above the neutral surface are given, respectively, by Eqs. (6.13) and (6.12). Since the moment of inertia of the cross section is

$$I = \frac{bh^3}{12} = \frac{(bh)(2c)^2}{12} = \frac{Ac^2}{3}$$

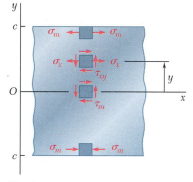

Fig. 8.5   Stress elements at selected points of a beam.

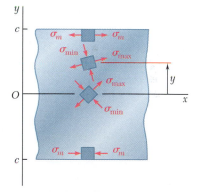

Fig. 8.6   Principal stress elements at selected points of a beam.

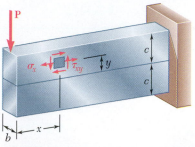

Fig. 8.7   Narrow rectangular cantilever beam supporting a single concentrated load.

where A is the cross-sectional area and c the half-depth of the beam,

$$\sigma_x = \frac{Pxy}{I} = \frac{Pxy}{\frac{1}{3}Ac^2} = 3\frac{P}{A}\frac{xy}{c^2} \tag{8.3}$$

and

$$\tau_{xy} = \frac{3}{2}\frac{P}{A}\left(1 - \frac{y^2}{c^2}\right) \tag{8.4}$$

Using the methods of Sec. 7.1B or Sec. 7.2, σ_{max} can be determined at any point of the beam. Figure 8.8 shows the results of the computation of the ratios σ_{max}/σ_m and σ_{min}/σ_m in two sections of the beam, corresponding respectively to $x = 2c$ and $x = 8c$. In each section, these ratios have been determined at 11 different points, and the orientation of the principal axes has been indicated at each point.[†]

It is clear that σ_{max} is smaller than σ_m in both of the two sections in Fig. 8.8. If it does exceed σ_m elsewhere, it is in sections close to load **P**, where σ_m is small compared to τ_m.[‡] But for sections close to load **P**, Saint-Venant's principle does not apply, and Eqs. (8.3) and (8.4) cease to be

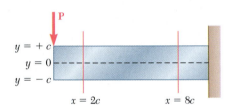

| y/c | x = 2c | | x = 8c | |
	σ_{min}/σ_m	σ_{max}/σ_m	σ_{min}/σ_m	σ_{max}/σ_m
1.0	0	1.000	0	1.000
0.8	−0.010	0.810	−0.001	0.801
0.6	−0.040	0.640	−0.003	0.603
0.4	−0.090	0.490	−0.007	0.407
0.2	−0.160	0.360	−0.017	0.217
0	−0.250	0.250	−0.063	0.063
−0.2	−0.360	0.160	−0.217	0.017
−0.4	−0.490	0.090	−0.407	0.007
−0.6	−0.640	0.040	−0.603	0.003
−0.8	−0.810	0.010	−0.801	0.001
−1.0	−1.000	0	−1.000	0

Fig. 8.8 Distribution of principal stresses in two transverse sections of a rectangular cantilever beam supporting a single concentrated load.

[†]See Prob. 8.C2, which refers to a program that can be written to obtain the results in Fig. 8.8.

[‡]As will be verified in Prob. 8.C2, σ_{max} exceeds σ_m if $x \le 0.544c$.

valid—except in the very unlikely case of a load distributed parabolically over the end section (see. Sec. 6.2), where advanced methods of analysis are used to account for the effect of stress concentrations. It can thus be concluded that, for beams of rectangular cross section, and within the scope of the theory presented in this text, the maximum normal stress can be obtained from Eq. (8.1).

In Fig. 8.8, the directions of the principal axes are found at 11 points in both of the sections considered. If this analysis is extended to a larger number of sections and points in each section, it possible to draw two orthogonal systems of curves on the side of the beam (Fig. 8.9). One

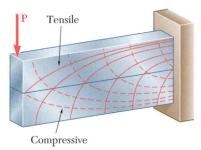

Fig. 8.9 Stress trajectories in a rectangular cantilevered beam supporting a single concentrated load.

system consists of curves tangent to the principal axes corresponding to σ_{max} and the other to σ_{min}. These curves are known as the *stress trajectories*. A trajectory of the first group (solid lines) defines the direction of the largest tensile stress at each of its points, while the second group (dashed lines) defines the direction of the largest compressive stress.[†]

The conclusion we have reached for beams of rectangular cross section, that the maximum normal stress in the beam can be obtained from Eq. (8.1), remains valid for many beams of nonrectangular cross section. However, when the width of the cross section varies so that large shearing stresses τ_{xy} occur at points close to the surface of the beam (where σ_x is also large), the principal stress σ_{max} may be larger than σ_m at such points. This is a distinct possibility when selecting W-beams or S-beams, where we should calculate the principal stress σ_{max} at the junctions b and d of the web with the flanges of the beam (Fig. 8.10). This is done by determining σ_x and τ_{xy} at that point from Eqs. (8.1) and (8.2), and by using either of the methods of analysis in Chap. 7 to obtain σ_{max} (see Sample Prob. 8.1). An alternative procedure for selecting an acceptable section uses the approximation $\tau_{max} = V/A_{web}$ [Eq. (6.11)]. This leads to a slightly larger and conservative value of the principal stress σ_{max} at the junction of the web with the flanges of the beam (see Sample Prob. 8.2).

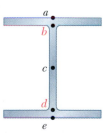

Fig. 8.10 Key stress analysis locations in I-shaped beams.

[†]A brittle material, such as concrete, fails in tension along planes that are perpendicular to the tensile-stress trajectories. Thus, to be effective, steel reinforcing bars should be placed so that they intersect these planes. On the other hand, stiffeners attached to the web of a plate girder are effective in preventing buckling only if they intersect planes perpendicular to the compressive-stress trajectories.

8.2 DESIGN OF TRANSMISSION SHAFTS

The design of transmission shafts in Sec. 3.4 considered only the stresses due to torques exerted on the shafts. However, if the power is transferred to and from the shaft by means of gears or sprocket wheels (Fig. 8.11*a*), the forces on the gear teeth or sprockets are equivalent to force-couple systems applied at the centers of the corresponding cross sections (Fig. 8.11*b*). This means that the shaft is subjected to both a transverse and a torsional load.

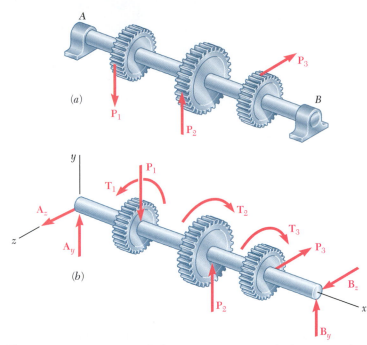

Fig. 8.11 Loadings on gear-shaft systems. (*a*) Forces applied to gear teeth. (*b*) Free-body diagram of shaft, with gear forces replaced by equivalent force-couple systems applied to shaft.

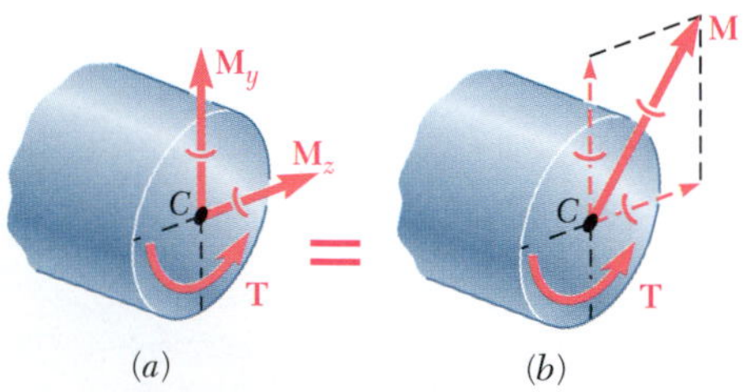

Fig. 8.12 (*a*) Torque and bending couples acting on shaft cross section. (*b*) Bending couples replaced by their resultant **M**.

The shearing stresses produced in the shaft by the transverse loads are usually much smaller than those produced by the torques and will be neglected in this analysis.[†] However, the normal stresses due to transverse loads may be quite large, and their contribution to the maximum shearing stress $\tau_{\max}$ should be taken into account.

Consider the cross section of the shaft at some point *C*. The torque **T** and the bending couples $\mathbf{M}_y$ and $\mathbf{M}_z$ acting in a horizontal and a vertical plane are represented by the couple vectors shown (Fig. 8.12*a*). Since any diameter of the section is a principal axis of inertia for the section, we can replace $\mathbf{M}_y$ and $\mathbf{M}_z$ by their resultant **M** (Fig. 8.12*b*) in order to compute the normal stresses σ_x. Thus σ_x is maximum at the end of the diameter

[†]For an application where the shearing stresses produced by the transverse loads must be considered, see Probs. 8.21 and 8.22.

perpendicular to the vector representing **M** (Fig. 8.13). Recalling that the values of the normal stresses at that point are $\sigma_m = Mc/I$ and zero and the shearing stress is $\tau_m = Tc/J$, plot the corresponding values as points X and Y on a Mohr's circle diagram (Fig. 8.14). The maximum shearing stress is found to be

$$\tau_{max} = R = \sqrt{\left(\frac{\sigma_m}{2}\right)^2 + (\tau_m)^2} = \sqrt{\left(\frac{Mc}{2I}\right)^2 + \left(\frac{Tc}{J}\right)^2}$$

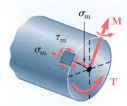

Fig. 8.13 Maximum stress element.

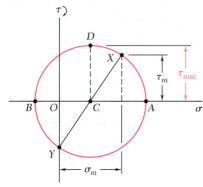

Fig. 8.14 Mohr's circle for shaft loading.

Recalling that $2I = J$ for a circular or annular cross section,

$$\tau_{max} = \frac{c}{J}\sqrt{M^2 + T^2} \tag{8.5}$$

It follows that the minimum allowable value of the ratio J/c for the cross section of the shaft is

$$\frac{J}{c} = \frac{\left(\sqrt{M^2 + T^2}\right)_{max}}{\tau_{all}} \tag{8.6}$$

where the numerator in the right-hand member represents the maximum value of $\sqrt{M^2 + T^2}$ in the shaft and τ_{all} is the allowable shearing stress. Expressing the bending moment M in terms of its components in the two coordinate planes, we obtain:

$$\frac{J}{c} = \frac{\left(\sqrt{M_y^2 + M_z^2 + T^2}\right)_{max}}{\tau_{all}} \tag{8.7}$$

Equations (8.6) and (8.7) can be used to design both solid and hollow circular shafts and should be compared to Eq. (3.21), which was obtained under the assumption of torsional loading only.

The maximum value of $\sqrt{M_y^2 + M_z^2 + T^2}$ is easier to find if both bending-moment diagrams corresponding to M_y and M_z and a third diagram representing the values of T along the shaft are drawn (see Sample Prob. 8.3).

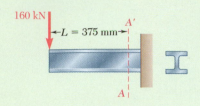

Sample Problem 8.1

A 160-kN force is applied as shown at the end of a W200 × 52 rolled-steel beam. Neglecting the effect of fillets and of stress concentrations, determine whether the normal stresses in the beam satisfy a design specification that they be equal to or less than 150 MPa at section *A–A′*.

STRATEGY: To determine the maximum normal stress, you should perform a beam stress analysis at the surface of the flange as well as at the junction of the web and flange. A Mohr's circle analysis will also be necessary at the web-flange junction to determine this maximum normal stress.

MODELING and ANALYSIS:

Shear and Bending Moment. Referring to Fig. 1, at section *A–A′*, we have

$$M_A = (160 \text{ kN})(0.375 \text{ m}) = 60 \text{ kN·m}$$
$$V_A = 160 \text{ kN}$$

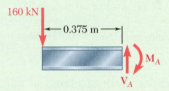

Fig. 1 Free-body diagram of beam, with section at *A–A′*

Normal Stresses on Transverse Plane. Referring to the table of *Properties of Rolled-Steel Shapes* in Appendix C to obtain the data shown, determine the stresses σ_a and σ_b (Fig. 2).

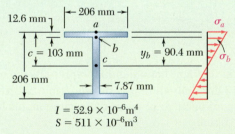

$$I = 52.9 \times 10^{-6} \text{m}^4$$
$$S = 511 \times 10^{-6} \text{m}^3$$

Fig. 2 Cross-section dimensions and normal stress distribution.

At point *a*,

$$\sigma_a = \frac{M_A}{S} = \frac{60 \text{ kN·m}}{511 \times 10^{-6} \text{ m}^3} = 117.4 \text{ MPa}$$

At point *b*,

$$\sigma_b = \sigma_a \frac{y_b}{c} = (117.4 \text{ MPa})\frac{90.4 \text{ mm}}{103 \text{ mm}} = 103.0 \text{ MPa}$$

(continued)

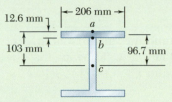

Fig. 3 Dimensions to evaluate Q at point b.

Note that all normal stresses on the transverse plane are less than 150 MPa.

Shearing Stresses on Transverse Plane. Referring to Fig. 3, we obtain the data necessary to evaluate Q and then determine the stresses τ_a and τ_b.

At point a,

$$Q = 0 \qquad \tau_a = 0$$

At point b,

$$Q = (206 \times 12.6)(96.7) = 251.0 \times 10^3 \text{ mm}^3 = 251.0 \times 10^{-6} \text{ m}^3$$

$$\tau_b = \frac{V_A Q}{It} = \frac{(160 \text{ kN})(251.0 \times 10^{-6} \text{ m}^3)}{(52.9 \times 10^{-6} \text{ m}^4)(0.00787 \text{ m})} = 96.5 \text{ MPa}$$

Principal Stress at Point *b*. The state of stress at point b consists of the normal stress $\sigma_b = 103.0$ MPa and the shearing stress $\tau_b = 96.5$ MPa. Draw Mohr's circle (Fig. 4) and find

$$\sigma_{\max} = \frac{1}{2}\sigma_b + R = \frac{1}{2}\sigma_b + \sqrt{\left(\frac{1}{2}\sigma_b\right)^2 + \tau_b^2}$$

$$= \frac{103.0}{2} + \sqrt{\left(\frac{103.0}{2}\right)^2 + (96.5)^2}$$

$$\sigma_{\max} = 160.9 \text{ MPa}$$

The specification, $\sigma_{\max} \leq 150$ MPa, is *not* satisfied ◄

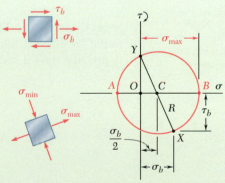

Fig. 4 Stress element for coordinate and principal orientations at point b; Mohr's circle for point b.

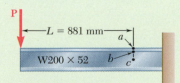

Fig. 5 Condition where maximum principal stress at point a begins to exceed that at point b.

REFLECT and THINK: For this beam and loading, the principal stress at point b is 36% larger than the normal stress at point a. For $L \geq 881$ mm (Fig. 5), the maximum normal stress would occur at point a.

(continued)

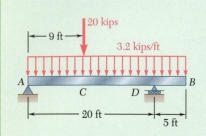

20 kips

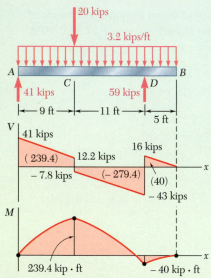

Fig. 1 Free-body diagram of beam; shear and bending moment diagrams.

Sample Problem 8.2

The overhanging beam AB supports a uniformly distributed load of 3.2 kips/ft and a concentrated load of 20 kips at C. Knowing that the grade of steel to be used has $\sigma_{all} = 24$ ksi and $\tau_{all} = 14.5$ ksi, select the wide-flange shape that should be used.

STRATEGY: Draw the shear and bending-moment diagrams to determine their maximum values. From the maximum bending moment, you can find the required section modulus and use this to select the lightest available wide-flange shape. You can then check to ensure that the maximum shearing stress in the web and the maximum principal stress at the web-flange junction do not exceed the given allowable stresses.

MODELING and ANALYSIS:

Reactions at _A_ and _D_. Draw the free-body diagram (Fig. 1) of the beam. From the equilibrium equations $\Sigma M_D = 0$ and $\Sigma M_A = 0$, the values of R_A and R_D are as shown.

Shear and Bending-Moment Diagrams. Using the methods discussed in Secs. 5.1 and 5.2, draw the diagrams (Fig. 1) and observe that

$$|M|_{max} = 239.4 \text{ kip·ft} = 2873 \text{ kip·in.} \qquad |V|_{max} = 43 \text{ kips}$$

Section Modulus. For $|M|_{max} = 2873$ kip·in. and $\sigma_{all} = 24$ ksi, the minimum acceptable section modulus of the rolled-steel shape is

$$S_{min} = \frac{|M|_{max}}{\sigma_{all}} = \frac{2873 \text{ kip·in.}}{24 \text{ ksi}} = 119.7 \text{ in}^3$$

Selection of Wide-Flange Shape. Choose from the table of *Properties of Rolled-Steel Shapes* in Appendix C the lightest shapes of a given depth that have a section modulus larger than S_{min}.

Shape	S (in³)
W24 × 68	154
W21 × 62	127
W18 × 76	146
W16 × 77	134
W14 × 82	123
W12 × 96	131

The lightest shape available is   W21 × 62 ◄

(continued)

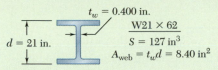

Fig. 2 I-shape cross section properties.

Shearing Stress. For the beam design, assume that the maximum shear is uniformly distributed over the web area of a W21 × 62 (Fig. 2). Write

$$\tau_m = \frac{V_{max}}{A_{web}} = \frac{43 \text{ kips}}{8.40 \text{ in}^2} = 5.12 \text{ ksi} < 14.5 \text{ ksi} \qquad (\text{OK})$$

Principal Stress at Point *b*. The maximum principal stress at point *b* in the critical section where *M* is maximum should not exceed $\sigma_{all} = 24$ ksi. Referring to Fig. 3, we write

$$\sigma_a = \frac{M_{max}}{S} = \frac{2873 \text{ kip·in.}}{127 \text{ in}^3} = 22.6 \text{ ksi}$$

$$\sigma_b = \sigma_a \frac{y_b}{c} = (22.6 \text{ ksi}) \frac{9.88 \text{ in.}}{10.50 \text{ in.}} = 21.3 \text{ ksi}$$

Conservatively, $\tau_b = \dfrac{V}{A_{web}} = \dfrac{12.2 \text{ kips}}{8.40 \text{ in}^2} = 1.45 \text{ ksi}$

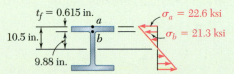

Fig. 3 Key stress analysis locations and normal stress distribution.

Draw Mohr's circle (Fig. 4) and find

$$\sigma_{max} = \tfrac{1}{2}\sigma_b + R = \frac{21.3 \text{ ksi}}{2} + \sqrt{\left(\frac{21.3 \text{ ksi}}{2}\right)^2 + (1.45 \text{ ksi})^2}$$

$$\sigma_{max} = 21.4 \text{ ksi} \leq 24 \text{ ksi} \quad (\text{OK}) \blacktriangleleft$$

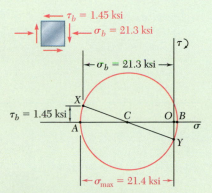

Fig. 4 Stress element at point *b* and Mohr's circle for point *b*.

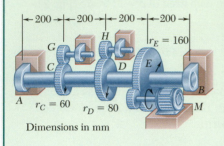

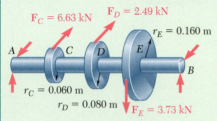

Fig. 1 Free-body diagram of shaft AB and its gears.

Sample Problem 8.3

The solid shaft AB rotates at 480 rpm and transmits 30 kW from the motor M to machine tools connected to gears G and H; 20 kW is taken off at gear G and 10 kW at gear H. Knowing that $\tau_{all} = 50$ MPa, determine the smallest permissible diameter for shaft AB.

STRATEGY: After determining the forces and couples exerted on the shaft, you can obtain its bending-moment and torque diagrams. Using these diagrams to aid in identifying the critical transverse section, you can then determine the required shaft diameter.

MODELING:

Draw the free-body diagram of the shaft and gears (Fig. 1). Observing that $f = 480$ rpm $= 8$ Hz, the torque exerted on gear E is

$$T_E = \frac{P}{2\pi f} = \frac{30\text{ kW}}{2\pi(8\text{ Hz})} = 597\text{ N·m}$$

The corresponding tangential force acting on the gear is

$$F_E = \frac{T_E}{r_E} = \frac{597\text{ N·m}}{0.16\text{ m}} = 3.73\text{ kN}$$

A similar analysis of gears C and D yields

$$T_C = \frac{20\text{ kW}}{2\pi(8\text{ Hz})} = 398\text{ N·m} \qquad F_C = 6.63\text{ kN}$$

$$T_D = \frac{10\text{ kW}}{2\pi(8\text{ Hz})} = 199\text{ N·m} \qquad F_D = 2.49\text{ kN}$$

Now replace the forces on the gears by equivalent force-couple systems as shown in Fig. 2.

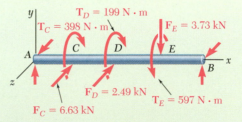

Fig. 2 Free-body diagram of shaft AB, with gear forces replaced by equivalent force-couple systems.

(continued)

ANALYSIS:

Bending-Moment and Torque Diagrams (Fig. 3)

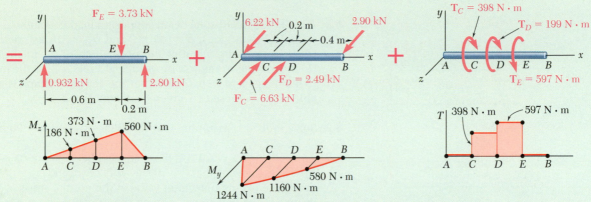

Fig. 3   Analysis of free-body diagram of shaft *AB* alone with equivalent force-couple loads is equivalent to superposition of bending moments from vertical loads, horizontal loads, and applied torques.

Critical Transverse Section.   By computing $\sqrt{M_y^2 + M_z^2 + T^2}$ at all potentially critical sections (Fig. 4), the maximum value occurs just to the right of *D*:

$$\sqrt{M_y^2 + M_z^2 + T^2}_{\max} = \sqrt{(1160)^2 + (373)^2 + (597)^2} = 1357 \text{ N·m}$$

Diameter of Shaft.   For $\tau_{\text{all}} = 50$ MPa, Eq. (7.32) yields

$$\frac{J}{c} = \frac{\sqrt{M_y^2 + M_z^2 + T^2}_{\max}}{\tau_{\text{all}}} = \frac{1357 \text{ N·m}}{50 \text{ MPa}} = 27.14 \times 10^{-6} \text{ m}^3$$

For a solid circular shaft of radius *c*,

$$\frac{J}{c} = \frac{\pi}{2} c^3 = 27.14 \times 10^{-6} \qquad c = 0.02585 \text{ m} = 25.85 \text{ mm}$$

$$\text{Diameter} = 2c = 51.7 \text{ mm} \quad \blacktriangleleft$$

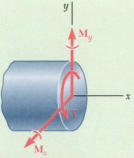

Fig. 4   Bending moment components and torque at critical section.

Problems

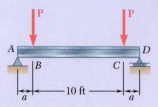

Fig. P8.1

8.1 A W10 × 39 rolled-steel beam supports a load **P** as shown. Knowing that $P = 45$ kips, $a = 10$ in., and $\sigma_{all} = 18$ ksi, determine (*a*) the maximum value of the normal stress σ_m in the beam, (*b*) the maximum value of the principal stress σ_{max} at the junction of the flange and web, (*c*) whether the specified shape is acceptable as far as these two stresses are concerned.

8.2 Solve Prob. 8.1, assuming that $P = 22.5$ kips and $a = 20$ in.

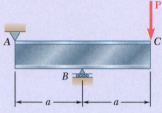

Fig. P8.3

8.3 An overhanging W920 × 449 rolled-steel beam supports a load **P** as shown. Knowing that $P = 700$ kN, $a = 2.5$ m, and $\sigma_{all} = 100$ MPa, determine (*a*) the maximum value of the normal stress σ_m in the beam, (*b*) the maximum value of the principal stress σ_{max} at the junction of the flange and web, (*c*) whether the specified shape is acceptable as far as these two stresses are concerned.

8.4 Solve Prob. 8.3, assuming that $P = 850$ kN and $a = 2.0$ m.

8.5 and 8.6 (*a*) Knowing that $\sigma_{all} = 160$ MPa and $\tau_{all} = 100$ MPa, select the most economical metric wide-flange shape that should be used to support the loading shown. (*b*) Determine the values to be expected for σ_m, τ_m, and the principal stress σ_{max} at the junction of a flange and the web of the selected beam.

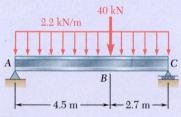

Fig. P8.5

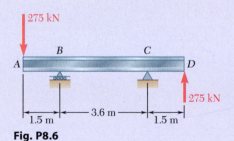

Fig. P8.6

8.7 and 8.8 (*a*) Knowing that $\sigma_{all} = 24$ ksi and $\tau_{all} = 14.5$ ksi, select the most economical wide-flange shape that should be used to support the loading shown. (*b*) Determine the values to be expected for σ_m, τ_m, and the principal stress σ_{max} at the junction of a flange and the web of the selected beam.

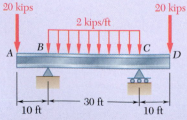

Fig. P8.7

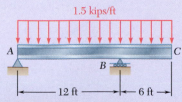

Fig. P8.8

8.9 through 8.14 Each of the following problems refers to a rolled-steel shape selected in a problem of Chap. 5 to support a given loading at a minimal cost while satisfying the requirement $\sigma_m \leq \sigma_{all}$. For the selected design, determine (a) the actual value of σ_m in the beam, (b) the maximum value of the principal stress σ_{max} at the junction of a flange and the web.

8.9 Loading of Prob. 5.73 and selected W530 × 92 shape.

8.10 Loading of Prob. 5.74 and selected W250 × 28.4 shape.

8.11 Loading of Prob. 5.75 and selected S12 × 31.8 shape.

8.12 Loading of Prob. 5.76 and selected S15 × 42.9 shape.

8.13 Loading of Prob. 5.77 and selected S510 × 98.2 shape.

8.14 Loading of Prob. 5.78 and selected S460 × 81.4 shape.

8.15 Determine the smallest allowable diameter of the solid shaft *ABCD*, knowing that $\tau_{all} = 60$ MPa and that the radius of disk *B* is *r* = 80 mm.

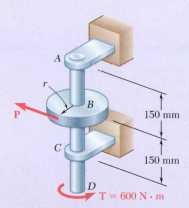

Fig. P8.15 and *P8.16*

8.16 Determine the smallest allowable diameter of the solid shaft *ABCD*, knowing that $\tau_{all} = 60$ MPa and that the radius of disk *B* is *r* = 120 mm.

8.17 Using the notation of Sec. 8.2 and neglecting the effect of shearing stresses caused by transverse loads, show that the maximum normal stress in a circular shaft can be expressed as follows:

$$\sigma_{max} = \frac{c}{J}\left[(M_y^2 + M_z^2)^{\frac{1}{2}} + (M_y^2 + M_z^2 + T^2)^{\frac{1}{2}} \right]_{max}$$

8.18 The 4-kN force is parallel to the *x* axis, and the force **Q** is parallel to the *z* axis. The shaft *AD* is hollow. Knowing that the inner diameter is half the outer diameter and that $\tau_{all} = 60$ MPa, determine the smallest permissible outer diameter of the shaft.

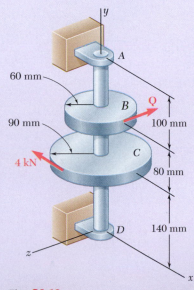

Fig. *P8.18*

8.19 The vertical force $\mathbf{P}_1$ and the horizontal force $\mathbf{P}_2$ are applied as shown to disks welded to the solid shaft AD. Knowing that the diameter of the shaft is 1.75 in. and that $\tau_{\text{all}} = 8$ ksi, determine the largest permissible magnitude of the force $\mathbf{P}_2$.

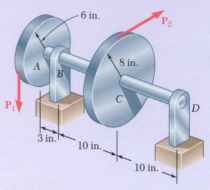

Fig. P8.19

8.20 The two 500-lb forces are vertical and the force $\mathbf{P}$ is parallel to the z axis. Knowing that $\tau_{\text{all}} = 8$ ksi, determine the smallest permissible diameter of the solid shaft AE.

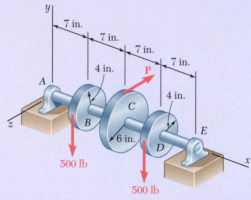

Fig. P8.20

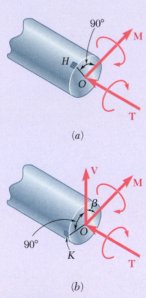

Fig. P8.21

8.21 It was stated in Sec. 8.2 that the shearing stresses produced in a shaft by the transverse loads are usually much smaller than those produced by the torques. In the preceding problems their effect was ignored, and it was assumed that the maximum shearing stress in a given section occurred at point H (Fig. P8.21a) and was equal to the expression obtained in Eq. (8.5), namely,

$$\tau_H = \frac{c}{J} \sqrt{M^2 + T^2}$$

Show that the maximum shearing stress at point K (Fig. P8.21b), where the effect of the shear V is greatest, can be expressed as

$$\tau_K = \frac{c}{J} \sqrt{(M\cos\beta)^2 + \left(\frac{2}{3}cV + T\right)^2}$$

where β is the angle between the vectors $\mathbf{V}$ and $\mathbf{M}$. It is clear that the effect of the shear V cannot be ignored when $\tau_K \geq \tau_H$. (*Hint:* Only the component of $\mathbf{M}$ along $\mathbf{V}$ contributes to the shearing stress at K.)

8.22 Assuming that the magnitudes of the forces applied to disks A and C of Prob. 8.19 are, respectively, $P_1 = 1080$ lb and $P_2 = 810$ lb, and using the expressions given in Prob. 8.21, determine the values of τ_H and τ_K in a section (a) just to the left of B, (b) just to the left of C.

8.23 The solid shaft AB rotates at 600 rpm and transmits 80 kW from the motor M to a machine tool connected to gear F. Knowing that $\tau_{all} = 60$ MPa, determine the smallest permissible diameter of shaft AB.

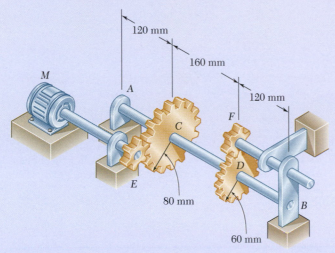

Fig. P8.23

8.24 Solve Prob. 8.23, assuming that shaft AB rotates at 720 rpm.

8.25 The solid shafts ABC and DEF and the gears shown are used to transmit 20 hp from the motor M to a machine tool connected to shaft DEF. Knowing that the motor rotates at 240 rpm and that $\tau_{all} = 7.5$ ksi, determine the smallest permissible diameter of (a) shaft ABC, (b) shaft DEF.

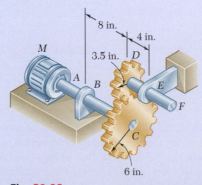

Fig. P8.25

8.26 Solve Prob. 8.25, assuming that the motor rotates at 360 rpm.

8.27 The solid shaft *ABC* and the gears shown are used to transmit 10 kW from the motor *M* to a machine tool connected to gear *D*. Knowing that the motor rotates at 240 rpm and that $\tau_{all} = 60$ MPa, determine the smallest permissible diameter of shaft *ABC*.

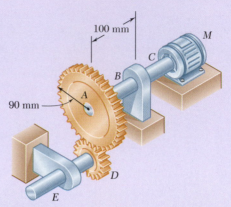

Fig. P8.27

8.28 Assuming that shaft *ABC* of Prob. 8.27 is hollow and has an outer diameter of 50 mm, determine the largest permissible inner diameter of the shaft.

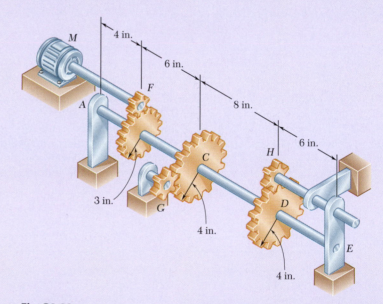

Fig. P8.29

8.29 The solid shaft *AE* rotates at 600 rpm and transmits 60 hp from the motor *M* to machine tools connected to gears *G* and *H*. Knowing that $\tau_{all} = 8$ ksi and that 40 hp is taken off at gear *G* and 20 hp is taken off at gear *H*, determine the smallest permissible diameter of shaft *AE*.

8.30 Solve Prob. 8.29, assuming that 30 hp is taken off at gear *G* and 30 hp is taken off at gear *H*.

8.3 STRESSES UNDER COMBINED LOADS

In Chaps. 1 and 2 you learned to determine the stresses caused by a centric axial load. In Chap. 3, you analyzed the distribution of stresses in a cylindrical member subjected to a twisting couple. In Chap. 4, you determined the stresses caused by bending couples and, in Chaps. 5 and 6, the stresses produced by transverse loads. As you will see presently, you can combine the knowledge you have acquired to determine the stresses in slender structural members or machine components under fairly general loading conditions. For example, the bent member *ABDE* of circular cross section is subjected to several forces (Fig. 8.15). In order to determine the stresses at points *H* or *K*, we first pass a section through these points and determine the force-couple system at the centroid *C* of the section that is required to maintain the equilibrium of portion *ABC*.[†] This system represents the internal forces in the section and consists of three force components and three couple vectors that are assumed to be directed as shown in Fig. 8.16.

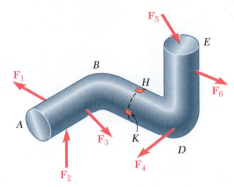

Fig. 8.15 Member *ABDE* subjected to several forces.

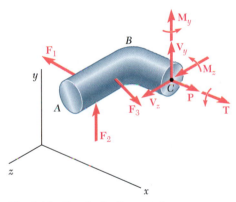

Fig. 8.16 Free-body diagram of segment *ABC* to determine the internal forces and couples at cross section *C*.

Force **P** is a centric axial force that produces normal stresses in the section. The couple vectors $\mathbf{M}_y$ and $\mathbf{M}_z$ cause the member to bend and also produce normal stresses in the section. These have been grouped in Fig. 8.17a, and the sums σ_x of the normal stresses produced at points *H* and *K* are shown in Fig. 8.18a. These stresses can be determined as shown in Sec. 4.9.

On the other hand, the twisting couple **T** and the shearing forces $\mathbf{V}_y$ and $\mathbf{V}_z$ as shown in Fig. 8.17b produce shearing stresses in the section. The sums τ_{xy} and τ_{xz} of the components of the shearing stresses produced at points *H* and *K* are shown in Fig. 8.18b and can be determined as indicated

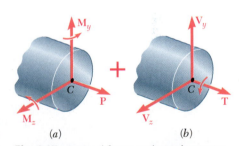

Fig. 8.17 Internal forces and couple vectors separated into (a) those causing normal stresses and (b) those causing shearing stresses.

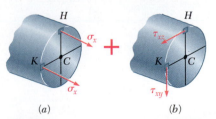

Fig. 8.18 Normal and shearing stresses at points *H* and *K*.

[†]The force-couple system at *C* can also be defined as *equivalent to the forces acting on the portion of the member located to the right of the section* (see Concept Application 8.1).

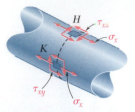

Fig. 8.19   Elements at points *H* and *K* showing combined stresses.

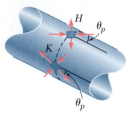

Fig. 8.20   Elements at points *H* and *K* showing principal stresses.

in Secs. 3.1C and 6.1B.[†] The normal and shearing stresses shown in parts *a* and *b* are now combined and displayed at points *H* and *K* on the surface of the member (Fig. 8.19).

The principal stresses and the orientation of the principal planes at points *H* and *K* are determined from σ_x, τ_{xy}, and τ_{xz} at each of these points by one of the methods presented in Chap. 7 (Fig. 8.20). The maximum shearing stress at each of these points and the corresponding planes can be found in a similar way.

The results in this section are valid only if the conditions of applicability of the superposition principle (Sec. 2.5) and of Saint-Venant's principle (Sec. 2.10) are met:

1. The stresses involved must not exceed the proportional limit of the material.
2. The deformations due to one of the loadings must not affect the determination of the stresses due to the others.
3. The section used in your analysis must not be too close to the points of application of the given forces.

The first of these requirements shows that the method presented here cannot be applied to plastic deformations.

[†]Note that your present knowledge allows you to determine the effect of the twisting couple **T** only in circular shafts, members with a rectangular cross section (Sec. 3.9), or thin-walled hollow members (Sec. 3.10).

Concept Application 8.1

Two forces **P**₁ and **P**₂, with a magnitude of $P_1 = 15$ kN and $P_2 = 18$ kN, are applied as shown in Fig. 8.21a to the end *A* of bar *AB*, which is welded to a cylindrical member *BD* of radius $c = 20$ mm. Knowing that the distance from *A* to the axis of member *BD* is $a = 50$ mm and assuming that all stresses remain below the proportional limit of the material, determine (*a*) the normal and shearing stresses at point *K* of the transverse section of member *BD* located at a distance $b = 60$ mm from end *B*, (*b*) the principal axes and principal stresses at *K*, and (*c*) the maximum shearing stress at *K*.

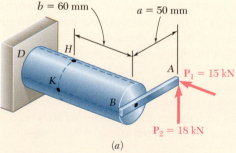

(*a*)

Fig. 8.21   Cylindrical member under combined loading. (*a*) Dimensions and loading.

(continued)

Internal Forces in Given Section. Replace the forces $\mathbf{P}_1$ and $\mathbf{P}_2$ by an equivalent system of forces and couples applied at the center C of the section containing point K (Fig. 8.21b). This system represents the internal forces in the section and consists of the following forces and couples:

1. A centric axial force $\mathbf{F}$ equal to the force $\mathbf{P}_1$ with the magnitude

$$F = P_1 = 15 \text{ kN}$$

2. A shearing force $\mathbf{V}$ equal to the force $\mathbf{P}_2$ with the magnitude

$$V = P_2 = 18 \text{ kN}$$

3. A twisting couple $\mathbf{T}$ of torque T equal to the moment of $\mathbf{P}_2$ about the axis of member BD:

$$T = P_2 a = (18 \text{ kN})(50 \text{ mm}) = 900 \text{ N·m}$$

4. A bending couple $\mathbf{M}_y$ of moment M_y equal to the moment of $\mathbf{P}_1$ about a vertical axis through C:

$$M_y = P_1 a = (15 \text{ kN})(50 \text{ mm}) = 750 \text{ N·m}$$

5. A bending couple $\mathbf{M}_z$ of moment M_z equal to the moment of $\mathbf{P}_2$ about a transverse, horizontal axis through C:

$$M_z = P_2 b = (18 \text{ kN})(60 \text{ mm}) = 1080 \text{ N·m}$$

The results are shown in Fig. 8.21c.

a. Normal and Shearing Stresses at Point K. Each of the forces and couples shown in Fig. 8.21c produce a normal or shear stress at point K. Compute each of these stresses seperately and then add the normal stresses and add the shearing stresses.

Geometric Properties of the Section For the given data, we have

$$A = \pi c^2 = \pi(0.020 \text{ m})^2 = 1.257 \times 10^{-3} \text{ m}^2$$

$$I_y = I_z = \tfrac{1}{4}\pi c^4 = \tfrac{1}{4}\pi(0.020 \text{ m})^4 = 125.7 \times 10^{-9} \text{ m}^4$$

$$J_C = \tfrac{1}{2}\pi c^4 = \tfrac{1}{2}\pi(0.020 \text{ m})^4 = 251.3 \times 10^{-9} \text{ m}^4$$

Also determine the first moment Q and the width t of the area of the cross section located above the z axis. Recall that $\bar{y} = 4c/3\pi$ for a semicircle of radius c, giving

$$Q = A'\bar{y} = \left(\frac{1}{2}\pi c^2\right)\left(\frac{4c}{3\pi}\right) = \frac{2}{3}c^3 = \frac{2}{3}(0.020 \text{ m})^3$$

$$= 5.33 \times 10^{-6} \text{ m}^3$$

and

$$t = 2c = 2(0.020 \text{ m}) = 0.040 \text{ m}$$

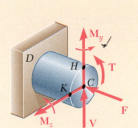

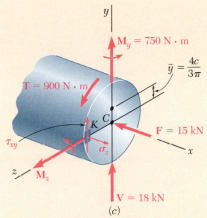

Fig. 8.21 (*Cont.*) (*b*) Internal forces and couples at section containing points H and K. (*c*) Values of forces and couples that produce stresses at point K, as well as the dimension needed to compute the first moment of area.

(*continued*)

Normal Stresses. Normal stresses are produced at K by the centric force $\mathbf{F}$ and the bending couple $\mathbf{M}_y$. However, the couple $\mathbf{M}_z$ does not produce any stress at K, since K is located on the neutral axis corresponding to that couple. Determining each sign from Fig. 8.21c gives

$$\sigma_x = -\frac{F}{A} + \frac{M_y c}{I_y} = -11.9\,\text{MPa} + \frac{(750\,\text{N·m})(0.020\,\text{m})}{125.7 \times 10^{-9}\,\text{m}^4}$$

$$= -11.9\,\text{MPa} + 119.3\,\text{MPa}$$

$$\sigma_x = +107.4\,\text{MPa}$$

Shearing Stresses. The shearing stress $(\tau_{xy})_V$ is due to the vertical shear $\mathbf{V}$, and the shearing stress $(\tau_{xy})_{\text{twist}}$ is caused by the torque $\mathbf{T}$. Using the values for Q, t, I_z, and J_C,

$$(\tau_{xy})_V = +\frac{VQ}{I_z t} = +\frac{(18 \times 10^3\,\text{N})(5.33 \times 10^{-6}\,\text{m}^3)}{(125.7 \times 10^{-9}\,\text{m}^4)(0.040\,\text{m})}$$

$$= +19.1\,\text{MPa}$$

$$(\tau_{xy})_{\text{twist}} = -\frac{Tc}{J_C} = -\frac{(900\,\text{N·m})(0.020\,\text{m})}{251.3 \times 10^{-9}\,\text{m}^4} = -71.6\,\text{MPa}$$

Adding these provides τ_{xy} at point K.

$$\tau_{xy} = (\tau_{xy})_V + (\tau_{xy})_{\text{twist}} = +19.1\,\text{MPa} - 71.6\,\text{MPa}$$

$$\tau_{xy} = -52.5\,\text{MPa}$$

In Fig. 8.21d, the normal stress σ_x and the shearing stresses τ_{xy} are acting on a square element located at K on the surface of the cylindrical member. Note that shearing stresses acting on the longitudinal sides of the element also are included.

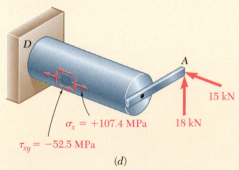

$\sigma_x = +107.4\,\text{MPa}$

$\tau_{xy} = -52.5\,\text{MPa}$

$15\,\text{kN}$

$18\,\text{kN}$

(d)

Fig. 8.21 (*Cont.*) (*d*) Element showing combined stresses at point K.

(continued)

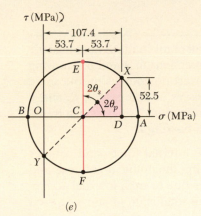

τ (MPa)

Fig. 8.21 (*Cont.*) (*e*) Mohr's circle for stresses at point *K*.

b. Principal Planes and Principal Stresses at Point *K*. Either of the two methods from Chap. 7 can be used to determine the principal planes and principal stresses at *K*. Selecting Mohr's circle, plot point *X* with coordinates $\sigma_x = +107.4$ MPa and $-\tau_{xy} = +52.5$ MPa and point *Y* with coordinates $\sigma_y = 0$ and $+\tau_{xy} = -52.5$ MPa and draw the circle with the diameter *XY* (Fig. 8.21*e*). Observing that

$$OC = CD = \tfrac{1}{2}(107.4) = 53.7 \text{ MPa} \qquad DX = 52.5 \text{ MPa}$$

we determine the orientation of the principal planes:

$$\tan 2\theta_p = \frac{DX}{CD} = \frac{52.5}{53.7} = 0.97765 \qquad 2\theta_p = 44.4° \downarrow$$

$$\theta_p = 22.2° \downarrow$$

The radius of the circle is

$$R = \sqrt{(53.7)^2 + (52.5)^2} = 75.1 \text{ MPa}$$

and the principal stresses are

$$\sigma_{\max} = OC + R = 53.7 + 75.1 = 128.8 \text{ MPa}$$
$$\sigma_{\min} = OC - R = 53.7 - 75.1 = -21.4 \text{ MPa}$$

The results are shown in Fig. 8.21*f*.

c. Maximum Shearing Stress at Point *K*. This stress corresponds to points *E* and *F* in Fig. 8.21*e*.

$$\tau_{\max} = CE = R = 75.1 \text{ MPa}$$

Observing that $2\theta_s = 90° - 2\theta_p = 90° - 44.4° = 45.6°$, the planes of maximum shearing stress form an angle $\theta_s = 22.8° \nwarrow$ with the horizontal. The corresponding element is shown in Fig. 8.21*g*. Note that the normal stresses acting on this element are represented by *OC* in Fig. 8.21*e* and are equal to +53.7 MPa.

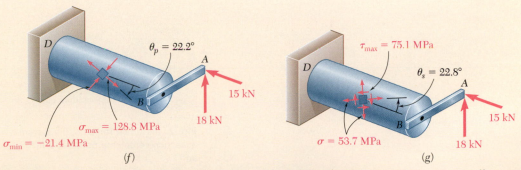

Fig. 8.21 (*Cont.*) (*f*) Principal stress element at point *K*. (*g*) Maximum shearing stress element at point *K*.

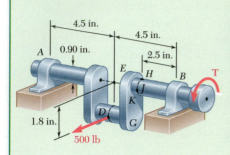

Sample Problem 8.4

A horizontal 500-lb force acts at point D of crankshaft AB held in static equilibrium by a twisting couple T and reactions at A and B. Knowing that the bearings are self-aligning and exert no couples on the shaft, determine the normal and shearing stresses at points H, J, K, and L located at the ends of the vertical and horizontal diameters of a transverse section located 2.5 in. to the left of bearing B.

STRATEGY: Begin by determining the internal forces and couples acting on the transverse section containing the points of interest, and then evaluate the stresses at these points due to each internal action. Combining these results will provide the total state of stress at each point.

MODELING:

Draw the free-body diagram of the crankshaft (Fig. 1). Find A = B = 250 lb

$$+\curvearrowleft\Sigma M_x = 0: \qquad -(500 \text{ lb})(1.8 \text{ in.}) + T = 0 \qquad T = 900 \text{ lb·in.}$$

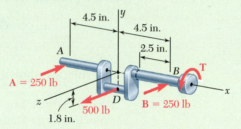

Fig. 1 Free-body diagram of crankshaft.

ANALYSIS:

Internal Forces in Transverse Section. Replace reaction **B** and the twisting couple **T** by an equivalent force-couple system at the center C of the transverse section containing H, J, K, and L. (Fig. 2.)

$$V = B = 250 \text{ lb} \qquad T = 900 \text{ lb·in.}$$

$$M_y = (250 \text{ lb})(2.5 \text{ in.}) = 625 \text{ lb·in.}$$

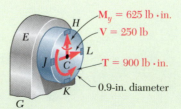

Fig. 2 Resultant force-couple system at section containing points H, J, K, and L.

(continued)

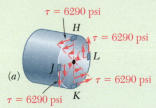

(a)

Fig. 3 Shearing stresses resulting from torque *T*.

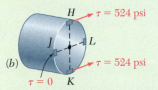

(b)

Fig. 4 Shearing stresses resulting from shearing force *V*.

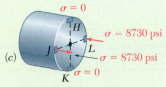

(c)

Fig. 5 Normal stresses resulting from bending couple M_y.

The geometric properties of the 0.9-in.-diameter section are

$$A = \pi(0.45 \text{ in.})^2 = 0.636 \text{ in}^2 \quad I = \tfrac{1}{4}\pi(0.45 \text{ in.})^4 = 32.2 \times 10^{-3} \text{ in}^4$$

$$J = \tfrac{1}{2}\pi(0.45 \text{ in.})^4 = 64.4 \times 10^{-3} \text{ in}^4$$

Stresses Produced by Twisting Couple T.

Using Eq. (3.10), determine the shearing stresses at points *H*, *J*, *K*, and *L* and show them in Fig. 3.

$$\tau = \frac{Tc}{J} = \frac{(900 \text{ lb·in.})(0.45 \text{ in.})}{64.4 \times 10^{-3} \text{ in}^4} = 6290 \text{ psi}$$

Stresses Produced by Shearing Force V.

The shearing force **V** produces no shearing stresses at points *J* and *L*. At points *H* and *K*, compute *Q* for a semicircle about a vertical diameter and then determine the shearing stress produced by the shear force *V* = 250 lb. These stresses are shown in Fig. 4.

$$Q = \left(\frac{1}{2}\pi c^2\right)\left(\frac{4c}{3\pi}\right) = \frac{2}{3}c^3 = \frac{2}{3}(0.45 \text{ in.})^3 = 60.7 \times 10^{-3} \text{ in}^3$$

$$\tau = \frac{VQ}{It} = \frac{(250 \text{ lb})(60.7 \times 10^{-3} \text{ in}^3)}{(32.2 \times 10^{-3} \text{ in}^4)(0.9 \text{ in.})} = 524 \text{ psi}$$

Stresses Produced by the Bending Couple M_y.

Since the bending couple **M**$_y$ acts in a horizontal plane, it produces no stresses at *H* and *K*. Use Eq. (4.15) to determine the normal stresses at points *J* and *L* and show them in Fig. 5.

$$\sigma = \frac{|M_y|c}{I} = \frac{(625 \text{ lb·in.})(0.45 \text{ in.})}{32.2 \times 10^{-3} \text{ in}^4} = 8730 \text{ psi}$$

Summary.

Add the stresses shown to obtain the total normal and shearing stresses at points *H*, *J*, *K*, and *L* (Fig. 6).

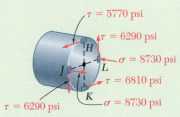

Fig. 6 Stress components at points *H*, *J*, *K*, and *L* from combining all loads.

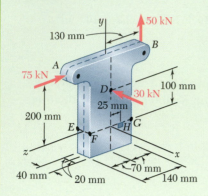

Sample Problem 8.5

Three forces are applied as shown at points A, B, and D of a short steel post. Knowing that the horizontal cross section of the post is a 40×140-mm rectangle, determine the principal stresses, principal planes, and maximum shearing stress at point H.

STRATEGY: Begin by determining the forces and couples acting on the section containing the point of interest, and then use these to calculate the normal and shearing stresses acting at the point. Using Mohr's circle, these stresses can then be transformed to obtain the principal stresses, principal planes, and maximim shearing stress.

MODELING and ANALYSIS:

Internal Forces in Section *EFG*.   Replace the three applied forces by an equivalent force-couple system at the center C of the rectangular section EFG (Fig. 1).

$$V_x = -30 \text{ kN} \qquad P = 50 \text{ kN} \qquad V_z = -75 \text{ kN}$$

$$M_x = (50 \text{ kN})(0.130 \text{ m}) - (75 \text{ kN})(0.200 \text{ m}) = -8.5 \text{ kN·m}$$

$$M_y = 0 \qquad M_z = (30 \text{ kN})(0.100 \text{ m}) = 3 \text{ kN·m}$$

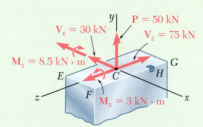

Fig. 1   Equivalent force-couple system at section containing points E, F, G, and H.

Note that there is no twisting couple about the y axis. The geometric properties of the rectangular section are

$$A = (0.040 \text{ m})(0.140 \text{ m}) = 5.6 \times 10^{-3} \text{ m}^2$$

$$I_x = \tfrac{1}{12}(0.040 \text{ m})(0.140 \text{ m})^3 = 9.15 \times 10^{-6} \text{ m}^4$$

$$I_z = \tfrac{1}{12}(0.140 \text{ m})(0.040 \text{ m})^3 = 0.747 \times 10^{-6} \text{ m}^4$$

(continued)

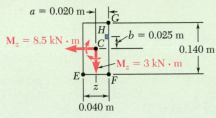

Fig. 2 Dimensions and bending couples used to determine normal stresses.

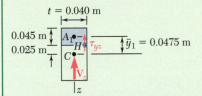

Fig. 3 Dimensions and shearing force used to determine the transverse shearing stress.

Normal Stress at *H*. The normal stresses σ_y are produced by the centric force **P** and by the bending couples $\mathbf{M}_x$ and $\mathbf{M}_z$. The sign of each stress is determined by carefully examining the force-couple system at *C* (Fig. 2).

$$\sigma_y = +\frac{P}{A} + \frac{|M_z|a}{I_z} - \frac{|M_x|b}{I_x}$$

$$= \frac{50\ \text{kN}}{5.6 \times 10^{-3}\ \text{m}^2} + \frac{(3\ \text{kN·m})(0.020\ \text{m})}{0.747 \times 10^{-6}\ \text{m}^4} - \frac{(8.5\ \text{kN·m})(0.025\ \text{m})}{9.15 \times 10^{-6}\ \text{m}^4}$$

$$\sigma_y = 8.93\ \text{MPa} + 80.3\ \text{MPa} - 23.2\ \text{MPa} \qquad \sigma_y = 66.0\ \text{MPa} \quad \blacktriangleleft$$

Shearing Stress at *H*. Considering the shearing force $\mathbf{V}_x$, we note that $Q = 0$ with respect to the *z* axis, since *H* is on the edge of the cross section. Thus, $\mathbf{V}_x$ produces no shearing stress at *H*. The shearing force $\mathbf{V}_z$ does produce a shearing stress at *H* (Fig. 3).

$$Q = A_1\bar{y}_1 = [(0.040\ \text{m})(0.045\ \text{m})](0.0475\ \text{m}) = 85.5 \times 10^{-6}\ \text{m}^3$$

$$\tau_{yz} = \frac{V_zQ}{I_xt} = \frac{(75\ \text{kN})(85.5 \times 10^{-6}\ \text{m}^3)}{(9.15 \times 10^{-6}\ \text{m}^4)(0.040\ \text{m})} \qquad \tau_{yz} = 17.52\ \text{MPa} \quad \blacktriangleleft$$

Principal Stresses, Principal Planes, and Maximum Shearing Stress at *H*. Draw Mohr's circle for the stresses at point *H* (Fig. 4).

$$\tan 2\theta_p = \frac{17.52}{33.0} \qquad 2\theta_p = 27.96° \qquad \theta_p = 13.98° \quad \blacktriangleleft$$

$$R = \sqrt{(33.0)^2 + (17.52)^2} = 37.4\ \text{MPa} \qquad \tau_{\text{max}} = 37.4\ \text{MPa} \quad \blacktriangleleft$$

$$\sigma_{\text{max}} = OA = OC + R = 33.0 + 37.4 \qquad \sigma_{\text{max}} = 70.4\ \text{MPa} \quad \blacktriangleleft$$

$$\sigma_{\text{min}} = OB = OC - R = 33.0 - 37.4 \qquad \sigma_{\text{min}} = -7.4\ \text{MPa} \quad \blacktriangleleft$$

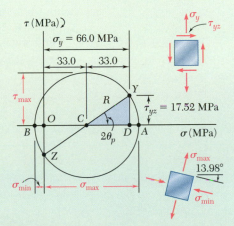

Fig. 4 Mohr's circle at point *H* used for finding principal stresses and maximum shearing stress and their orientation.

Problems

8.31 Two 1.2-kip forces are applied to an L-shaped machine element *AB* as shown. Determine the normal and shearing stresses at (*a*) point *a*, (*b*) point *b*, (*c*) point *c*.

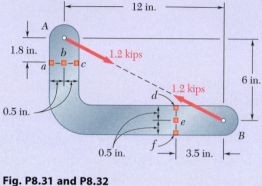

Fig. P8.31 and P8.32

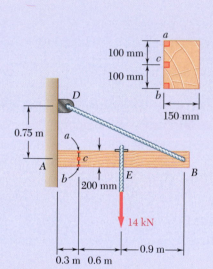

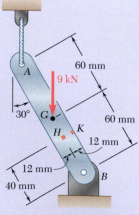

Fig. P8.33

8.32 Two 1.2-kip forces are applied to an L-shaped machine element *AB* as shown. Determine the normal and shearing stresses at (*a*) point *d*, (*b*) point *e*, (*c*) point *f*.

8.33 The cantilever beam *AB* has a rectangular cross section of 150 × 200 mm. Knowing that the tension in the cable *BD* is 10.4 kN and neglecting the weight of the beam, determine the normal and shearing stresses at the three points indicated.

8.34 through 8.36 Member *AB* has a uniform rectangular cross section of 10 × 24 mm. For the loading shown, determine the normal and shearing stresses at (*a*) point *H*, (*b*) point *K*.

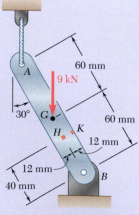

Fig. P8.34

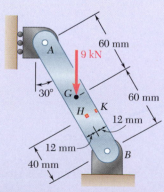

Fig. P8.35

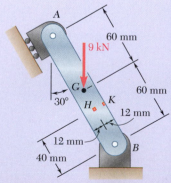

Fig. P8.36

8.37 A 1.5-kip force and a 9-kip·in. couple are applied at the top of the 2.5-in.-diameter cast-iron post shown. Determine the normal and shearing stresses at (*a*) point *H*, (*b*) point *K*.

8.38 Two forces are applied to the pipe *AB* as shown. Knowing that the pipe has inner and outer diameters equal to 35 and 42 mm, respectively, determine the normal and shearing stresses at (*a*) point *a*, (*b*) point *b*.

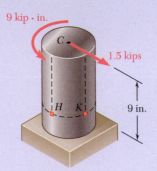

Fig. P8.37

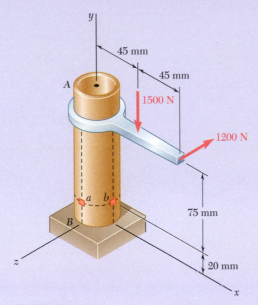

Fig. P8.38

8.39 Several forces are applied to the pipe assembly shown. Knowing that the pipe has inner and outer diameters equal to 1.61 in. and 1.90 in., respectively, determine the normal and shearing stresses at (*a*) point *H*, (*b*) point *K*.

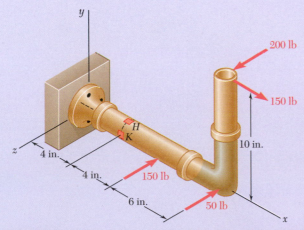

Fig. P8.39

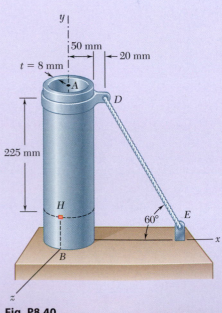

Fig. P8.40

8.40 The steel pile *AB* has a 100-mm outer diameter and an 8-mm wall thickness. Knowing that the tension in the cable is 40 kN, determine the normal and shearing stresses at point *H*.

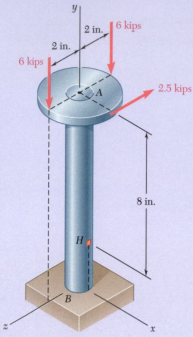

Fig. P8.41

8.41 Three forces are applied to a 4-in.-diameter plate that is attached to the solid 1.8-in.-diameter shaft *AB*. At point *H*, determine (*a*) the principal stresses and principal planes, (*b*) the maximum shearing stress.

8.42 The steel pipe *AB* has a 72-mm outer diameter and a 5-mm wall thickness. Knowing that the arm *CDE* is rigidly attached to the pipe, determine the principal stresses, principal planes, and the maximum shearing stress at point *H*.

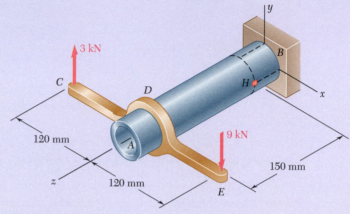

Fig. P8.42

8.43 A 13-kN force is applied as shown to the 60-mm-diameter cast-iron post *ABD*. At point *H*, determine (*a*) the principal stresses and principal planes, (*b*) the maximum shearing stress.

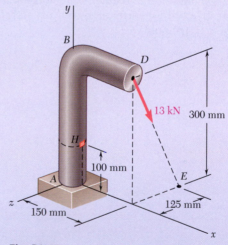

Fig. P8.43

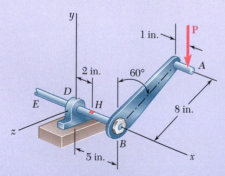

Fig. P8.44

8.44 A vertical force **P** of magnitude 60 lb is applied to the crank at point *A*. Knowing that the shaft *BDE* has a diameter of 0.75 in., determine the principal stresses and the maximum shearing stress at point *H* located at the top of the shaft, 2 in. to the right of support *D*.

8.45 Three forces are applied to the bar shown. Determine the normal and shearing stresses at (*a*) point *a*, (*b*) point *b*, (*c*) point *c*.

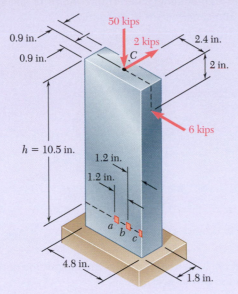

Fig. P8.45

8.46 Solve Prob. 8.45, assuming that *h* = 12 in.

8.47 Three forces are applied to the bar shown. Determine the normal and shearing stresses at (*a*) point *a*, (*b*) point *b*, (*c*) point *c*.

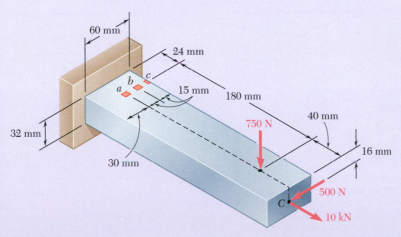

Fig. P8.47

8.48 Solve Prob. 8.47, assuming that the 750-N force is directed vertically upward.

8.49 Two forces are applied to the small post *BD* as shown. Knowing that the vertical portion of the post has a cross section of 1.5 × 2.4 in., determine the principal stresses, principal planes, and maximum shearing stress at point *H*.

8.50 Solve Prob. 8.49, assuming that the magnitude of the 6000-lb force is reduced to 1500 lb.

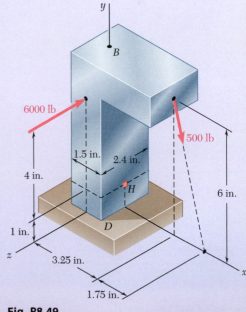

Fig. P8.49

8.51 Three forces are applied to the machine component ABD as shown. Knowing that the cross section containing point H is a 20 × 40-mm rectangle, determine the principal stresses and the maximum shearing stress at point H.

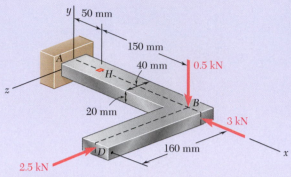

Fig. P8.51

8.52 Solve Prob. 8.51, assuming that the magnitude of the 2.5-kN force is increased to 10 kN.

8.53 Three steel plates, each 13 mm thick, are welded together to form a cantilever beam. For the loading shown, determine the normal and shearing stresses at points a and b.

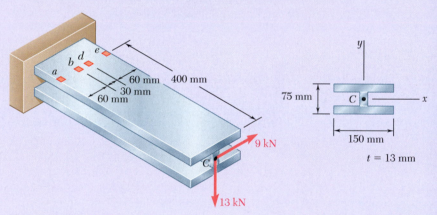

Fig. P8.53 and *P8.54*

8.54 Three steel plates, each 13 mm thick, are welded together to form a cantilever beam. For the loading shown, determine the normal and shearing stresses at points d and e.

8.55 Two forces P_1 and P_2 are applied as shown in directions perpendicular to the longitudinal axis of a W310 × 60 beam. Knowing that $P_1 = 25$ kN and $P_2 = 24$ kN, determine the principal stresses and the maximum shearing stress at point a.

8.56 Two forces P_1 and P_2 are applied as shown in directions perpendicular to the longitudinal axis of a W310 × 60 beam. Knowing that $P_1 = 25$ kN and $P_2 = 24$ kN, determine the principal stresses and the maximum shearing stress at point b.

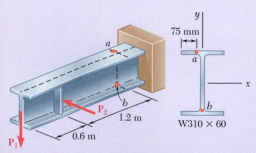

Fig. P8.55 and *P8.56*

8.57 Four forces are applied to a W8 × 28 rolled-steel beam as shown. Determine the principal stresses and maximum shearing stress at point *a*.

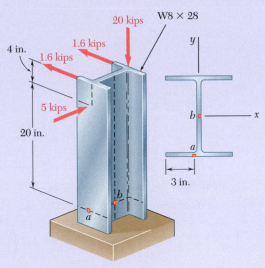

Fig. P8.57 and P8.58

8.58 Four forces are applied to a W8 × 28 rolled-steel beam as shown. Determine the principal stresses and maximum shearing stress at point *b*.

8.59 A force **P** is applied to a cantilever beam by means of a cable attached to a bolt located at the center of the free end of the beam. Knowing that **P** acts in a direction perpendicular to the longitudinal axis of the beam, determine (*a*) the normal stress at point *a* in terms of *P*, *b*, *h*, *l*, and β, (*b*) the values of β for which the normal stress at *a* is zero.

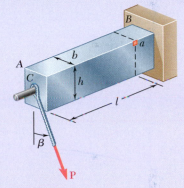

Fig. P8.59

8.60 A vertical force **P** is applied at the center of the free end of cantilever beam *AB*. (*a*) If the beam is installed with the web vertical ($\beta = 0$) and with its longitudinal axis *AB* horizontal, determine the magnitude of the force **P** for which the normal stress at point *a* is +120 MPa. (*b*) Solve part *a*, assuming that the beam is installed with $\beta = 3°$.

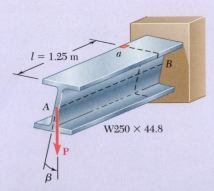

Fig. P8.60

*8.61 A 5-kN force **P** is applied to a wire that is wrapped around bar *AB* as shown. Knowing that the cross section of the bar is a square of side $d = 40$ mm, determine the principal stresses and the maximum shearing stress at point *a*.

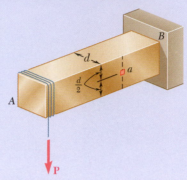

Fig. P8.61

*8.62 Knowing that the structural tube shown has a uniform wall thickness of 0.3 in., determine the principal stresses, principal planes, and maximum shearing stress at (*a*) point *H*, (*b*) point *K*.

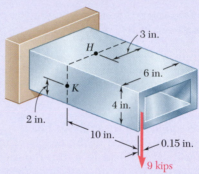

Fig. P8.62

*8.63 The structural tube shown has a uniform wall thickness of 0.3 in. Knowing that the 15-kip load is applied 0.15 in. above the base of the tube, determine the shearing stress at (*a*) point *a*, (*b*) point *b*.

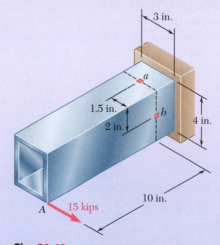

Fig. P8.63

*8.64 For the tube and loading of Prob. 8.63, determine the principal stresses and the maximum shearing stress at point *b*.

Review and Summary

Normal and Shearing Stresses in a Beam

The two fundamental relationships for the normal stress σ_x and the shearing stress τ_{xy} at any given point of a cross section of a prismatic beam are

$$\sigma_x = -\frac{My}{I} \qquad \text{(8.1)}$$

and

$$\tau_{xy} = -\frac{VQ}{It} \qquad \text{(8.2)}$$

where V = shear in the section

M = bending moment in the section

y = distance of the point from the neutral surface

I = centroidal moment of inertia of the cross section

Q = first moment about the neutral axis of the portion of the cross section located above the given point

t = width of the cross section at the given point

Principal Planes and Principal Stresses in a Beam

Using one of the methods of Ch. 7 for the transformation of stresses, the principal planes and principal stresses were obtained at various points (Fig. 8.22).

We investigated the distribution of principal stresses in a narrow rectangular cantilever beam subjected to a concentrated load **P** at its free end, and found that in any given transverse section—except close to the point of application of the load—the maximum principal stress $\sigma_{\max}$ did not exceed the determination of the maximum normal stress σ_m occurring at the surface of the beam.

While this is true for many beams of nonrectangular cross section, it may not hold for W-beams or S-beams, where $\sigma_{\max}$ at the junctions b and d of the web with the flanges of the beam (Fig. 8.23) may exceed the value of σ_m occurring at points a and e. Therefore, the design of a rolled-steel beam should include the determination of the maximum principal stress at these points.

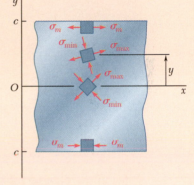

Fig. 8.22 Principal stress elements at selected points of beam.

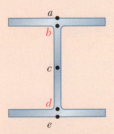

Fig. 8.23 Key locations for determination of principal stresses in I-shaped beams.

Design of Transmission Shafts Under Transverse Loads

The design of *transmission shafts* subjected to *transverse loads* and torques should include consideration of both the normal stresses due to the bending moment M and the shearing stresses due to the torque T. At any given transverse section of a cylindrical shaft (either solid or hollow), the minimum allowable value of the ratio J/c for the cross section is:

$$\frac{J}{c} = \frac{\left(\sqrt{M^2 + T^2}\right)_{\text{max}}}{\tau_{\text{all}}} \tag{8.6}$$

Stresses Under General Loading Conditions

In preceding chapters, you learned to determine the stresses in prismatic members caused by axial loadings (Chaps. 1 and 2), torsion (Chap. 3), bending (Chap. 4), and transverse loadings (Chaps. 5 and 6). In the second part of this chapter (Sec. 8.3), we combined this knowledge to determine stresses under more general loading conditions.

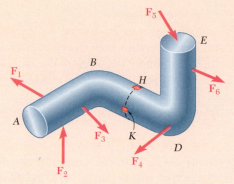

Fig. 8.24 Member *ABCD* subjected to several loads.

For instance, to determine the stresses at point H or K of the bent member shown in Fig. 8.24, a section is passed through these points and the applied loads are replaced by an equivalent force-couple system at the centroid C of the section (Fig. 8.25). The normal and shearing stresses produced at H or K by each of the forces and couples applied at C are determined and then combined to obtain the resulting normal stress σ_x and the resulting shearing stresses τ_{xy} and τ_{xz} at H or K. The principal stresses, the orientation of the principal planes, and the maximum shearing stress at point H or K are then determined using one of the methods presented in Chap. 7.

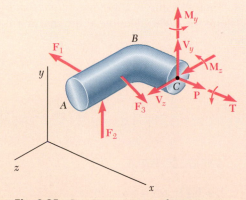

Fig. 8.25 Free-body diagram of segment *ABC* to determine the internal forces and couples at cross section *C*.

Review Problems

8.65 (a) Knowing that $\sigma_{all} = 24$ ksi and $\tau_{all} = 14.5$ ksi, select the most economical wide-flange shape that should be used to support the loading shown. (b) Determine the values to be expected for σ_m, τ_m, and the principal stress σ_{max} at the junction of a flange and the web of the selected beam.

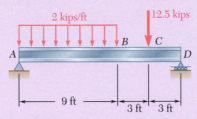

Fig. P8.65

8.66 Neglecting the effect of fillets and of stress concentrations, determine the smallest permissible diameters of the solid rods BC and CD. Use $\tau_{all} = 60$ MPa.

8.67 Knowing that rods BC and CD are of diameter 24 mm and 36 mm, respectively, determine the maximum shearing stress in each rod. Neglect the effect of fillets and of stress concentrations.

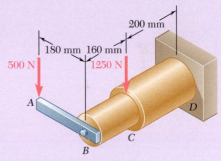

Fig. P8.66 and *P8.67*

8.68 The solid shaft AB rotates at 450 rpm and transmits 20 kW from the motor M to machine tools connected to gears F and G. Knowing that $\tau_{all} = 55$ MPa and assuming that 8 kW is taken off at gear F and 12 kW is taken off at gear G, determine the smallest permissible diameter of shaft AB.

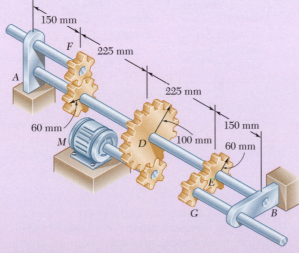

Fig. P8.68

8.69 A 6-kip force is applied to the machine element *AB* as shown. Knowing that the uniform thickness of the element is 0.8 in., determine the normal and shearing stresses at (*a*) point *a*, (*b*) point *b*, (*c*) point *c*.

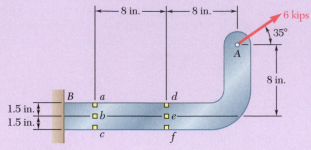

Fig. P8.69

8.70 A thin strap is wrapped around a solid rod of radius *c* = 20 mm as shown. Knowing that *l* = 100 mm and *F* = 5 kN, determine the normal and shearing stresses at (*a*) point *H*, (*b*) point *K*.

8.71 A close-coiled spring is made of a circular wire of radius *r* that is formed into a helix of radius *R*. Determine the maximum shearing stress produced by the two equal and opposite forces **P** and **P′**. (*Hint:* First determine the shear **V** and the torque **T** in a transverse cross section.)

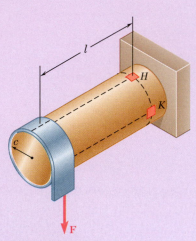

Fig. P8.70

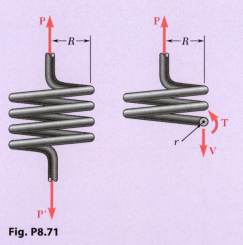

Fig. P8.71

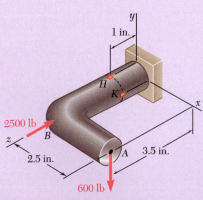

Fig. P8.72

8.72 Forces are applied at points *A* and *B* of the solid cast-iron bracket shown. Knowing that the bracket has a diameter of 0.8 in., determine the principal stresses and the maximum shearing stress at (*a*) point *H*, (*b*) point *K*.

8.73 Knowing that the bracket AB has a uniform thickness of $\frac{5}{8}$ in., determine (*a*) the principal planes and principal stresses at point K, (*b*) the maximum shearing stress at point K.

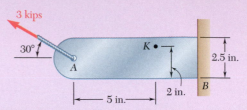

Fig. P8.73

8.74 For the post and loading shown, determine the principal stresses, principal planes, and maximum shearing stress at point H.

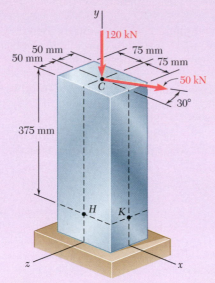

Fig. P8.74

8.75 Knowing that the structural tube shown has a uniform wall thickness of 0.25 in., determine the normal and shearing stresses at the three points indicated.

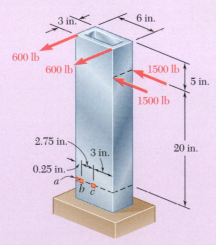

Fig. P8.75

8.76 The cantilever beam AB will be installed so that the 60-mm side forms an angle β between 0 and 90° with the vertical. Knowing that the 600-N vertical force is applied at the center of the free end of the beam, determine the normal stress at point a when (*a*) $\beta = 0$, (*b*) $\beta = 90°$. (*c*) Also, determine the value of β for which the normal stress at point a is a maximum and the corresponding value of that stress.

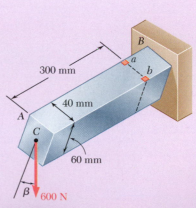

Fig. P8.76

Computer Problems

The following problems are designed to be solved with a computer.

8.C1 Let us assume that the shear V and the bending moment M have been determined in a given section of a rolled-steel beam. Write a computer program to calculate in that section, from the data available in Appendix C, (a) the maximum normal stress σ_m, (b) the principal stress σ_{max} at the junction of a flange and the web. Use this program to solve parts a and b of the following problems:

(1) Prob. 8.1 (Use $V = 45$ kips and $M = 450$ kip·in.)
(2) Prob. 8.2 (Use $V = 22.5$ kips and $M = 450$ kip·in.)
(3) Prob. 8.3 (Use $V = 700$ kN and $M = 1750$ kN·m)
(4) Prob. 8.4 (Use $V = 850$ kN and $M = 1700$ kN·m)

8.C2 A cantilever beam AB with a rectangular cross section of width b and depth $2c$ supports a single concentrated load **P** at its end A. Write a computer program to calculate, for any values of x/c and y/c, (a) the ratios σ_{max}/σ_m and σ_{min}/σ_m, where σ_{max} and σ_{min} are the principal stresses at point $K(x, y)$ and σ_m the maximum normal stress in the same transverse section, (b) the angle θ_p that the principal planes at K form with a transverse and a horizontal plane through K. Use this program to check the values shown in Fig. 8.8 and to verify that σ_{max} exceeds σ_m if $x \leq 0.544c$, as indicated in the second footnote on page 560.

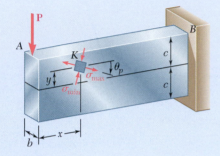

Fig. P8.C2

8.C3 Disks $D_1, D_2, \ldots, D_n$ are attached as shown in Fig. 8.C3 to the solid shaft AB of length L, uniform diameter d, and allowable shearing stress τ_{all}. Forces $\mathbf{P}_1, \mathbf{P}_2, \ldots, \mathbf{P}_n$ of known magnitude (except for one of them) are applied to the disks, either at the top or bottom of its vertical diameter, or at the left or right end of its horizontal diameter. Denoting by r_i the radius of disk D_i and by c_i its distance from the support at A, write a computer program to calculate (a) the magnitude of the unknown force $\mathbf{P}_i$, (b) the smallest permissible value of the diameter d of shaft AB. Use this program to solve Prob. 8.18.

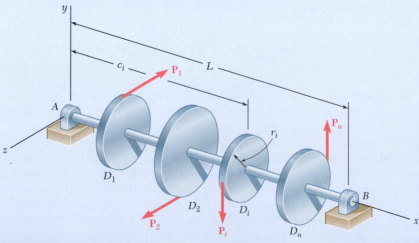

Fig. P8.C3

8.C4 The solid shaft AB of length L, uniform diameter d, and allowable shearing stress τ_{all} rotates at a given speed expressed in rpm (Fig. 8.C4). Gears $G_1, G_2, \ldots, G_n$ are attached to the shaft and each of these gears meshes with another gear (not shown), either at the top or bottom of its vertical diameter, or at the left or right end of its horizontal diameter. One of these gears is connected to a motor and the rest of them to various machine tools. Denoting by r_i the radius of disk G_i, by c_i its distance from the support at A, and by P_i the power transmitted to that gear ($+$ sign) or taken off that gear ($-$sign), write a computer program to calculate the smallest permissible value of the diameter d of shaft AB. Use this program to solve Probs. 8.27 and 8.68.

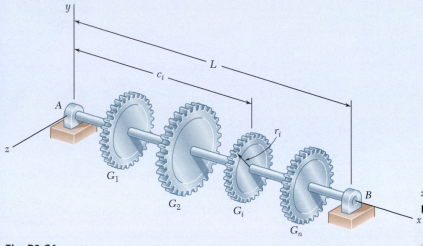

Fig. P8.C4

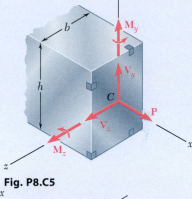

Fig. P8.C5

8.C5 Write a computer program that can be used to calculate the normal and shearing stresses at points with given coordinates y and z located on the surface of a machine part having a rectangular cross section. The internal forces are known to be equivalent to the force-couple system shown. Write the program so that the loads and dimensions can be expressed in either SI or U.S. customary units. Use this program to solve (a) Prob. 8.45b, (b) Prob. 8.47a.

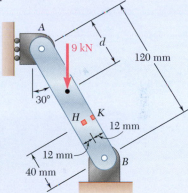

Fig. P8.C6

8.C6 Member AB has a rectangular cross section of 10×24 mm. For the loading shown, write a computer program that can be used to determine the normal and shearing stresses at points H and K for values of d from 0 to 120 mm, using 15-mm increments. Use this program to solve Prob. 8.35.

*8.C7 The structural tube shown has a uniform wall thickness of 0.3 in. A 9-kip force is applied at a bar (not shown) that is welded to the end of the tube. Write a computer program that can be used to determine, for any given value of c, the principal stresses, principal planes, and maximum shearing stress at point H for values of d from -3 in. to 3 in., using one-inch increments. Use this program to solve Prob. 8.62a.

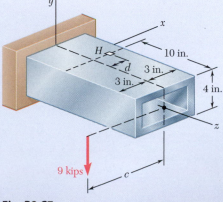

Fig. P8.C7

9

Deflection of Beams

In addition to strength considerations, the design of this bridge is also based on deflection evaluations.

Objectives

In this chapter, you will:

- **Develop** the governing differential equation for the elastic curve, the basis for the several techniques considered in this chapter for determining beam deflections.

- **Use** direct integration to obtain slope and deflection equations for beams of simple constraints and loadings.

- **Use** singularity functions to determine slope and deflection equations for beams of more complex constraints and loadings.

- **Use** the method of superposition to determine slope and deflection in beams by combining tabulated formulae.

- **Use** the moment-area theorems as an alternate technique to determine slope and deflection at specific points in a beam.

- **Apply** direct integration, singularity functions, superposition, and the moment-area theorems to analyze statically indeterminate beams.

Introduction

In the preceding chapter we learned to design beams for strength. This chapter discusses another aspect in the design of beams: the determination of the *deflection*. The *maximum deflection* of a beam under a given load is of particular interest, since the design specifications of a beam will generally include a maximum allowable value for its deflection. A knowledge of deflections is also required to analyze *indeterminate beams*, in which the number of reactions at the supports exceeds the number of equilibrium equations available to determine unknowns.

Recall from Sec. 4.2 that a prismatic beam subjected to pure bending is bent into a circular arc and, within the elastic range, the curvature of the neutral surface is

$$\frac{1}{\rho} = \frac{M}{EI} \tag{4.21}$$

where M is the bending moment, E is the modulus of elasticity, and I is the moment of inertia of the cross section about its neutral axis.

When a beam is subjected to a transverse loading, Eq. (4.21) remains valid for any transverse section, provided that Saint-Venant's principle applies. However, both the bending moment and the curvature of the neutral surface vary from section to section. Denoting by x the distance from the left end of the beam, we write

$$\frac{1}{\rho} = \frac{M(x)}{EI} \tag{9.1}$$

Knowing the curvature at various points of the beam will help us to draw some general conclusions about the deformation of the beam under loading (Sec. 9.1).

To determine the slope and deflection of the beam at any given point, the second-order linear differential equation, which governs the *elastic curve* characterizing the shape of the deformed beam (Sec. 9.1A), is given as

$$\frac{d^2y}{dx^2} = \frac{M(x)}{EI}$$

If the bending moment can be represented for all values of x by a single function $M(x)$, as shown in Fig. 9.1, the slope $\theta = dy/dx$ and the

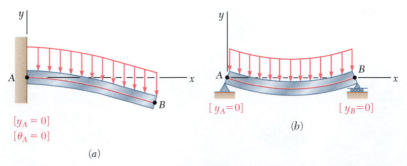

Fig. 9.1 Situations where bending moment can be expressed by a single function $M(x)$. (*a*) Uniformly-loaded cantilever beam. (*b*) Uniformly-loaded simply supported beam.

deflection y at any point of the beam can be obtained through two successive integrations. The two constants of integration introduced in the process are determined from the boundary conditions.

However, if different analytical functions are required to represent the bending moment in various portions of the beam, different differential equations are also required, leading to different functions defining the elastic curve in various portions of the beam. For the beam and loading of Fig. 9.2, for example, two differential equations are

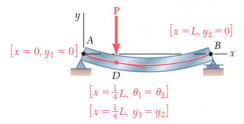

Fig. 9.2 Situation where two sets of equations are required.

required: one for the portion AD and the other for the portion DB. The first equation yields functions θ_1 and y_1, and the second functions θ_2 and y_2. Altogether, four constants of integration must be determined; two will be obtained with the deflection being zero at A and B, and the other two by expressing that the portions AD and DB have the same slope and the same deflection at D.

Sec. 9.1B shows that, in a beam supporting a distributed load $w(x)$, the elastic curve can be obtained directly from $w(x)$ through four successive integrations. The constants introduced in this process are determined from the boundary values of V, M, θ, and y.

Section 9.2 discusses *statically indeterminate beams* where the reactions at the supports involve four or more unknowns. The three equilibrium equations must be supplemented with equations obtained from the boundary conditions that are imposed by the supports.

Determining the elastic curve when several functions are required for the bending moment M can be quite complex, since it requires matching slopes and deflections at every transition point. Section 9.3 uses *singularity functions* to simplify the determination of θ and y at any point of the beam.

The *method of superposition* consists of separately determining and then adding the slope and deflection caused by the various loads applied to a beam (Sec. 9.4). This procedure can be facilitated by the use of the table in Appendix D, which gives the slopes and deflections of beams for various loadings and types of support.

In Sec. 9.5, certain geometric properties of the elastic curve are used to determine the deflection and slope of a beam at a given point. Instead of expressing the bending moment as a function $M(x)$ and integrating it analytically, a diagram representing a variation of M/EI over the length of the beam is drawn, and two moment-area theorems are derived. The *first moment-area theorem* enables the calculation of the angle between the tangents to the beam at two different points. The *second moment-area theorem* is used to calculate the vertical distance from a point on the beam to a tangent through a second point.

The moment-area theorems are used in Sec. 9.5B to determine the slope and deflection at selected points of cantilever beams and beams with symmetric loads. Section 9.5C shows that the areas and moments of areas defined by the *M/EI* diagram can be determined more easily if we draw the *bending-moment diagram by parts*. This method is particularly effective for *beams of variable cross section*.

Beams with unsymmetric loads and overhanging beams are considered in Sec. 9.6A. Since in beams with unsymmetric loads the maximum deflection does not occur at the center of a beam, Sec. 9.6B shows how to locate the point where the tangent is horizontal in order to determine the *maximum deflection*. Section 9.6C is devoted to the solution of problems involving *statically indeterminate beams*.

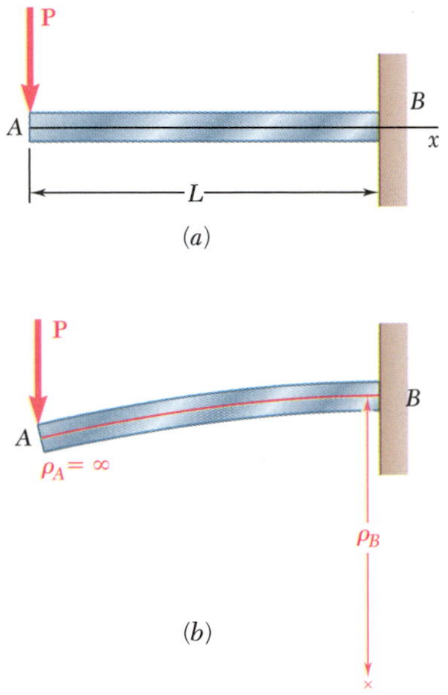

9.1 DEFORMATION UNDER TRANSVERSE LOADING

Recall that Eq. (4.21) relates the curvature of the neutral surface to the bending moment in a beam in pure bending. This equation is valid for any given transverse section of a beam subjected to a transverse loading, provided that Saint-Venant's principle applies. However, both the bending moment and the curvature of the neutral surface vary from section to section. Denoting by x the distance of the section from the left end of the beam,

$$\boxed{\frac{1}{\rho} = \frac{M(x)}{EI}} \tag{9.1}$$

Consider, for example, a cantilever beam AB of length L subjected to a concentrated load **P** at its free end A (Fig. 9.3*a*). We have $M(x) = -Px$, and substituting into Eq. (9.1) gives

$$\frac{1}{\rho} = -\frac{Px}{EI}$$

which shows that the curvature of the neutral surface varies linearly with x from zero at A, where ρ_A itself is infinite, to $-PL/EI$ at B, where $|\rho_B| = EI/PL$ (Fig. 9.3*b*).

Now consider the overhanging beam AD of Fig. 9.4*a* that supports two concentrated loads. From the free-body diagram of the beam (Fig. 9.4*b*), the reactions at the supports are $R_A = 1$ kN and $R_C = 5$ kN. The corresponding

Fig. 9.3 (*a*) Cantilever beam with concentrated load. (*b*) Deformed beam showing curvature at ends.

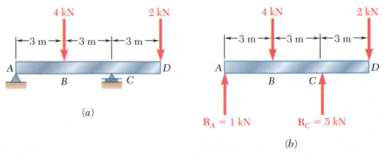

(*a*)

(*b*)

Fig. 9.4 (*a*) Overhanging beam with two concentrated loads. (*b*) Free-body diagram showing reaction forces.

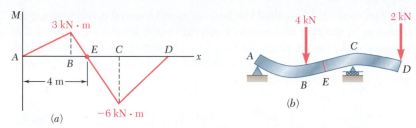

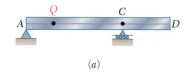

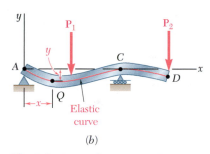

Fig. 9.5 Beam of Fig. 9.4. (*a*) Bending-moment diagram. (*b*) Deformed shape.

bending-moment diagram is shown in Fig. 9.5*a*. Note from the diagram that *M* and the curvature of the beam are both zero at each end and at a point *E* located at $x = 4$ m. Between *A* and *E*, the bending moment is positive, and the beam is concave upward. Between *E* and *D*, the bending moment is negative and the beam is concave downward (Fig. 9.5*b*). The largest value of the curvature (i.e., the smallest value of the radius of curvature) occurs at support *C*, where |*M*| is maximum.

The shape of the deformed beam is obtained from the information about its curvature. However, the analysis and design of a beam usually requires more precise information on the *deflection* and the *slope* at various points. Of particular importance is the maximum deflection of the beam. Equation (9.1) will be used in the next section to find the relationship between the deflection *y* measured at a given point *Q* on the axis of the beam and the distance *x* of that point from some fixed origin (Fig. 9.6). This relationship is the equation of the *elastic curve,* into which the axis of the beam is transformed under the given load (Fig. 9.6*b*).[†]

Fig. 9.6 Beam of Fig. 9.4. (*a*) Undeformed. (*b*) Deformed.

9.1A Equation of The Elastic Curve

Recall from elementary calculus that the curvature of a plane curve at a point $Q(x,y)$ is

$$\frac{1}{\rho} = \frac{\dfrac{d^2y}{dx^2}}{\left[1 + \left(\dfrac{dy}{dx}\right)^2\right]^{3/2}} \tag{9.2}$$

where dy/dx and d^2y/dx^2 are the first and second derivatives of the function $y(x)$ represented by that curve. For the elastic curve of a beam, however, the slope dy/dx is very small, and its square is negligible compared to unity. Therefore,

$$\frac{1}{\rho} = \frac{d^2y}{dx^2} \tag{9.3}$$

Substituting for $1/\rho$ from Eq. (9.3) into Eq. (9.1),

$$\frac{d^2y}{dx^2} = \frac{M(x)}{EI} \tag{9.4}$$

This equation is a second-order linear differential equation; it is the governing differential equation for the elastic curve.

[†]In this chapter, *y* represents a vertical displacement. It was used in previous chapters to represent the distance of a given point in a transverse section from the neutral axis of that section.

The product EI is called the *flexural rigidity,* and if it varies along the beam, as in the case of a beam of varying depth, it must be expressed as a function of x before integrating Eq. (9.4). However, for a prismatic beam, the flexural rigidity is constant. Multiply both members of Eq. (9.4) by EI and integrate in x to obtain

$$EI\frac{dy}{dx} = \int_0^x M(x)\,dx + C_1 \tag{9.5a}$$

where C_1 is a constant of integration. Denoting by $\theta(x)$ the angle, measured in radians, that the tangent to the elastic curve at Q forms with the horizontal (Fig. 9.7), and recalling that this angle is very small,

$$\frac{dy}{dx} = \tan\theta \simeq \theta(x)$$

Thus, Eq. (9.5a) in the alternative form is

$$EI\,\theta(x) = \int_0^x M(x)\,dx + C_1 \tag{9.5b}$$

Integrating Eq. (9.5) in x,

$$EI\,y = \int_0^x \left[\int_0^x M(x)\,dx + C_1\right] dx + C_2$$

$$EI\,y = \int_0^x dx \int_0^x M(x)\,dx + C_1 x + C_2 \tag{9.6}$$

where C_2 is a second constant and where the first term in the right-hand member represents the function of x obtained by integrating the bending moment $M(x)$ twice in x. Although the constants C_1 and C_2 are as yet undetermined, Eq. (9.6) defines the deflection of the beam at any given point Q, and Eqs. (9.5a) or (9.5b) similarly define the slope of the beam at Q.

The constants C_1 and C_2 are determined from the *boundary conditions* or, more precisely, from the conditions imposed on the beam by its supports. Limiting this analysis to *statically determinate beams,* which are supported so that the reactions at the supports can be obtained by the methods of statics, only three types of beams need to be considered here (Fig. 9.8): (*a*) the *simply supported beam,* (*b*) the *overhanging beam,* and (*c*) the *cantilever beam.*

In Figs. 9.8*a* and *b*, the supports consist of a pin and bracket at A and a roller at B and require that the deflection be zero at each of these points. Letting $x = x_A$, $y = y_A = 0$ in Eq. (9.6) and then setting $x = x_B$, $y = y_B = 0$ in the same equation, two equations are obtained that can be solved for C_1 and C_2. For the cantilever beam (Fig. 9.8*c*), both the deflection and the slope at A must be zero. Letting $x = x_A$, $y = y_A = 0$ in Eq. (9.6) and $x = x_A$, $\theta = \theta_A = 0$ in Eq. (9.5b), two equations are again obtained that can be solved for C_1 and C_2.

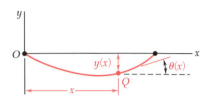

Fig. 9.7 Slope $\theta(x)$ of tangent to the elastic curve.

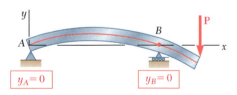

(*a*) Simply supported beam

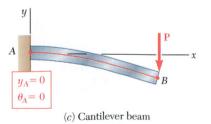

(*b*) Overhanging beam

(*c*) Cantilever beam

Fig. 9.8 Known boundary conditions for statically determinate beams.

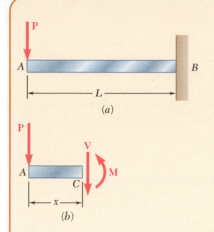

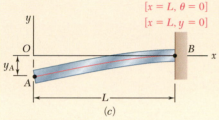

Fig. 9.9 (a) Cantilever beam with end load. (b) Free-body diagram of section *AC*. (c) Deformed shape and boundary conditions.

Concept Application 9.1

The cantilever beam *AB* is of uniform cross section and carries a load **P** at its free end A (Fig. 9.9a). Determine the equation of the elastic curve and the deflection and slope at *A*.

Using the free-body diagram of the portion *AC* of the beam (Fig. 9.9b), where *C* is located at a distance *x* from end *A*,

$$M = -Px \tag{1}$$

Substituting for *M* into Eq. (9.4) and multiplying both members by the constant *EI* gives

$$EI \frac{d^2y}{dx^2} = -Px$$

Integrating in *x*,

$$EI \frac{dy}{dx} = -\tfrac{1}{2}Px^2 + C_1 \tag{2}$$

Now observe the fixed end *B* where $x = L$ and $\theta = dy/dx = 0$ (Fig. 9.9c). Substituting these values into Eq. (2) and solving for C_1 gives

$$C_1 = \tfrac{1}{2}PL^2$$

which we carry back into Eq. (2):

$$EI \frac{dy}{dx} = -\tfrac{1}{2}Px^2 + \tfrac{1}{2}PL^2 \tag{3}$$

Integrating both members of Eq. (3),

$$EI\,y = -\tfrac{1}{6}Px^3 + \tfrac{1}{2}PL^2x + C_2 \tag{4}$$

But at *B*, $x = L$, $y = 0$. Substituting into Eq. (4),

$$0 = -\tfrac{1}{6}PL^3 + \tfrac{1}{2}PL^3 + C_2$$
$$C_2 = -\tfrac{1}{3}PL^3$$

Carrying the value of C_2 back into Eq. (4), the equation of the elastic curve is

$$EI\,y = -\tfrac{1}{6}Px^3 + \tfrac{1}{2}PL^2x - \tfrac{1}{3}PL^3$$

or

$$y = \frac{P}{6EI}(-x^3 + 3L^2x - 2L^3) \tag{5}$$

The deflection and slope at *A* are obtained by letting $x = 0$ in Eqs. (3) and (5).

$$y_A = -\frac{PL^3}{3EI} \quad \text{and} \quad \theta_A = \left(\frac{dy}{dx}\right)_A = \frac{PL^2}{2EI}$$

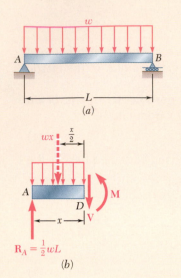

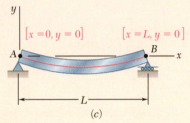

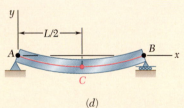

Fig. 9.10 (a) Simply supported beam with a uniformly distributed load. (b) Free-body diagram of segment AD. (c) Boundary conditions. (d) Point of maximum deflection.

Concept Application 9.2

The simply supported prismatic beam AB carries a uniformly distributed load w per unit length (Fig. 9.10a). Determine the equation of the elastic curve and the maximum deflection of the beam.

Draw the free-body diagram of the portion AD of the beam (Fig. 9.10b) and take moments about D for

$$M = \tfrac{1}{2}wLx - \tfrac{1}{2}wx^2 \tag{1}$$

Substituting for M into Eq. (9.4) and multiplying both members of this equation by the constant EI gives

$$EI\frac{d^2y}{dx^2} = -\frac{1}{2}wx^2 + \frac{1}{2}wLx \tag{2}$$

Integrating twice in x,

$$EI\frac{dy}{dx} = -\frac{1}{6}wx^3 + \frac{1}{4}wLx^2 + C_1 \tag{3}$$

$$EI\,y = -\frac{1}{24}wx^4 + \frac{1}{12}wLx^3 + C_1x + C_2 \tag{4}$$

Observing that y = 0 at both ends of the beam (Fig. 9.10c), let x = 0 and y = 0 in Eq. (4) and obtain $C_2 = 0$. Then make x = L and y = 0 in the same equation, so

$$0 = -\tfrac{1}{24}wL^4 + \tfrac{1}{12}wL^4 + C_1L$$

$$C_1 = -\tfrac{1}{24}wL^3$$

Carrying the values of C_1 and C_2 back into Eq. (9.4), the elastic curve is

$$EI\,y = -\tfrac{1}{24}wx^4 + \tfrac{1}{12}wLx^3 - \tfrac{1}{24}wL^3x$$

or

$$y = \frac{w}{24EI}(-x^4 + 2Lx^3 - L^3x) \tag{5}$$

Substituting the value for C_1 into Eq. (3), we check that the slope of the beam is zero for x = L/2 and thus that the elastic curve has a minimum at the midpoint C (Fig. 9.10d). Letting x = L/2 in Eq. (5),

$$y_C = \frac{w}{24EI}\left(-\frac{L^4}{16} + 2L\frac{L^3}{8} - L^3\frac{L}{2}\right) = -\frac{5wL^4}{384EI}$$

The maximum deflection (the maximum absolute value) is

$$|y|_{max} = \frac{5wL^4}{384EI}$$

In both concept applications considered so far, only one free-body diagram was required to determine the bending moment in the beam. As a result, a single function of x was used to represent M throughout the beam. However, concentrated loads, reactions at supports, or discontinuities in a distributed load make it necessary to divide the beam into several portions and to represent the bending moment by a different function $M(x)$ in each. As an example, Photo 9.1 shows an elevated roadway supported by beams, which in turn are subjected to concentrated loads from vehicles crossing the bridge. Each of the functions $M(x)$ leads to a different expression for the slope $\theta(x)$ and the deflection $y(x)$. Since each expression must contain two constants of integration, a large number of constants will have to be determined. As shown in the following concept application, the required additional boundary conditions can be obtained by observing that, while the shear and bending moment can be discontinuous at several points in a beam, the *deflection* and the *slope* of the beam *cannot be discontinuous* at any point.

Photo 9.1   A different function $M(x)$ is required in each portion of the beams when a vehicle crosses the bridge.

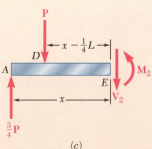

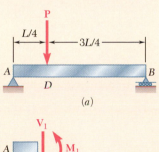

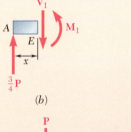

Fig. 9.11   (a) Simply supported beam with transverse load P. (b) Free-body diagram of portion AE to find moment left of load P. (c) Free-body diagram of portion AE to find moment right of load P.

Concept Application 9.3

For the prismatic beam and load shown (Fig. 9.11a), determine the slope and deflection at point D.

Divide the beam into two portions, AD and DB, and determine the function $y(x)$ that defines the elastic curve for each of these portions.

1. From A to D ($x < L/4$).   Draw the free-body diagram of a portion of beam AE of length $x < L/4$ (Fig. 9.11b). Take moments about E to obtain

$$M_1 = \frac{3P}{4}x \tag{1}$$

and recalling Eq. (9.4), we write

$$EI\frac{d^2y_1}{dx^2} = \frac{3}{4}Px \tag{2}$$

where $y_1(x)$ is the function that defines the elastic curve *for portion AD of the beam*. Integrating in x,

$$EI\,\theta_1 = EI\frac{dy_1}{dx} = \frac{3}{8}Px^2 + C_1 \tag{3}$$

$$EI\,y_1 = \frac{1}{8}Px^3 + C_1x + C_2 \tag{4}$$

2. From D to B ($x > L/4$).   Now draw the free-body diagram of a portion of beam AE of the length $x > L/4$ (Fig. 9.11c) and write

$$M_2 = \frac{3P}{4}x - P\left(x - \frac{L}{4}\right) \tag{5}$$

(continued)

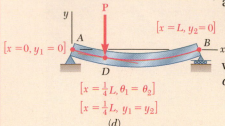

$[x = 0, y_1 = 0]$

$[x = L, y_2 = 0]$

$[x = \frac{1}{4}L, \theta_1 = \theta_2]$

$[x = \frac{1}{4}L, y_1 = y_2]$

(d)

Fig. 9.11   (*cont.*) (*d*) Boundary conditions.

and recalling Eq. (9.4) and rearranging terms, we have

$$EI \frac{d^2 y_2}{dx^2} = -\frac{1}{4}Px + \frac{1}{4}PL \tag{6}$$

where $y_2(x)$ is the function that defines the elastic curve *for portion DB of the beam*. Integrating in x,

$$EI\theta_2 = EI\frac{dy_2}{dx} = -\frac{1}{8}Px^2 + \frac{1}{4}PLx + C_3 \tag{7}$$

$$EIy_2 = -\frac{1}{24}Px^3 + \frac{1}{8}PLx^2 + C_3 x + C_4 \tag{8}$$

Determination of the Constants of Integration.   The conditions satisfied by the constants of integration are summarized in Fig. 9.11*d*. At the support *A*, where the deflection is defined by Eq. (4), $x = 0$ and $y_1 = 0$. At the support *B*, where the deflection is defined by Eq. (8), $x = L$ and $y_2 = 0$. Also, the fact that there can be no sudden change in deflection or in slope at point *D* requires that $y_1 = y_2$ and $\theta_1 = \theta_2$ when $x = L/4$. Therefore,

$[x = 0, y_1 = 0]$, Eq. (4): $0 = C_2$ $\hspace{2cm}$ (9)

$[x = L, y_2 = 0]$, Eq. (8): $0 = \frac{1}{12}PL^3 + C_3 L + C_4$ $\hspace{1cm}$ (10)

$[x = L/4, \theta_1 = \theta_2]$, Eqs. (3) and (7):

$$\frac{3}{128}PL^2 + C_1 = \frac{7}{128}PL^2 + C_3 \tag{11}$$

$[x = L/4, y_1 = y_2]$, Eqs. (4) and (8):

$$\frac{PL^3}{512} + C_1\frac{L}{4} = \frac{11PL^3}{1536} + C_3\frac{L}{4} + C_4 \tag{12}$$

Solving these equations simultaneously,

$$C_1 = -\frac{7PL^2}{128}, \quad C_2 = 0, \quad C_3 = -\frac{11PL^2}{128}, \quad C_4 = \frac{PL^3}{384}$$

Substituting for C_1 and C_2 into Eqs. (3) and (4), $x \le L/4$ is

$$EI\theta_1 = \frac{3}{8}Px^2 - \frac{7PL^2}{128} \tag{13}$$

$$EIy_1 = \frac{1}{8}Px^3 - \frac{7PL^2}{128}x \tag{14}$$

Letting $x = L/4$ in each of these equations, the slope and deflection at point *D* are

$$\theta_D = -\frac{PL^2}{32EI} \quad \text{and} \quad y_D = -\frac{3PL^3}{256EI}$$

Note that since $\theta_D \ne 0$, the deflection at *D* is *not* the maximum deflection of the beam.

*9.1B Determination of the Elastic Curve from the Load Distribution

Section. 9.1A showed that the equation of the elastic curve can be obtained by integrating twice the differential equation

$$\frac{d^2y}{dx^2} = \frac{M(x)}{EI} \qquad (9.4)$$

where $M(x)$ is the bending moment in the beam. Now recall from Sec. 5.2 that, when a beam supports a distributed load $w(x)$, we have $dM/dx = V$ and $dV/dx = -w$ at any point of the beam. Differentiating both members of Eq. (9.4) with respect to x and assuming EI to be constant,

$$\frac{d^3y}{dx^3} = \frac{1}{EI}\frac{dM}{dx} = \frac{V(x)}{EI} \qquad (9.7)$$

and differentiating again,

$$\frac{d^4y}{dx^4} = \frac{1}{EI}\frac{dV}{dx} = -\frac{w(x)}{EI}$$

Thus, when a prismatic beam supports a distributed load $w(x)$, its elastic curve is governed by the fourth-order linear differential equation

$$\frac{d^4y}{dx^4} = -\frac{w(x)}{EI} \qquad (9.8)$$

Multiply both members of Eq. (9.8) by the constant EI and integrate four times to obtain

$$EI\frac{d^4y}{dx^4} = -w(x)$$

$$EI\frac{d^3y}{dx^3} = V(x) = -\int w(x)\,dx + C_1$$

$$EI\frac{d^2y}{dx^2} = M(x) = -\int dx \int w(x)\,dx + C_1x + C_2 \qquad (9.9)$$

$$EI\frac{dy}{dx} = EI\,\theta(x) = -\int dx \int dx \int w(x)\,dx + \frac{1}{2}C_1x^2 + C_2x + C_3$$

$$EIy(x) = -\int dx \int dx \int dx \int w(x)\,dx + \frac{1}{6}C_1x^3 + \frac{1}{2}C_2x^2 + C_3x + C_4$$

The four constants of integration are determined from the boundary conditions. These conditions include (a) the conditions imposed on the deflection or slope of the beam by its supports (see. Sec. 9.1A) and (b) the condition that V and M be zero at the free end of a cantilever beam or that M be zero at both ends of a simply supported beam (see. Sec. 5.2). This has been illustrated in Fig. 9.12.

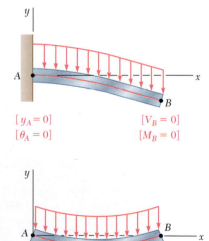

$[y_A = 0]$ $[V_B = 0]$
$[\theta_A = 0]$ $[M_B = 0]$

$[y_A = 0]$ $[y_B = 0]$
$[M_A = 0]$ $[M_B = 0]$

Fig. 9.12 Boundary conditions for (a) cantilever beam (b) simply supported beam.

This method can be used effectively with cantilever or simply supported beams carrying a distributed load. In the case of overhanging beams, the reactions at the supports cause discontinuities in the shear (i.e., in the third derivative of y,) and different functions are required to define the elastic curve over the entire beam.

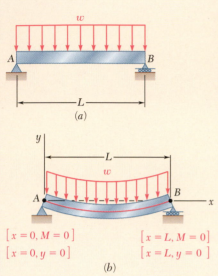

$[x = 0, M = 0]$ $[x = L, M = 0]$
$[x = 0, y = 0]$ $[x = L, y = 0]$

(b)

Fig. 9.13 (a) Simply supported beam with a uniformly distributed load. (b) Boundary conditions.

Concept Application 9.4

The simply supported prismatic beam AB carries a uniformly distributed load w per unit length (Fig. 9.13a). Determine the equation of the elastic curve and the maximum deflection of the beam. (This is the same beam and load as in Concept Application 9.2.)

Since $w = $ constant, the first three of Eqs. (9.9) yield

$$EI \frac{d^4y}{dx^4} = -w$$

$$EI \frac{d^3y}{dx^3} = V(x) = -wx + C_1$$

$$EI \frac{d^2y}{dx^2} = M(x) = -\frac{1}{2}wx^2 + C_1x + C_2 \tag{1}$$

Noting that the boundary conditions require that $M = 0$ at both ends of the beam (Fig. 9.13b), let $x = 0$ and $M = 0$ in Eq. (1) and obtain $C_2 = 0$. Then make $x = L$ and $M = 0$ in the same equation and obtain $C_1 = \frac{1}{2}wL$.

Carry the values of C_1 and C_2 back into Eq. (1) and integrate twice to obtain

$$EI \frac{d^2y}{dx^2} = -\frac{1}{2}wx^2 + \frac{1}{2}wLx$$

$$EI \frac{dy}{dx} = -\frac{1}{6}wx^3 + \frac{1}{4}wLx^2 + C_3$$

$$EIy = -\frac{1}{24}wx^4 + \frac{1}{12}wLx^3 + C_3x + C_4 \tag{2}$$

But the boundary conditions also require that $y = 0$ at both ends of the beam. Letting $x = 0$ and $y = 0$ in Eq. (2), $C_4 = 0$. Letting $x = L$ and $y = 0$ in the same equation gives

$$0 = -\tfrac{1}{24}wL^4 + \tfrac{1}{12}wL^4 + C_3L$$
$$C_3 = -\tfrac{1}{24}wL^3$$

Carrying the values of C_3 and C_4 back into Eq. (2) and dividing both members by EI, the equation of the elastic curve is

$$y = \frac{w}{24EI}(-x^4 + 2Lx^3 - L^3x) \tag{3}$$

The maximum deflection is obtained by making $x = L/2$ in Eq. (3).

$$|y|_{max} = \frac{5wL^4}{384EI}$$

9.2   STATICALLY INDETERMINATE BEAMS

In the preceding sections, our analysis was limited to statically determinate beams. Now consider the prismatic beam AB (Fig. 9.14a), which has a fixed end at A and is supported by a roller at B. Drawing the free-body diagram of the beam (Fig. 9.14b), the reactions involve four unknowns, with only three equilibrium equations:

$$\Sigma F_x = 0 \quad \Sigma F_y = 0 \quad \Sigma M_A = 0 \qquad \textbf{(9.10)}$$

Since only A_x can be determined from these equations, the beam is *statically indeterminate*.

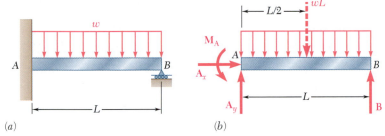

(a)   (b)

Fig. 9.14   (a) Statically indeterminate beam with a uniformly distributed load. (b) Free-body diagram with four unknown reactions.

Recall from Chaps. 2 and 3 that, in a statically indeterminate problem, the reactions can be obtained by considering the *deformations* of the structure. Therefore, we proceed with the computation of the slope and deformation along the beam. Following the method used in Sec. 9.1A, the bending moment $M(x)$ at any given point AB is expressed in terms of the distance x from A, the given load, and the unknown reactions. Integrating in x, expressions for θ and y are found. These contain two additional unknowns: the constants of integration C_1 and C_2. Altogether, six equations are available to determine the reactions and constants C_1 and C_2; they are the three equilibrium equations of Eq. (9.10) and the three equations expressing that the boundary conditions are satisfied (i.e., that the slope and deflection at A are zero and that the deflection at B is zero (Fig. 9.15)). Thus, the reactions at the supports can be determined, and the equation of the elastic curve can be obtained.

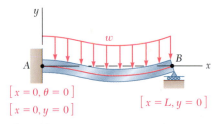

Fig. 9.15   Boundary conditions for beam of Fig. 9.14.

Concept Application 9.5

Determine the reactions at the supports for the prismatic beam of Fig. 9.14a.

Equilibrium Equations.   From the free-body diagram of Fig. 9.14b,

$$\xrightarrow{+} \Sigma F_x = 0: \quad A_x = 0$$

$$+\uparrow \Sigma F_y = 0: \quad A_y + B - wL = 0 \qquad \textbf{(1)}$$

$$+\circlearrowleft \Sigma M_A = 0: \quad M_A + BL - \tfrac{1}{2}wL^2 = 0$$

(continued)

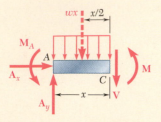

Fig. 9.16 Free-body diagram of beam portion *AC*.

Equation of Elastic Curve. Draw the free-body diagram of a portion of beam *AC* (Fig. 9.16) to obtain

$$+\curvearrowleft \Sigma M_C = 0: \quad M + \tfrac{1}{2}wx^2 + M_A - A_y x = 0 \tag{2}$$

Solving Eq. (2) for *M* and carrying into Eq. (9.4),

$$EI\frac{d^2y}{dx^2} = -\frac{1}{2}wx^2 + A_y x - M_A$$

Integrating in *x* gives

$$EI\,\theta = EI\frac{dy}{dx} = -\frac{1}{6}wx^3 + \frac{1}{2}A_y x^2 - M_A x + C_1 \tag{3}$$

$$EI\,y = -\frac{1}{24}wx^4 + \frac{1}{6}A_y x^3 - \frac{1}{2}M_A x^2 + C_1 x + C_2 \tag{4}$$

Referring to the boundary conditions indicated in Fig. 9.15, $x = 0$, $\theta = 0$ in Eq. (3), $x = 0$, $y = 0$ in Eq. (4), and conclude that $C_1 = C_2 = 0$. Thus, Eq. (4) is rewritten as

$$EI\,y = -\tfrac{1}{24}wx^4 + \tfrac{1}{6}A_y x^3 - \tfrac{1}{2}M_A x^2 \tag{5}$$

But the third boundary condition requires that $y = 0$ for $x = L$. Carrying these values into Eq. (5),

$$0 = -\tfrac{1}{24}wL^4 + \tfrac{1}{6}A_y L^3 - \tfrac{1}{2}M_A L^2$$

or

$$3M_A - A_y L + \tfrac{1}{4}wL^2 = 0 \tag{6}$$

Solving this equation simultaneously with the three equilibrium equations of Eq. (1), the reactions at the supports are

$$A_x = 0 \qquad A_y = \tfrac{5}{8}wL \qquad M_A = \tfrac{1}{8}wL^2 \qquad B = \tfrac{3}{8}wL$$

In the previous Concept Application, there was one redundant reaction (i.e., one more than could be determined from the equilibrium equations alone). The corresponding beam is *statically indeterminate to the first degree*. Another example of a beam indeterminate to the first degree is provided in Sample Prob. 9.3. If the beam supports are such that two reactions are redundant (Fig. 9.17*a*), the beam is *indeterminate to the second degree*. While there are now five unknown reactions (Fig. 9.17*b*), four equations can be obtained from the boundary conditions (Fig. 9.17*c*). Thus, seven equations are available to determine the five reactions and the two constants of integration.

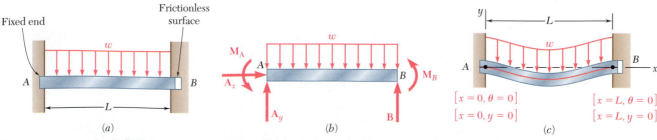

Fig. 9.17 (*a*) Beam statically indeterminate to the second degree. (*b*) Free-body diagram. (*c*) Boundary conditions.

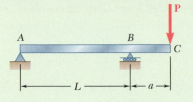

Sample Problem 9.1

The overhanging steel beam ABC carries a concentrated load **P** at end C. For portion AB of the beam, (a) derive the equation of the elastic curve, (b) determine the maximum deflection, (c) evaluate $y_{\max}$ for the following data:

$$W14 \times 68 \qquad I = 722 \text{ in}^4 \qquad E = 29 \times 10^6 \text{ psi}$$

$$P = 50 \text{ kips} \qquad L = 15 \text{ ft} = 180 \text{ in.} \qquad a = 4 \text{ ft} = 48 \text{ in.}$$

STRATEGY: You should begin by determining the bending-moment equation for the portion of interest. Substituting this into the differential equation of the elastic curve, integrating twice, and applying the boundary conditions, you can then obtain the equation of the elastic curve. Use this equation to find the desired deflections.

MODELING: Using the free-body diagram of the entire beam (Fig. 1) gives the reactions: $\mathbf{R}_A = Pa/L \downarrow$ $\mathbf{R}_B = P(1 + a/L) \uparrow$. The free-body diagram of the portion of beam AD of length x (Fig. 1) gives

$$M = -P\frac{a}{L}x \qquad (0 < x < L)$$

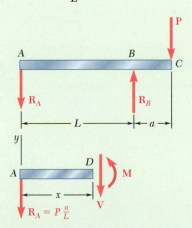

Fig. 1 Free-body diagrams of beam and portion AD.

ANALYSIS:

Differential Equation of the Elastic Curve. Using Eq. (9.4) gives

$$EI\frac{d^2y}{dx^2} = -P\frac{a}{L}x$$

Noting that the flexural rigidity EI is constant, integrate twice and find

$$EI\frac{dy}{dx} = -\frac{1}{2}P\frac{a}{L}x^2 + C_1 \tag{1}$$

$$EIy = -\frac{1}{6}P\frac{a}{L}x^3 + C_1x + C_2 \tag{2}$$

(continued)

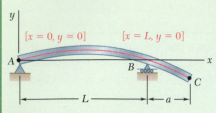

Fig. 2 Boundary conditions.

Determination of Constants. For the boundary conditions shown (Fig. 2),

$[x = 0, y = 0]$:   From Eq. (2), $C_2 = 0$

$[x = L, y = 0]$:   Again using Eq. (2),

$$EI(0) = -\frac{1}{6}P\frac{a}{L}L^3 + C_1L \quad C_1 = +\frac{1}{6}PaL$$

a. Equation of the Elastic Curve.
Substituting for C_1 and C_2 into Eqs. (1) and (2),

$$EI\frac{dy}{dx} = -\frac{1}{2}P\frac{a}{L}x^2 + \frac{1}{6}PaL \qquad \frac{dy}{dx} = \frac{PaL}{6EI}\left[1 - 3\left(\frac{x}{L}\right)^2\right] \quad \textbf{(3)}$$

$$EIy = -\frac{1}{6}P\frac{a}{L}x^3 + \frac{1}{6}PaLx \qquad y = \frac{PaL^2}{6EI}\left[\frac{x}{L} - \left(\frac{x}{L}\right)^3\right] \quad \textbf{(4)} \blacktriangleleft$$

b. Maximum Deflection in Portion AB.
The maximum deflection y_{max} occurs at point E where the slope of the elastic curve is zero (Fig. 3). Setting $dy/dx = 0$ in Eq. (3), the abscissa x_m of point E is

$$0 = \frac{PaL}{6EI}\left[1 - 3\left(\frac{x_m}{L}\right)^2\right] \quad x_m = \frac{L}{\sqrt{3}} = 0.577L$$

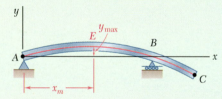

Fig. 3 Deformed elastic curve with location of maximum deflection.

Substitute $x_m/L = 0.577$ into Eq. (4):

$$y_{max} = \frac{PaL^2}{6EI}[(0.577) - (0.577)^3] \quad y_{max} = 0.0642\frac{PaL^2}{EI} \blacktriangleleft$$

c. Evaluation of y_{max}.
For the data given, the value of y_{max} is

$$y_{max} = 0.0642\frac{(50 \text{ kips})(48 \text{ in.})(180 \text{ in.})^2}{(29 \times 10^6 \text{ psi})(722 \text{ in}^4)} \quad y_{max} = 0.238 \text{ in.} \blacktriangleleft$$

REFLECT and THINK: Because the maximum deflection is positive, it is upward. As a check, we see that this is consistent with the deflected shape anticipated for this loading (Fig. 3).

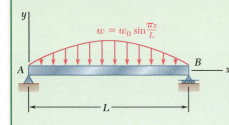

Sample Problem 9.2

For the beam and loading shown determine (*a*) the equation of the elastic curve, (*b*) the slope at end *A*, (*c*) the maximum deflection.

STRATEGY: Determine the elastic curve directly from the load distribution using Eq. (9.8), applying the appropriate boundary conditions. Use this equation to find the desired slope and deflection.

MODELING and ANALYSIS:

Differential Equation of the Elastic Curve.   From Eq. (9.8),

$$EI\frac{d^4y}{dx^4} = -w(x) = -w_0 \sin\frac{\pi x}{L} \tag{1}$$

Integrate Eq. (1) twice:

$$EI\frac{d^3y}{dx^3} = V = +w_0\frac{L}{\pi}\cos\frac{\pi x}{L} + C_1 \tag{2}$$

$$EI\frac{d^2y}{dx^2} = M = +w_0\frac{L^2}{\pi^2}\sin\frac{\pi x}{L} + C_1 x + C_2 \tag{3}$$

Boundary Conditions: Refer to Fig. 1.

$[x = 0, M = 0]$:   From Eq. (3), $C_2 = 0$

$[x = L, M = 0]$:   Again using Eq. (3),

$$0 = w_0\frac{L^2}{\pi^2}\sin\pi + C_1 L \quad C_1 = 0$$

Thus,

$$EI\frac{d^2y}{dx^2} = +w_0\frac{L^2}{\pi^2}\sin\frac{\pi x}{L} \tag{4}$$

Integrate Eq. (4) twice:

$$EI\frac{dy}{dx} = EI\theta = -w_0\frac{L^3}{\pi^3}\cos\frac{\pi x}{L} + C_3 \tag{5}$$

$$EI\, y = -w_0\frac{L^4}{\pi^4}\sin\frac{\pi x}{L} + C_3 x + C_4 \tag{6}$$

Boundary Conditions: Refer to Fig. 1.

$[x = 0, y = 0]$:   Using Eq. (6), $C_4 = 0$

$[x = L, y = 0]$:   Again using Eq. (6), $C_3 = 0$

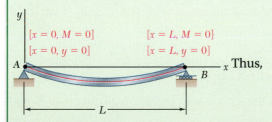

Fig. 1   Boundary conditions.

(continued)

a. Equation of Elastic Curve.

$$EIy = -w_0 \frac{L^4}{\pi^4} \sin \frac{\pi x}{L}$$ ◄

b. Slope at End A.   Refer to Fig. 2. For $x = 0$,

$$EI\theta_A = -w_0 \frac{L^3}{\pi^3} \cos 0 \qquad \theta_A = \frac{w_0 L^3}{\pi^3 EI} \searrow$$ ◄

c. Maximum Deflection.   Referring to Fig. 2, for $x = \frac{1}{2}L$,

$$ELy_{max} = -w_0 \frac{L^4}{\pi^4} \sin \frac{\pi}{2} \qquad y_{max} = \frac{w_0 L^4}{\pi^4 EI} \downarrow$$ ◄

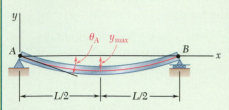

Fig. 2 Deformed elastic curve showing slope at A and maximum deflection.

REFLECT and THINK: As a check, we observe that the directions of the slope at end A and the maximum deflection are consistent with the deflected shape anticipated for this loading (Fig. 1).

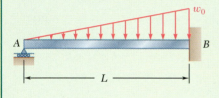

Sample Problem 9.3

For the uniform beam AB (a) determine the reaction at A, (b) derive the equation of the elastic curve, (c) determine the slope at A. (Note that the beam is statically indeterminate to the first degree.)

STRATEGY: The beam is statically indeterminate to the first degree. Treating the reaction at A as the redundant, write the bending-moment equation as a function of this redundant reaction and the existing load. After substituting the bending-moment equation into the differential equation of the elastic curve, integrating twice, and applying the boundary conditions, the reaction can be determined. Use the equation for the elastic curve to find the desired slope.

MODELING: Using the free body shown in Fig. 1, obtain the bending moment diagram:

$$+\!\!\downarrow\!\Sigma M_D = 0: \qquad R_A x - \frac{1}{2}\left(\frac{w_0 x^2}{L}\right)\frac{x}{3} - M = 0 \qquad M = R_A x - \frac{w_0 x^3}{6L}$$

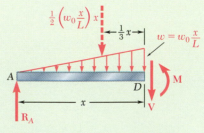

Fig. 1   Free-body diagram of portion AD of beam.

(continued)

ANALYSIS:

Differential Equation of the Elastic Curve.
Use Eq. (9.4) for

$$EI\frac{d^2y}{dx^2} = R_Ax - \frac{w_0x^3}{6L}$$

Noting that the flexural rigidity EI is constant, integrate twice and find

$$EI\frac{dy}{dx} = EI\theta = \frac{1}{2}R_Ax^2 - \frac{w_0x^4}{24L} + C_1 \tag{1}$$

$$EIy = \frac{1}{6}R_Ax^3 - \frac{w_0x^5}{120L} + C_1x + C_2 \tag{2}$$

Boundary Conditions.
The three boundary conditions that must be satisfied are shown in Fig. 2.

$$[x = 0, y = 0]: \quad C_2 = 0 \tag{3}$$

$$[x = L, \theta = 0]: \quad \frac{1}{2}R_AL^2 - \frac{w_0L^3}{24} + C_1 = 0 \tag{4}$$

$$[x = L, y = 0]: \quad \frac{1}{6}R_AL^3 - \frac{w_0L^4}{120} + C_1L + C_2 = 0 \tag{5}$$

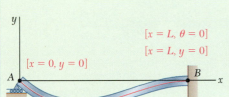

$[x = L, \theta = 0]$
$[x = L, y = 0]$
$[x = 0, y = 0]$

Fig. 2 Boundary conditions.

a. Reaction at A.
Multiplying Eq. (4) by L, subtracting Eq. (5) member by member from the equation obtained, and noting that $C_2 = 0$, give

$$\tfrac{1}{3}R_AL^3 - \tfrac{1}{30}w_0L^4 = 0 \qquad \mathbf{R_A} = \tfrac{1}{10}w_0L\uparrow \quad \blacktriangleleft$$

The reaction is independent of E and I. Substituting $R_A = \frac{1}{10}w_0L$ into Eq. (4),

$$\tfrac{1}{2}(\tfrac{1}{10}w_0L)L^2 - \tfrac{1}{24}w_0L^3 + C_1 = 0 \qquad C_1 = -\tfrac{1}{120}w_0L^3$$

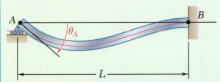

Fig. 3 Deformed elastic curve showing slope at A.

b. Equation of the Elastic Curve.
Substituting for R_A, C_1, and C_2 into Eq. (2),

$$EIy = \frac{1}{6}\left(\frac{1}{10}w_0L\right)x^3 - \frac{w_0x^5}{120L} - \left(\frac{1}{120}w_0L^3\right)x$$

$$y = \frac{w_0}{120EIL}(-x^5 + 2L^2x^3 - L^4x) \quad \blacktriangleleft$$

c. Slope at A (Fig. 3).
Differentiate the equation of the elastic curve with respect to x:

$$\theta = \frac{dy}{dx} = \frac{w_0}{120EIL}(-5x^4 + 6L^2x^2 - L^4)$$

Making $x = 0$, $\qquad \theta_A = -\frac{w_0L^3}{120EI} \qquad \theta_A = \frac{w_0L^3}{120EI} \quad \blacktriangleleft$

Problems

In the following problems assume that the flexural rigidity EI of each beam is constant.

9.1 through 9.4 For the loading shown, determine (*a*) the equation of the elastic curve for the cantilever beam AB, (*b*) the deflection at the free end, (*c*) the slope at the free end.

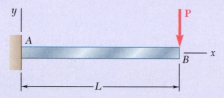

Fig. P9.1

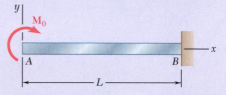

Fig. P9.2

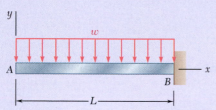

Fig. P9.3

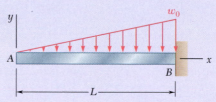

Fig. P9.4

9.5 and 9.6 For the cantilever beam and loading shown, determine (*a*) the equation of the elastic curve for portion AB of the beam, (*b*) the deflection at B, (*c*) the slope at B.

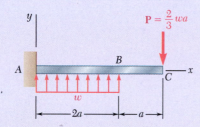

Fig. P9.5

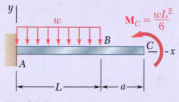

Fig. P9.6

9.7 For the beam and loading shown, determine (*a*) the equation of the elastic curve for portion AB of the beam, (*b*) the deflection at midspan, (*c*) the slope at B.

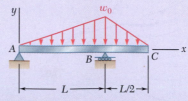

Fig. P9.7

9.8 For the beam and loading shown, determine (*a*) the equation of the elastic curve for portion *AB* of the beam, (*b*) the slope at *A*, (*c*) the slope at *B*.

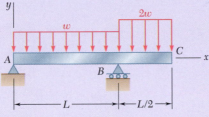

Fig. P9.8

9.9 Knowing that beam *AB* is a W10 × 33 rolled shape and that $w_0 = 3$ kips/ft, $L = 12$ ft, and $E = 29 × 10^6$ psi, determine (*a*) the slope at *A*, (*b*) the deflection at *C*.

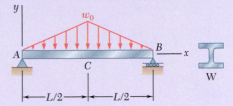

Fig. P9.9

9.10 Knowing that beam *AB* is an S200 × 34 rolled shape and that $P = 60$ kN, $L = 2$ m, and $E = 200$ GPa, determine (*a*) the slope at *A*, (*b*) the deflection at *C*.

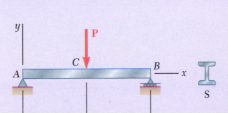

Fig. P9.10

9.11 For the beam and loading shown, (*a*) express the magnitude and location of the maximum deflection in terms of w_0, L, E, and I. (*b*) Calculate the value of the maximum deflection, assuming that beam *AB* is a W18 × 50 rolled shape and that $w_0 = 4.5$ kips/ft, $L = 18$ ft, and $E = 29 × 10^6$ psi.

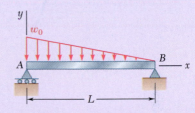

Fig. P9.11

9.12 (*a*) Determine the location and magnitude of the maximum absolute deflection in *AB* between *A* and the center of the beam. (*b*) Assuming that beam *AB* is a W460 × 113, $M_0 = 224$ kN·m, and $E = 200$ GPa, determine the maximum allowable length *L* of the beam if the maximum deflection is not to exceed 1.2 mm.

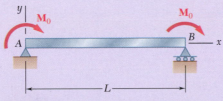

Fig. P9.12

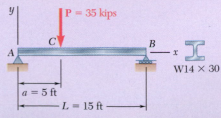

Fig. P9.13

9.13 For the beam and loading shown, determine the deflection at point *C*. Use $E = 29 × 10^6$ psi.

619

9.14 Knowing that beam AE is a W360 × 101 rolled shape and that $M_0 = 310$ kN·m, $L = 2.4$ m, $a = 0.5$ m, and $E = 200$ GPa, determine (*a*) the equation of the elastic curve for portion BD, (*b*) the deflection at point C.

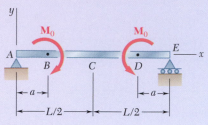

Fig. P9.14

9.15 For the beam and loading shown, knowing that $a = 2$ m, $w = 50$ kN/m, and $E = 200$ GPa, determine (*a*) the slope at support A, (*b*) the deflection at point C.

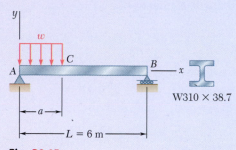

Fig. P9.15

9.16 Knowing that beam AE is an S200 × 27.4 rolled shape and that $P = 17.5$ kN, $L = 2.5$ m, $a = 0.8$ m, and $E = 200$ GPa, determine (*a*) the equation of the elastic curve for portion BD, (*b*) the deflection at the center C of the beam.

9.17 For the beam and loading shown, determine (*a*) the equation of the elastic curve, (*b*) the deflection at the free end.

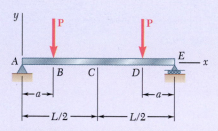

Fig. P9.16

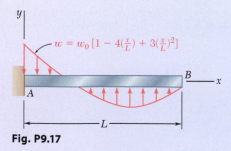

$w = w_0 \left[1 - 4\left(\frac{x}{L}\right) + 3\left(\frac{x}{L}\right)^2\right]$

Fig. P9.17

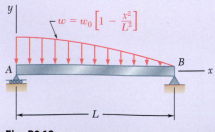

$w = w_0 \left[1 - \frac{x^2}{L^2}\right]$

Fig. P9.18

9.18 For the beam and loading shown, determine (*a*) the equation of the elastic curve, (*b*) the slope at end A, (*c*) the deflection at the midpoint of the span.

9.19 through 9.22 For the beam and loading shown, determine the reaction at the roller support.

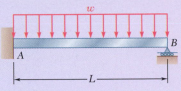

Fig. P9.19

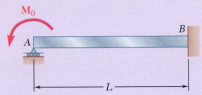

Fig. P9.20

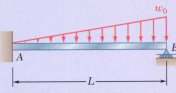

Fig. *P9.21*

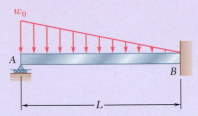

Fig. *P9.22*

9.23 For the beam shown, determine the reaction at the roller support when $w_0 = 6$ kips/ft.

9.24 For the beam shown, determine the reaction at the roller support when $w_0 = 15$ kN/m.

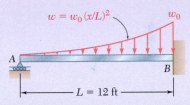

$w = w_0(x/L)^2$

Fig. P9.23

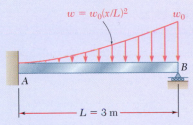

$w = w_0(x/L)^2$

Fig. P9.24

9.25 through 9.28 Determine the reaction at the roller support and draw the bending moment diagram for the beam and loading shown.

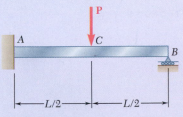

Fig. P9.25

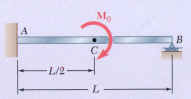

Fig. P9.26

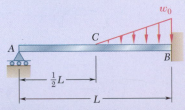

Fig. P9.27

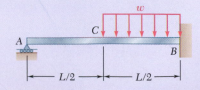

Fig. P9.28

9.29 and 9.30 Determine the reaction at the roller support and the deflection at point C.

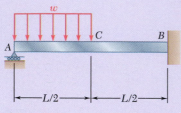

Fig. *P9.29*

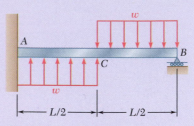

Fig. P9.30

9.31 and 9.32 Determine the reaction at the roller support and the deflection at point D if a is equal to $L/3$.

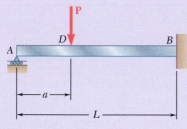

Fig. *P9.31*

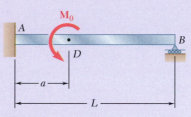

Fig. P9.32

9.33 and 9.34 Determine the reaction at A and draw the bending moment diagram for the beam and loading shown.

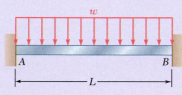

Fig. P9.33

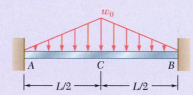

Fig. P9.34

*9.3 SINGULARITY FUNCTIONS TO DETERMINE SLOPE AND DEFLECTION

The integration method provides a convenient and effective way of determining the slope and deflection at any point of a prismatic beam, *as long as the bending moment can be represented by a single analytical function M(x)*. However, when the loading of the beam needs two different functions to represent the bending moment over the entire length, as in Concept Application 9.3 (Fig. 9.11*a*), four constants of integration are required. An equal number of equations, expressing continuity conditions at point *D* as well as boundary conditions at supports *A* and *B*, must be used to determine these constants. If three or more functions are needed for the bending moment, additional constants and a corresponding number of additional equations are required, resulting in rather lengthy computations. Such is the case for the beam shown in Photo 9.2. This section simplifies computations in situations like this through the use of the singularity functions discussed in Sec. 5.4.

Photo 9.2 In this roof structure, each of the open-web joists applies a concentrated load to the beam that supports it.

Consider again the beam and loading of Concept Application 9.3 (Fig. 9.11) and draw the free-body diagram of that beam (Fig. 9.18). Use the appropriate singularity function (Sec. 5.4) to represent the contribution to the shear of the concentrated load **P**, and write

$$V(x) = \frac{3P}{4} - P\langle x - \tfrac{1}{4}L\rangle^0$$

Integrate in *x* and recall from Sec. 5.4 that, in the absence of any concentrated couple, the expression for the bending moment does not contain a constant term, so

$$M(x) = \frac{3P}{4}x - P\langle x - \tfrac{1}{4}L\rangle \qquad (9.11)$$

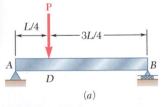

Fig. 9.11 (*repeated*) Simply supported beam with transverse load **P**.

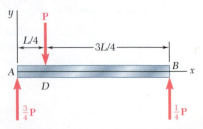

Fig. 9.18 Free-body diagram for beam of Fig. 9.11.

Substituting for $M(x)$ from Eq. (9.11) into Eq. (9.4),

$$EI\frac{d^2y}{dx^2} = \frac{3P}{4}x - P\langle x - \tfrac{1}{4}L\rangle \qquad \textbf{(9.12)}$$

and, integrating in x,

$$EI\theta = EI\frac{dy}{dx} = \frac{3}{8}Px^2 - \frac{1}{2}P\langle x - \tfrac{1}{4}L\rangle^2 + C_1 \qquad \textbf{(9.13)}$$

$$EI\,y = \frac{1}{8}Px^3 - \frac{1}{6}P\langle x - \tfrac{1}{4}L\rangle^3 + C_1 x + C_2 \qquad \textbf{(9.14)}^{\dagger}$$

The constants C_1 and C_2 can be determined from the boundary conditions shown in Fig. 9.19. Letting $x = 0$, $y = 0$ in Eq. (9.14),

$$0 = 0 - \frac{1}{6}P\langle 0 - \tfrac{1}{4}L\rangle^3 + 0 + C_2$$

which reduces to $C_2 = 0$, since any bracket containing a negative quantity is equal to zero. Letting now $x = L$, $y = 0$, and $C_2 = 0$ in Eq. (9.14),

$$0 = \frac{1}{8}PL^3 - \frac{1}{6}P\langle \tfrac{3}{4}L\rangle^3 + C_1 L$$

Since the quantity between brackets is positive, the brackets can be replaced by ordinary parentheses. Solving for C_1 gives

$$C_1 = -\frac{7PL^2}{128}$$

The expressions obtained for the constants C_1 and C_2 are the same found in Concept Application 9.3. But the need for additional constants C_3 and C_4 has been eliminated, and the equations expressing that the slope and the deflection are continuous at point D are not needed.

$[x = 0, y = 0]$ $[x = L, y = 0]$

Fig. 9.19 Boundary conditions for beam of Fig. 9.11.

[†]The continuity conditions for the slope and deflection at D are "built-in" in Eqs. (9.13) and (9.14). Indeed, the difference between the expressions for the slope θ_1 in AD and the slope θ_2 in DB is represented by the term $-\frac{1}{2}P\langle x - \tfrac{1}{4}L\rangle^2$ in Eq. (9.13), and this term is equal to zero at D. Similarly, the difference between the expressions for the deflection y_1 in AD and the deflection y_2 in DB is represented by $-\frac{1}{6}P\langle x - \tfrac{1}{4}L\rangle^3$ in Eq. (9.14), and this term is also equal to zero at D.

Concept Application 9.6

For the beam and loading shown (Fig. 9.20a) and using singularity functions, (a) express the slope and deflection as functions of the distance x from the support at A, (b) determine the deflection at the midpoint D. Use $E = 200$ GPa and $I = 6.87 \times 10^{-6}$ m^4.

(a) The beam is loaded and supported in the same manner as the beam of Concept Application 5.5. Recall that the distributed load was

(continued)

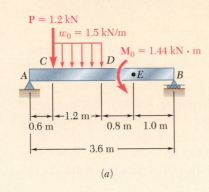

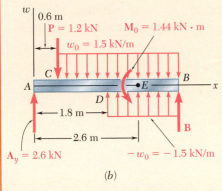

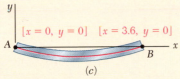

Fig. 9.20 (a) Simply supported beam with multiple loads. (b) Free-body diagram of beam showing equivalent loading system. (c) Boundary conditions.

replaced by the two equivalent open-ended loads shown in Fig. 9.20b. The expressions for the shear and bending moment are

$$V(x) = -1.5\langle x - 0.6\rangle^1 + 1.5\langle x - 1.8\rangle^1 + 2.6 - 1.2\langle x - 0.6\rangle^0$$

$$M(x) = -0.75\langle x - 0.6\rangle^2 + 0.75\langle x - 1.8\rangle^2$$
$$+ 2.6x - 1.2\langle x - 0.6\rangle^1 - 1.44\langle x - 2.6\rangle^0$$

Integrating the last expression twice,

$$EI\theta = -0.25\langle x - 0.6\rangle^3 + 0.25\langle x - 1.8\rangle^3$$
$$+ 1.3x^2 - 0.6\langle x - 0.6\rangle^2 - 1.44\langle x - 2.6\rangle^1 + C_1 \quad \textbf{(1)}$$

$$EIy = -0.0625\langle x - 0.6\rangle^4 + 0.0625\langle x - 1.8\rangle^4 + 0.4333x^3$$
$$- 0.2\langle x - 0.6\rangle^3 - 0.72\langle x - 2.6\rangle^2 + C_1x + C_2 \quad \textbf{(2)}$$

The constants C_1 and C_2 can be determined from the boundary conditions shown in Fig. 9.20c. Letting $x = 0$, $y = 0$ in Eq. (2) and noting that all the brackets contain negative quantities and, therefore, are equal to zero, we conclude that $C_2 = 0$. Letting $x = 3.6$, $y = 0$, and $C_2 = 0$ in Eq. (2) gives

$$0 = -0.0625\langle 3.0\rangle^4 + 0.0625\langle 1.8\rangle^4$$
$$+ 0.4333(3.6)^3 - 0.2\langle 3.0\rangle^3 - 0.72\langle 1.0\rangle^2 + C_1(3.6) + 0$$

Since all the quantities between brackets are positive, the brackets can be replaced by ordinary parentheses. Solving for C_1, we find $C_1 = -2.692$.

(b) Substituting for C_1 and C_2 into Eq. (2) and making $x = x_D = 1.8$ m, we find that the deflection at point D is defined by the relation

$$EIy_D = -0.0625\langle 1.2\rangle^4 + 0.0625\langle 0\rangle^4$$
$$+ 0.4333(1.8)^3 - 0.2\langle 1.2\rangle^3 - 0.72\langle -0.8\rangle^2 - 2.692(1.8)$$

The last bracket contains a negative quantity and, therefore, is equal to zero. All the other brackets contain positive quantities and can be replaced by ordinary parentheses.

$$EIy_D = -0.0625(1.2)^4 + 0.0625(0)^4$$
$$+ 0.4333(1.8)^3 - 0.2(1.2)^3 - 0 - 2.692(1.8) = -2.794$$

Recalling the given numerical values of E and I,

$$(200 \text{ GPa})(6.87 \times 10^{-6} \text{ m}^4)y_D = -2.794 \text{ kN·m}^3$$
$$y_D = -13.64 \times 10^{-3} \text{ m} = -2.03 \text{ mm}$$

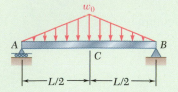

Sample Problem 9.4

For the prismatic beam and loading shown, determine (*a*) the equation of the elastic curve, (*b*) the slope at *A*, (*c*) the maximum deflection.

STRATEGY: You can begin by determining the bending-moment equation of the beam, using a singularity function for any transition in loading. Substituting this into the differential equation of the elastic curve, integrating twice, and applying the boundary conditions, you can then obtain the equation of the elastic curve. Use this equation to find the desired slope and deflection.

MODELING: The equation defining the bending moment of the beam was obtained in Sample Prob. 5.9. Using the modified loading diagram shown in Fig. 1, we had [Eq. (3)]:

$$M(x) = -\frac{w_0}{3L}x^3 + \frac{2w_0}{3L}\left\langle x - \tfrac{1}{2}L \right\rangle^3 + \tfrac{1}{4}w_0 L x$$

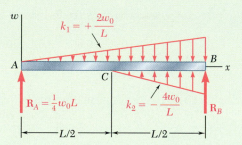

Fig. 1 Free-body diagram showing modified loading.

ANALYSIS:

a. Equation of the Elastic Curve. Using Eq. (9.4),

$$EI\frac{d^2y}{dx^2} = -\frac{w_0}{3L}x^3 + \frac{2w_0}{3L}\left\langle x - \tfrac{1}{2}L \right\rangle^3 + \tfrac{1}{4}w_0 L x \tag{1}$$

and, integrating twice in *x*,

$$EI\,\theta = -\frac{w_0}{12L}x^4 + \frac{w_0}{6L}\left\langle x - \tfrac{1}{2}L \right\rangle^4 + \frac{w_0 L}{8}x^2 + C_1 \tag{2}$$

$$EI\,y = -\frac{w_0}{60L}x^5 + \frac{w_0}{30L}\left\langle x - \tfrac{1}{2}L \right\rangle^5 + \frac{w_0 L}{24}x^3 + C_1 x + C_2 \tag{3}$$

(continued)

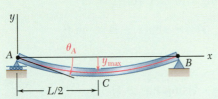

Fig. 2 Boundary conditions.

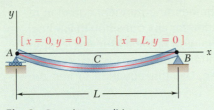

Fig. 3 Deformed elastic curve showing slope at A and maximum deflection at C.

Boundary Conditions. Referring to Fig. 2,

$[x = 0, y = 0]$: Using Eq. (3) and noting that each bracket $\langle \ \rangle$ contains a negative quantity and is equal to zero, $C_2 = 0$.

$[x = L, y = 0]$: Again using Eq. (3),

$$0 = -\frac{w_0 L^4}{60} + \frac{w_0}{30L}\left(\frac{L}{2}\right)^5 + \frac{w_0 L^4}{24} + C_1 L \qquad C_1 = -\frac{5}{192} w_0 L^3$$

Substituting C_1 and C_2 into Eqs. (2) and (3),

$$EI\,\theta = -\frac{w_0}{12L} x^4 + \frac{w_0}{6L}\left\langle x - \tfrac{1}{2}L\right\rangle^4 + \frac{w_0 L}{8} x^2 - \frac{5}{192} w_0 L^3 \qquad \textbf{(4)}$$

$$EI\,y = -\frac{w_0}{60L} x^5 + \frac{w_0}{30L}\left\langle x - \tfrac{1}{2}L\right\rangle^5 + \frac{w_0 L}{24} x^3 - \frac{5}{192} w_0 L^3 x \qquad \textbf{(5)} \ \blacktriangleleft$$

b. Slope at A (Fig. 3). Substituting $x = 0$ into Eq. (4),

$$EI\,\theta_A = -\frac{5}{192} w_0 L^3 \qquad \theta_A = \frac{5w_0 L^3}{192EI} \ \measuredangle \quad \blacktriangleleft$$

c. Maximum Deflection (Fig. 3). Because of the symmetry of the supports and loading, the maximum deflection occurs at point C where $x = \frac{1}{2}L$. Substituting into Eq. (5),

$$EI\,y_{max} = w_0 L^4\left[-\frac{1}{60(32)} + 0 + \frac{1}{24(8)} - \frac{5}{192(2)}\right] = -\frac{w_0 L^4}{120}$$

$$y_{max} = \frac{w_0 L^4}{120EI} \downarrow \quad \blacktriangleleft$$

Sample Problem 9.5

The rigid bar *DEF* is welded at point *D* to the uniform steel beam *AB*. For the loading shown, determine (*a*) the equation of the elastic curve of the beam, (*b*) the deflection at the midpoint *C* of the beam. Use $E = 29 \times 10^6$ psi.

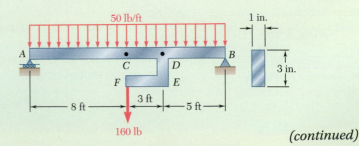

(continued)

STRATEGY: Begin by determining the bending-moment equation of the beam ADB, using a singularity function for any transition in loading. Substituting this into the differential equation of the elastic curve, integrating twice, and applying the boundary conditions, you can then obtain the equation of the elastic curve. Use this equation to find the desired deflection.

MODELING: The equation defining the bending moment of the beam was obtained in Sample Prob. 5.10. Using the modified loading diagram shown in Fig. 1 and expressing x in feet, [Eq. (3)] is

$$M(x) = -25x^2 + 480x - 160\langle x - 11 \rangle^1 - 480\langle x - 11 \rangle^0 \quad \text{lb·ft}$$

ANALYSIS:

a. Equation of the Elastic Curve. Using Eq. (9.4),

$$EI(d^2y/dx^2) = -25x^2 + 480x - 160\langle x - 11 \rangle^1 - 480\langle x - 11 \rangle^0 \quad \text{lb·ft} \quad \textbf{(1)}$$

and, integrating twice in x,

$$EI\,\theta = -8.333x^3 + 240x^2 - 80\langle x - 11 \rangle^2 - 480\langle x - 11 \rangle^1 + C_1 \quad \text{lb·ft}^2 \quad \textbf{(2)}$$

$$EI\,y = -2.083x^4 + 80x^3 - 26.67\langle x - 11 \rangle^3 - 240\langle x - 11 \rangle^2$$
$$+ C_1x + C_2 \quad \text{lb·ft}^3 \quad \textbf{(3a)}$$

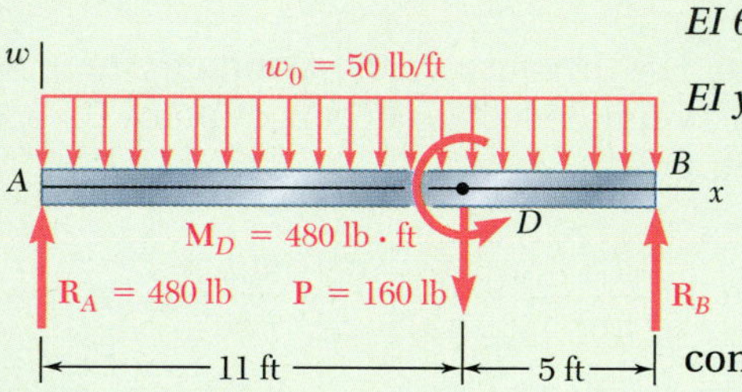

Boundary Conditions. Referring to Fig. 2,

$[x = 0, y = 0]$: Using Eq. (3) and noting that each bracket $\langle \; \rangle$ contains a negative quantity and, thus, is equal to zero, we find $C_2 = 0$.

$[x = 16 \text{ ft}, y = 0]$: Again using Eq. (3), each bracket contains a positive quantity and can be replaced by a parenthesis:

$$0 = -2.083(16)^4 + 80(16)^3 - 26.67(5)^3 - 240(5)^2 + C_1(16)$$
$$C_1 = -11.36 \times 10^3$$

Substituting the values found for C_1 and C_2 into Eq. (3) gives

$$EI\,y = -2.083x^4 + 80x^3 - 26.67\langle x - 11 \rangle^3 - 240\langle x - 11 \rangle^2$$
$$- 11.36 \times 10^3 x \quad \text{lb·ft}^3 \quad \textbf{(3b)} \quad \blacktriangleleft$$

To determine EI, recall that $E = 29 \times 10^6$ psi and compute

$$I = \tfrac{1}{12}bh^3 = \tfrac{1}{12}(1 \text{ in.})(3 \text{ in.})^3 = 2.25 \text{ in}^4$$
$$EI = (29 \times 10^6 \text{ psi})(2.25 \text{ in}^4) = 65.25 \times 10^6 \text{ lb·in}^2$$

However, since all previous computations have been carried out with feet as the unit of length,

$$EI = (65.25 \times 10^6 \text{ lb·in}^2)(1 \text{ ft}/12 \text{ in.})^2 = 453.1 \times 10^3 \text{ lb·ft}^2$$

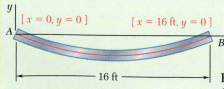

Fig. 1 Free-body diagram of showing equivalent force-couple system.

Fig. 2 Boundary conditions.

(continued)

b. Deflection at Midpoint C. (Fig. 3).   Making $x = 8$ ft in Eq. (3b),

$$EI\, y_C = -2.083(8)^4 + 80(8)^3 - 26.67\langle -3 \rangle^3 - 240\langle -3 \rangle^2 - 11.36 \times 10^3(8)$$

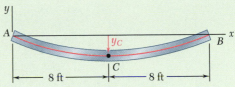

Fig. 3   Deformed elastic curve
showing displacement at midpoint C.

Noting that each bracket is equal to zero and substituting for EI its numerical value gives

$$(453.1 \times 10^3 \text{ lb·ft}^2)\, y_C = -58.45 \times 10^3 \text{ lb·ft}^3$$

and solving for y_C: $y_C = -0.1290$ ft $y_C = -1.548$ in. ◀

REFLECT and THINK: Note that the deflection obtained at midpoint C is *not* the maximum deflection.

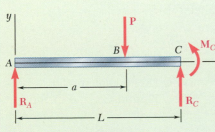

Fig. 1   Free-body diagram.

Sample Problem 9.6

For the uniform beam ABC, (a) express the reaction at A in terms of P, L, a, E, and I, (b) determine the reaction at A and the deflection under the load when $a = L/2$.

STRATEGY: The beam is statically indeterminate to the first degree. Using singularity functions, you can write the bending-moment equation for the beam, including the unknown reaction at A as part of the expression. After substituting this equation into the differential equation of the elastic curve, integrating twice, and applying the boundary conditions, the reaction at A can be determined, followed by the determination of the desired deflection.

MODELING:

Reactions.   For the given vertical load **P** the reactions are as shown in Fig. 1. We note that they are statically indeterminate.

Shear and Bending Moment.   Using a step function to represent the contribution of **P** to the shear,

$$V(x) = R_A - P\langle x - a \rangle^0$$

Integrating in x, the bending moment is

$$M(x) = R_A x - P\langle x - a \rangle^1$$

(continued)

ANALYSIS:

Equation of the Elastic Curve. Using Eq. (9.4),

$$EI\frac{d^2y}{dx^2} = R_A x - P\langle x - a \rangle^1$$

Integrating twice in x,

$$EI\frac{dy}{dx} = EI\theta = \frac{1}{2}R_A x^2 - \frac{1}{2}P\langle x - a \rangle^2 + C_1$$

$$EI\,y = \frac{1}{6}R_A x^3 - \frac{1}{6}P\langle x - a \rangle^3 + C_1 x + C_2$$

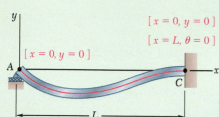

$[x = 0, y = 0]$
$[x = L, \theta = 0]$
$[x = 0, y = 0]$

Fig. 2 Boundary conditions.

Boundary Conditions Referring to Fig. 2 and noting that the bracket $\langle x - a \rangle$ is equal to zero for $x = 0$ and to $(L - a)$ for $x = L$,

$$[x = 0, y = 0]: \quad C_2 = 0 \tag{1}$$

$$[x = L, \theta = 0]: \quad \tfrac{1}{2}R_A L^2 - \tfrac{1}{2}P(L - a)^2 + C_1 = 0 \tag{2}$$

$$[x = L, y = 0]: \quad \tfrac{1}{6}R_A L^3 - \tfrac{1}{6}P(L - a)^3 + C_1 L + C_2 = 0 \tag{3}$$

a. Reaction at A. Multiplying Eq. (2) by L, subtracting Eq. (3) member by member from the equation, and noting that $C_2 = 0$, we obtain

$$\frac{1}{3}R_A L^3 - \frac{1}{6}P(L - a)^2[3L - (L - a)] = 0$$

$$\mathbf{R}_A = P\left(1 - \frac{a}{L}\right)^2\left(1 + \frac{a}{2L}\right)\uparrow \quad \blacktriangleleft$$

The reaction is independent of E and I.

b. Reaction at A and Deflection at B when $a = \frac{1}{2}L$ (Fig. 3). Making $a = \frac{1}{2}L$ in the expression obtained for R_A,

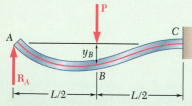

Fig. 3 Deformed elastic curve showing deflection at B.

$$R_A = P(1 - \tfrac{1}{2})^2(1 + \tfrac{1}{4}) = 5P/16 \qquad \mathbf{R}_A = \frac{5}{16}P\uparrow \quad \blacktriangleleft$$

Substituting $a = L/2$ and $R_A = 5P/16$ into Eq. (2) and solving for C_1, $C_1 = -PL^2/32$. Making $x = L/2$, $C_1 = -PL^2/32$, and $C_2 = 0$ in the expression obtained for y,

$$y_B = -\frac{7PL^3}{768EI} \qquad\qquad y_B = \frac{7PL^3}{768EI}\downarrow \quad \blacktriangleleft$$

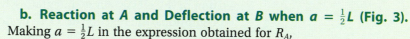

REFLECT and THINK: Note that the deflection obtained at B is *not* the maximum deflection.

Problems

Use singularity functions to solve the following problems and assume that the flexural rigidity *EI* of each beam is constant.

9.35 and 9.36 For the beam and loading shown, determine (*a*) the equation of the elastic curve, (*b*) the slope at end *A*, (*c*) the deflection at point *C*.

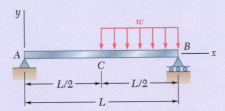

Fig. P9.35

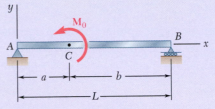

Fig. P9.36

9.37 and 9.38 For the beam and loading shown, determine the deflection at (*a*) point *B*, (*b*) point *C*, (*c*) point *D*.

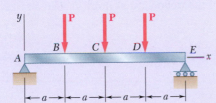

Fig. P9.37

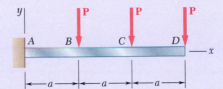

Fig. P9.38

9.39 and 9.40 For the beam and loading shown, determine (*a*) the deflection at end *A*, (*b*) the deflection at point *C*, (*c*) the slope at end *D*.

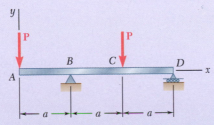

Fig. P9.39

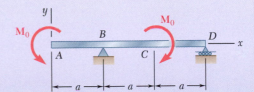

Fig. P9.40

9.41 and 9.42 For the beam and loading shown, determine (*a*) the equation of the elastic curve, (*b*) the deflection at the midpoint *C*.

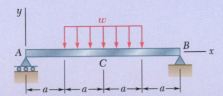

Fig. P9.41

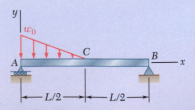

Fig. P9.42

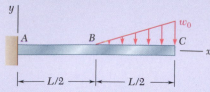

Fig. P9.43

9.43 For the beam and loading shown, determine (*a*) the equation of the elastic curve, (*b*) the deflection at point *B*, (*c*) the deflection at point *C*.

9.44 For the beam and loading shown, determine (*a*) the equation of the elastic curve, (*b*) the deflection at point *B*, (*c*) the deflection at point *D*.

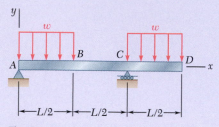

Fig. P9.44

9.45 For the timber beam and loading shown, determine (*a*) the slope at end *A*, (*b*) the deflection at the midpoint *C*. Use *E* = 12 GPa.

9.46 For the beam and loading shown, determine (*a*) the slope at end *A*, (*b*) the deflection at point *B*. Use *E* = 29 × 10⁶ psi.

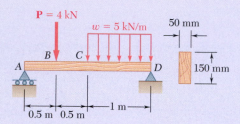

Fig. P9.45

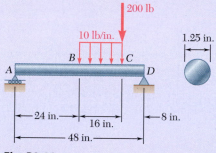

Fig. P9.46

9.47 For the beam and loading shown, determine (*a*) the slope at end *A*, (*b*) the deflection at point *C*. Use *E* = 29 × 10⁶ psi.

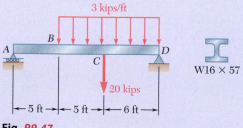

Fig. P9.47

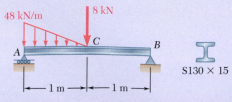

Fig. P9.48

9.48 For the beam and loading shown, determine (*a*) the slope at end *A*, (*b*) the deflection at the midpoint *C*. Use *E* = 200 GPa.

9.49 and 9.50 For the beam and loading shown, determine (a) the reaction at the roller support, (b) the deflection at point C.

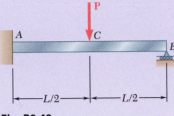

Fig. P9.49

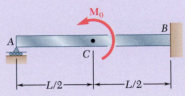

Fig. P9.50

9.51 and 9.52 For the beam and loading shown, determine (a) the reaction at the roller support, (b) the deflection at point B.

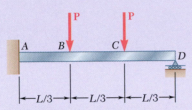

Fig. P9.51

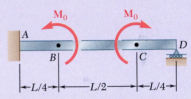

Fig. *P9.52*

9.53 For the beam and loading shown, determine (a) the reaction at point C, (b) the deflection at point B. Use $E = 200$ GPa.

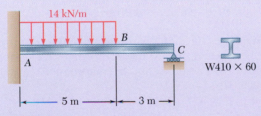

Fig. P9.53

W410 × 60

9.54 For the beam shown and knowing that $P = 40$ kN, determine (a) the reaction at point E, (b) the deflection at point C. Use $E = 200$ GPa.

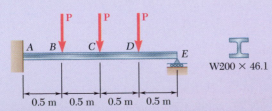

W200 × 46.1

Fig. P9.54

9.55 and 9.56 For the beam and loading shown, determine (*a*) the reaction at point *A*, (*b*) the deflection at point *C*. Use $E = 29 \times 10^6$ psi.

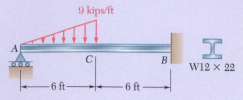

9 kips/ft
A
C
B
6 ft
6 ft
W12 × 22

Fig. *P9.55*

w = 4.5 kips/ft
A
B
C
D
E
2.5 ft
2.5 ft
2.5 ft
2.5 ft
W14 × 22

Fig. P9.56

9.57 For the beam and loading shown, determine (*a*) the reaction at point *A*, (*b*) the deflection at point *D*.

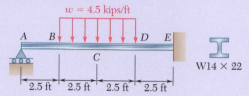

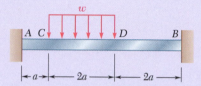

w
A C
D
B
a
2a
2a

Fig. P9.57

P
A
B
C
D
L/3
L/2
L/2

Fig. P9.58

9.58 For the beam and loading shown, determine (*a*) the reaction at point *A*, (*b*) the deflection at midpoint *C*.

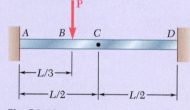

9.59 through 9.62 For the beam and loading indicated, determine the magnitude and location of the largest downward deflection.

9.59 Beam and loading of Prob. 9.45.
9.60 Beam and loading of Prob. 9.46.
9.61 Beam and loading of Prob. 9.47.
9.62 Beam and loading of Prob. 9.48.

9.63 The rigid bars *BF* and *DH* are welded to the rolled-steel beam *AE* as shown. Determine for the loading shown (*a*) the deflection at point *B*, (*b*) the deflection at midpoint *C* of the beam. Use $E = 200$ GPa.

0.5 m
0.3 m
0.3 m
0.5 m
A
B
C
D
E
F
H
G
0.4 m
0.15 m
100 kN
W100 × 19.3

Fig. *P9.63*

30 kN/m
A
C
D
B
F
E
50 kN
2.4 m
1.2 m
1.2 m
W460 × 52

Fig. *P9.64*

9.64 The rigid bar *DEF* is welded at point *D* to the rolled-steel beam *AB*. For the loading shown, determine (*a*) the slope at point *A*, (*b*) the deflection at midpoint *C* of the beam. Use $E = 200$ GPa.

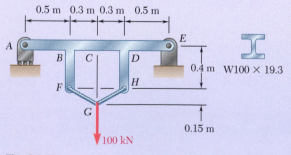

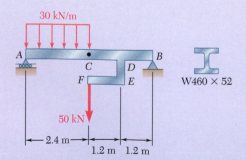

9.4 METHOD OF SUPERPOSITION

9.4A Statically Determinate Beams

When a beam is subjected to several concentrated or distributed loads, it is convenient to compute separately the slope and deflection caused by each of the given loads. The slope and deflection due to the combined loads are obtained by applying the principle of superposition (Sec. 2.5) and adding the values of the slope or deflection corresponding to the various loads.

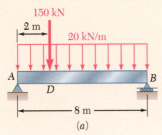

Fig. 9.21 (a) Simply supported beam having distributed and concentrated loads.

Concept Application 9.7

Determine the slope and deflection at D for the beam and loading shown (Fig. 9.21a), knowing that the flexural rigidity of the beam is $EI = 100$ MN·m².

The slope and deflection at any point of the beam can be obtained by superposing the slopes and deflections caused by the concentrated load and by the distributed load (Fig. 9.21b).

Since the concentrated load in Fig. 9.21c is applied at quarter span, the results for the beam and loading of Concept Application 9.3 can be used to write

$$(\theta_D)_P = -\frac{PL^2}{32EI} = -\frac{(150 \times 10^3)(8)^2}{32(100 \times 10^6)} = -3 \times 10^{-3} \text{ rad}$$

$$(y_D)_P = -\frac{3PL^3}{256EI} = -\frac{3(150 \times 10^3)(8)^3}{256(100 \times 10^6)} = -9 \times 10^{-3} \text{ m}$$

$$= -9 \text{ mm}$$

On the other hand, recalling the equation of the elastic curve obtained for a uniformly distributed load in Concept Application 9.2, the deflection in Fig. 9.21d is

$$y = \frac{w}{24EI}(-x^4 + 2Lx^3 - L^3x) \tag{1}$$

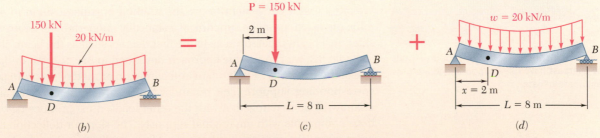

Fig. 9.21 (b) The beam's loading can be obtained by superposing deflections due to (c) the concentrated load and (d) the distributed load.

Differentiating with respect to x gives

$$\theta = \frac{dy}{dx} = \frac{w}{24EI}(-4x^3 + 6Lx^2 - L^3) \tag{2}$$

(continued)

Making $w = 20$ kN/m, $x = 2$ m, and $L = 8$ m in Eqs. (1) and (2), we obtain

$$(\theta_D)_w = \frac{20 \times 10^3}{24(100 \times 10^6)}(-352) = -2.93 \times 10^{-3} \, \text{rad}$$

$$(y_D)_w = \frac{20 \times 10^3}{24(100 \times 10^6)}(-912) = -7.60 \times 10^{-3} \, \text{m}$$

$$= -7.60 \, \text{mm}$$

Combining the slopes and deflections produced by the concentrated and the distributed loads,

$$\theta_D = (\theta_D)_P + (\theta_D)_w = -3 \times 10^{-3} - 2.93 \times 10^{-3}$$

$$= -5.93 \times 10^{-3} \, \text{rad}$$

$$y_D = (y_D)_P + (y_D)_w = -9 \, \text{mm} - 7.60 \, \text{mm} = -16.60 \, \text{mm}$$

To facilitate the work of practicing engineers, most structural and mechanical engineering handbooks include tables giving the deflections and slopes of beams for various loadings and types of support. Such a table is found in Appendix D. The slope and deflection of the beam of Fig. 9.21a could have been determined from that table. Indeed, using the information given under cases 5 and 6, we could have expressed the deflection of the beam for any value $x \leq L/4$. Taking the derivative of the expression obtained in this way would have yielded the slope of the beam over the same interval. We also note that the slope at both ends of the beam can be obtained by simply adding the corresponding values given in the table. However, the maximum deflection of the beam of Fig. 9.21a *cannot* be obtained by adding the maximum deflections of cases 5 and 6, since these deflections occur at different points of the beam.[†]

9.4B Statically Indeterminate Beams

We often find it convenient to use the method of superposition to determine the reactions at the supports of a statically indeterminate beam. Considering a beam indeterminate to the first degree, such as the beam shown in Photo 9.3, we can use the approach described in Sec. 9.2. We designate one of the reactions as redundant and eliminate or modify accordingly the corresponding support. The redundant reaction is then treated as an unknown load that, together with the other loads, must produce deformations compatible with the original supports. The slope or deflection at the point where the support has been modified or eliminated is obtained by computing the deformations caused by both the given loads and the redundant reaction and by superposing the results. Once the reactions at the supports are found, the slope and deflection can be determined.

Photo 9.3 The continuous beams supporting this highway overpass have three supports and are thus statically indeterminate.

[†]An approximate value of the maximum deflection of the beam can be obtained by plotting the values of y corresponding to various values of x. The determination of the exact location and magnitude of the maximum deflection would require setting equal to zero the expression obtained for the slope of the beam and solving this equation for x.

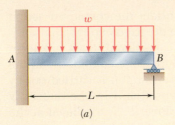

Fig. 9.22 (a) Statically indeterminate beam with a uniformly distributed load.

Concept Application 9.8

Determine the reactions at the supports for the prismatic beam and loading shown in Fig. 9.22a. (This is the same beam and loading as in Concept Application 9.5.)

We consider the reaction at B as redundant and release the beam from the support. The reaction $\mathbf{R}_B$ is now considered as an unknown load (Fig. 9.22b) and will be determined from the condition that the deflection of the beam at B must be zero. The solution is carried out by considering separately the deflection $(y_B)_w$ caused at B by the uniformly distributed load w (Fig. 9.22c) and the deflection $(y_B)_R$ produced at the same point by the redundant reaction $\mathbf{R}_B$ (Fig. 9.22d).

From the table of Appendix D (cases 2 and 1),

$$(y_B)_w = -\frac{wL^4}{8EI} \qquad (y_B)_R = +\frac{R_B L^3}{3EI}$$

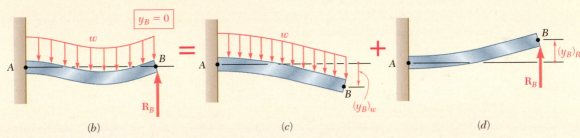

Fig. 9.22 (b) Analyze the indeterminate beam by superposing two determinate cantilever beams, subjected to (c) a uniformly distributed load, (d) the redundant reaction.

Writing that the deflection at B is the sum of these two quantities and that it must be zero,

$$y_B = (y_B)_w + (y_B)_R = 0$$

$$y_B = -\frac{wL^4}{8EI} + \frac{R_B L^3}{3EI} = 0$$

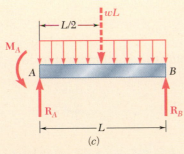

Fig. 9.22 (e) Free-body diagram of indeterminate beam.

and, solving for R_B, $\qquad R_B = \frac{3}{8}wL \qquad \mathbf{R}_B = \frac{3}{8}wL \uparrow$

Drawing the free-body diagram of the beam (Fig. 9.22e) and writing the corresponding equilibrium equations,

$$+\uparrow \Sigma F_y = 0: \qquad\qquad R_A + R_B - wL = 0 \qquad\qquad (1)$$

$$R_A = wL - R_B = wL - \tfrac{3}{8}wL = \tfrac{5}{8}wL$$

$$\mathbf{R}_A = \tfrac{5}{8}wL \uparrow$$

$$+\!\!\curvearrowleft \Sigma M_A = 0: \qquad M_A + R_B L - (wL)(\tfrac{1}{2}L) = 0 \qquad (2)$$

$$M_A = \tfrac{1}{2}wL^2 - R_B L = \tfrac{1}{2}wL^2 - \tfrac{3}{8}wL^2 = \tfrac{1}{8}wL^2$$

$$\mathbf{M}_A = \tfrac{1}{8}wL^2 \curvearrowleft$$

(continued)

Alternative Solution. We may consider the couple exerted at the fixed end A as redundant and replace the fixed end by a pin-and-bracket support. The couple $\mathbf{M}_A$ is now considered as an unknown load (Fig. 9.22f) and will be determined from the condition that the slope of the beam at A must be zero. The solution is carried out by considering separately the slope $(\theta_A)_w$ caused at A by the uniformly distributed load w (Fig. 9.22g) and the slope $(\theta_A)_M$ produced at the same point by the unknown couple $\mathbf{M}_A$ (Fig 9.22h).

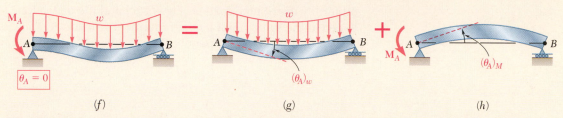

(f) (g) (h)

Fig. 9.22 (f) Analyze the indeterminate beam by superposing two determinate simply supported beams, subjected to (g) a uniformly distributed load, (h) the redundant reaction.

Using the table of Appendix D (cases 6 and 7) and noting that A and B must be interchanged in case 7,

$$(\theta_A)_w = -\frac{wL^3}{24EI} \qquad (\theta_A)_M = \frac{M_A L}{3EI}$$

Writing that the slope at A is the sum of these two quantities and that it must be zero gives

$$\theta_A = (\theta_A)_w + (\theta_A)_M = 0$$

$$\theta_A = -\frac{wL^3}{25EI} + \frac{M_A L}{3EI} = 0$$

where M_A is

$$M_A = \tfrac{1}{8}wL^2 \qquad \mathbf{M}_A = \tfrac{1}{8}wL^2 \,\text{\Large$\uparrow$}$$

The values of R_A and R_B are found by using the equilibrium equations (1) and (2).

The beam considered in the preceding Concept Application was indeterminate to the first degree. In the case of a beam indeterminate to the second degree (see Sec. 9.2), two reactions must be designated as redundant, and the corresponding supports must be eliminated or modified accordingly. The redundant reactions are then treated as unknown loads that, simultaneously and together with the other loads, must produce deformations that are compatible with the original supports. (See Sample Prob. 9.9.)

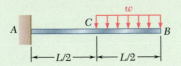

Sample Problem 9.7

For the beam and loading shown, determine the slope and deflection at point B.

STRATEGY: Using the method of superposition, you can model the given problem using a summation of beam load cases for which deflection formulae are readily available.

MODELING: Through the principle of superposition, the given loading can be obtained by superposing the loadings shown in the following picture equation of Fig. 1. The beam AB is the same in each part of the figure.

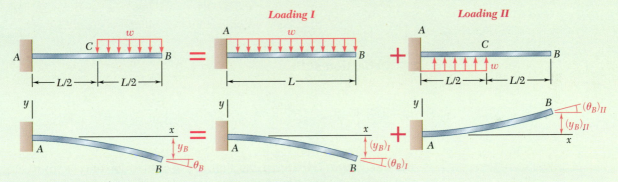

Fig. 1 Actual loading is equivalent to the superposition of two distributed loads.

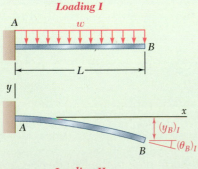

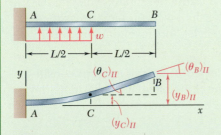

Fig. 2 Deformation details of the superposed loadings I and II.

ANALYSIS: For each of the loadings I and II (detailed further in Fig. 2), determine the slope and deflection at B by using the table of *Beam Deflections and Slopes* in Appendix D.

Loading I

$$(\theta_B)_\mathrm{I} = -\frac{wL^3}{6EI} \qquad (y_B)_\mathrm{I} = -\frac{wL^4}{8EI}$$

Loading II

$$(\theta_C)_\mathrm{II} = +\frac{w(L/2)^3}{6EI} = +\frac{wL^3}{48EI} \qquad (y_C)_\mathrm{II} = +\frac{w(L/2)^4}{8EI} = +\frac{wL^4}{128EI}$$

(continued)

In portion CB, the bending moment for loading II is zero. Thus, the elastic curve is a straight line.

$$(\theta_B)_{II} = (\theta_C)_{II} = +\frac{wL^3}{48EI} \qquad (y_B)_{II} = (y_C)_{II} + (\theta_C)_{II}\left(\frac{L}{2}\right)$$

$$= \frac{wL^4}{128EI} + \frac{wL^3}{48EI}\left(\frac{L}{2}\right) = +\frac{7wL^4}{384EI}$$

Slope at Point B

$$\theta_B = (\theta_B)_I + (\theta_B)_{II} = -\frac{wL^3}{6EI} + \frac{wL^3}{48EI} = -\frac{7wL^3}{48EI} \qquad \theta_B = \frac{7wL^3}{48EI} \; \triangleleft \; \blacktriangleleft$$

Deflection at B

$$y_B = (y_B)_I + (y_B)_{II} = -\frac{wL^4}{8EI} + \frac{7wL^4}{384EI} = -\frac{41wL^4}{384EI} \qquad y_B = \frac{41wL^4}{384EI} \downarrow \; \blacktriangleleft$$

REFLECT and THINK: Note that the formulae for one beam case can sometimes be extended to obtain the desired deflection of another case, as you saw here for loading II.

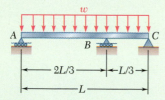

Sample Problem 9.8

For the uniform beam and loading shown, determine (*a*) the reaction at each support, (*b*) the slope at end *A*.

STRATEGY: The beam is statically indeterminate to the first degree. Strategically selecting the reaction at *B* as the redundant, you can use the method of superposition to model the given problem by using a summation of load cases for which deflection formulae are readily available.

MODELING: The reaction $\mathbf{R}_B$ is selected as redundant and considered as an unknown load. Applying the principle of superposition, the deflections due to the distributed load and to the reaction $\mathbf{R}_B$ are considered separately as shown in Fig. 1.

(continued)

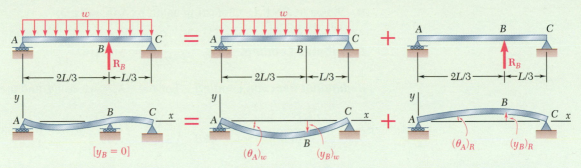

Fig. 1 Indeterminate beam modeled as superposition of two determinate simply supported beams with reaction at B chosen as redundant.

ANALYSIS: For each loading case, the deflection at point B is found by using the table of *Beam Deflections* and *Slopes* in Appendix D.

Distributed Loading. Use case 6, Appendix D:

$$y = -\frac{w}{24EI}(x^4 - 2Lx^3 + L^3x)$$

At point B, $x = \frac{2}{3}L$:

$$(y_B)_w = -\frac{w}{24EI}\left[\left(\frac{2}{3}L\right)^4 - 2L\left(\frac{2}{3}L\right)^3 + L^3\left(\frac{2}{3}L\right)\right] = -0.01132\frac{wL^4}{EI}$$

Redundant Reaction Loading. From case 5, Appendix D, with $a = \frac{2}{3}L$ and $b = \frac{1}{3}L$,

$$(y_B)_R = -\frac{Pa^2b^2}{3EIL} = +\frac{R_B}{3EIL}\left(\frac{2}{3}L\right)^2\left(\frac{L}{3}\right)^2 = 0.01646\frac{R_BL^3}{EI}$$

a. Reactions at Supports. Recalling that $y_B = 0$,

$$y_B = (y_B)_w + (y_B)_R$$

$$0 = -0.01132\frac{wL^4}{EI} + 0.01646\frac{R_BL^3}{EI} \qquad \mathbf{R}_B = 0.688wL \uparrow \quad \blacktriangleleft$$

Since the reaction R_B is now known, use the methods of statics to determine the other reactions (Fig. 2):

$$\mathbf{R}_A = 0.271wL \uparrow \quad \mathbf{R}_C = 0.0413wL \uparrow \quad \blacktriangleleft$$

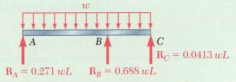

Fig. 2 Free-body diagram of beam with calculated reactions.

(continued)

b. Slope at End *A*.   Referring again to Appendix D,

Distributed Loading. $(\theta_A)_w = -\dfrac{wL^3}{24EI} = -0.04167\dfrac{wL^3}{EI}$

Redundant Reaction Loading.   For $P = -R_B = -0.688wL$ and $b = \frac{1}{3}L$,

$(\theta_A)_R = -\dfrac{Pb(L^2 - b^2)}{6EIL} = +\dfrac{0.688wL}{6EIL}\left(\dfrac{L}{3}\right)\left[L^2 - \left(\dfrac{L}{3}\right)^2\right] \quad (\theta_A)_R = 0.03398\dfrac{wL^3}{EI}$

Finally, $\theta_A = (\theta_A)_w + (\theta_A)_R$

$$\theta_A = -0.04167\dfrac{wL^3}{EI} + 0.03398\dfrac{wL^3}{EI} = -0.00769\dfrac{wL^3}{EI}$$

$$\theta_A = 0.00769\dfrac{wL^3}{EI} \;\blacktriangleleft$$

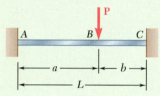

Sample Problem 9.9

For the beam and loading shown, determine the reaction at the fixed support *C*.

STRATEGY: The beam is statically indeterminate to the second degree. Strategically selecting the reactions at *C* as redundants, you can use the method of superposition and model the given problem by using a summation of load cases for which deflection formulae are readily available.

MODELING: Assuming the axial force in the beam to be zero, the beam *ABC* is indeterminate to the second degree, and we choose two reaction components as redundants: the vertical force $\mathbf{R}_C$ and the couple $\mathbf{M}_C$. The deformations caused by the given load $\mathbf{P}$, the force $\mathbf{R}_C$, and the couple $\mathbf{M}_C$ are considered separately, as shown in Fig. 1.

ANALYSIS: For each load, the slope and deflection at point *C* is found by using the table of *Beam Deflections and Slopes* in Appendix D.

Load *P*.   For this load, portion *BC* of the beam is straight.

$$(\theta_C)_P = (\theta_B)_P = -\dfrac{Pa^2}{2EI} \qquad (y_C)_P = (y_B)_P + (\theta_B)_P b$$

$$= -\dfrac{Pa^3}{3EI} - \dfrac{Pa^2}{2EI}b = -\dfrac{Pa^2}{6EI}(2a + 3b)$$

(continued)

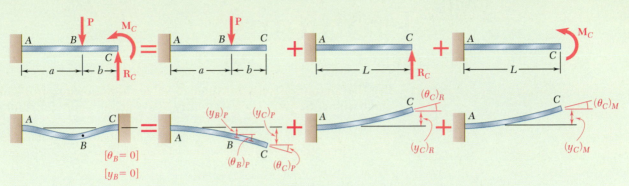

Fig. 1 Indeterminate beam modeled as the superposition of three determinate cases, including one for each of the two redundant reactions.

Force R_C $(\theta_C)_R = +\dfrac{R_C L^2}{2EI}$ $(y_C)_R = +\dfrac{R_C L^3}{3EI}$

Couple M_C $(\theta_C)_M = +\dfrac{M_C L}{EI}$ $(y_C)_M = +\dfrac{M_C L^2}{2EI}$

Boundary Conditions.   At end C, the slope and deflection must be zero:

$[x = L, \theta_C = 0]$: $\theta_C = (\theta_C)_P + (\theta_C)_R + (\theta_C)_M$

$$0 = -\frac{Pa^2}{2EI} + \frac{R_C L^2}{2EI} + \frac{M_C L}{EI} \tag{1}$$

$[x = L, y_C = 0]$: $y_C = (y_C)_P + (y_C)_R + (y_C)_M$

$$0 = -\frac{Pa^2}{6EI}(2a + 3b) + \frac{R_C L^3}{3EI} + \frac{M_C L^2}{2EI} \tag{2}$$

Reaction Components at C.   Solve Eqs. (1) and (2) simultaneously:

$$R_C = +\frac{Pa^2}{L^3}(a + 3b) \qquad \mathbf{R_C} = \frac{Pa^2}{L^3}(a + 3b) \uparrow \ \blacktriangleleft$$

$$M_C = -\frac{Pa^2 b}{L^2} \qquad\qquad \mathbf{M_C} = \frac{Pa^2 b}{L^2} \downarrow \ \blacktriangleleft$$

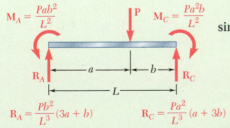

$M_A = \dfrac{Pab^2}{L^2}$

$R_A = \dfrac{Pb^2}{L^3}(3a + b)$ $R_C = \dfrac{Pa^2}{L^3}(a + 3b)$

Fig. 2 Free-body diagram showing the reaction results.

The methods of statics are used to determine the reaction at A, shown in Fig. 2.

REFLECT and THINK: Note that an alternate strategy that could have been used in this particular problem is to treat the couple reactions at the ends as redundant. The application of superposition would then have involved a simply-supported beam, for which deflection formulae are also readily available.

Problems

Use the method of superposition to solve the following problems and assume that the flexural rigidity EI of each beam is constant.

9.65 through 9.68 For the cantilever beam and loading shown, determine the slope and deflection at the free end.

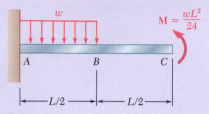

Fig. P9. 65

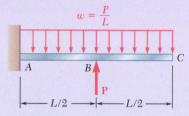

Fig. P9. 66

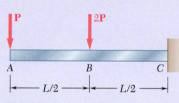

Fig. P9.67

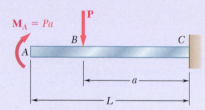

Fig. P9.68

9.69 through 9.72 For the beam and loading shown, determine (*a*) the deflection at point *C*, (*b*) the slope at end *A*.

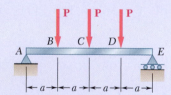

Fig. *P9.69*

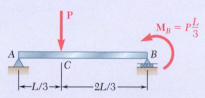

Fig. *P9.70*

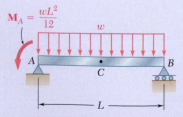

Fig. P9.71

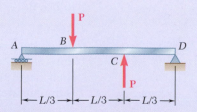

Fig. P9.72

9.73 For the cantilever beam and loading shown, determine the slope and deflection at end C. Use $E = 200$ GPa.

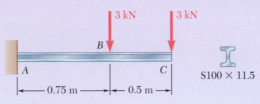

Fig. P9.73 and *P9.74*

9.74 For the cantilever beam and loading shown, determine the slope and deflection at point B. Use $E = 200$ GPa.

9.75 For the cantilever beam and loading shown, determine the slope and deflection at end A. Use $E = 29 \times 10^6$ psi.

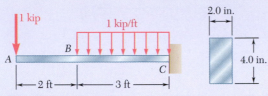

Fig. P9.75 and P9.76

9.76 For the cantilever beam and loading shown, determine the slope and deflection at point B. Use $E = 29 \times 10^6$ psi.

9.77 and 9.78 For the beam and loading shown, determine (*a*) the slope at end A, (*b*) the deflection at point C. Use $E = 200$ GPa.

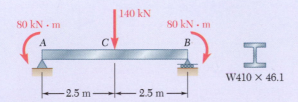

Fig. P9.77

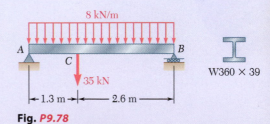

Fig. *P9.78*

9.79 and 9.80 For the uniform beam shown, determine (*a*) the reaction at A, (*b*) the reaction at B.

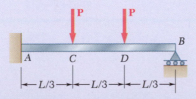

Fig. P9.79

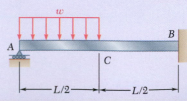

Fig. P9.80

9.81 and 9.82 For the uniform beam shown, determine the reaction at each of the three supports.

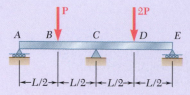

Fig. P9.81

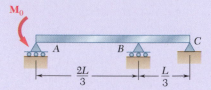

Fig. P9.82

9.83 and 9.84 For the beam shown, determine the reaction at *B*.

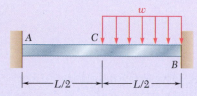

Fig. P9.83

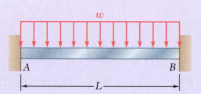

Fig. P9.84

9.85 Beam *DE* rests on the cantilever beam *AC* as shown. Knowing that a square rod of side 10 mm is used for each beam, determine the deflection at end *C* if the 25-N · m couple is applied (*a*) to end *E* of the beam *DE*, (*b*) to end *C* of the beam *AC*. Use *E* = 200 GPa.

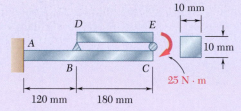

Fig. P9.85

9.86 Beam *AD* rests on beam *EF* as shown. Knowing that a W12 × 26 rolled-steel shape is used for each beam, determine for the loading shown the deflection at points *B* and *C*. Use $E = 29 \times 10^6$ psi.

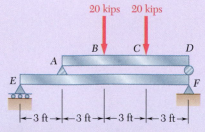

Fig. P9.86

9.87 The two beams shown have the same cross section and are joined by a hinge at C. For the loading shown, determine (*a*) the slope at point A, (*b*) the deflection at point B. Use $E = 29 \times 10^6$ psi.

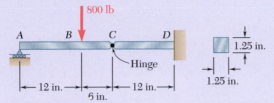

Fig. P9.87

9.88 A central beam BD is joined at hinges to two cantilever beams AB and DE. All beams have the cross section shown. For the loading shown, determine the largest w so that the deflection at C does not exceed 3 mm. Use $E = 200$ GPa.

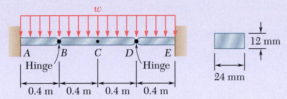

Fig. P9.88

9.89 For the loading shown, and knowing that beams AB and DE have the same flexural rigidity, determine the reaction (*a*) at B, (*b*) at E.

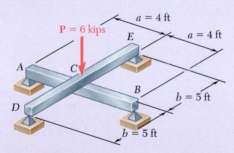

Fig. P9.89

9.90 Before the load **P** was applied, a gap, $\delta_0 = 0.5$ mm, existed between the cantilever beam AC and the support at B. Knowing that $E = 200$ GPa, determine the magnitude of **P** for which the deflection at C is 1 mm.

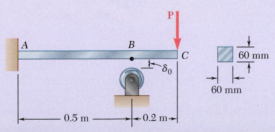

Fig. P9.90

9.91 Knowing that the rod ABC and the wire BD are both made of steel, determine (a) the deflection at B, (b) the reaction at A. Use $E = 200$ GPa.

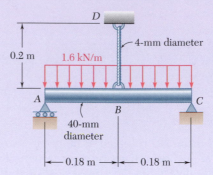

Fig. P9.91

9.92 Before the 2-kip/ft load is applied, a gap, $\delta_0 = 0.8$ in., exists between the W16 × 40 beam and the support at C. Knowing that $E = 29 \times 10^6$ psi, determine the reaction at each support after the uniformly distributed load is applied.

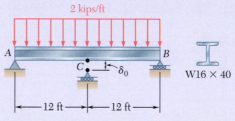

Fig. P9.92

9.93 A $\frac{7}{8}$-in.-diameter rod BC is attached to the lever AB and to the fixed support at C. Lever AB has a uniform cross section $\frac{3}{8}$ in. thick and 1 in. deep. For the loading shown, determine the deflection of point A. Use $E = 29 \times 10^6$ psi and $G = 11.2 \times 10^6$ psi.

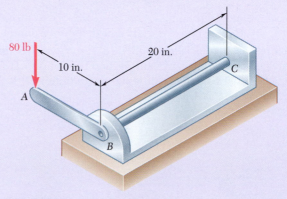

Fig. P9.93

9.94 A 16-mm-diameter rod has been bent into the shape shown. Determine the deflection of end C after the 200-N force is applied. Use $E = 200$ GPa and $G = 80$ GPa.

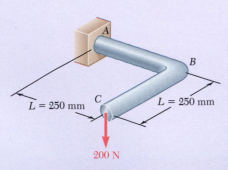

Fig. P9.94

*9.5   MOMENT-AREA THEOREMS

*9.5A   General Principles

In Sec. 9.1 through Sec. 9.3 we used a mathematical method based on the integration of a differential equation to determine the deflection and slope of a beam at any given point. The bending moment was expressed as a function $M(x)$ of the distance x measured along the beam, and two successive integrations led to the functions $\theta(x)$ and $y(x)$ representing, respectively, the slope and deflection at any point of the beam. In this section you will see how geometric properties of the elastic curve can be used to determine the deflection and slope of a beam at a specific point (Photo 9.4).

Photo 9.4   The maximum deflection of each beam supporting the floors of a building should be taken into account in the design process.

First Moment-Area Theorem.   Consider a beam AB subjected to some arbitrary loading (Fig. 9.23a). Draw the diagram representing the variation along the beam of M/EI obtained by dividing the bending moment M by the flexural rigidity EI (Fig. 9.23b). Except for a difference in the scales of ordinates, this diagram is the same as the bending-moment diagram if the flexural rigidity of the beam is constant.

Recalling Eq. (9.4) and that $dy/dx = \theta$,

$$\frac{d\theta}{dx} = \frac{d^2y}{dx^2} = \frac{M}{EI}$$

or

$$d\theta = \frac{M}{EI}\,dx \qquad\qquad \textbf{(9.15)}^{\dagger}$$

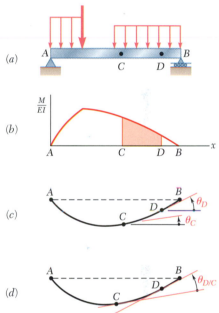

Fig. 9.23   First moment-area theorem. (a) Beam subjected to arbitrary load. (b) Plot of M/EI curve. (c) Elastic curve showing slope at C and D. (d) Elastic curve showing slope at D with respect to C.

[†]This relationship also can be determined by noting that the angle $d\theta$ formed by the tangents to the elastic curve at P and P' (Fig. 9.24) is also the angle formed by the corresponding normals to that curve. Thus $d\theta = ds/\rho$, where ds is the length of the arc PP' and ρ is the radius of curvature at P. Substituting for $1/\rho$ from Eq. (4.21) and noting that since the slope at P is very small, ds is equal in first approximation to the horizontal distance dx between P and P', we will again obtain Eq. (9.15).

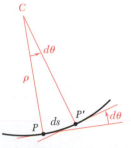

Fig. 9.24   Geometry of the elastic curve used to define the slope at point P' with respect to P.

Next consider two arbitrary points C and D on the beam and integrate both members of Eq. (9.15) from C to D:

$$\int_{\theta_C}^{\theta_D} d\theta = \int_{x_C}^{x_D} \frac{M}{EI}\, dx$$

or

$$\theta_D - \theta_C = \int_{x_C}^{x_D} \frac{M}{EI}\, dx \tag{9.16}$$

where θ_C and θ_D indicate the slope at C and D (Fig. 9.24c). But the right-hand member of Eq. (9.16) represents the area under the M/EI diagram between C and D, while the left-hand member is the angle between the tangents to the elastic curve at C and D (Fig. 9.23d). This angle is given as

$$\boxed{\theta_{D/C} = \text{area under } M/EI \text{ diagram between } C \text{ and } D} \tag{9.17}$$

This is the *first moment-area theorem.*

Note that $\theta_{D/C}$ and the area under the M/EI diagram have the same sign. This positive area (i.e., located above the x axis) corresponds to a counterclockwise rotation of the tangent to the elastic curve moving from C to D, and a negative area corresponds to a clockwise rotation.

Second Moment-Area Theorem.

Now consider two points P and P' located between C and D at a distance dx from each other (Fig. 9.25). The tangents to the elastic curve drawn at P and P' intercept a segment with a length dt on the vertical through point C. Since the slope θ at P and the angle $d\theta$ formed by the tangents at P and P' are both small quantities, dt is assumed to be equal to the arc of the circle of radius x_1 subtending the angle $d\theta$. Therefore,

$$dt = x_1\, d\theta$$

or substituting for $d\theta$ from Eq. (9.15),

$$dt = x_1 \frac{M}{EI}\, dx \tag{9.18}$$

Now integrate Eq. (9.18) from C to D. As point P describes the elastic curve from C to D, the tangent at P sweeps the vertical through C from C to E. Thus, the integral of the left-hand member is equal to the vertical distance from C to the tangent at D. This distance is denoted by $t_{C/D}$ and is called the *tangential deviation of C with respect to D.* Therefore,

$$t_{C/D} = \int_{x_C}^{x_D} x_1 \frac{M}{EI}\, dx \tag{9.19}$$

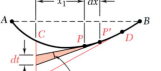

Fig. 9.25 Geometry used to determine the tangential deviation of C with respect to D.

Observe that $(M/EI)\,dx$ represents an element of area under the (M/EI) diagram, and $x_1\,(M/EI)\,dx$ is the first moment of that element with respect to a vertical axis through C (Fig. 9.26). The right-hand member in Eq. (9.19) represents the first moment with respect to that axis of the area located under the M/EI diagram between C and D.

We can, therefore, state the *second moment-area theorem* as follows: *The tangential deviation $t_{C/D}$ of C with respect to D is equal to the first moment with respect to a vertical axis through C of the area under the (M/EI) diagram between C and D.*

Recalling that the first moment of an area with respect to an axis is equal to the product of the area and the distance from its centroid to that axis, the second moment-area theorem is expressed as:

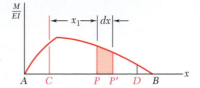

Fig. 9.26   The expression $x_1(M/EI)dx$ is the first moment of the shaded area with respect to C.

$$t_{C/D} = (\text{area between } C \text{ and } D)\,\bar{x}_1 \qquad \textbf{(9.20)}$$

where the area refers to the area under the M/EI diagram and where $\bar{x}_1$ is the distance from the centroid of the area to the vertical axis through C (Fig. 9.27a).

Remember to distinguish between the tangential deviation of C with respect to D ($t_{C/D}$) and the tangential deviation of D with respect to C ($t_{D/C}$). The tangential deviation $t_{D/C}$ represents the vertical distance from D to the tangent to the elastic curve at C and is obtained by multiplying the area under the (M/EI) diagram by the distance $\bar{x}_2$ from its centroid to the vertical axis through D (Fig. 9.27b):

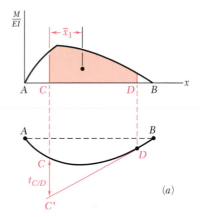

$$t_{D/C} = (\text{area between } C \text{ and } D)\,\bar{x}_2 \qquad \textbf{(9.21)}$$

Note that if an area under the M/EI diagram is located above the x axis, its first moment with respect to a vertical axis is positive. If it is located below the x axis, its first moment is negative. As shown in Figure 9.27, a point with a *positive* tangential deviation is located *above* the corresponding tangent. A point with a *negative* tangential deviation is located *below* that tangent.

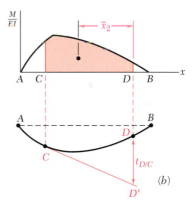

Fig. 9.27   Second moment-area theorem illustrated. (*a*) Evaluating $t_{C/D}$. (*b*) Evaluating $t_{D/C}$.

*9.5B   Cantilever Beams and Beams with Symmetric Loadings

Recall that the first moment-area theorem defines the angle $\theta_{D/C}$ *between* the tangents at two points C and D of the elastic curve. The angle θ_D that the tangent at D forms with the horizontal (i.e., the slope at D) can be obtained only if the slope at C is known. Similarly, the second moment-area theorem defines the vertical distance of one point of the elastic curve from the tangent at another point. Therefore, the tangential deviation $t_{D/C}$ helps to locate point D only if the tangent at C is known. Thus, the two moment-area theorems can be applied effectively to determine slopes and deflections only if a certain *reference tangent* to the elastic curve has been determined.

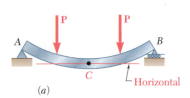

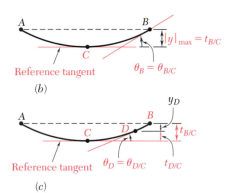

(a)

(b)

(c)

Fig. 9.29 Application of moment-area method to simply supported beams with symmetric loads. (a) Beam and loads. (b) Maximum deflection and slope at point B. (c) Deflection and slope at arbitrary point D.

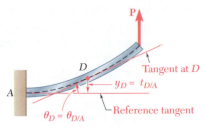

Fig. 9.28 Application of moment-area method to cantilever beams.

In *cantilever beams* (Fig. 9.28), the tangent to the elastic curve at the fixed end A is known and can be used as the reference tangent. Since $\theta_A = 0$, the slope of the beam at any point D is $\theta_D = \theta_{D/A}$ and can be obtained using the first moment-area theorem. On the other hand, the deflection y_D of point D is equal to the tangential deviation $t_{D/A}$ measured from the horizontal reference tangent at A and can be obtained using the second moment-area theorem.

In a simply supported beam AB with a *symmetric load* (Fig. 9.29a) or an overhanging symmetric beam with a symmetric load (see Sample Prob. 9.11), the tangent at the center C of the beam must be horizontal (by reason of symmetry) and can be used as the reference tangent (Fig. 9.29b). Since $\theta_C = 0$, the slope at the support B is $\theta_B = \theta_{B/C}$ and can be obtained using the first moment-area theorem. Also, $|y|_{max}$ is equal to the tangential deviation $t_{B/C}$ and can be obtained with the second moment-area theorem. The slope at any other point D of the beam (Fig. 9.29c) is found in a similar way, and the deflection at D is $y_D = t_{D/C} - t_{B/C}$.

Concept Application 9.9

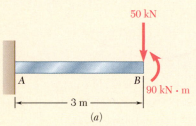

Fig. 9.30 (a) Cantilevered beam with end loads.

Determine the slope and deflection at end B of the prismatic cantilever beam AB when it is loaded as shown (Fig. 9.30a), knowing that the flexural rigidity of the beam is $EI = 10$ MN·m^2.

Draw the free-body diagram of the beam (Fig. 9.30b). Summing vertical components and moments about A, the reaction at the fixed end A consists of a 50 kN upward vertical force $\mathbf{R}_A$ and a 60 kN·m counterclockwise couple $\mathbf{M}_A$. Next, draw the bending-moment diagram (Fig. 9.30c) and determine from similar triangles the distance x_D from end A to point D of the beam where $M = 0$:

$$\frac{x_D}{60} = \frac{3 - x_D}{90} = \frac{3}{150} \qquad x_D = 1.2 \text{ m}$$

Dividing the values obtained for M by the flexural rigidity EI, draw the M/EI diagram (Fig. 9.30d) and compute the areas corresponding respectively to the segments AD and DB, assigning a positive sign to

(continued)

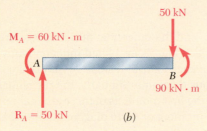

$M_A = 60 \text{ kN} \cdot \text{m}$

$R_A = 50 \text{ kN}$

(b)

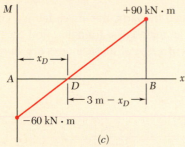

(c)

Fig. 9.30 (b) Free-body diagram with reactions. (c) Moment diagram.

the area located above the x axis and a negative sign to the area located below that axis. Use the first moment-area theorem to obtain

$$\theta_{B/A} = \theta_B - \theta_A = \text{area from } A \text{ to } B = A_1 + A_2$$

$$= -\tfrac{1}{2}(1.2 \text{ m})(6 \times 10^{-3} \text{ m}^{-1}) + \tfrac{1}{2}(1.8 \text{ m})(9 \times 10^{-3} \text{ m}^{-1})$$

$$= -3.6 \times 10^{-3} + 8.1 \times 10^{-3}$$

$$= +4.5 \times 10^{-3} \text{ rad}$$

and, since $\theta_A = 0$,

$$\theta_B = +4.5 \times 10^{-3} \text{ rad}$$

Using the second moment-area theorem, the tangential deviation $t_{B/A}$ is equal to the first moment about a vertical axis through B of the total area between A and B. The moment of each partial area is the product of that area and the distance from its centroid to the axis through B:

$$t_{B/A} = A_1(2.6 \text{ m}) + A_2(0.6 \text{ m})$$

$$= (-3.6 \times 10^{-3})(2.6 \text{ m}) + (8.1 \times 10^{-3})(0.6 \text{ m})$$

$$= -9.36 \text{ mm} + 4.86 \text{ mm} = -4.50 \text{ mm}$$

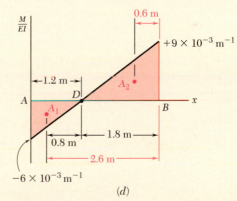

(d)

Fig. 9.30 (d) Plot of M/EI showing locations of area centroids.

Since the reference tangent at A is horizontal, the deflection at B is equal to $t_{B/A}$, so

$$y_B = t_{B/A} = -4.50 \text{ mm}$$

The deflected beam is shown in Fig. 9.30e.

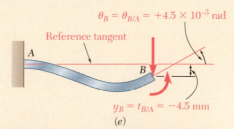

(e)

Fig. 9.30 (e) Deflected beam showing slope and deflection results at end B.

*9.5C   Bending-Moment Diagrams by Parts

In many applications, the angle $\theta_{D/C}$ and the tangential deviation $t_{D/C}$ are easier to determine if the effect of each load is evaluated independently. A separate M/EI diagram is drawn for each load, and angle $\theta_{D/C}$ is obtained by adding the areas under the various diagrams. Similarly, the tangential deviation $t_{D/C}$ is obtained by adding the first moments of these areas about a vertical axis through D. A bending-moment or M/EI diagram plotted this way is said to be *drawn by parts*.

When an M/EI diagram is drawn by parts, the areas defined consist of simple geometric shapes, such as rectangles, triangles, and parabolic spandrels. The areas and centroids of some of these shapes are shown in Fig. 9.31.

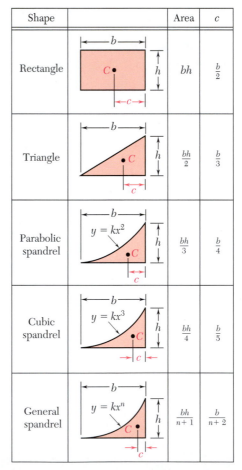

Shape		Area	c
Rectangle		bh	$\frac{b}{2}$
Triangle		$\frac{bh}{2}$	$\frac{b}{3}$
Parabolic spandrel	$y = kx^2$	$\frac{bh}{3}$	$\frac{b}{4}$
Cubic spandrel	$y = kx^3$	$\frac{bh}{4}$	$\frac{b}{5}$
General spandrel	$y = kx^n$	$\frac{bh}{n+1}$	$\frac{b}{n+2}$

Fig. 9.31   Areas and centroids of common shapes.

Concept Application 9.10

Determine the slope and deflection at end B of the prismatic beam of Concept Application 9.9, drawing the bending-moment diagram by parts.

The given load is replaced by the two equivalent loads shown in Fig. 9.32a, and the corresponding bending-moment and M/EI diagrams are drawn from right to left, starting at the free end B.

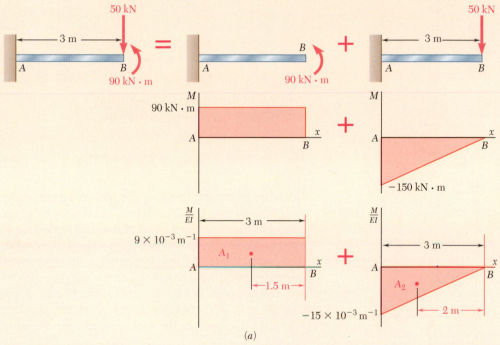

(a)

Fig. 9.32 (a) Superposition of loads and their resulting bending-moment and M/EI diagrams.

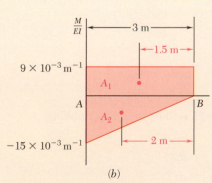

(b)

Fig. 9.32 M/EI diagrams combined into a single drawing.

Applying the first moment-area theorem and recalling that $\theta_A = 0$

$$\theta_B = \theta_{B/A} = A_1 + A_2$$

$$= (9 \times 10^{-3}\,\text{m}^{-1})(3\,\text{m}) - \tfrac{1}{2}(15 \times 10^{-3}\,\text{m}^{-1})(3\,\text{m})$$

$$= 27 \times 10^{-3} - 22.5 \times 10^{-3} = 4.5 \times 10^{-3}\,\text{rad}$$

Applying the second moment-area theorem, compute the first moment of each area about a vertical axis through B and write

$$y_B = t_{B/A} = A_1(1.5\,\text{m}) + A_2(2\,\text{m})$$

$$= (27 \times 10^{-3})(1.5\,\text{m}) - (22.5 \times 10^{-3})(2\,\text{m})$$

$$= 40.5\,\text{mm} - 45\,\text{mm} = -4.5\,\text{mm}$$

It practice, it is convenient to combine the two portions of the M/EI diagram into a single drawing (Fig. 9.32b).

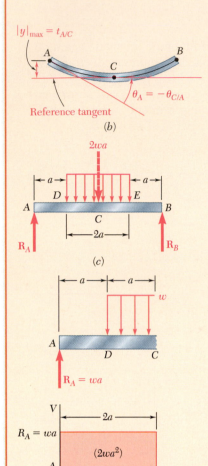

$|y|_{max} = t_{A/C}$

Reference tangent

$\theta_A = -\theta_{C/A}$

(b)

$2wa$

A D C E B

$2a$

R_A R_B

(c)

w

A D C

$R_A = wa$

V

$2a$

$R_A = wa$

$(2wa^2)$

A D C x

$(-\frac{1}{2}wa^2)$

$-wa$

a

$\frac{M}{EI}$

$\frac{4a}{3}$ $\frac{2wa^2}{EI}$

A A_1 C x

D A_2

$-\frac{wa^2}{2EI}$

$\frac{7a}{4}$ $\frac{1}{4}a$

a a

(d)

Fig. 9.33 (b) Elastic curve with maximum deflection and slope at point A shown. (c) Free-body diagram of the beam. (d) Shear and M/EI diagrams for the left half of the beam.

Concept Application 9.11

For the prismatic beam AB and the loading shown in Fig. 9.33a, determine the slope at a support and the maximum deflection.

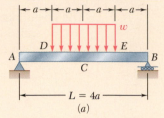

a a a a

w

A D C E B

$L = 4a$

(a)

Fig. 9.33 (a) Simply supported beam with symmetric distributed loading.

Sketch the deflected beam (Fig. 9.33b). Since the tangent at the center C of the beam is horizontal, it is used as the reference tangent, and $|y|_{max} = t_{A/C}$. But since $\theta_C = 0$,

$$\theta_{C/A} = \theta_C - \theta_A = -\theta_A \qquad \text{or} \qquad \theta_A = -\theta_{C/A}$$

The free-body diagram of the beam (Fig. 9.33c) shows

$$R_A = R_B = wa$$

Next, the shear and bending-moment diagrams are drawn for portion AC of the beam. These diagrams are drawn by parts, considering the effects of the reaction $\mathbf{R}_A$ and of the distributed load separately. However, for convenience the two parts of each diagram have been plotted together in Fig. 9.33d. Recall that when the distributed load is uniform, the corresponding parts of the shear and bending-moment diagrams are, respectively, linear and parabolic. The area and centroid of the triangle and of the parabolic spandrel are obtained by referring to Fig. 9.31. The areas of the triangle and spandrel are

$$A_1 = \frac{1}{2}(2a)\left(\frac{2wa^2}{EI}\right) = \frac{2wa^3}{EI}$$

and

$$A_2 = -\frac{1}{3}(a)\left(\frac{wa^2}{2EI}\right) = -\frac{wa^3}{6EI}$$

Applying the first moment-area theorem,

$$\theta_{C/A} = A_1 + A_2 = \frac{2wa^3}{EI} - \frac{wa^3}{6EI} = \frac{11wa^3}{6EI}$$

Recall from Figs. 9.33a and b that $a = \frac{1}{4}L$ and $\theta_A = -\theta_{C/A}$, making

$$\theta_A = -\frac{11wa^3}{6EI} = -\frac{11wL^3}{384EI}$$

Applying the second moment-area theorem

$$t_{A/C} = A_1\frac{4a}{3} + A_2\frac{7a}{4} = \left(\frac{2wa^3}{EI}\right)\frac{4a}{3} + \left(-\frac{wa^3}{6EI}\right)\frac{7a}{4} = \frac{19wa^4}{8EI}$$

and

$$|y|_{max} = t_{A/C} = \frac{19wa^4}{8EI} = \frac{19wL^4}{2048EI}$$

Sample Problem 9.10

Prismatic rods AD and DB are welded together to form the cantilever beam ADB. Knowing that the flexural rigidity is EI in portion AD of the beam and $2EI$ in portion DB, determine the slope and deflection at end A for the loading shown.

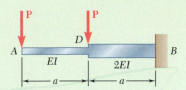

STRATEGY: To apply the moment-area theorems, you should first obtain the M/EI diagram for the beam. For a cantilever beam, it is convenient to place the reference tangent at the fixed end, since it is known to be horizontal.

MODELING and ANALYSIS:

(M/EI) Diagram. Referring to Fig. 1, draw the bending-moment diagram for the beam and then obtain the M/EI diagram by dividing the value of M at each point of the beam by the corresponding value of the flexural rigidity.

Reference Tangent. Referring to Fig. 2, choose the horizontal tangent at the fixed end B as the reference tangent. Since $\theta_B = 0$ and $y_B = 0$,

$$\theta_A = -\theta_{B/A} \quad y_A = t_{A/B}$$

Fig. 1 Free-body diagram and construction of the M/EI diagram.

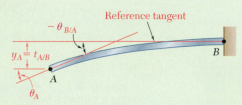

Fig. 2 Slope and deflection at end A related to reference tangent at fixed end B.

(continued)

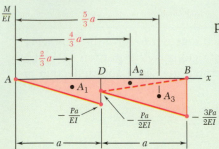

Fig. 3 Areas and centroids of moment-area diagram used to find slope and deflection.

Slope at A. Divide the M/EI diagram into the three triangular portions shown in Fig. 3.

$$A_1 = -\frac{1}{2}\frac{Pa}{EI}a = -\frac{Pa^2}{2EI}$$

$$A_2 = -\frac{1}{2}\frac{Pa}{2EI}a = -\frac{Pa^2}{4EI}$$

$$A_3 = -\frac{1}{2}\frac{3Pa}{2EI}a = -\frac{3Pa^2}{4EI}$$

Using the first moment-area theorem,

$$\theta_{B/A} = A_1 + A_2 + A_3 = -\frac{Pa^2}{2EI} - \frac{Pa^2}{4EI} - \frac{3Pa^2}{4EI} = -\frac{3Pa^2}{2EI}$$

$$\theta_A = -\theta_{B/A} = +\frac{3Pa^2}{2EI} \qquad \theta_A = \frac{3Pa^2}{2EI} \; \angle \quad \blacktriangleleft$$

Deflection at A. Using the second moment-area theorem,

$$y_A = t_{A/B} = A_1\left(\frac{2}{3}a\right) + A_2\left(\frac{4}{3}a\right) + A_3\left(\frac{5}{3}a\right)$$

$$= \left(-\frac{Pa^2}{2EI}\right)\frac{2a}{3} + \left(-\frac{Pa^2}{4EI}\right)\frac{4a}{3} + \left(-\frac{3Pa^2}{4EI}\right)\frac{5a}{3}$$

$$y_A = -\frac{23Pa^3}{12EI} \qquad\qquad y_A = \frac{23Pa^3}{12EI} \downarrow \quad \blacktriangleleft$$

REFLECT and THINK: This example demonstrates that the moment-area theorems can be just as easily used for nonprismatic beams as for prismatic beams.

Sample Problem 9.11

For the prismatic beam and loading shown, determine the slope and deflection at end E.

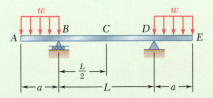

STRATEGY: To apply the moment-area theorems, you should first obtain the M/EI diagram for the beam. Due to the symmetry of both the beam and its loading, it is convenient to place the reference tangent at the mid-point since it is known to be horizontal.

(continued)

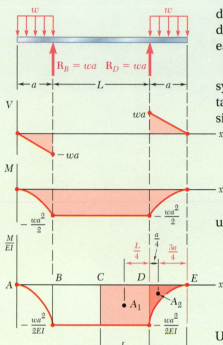

Fig. 1 Free-body diagram and construction of the moment-area diagram.

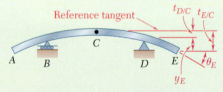

Fig. 2 Due to symmetry, reference tangent at midpoint C is horizontal. Shown are the slope and deflection at end E related to this reference tangent.

MODELING and ANALYSIS:

M/EI Diagram. From a free-body diagram of the beam (Fig. 1), determine the reactions and then draw the shear and bending-moment diagrams. Since the flexural rigidity of the beam is constant, divide each value of M by EI and obtain the M/EI diagram shown.

Reference Tangent. In Fig. 2, since the beam and its loads are symmetric with respect to the midpoint C, the tangent at C is horizontal and can be used as the reference tangent. Referring to Fig. 2 and since $\theta_C = 0$,

$$\theta_E = \theta_C + \theta_{E/C} = \theta_{E/C} \tag{1}$$

$$y_E = t_{E/C} - t_{D/C} \tag{2}$$

Slope at E. Referring to the M/EI diagram shown in Fig. 1 and using the first moment-area theorem,

$$A_1 = -\frac{wa^2}{2EI}\left(\frac{L}{2}\right) = -\frac{wa^2 L}{4EI}$$

$$A_2 = -\frac{1}{3}\left(\frac{wa^2}{2EI}\right)(a) = -\frac{wa^3}{6EI}$$

Using Eq. (1),

$$\theta_E = \theta_{E/C} = A_1 + A_2 = -\frac{wa^2 L}{4EI} - \frac{wa^3}{6EI}$$

$$\theta_E = -\frac{wa^2}{12EI}(3L + 2a) \qquad \theta_E = \frac{wa^2}{12EI}(3L + 2a) \ \searrow \ \blacktriangleleft$$

Deflection at E. Use the second moment-area theorem to write

$$t_{D/C} = A_1 \frac{L}{4} = \left(-\frac{wa^2 L}{4EI}\right)\frac{L}{4} = -\frac{wa^2 L^2}{16EI}$$

$$t_{E/C} = A_1\left(a + \frac{L}{4}\right) + A_2\left(\frac{3a}{4}\right)$$

$$= \left(-\frac{wa^2 L}{4EI}\right)\left(a + \frac{L}{4}\right) + \left(-\frac{wa^3}{6EI}\right)\left(\frac{3a}{4}\right)$$

$$= -\frac{wa^3 L}{4EI} - \frac{wa^2 L^2}{16EI} - \frac{wa^4}{8EI}$$

Use Eq. (2) to obtain

$$y_E = t_{E/C} - t_{D/C} = -\frac{wa^3 L}{4EI} - \frac{wa^4}{8EI}$$

$$y_E = -\frac{wa^3}{8EI}(2L + a) \qquad y_E = \frac{wa^3}{8EI}(2L + a) \downarrow \ \blacktriangleleft$$

Problems

Use the moment-area method to solve the following problems.

9.95 through 9.98 For the uniform cantilever beam and loading shown, determine (*a*) the slope at the free end, (*b*) the deflection at the free end.

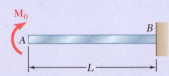

Fig. P9.95

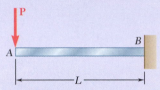

Fig. P9.96

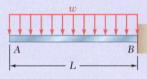

Fig. P9.97

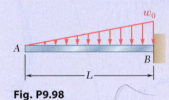

Fig. P9.98

9.99 and **9.100** For the uniform cantilever beam and loading shown, determine the slope and deflection at (*a*) point *B*, (*b*) point *C*.

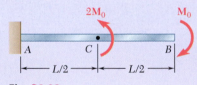

Fig. *P9.99*

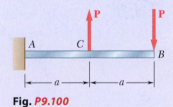

Fig. *P9.100*

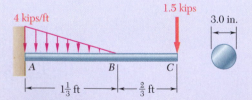

Fig. P9.101

9.101 For the cantilever beam and loading shown, determine (*a*) the slope at point *C*, (*b*) the deflection at point *C*. Use $E = 29 \times 10^6$ psi.

9.102 For the cantilever beam and loading shown, determine (*a*) the slope at point *A*, (*b*) the deflection at point *A*. Use $E = 200$ GPa.

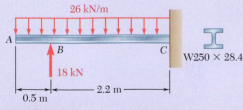

Fig. P9.102

9.103 Two C6 × 8.2 channels are welded back to back and loaded as shown. Knowing that $E = 29 \times 10^6$ psi, determine (a) the slope at point D, (b) the deflection at point D.

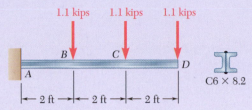

Fig. P9.103

9.104 For the cantilever beam and loading shown, determine (a) the slope at point A, (b) the deflection at point A. Use $E = 200$ GPa.

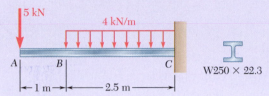

Fig. P9.104

9.105 For the cantilever beam and loading shown, determine (a) the slope at point A, (b) the deflection at point A.

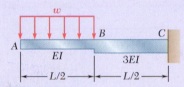

Fig. P9.105

9.106 For the cantilever beam and loading shown, determine the deflection and slope at end A caused by the moment $\mathbf{M}_0$.

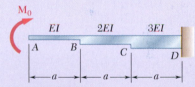

Fig. P9.106

9.107 Two cover plates are welded to the rolled-steel beam as shown. Using $E = 200$ GPa, determine (a) the slope at end A, (b) the deflection at end A.

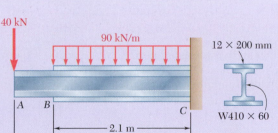

Fig. P9.107

9.108 Two cover plates are welded to the rolled-steel beam as shown. Using $E = 29 \times 10^6$ psi, determine (a) the slope at end C, (b) the deflection at end C.

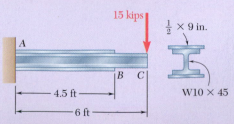

Fig. P9.108

9.109 through 9.114 For the prismatic beam and loading shown, determine (a) the slope at end A, (b) the deflection at the center C of the beam.

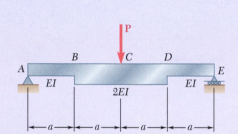

Fig. P9.109

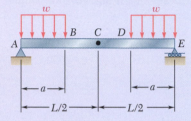

Fig. P9.110

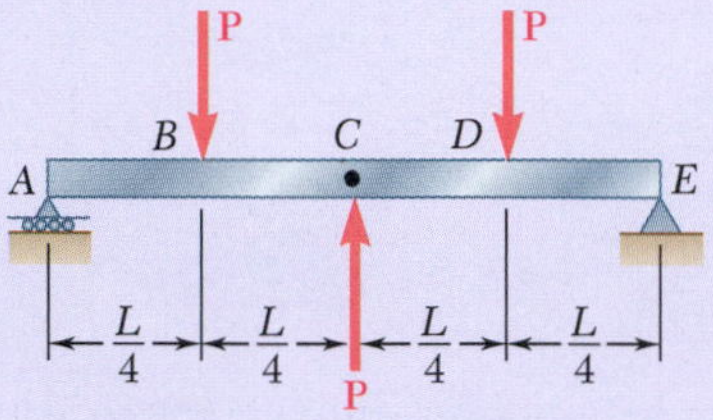

Fig. P9.111

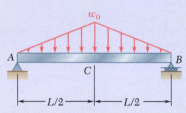

Fig. P9.112

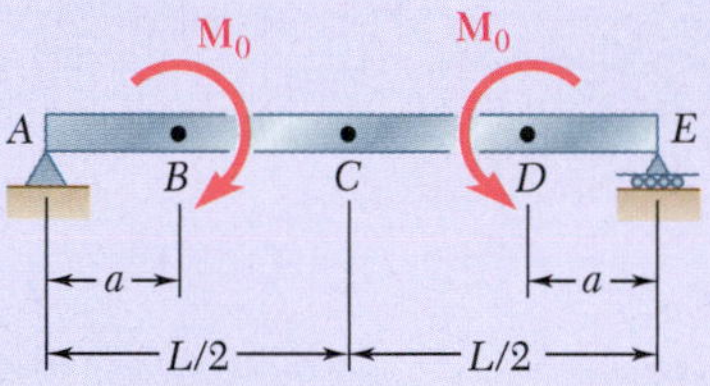

Fig. P9.113

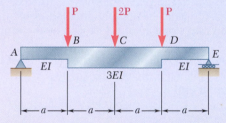

Fig. P9.114

9.115 and 9.116 For the beam and loading shown, determine (a) the slope at end A, (b) the deflection at the center C of the beam.

Fig. P9.115

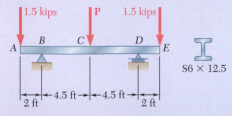

Fig. P9.116

9.117 Knowing that the magnitude of the load **P** is 7 kips, determine (a) the slope at end A, (b) the deflection at end A, (c) the deflection at midpoint C of the beam. Use $E = 29 \times 10^6$ psi.

Fig. P9.117

9.118 and 9.119 For the beam and loading shown, determine (*a*) the slope at end *A*, (*b*) the deflection at the midpoint of the beam. Use $E = 200$ GPa.

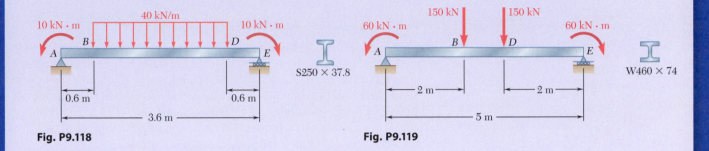

Fig. P9.118   **Fig. P9.119**

9.120 For the beam and loading shown and knowing that $w = 8$ kN/m, determine (*a*) the slope at end *A*, (*b*) the deflection at midpoint *C*. Use $E = 200$ GPa.

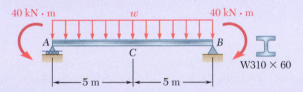

Fig. P9.120

9.121 For the beam and loading of Prob. 9.117, determine (*a*) the load **P** for which the deflection is zero at the midpoint *C* of the beam, (*b*) the corresponding deflection at end *A*. Use $E = 29 \times 10^6$ psi.

9.122 For the beam and loading of Prob. 9.120, determine the value of *w* for which the deflection is zero at the midpoint *C* of the beam. Use $E = 200$ GPa.

*9.123 A uniform rod *AE* is to be supported at two points *B* and *D*. Determine the distance *a* for which the slope at ends *A* and *E* is zero.

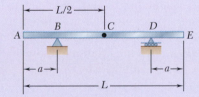

Fig. P9.123 and P9.124

*9.124 A uniform rod *AE* is to be supported at two points *B* and *D*. Determine the distance *a* from the ends of the rod to the points of support, if the downward deflections of points *A*, *C*, and *E* are to be equal.

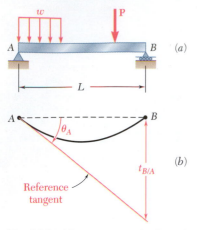

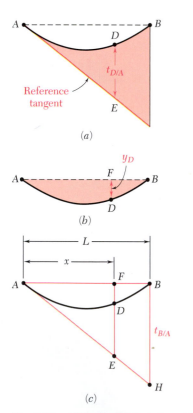

Fig. 9.34 (a) Unsymmetric loading. (b) Application of moment-area method to find slope at point A.

Fig. 9.36 (a) Tangential deviation of point D with respect to point A. (b) Deflection of point D. (c) Knowing HB through $t_{B/A}$, EF can be found by similar triangles.

*9.6 MOMENT-AREA THEOREMS APPLIED TO BEAMS WITH UNSYMMETRIC LOADINGS

*9.6A General Principles

When a simply supported or overhanging beam carries a symmetric load, the tangent at the center C of the beam is horizontal and can be used as the reference tangent (Sec. 9.6). When a simply supported or overhanging beam carries an unsymmetric load, it is not always possible to determine by inspection the point of the beam where the tangent is horizontal. Other means must be used to locate a reference tangent (i.e., a tangent of known slope for applying either of the two moment-area theorems).

It is usually convenient to select the reference tangent at one of the beam supports. For example, considering the tangent at the support A of the simply supported beam AB (Fig. 9.34), its slope can be determined by computing the tangential deviation $t_{B/A}$ of the support B with respect to A and dividing $t_{B/A}$ by the distance L between the supports. Recalling that the tangential deviation of a point located above the tangent is positive,

$$\theta_A = -\frac{t_{B/A}}{L} \qquad (9.22)$$

Once the slope of the reference tangent has been found, the slope θ_D of the beam at any point D (Fig. 9.35) can be determined by using the first moment-area theorem to obtain $\theta_{D/A}$, and then writing:

$$\theta_D = \theta_A + \theta_{D/A} \qquad (9.23)$$

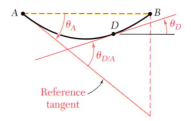

Fig. 9.35 Finding the tangential deviation between supports provides a convenient reference tangent for evaluating slopes.

The tangential deviation $t_{D/A}$ of D with respect to the support A can be obtained from the second moment-area theorem. Note that $t_{D/A}$ is equal to segment ED (Fig. 9.36a) and represents the vertical distance D *from the reference tangent*. On the other hand, the deflection y_D of point D represents the vertical distance of D *from the horizontal line AB* (Fig. 9.36b). Since y_D is equal in magnitude to the segment FD, it can be expressed as the difference between EF and ED (Fig. 9.36c). Observing from the similar triangles AFE and ABH that

$$\frac{EF}{x} = \frac{HB}{L} \qquad \text{or} \qquad EF = \frac{x}{L} t_{B/A}$$

and recalling the sign conventions used for deflections and tangential deviations,

$$y_D = ED - EF = t_{D/A} - \frac{x}{L} t_{B/A} \qquad (9.24)$$

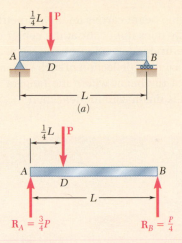

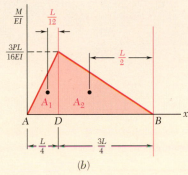

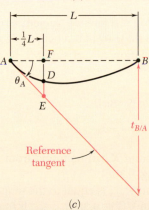

Fig. 9.37 (a) Simply supported beam with unsymmetric load. (b) Free-body diagram and *M/EI* diagram. (c) Reference tangent and geometry to determine slope and deflection at point D.

Concept Application 9.12

For the prismatic beam and loading shown (Fig. 9.37a), determine the slope and deflection at point D.

Reference Tangent at Support A.   Compute the reactions at the supports and draw the M/EI diagram (Fig. 9.37b). The tangential deviation $t_{B/A}$ of support B with respect to support A is found by applying the second moment-area theorem and computing the moments about a vertical axis through B of the areas A_1 and A_2.

$$A_1 = \frac{1}{2}\frac{L}{4}\frac{3PL}{16EI} = \frac{3PL^2}{128EI} \qquad A_2 = \frac{1}{2}\frac{3L}{4}\frac{3PL}{16EI} = \frac{9PL^2}{128EI}$$

$$t_{B/A} = A_1\left(\frac{L}{12} + \frac{3L}{4}\right) + A_2\left(\frac{L}{2}\right)$$

$$= \frac{3PL^2}{128EI}\frac{10L}{12} + \frac{9PL^2}{128EI}\frac{L}{2} = \frac{7PL^3}{128EI}$$

The slope of the reference tangent at A (Fig. 9.37c) is

$$\theta_A = -\frac{t_{B/A}}{L} = -\frac{7PL^2}{128EI}$$

Slope at D.   Applying the first moment-area theorem from A to D,

$$\theta_{D/A} = A_1 = \frac{3PL^2}{128EI}$$

Thus, the slope at D is

$$\theta_D = \theta_A + \theta_{D/A} = -\frac{7PL^2}{128EI} + \frac{3PL^2}{128EI} = -\frac{PL^2}{32EI}$$

Deflection at D.   The tangential deviation $DE = t_{D/A}$ is found by computing the moment of the area A_1 about a vertical axis through D:

$$DE = t_{D/A} = A_1\left(\frac{L}{12}\right) = \frac{3PL^2}{128EI}\frac{L}{12} = \frac{PL^3}{512EI}$$

The deflection at D is equal to the difference between the segments DE and EF (Fig. 9.37c). Thus,

$$y_D = DE - EF = t_{D/A} - \tfrac{1}{4}t_{B/A}$$

$$= \frac{PL^3}{512EI} - \frac{1}{4}\frac{7PL^3}{128EI}$$

$$y_D = -\frac{3PL^3}{256EI} = -0.01172PL^3/EI$$

*9.6B   Maximum Deflection

When a simply supported or overhanging beam carries an unsymmetric load, the maximum deflection generally does not occur at the center of the beam. As shown in Photo 9.5, the bridge is loaded by the truck at each axle location. To determine the maximum deflection of such a beam, it is first necessary to locate point K of the beam where the tangent is horizontal. The deflection at that point is the maximum deflection.

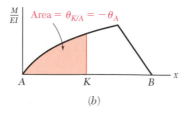

Photo 9.5   The deflections of the beams used for the bridge must be reviewed for different possible positions of the truck.

This analysis must begin by determining a reference tangent at one of the supports. If support A is selected, the slope θ_A of the tangent at A is obtained by computing the tangential deviation $t_{B/A}$ of support B with respect to A and dividing that quantity by the distance L between the two supports.

Since the slope θ_K at point K is zero (Fig. 9.38a),

$$\theta_{K/A} = \theta_K - \theta_A = 0 - \theta_A = -\theta_A$$

Recalling the first moment-area theorem, point K can be found from the M/EI diagram knowing that $\theta_{K/A} = -\theta_A$ (Fig. 9.38b).

Observing that the maximum deflection $|y|_{\max}$ is equal to the tangential deviation $t_{A/K}$ of support A with respect to K (Fig. 9.38a), $|y|_{\max}$ is found by computing the first moment with respect to the vertical axis through A of the area between A and K (Fig. 9.38b).

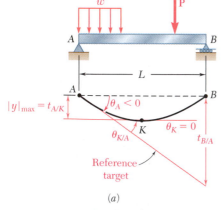

(a)

(b)

Fig. 9.38   Determination of maximum deflection using moment-area method. (a) The maximum deflection occurs at a point K where $\theta_K = 0$, which is where $\theta_{K/A} = -\theta_A$. (b) With point K so located, the maximum deflection is equal to the first moment of the shaded area with respect to A.

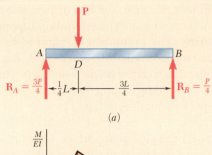

(a)

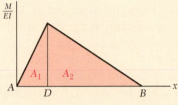

(b)

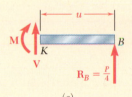

(c)

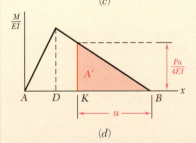

(d)

Fig. 9.39 (a) Free-body diagram. (b) M/EI diagram and geometry to determine the maximum deflection. (c) Free-body diagram of portion KB. (d) Maximum deflection is the first moment of the shaded area with respect to B.

Concept Application 9.13

Determine the maximum deflection of the beam of Concept Application 9.12. The free-body diagram is shown in Fig. 9.39a.

Determination of Point K Where Slope Is Zero. Recall that the slope at point D, where the load is applied, is negative. It follows that point K, where the slope is zero, is located between D and the support B (Fig. 9.39b). Our computations are simplified if the slope at K is related to the slope at B, rather than to the slope at A.

Since the slope at A has already been determined in Concept Application 9.12, the slope at B is obtained by

$$\theta_B = \theta_A + \theta_{B/A} = \theta_A + A_1 + A_2$$

$$\theta_B = -\frac{7PL^2}{128EI} + \frac{3PL^2}{128EI} + \frac{9PL^2}{128EI} = \frac{5PL^2}{128EI}$$

Observing that the bending moment at a distance u from end B is $M = \frac{1}{4}Pu$ (Fig. 9.39c), the area A' located between K and B under the M/EI diagram (Fig. 9.39d) is expressed as

$$A' = \frac{1}{2}\frac{Pu}{4EI}u = \frac{Pu^2}{8EI}$$

Use the first moment-area theorem to obtain

$$\theta_{B/K} = \theta_B - \theta_K = A'$$

and since $\theta_K = 0$, $\qquad \theta_B = A'$

Substituting the values obtained for θ_B and A',

$$\frac{5PL^2}{128EI} = \frac{Pu^2}{8EI}$$

and solving for u,

$$u = \frac{\sqrt{5}}{4}L = 0.559L$$

Thus, the distance from the support A to point K is

$$AK = L - 0.559L = 0.441L$$

Maximum Deflection. The maximum deflection $|y|_{max}$ is equal to the tangential deviation $t_{B/K}$ and thus to the first moment of area A' about a vertical axis through B (Fig. 9.39d).

$$|y|_{max} = t_{B/K} = A'\left(\frac{2u}{3}\right) = \frac{Pu^2}{8EI}\left(\frac{2u}{3}\right) = \frac{Pu^3}{12EI}$$

Substituting the value obtained for u,

$$|y|_{max} = \frac{P}{12EI}\left(\frac{\sqrt{5}}{4}L\right)^3 = 0.01456PL^3/EI$$

*9.6C Statically Indeterminate Beams

Reactions at the supports of a statically indeterminate beam can be determined using the moment-area method in much the same way that was described in Sec. 9.4. For a beam indeterminate to the first degree, one of the reactions is designated as redundant, and the corresponding support is eliminated or modified accordingly. The redundant reaction is then treated as an unknown load, which, together with the other loads, must produce deformations that are compatible with the original supports. This compatibility condition is usually expressed by writing that the tangential deviation of one support with respect to another either is zero or has a predetermined value.

Two separate free-body diagrams of the beam are drawn. One shows *the given loads and the corresponding reactions* at the supports that have not been eliminated; the other shows *the redundant reaction and the corresponding reactions* at the same supports (see Concept Application 9.14). An M/EI diagram is drawn for each of the two loadings, and the desired tangential deviations are obtained using the second moment-area theorem. Superposing the results, we obtain the required compatibility condition needed to determine the redundant reaction. The other reactions are obtained from the free-body diagram of beam.

Once the reactions at the supports are found, the slope and deflection are obtained using the moment-area method at any other point of the beam.

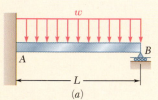

Fig. 9.40 (a) Statically indeterminate beam with a uniformly distributed load.

Concept Application 9.14

Determine the reaction at the supports for the prismatic beam and loading shown (Fig. 9.40a).

Consider the couple exerted at the fixed end A as redundant and replace the fixed end by a pin-and-bracket support. Couple $\mathbf{M}_A$ is now considered to be an unknown load (Fig. 9.40b) and will be determined from the condition that the tangent to the beam at A must be horizontal. Thus, this tangent must pass through the support B, and the tangential deviation $t_{B/A}$ of B with respect to A must be zero. The solution is

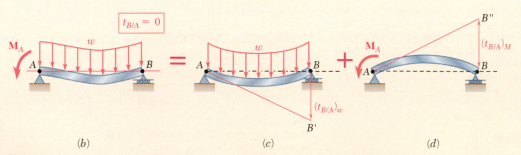

Fig. 9.40 (b) Analyze the indeterminate beam by superposing two determinate simply supported beams, subjected to (c) a uniformly distributed load, (d) the redundant reaction.

(continued)

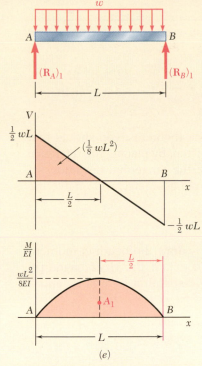

Fig. 9.40   (e) Free-body diagram of beam with distributed load, shear diagram, and *M/EI* diagram.

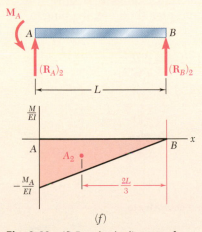

Fig. 9.40   (f) Free-body diagram of beam with redundant couple and *M/EI* diagram.

carried out by computing separately the tangential deviation $(t_{B/A})_w$ caused by the uniformly distributed load w (Fig. 9.40c) and the tangential deviation $(t_{B/A})_M$ produced by the unknown couple $\mathbf{M}_A$ (Fig. 9.40d).

Using the free-body diagram of the beam under the known distributed load w (Fig. 9.40e), determine the corresponding reactions at the supports A and B.

$$(\mathbf{R}_A)_1 = (\mathbf{R}_B)_1 = \tfrac{1}{2}wL \uparrow \tag{1}$$

Now draw the corresponding shear and M/EI diagrams (Fig. 9.40e). Observing that M/EI is represented by an arc of parabola and recalling the formula $A = \tfrac{2}{3}bh$ for the area under a parabola, the first moment of this area about a vertical axis through B is

$$(t_{B/A})_w = A_1\left(\frac{L}{2}\right) = \left(\frac{2}{3}L\frac{wL^2}{8EI}\right)\left(\frac{L}{2}\right) = \frac{wL^4}{24EI} \tag{2}$$

Using the free-body diagram of the beam when it is subjected to the unknown couple $\mathbf{M}_A$ (Fig. 9.40f), the corresponding reactions at A and B are

$$(\mathbf{R}_A)_2 = \frac{M_A}{L} \uparrow \qquad (\mathbf{R}_B)_2 = \frac{M_A}{L} \downarrow \tag{3}$$

Drawing the corresponding M/EI diagram (Fig. 9.40f), the second moment-area theorem is applied to obtain

$$(t_{B/A})_M = A_2\left(\frac{2L}{3}\right) = \left(-\frac{1}{2}L\frac{M_A}{EI}\right)\left(\frac{2L}{3}\right) = -\frac{M_AL^2}{3EI} \tag{4}$$

Combining the results obtained in Eqs. (2) and (4) and expressing that the resulting tangential deviation $t_{B/A}$ must be zero (Fig. 9.40b, c, d),

$$t_{B/A} = (t_{B/A})_w + (t_{B/A})_M = 0$$

$$\frac{wL^4}{24EI} - \frac{M_AL^2}{3EI} = 0$$

and solving for M_A,

$$M_A = +\tfrac{1}{8}wL^2 \qquad \mathbf{M}_A = \tfrac{1}{8}wL^2 \curvearrowleft$$

Substituting for M_A into Eq. (3), and recalling Eq. (1), the values of R_A and R_B are

$$R_A = (R_A)_1 + (R_A)_2 = \tfrac{1}{2}wL + \tfrac{1}{8}wL = \tfrac{5}{8}wL$$
$$R_B = (R_B)_1 + (R_B)_2 = \tfrac{1}{2}wL - \tfrac{1}{8}wL = \tfrac{3}{8}wL$$

In Concept Application 9.14, there was a single redundant reaction (i.e., the beam was *statically indeterminate to the first degree*). The *moment-area theorems* also can be used when there are additional redundant reactions, but it is necessary to write additional equations. Thus, for a beam that is *statically indeterminate to the second degree*, it would be necessary to select two redundant reactions and write two equations considering the *deformations* of the structure involved.

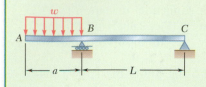

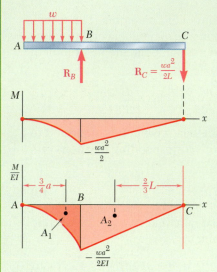

Fig. 1 Free-body, moment, and *M/EI* diagrams.

Sample Problem 9.12

For the beam and loading shown, (*a*) determine the deflection at end *A*, (*b*) evaluate y_A for the following data:

W 10 × 33: $I = 171 \text{ in}^4$ $E = 29 \times 10^6$ psi
$a = 3$ ft $= 36$ in. $L = 5.5$ ft $= 66$ in.
$w = 13.5$ kips/ft $= 1125$ lb/in.

STRATEGY: To apply the moment-area theorems, you should first obtain the *M/EI* diagram for the beam. Then, by placing the reference tangent at a support, you can evaluate the tangential deviations at other strategic points that, through simple geometry, will enable the determination of the desired deflection.

MODELING and ANALYSIS:

M/EI **Diagram.** Referring to Fig. 1, draw the bending-moment diagram. Since the flexural rigidity *EI* is constant, the *M/EI* diagram is as shown, which consists of a parabolic spandrel of area A_1 and a triangle of area A_2.

$$A_1 = \frac{1}{3}\left(-\frac{wa^2}{2EI}\right)a = -\frac{wa^3}{6EI}$$

$$A_2 = \frac{1}{2}\left(-\frac{wa^2}{2EI}\right)L = -\frac{wa^2L}{4EI}$$

Reference Tangent at *B*. The reference tangent is drawn at point *B* in Fig. 2. Using the second moment-area theorem, the tangential deviation of *C* with respect to *B* is

$$t_{C/B} = A_2\frac{2L}{3} = \left(-\frac{wa^2L}{4EI}\right)\frac{2L}{3} = -\frac{wa^2L^2}{6EI}$$

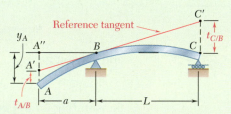

Fig. 2 Reference tangent and geometry to determine deflection at *A*.

From the similar triangles *A″A′B* and *CC′B*,

$$A''A' = t_{C/B}\left(\frac{a}{L}\right) = -\frac{wa^2L^2}{6EI}\left(\frac{a}{L}\right) = -\frac{wa^3L}{6EI}$$

Again using the second moment-area theorem,

$$t_{A/B} = A_1\frac{3a}{4} = \left(-\frac{wa^3}{6EI}\right)\frac{3a}{4} = -\frac{wa^4}{8EI}$$

(continued)

a. Deflection at End A

$$y_A = A''A' + t_{A/B} = -\frac{wa^3L}{6EI} - \frac{wa^4}{8EI} = -\frac{wa^4}{8EI}\left(\frac{4}{3}\frac{L}{a} + 1\right)$$

$$y_A = \frac{wa^4}{8EI}\left(1 + \frac{4}{3}\frac{L}{a}\right)\downarrow \quad \blacktriangleleft$$

b. Evaluation of y_A.   Substituting the data,

$$y_A = \frac{(1125\text{ lb/in.})(36\text{ in.})^4}{8(29 \times 10^6\text{ lb/in}^2)(171\text{ in}^4)}\left(1 + \frac{4}{3}\frac{66\text{ in.}}{36\text{ in.}}\right)$$

$$y_A = 0.1641\text{ in.}\downarrow \quad \blacktriangleleft$$

REFLECT and THINK: Note that an equally effective alternate strategy would be to draw a reference tangent at point C.

Sample Problem 9.13

For the beam and loading shown, determine the magnitude and location of the largest deflection. Use $E = 200$ GPa.

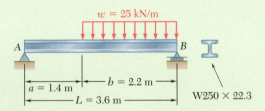

STRATEGY: To apply the moment-area theorems, you should first obtain the M/EI diagram for the beam. Then, by placing the reference tangent at a support, you can evaluate the tangential deviation at the other support that, through simple geometry and the further application of the moment-area theorems, will enable the determination of the maximum deflection.

MODELING: Use the free-body diagram of the entire beam in Fig. 1 to obtain

$$\mathbf{R}_A = 16.81\text{ kN}\uparrow \qquad \mathbf{R}_B = 38.2\text{ kN}\uparrow$$

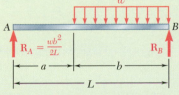

Fig. 1   Free-body diagram.

(continued)

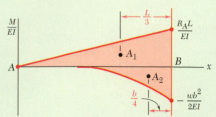

Fig. 2 Parts of M/EI diagram with centroid locations.

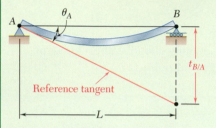

Fig. 3 Determination of θ_A through tangential deviation $t_{B/A}$.

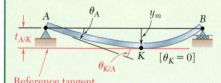

Fig. 4 Geometry to determine maximum deflection.

ANALYSIS:

***M/EI* Diagram.** Draw the M/EI diagram by parts (Fig. 2), considering the effects of the reaction $\mathbf{R}_A$ and of the distributed load separately. The areas of the triangle and of the spandrel are

$$A_1 = \frac{1}{2}\frac{R_A L}{EI}L = \frac{R_A L^2}{2EI} \qquad A_2 = \frac{1}{3}\left(-\frac{wb^2}{2EI}\right)b = -\frac{wb^3}{6EI}$$

Reference Tangent. As seen in Fig. 3, the tangent to the beam at support A is chosen as the reference tangent. Using the second moment-area theorem, the tangential deviation $t_{B/A}$ of support B with respect to support A is

$$t_{B/A} = A_1\frac{L}{3} + A_2\frac{b}{4} = \left(\frac{R_A L^2}{2EI}\right)\frac{L}{3} + \left(-\frac{wb^3}{6EI}\right)\frac{b}{4} = \frac{R_A L^3}{6EI} - \frac{wb^4}{24EI}$$

Slope at A

$$\theta_A = -\frac{t_{B/A}}{L} = -\left(\frac{R_A L^2}{6EI} - \frac{wb^4}{24EIL}\right) \tag{1}$$

Largest Deflection. As seen in Fig. 4, the largest deflection occurs at point K, where the slope of the beam is zero. Using Fig. 5, write

$$\theta_K = \theta_A + \theta_{K/A} = 0 \tag{2}$$

But

$$\theta_{K/A} = A_3 + A_4 = \frac{R_A x_m^2}{2EI} - \frac{w}{6EI}(x_m - a)^3 \tag{3}$$

Substitute for θ_A and $\theta_{K/A}$ from Eqs. (1) and (3) into Eq. (2):

$$-\left(\frac{R_A L^2}{6EI} - \frac{wb^4}{24EIL}\right) + \left[\frac{R_A x_m^2}{2EI} - \frac{w}{6EI}(x_m - a)^3\right] = 0$$

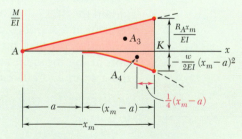

Fig. 5 M/EI diagram between Point A and the location of maximum deflection, point K.

(continued)

Substituting the numerical data gives

$$-29.53 \frac{10^3}{EI} + 8.405 x_m^2 \frac{10^3}{EI} - 4.167(x_m - 1.4)^3 \frac{10^3}{EI} = 0$$

Solving by trial and error for x_m, $x_m = 1.890$ m ◀

Computing the moments of A_3 and A_4 about a vertical axis through A gives

$$|y|_m = t_{A/K} = A_3 \frac{2x_m}{3} + A_4\left[a + \frac{3}{4}(x_m - a)\right]$$

$$= \frac{R_A x_m^3}{3EI} - \frac{wa}{6EI}(x_m - a)^3 - \frac{w}{8EI}(x_m - a)^4$$

Using the given data, $R_A = 16.81$ kN, and $I = 28.7 \times 10^{-6}$ m^4,

$$y_m = 6.44 \text{ mm} \downarrow \quad ◀$$

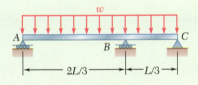

Sample Problem 9.14

For the uniform beam and loading shown, determine the reaction at B.

STRATEGY: Applying the superposition concept, you can model this statically indeterminate problem as a summation of the displacements for the given load and the redundant load cases. The redundant reaction can then be found by noting that a displacement associated with the two cases must be consistent with the geometry of the original beam.

MODELING: The beam is indeterminate to the first degree. Referring to Fig. 1, the reaction $\mathbf{R}_B$ is chosen as redundant, and the distributed load and redundant reaction load are considered separately.

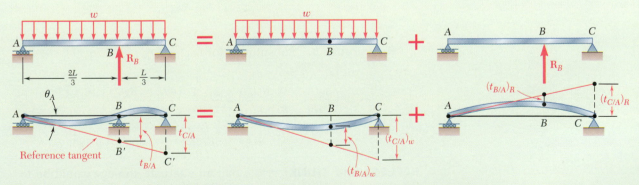

Fig. 1   Indeterminate beam modeled as superposition of two determinate beams with reaction at B chosen as redundant.

(continued)

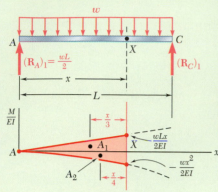

Fig. 2 Free-body and *M/EI* diagrams for beam with distributed load.

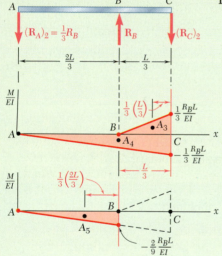

Fig. 3 Free-body and *M/EI* diagrams for beam with redundant reaction.

Next the tangent at *A* is selected as the reference tangent. From the similar triangles *ABB'* and *ACC'*,

$$\frac{t_{C/A}}{L} = \frac{t_{B/A}}{\frac{2}{3}L} \qquad t_{C/A} = \frac{3}{2}t_{B/A} \qquad \text{(1)}$$

For each loading, we draw the *M/EI* diagram and then determine the tangential deviations of *B* and *C* with respect to *A*.

ANALYSIS:

Distributed Loading (Fig. 2). Considering the *M/EI* diagram from end *A* to an arbitrary point *X*,

$$(t_{X/A})_w = A_1\frac{x}{3} + A_2\frac{x}{4} = \left(\frac{1}{2}\frac{wLx}{2EI}x\right)\frac{x}{3} + \left(-\frac{1}{3}\frac{wx^2}{2EI}x\right)\frac{x}{4} = \frac{wx^3}{24EI}(2L - x)$$

Letting $x = L$ and $x = \frac{2}{3}L$,

$$(t_{C/A})_w = \frac{wL^4}{24EI} \qquad (t_{B/A})_w = \frac{4}{243}\frac{wL^4}{EI}$$

Redundant Reaction Loading (Fig. 3).

$$(t_{C/A})_R = A_3\frac{L}{9} + A_4\frac{L}{3} = \left(\frac{1}{2}\frac{R_BL}{3EI}\frac{L}{3}\right)\frac{L}{9} + \left(-\frac{1}{2}\frac{R_BL}{3EI}L\right)\frac{L}{3} = -\frac{4}{81}\frac{R_BL^3}{EI}$$

$$(t_{B/A})_R = A_5\frac{2L}{9} = \left[-\frac{1}{2}\frac{2R_BL}{9EI}\left(\frac{2L}{3}\right)\right]\frac{2L}{9} = -\frac{4}{243}\frac{R_BL^3}{EI}$$

Combined Loading. Adding the results gives

$$t_{C/A} = \frac{wL^4}{24EI} - \frac{4}{81}\frac{R_BL^3}{EI} \qquad t_{B/A} = \frac{4}{243}\frac{(wL^4 - R_BL^3)}{EI}$$

Reaction at *B*. Substituting for $t_{C/A}$ and $t_{B/A}$ into Eq. (1),

$$\left(\frac{wL^4}{24EI} - \frac{4}{81}\frac{R_BL^3}{EI}\right) = \frac{3}{2}\left[\frac{4}{243}\frac{(wL^4 - R_BL^3)}{EI}\right]$$

$$R_B = 0.6875wL \qquad R_B = 0.688wL \uparrow \quad \blacktriangleleft$$

REFLECT and THINK: Note that an alternate strategy would be to determine the deflections at *B* for the given load and the redundant reaction and to set the sum equal to zero.

Problems

Use the moment-area method to solve the following problems.

9.125 through 9.128 For the prismatic beam and loading shown, determine (*a*) the deflection at point *D*, (*b*) the slope at end *A*.

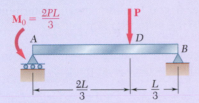

Fig. P9.125

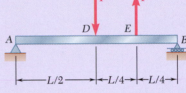

Fig. P9.126

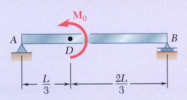

Fig. *P9.127*

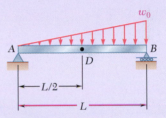

Fig. P9.128

9.129 and 9.130 For the beam and loading shown, determine (*a*) the slope at end *A*, (*b*) the deflection at point *D*. Use *E* = 200 GPa.

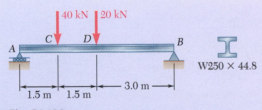

Fig. P9.129

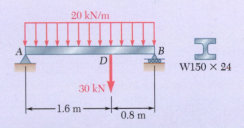

Fig. P9.130

9.131 For the timber beam and loading shown, determine (*a*) the slope at point *A*, (*b*) the deflection at point *C*. Use $E = 1.7 \times 10^6$ psi.

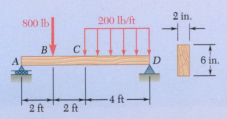

Fig. *P9.131*

9.132 For the beam and loading shown, determine (*a*) the slope at point *A*, (*b*) the deflection at point *E*. Use $E = 29 \times 10^6$ psi.

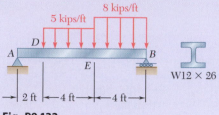

Fig. P9.132

9.133 For the beam and loading shown, determine (*a*) the slope at point *A*, (*b*) the deflection at point *A*.

9.134 For the beam and loading shown, determine (*a*) the slope at point *A*, (*b*) the deflection at point *D*.

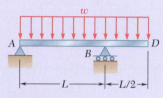

Fig. *P9.134*

9.135 Knowing that the beam *AB* is made of a solid steel rod of diameter *d* = 0.75 in., determine for the loading shown (*a*) the slope at point *D*, (*b*) the deflection at point *A*. Use $E = 29 \times 10^6$ psi.

9.136 Knowing that the beam *AD* is made of a solid steel bar, determine (*a*) the slope at point *B*, (*b*) the deflection at point *A*. Use $E = 200$ GPa.

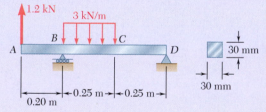

Fig. P9.136

9.137 For the beam and loading shown, determine (*a*) the slope at point *C*, (*b*) the deflection at point *D*. Use $E = 29 \times 10^6$ psi.

9.138 For the beam and loading shown, determine (*a*) the slope at point *B*, (*b*) the deflection at point *D*. Use $E = 200$ GPa.

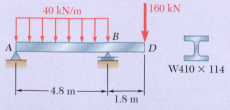

Fig. *P9.138*

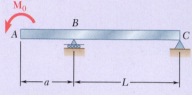

Fig. P9.133

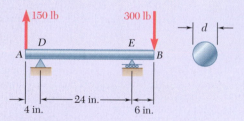

Fig. P9.135

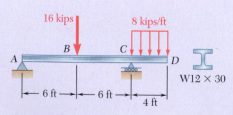

Fig. P9.137

9.139 For the beam and loading shown, determine (a) the slope at end A, (b) the slope at end B, (c) the deflection at the midpoint C.

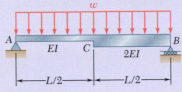

Fig. P9.139

9.140 For the beam and loading shown, determine the deflection (a) at point D, (b) at point E.

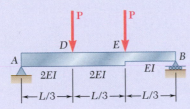

Fig. P9.140

9.141 through 9.144 For the beam and loading shown, determine the magnitude and location of the largest downward deflection.
   9.141 Beam and loading of Prob. 9.126
   9.142 Beam and loading of Prob. 9.128
   9.143 Beam and loading of Prob. 9.129
   9.144 Beam and loading of Prob. 9.132

9.145 For the beam and loading of Prob. 9.135, determine the largest upward deflection in span DE.

9.146 For the beam and loading of Prob. 9.138, determine the largest upward deflection in span AB.

9.147 through 9.150 For the beam and loading shown, determine the reaction at the roller support.

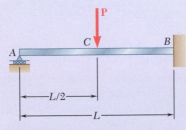

Fig. P9.147

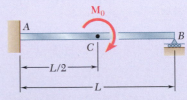

Fig. P9.148

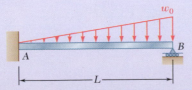

Fig. P9.149

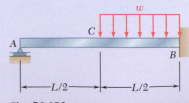

Fig. P9.150

9.151 and 9.152 For the beam and loading shown, determine the reaction at each support.

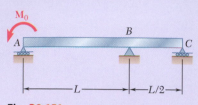

Fig. *P9.151*

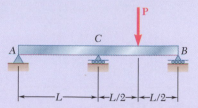

Fig. P9.152

9.153 A hydraulic jack can be used to raise point B of the cantilever beam ABC. The beam was originally straight, horizontal, and unloaded. A 20-kN load was then applied at point C, causing this point to move down. Determine (*a*) how much point B should be raised to return point C to its original position, (*b*) the final value of the reaction at B. Use $E = 200$ GPa.

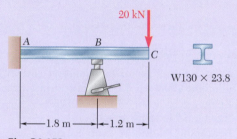

Fig. P9.153

9.154 Determine the reaction at the roller support and draw the bending-moment diagram for the beam and loading shown.

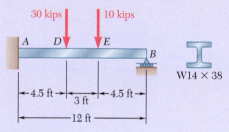

Fig. P9.154

9.155 For the beam and loading shown, determine the spring constant k for which the force in the spring is equal to one-third of the total load on the beam.

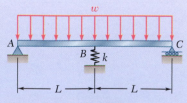

Fig. P9.155 and P9.156

9.156 For the beam and loading shown, determine the spring constant k for which the bending moment at B is $M_B = -wL^2/10$.

Review and Summary

Two approaches were used in this chapter to determine the slopes and deflections of beams under transverse loadings. A mathematical method based on the method of integration of a differential equation was used to get the slopes and deflections at any point along the beam. Then the *moment-area method* was used to find the slopes and deflections at a given point along the beam. Particular emphasis was placed on the computation of the maximum deflection of a beam under a given loading. These methods also were used to determine support reactions and deflections of *indeterminate beams*, where the number of reactions at the supports exceeds the number of equilibrium equations available to determine these unknowns.

Deformation Under Transverse Loading

The relationship of the curvature $1/\rho$ of the neutral surface and the bending moment M in a prismatic beam in pure bending can be applied to a beam under a transverse loading, but in this case both M and $1/\rho$ vary from section to section. Using the distance x from the left end of the beam,

$$\frac{1}{\rho} = \frac{M(x)}{EI} \tag{9.1}$$

This equation enables us to determine the radius of curvature of the neutral surface for any value of x and to draw some general conclusions regarding the shape of the deformed beam.

A relationship was found between the deflection y of a beam, measured at a given point Q, and the distance x of that point from some fixed origin (Fig. 9.41). The resulting equation defines the *elastic curve* of a beam. Expressing the curvature $1/\rho$ in terms of the derivatives of the function $y(x)$ and substituting into Eq. (9.1), we obtained the second-order linear differential equation

$$\frac{d^2y}{dx^2} = \frac{M(x)}{EI} \tag{9.4}$$

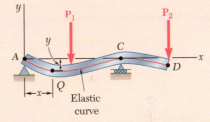

Fig. 9.41 Elastic curve for beam with transverse loads.

Integrating this equation twice, the expressions defining the slope $\theta(x) = dy/dx$ and the deflection $y(x)$ were obtained:

$$EI\frac{dy}{dx} = \int_0^x M(x)\,dx + C_1 \tag{9.5}$$

$$EI\,y = \int_0^x dx \int_0^x M(x)\,dx + C_1 x + C_2 \tag{9.6}$$

The product EI is known as the *flexural rigidity* of the beam. Two constants of integration C_1 and C_2 can be determined from the *boundary conditions* imposed on the beam by its supports (Fig. 9.42). The maximum deflection can be obtained by first determining the value of x for which the slope is zero and then computing the corresponding value of y.

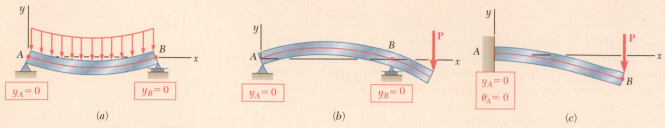

Fig. 9.42 Known boundary conditions for statically determinate beams. (*a*) Simply supported beam. (*b*) Overhanging beam. (*c*) Cantilever beam.

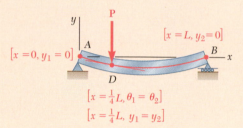

Fig. 9.43 Simply supported beam and boundary conditions, where two sets of functions are required due to the discontinuity in load at point *D*.

Elastic Curve Defined by Different Functions

When the load requires different analytical functions to represent the bending moment in various portions of the beam, multiple differential equations are required to represent the slope $\theta(x)$ and the deflection $y(x)$. For the beam and load considered in Fig. 9.43, two differential equations are required: one for the portion of beam *AD* and the other for the portion *DB*. The first equation yields the functions θ_1 and y_1, and the second the functions θ_2 and y_2. Altogether, four constants of integration must be determined: two by writing that the deflections at *A* and *B* are zero and two by expressing that the portions of beam *AD* and *DB* have the same slope and the same deflection at *D*.

For a beam supporting a distributed load $w(x)$, the elastic curve can be determined directly from $w(x)$ through four integrations yielding V, M, θ, and y (in that order). For the cantilever beam of Fig. 9.44*a* and the simply supported beam of Fig. 9.44*b*, four constants of integration can be determined from the four boundary conditions.

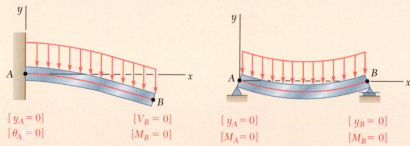

Fig. 9.44 Boundary conditions for beams carrying a distributed load. (*a*) Cantilever beam. (*b*) Simply supported beam.

Statically Indeterminate Beams

Statically indeterminate beams are supported such that the reactions at the supports involve four or more unknowns. Since only three equilibrium equations are available to determine these unknowns, they are supplemented with equations obtained from the boundary conditions imposed

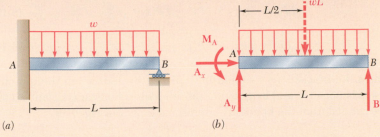

Fig. 9.45 (a) Statically indeterminate beam with a uniformly distributed load. (b) Free-body diagram with four unknown reactions.

by the supports. For the beam of Fig 9.45, the reactions at the supports involve four unknowns: M_A, A_x, A_y, and B. This beam is *indeterminate to the first degree.* (If five unknowns are involved, the beam is indeterminate to the *second degree.*) Expressing the bending moment $M(x)$ in terms of the four unknowns and integrating twice, the slope $\theta(x)$ and the deflection $y(x)$ are determined in terms of the same unknowns and the constants of integration C_1 and C_2. The six unknowns are obtained by solving the three equilibrium equations for the free body of Fig. 9.45b and the three equations expressing that $\theta = 0$, $y = 0$ for $x = 0$, and that $y = 0$ for $x = L$ (Fig. 9.46) simultaneously.

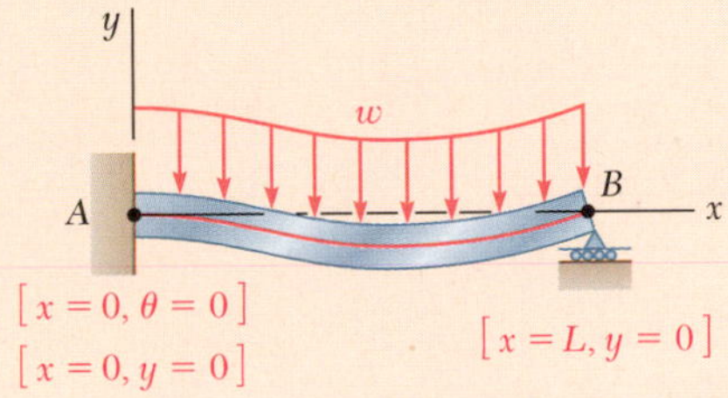

Fig. 9.46 Boundary conditions for beam of Fig. 9.45.

$$[\,x = 0, \theta = 0\,]$$
$$[\,x = 0, y = 0\,]$$
$$[\,x = L, y = 0\,]$$

Use of Singularity Functions

The integration method provides an effective way to determine the slope and deflection at any point of a prismatic beam, as long as the bending moment M can be represented by a single analytical function. However, when several functions are required to represent M over the entire length of the beam, the use of *singularity functions* considerably simplifies the determination of θ and y at any point of the beam. Considering the beam of Fig. 9.47 and drawing its free-body diagram (Fig. 9.48), the shear at any point of the beam is

$$V(x) = \frac{3P}{4} - P\langle x - \tfrac{1}{4}L \rangle^0$$

where the step function $\langle x - \tfrac{1}{4}L \rangle^0$ is equal to zero when the quantity inside the brackets $\langle \; \rangle$ is negative and otherwise is equal to one. Integrating three times,

$$M(x) = \frac{3P}{4}x - P\langle x - \tfrac{1}{4}L \rangle \qquad \textbf{(9.11)}$$

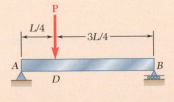

Fig. 9.47 Simply supported beam with concentrated load.

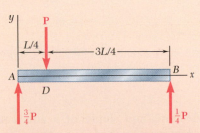

Fig. 9.48 Free-body diagram for beam of Fig. 9.47.

$$EI\,\theta = EI\frac{dy}{dx} = \tfrac{3}{8}Px^2 - \tfrac{1}{2}P\langle x - \tfrac{1}{4}L\rangle^2 + C_1 \qquad \textbf{(9.13)}$$

$$EI\,y = \tfrac{1}{8}Px^3 - \tfrac{1}{6}P\langle x - \tfrac{1}{4}L\rangle^3 + C_1x + C_2 \qquad \textbf{(9.14)}$$

where the brackets $\langle\ \rangle$ should be replaced by zero when the quantity inside is negative and by parentheses otherwise. Constants C_1 and C_2 are determined from the boundary conditions shown in Fig. 9.49.

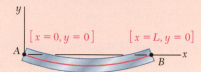

Fig. 9.49 Boundary conditions for simply supported beam.

Method of Superposition

The *method of superposition* separately determines and then adds the slope and deflection caused by the various loads applied to a beam. This procedure is made easier using the table of Appendix D, which gives the slopes and deflections of beams for various loadings and types of support.

Statically Indeterminate Beams by Superposition

The method of superposition can be effective for analyzing *statically indeterminate beams*. For example, the beam of Fig. 9.50 involves four unknown reactions and is indeterminate to the *first degree;* the reaction at B is chosen as *redundant*, and the beam is released from that support. Treating the reaction $\mathbf{R}_B$ as an unknown load and considering the deflections caused at B by the given distributed load and by $\mathbf{R}_B$ separately, the sum of these deflections is zero (Fig. 9.51). For a beam indeterminate to the *second degree* (i.e., with reactions at the supports involving five unknowns), two reactions are redundant, and the corresponding supports must be eliminated or modified accordingly.

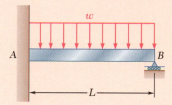

Fig. 9.50 Indeterminate beam with uniformly distributed load.

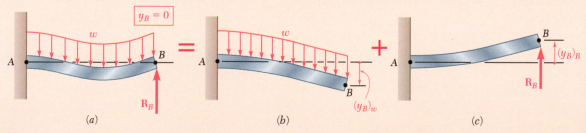

Fig. 9.51 (*a*) Analyze indeterminate beam by superposing two determinate beams, with (*b*) a uniformly distributed load, (*c*) the redundant reaction.

First Moment-Area Theorem

Deflections and slopes of beams can also be determined using the *moment-area method*. The *moment-area theorems* were developed by drawing a diagram representing the variation along the beam of the quantity M/EI, which is obtained by dividing the bending moment M by the flexural rigidity EI (Fig. 9.52). The *first moment-area theorem* is stated as: *The area under the* (M/EI) *diagram between two points is equal to the angle between the tangents to the elastic curve drawn at these points.* Considering tangents at C and D,

$$\theta_{D/C} = \text{area under } (M/EI) \text{ diagram between } C \text{ and } D \qquad \textbf{(9.17)}$$

Second Moment-Area Theorem

Again using the M/EI diagram and a sketch of the deflected beam (Fig. 9.53), a tangent at point D is drawn and the vertical distance $t_{C/D}$, which is called the tangential deviation of C with respect to D, is considered. The *second moment-area theorem* is stated as: *The tangential deviation $t_{C/D}$ of C with respect to D is equal to the first moment with respect to a vertical axis through C of the area under the M/EI diagram between C and D.* It is important to distinguish between the tangential deviation of C with respect to D (Fig. 9.53a), which is

$$t_{C/D} = (\text{area between } C \text{ and } D)\,\bar{x}_1 \qquad \textbf{(9.20)}$$

and the tangential deviation of D with respect to C (Fig. 9.53b), which is

$$t_{D/C} = (\text{area between } C \text{ and } D)\,\bar{x}_2 \qquad \textbf{(9.21)}$$

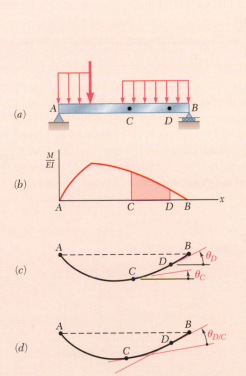

Fig. 9.52 First moment-area theorem Illustrated. (*a*) Beam subjected to arbitrary load. (*b*) M/EI diagram. (*c*) Elastic curve showing slope at C and D. (*d*) Elastic curve showing slope at D with respect to C.

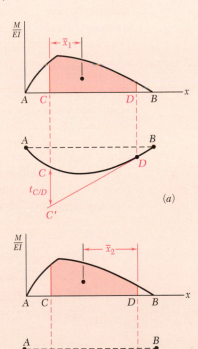

Fig. 9.53 Second moment-area theorem illustrated. (*a*) Evaluating $t_{C/D}$. (*b*) Evaluating $t_{D/C}$.

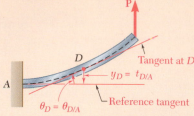

Fig. 9.54 Application of moment-area method to cantilever beams.

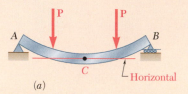

(a)

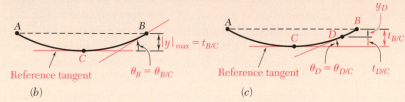

(b)   (c)

Fig. 9.55 Application of moment-area method to simply supported beams with symmetric loadings. (a) Beam and loadings. (b) Maximum deflection and slope at point B. (c) Deflection and slope at arbitrary point D.

Cantilever Beams
Beams with Symmetric Loadings

To determine the slope and deflection at points of *cantilever* beams, the tangent at the fixed support is horizontal (Fig. 9.54). For *symmetrically loaded beams*, the tangent is horizontal at the midpoint C of the beam (Fig. 9.55). Using the horizontal tangent as a *reference tangent,* slopes and deflections are determined by using, respectively, the first and second moment-area theorems. To find a deflection that is not a tangential deviation (Fig. 9.55c), it is first necessary to determine which tangential deviations can be combined to obtain the desired deflection.

Bending-Moment Diagram by Parts

In many cases, the application of the moment-area theorems is simplified if the effect of each load is considered separately. To do this, we draw the M/EI *diagram by parts* with a separate M/EI diagram for each load. The areas and the moments of areas under the several diagrams are added to determine slopes and tangential deviations for the original beam and loading.

Unsymmetric Loadings

The moment-area method is also used to analyze beams with *unsymmetric loadings.* Observing that the location of a horizontal tangent is usually not obvious, a reference tangent is selected at one of the beam supports, since the slope of that tangent is easily determined. For the beam and loading shown in Fig. 9.56, the slope of the tangent at A is obtained by computing the tangential deviation $t_{B/A}$ and dividing it by the distance L between supports A and B. Then, using both moment-area theorems and simple geometry, the slope and deflection are determined at any point of the beam.

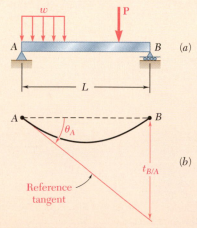

Fig. 9.56 Application of moment-area method to unsymmetrically loaded beam establishes a reference tangent at a support.

Maximum Deflection

The *maximum deflection* of an unsymmetrically loaded beam generally does not occur at midspan. The approach indicated in the preceding paragraph was used to determine point K where the maximum deflection occurs and the magnitude of that deflection. Observing that the slope at K is zero (Fig. 9.57), $\theta_{K/A} = -\theta_A$. Recalling the first moment-area theorem, the location of K is found by determining an area under the M/EI diagram equal to $\theta_{K/A}$. The maximum deflection is then obtained by computing the tangential deviation $t_{A/K}$.

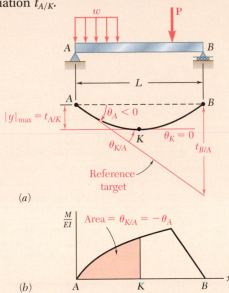

Fig. 9.57 Determination of maximum deflection using moment-area method.

Statically Indeterminate Beams

The moment-area method can be used for the analysis of *statically indeterminate beams*. Since the reactions for the beam and loading shown in Fig. 9.58 cannot be determined by statics alone, one of the reactions of the beam is designated as redundant ($\mathbf{M}_A$ in Fig. 9.59a), and the redundant reaction is considered to be an unknown load. The tangential deviation of B with respect to A is considered separately for the distributed load (Fig. 9.59b) and for the redundant reaction (Fig. 9.59c). Expressing that under the combined action of the distributed load and of the couple $\mathbf{M}_A$ the tangential deviation of B with respect to A must be zero,

$$t_{B/A} = (t_{B/A})_w + (t_{B/A})_M = 0$$

From this equation, the magnitude of the redundant reaction $\mathbf{M}_A$ can be found.

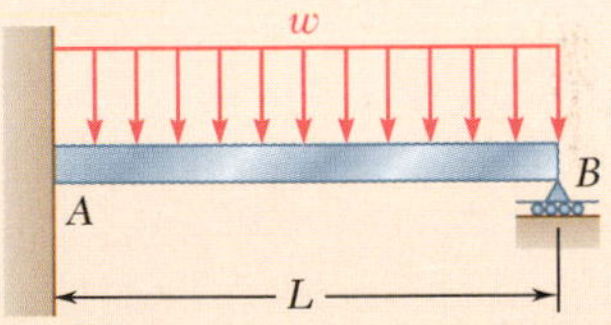

Fig. 9.58 Statically indeterminate beam.

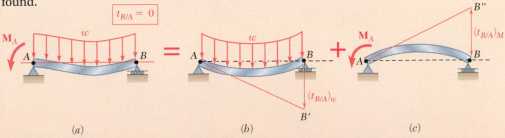

Fig. 9.59 Modeling the indeterminate beam as the superposition of two determinate cases.

Review Problems

9.157 For the loading shown, determine (*a*) the equation of the elastic curve for the cantilever beam *AB*, (*b*) the deflection at the free end, (*c*) the slope at the free end.

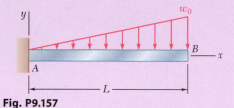

Fig. P9.157

9.158 (*a*) Determine the location and magnitude of the maximum deflection of beam *AB*. (*b*) Assuming that beam *AB* is a W360 × 64, *L* = 3.5 m, and *E* = 200 GPa, calculate the maximum allowable value of the applied moment M_0 if the maximum deflection is not to exceed 1 mm.

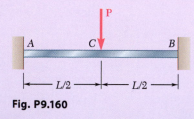

Fig. P9.158

9.159 For the beam and loading shown, determine (*a*) the equation of the elastic curve, (*b*) the slope at end *A*, (*c*) the deflection at the midpoint of the span.

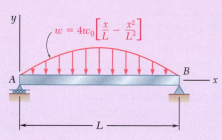

$$w = 4w_0 \left[\frac{x}{L} - \frac{x^2}{L^2} \right]$$

Fig. *P9.159*

9.160 Determine the reaction at *A* and draw the bending moment diagram for the beam and loading shown.

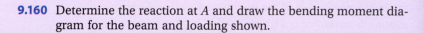

Fig. P9.160

9.161 For the beam and loading shown, determine (*a*) the slope at end *A*, (*b*) the deflection at point *C*. Use *E* = 200 GPa.

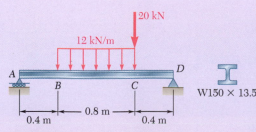

Fig. P9.161

9.162 For the beam and loading shown, determine (*a*) the reaction at point *C*, (*b*) the deflection at point *B*. Use $E = 29 \times 10^6$ psi.

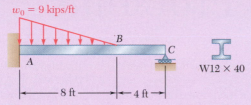

Fig. P9.162

9.163 Beam *CE* rests on beam *AB* as shown. Knowing that a W10 × 30 rolled-steel shape is used for each beam, determine for the loading shown the deflection at point *D*. Use $E = 29 \times 10^6$ psi.

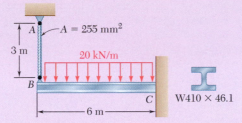

Fig. P9.163

9.164 The cantilever beam *BC* is attached to the steel cable *AB* as shown. Knowing that the cable is initially taut, determine the tension in the cable caused by the distributed load shown. Use $E = 200$ GPa.

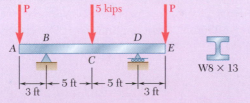

Fig. P9.164

9.165 For the cantilever beam and loading shown, determine (*a*) the slope at point *A*, (*b*) the deflection at point *A*. Use $E = 200$ GPa.

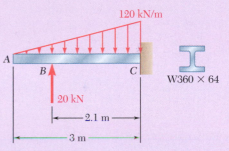

Fig. P9.165

9.166 Knowing that $P = 4$ kips, determine (*a*) the slope at end *A*, (*b*) the deflection at the midpoint *C* of the beam. Use $E = 29 \times 10^6$ psi.

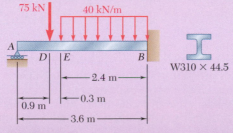

Fig. P9.166

9.167 For the beam and loading shown, determine (*a*) the slope at point *A*, (*b*) the deflection at point *D*.

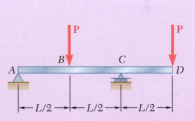

Fig. P9.167

9.168 Determine the reaction at the roller support and draw the bending-moment diagram for the beam and loading shown.

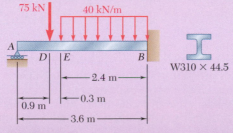

Fig. P9.168

Computer Problems

The following problems are designed to be solved with a computer.

9.C1 Several concentrated loads can be applied to the cantilever beam AB. Write a computer program to calculate the slope and deflection of beam AB from $x = 0$ to $x = L$, using given increments Δx. Apply this program with increments $\Delta x = 50$ mm to the beam and loading of Prob. 9.73 and Prob. 9.74.

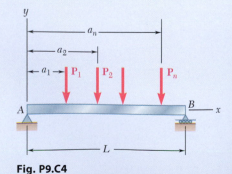

Fig. P9.C1

9.C2 The 22-ft beam AB consists of a W21 × 62 rolled-steel shape and supports a 3.5-kip/ft distributed load as shown. Write a computer program and use it to calculate for values of a from 0 to 22 ft, using 1-ft increments, (a) the slope and deflection at D, (b) the location and magnitude of the maximum deflection. Use $E = 29 \times 10^6$ psi.

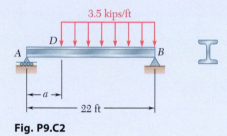

3.5 kips/ft

Fig. P9.C2

9.C3 The cantilever beam AB carries the distributed loads shown. Write a computer program to calculate the slope and deflection of beam AB from $x = 0$ to $x = L$ using given increments Δx. Apply this program with increments $\Delta x = 100$ mm, assuming that $L = 2.4$ m, $w = 36$ kN/m, and (a) $a = 0.6$ m, (b) $a = 1.2$ m, (c) $a = 1.8$ m. Use $E = 200$ GPa.

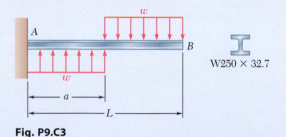

W250 × 32.7

Fig. P9.C3

9.C4 The simple beam AB is of constant flexural rigidity EI and carries several concentrated loads as shown. Using the *Method of Integration*, write a computer program that can be used to calculate the slope and deflection at points along the beam from $x = 0$ to $x = L$ using given increments Δx. Apply this program to the beam and loading of (a) Prob. 9.13 with $\Delta x = 1$ ft, (b) Prob. 9.16 with $\Delta x = 0.05$ m, (c) Prob. 9.129 with $\Delta x = 0.25$ m.

Fig. P9.C4

9.C5 The supports of beam AB consist of a fixed support at end A and a roller support located at point D. Write a computer program that can be used to calculate the slope and deflection at the free end of the beam for values of a from 0 to L using given increments Δa. Apply this program to calculate the slope and deflection at point B for each of the following cases:

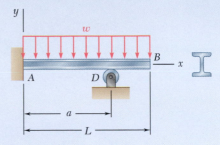

	L	ΔL	w	E	Shape
(a)	12 ft	0.5 ft	1.6 k/ft	29×10^6 psi	W16 × 57
(b)	3 m	0.2 m	18 kN/m	200 GPa	W460 × 113

Fig. P9.C5

9.C6 For the beam and loading shown, use the *Moment-Area Method* to write a computer program to calculate the slope and deflection at points along the beam from $x = 0$ to $x = L$ using given increments Δx. Apply this program to calculate the slope and deflection at each concentrated load for the beam of (a) Prob. 9.77 with $\Delta x = 0.5$ m, (b) Prob. 9.119 with $\Delta x = 0.5$ m.

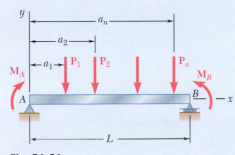

Fig. P9.C6

9.C7 Two 52-kN loads are maintained 2.5 m apart as they are moved slowly across beam AB. Write a computer program to calculate the deflection at the midpoint C of the beam for values of x from 0 to 9 m, using 0.5-m increments. Use $E = 200$ GPa.

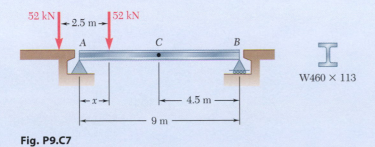

Fig. P9.C7

9.C8 A uniformly distributed load w and several distributed loads P_i may be applied to beam AB. Write a computer program to determine the reaction at the roller support and apply this program to the beam and loading of (a) Prob. 9.53a, (b) Prob. 9.154.

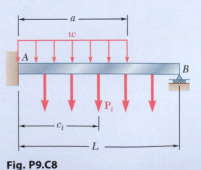

Fig. P9.C8

10

Columns

The curved pedestrian bridge is supported by a series of columns. The analysis and design of members supporting axial compressive loads will be discussed in this chapter.

Objectives

In this chapter, you will:

- **Describe** the behavior of columns in terms of stability
- **Develop** Euler's formula for columns, using effective lengths to account for different end conditions
- **Develop** the secant formula for analysis of eccentrically loaded columns
- **Use** allowable-stress design for columns made of steel, aluminum, and wood
- **Provide** the basis for using load and resistance factor design for steel columns
- **Present** two design approaches to use for eccentrically loaded columns: the allowable-stress method and the interaction method

Introduction

In the preceding chapters, we had two primary concerns: (1) the strength of the structure, i.e., its ability to support a specified load without experiencing excessive stress; (2) the ability of the structure to support a specified load without undergoing unacceptable deformations. This chapter is concerned with the stability of the structure (its ability to support a given load without experiencing a sudden change in configuration). This discussion is focused on columns, that is, the analysis and design of vertical prismatic members supporting axial loads.

In Sec. 10.1, the stability of a simplified model is discussed, where the column consists of two rigid rods connected by a pin and a spring and supports a load **P.** If its equilibrium is disturbed, this system will return to its original equilibrium position as long as P does not exceed a certain value P_{cr}, called the *critical load.* This is a *stable* system. However, if $P > P_{cr}$, the system moves away from its original position and settles in a new position of equilibrium. This system is said to be *unstable.*

In Sec. 10.1A, the *stability of elastic columns* considers a pin-ended column subjected to a centric axial load. *Euler's formula* for the critical load of the column is derived, and the corresponding critical normal stress in the column is determined. Applying a factor of safety to the critical load, we obtain the allowable load that can be safely applied to a pin-ended column.

In Sec. 10.1B, the analysis of the stability of columns with different end conditions is considered by learning how to determine the *effective length* of a column.

Columns supporting eccentric axial loads are discussed in Sec. 10.2. These columns have transverse deflections for all magnitudes of the load. An equation for the maximum deflection under a given load is developed and used to determine the maximum normal stress in the column. Finally, the *secant formula* relating the average and maximum stresses in a column is developed.

In the first sections of the chapter, each column is assumed to be a straight, homogeneous prism. In the last part of the chapter, real columns are designed and analyzed using empirical formulas set forth by professional organizations. In Sec. 10.3A, design equations are presented for the allowable stress in columns made of steel, aluminum, or wood that are subjected to a centric load. Section 10.3B describes an alternative approach for steel columns, the load and resistance factor design method. The design of columns under an eccentric axial load is covered in Sec. 10.4.

10.1 STABILITY OF STRUCTURES

Consider the design of a column AB of length L to support a given load **P** (Fig. 10.1). The column is pin-connected at both ends, and **P** is a centric axial load. If the cross-sectional area A is selected so that the value $\sigma = P/A$ of the stress on a transverse section is less than the allowable stress σ_{all} for the material used and the deformation $\delta = PL/AE$ falls within the given specifications, we might conclude that the column has been properly designed. However, it may happen that as the load is applied, the column *buckles* (Fig. 10.2). Instead of remaining straight, it suddenly

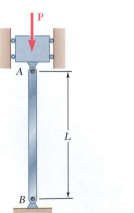

Fig. 10.1 Pinned-ended axially loaded column.

Fig. 10.2 Buckled pin-ended column.

Photo 10.1 Laboratory test showing a buckled column.

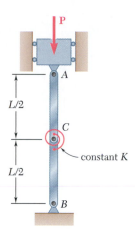

Fig. 10.3 Model column made of two rigid rods joined by a torsional spring at *C*.

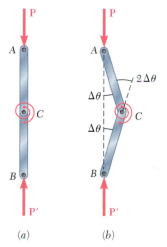

Fig. 10.4 Free-body diagram of model column (*a*) perfectly aligned (*b*) point *C* moved slightly out of alignment.

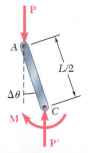

Fig. 10.5 Free-body diagram of rod *AC* in unaligned position.

becomes sharply curved such as shown in Photo 10.1. Clearly, a column that buckles under the load it is to support is not properly designed.

Before getting into the actual discussion of the stability of elastic columns, some insight will be gained on the problem by considering a simplified model consisting of two rigid rods *AC* and *BC* connected at *C* by a pin and a torsional spring of constant *K* (Fig. 10.3).

If the two rods and forces **P** and **P′** are perfectly aligned, the system will remain in the position of equilibrium shown in Fig.10.4*a* as long as it is not disturbed. But suppose we move *C* slightly to the right so that each rod forms a small angle $\Delta\theta$ with the vertical (Fig. 10.4*b*). Will the system return to its original equilibrium position, or will it move further away? In the first case, the system is *stable;* in the second, it is *unstable.*

To determine whether the two-rod system is stable or unstable, consider the forces acting on rod *AC* (Fig. 10.5). These forces consist of the couple formed by **P** and **P′** of moment $P(L/2)\sin\Delta\theta$, which tends to move the rod away from the vertical, and the couple **M** exerted by the spring, which tends to bring the rod back into its original vertical position. Since the angle of deflection of the spring is $2\,\Delta\theta$, the moment of couple **M** is $M = K(2\,\Delta\theta)$. If the moment of the second couple is larger than the moment of the first couple, the system tends to return to its original equilibrium position; the system is stable. If the moment of the first couple is larger than the moment of the second couple, the system tends to move

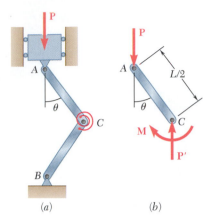

Fig. 10.6 (a) Model column in buckled position, (b) free-body diagram of rod AC.

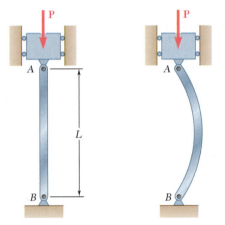

Fig. 10.1 (repeated).   **Fig. 10.2** (repeated).

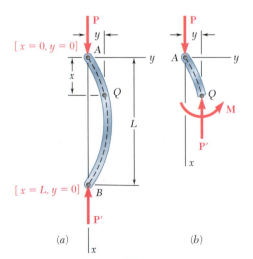

Fig. 10.7 Free-body diagrams of (a) buckled column and (b) portion AQ.

away from its original equilibrium position; the system is unstable. The load when the two couples balance each other is called the *critical load*, P_{cr}, which is given as

$$P_{cr}(L/2)\sin\Delta\theta = K(2\Delta\theta) \tag{10.1}$$

or since $\sin\Delta\theta \approx \Delta\theta$, when the displacement of C is very small (at the immediate onset of buckling),

$$P_{cr} = 4K/L \tag{10.2}$$

Clearly, the system is stable for $P < P_{cr}$ and unstable for $P > P_{cr}$.

Assume that a load $P > P_{cr}$ has been applied to the two rods of Fig. 10.3 and the system has been disturbed. Since $P > P_{cr}$, the system will move further away from the vertical and, after some oscillations, will settle into a new equilibrium position (Fig. 10.6a). Considering the equilibrium of the free body AC (Fig. 10.6b), an equation similar to Eq. (10.1) but involving the finite angle θ, is

$$P(L/2)\sin\theta = K(2\theta)$$

or

$$\frac{PL}{4K} = \frac{\theta}{\sin\theta} \tag{10.3}$$

The value of θ corresponding to the equilibrium position in Fig. 10.6 is obtained by solving Eq. (10.3) by trial and error. But for any positive value of θ, $\sin\theta < \theta$. Thus, Eq. (10.3) yields a value of θ different from zero only when the left-hand member of the equation is larger than one. Recalling Eq. (10.2), this is true only if $P > P_{cr}$. But, if $P < P_{cr}$, the second equilibrium position shown in Fig. 10.6 would not exist, and the only possible equilibrium position would be the one corresponding to $\theta = 0$. Thus, for $P < P_{cr}$, the position where $\theta = 0$ must be stable.

This observation applies to structures and mechanical systems in general and is used in the next section for the stability of elastic columns.

10.1A   Euler's Formula for Pin-Ended Columns

Returning to the column AB considered in the preceding section (Fig. 10.1), we propose to determine the critical value of the load **P,** i.e., the value P_{cr} of the load for which the position shown in Fig. 10.1 ceases to be stable. If $P > P_{cr}$, the slightest misalignment or disturbance will cause the column to buckle into a curved shape, as shown in Fig. 10.2.

This approach determines the conditions under which the configuration of Fig. 10.2 is possible. Since a column is like a beam placed in a vertical position and subjected to an axial load, we proceed as in Chap. 9 and denote by x the distance from end A of the column to a point Q of its elastic curve and by y the deflection of that point (Fig. 10.7a). The x axis is vertical and directed downward, and the y axis is horizontal and directed to the right. Considering the equilibrium of the free body AQ (Fig. 10.7b), the bending moment at Q is $M = -Py$. Substituting this value for M in Eq. (9.4) gives

$$\frac{d^2y}{dx^2} = \frac{M}{EI} = -\frac{P}{EI}y \qquad \textbf{(10.4)}$$

or transposing the last term,

$$\frac{d^2y}{dx^2} + \frac{P}{EI}y = 0 \qquad \textbf{(10.5)}$$

This equation is a linear, homogeneous differential equation of the second order with constant coefficients. Setting

$$p^2 = \frac{P}{EI} \qquad \textbf{(10.6)}$$

Eq. (10.5) is rewritten as

$$\frac{d^2y}{dx^2} + p^2y = 0 \qquad \textbf{(10.7)}$$

which is the same as the differential equation for simple harmonic motion, except the independent variable is now the distance x instead of the time t. The general solution of Eq. (10.7) is

$$y = A \sin px + B \cos px \qquad \textbf{(10.8)}$$

and is easily checked by calculating d^2y/dx^2 and substituting for y and d^2y/dx^2 into Eq. (10.7).

Recalling the boundary conditions that must be satisfied at ends A and B of the column (Fig. 10.7a), make $x = 0$, $y = 0$ in Eq. (10.8), and find that $B = 0$. Substituting $x = L$, $y = 0$, obtain

$$A \sin pL = 0 \qquad \textbf{(10.9)}$$

This equation is satisfied if either $A = 0$ or $\sin pL = 0$. If the first of these conditions is satisfied, Eq. (10.8) reduces to $y = 0$ and the column is straight (Fig. 10.1). For the second condition to be satisfied, $pL = n\pi$, or substituting for p from (10.6) and solving for P,

$$P = \frac{n^2\pi^2EI}{L^2} \qquad \textbf{(10.10)}$$

The smallest value of P defined by Eq. (10.10) is that corresponding to $n = 1$. Thus,

$$P_{cr} = \frac{\pi^2EI}{L^2} \qquad \textbf{(10.11a)}$$

This expression is known as *Euler's formula,* after the Swiss mathematician Leonhard Euler (1707–1783). Substituting this expression for P into Eq. (10.6), the value for p into Eq. (10.8), and recalling that $B = 0$,

$$y = A \sin\frac{\pi x}{L} \qquad \textbf{(10.12)}$$

which is the equation of the elastic curve after the column has buckled (Fig. 10.2). Note that the maximum deflection $y_m = A$ is indeterminate.

This is because the differential Eq. (10.5) is a linearized approximation of the governing differential equation for the elastic curve.[†]

If $P < P_{cr}$, the condition $\sin pL = 0$ cannot be satisfied, and the solution of Eq. (10.12) does not exist. Then we must have $A = 0$, and the only possible configuration for the column is a straight one. Thus, for $P < P_{cr}$ the straight configuration of Fig. 10.1 is stable.

In a column with a circular or square cross section, the moment of inertia I is the same about any centroidal axis, and the column is as likely to buckle in one plane as another (except for the restraints that can be imposed by the end connections). For other cross-sectional shapes, the critical load should be found by making $I = I_{min}$ in Eq. (10.11a). If it occurs, buckling will take place in a plane perpendicular to the corresponding principal axis of inertia.

The stress corresponding to the critical load is the *critical stress* σ_{cr}. Recalling Eq. (10.11a) and setting $I = Ar^2$, where A is the cross-sectional area and r its radius of gyration gives

$$\sigma_{cr} = \frac{P_{cr}}{A} = \frac{\pi^2 EAr^2}{AL^2}$$

or

$$\sigma_{cr} = \frac{\pi^2 E}{(L/r)^2} \qquad \textbf{(10.13a)}$$

The quantity L/r is the *slenderness ratio* of the column. The minimum value of the radius of gyration r should be used to obtain the slenderness ratio and the critical stress in a column.

Equation (10.13a) shows that the critical stress is proportional to the modulus of elasticity of the material and inversely proportional to the square of the slenderness ratio of the column. The plot of σ_{cr} versus L/r is shown in Fig. 10.8 for structural steel, assuming $E = 200$ GPa and $\sigma_Y = 250$ MPa. Keep in mind that no factor of safety has been used in plotting σ_{cr}. Also, if σ_{cr} obtained from Eq. (10.13a) or from the curve of Fig. 10.8 is larger than the yield strength σ_Y, this value is of no interest, since the column will yield in compression and cease to be elastic before it has a chance to buckle.

The analysis of the behavior of a column has been based on the assumption of a perfectly aligned centric load. In practice, this is seldom the case, and in Sec. 10.2, the effect of eccentric loading is taken into account. This approach leads to a smoother transition from the buckling failure of long, slender columns to the compression failure of short, stubby columns. It also provides a more realistic view of the relationship between the slenderness ratio of a column and the load that causes it to fail.

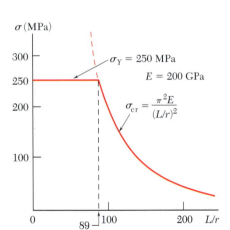

Fig. 10.8 Plot of critical stress.

[†]Recall that $d^2y/dx^2 = M/EI$ was obtained in Sec. 9.1A by assuming that the slope dy/dx of the beam could be neglected and that the exact expression in Eq. (9.3) for the curvature of the beam could be replaced by $1/\rho = d^2y/dx^2$.

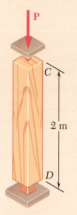

Fig. 10.9 Pin-ended wood column of square cross section.

Concept Application 10.1

A 2-m-long pin-ended column with a square cross section is to be made of wood (Fig 10.9). Assuming $E = 13$ GPa, $\sigma_{all} = 12$ MPa, and using a factor of safety of 2.5 to calculate Euler's critical load for buckling, determine the size of the cross section if the column is to safely support (*a*) a 100-kN load, (*b*) a 200-kN load.

a. For the 100-kN Load. Use the given factor of safety to obtain

$$P_{cr} = 2.5(100 \text{ kN}) = 250 \text{ kN} \qquad L = 2 \text{ m} \qquad E = 13 \text{ GPa}$$

Use Euler's formula, Eq. (10.11a), and solve for *I*:

$$I = \frac{P_{cr}L^2}{\pi^2 E} = \frac{(250 \times 10^3 \text{ N})(2 \text{ m})^2}{\pi^2(13 \times 10^9 \text{ Pa})} = 7.794 \times 10^{-6} \text{ m}^4$$

Recalling that, for a square of side *a*, $I = a^4/12$, write

$$\frac{a^4}{12} = 7.794 \times 10^{-6} \text{ m}^4 \qquad a = 98.3 \text{ mm} \approx 100 \text{ mm}$$

Check the value of the normal stress in the column:

$$\sigma = \frac{P}{A} = \frac{100 \text{ kN}}{(0.100 \text{ m})^2} = 10 \text{ MPa}$$

Since σ is smaller than the allowable stress, a 100×100-mm cross section is acceptable.

b. For the 200-kN Load. Solve Eq. (10.11a) again for *I*, but make $P_{cr} = 2.5(200) = 500$ kN to obtain

$$I = 15.588 \times 10^{-6} \text{ m}^4$$

$$\frac{a^4}{12} = 15.588 \times 10^{-6} \qquad a = 116.95 \text{ mm}$$

The value of the normal stress is

$$\sigma = \frac{P}{A} = \frac{200 \text{ kN}}{(0.11695 \text{ m})^2} = 14.62 \text{ MPa}$$

Since this is larger than the allowable stress, the dimension obtained is not acceptable, and the cross section must be selected on the basis of its resistance to compression.

$$A = \frac{P}{\sigma_{all}} = \frac{200 \text{ kN}}{12 \text{ MPa}} = 16.67 \times 10^{-3} \text{ m}^2$$

$$a^2 = 16.67 \times 10^{-3} \text{ m}^2 \qquad a = 129.1 \text{ mm}$$

A 130×130-mm cross section is acceptable.

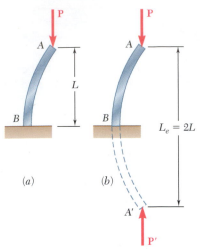

Fig. 10.10 Effective length of a fixed-free column of length L is equivalent to a pin-ended column of length 2L.

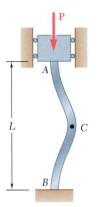

Fig. 10.11 Column with fixed ends.

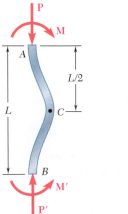

Fig. 10.12 Free-body diagram of buckled fixed-ended column.

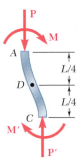

Fig. 10.13 Free-body diagram of upper half of fixed-ended column.

10.1B Euler's Formula for Columns with Other End Conditions

Euler's formula (10.11) was derived in the preceding section for a column that was pin-connected at both ends. Now the critical load P_{cr} will be determined for columns with different end conditions.

A column with one free end A supporting a load **P** and one fixed end B (Fig. 10.10a) behaves as the upper half of a pin-connected column (Fig. 10.10b). The critical load for the column of Fig. 10.10a is thus the same as for the pin-ended column of Fig. 10.10b and can be obtained from Euler's formula Eq. (10.11a) by using a column length equal to twice the actual length L. We say that the *effective length* L_e of the column of Fig. 10.10 is equal to $2L$, and substitute $L_e = 2L$ in Euler's formula:

$$P_{cr} = \frac{\pi^2 EI}{L_e^2} \qquad \textbf{(10.11b)}$$

The critical stress is

$$\sigma_{cr} = \frac{\pi^2 E}{(L_e/r)^2} \qquad \textbf{(10.13b)}$$

The quantity L_e/r is called the *effective slenderness ratio* of the column and for Fig. 10.10a is equal to $2L/r$.

Now consider a column with two fixed ends A and B supporting a load **P** (Fig. 10.11). The symmetry of the supports and the load about a horizontal axis through the midpoint C requires that the shear at C and the horizontal components of the reactions at A and B be zero (Fig. 10.12a). Thus, the restraints imposed on the upper half AC of the column by the support at A and by the lower half CB are identical (Fig. 10.13). Portion AC must be symmetric about its midpoint D, and this point must be a point of inflection where the bending moment is zero. The bending moment at the midpoint E of the lower half of the column also must be zero (Fig. 10.14a). Since the bending moment at the ends of a pin-ended column is zero, portion DE of the column in Fig. 10.13a must behave like a pin-ended column (Fig. 10.14b). Thus, the effective length of a column with two fixed ends is $L_e = L/2$.

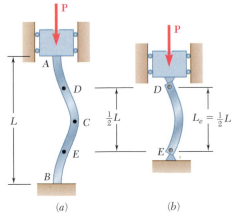

Fig. 10.14 Effective length of a fixed-ended column of length L is equivalent to a pin-ended column of length L/2.

In a column with one fixed end B and one pin-connected end A supporting a load **P** (Fig. 10.15), the differential equation of the elastic curve must be solved to determine the effective length. From the free-body diagram of the entire column (Fig. 10.16), a transverse force **V** is exerted at end A, in addition to the axial load **P**, and **V** is statically indeterminate. Considering the free-body diagram of a portion AQ of the column (Fig. 10.17), the bending moment at Q is

$$M = -Py - Vx$$

Substituting this value into Eq. (9.4) of Sec. 9.1A,

$$\frac{d^2y}{dx^2} = \frac{M}{EI} = -\frac{P}{EI}y - \frac{V}{EI}x$$

Transposing the term containing y and setting

$$p^2 = \frac{P}{EI} \qquad \textbf{(10.6)}$$

as in Sec. 10.1A gives

$$\frac{d^2y}{dx^2} + p^2y = -\frac{V}{EI}x \qquad \textbf{(10.14)}$$

This is a linear, nonhomogeneous differential equation of the second order with constant coefficients. Observing that the left-hand members of Eqs. (10.7) and (10.14) are identical, the general solution of Eq. (10.14) can be obtained by adding a particular solution of Eq. (10.14) to the solution of Eq. (10.8) obtained for Eq. (10.7). Such a particular solution is

$$y = -\frac{V}{p^2EI}x$$

or recalling Eq. (10.6),

$$y = -\frac{V}{P}x \qquad \textbf{(10.15)}$$

Adding the solutions of Eq. (10.8) and (10.15), the general solution of Eq. (10.14) is

$$y = A\sin px + B\cos px - \frac{V}{P}x \qquad \textbf{(10.16)}$$

The constants A and B and the magnitude V of the unknown transverse force **V** are obtained from the boundary conditions in Fig. (10.16). Making $x = 0$, $y = 0$ in Eq. (10.16), $B = 0$. Making $x = L$, $y = 0$, gives

$$A\sin pL = \frac{V}{P}L \qquad \textbf{(10.17)}$$

Taking the derivative of Eq. (10.16), with $B = 0$,

$$\frac{dy}{dx} = Ap\cos px - \frac{V}{P}$$

and making $x = L$, $dy/dx = 0$,

$$Ap\cos pL = \frac{V}{P} \qquad \textbf{(10.18)}$$

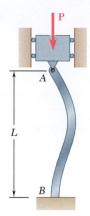

Fig. 10.15 Column with fixed-pinned end conditions.

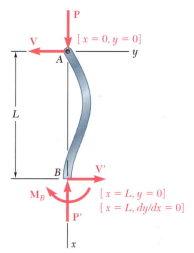

Fig. 10.16 Free-body diagram of buckled fixed-pinned column.

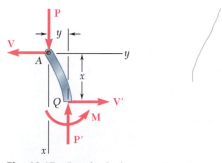

Fig. 10.17 Free-body diagram of portion AQ of buckled fixed-pinned column.

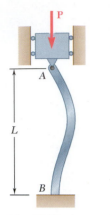

Fig. 10.15   (repeated).

Dividing Eq. (10.17) by Eq. (10.18) member by member, a solution like Eq. (10.16) can exist only if

$$\tan pL = pL \tag{10.19}$$

Solving this equation by trial and error, the smallest value of pL that satisfies Eq. (10.19) is

$$pL = 4.4934 \tag{10.20}$$

Carrying the value of p from Eq. (10.20) into Eq. (10.6) and solving for P, the critical load for the column of Fig. 10.15 is

$$P_{cr} = \frac{20.19EI}{L^2} \tag{10.21}$$

The effective length of the column is obtained by equating the right-hand members of Eqs. (10.11b) and (10.21):

$$\frac{\pi^2 EI}{L_e^2} = \frac{20.19EI}{L^2}$$

Solving for L_e, the effective length of a column with one fixed end and one pin-connected end is $L_e = 0.699L \approx 0.7L$.

The effective lengths corresponding to the various end conditions are shown in Fig. 10.18.

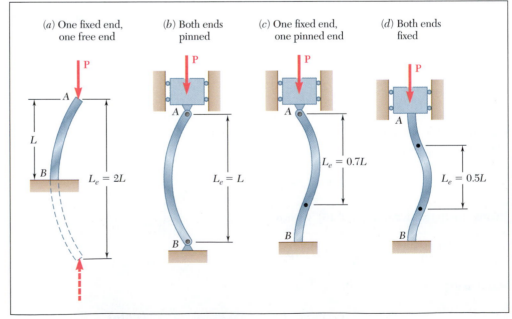

Fig. 10.18   Effective length of column for various end conditions.

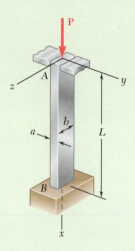

Sample Problem 10.1

An aluminum column with a length of L and a rectangular cross section has a fixed end B and supports a centric load at A. Two smooth and rounded fixed plates restrain end A from moving in one of the vertical planes of symmetry of the column but allow it to move in the other plane. (*a*) Determine the ratio a/b of the two sides of the cross section corresponding to the most efficient design against buckling. (*b*) Design the most efficient cross section for the column, knowing that $L = 20$ in., $E = 10.1 \times 10^6$ psi, $P = 5$ kips, and a factor of safety of 2.5 is required.

STRATEGY: The most efficient design is that for which the critical stresses corresponding to the two possible buckling modes are equal. This occurs if the two critical stresses obtained from Eq. (10.13b) are the same. Thus for this problem, the two effective slenderness ratios in this equation must be equal to solve part *a*. Use Fig. 10.18 to determine the effective lengths. The design data can then be used with Eq. (10.13b) to size the cross section for part *b*.

MODELING:

Buckling in *xy* Plane. Referring to Fig. 10.18*c*, the effective length of the column with respect to buckling in this plane is $L_e = 0.7L$. The radius of gyration r_z of the cross section is obtained by

$$I_z = \tfrac{1}{12}ba^3 \quad A = ab$$

and since $I_z = Ar_z^2$, $\qquad r_z^2 = \dfrac{I_z}{A} = \dfrac{\tfrac{1}{12}ba^3}{ab} = \dfrac{a^2}{12} \qquad r_z = a/\sqrt{12}$

The effective slenderness ratio of the column with respect to buckling in the *xy* plane is

$$\frac{L_e}{r_z} = \frac{0.7L}{a/\sqrt{12}} \tag{1}$$

Buckling in *xz* Plane. Referring to Fig. 10.18*a*, the effective length of the column with respect to buckling in this plane is $L_e = 2L$, and the corresponding radius of gyration is $r_y = b/\sqrt{12}$. Thus,

$$\frac{L_e}{r_y} = \frac{2L}{b/\sqrt{12}} \tag{2}$$

(continued)

ANALYSIS:

a. Most Efficient Design.   The most efficient design is when the critical stresses corresponding to the two possible modes of buckling are equal. Referring to Eq. (10.13b), this is the case if the two values obtained above for the effective slenderness ratio are equal.

$$\frac{0.7L}{a/\sqrt{12}} = \frac{2L}{b/\sqrt{12}}$$

and solving for the ratio a/b, $\dfrac{a}{b} = \dfrac{0.7}{2}$ $\dfrac{a}{b} = 0.35$ ◄

b. Design for Given Data.   Since $F.S. = 2.5$ is required,

$$P_{cr} = (F.S.)P = (2.5)(5 \text{ kips}) = 12.5 \text{ kips}$$

Using $a = 0.35b$,

$$A = ab = 0.35b^2 \quad \text{and} \quad \sigma_{cr} = \frac{P_{cr}}{A} = \frac{12{,}500 \text{ lb}}{0.35b^2}$$

Making $L = 20$ in. in Eq. (2), $L_e/r_y = 138.6/b$. Substituting for E, L_e/r, and σ_{cr} into Eq. (10.13b) gives

$$\sigma_{cr} = \frac{\pi^2 E}{(L_e/r)^2} \qquad \frac{12{,}500 \text{ lb}}{0.35b^2} = \frac{\pi^2(10.1 \times 10^6 \text{ psi})}{(138.6/b)^2}$$

$b = 1.620$ in. $a = 0.35b = 0.567$ in.   ◄

REFLECT and THINK: The calculated critical Euler buckling stress can never be taken to exceed the yield strength of the material. In this problem, you can readily determine that the critical stress $\sigma_{cr} = 13.6$ ksi; though the specific alloy was not given, this stress is less than the tensile yield strength σ_y values for all aluminum alloys listed in Appendix B.

Problems

10.1 Knowing that the spring at A is of constant k and that the bar AB is rigid, determine the critical load P_{cr}.

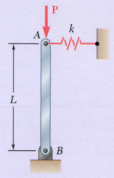

Fig. P10.1

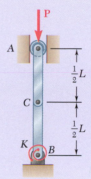

Fig. P10.2

10.2 Two rigid bars AC and BC are connected by a pin at C as shown. Knowing that the torsional spring at B is of constant K, determine the critical load P_{cr} for the system.

10.3 and 10.4 Two rigid bars AC and BC are connected as shown to a spring of constant k. Knowing that the spring can act in either tension or compression, determine the critical load P_{cr} for the system.

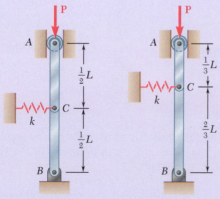

Fig. P10.3   **Fig. P10.4**

10.5 The steel rod BC is attached to the rigid bar AB and to the fixed support at C. Knowing that $G = 11.2 \times 10^6$ psi, determine the diameter of rod BC for which the critical load P_{cr} of the system is 80 lb.

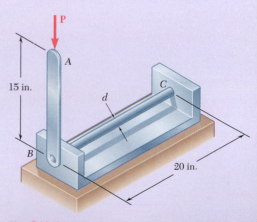

Fig. P10.5

10.6 The rigid rod AB is attached to a hinge at A and to two springs, each of constant $k = 2$ kips/in., that can act in either tension or compression. Knowing that $h = 2$ ft, determine the critical load.

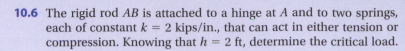

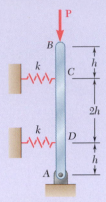

Fig. P10.6

10.7 The rigid bar AD is attached to two springs of constant k and is in equilibrium in the position shown. Knowing that the equal and opposite loads P and P' remain horizontal, determine the magnitude P_{cr} of the critical load for the system.

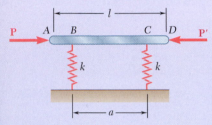

Fig. P10.7

10.8 A frame consists of four L-shaped members connected by four torsional springs, each of constant K. Knowing that equal loads **P** are applied at points A and D as shown, determine the critical value P_{cr} of the loads applied to the frame.

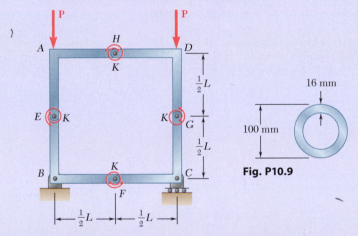

Fig. P10.8

10.9 Determine the critical load of a pin-ended steel tube that is 5 m long and has a 100-mm outer diameter and a 16-mm wall thickness. Use $E = 200$ GPa.

Fig. P10.9

10.10 Determine the critical load of a pin-ended wooden stick that is 3 ft long and has a $\frac{3}{16} \times 1\frac{1}{4}$-in. rectangular cross section. Use $E = 1.6 \times 10^6$ psi.

10.11 A column of effective length L can be made by gluing together identical planks in either of the arrangements shown. Determine the ratio of the critical load using the arrangement a to the critical load using the arrangement b.

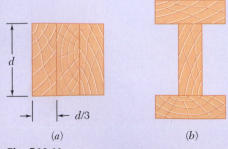

Fig. P10.11

10.12 A compression member of 1.5-m effective length consists of a solid 30-mm-diameter brass rod. In order to reduce the weight of the member by 25%, the solid rod is replaced by a hollow rod of the cross section shown. Determine (a) the percent reduction in the critical load, (b) the value of the critical load for the hollow rod. Use $E = 200$ GPa.

10.13 Determine the radius of the round strut so that the round and square struts have the same cross-sectional area and compute the critical load of each strut. Use $E = 200$ GPa.

Fig. P10.12

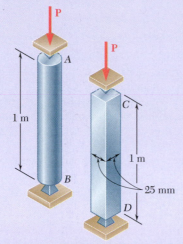
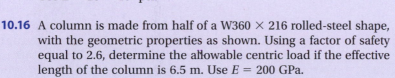

Fig. P10.13 and P10.14

10.14 Determine (a) the critical load for the square strut, (b) the radius of the round strut for which both struts have the same critical load. (c) Express the cross-sectional area of the square strut as a percentage of the cross-sectional area of the round strut. Use $E = 200$ GPa.

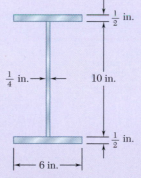

Fig. P10.15

10.15 A column with the cross section shown has a 13.5-ft effective length. Using a factor of safety equal to 2.8, determine the allowable centric load that can be applied to the column. Use $E = 29 \times 10^6$ psi.

10.16 A column is made from half of a W360 × 216 rolled-steel shape, with the geometric properties as shown. Using a factor of safety equal to 2.6, determine the allowable centric load if the effective length of the column is 6.5 m. Use $E = 200$ GPa.

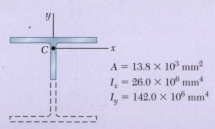

$A = 13.8 \times 10^3$ mm^2
$I_x = 26.0 \times 10^6$ mm^4
$I_y = 142.0 \times 10^6$ mm^4

Fig. P10.16

10.17 A column of 22-ft effective length is made by welding two 9 × 0.5-in. plates to a W8 × 35 as shown. Determine the allowable centric load if a factor of safety of 2.3 is required. Use $E = 29 \times 10^6$ psi.

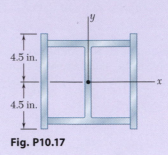

4.5 in.

4.5 in.

Fig. P10.17

10.18 A single compression member of 8.2-m effective length is obtained by connecting two C200 × 17.1 steel channels with lacing bars as shown. Knowing that the factor of safety is 1.85, determine the allowable centric load for the member. Use $E = 200$ GPa and $d = 100$ mm.

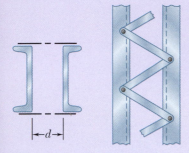

$\leftarrow d \rightarrow$

Fig. P10.18

10.19 Knowing that $P = 5.2$ kN, determine the factor of safety for the structure shown. Use $E = 200$ GPa and consider only buckling in the plane of the structure.

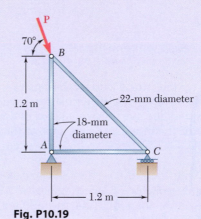

P

70°

B

1.2 m

22-mm diameter

18-mm diameter

A

C

1.2 m

Fig. P10.19

10.20 Members AB and CD are 30-mm-diameter steel rods, and members BC and AD are 22-mm-diameter steel rods. When the turnbuckle is tightened, the diagonal member AC is put in tension. Knowing that a factor of safety with respect to buckling of 2.75 is required, determine the largest allowable tension in AC. Use $E = 200$ GPa and consider only buckling in the plane of the structure.

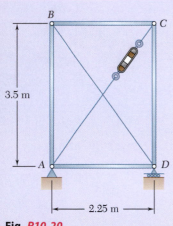

B   C

3.5 m

A   D

2.25 m

Fig. P10.20

10.21 The uniform brass bar AB has a rectangular cross section and is supported by pins and brackets as shown. Each end of the bar can rotate freely about a horizontal axis through the pin, but rotation about a vertical axis is prevented by the brackets. (a) Determine the ratio b/d for which the factor of safety is the same about the horizontal and vertical axes. (b) Determine the factor of safety if $P = 1.8$ kips, $L = 7$ ft, $d = 1.5$ in., and $E = 29 \times 10^6$ psi.

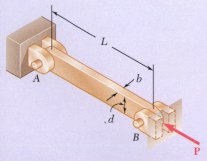

L

A

b

d

B

P

Fig. P10.21

10.22 A 1-in.-square aluminum strut is maintained in the position shown by a pin support at A and by sets of rollers at B and C that prevent rotation of the strut in the plane of the figure. Knowing that $L_{AB} = 3$ ft, determine (a) the largest values of L_{BC} and L_{CD} that can be used if the allowable load $\mathbf{P}$ is to be as large as possible, (b) the magnitude of the corresponding allowable load. Consider only buckling in the plane of the figure and use $E = 10.4 \times 10^6$ psi.

10.23 A 1-in.-square aluminum strut is maintained in the position shown by a pin support at A and by sets of rollers at B and C that prevent rotation of the strut in the plane of the figure. Knowing that $L_{AB} = 3$ ft, $L_{BC} = 4$ ft, and $L_{CD} = 1$ ft, determine the allowable load $\mathbf{P}$ using a factor of safety with respect to buckling of 3.2. Consider only buckling in the plane of the figure and use $E = 10.4 \times 10^6$ psi.

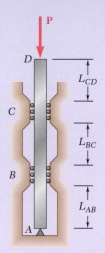

Fig. P10.22 and P10.23

10.24 Column ABC has a uniform rectangular cross section with $b = 12$ mm and $d = 22$ mm. The column is braced in the xz plane at its midpoint C and carries a centric load $\mathbf{P}$ of magnitude 3.8 kN. Knowing that a factor of safety of 3.2 is required, determine the largest allowable length L. Use $E = 200$ GPa.

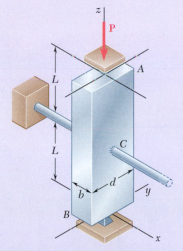

Fig. P10.24 and P10.25

10.25 Column ABC has a uniform rectangular cross section and is braced in the xz plane at its midpoint C. (a) Determine the ratio b/d for which the factor of safety is the same with respect to buckling in the xz and yz planes. (b) Using the ratio found in part a, design the cross section of the column so that the factor of safety will be 3.0 when $P = 4.4$ kN, $L = 1$ m, and $E = 200$ GPa.

10.26 Column AB carries a centric load $\mathbf{P}$ of magnitude 15 kips. Cables BC and BD are taut and prevent motion of point B in the xz plane. Using Euler's formula and a factor of safety of 2.2, and neglecting the tension in the cables, determine the maximum allowable length L. Use $E = 29 \times 10^6$ psi.

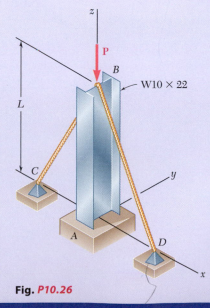

Fig. P10.26

10.27 Each of the five struts shown consists of a solid steel rod. (a) Knowing that the strut of Fig. (1) is of a 20-mm diameter, determine the factor of safety with respect to buckling for the loading shown. (b) Determine the diameter of each of the other struts for which the factor of safety is the same as the factor of safety obtained in part a. Use $E = 200$ GPa.

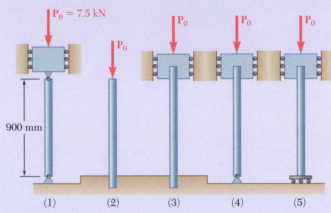

Fig. P10.27

10.28 A rigid block of mass m can be supported in each of the four ways shown. Each column consists of an aluminum tube that has a 44-mm outer diameter and a 4-mm wall thickness. Using $E = 70$ GPa and a factor of safety of 2.8, determine the allowable mass for each support condition.

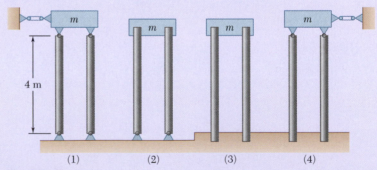

Fig. P10.28

*10.2   ECCENTRIC LOADING AND THE SECANT FORMULA

In this section, column buckling is approached by observing that the load **P** applied to a column is never perfectly centric. The eccentricity of the load is the distance between the line of action **P** and the axis of the column (Fig. 10.19*a*). The given eccentric load is replaced by a centric force **P** and a couple $\mathbf{M}_A$ of moment $M_A = Pe$ (Fig. 10.19*b*). Regardless of how small the

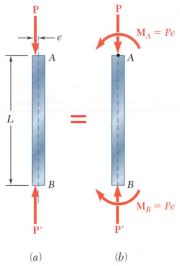

(a)   (b)

Fig. 10.19   (*a*) Column with an eccentric load (*b*) modeled as a column with an equivalent centric force-couple load.

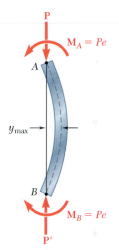

Fig. 10.20   Free-body diagram of an eccentrically loaded column.

load **P** and the eccentricity *e* are, the couple $\mathbf{M}_A$ will cause some bending of the column (Fig. 10.20). As the eccentric load increases, both the couple $\mathbf{M}_A$ and the axial force **P** increase, and both cause the column to bend further. Viewed in this way, buckling is not a question of determining how long the column can remain straight and stable under an increasing load, but how much the column can be permitted to bend if the allowable stress is not exceeded and the deflection $y_{\max}$ is not excessive.

We first write and solve the differential equation of the elastic curve, proceeding in the same manner as we did earlier in Secs. 10.1A and B. Drawing the free-body diagram of a portion *AQ* of the column and choosing the coordinate axes as shown (Fig. 10.21), we find that the bending moment at *Q* is

$$M = -Py - M_A = -Py - Pe \qquad \textbf{(10.22)}$$

Substituting the value of *M* into Eq. (9.4) gives

$$\frac{d^2y}{dx^2} = \frac{M}{EI} = -\frac{P}{EI}y - \frac{Pe}{EI}$$

Transposing the term containing *y* and setting

$$p^2 = \frac{P}{EI} \qquad \textbf{(10.6)}$$

as done earlier gives

$$\frac{d^2y}{dx^2} + p^2 y = -p^2 e \qquad \textbf{(10.23)}$$

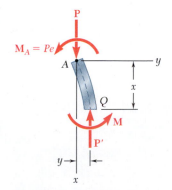

Fig. 10.21   Free-body diagram of portion *AQ* of an eccentrically loaded column.

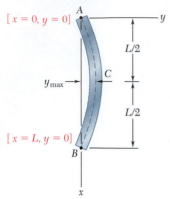

Fig. 10.22 Boundary conditions for an eccentrically loaded column.

Since the left-hand member of Eq. (10.23) is the same as Eq. (10.7), the general solution of Eq. (10.23) is rewritten as

$$y = A \sin px + B \cos px - e \qquad (10.24)$$

where the last term is a particular solution.

Constants A and B are obtained from the boundary conditions shown in Fig. 10.22. Making $x = 0$, $y = 0$ in Eq. (10.24), we have

$$B = e$$

Making $x = L$, $y = 0$ gives

$$A \sin pL = e(1 - \cos pL) \qquad (10.25)$$

Recalling that

$$\sin pL = 2 \sin \frac{pL}{2} \cos \frac{pL}{2}$$

and

$$1 - \cos pL = 2 \sin^2 \frac{pL}{2}$$

and substituting into Eq. (10.25) after reductions gives

$$A = e \tan \frac{pL}{2}$$

Substituting for A and B into Eq. (10.24), the equation of the elastic curve is

$$y = e \left(\tan \frac{pL}{2} \sin px + \cos px - 1 \right) \qquad (10.26)$$

The maximum deflection is obtained by setting $x = L/2$ in Eq. (10.26).

$$y_{\max} = e \left(\tan \frac{pL}{2} \sin \frac{pL}{2} + \cos \frac{pL}{2} - 1 \right)$$

$$= e \left(\frac{\sin^2 \frac{pL}{2} + \cos^2 \frac{pL}{2}}{\cos \frac{pL}{2}} - 1 \right)$$

$$y_{\max} = e \left(\sec \frac{pL}{2} - 1 \right) \qquad (10.27)$$

Recalling Eq. (10.6), we have

$$y_{\max} = e \left[\sec \left(\sqrt{\frac{P}{EI}} \frac{L}{2} \right) - 1 \right] \qquad (10.28)$$

The expression obtained indicates that $y_{\max}$ becomes infinite when

$$\sqrt{\frac{P}{EI}} \frac{L}{2} = \frac{\pi}{2} \qquad (10.29)$$

While the deflection does not actually become infinite, it nevertheless becomes unacceptably large, and P should not be allowed to reach the critical value which satisfies Eq. (10.29). Solving Eq. (10.29) for P,

$$P_{cr} = \frac{\pi^2 EI}{L^2} \qquad \textbf{(10.30)}$$

which also was obtained in Sec. 10.1A for a column under a centric load. Solving Eq. (10.30) for EI and substituting into Eq. (10.28), the maximum deflection in the alternative form is

$$y_{max} = e\left(\sec \frac{\pi}{2} \sqrt{\frac{P}{P_{cr}}} - 1 \right) \qquad \textbf{(10.31)}$$

The maximum stress σ_{max} occurs in the section of the column where the bending moment is maximum (i.e., the transverse section through the midpoint C) and can be obtained by adding the normal stresses due to the axial force and the bending couple exerted on that section (see Sec. 4.7). Thus,

$$\sigma_{max} = \frac{P}{A} + \frac{M_{max}c}{I} \qquad \textbf{(10.32)}$$

From the free-body diagram of portion AC (Fig. 10.23),

$$M_{max} = Py_{max} + M_A = P(y_{max} + e)$$

Substituting this into Eq. (10.32) and recalling that $I = Ar^2$,

$$\sigma_{max} = \frac{P}{A}\left[1 + \frac{(y_{max} + e)c}{r^2} \right] \qquad \textbf{(10.33)}$$

Substituting the value obtained in Eq. (10.28) for y_{max}:

$$\sigma_{max} = \frac{P}{A}\left[1 + \frac{ec}{r^2} \sec\left(\sqrt{\frac{P}{EI}} \frac{L}{2} \right) \right] \qquad \textbf{(10.34)}$$

An alternative form of σ_{max} is obtained by substituting from (10.31) into Eq. (10.33) for y_{max}. Thus,

$$\sigma_{max} = \frac{P}{A}\left(1 + \frac{ec}{r^2} \sec \frac{\pi}{2} \sqrt{\frac{P}{P_{cr}}} \right) \qquad \textbf{(10.35)}$$

This equation can be used with any end conditions, as long as the appropriate value is used for the critical load (see Sec. 10.1B).

Since σ_{max} does not vary linearly with load P, the principle of superposition does not apply to the determination of stress due to the simultaneous application of several loads. The resultant load must be computed first, and then Eqs. (10.34) or (10.35) can be used to find the corresponding stress. For the same reason, any given factor of safety should be applied to the load—not to the stress—when using the *second formula*.

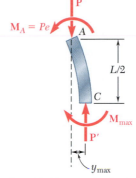

Fig. 10.23 Free-body diagram of upper half of eccentrically loaded column.

Making $I = Ar^2$ in Eq. (10.34) and solving for the ratio P/A in front of the bracket gives

$$\frac{P}{A} = \frac{\sigma_{max}}{1 + \dfrac{ec}{r^2}\sec\left(\dfrac{1}{2}\sqrt{\dfrac{P}{EA}}\dfrac{L_e}{r}\right)} \qquad (10.36)$$

where the effective length is applied for various end conditions. This equation is *the secant formula*. It defines the force per unit area, P/A, that causes a specified maximum stress σ_{max} in a column with a given effective slenderness ratio L_e/r for a ratio ec/r^2, where e is the eccentricity of the applied load. Since P/A appears in both members, it is necessary to solve a transcendental equation by trial and error to obtain the value of P/A corresponding to a given column and loading condition.

Equation (10.36) is used to draw the curves shown in Fig. 10.24a and b for a steel column, assuming the values of E and σ_Y shown. These curves make it possible to determine the load per unit area P/A, which causes the column to yield for given values of the ratios L_e/r and ec/r^2.

For small values of L_e/r, the secant is almost equal to 1 in Eq. (10.36), and thus P/A can be assumed equal to

$$\frac{P}{A} = \frac{\sigma_{max}}{1 + \dfrac{ec}{r^2}} \qquad (10.37)$$

This value can be obtained by neglecting the effect of the lateral deflection of the column and using the method of Sec. 4.8. On the other hand, Fig. 10.24 shows that, for large values of L_e/r, the curves corresponding to the ratio ec/r^2 get very close to Euler's curve given in Eq. (10.13b). Thus, the effect of the eccentricity of the load on P/A becomes negligible. The secant formula is mainly used for intermediate values of L_e/r. However, to use it effectively, the eccentricity e of the load should be known, but unfortunately, this quantity is seldom known with any degree of precision.

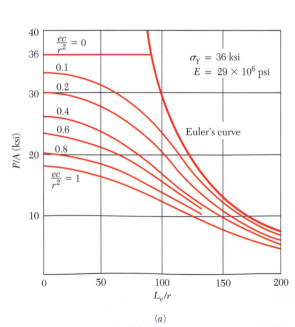

(a)

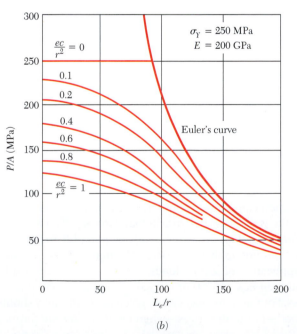

(b)

Fig. 10.24 Secant formula plots for buckling in eccentrically loaded columns. (a) U.S. customary units. (b) Metric units.

Sample Problem 10.2

The uniform column *AB* consists of an 8-ft section of structural tubing with the cross section shown. (*a*) Using Euler's formula and a factor of safety of 2, determine the allowable centric load for the column and the corresponding normal stress. (*b*) Assuming that the allowable load found in part *a* is applied at a point 0.75 in. from the geometric axis of the column, determine the horizontal deflection of the top of the column and the maximum normal stress in the column. Use $E = 29 \times 10^6$ psi.

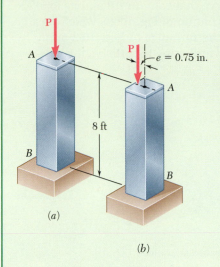

(*a*)

(*b*)

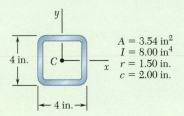

$A = 3.54 \text{ in}^2$
$I = 8.00 \text{ in}^4$
$r = 1.50$ in.
$c = 2.00$ in.

STRATEGY: For part *a*, use the factor of safety with Euler's formula to determine the allowable centric load. For part *b*, use Eqs. (10.31) and (10.35) to find the horizontal deflection and maximum normal stress in the column, respectively.

MODELING:

Effective Length.   Since the column has one end fixed and one end free, its effective length is

$$L_e = 2(8 \text{ ft}) = 16 \text{ ft} = 192 \text{ in.}$$

Critical Load.   Using Euler's formula,

$$P_{cr} = \frac{\pi^2 EI}{L_e^2} = \frac{\pi^2 (29 \times 10^6 \text{ psi})(8.00 \text{ in}^4)}{(192 \text{ in.})^2} \qquad P_{cr} = 62.1 \text{ kips}$$

ANALYSIS:

a. Allowable Load and Stress.   For a factor of safety of 2,

$$P_{all} = \frac{P_{cr}}{F.S.} = \frac{62.1 \text{ kips}}{2} \qquad \blacktriangleleft \ P_{all} = 31.1 \text{ kips} \ \blacktriangleleft$$

and

$$\sigma = \frac{P_{all}}{A} = \frac{31.1 \text{ kips}}{3.54 \text{ in}^2} \qquad \sigma = 8.79 \text{ ksi} \ \blacktriangleleft$$

(continued)

b. Eccentric Load (Fig. 1). Observe that column AB and its load are identical to the upper half of the column of Fig. 10.20, which was used for the secant formulas. Thus, the formulas of Sec. 10.2 apply directly here. Recalling that $P_{all}/P_{cr} = \frac{1}{2}$ and using Eq. (10.31), the horizontal deflection of point A is

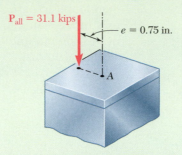

$P_{all} = 31.1$ kips

$e = 0.75$ in.

A

Fig. 1 Allowable load applied at assumed eccentricity.

$$y_m = e\left[\sec\left(\frac{\pi}{2}\sqrt{\frac{P}{P_{cr}}}\right) - 1\right] = (0.75 \text{ in.})\left[\sec\left(\frac{\pi}{2\sqrt{2}}\right) - 1\right]$$

$$= (0.75 \text{ in.})(2.252 - 1) \qquad y_m = 0.939 \text{ in.} \blacktriangleleft$$

This result is illustrated in Fig. 2.

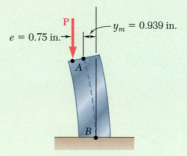

P

$e = 0.75$ in.

$y_m = 0.939$ in.

A

B

Fig. 2 Deflection of eccentrically loaded column.

The maximum normal stress is obtained from Eq. (10.35) as

$$\sigma_m = \frac{P}{A}\left[1 + \frac{ec}{r^2}\sec\left(\frac{\pi}{2}\sqrt{\frac{P}{P_{cr}}}\right)\right]$$

$$= \frac{31.1 \text{ kips}}{3.54 \text{ in}^2}\left[1 + \frac{(0.75 \text{ in.})(2 \text{ in.})}{(1.50 \text{ in.})^2}\sec\left(\frac{\pi}{2\sqrt{2}}\right)\right]$$

$$= (8.79 \text{ ksi})[1 + 0.667(2.252)] \qquad \sigma_m = 22.0 \text{ ksi} \blacktriangleleft$$

Problems

10.29 An axial load **P** = 15 kN is applied at point D that is 4 mm from the geometric axis of the square aluminum bar BC. Using E = 70 GPa, determine (a) the horizontal deflection of end C, (b) the maximum stress in the column.

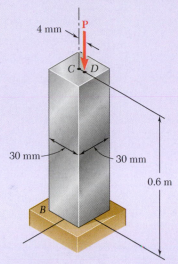

Fig. P10.29

10.30 An axial load **P** is applied to the 32-mm-diameter steel rod AB as shown. For P = 37 kN and e = 1.2 mm, determine (a) the deflection at the midpoint C of the rod, (b) the maximum stress in the rod. Use E = 200 GPa.

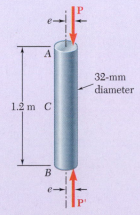

Fig. P10.30

10.31 The line of action of the 310-kN axial load is parallel to the geometric axis of the column AB and intersects the x axis at $x = e$. Using E = 200 GPa, determine (a) the eccentricity e when the deflection of the midpoint C of the column is 9 mm, (b) the maximum stress in the column.

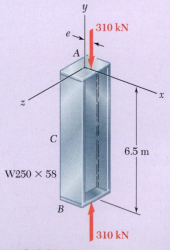

Fig. P10.31

10.32 An axial load **P** is applied to the 1.375-in. diameter steel rod *AB* as shown. When $P = 21$ kips, it is observed that the horizontal deflection at midpoint *C* is 0.03 in. Using $E = 29 \times 10^6$ psi, determine (*a*) the eccentricity *e* of the load, (*b*) the maximum stress in the rod.

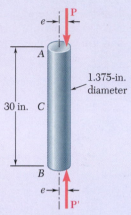

Fig. P10.32

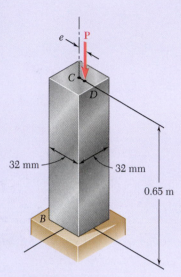

Fig. *P10.33*

10.33 An axial load **P** is applied to the 32-mm-square aluminum bar *BC* as shown. When $P = 24$ kN, the horizontal deflection at end *C* is 4 mm. Using $E = 70$ GPa, determine (*a*) the eccentricity *e* of the load, (*b*) the maximum stress in the bar.

10.34 The axial load **P** is applied at a point located on the *x* axis at a distance *e* from the geometric axis of the rolled-steel column *BC*. When $P = 82$ kips, the horizontal deflection of the top of the column is 0.20 in. Using $E = 29 \times 10^6$ psi, determine (*a*) the eccentricity *e* of the load, (*b*) the maximum stress in the column.

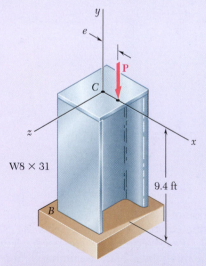

Fig. P10.34

10.35 An axial load **P** is applied at point *D* that is 0.25 in. from the geometric axis of the square aluminum bar *BC*. Using $E = 10.1 \times 10^6$ psi, determine (*a*) the load **P** for which the horizontal deflection of end *C* is 0.50 in., (*b*) the corresponding maximum stress in the column.

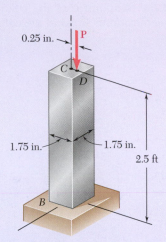

Fig. P10.35

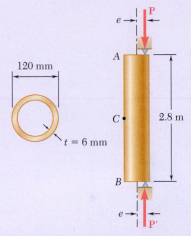

Fig. P10.36

10.36 A brass pipe having the cross section shown has an axial load **P** applied 5 mm from its geometric axis. Using $E = 120$ GPa, determine (*a*) the load **P** for which the horizontal deflection at the midpoint *C* is 5 mm, (*b*) the corresponding maximum stress in the column.

10.37 Solve Prob. 10.36, assuming that the axial load **P** is applied 10 mm from the geometric axis of the column.

10.38 The line of action of the axial load **P** is parallel to the geometric axis of the column *AB* and intersects the *x* axis at *x* = 0.8 in. Using $E = 29 \times 10^6$ psi, determine (*a*) the load **P** for which the horizontal deflection at the end *C* is 0.5 in., (*b*) the corresponding maximum stress in the column.

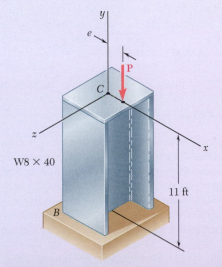

Fig. P10.38

10.39 The line of action of the axial load **P** is parallel to the geometric axis of the column and applied at a point located on the x axis at a distance $e = 12$ mm from the geometric axis of the W310 × 60 rolled-steel column BC. Assuming that $L = 7.0$ m and using $E = 200$ GPa, determine (a) the load **P** for which the horizontal deflection of the midpoint C of the column is 15 mm, (b) the corresponding maximum stress in the column.

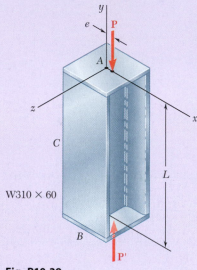

Fig. P10.39

10.40 Solve Prob. 10.39, assuming that L is 9.0 m.

10.41 The steel bar AB has a $\frac{3}{8} \times \frac{3}{8}$-in. square cross section and is held by pins that are a fixed distance apart and are located at a distance $e = 0.03$ in. from the geometric axis of the bar. Knowing that at temperature T_0 the pins are in contact with the bar and that the force in the bar is zero, determine the increase in temperature for which the bar will just make contact with point C if $d = 0.01$ in. Use $E = 29 \times 10^6$ psi and a coefficient of thermal expansion $\alpha = 6.5 \times 10^{-6}/°$F.

10.42 For the bar of Prob. 10.41, determine the required distance d for which the bar will just make contact with point C when the temperature increases by 120 °F.

10.43 A 3.5-m-long steel tube having the cross section and properties shown is used as a column. For the grade of steel used $\sigma_Y = 250$ MPa and $E = 200$ GPa. Knowing that a factor of safety of 2.6 with respect to permanent deformation is required, determine the allowable load **P** when the eccentricity e is (a) 15 mm, (b) 7.5 mm. (*Hint:* Since the factor of safety must be applied to the load **P**, not to the stress, use Fig. 10.24 to determine P_Y).

10.44 Solve Prob. 10.43, assuming that the length of the tube is increased to 5 m.

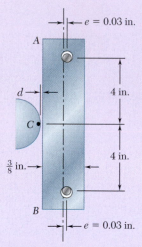

Fig. P10.41

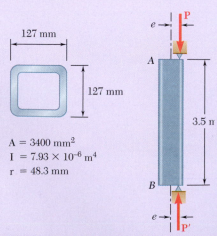

$A = 3400$ mm^2
$I = 7.93 \times 10^{-6}$ m^4
$r = 48.3$ mm

Fig. P10.43

10.45 An axial load **P** is applied to the W8 × 28 rolled-steel column *BC* that is free at its top *C* and fixed at its base *B*. Knowing that the eccentricity of the load is *e* = 0.6 in. and that for the grade of steel used σ_Y = 36 ksi and *E* = 29 × 10⁶ psi, determine (*a*) the magnitude of *P* of the allowable load when a factor of safety of 2.5 with respect to permanent deformation is required, (*b*) the ratio of the load found in part *a* to the magnitude of the allowable centric load for the column. (See hint of Prob. 10.43.)

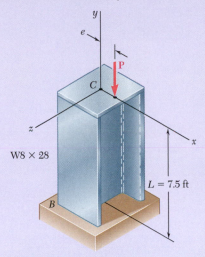

Fig. 10.45 and *10.46*

10.46 An axial load **P** of magnitude 50 kips is applied at a point located on the *x* axis at a distance *e* = 0.25 in. from the geometric axis of the W8 × 28 rolled-steel column *BC*. Knowing that the column is free at its top *C* and fixed at its base *B* and that σ_Y = 36 ksi and *E* = 29 × 10⁶ psi, determine the factor of safety with respect to yield. (See hint of Prob. 10.43.)

10.47 A 100-kN axial load **P** is applied to the W150 × 18 rolled-steel column *BC* that is free at its top *C* and fixed at its base *B*. Knowing that the eccentricity of the load is *e* = 6 mm, determine the largest permissible length *L* if the allowable stress in the column is 80 MPa. Use *E* = 200 GPa.

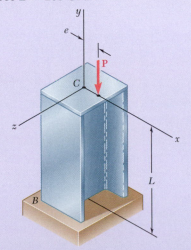

Fig. 10.47

10.48 A 26-kip axial load **P** is applied to a W6 × 12 rolled-steel column *BC* that is free at its top *C* and fixed at its base *B*. Knowing that the eccentricity of the load is *e* = 0.25 in., determine the largest permissible length *L* if the allowable stress in the column is 14 ksi. Use $E = 29 \times 10^6$ psi.

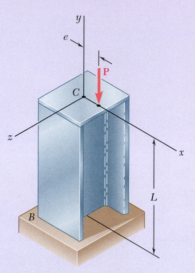

Fig. 10.48

10.49 Axial loads of magnitude *P* = 135 kips are applied parallel to the geometric axis of the W10 × 54 rolled-steel column *AB* and intersect the *x* axis at a distance *e* from the geometric axis. Knowing that σ_{all} = 12 ksi and $E = 29 \times 10^6$ psi, determine the largest permissible length *L* when (*a*) *e* = 0.25 in., (*b*) *e* = 0.5 in.

10.50 Axial loads of magnitude *P* = 84 kN are applied parallel to the geometric axis of the W200 × 22.5 rolled-steel column *AB* and intersect the *x* axis at a distance *e* from the geometric axis. Knowing that σ_{all} = 75 MPa and *E* = 200 GPa, determine the largest permissible length *L* when (*a*) *e* = 5 mm, (*b*) *e* = 12 mm.

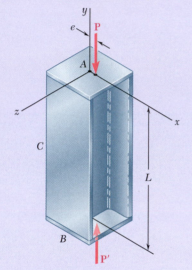

Fig. 10.49

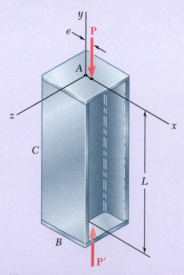

Fig. 10.50

10.51 An axial load of magnitude $P = 220$ kN is applied at a point located on the x axis at a distance $e = 6$ mm from the geometric axis of the wide-flange column BC. Knowing that $E = 200$ GPa, choose the lightest W200 shape that can be used if $\sigma_{all} = 120$ MPa.

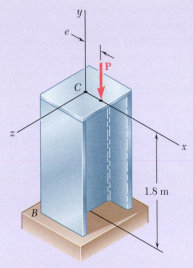

Fig. 10.51

10.52 Solve Prob. 10.51, assuming that the magnitude of the axial load is $P = 345$ kN.

10.53 A 12-kip axial load is applied with an eccentricity $e = 0.375$ in. to the circular steel rod BC that is free at its top C and fixed at its base B. Knowing that the stock of rods available for use have diameters in increments of $\frac{1}{8}$ in. from 1.5 in. to 3.0 in., determine the lightest rod that can be used if $\sigma_{all} = 15$ ksi. Use $E = 29 \times 10^6$ psi.

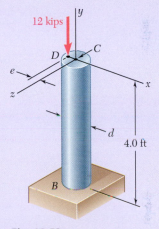

Fig. 10.53

10.54 Solve Prob. 10.53, assuming that the 12-kip axial load will be applied to the rod with an eccentricity $e = \frac{1}{2}d$.

10.55 Axial loads of magnitude $P = 175$ kN are applied parallel to the geometric axis of a W250 × 44.8 rolled-steel column AB and intersect the x axis at a distance $e = 12$ mm from its geometric axis. Knowing that $\sigma_Y = 250$ MPa and $E = 200$ GPa, determine the factor of safety with respect to yield. (*Hint:* Since the factor of safety must be applied to the load **P**, not to the stresses, use Fig. 10.24 to determine P_Y.)

10.56 Solve Prob. 10.55, assuming that $e = 16$ mm and $P = 155$ kN.

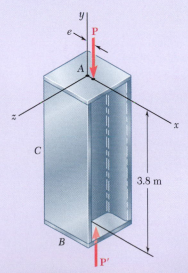

Fig. 10.55

10.3 CENTRIC LOAD DESIGN

The preceding sections determined the critical load of a column by using Euler's formula and investigated the deformations and stresses in eccentrically loaded columns by using the secant formula. In each case, all stresses remained below the proportional limit, and the column was initially a straight, homogeneous prism. Real columns fall short of such an idealization, and in practice the design of columns is based on empirical formulas that reflect the results of numerous laboratory tests.

Over the last century, many steel columns have been tested by applying to them a centric axial load and increasing the load until failure occurred. The results of such tests are represented in Fig. 10.25 where a point has been plotted with its ordinate equal to the normal stress σ_{cr} at failure and its abscissa is equal to the corresponding effective slenderness ratio L_e/r. Although there is considerable scatter in the test results, regions corresponding to three types of failure can be observed.

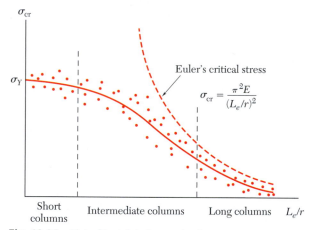

Fig. 10.25 Plot of test data for steel columns.

- For long columns, where L_e/r is large, failure is closely predicted by Euler's formula, and the value of σ_{cr} depends on the modulus of elasticity E of the steel used—but not on its yield strength σ_Y.
- For very short columns and compression blocks, failure essentially occurs as a result of yield, and $\sigma_{cr} \approx \sigma_Y$.
- For columns of intermediate length, failure is dependent on both σ_Y and E. In this range, column failure is an extremely complex phenomenon, and test data is used extensively to guide the development of specifications and design formulas.

Empirical formulas for an allowable or critical stress given in terms of the effective slenderness ratio were first introduced over a century ago. Since then, they have undergone a process of refinement and improvement. Typical empirical formulas used to approximate test data are shown

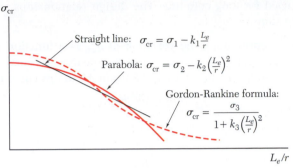

σ_{cr}

Straight line: $\sigma_{cr} = \sigma_1 - k_1 \frac{L_e}{r}$

Parabola: $\sigma_{cr} = \sigma_2 - k_2 \left(\frac{L_e}{r}\right)^2$

Gordon-Rankine formula:

$$\sigma_{cr} = \frac{\sigma_3}{1 + k_3 \left(\frac{L_e}{r}\right)^2}$$

L_e/r

Fig. 10.26   Plots of empirical formulas for critical stresses.

in Fig. 10.26. It is not always possible to use a single formula for all values of L_e/r. Most design specifications use different formulas—each with a definite range of applicability. In each case we must check that the equation used is applicable for the value of L_e/r for the column involved. Furthermore, it must determined whether the equation provides the critical stress for the column, to which the appropriate factor of safety must be applied, or if it provides an allowable stress.

Photo 10.2 shows examples of columns that are designed using such design specification formulas. The design formulas for three different materials using *Allowable Stress Design* are presented next, followed by formulas for the design of steel columns based on *Load and Resistance Factor Design*.[†]

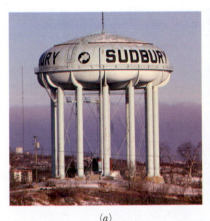

(a)   (b)

Photo 10.2   (a) The water tank is supported by steel columns. (b) The house under construction is framed with wood columns.

10.3A   Allowable Stress Design

Structural Steel.   The most commonly used formulas for allowable stress design of steel columns under a centric load are found in the *Specification for Structural Steel Buildings* of the American Institute of Steel Construction.[‡] An exponential expression is used to predict σ_{all} for columns of short and intermediate lengths, and an Euler-based

[†]In specific design formulas, the letter L always refers to the effective length of the column.

[‡]*Manual of Steel Construction*, 14th ed., American Institute of Steel Construction, Chicago, 2011.

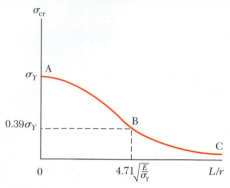

Fig. 10.27 Design curve for columns recommended by the American Institute of Steel Construction.

relation is used for long columns. The design relationships are developed in two steps.

1. A curve representing the variation of σ_{cr} as a function of L/r is obtained (Fig. 10.27). It is important to note that this curve does not incorporate any factor of safety.[§] Portion AB of this curve is

$$\sigma_{cr} = \left[0.658^{(\sigma_Y/\sigma_e)} \right] \sigma_Y \qquad \text{(10.38)}$$

where

$$\sigma_e = \frac{\pi^2 E}{(L/r)^2} \qquad \text{(10.39)}$$

Portion BC is

$$\sigma_{cr} = 0.877 \sigma_e \qquad \text{(10.40)}$$

When $L/r = 0$, $\sigma_{cr} = \sigma_Y$ in Eq. (10.38). At point B, Eq. (10.38) intersects Eq. (10.40). The slenderness L/r at the junction between the two equations is

$$\frac{L}{r} = 4.71 \sqrt{\frac{E}{\sigma_Y}} \qquad \text{(10.41)}$$

If L/r is smaller than the value from Eq. (10.41), σ_{cr} is determined from Eq. (10.38). If L/r is greater, σ_{cr} is determined from Eq. (10.40). At the slenderness L/r specified in Eq. (10.41), the stress $\sigma_e = 0.44\,\sigma_Y$. Using Eq. (10.40), $\sigma_{cr} = 0.877\,(0.44\,\sigma_Y) = 0.39\,\sigma_Y$.

2. A factor of safety must be used for the final design. The factor of safety given by the specification is 1.67. Thus,

$$\sigma_{all} = \frac{\sigma_{cr}}{1.67} \qquad \text{(10.42)}$$

These equations can be used with SI or U.S. customary units.

By using Eqs. (10.38), (10.40), (10.41), and (10.42), the allowable axial stress can be determined for a given grade of steel and any given value of L/r. The procedure is to compute L/r at the intersection between the two equations from Eq. (10.41). For smaller given values of L/r, use Eqs. (10.38) and (10.42) to calculate σ_{all}, and if greater, use Eqs. (10.40) and (10.42). Figure 10.28 provides an example of how σ_e varies as a function of L/r for different grades of structural steel.

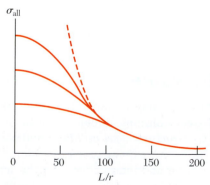

Fig. 10.28 Steel column design curves for different grades of steel.

[§]In the *Specification for Structural Steel Buildings*, the symbol F is used for stresses.

Concept Application 10.2

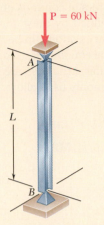

$P = 60$ kN

A

L

B

Fig. 10.29 Centrically loaded S100 × 11.5 rolled steel member.

Determine the longest unsupported length L for which the S100 × 11.5 rolled-steel compression member AB can safely carry the centric load shown (Fig. 10.29). Assume $\sigma_Y = 250$ MPa and $E = 200$ GPa.

From Appendix C, for an S100 × 11.5 shape,

$$A = 1460 \text{ mm}^2 \quad r_x = 41.7 \text{ mm} \quad r_y = 14.6 \text{ mm}$$

If the 60-kN load is to be safely supported,

$$\sigma_{\text{all}} = \frac{P}{A} = \frac{60 \times 10^3 \text{ N}}{1460 \times 10^{-6} \text{ m}^2} = 41.1 \times 10^6 \text{ Pa}$$

To compute the critical stress σ_{cr}, we start by assuming that L/r is larger than the slenderness specified by Eq. (10.41). We then use Eq. (10.40) with Eq. (10.39) and write

$$\sigma_{\text{cr}} = 0.877 \, \sigma_e = 0.877 \frac{\pi^2 E}{(L/r)^2}$$

$$= 0.877 \frac{\pi^2 (200 \times 10^9 \text{ Pa})}{(L/r)^2} = \frac{1.731 \times 10^{12} \text{ Pa}}{(L/r)^2}$$

Using this expression in Eq. (10.42),

$$\sigma_{\text{all}} = \frac{\sigma_{\text{cr}}}{1.67} = \frac{1.037 \times 10^{12} \text{ Pa}}{(L/r)^2}$$

Equating this expression to the required value of σ_{all} gives

$$\frac{1.037 \times 10^{12} \text{ Pa}}{(L/r)^2} = 41.1 \times 10^6 \text{ Pa} \qquad L/r = 158.8$$

The slenderness ratio from Eq. (10.41) is

$$\frac{L}{r} = 4.71 \sqrt{\frac{200 \times 10^9}{250 \times 10^6}} = 133.2$$

Our assumption that L/r is greater than this slenderness ratio is correct. Choosing the smaller of the two radii of gyration:

$$\frac{L}{r_y} = \frac{L}{14.6 \times 10^{-3} \text{ m}} = 158.8 \quad L = 2.32 \text{ m}$$

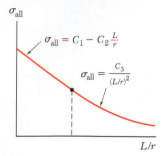

Fig. 10.30 Design curve for columns recommended by the Aluminum Association.

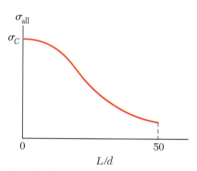

Fig. 10.31 Design curve for columns recommended by the American Forest & Paper Association.

Aluminum. Many aluminum alloys are used in structures and machines. For most columns, the specifications of the Aluminum Association[†] provide two formulas for the allowable stress in columns under centric loading. The variation of σ_{all} with L/r defined by these formulas is shown in Fig. 10.30. For short columns, a linear relationship between σ_{all} and L/r is used. For long columns, an Euler-type equation is used. Specific formulas for the design of buildings and similar structures are given in both SI and U.S. customary units for two commonly used alloys.

Alloy 6061-T6:

$$L/r < 66: \quad \sigma_{all} = [20.3 - 0.127(L/r)]\ \text{ksi} \qquad \textbf{(10.43a)}$$

$$= [140 - 0.874(L/r)]\ \text{MPa} \qquad \textbf{(10.43b)}$$

$$L/r \geq 66: \quad \sigma_{all} = \frac{51{,}400\ \text{ksi}}{(L/r)^2} = \frac{354 \times 10^3\ \text{MPa}}{(L/r)^2} \quad \textbf{(10.44a, b)}$$

Alloy 2014-T6:

$$L/r < 55: \quad \sigma_{all} = [30.9 - 0.229(L/r)]\ \text{ksi} \qquad \textbf{(10.45a)}$$

$$= [213 - 1.577(L/r)]\ \text{MPa} \qquad \textbf{(10.45b)}$$

$$L/r \geq 55: \quad \sigma_{all} = \frac{55{,}400\ \text{ksi}}{(L/r)^2} = \frac{382 \times 10^3\ \text{MPa}}{(L/r)^2} \quad \textbf{(10.46a, b)}$$

Wood. For the design of wood columns, the specifications of the American Forest & Paper Association[‡] provide a single equation to obtain the allowable stress for short, intermediate, and long columns under centric loading. For a column with a *rectangular* cross section of sides b and d, where $d < b$, the variation of σ_{all} with L/d is shown in Fig. 10.31.

For solid columns made from a single piece of wood or by gluing laminations together, the allowable stress σ_{all} is

$$\sigma_{all} = \sigma_C C_P \qquad \textbf{(10.47)}$$

where σ_C is the adjusted allowable stress for compression parallel to the grain.[§] Adjustments for σ_C are included in the specifications to account for different variations (such as in the load duration). The column stability factor C_P accounts for the column length and is defined by

$$C_P = \frac{1 + (\sigma_{CE}/\sigma_C)}{2c} - \sqrt{\left[\frac{1 + (\sigma_{CE}/\sigma_C)}{2c}\right]^2 - \frac{\sigma_{CE}/\sigma_C}{c}} \qquad \textbf{(10.48)}$$

The parameter c accounts for the type of column, and it is equal to 0.8 for sawn lumber columns and 0.90 for glued laminated wood columns. The value of σ_{CE} is defined as

$$\sigma_{CE} = \frac{0.822E}{(L/d)^2} \qquad \textbf{(10.49)}$$

where E is an adjusted modulus of elasticity for column buckling. Columns in which L/d exceeds 50 are not permitted by the *National Design Specification for Wood Construction*.

[†]*Specifications for Aluminum Structures*, Aluminum Association, Inc., Washington, D.C., 2010.

[‡]*National Design Specification for Wood Construction*, American Forest & Paper Association, American Wood Council, Washington, D.C., 2012.

[§]In the *National Design Specification for Wood Construction*, the symbol F is used for stresses.

Concept Application 10.3

Knowing that column AB (Fig. 10.32) has an effective length of 14 ft and must safely carry a 32-kip load, design the column using a square glued laminated cross section. The adjusted modulus of elasticity for the wood is $E = 800 \times 10^3$ psi, and the adjusted allowable stress for compression parallel to the grain is $\sigma_C = 1060$ psi.

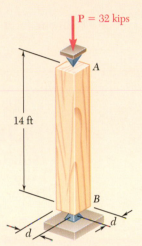

Fig. 10.32 Centrically loaded wood column.

Note that $c = 0.90$ for glued laminated wood columns. Computing the value of σ_{CE}, using Eq. (10.49), gives

$$\sigma_{CE} = \frac{0.822E}{(L/d)^2} = \frac{0.822(800 \times 10^3 \text{ psi})}{(168 \text{ in.}/d)^2} = 23.299d^2 \text{ psi}$$

Equation (10.48) is used to express the column stability factor in terms of d, with $(\sigma_{CE}/\sigma_C) = (23.299d^2/1.060 \times 10^3) = 21.98 \times 10^{-3}\,d^2$,

$$C_P = \frac{1 + (\sigma_{CE}/\sigma_C)}{2c} - \sqrt{\left[\frac{1 + (\sigma_{CE}/\sigma_C)}{2c}\right]^2 - \frac{\sigma_{CE}/\sigma_C}{c}}$$

$$= \frac{1 + 21.98 \times 10^{-3}\,d^2}{2(0.90)} - \sqrt{\left[\frac{1 + 21.98 \times 10^{-3}\,d^2}{2(0.90)}\right]^2 - \frac{21.98 \times 10^{-3}\,d^2}{0.90}}$$

(continued)

Since the column must carry 32 kips, Eq. (10.47) gives

$$\sigma_{\text{all}} = \frac{32 \text{ kips}}{d^2} = \sigma_C C_P = 1.060 C_P$$

Solving this equation for C_P and substituting the value into the previous equation, we obtain

$$\frac{30.19}{d^2} = \frac{1 + 21.98 \times 10^{-3} d^2}{2(0.90)} - \sqrt{\left[\frac{1 + 21.98 \times 10^{-3} d^2}{2(0.90)}\right]^2 - \frac{21.98 \times 10^{-3} d^2}{0.90}}$$

Solving for d by trial and error yields $d = 6.45$ in.

10.3B   Load and Resistance Factor Design

*Structural Steel   Section 1.5D gave an alternative method of design based on determining the load when the structure ceases to be useful. Design is based on the inequality given by

$$\gamma_D P_D + \gamma_L P_L \leq \phi P_U \qquad \textbf{(1.27)}$$

The design of steel columns under a centric load using Load and Resistance Factor Design with the American Institute of Steel Construction Specification is similar to that for the Allowable Stress Design. Using the critical stress σ_{cr}, the ultimate load P_U is

$$P_U = \sigma_{cr} A \qquad \textbf{(10.50)}$$

The critical stress σ_{cr} is determined using Eq. (10.41) for the slenderness at the junction between Eqs. (10.38) and Eq. (10.40). If the specified slenderness L/r is smaller than in Eq. (10.41), Eq. (10.38) governs. If it is larger, Eq. (10.40) governs. The equations can be used with SI or U.S. customary units.

By using Eq. (10.50) with Eq. (1.27), it can be determined if the design is acceptable. First calculate the slenderness ratio from Eq. (10.41). For values of L/r smaller than this slenderness, the *ultimate load P_U* used with Eq. (1.27) is obtained from Eq. (10.50), where σ_{cr} is determined from Eq. (10.38). For values of L/r larger than this slenderness, the *ultimate load P_U* is found by using Eq. (10.50) with Eq. (10.40). The Load and Resistance Factor Design Specification of the American Institute of Steel Construction specifies that the *resistance factor ϕ* is 0.90.

Note: The design formulas presented throughout Sec. 10.3 are examples of different design approaches. These equations do not provide all of the requirements needed for many designs, and the student should refer to the appropriate design specifications before attempting actual designs.

W10 × 39
$A = 11.5 \text{ in}^2$
$r_x = 4.27 \text{ in.}$
$r_y = 1.98 \text{ in.}$

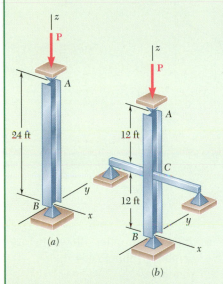

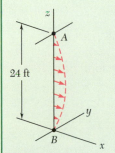

Fig. 1   Centrically loaded column
(a) unbraced, (b) braced.

Fig. 2   Buckled shape for
unbraced column.

Sample Problem 10.3

Column AB consists of a $W10 \times 39$ rolled-steel shape made of a grade of steel for which $\sigma_Y = 36$ ksi and $E = 29 \times 10^6$ psi. Determine the allowable centric load $\mathbf{P}$ (a) if the effective length of the column is 24 ft in all directions, (b) if bracing is provided to prevent the movement of the midpoint C in the xz plane. (Assume that the movement of point C in the yz plane is not affected by the bracing.)

STRATEGY: The allowable centric load for part a is determined from the governing allowable stress design equation for steel, Eq. (10.38) or Eq. (10.40), based on buckling associated with the axis with a smaller radius of gyration since the effective lengths are the same. In part b, it is necessary to determine the effective slenderness ratios for both axes, including the reduced effective length due to the bracing. The larger slenderness ratio governs the design.

MODELING: First compute the slenderness ratio from Eq. (10.41) corresponding to the given yield strength $\sigma_Y = 36$ ksi.

$$\frac{L}{r} = 4.71 \sqrt{\frac{29 \times 10^6}{36 \times 10^3}} = 133.7$$

ANALYSIS:

a. Effective Length = 24 ft.   The column is shown in Fig. 1a. Knowing that $r_y < r_x$, buckling takes place in the xz plane (Fig. 2). For $L = 24$ ft and $r = r_y = 1.98$ in., the slenderness ratio is

$$\frac{L}{r_y} = \frac{(24 \times 12) \text{ in.}}{1.98 \text{ in.}} = \frac{288 \text{ in.}}{1.98 \text{ in.}} = 145.5$$

Since $L/r > 133.7$, Eq. (10.39) in Eq. (10.40) is used to determine

$$\sigma_{cr} = 0.877\sigma_e = 0.877 \frac{\pi^2 E}{(L/r)^2} = 0.877 \frac{\pi^2 (29 \times 10^3 \text{ ksi})}{(145.5)^2} = 11.86 \text{ ksi}$$

The allowable stress is determined using Eq. (10.42)

$$\sigma_{all} = \frac{\sigma_{cr}}{1.67} = \frac{11.86 \text{ ksi}}{1.67} = 7.10 \text{ ksi}$$

and

$$P_{all} = \sigma_{all}A = (7.10 \text{ ksi})(11.5 \text{ in}^2) = 81.7 \text{ kips} \blacktriangleleft$$

(continued)

b. Bracing at Midpoint C. The column is shown in Fig. 1*b*. Since bracing prevents movement of point *C* in the *xz* plane but not in the *yz* plane, the slenderness ratio corresponding to buckling in each plane (Fig. 3) is computed to determine which is larger.

xz Plane: Effective length = 12 ft = 144 in., $r = r_y$ = 1.98 in.

$$L/r = (144 \text{ in.})/(1.98 \text{ in.}) = 72.7$$

yz Plane: Effective length = 24 ft = 288 in., $r = r_x$ = 4.27 in.

$$L/r = (288 \text{ in.})/(4.27 \text{ in.}) = 67.4$$

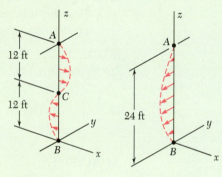

Buckling in *xz* plane   Buckling in *yz* plane

Fig. 3 Buckled shapes for braced column.

Since the larger slenderness ratio corresponds to a smaller allowable load, we choose L/r = 72.7. Since this is smaller than L/r = 133.7, Eqs. (10.39) and (10.38) are used to determine σ_{cr}:

$$\sigma_e = \frac{\pi^2 E}{(L/r)^2} = \frac{\pi^2 (29 \times 10^3 \text{ ksi})}{(72.7)^2} = 54.1 \text{ ksi}$$

$$\sigma_{cr} = \left[0.658^{(\sigma_Y/\sigma_e)} \right] F_Y = \left[0.658^{(36 \text{ ksi}/54.1 \text{ ksi})} \right] 36 \text{ ksi} = 27.3 \text{ ksi}$$

The allowable stress using Eq. (10.42) and the allowable load are

$$\sigma_{\text{all}} = \frac{\sigma_{cr}}{1.67} = \frac{27.3 \text{ ksi}}{1.67} = 16.32 \text{ ksi}$$

$$P_{\text{all}} = \sigma_{\text{all}} A = (16.32 \text{ ksi})(11.5 \text{ in}^2) \qquad P_{\text{all}} = 187.7 \text{ kips} \blacktriangleleft$$

REFLECT and THINK: This sample problem shows the benefit of using bracing to reduce the effective length for buckling about the weak axis when a column has significantly different radii of gyration, which is typical for steel wide-flange columns.

Sample Problem 10.4

Using the aluminum alloy 2014-T6 for the circular rod shown, determine the smallest diameter that can be used to support the centric load $P = 60$ kN if (a) $L = 750$ mm, (b) $L = 300$ mm.

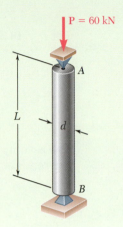

STRATEGY: Use the aluminum allowable stress equations to design the column, i.e., to determine the smallest diameter that can be used. Since there are two design equations based on L/r, it is first necessary to assume which governs. Then check the assumption.

MODELING: For the cross section of the solid circular rod shown in Fig. 1,

$$I = \frac{\pi}{4}c^4 \qquad A = \pi c^2 \qquad r = \sqrt{\frac{I}{A}} = \sqrt{\frac{\pi c^4/4}{\pi c^2}} = \frac{c}{2}$$

Fig. 1 Cross section of aluminum column.

ANALYSIS:

a. Length of 750 mm. Since the diameter of the rod is not known, L/r must be assumed. Assume that $L/r > 55$ and use Eq. (10.46). For the centric load **P**, $\sigma = P/A$ and write

$$\frac{P}{A} = \sigma_{\text{all}} = \frac{382 \times 10^3 \text{ MPa}}{(L/r)^2}$$

(continued)

$$\frac{60 \times 10^3 \, \text{N}}{\pi c^2} = \frac{382 \times 10^9 \, \text{Pa}}{\left(\dfrac{0.750 \, \text{m}}{c/2}\right)^2}$$

$$c^4 = 112.5 \times 10^{-9} \, \text{m}^4 \qquad c = 18.31 \, \text{mm}$$

For $c = 18.31$ mm, the slenderness ratio is

$$\frac{L}{r} = \frac{L}{c/2} = \frac{750 \, \text{mm}}{(18.31 \, \text{mm})/2} = 81.9 > 55$$

The assumption that L/r is greater than 55 is correct. For $L = 750$ mm, the required diameter is

$$d = 2c = 2(18.31 \, \text{mm}) \qquad d = 36.6 \, \text{mm} \quad \blacktriangleleft$$

b. Length of 300 mm. Assume that $L/r > 55$. Using Eq. (10.46) and following the procedure used in part a, $c = 11.58$ mm and $L/r = 51.8$. Since L/r is less than 55, this assumption is wrong. Now assume that $L/r < 55$ and use Eq. (10.45b) for the design of this rod.

$$\frac{P}{A} = \sigma_{\text{all}} = \left[213 - 1.577 \left(\frac{L}{r} \right) \right] \text{MPa}$$

$$\frac{60 \times 10^3 \, \text{N}}{\pi c^2} = \left[213 - 1.577 \left(\frac{0.3 \, \text{m}}{c/2} \right) \right] 10^6 \, \text{Pa}$$

$$c = 11.95 \, \text{mm}$$

For $c = 11.95$ mm, the slenderness ratio is

$$\frac{L}{r} = \frac{L}{c/2} = \frac{300 \, \text{mm}}{(11.95 \, \text{mm})/2} = 50.2$$

The second assumption that $L/r < 55$ is correct. For $L = 300$ mm, the required diameter is

$$d = 2c = 2(11.95 \, \text{mm}) \qquad d = 23.9 \, \text{mm} \quad \blacktriangleleft$$

Problems

10.57 Using allowable stress design, determine the allowable centric load for a column of 6-m effective length that is made from the following rolled-steel shape: (*a*) W200 × 35.9, (*b*) W200 × 86. Use $\sigma_Y = 250$ MPa and $E = 200$ GPa.

10.58 A W8 × 31 rolled-steel shape is used for a column of 21-ft effective length. Using allowable stress design, determine the allowable centric load if the yield strength of the grade of steel used is (*a*) $\sigma_Y = 36$ ksi, (*b*) $\sigma_Y = 50$ ksi. Use $E = 29 \times 10^6$ psi.

10.59 A rectangular structural tube having the cross section shown is used as a column of 5-m effective length. Knowing that $\sigma_Y = 250$ MPa and $E = 200$ GPa, use allowable stress design to determine the largest centric load that can be applied to the steel column.

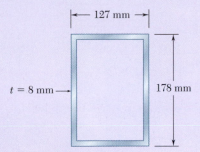

Fig. P10.59

10.60 A column having a 3.5-m effective length is made of sawn lumber with a 114 × 140-mm cross section. Knowing that for the grade of wood used the adjusted allowable stress for compression parallel to the grain is $\sigma_C = 7.6$ MPa and the adjusted modulus $E = 2.8$ GPa, determine the maximum allowable centric load for the column.

10.61 A sawn lumber column with a 7.5 × 5.5-in. cross section has an 18-ft effective length. Knowing that for the grade of wood used the adjusted allowable stress for compression parallel to the grain is $\sigma_C = 1200$ psi and that the adjusted modulus $E = 470 \times 10^3$ psi, determine the maximum allowable centric load for the column.

10.62 Bar *AB* is free at its end *A* and fixed at its base *B*. Determine the allowable centric load **P** if the aluminum alloy is (*a*) 6061-T6, (*b*) 2014-T6.

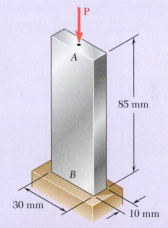

10.63 A compression member has the cross section shown and an effective length of 5 ft. Knowing that the aluminum alloy used is 2014-T6, determine the allowable centric load.

Fig. P10.62

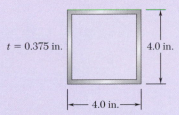

Fig. P10.63

10.64 A compression member has the cross section shown and an effective length of 5 ft. Knowing that the aluminum alloy used is 6061-T6, determine the allowable centric load.

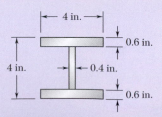

Fig. P10.64

Fig. P10.65

Fig. P10.66

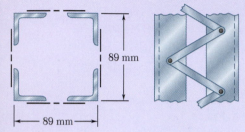

Fig. P10.67

10.65 A compression member of 8.2-ft effective length is obtained by bolting together two L5 × 3 × ½-in. steel angles as shown. Using allowable stress design, determine the allowable centric load for the column. Use $\sigma_Y = 36$ ksi and $E = 29 \times 10^6$ psi.

10.66 A compression member of 9-m effective length is obtained by welding two 10-mm-thick steel plates to a W250 × 80 rolled-steel shape as shown. Knowing that $\sigma_Y = 345$ MPa and $E = 200$ GPa and using allowable stress design, determine the allowable centric load for the compression member.

10.67 A column of 6.4-m effective length is obtained by connecting four L89 × 89 × 9.5-mm steel angles with lacing bars as shown. Using allowable stress design, determine the allowable centric load for the column. Use $\sigma_Y = 345$ MPa and $E = 200$ GPa.

10.68 A column of 21-ft effective length is obtained by connecting C10 × 20 steel channels with lacing bars as shown. Using allowable stress design, determine the allowable centric load for the column. Use $\sigma_Y = 36$ ksi and $E = 29 \times 10^6$ psi.

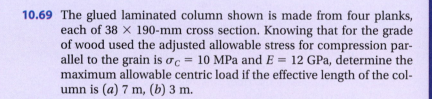

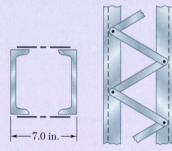

Fig. P10.68

10.69 The glued laminated column shown is made from four planks, each of 38 × 190-mm cross section. Knowing that for the grade of wood used the adjusted allowable stress for compression parallel to the grain is $\sigma_C = 10$ MPa and $E = 12$ GPa, determine the maximum allowable centric load if the effective length of the column is (a) 7 m, (b) 3 m.

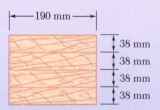

Fig. P10.69

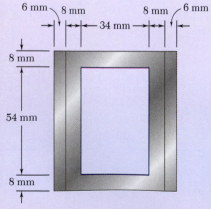

Fig. P10.70

10.70 An aluminum structural tube is reinforced by bolting two plates to it as shown for use as a column of 1.7-m effective length. Knowing that all material is aluminum alloy 2014-T6, determine the maximum allowable centric load.

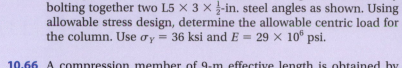

10.71 The glued laminated column shown is free at its top A and fixed at its base B. Using wood that has an adjusted allowable stress for compression parallel to the grain $\sigma_C = 9.2$ MPa and an adjusted modulus of elasticity $E = 5.7$ GPa, determine the smallest cross section that can support a centric load of 62 kN.

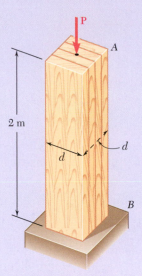

Fig. P10.71

Fig. P10.72

10.72 An 18-kip centric load is applied to a rectangular sawn lumber column of 22-ft effective length. Using lumber for which the adjusted allowable stress for compression parallel to the grain is $\sigma_C = 1050$ psi and the adjusted modulus is $E = 440 \times 10^3$ psi, determine the smallest cross section that can be used. Use $b = 2d$.

10.73 A laminated column of 2.1-m effective length is to be made by gluing together wood pieces of 25×150-mm cross section. Knowing that for the grade of wood used the adjusted allowable stress for compression parallel to the grain is $\sigma_C = 7.7$ MPa and the adjusted modulus is $E = 5.4$ GPa, determine the number of wood pieces that must be used to support the concentric load shown when (a) $P = 52$ kN, (b) $P = 108$ kN.

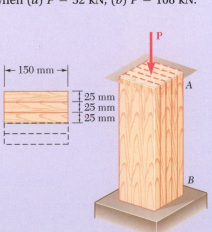

Fig. P10.73

10.74 For a rod made of aluminum alloy 2014-T6, select the smallest square cross section that can be used if the rod is to carry a 55-kip centric load.

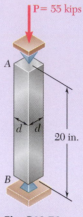

P= 55 kips

A

d   d   20 in.

B

Fig. P10.74

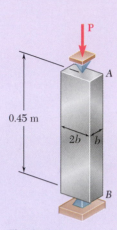

P

0.45 m

2b   b

A

B

Fig. P10.75

10.75 A 72-kN centric load must be supported by an aluminum column as shown. Using the aluminum alloy 6061-T6, determine the minimum dimension b that can be used.

10.76 An aluminum tube of 90-mm outer diameter is to carry a centric load of 120 kN. Knowing that the stock of tubes available for use are made of alloy 2014-T6 and with wall thicknesses in increments of 3 mm from 6 mm to 15 mm, determine the lightest tube that can be used.

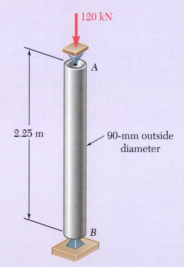

120 kN

A

2.25 m   90-mm outside
                diameter

B

Fig. P10.76

10.77 A column of 4.6-m effective length must carry a centric load of 525 kN. Knowing that $\sigma_Y = 345$ MPa and $E = 200$ GPa, use allowable stress design to select the wide-flange shape of 200-mm nominal depth that should be used.

10.78 A column of 22.5-ft effective length must carry a centric load of 288 kips. Using allowable stress design, select the wide-flange shape of 14-in. nominal depth that should be used. Use $\sigma_Y = 50$ ksi and $E = 29 \times 10^6$ psi.

10.79 A column of 17-ft effective length must carry a centric load of 235 kips. Using allowable stress design, select the wide-flange shape of 10-in. nominal depth that should be used. Use $\sigma_Y = 36$ ksi and $E = 29 \times 10^6$ psi.

10.80 A centric load **P** must be supported by the steel bar AB. Using allowable stress design, determine the smallest dimension d of the cross section that can be used when (*a*) $P = 108$ kN, (*b*) $P = 166$ kN. Use $\sigma_Y = 250$ MPa and $E = 200$ GPa.

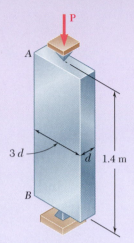

Fig. P10.80

10.81 A square steel tube having the cross section shown is used as a column of 26-ft effective length to carry a centric load of 65 kips. Knowing that the tubes available for use are made with wall thicknesses ranging from $\frac{1}{4}$ in. to $\frac{3}{4}$ in. in increments of $\frac{1}{16}$ in., use allowable stress design to determine the lightest tube that can be used. Use $\sigma_Y = 36$ ksi and $E = 29 \times 10^6$ psi.

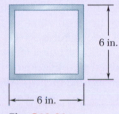

6 in.

6 in.

Fig. P10.81

10.82 Solve Prob. 10.81, assuming that the effective length of the column is decreased to 20 ft.

10.83 Two 89×64-mm angles are bolted together as shown for use as a column of 2.4-m effective length to carry a centric load of 180 kN. Knowing that the angles available have thicknesses of 6.4 mm, 9.5 mm, and 12.7 mm, use allowable stress design to determine the lightest angles that can be used. Use $\sigma_Y = 250$ MPa and $E = 200$ GPa.

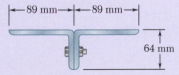

89 mm — 89 mm

64 mm

Fig. P10.83

10.84 Two 89×64-mm angles are bolted together as shown for use as a column of 2.4-m effective length to carry a centric load of 325 kN. Knowing that the angles available have thicknesses of 6.4 mm, 9.5 mm, and 12.7 mm, use allowable stress design to determine the lightest angles that can be used. Use $\sigma_Y = 250$ MPa and $E = 200$ GPa.

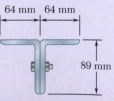

64 mm   64 mm

89 mm

Fig. P10.84

*10.85 A rectangular steel tube having the cross section shown is used as a column of 14.5-ft effective length. Knowing that $\sigma_Y = 36$ ksi and $E = 29 \times 10^6$ psi, use load and resistance factor design to determine the largest centric live load that can be applied if the centric dead load is 54 kips. Use a dead load factor $\gamma_D = 1.2$, a live load factor $\gamma_L = 1.6$ and the resistance factor $\phi = 0.90$.

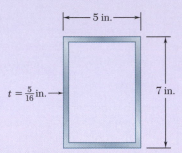

5 in.

$t = \frac{5}{16}$ in.

7 in.

Fig. P10.85

*10.86 A column with a 5.8-m effective length supports a centric load, with ratio of dead to live load equal to 1.35. The dead load factor is $\gamma_D = 1.2$, the live load factor $\gamma_L = 1.6$, and the resistance factor $\phi = 0.90$. Use load and resistance factor design to determine the allowable centric dead and live loads if the column is made of the following rolled-steel shape: (a) W250 × 67, (b) W360 × 101. Use $\sigma_Y = 345$ MPa and $E = 200$ GPa.

*10.87 A steel column of 5.5-m effective length must carry a centric dead load of 310 kN and a centric live load of 375 kN. Knowing that $\sigma_Y = 250$ MPa and $E = 200$ GPa, use load and resistance factor design to select the wide-flange shape of 310-mm nominal depth that should be used. The dead load factor $\gamma_D = 1.2$, the live load factor $\gamma_L = 1.6$, and the resistance factor $\phi = 0.90$.

*10.88 The steel tube having the cross section shown is used as a column of 15-ft effective length to carry a centric dead load of 51 kips and a centric live load of 58 kips. Knowing that the tubes available for use are made with wall thicknesses in increments of $\frac{1}{16}$ in. from $\frac{3}{16}$ in. to $\frac{3}{8}$ in., use load and resistance factor design to determine the lightest tube that can be used. Use $\sigma_Y = 36$ ksi and $E = 29 \times 10^6$ psi. The dead load factor $\gamma_D = 1.2$, the live load factor $\gamma_L = 1.6$, and the resistance factor $\phi = 0.90$.

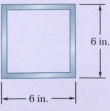

6 in.

6 in.

Fig. P10.88

10.4 ECCENTRIC LOAD DESIGN

In this section, the design of columns subjected to an eccentric load is considered. The empirical formulas developed in the preceding section for columns under a centric load can be modified and used when load **P** is applied to the column with a known eccentricity e.

Recall from Sec. 4.7 that an eccentric axial load **P** applied in a plane of symmetry can be replaced by an equivalent system consisting of a centric load **P** and a couple **M** of moment $M = Pe$, where e is the distance from the line of action of the load to the longitudinal axis of the column (Fig. 10.33). The normal stresses exerted on a transverse section of the column are obtained by superposing the stresses due to the centric load **P** and the couple **M** (Fig. 10.34), provided that the section is not too close

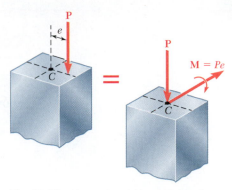

Fig. 10.33 Eccentric axial load replaced with an equivalent centric load and couple.

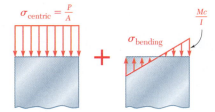

Fig. 10.34 Normal stress of an eccentrically loaded column is the superposition of centric axial and bending stresses.

to either end of the column and as long as the stresses do not exceed the proportional limit of the material. The normal stress due to the eccentric load **P** can be expressed as

$$\sigma = \sigma_{\text{centric}} + \sigma_{\text{bending}} \tag{10.51}$$

The maximum compressive stress in the column is

$$\sigma_{\max} = \frac{P}{A} + \frac{Mc}{I} \tag{10.52}$$

In a properly designed column, the maximum stress given in Eq. (10.52) should not exceed the allowable stress for the column. Two alternative approaches can be used to satisfy this requirement: the *allowable-stress method* and the *interaction method*.

Allowable-Stress Method. This method is based on the assumption that the allowable stress for an eccentrically loaded column is the same as if the column were centrically loaded. Therefore, $\sigma_{\max} \leq \sigma_{\text{all}}$, where σ_{all} is the allowable stress under a centric load. Substituting for $\sigma_{\max}$ from Eq. (10.52) gives

$$\frac{P}{A} + \frac{Mc}{I} \leq \sigma_{\text{all}} \tag{10.53}$$

The allowable stress is determined using the same equations in Sec. 10.3. For a given material, these equations express σ_{all} as a function of the slenderness ratio of the column. Engineering codes require that the largest value of the slenderness ratio of the column be used to determine the allowable stress, whether it corresponds to the actual plane of bending or not. This requirement sometimes results in an overly conservative design.

Concept Application 10.4

A column with a 2-in.-square cross section and 28-in. effective length is made of the aluminum alloy 2014-T6 (Fig. 10.35). Using the allowable-stress method, determine the maximum load P that can be safely supported with an eccentricity of 0.8 in.

Compute the radius of gyration r using the given data:

$$A = (2 \text{ in.})^2 = 4 \text{ in}^2 \qquad I = \tfrac{1}{12}(2 \text{ in.})^4 = 1.333 \text{ in}^4$$

$$r = \sqrt{\frac{I}{A}} = \sqrt{\frac{1.333 \text{ in}^4}{4 \text{ in}^2}} = 0.5774 \text{ in.}$$

Next, compute $L/r = (28 \text{ in.})/(0.5774 \text{ in.}) = 48.50$.

Since $L/r < 55$, use Eq. (10.45a) to determine the allowable stress for the aluminum column subjected to a centric load.

$$\sigma_{\text{all}} = [30.9 - 0.229(48.50)] = 19.79 \text{ ksi}$$

Now use Eq. (10.53) with $M = Pe$ and $c = \tfrac{1}{2}(2 \text{ in.}) = 1 \text{ in.}$ to determine the allowable load:

$$\frac{P}{4 \text{ in}^2} + \frac{P(0.8 \text{ in.})(1 \text{ in.})}{1.333 \text{ in}^4} \le 19.79 \text{ ksi}$$

$$P \le 23.3 \text{ kips}$$

The maximum load that can be safely applied is $P = 23.3$ kips.

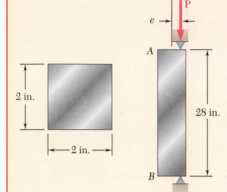

Fig. 10.35 Column subjected to eccentric axial load.

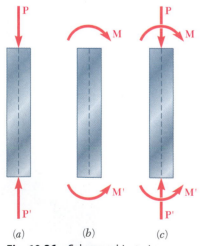

Fig. 10.36 Column subjected to (a) centric axial load, (b) pure bending, (c) eccentric load.

Interaction Method. Recall that the allowable stress for a column subjected to a centric load (Fig. 10.36a) is generally smaller than the allowable stress for a column in pure bending (Fig. 10.36b), since it takes into account the possibility of buckling. Therefore, when the allowable-stress method is used to design an eccentrically loaded column such that the sum of the stresses due to the centric load **P** and the bending couple **M** (Fig. 10.36c) does not exceed the allowable stress for a centrically loaded column, the resulting design is often overly conservative. An improved method of design can be developed by rewriting Eq. (10.53) in the form

$$\frac{P/A}{\sigma_{\text{all}}} + \frac{Mc/I}{\sigma_{\text{all}}} \le 1 \qquad (10.54)$$

and substituting for σ_{all} in the first and second terms the allowable stresses, that correspond, respectively, to the allowable stresses obtained for the centric load of Fig. 10.36a and the pure bending of Fig. 10.36b. Thus,

$$\frac{P/A}{(\sigma_{\text{all}})_{\text{centric}}} + \frac{Mc/I}{(\sigma_{\text{all}})_{\text{bending}}} \le 1 \qquad (10.55)$$

This is known as an *interaction formula*.

When $M = 0$, the use of Eq. (10.55) results in the design of a centrically loaded column by the method of Sec. 10.3. On the other hand, when $P = 0$, this equation results in the design of a beam in pure bending by the method of Sec. 5.3. When P and M are both different from zero, the interaction formula results in a design that takes into account the capacity of the member to resist bending as well as axial loading. In all cases, $(\sigma_{all})_{centric}$ is determined by using the largest slenderness ratio of the column, regardless of the plane in which bending takes place.[†]

Concept Application 10.5

Use the interaction method to determine the maximum load P that can be safely supported by the column of Concept Application 10.4 with an eccentricity of 0.8 in. The allowable stress in bending is 24 ksi.

The value of $(\sigma_{all})_{centric}$ has been determined and thus

$$(\sigma_{all})_{centric} = 19.79 \text{ ksi} \qquad (\sigma_{all})_{bending} = 24 \text{ ksi}$$

Substituting these values into Eq. (10.55),

$$\frac{P/A}{19.79 \text{ ksi}} + \frac{Mc/I}{24 \text{ ksi}} \leq 1.0$$

Use the numerical data from Concept Application 10.4 to write

$$\frac{P/4}{19.79 \text{ ksi}} + \frac{P(0.8)(1.0)/1.333}{24 \text{ ksi}} \leq 1.0$$

$$P \leq 26.6 \text{ kips}$$

The maximum load that can be safely applied is $P = 26.6$ kips.

When the eccentric load **P** is not applied in a plane of symmetry, it causes bending about both of the principal axes of the cross section of the column. The eccentric load **P** then can be replaced by a centric load **P** and two couples $\mathbf{M}_x$ and $\mathbf{M}_z$, as shown in Fig. 10.37. The interaction formula to be used is

$$\frac{P/A}{(\sigma_{all})_{centric}} + \frac{|M_x|z_{max}/I_x}{(\sigma_{all})_{bending}} + \frac{|M_z|x_{max}/I_z}{(\sigma_{all})_{bending}} \leq 1 \qquad \textbf{(10.56)}$$

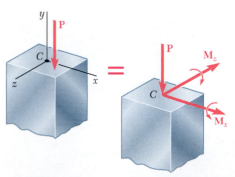

Fig. 10.37   Equivalent centric load and couples for an eccentric axial load causing biaxial bending.

[†]This procedure is required by all major codes for the design of steel, aluminum, and timber compression members. In addition, many specifications call for the use of an additional factor in the second term of Eq. (10.55). This factor takes into account the additional stresses resulting from the deflection of the column due to bending.

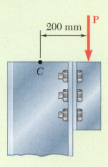

Sample Problem 10.5

Using the allowable-stress method, determine the largest load **P** that can be safely carried by a W310 × 74 steel column of 4.5-m effective length. Use $E = 200$ GPa and $\sigma_Y = 250$ MPa.

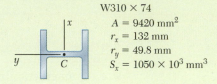

W310 × 74
$A = 9420$ mm^2
$r_x = 132$ mm
$r_y = 49.8$ mm
$S_x = 1050 \times 10^3$ mm^3

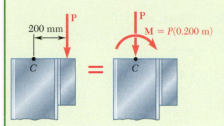

Fig. 1 Eccentric loading replaced by centric force-couple system acting at column's centroid.

STRATEGY: Determine the allowable centric stress for the column, using the governing allowable stress design equation for steel, Eq. (10.38) or Eq. (10.40) with Eq. (10.42). Then use Eq. (10.53) to calculate the load **P**.

MODELING and ANALYSIS:

The largest slenderness ratio of the column is $L/r_y = (4.5$ m$)/$ $(0.0498$ m$) = 90.4$. Using Eq. (10.41) with $E = 200$ GPa and $\sigma_Y = 250$ MPa, the slenderness ratio at the junction between the two equations for σ_{cr} is $L/r = 133.2$. Thus, Eqs. (10.38) and (10.39) are used, and $\sigma_{cr} = 162.2$ MPa. Using Eq. (10.42), the allowable stress is

$$(\sigma_{\text{all}})_{\text{centric}} = 162.2/1.67 = 97.1 \text{ MPa}$$

For the given column, replacing the eccentric loading with a centric force-couple system acting at the centroid (Fig. 1), we write

$$\frac{P}{A} = \frac{P}{9.42 \times 10^{-3} \text{ m}^2} \qquad \frac{Mc}{I} = \frac{M}{S} = \frac{P(0.200 \text{ m})}{1.050 \times 10^{-3} \text{ m}^3}$$

Substituting into Eq. (10.53), we obtain

$$\frac{P}{A} + \frac{Mc}{I} \leq \sigma_{\text{all}}$$

$$\frac{P}{9.42 \times 10^{-3} \text{ m}^2} + \frac{P(0.200 \text{ m})}{1.050 \times 10^{-3} \text{ m}^3} \leq 97.1 \text{ MPa} \qquad P \leq 327 \text{ kN}$$

The largest allowable load **P** is $\qquad$ **P** $= 327$ kN $\downarrow$ ◀

Sample Problem 10.6

Using the interaction method, solve Sample Prob. 10.5. Assume $(\sigma_{all})_{bending} = 150$ MPa.

STRATEGY: Use the allowable centric stress found in Sample Problem 10.5 to calculate the load **P**.

MODELING and ANALYSIS:

Using Eq. (10.55),

$$\frac{P/A}{(\sigma_{all})_{centric}} + \frac{Mc/I}{(\sigma_{all})_{bending}} \leq 1$$

Substituting the allowable bending stress and the allowable centric stress found in Sample Prob. 10.5, as well as the other given data, we obtain

$$\frac{P/(9.42 \times 10^{-3}\,\text{m}^2)}{97.1 \times 10^6\,\text{Pa}} + \frac{P(0.200\,\text{m})/(1.050 \times 10^{-3}\,\text{m}^3)}{150 \times 10^6\,\text{Pa}} \leq 1$$

$$P \leq 423\,\text{kN}$$

The largest allowable load **P** is   **P = 423 kN** ↓ ◀

Sample Problem 10.7

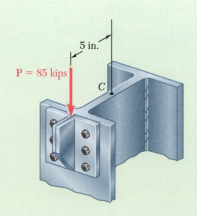

A steel column with an effective length of 16 ft is loaded eccentrically as shown. Using the interaction method, select the wide-flange shape of 8-in. nominal depth that should be used. Assume $E = 29 \times 10^6$ psi and $\sigma_Y = 36$ ksi, and use an allowable stress in bending of 22 ksi.

STRATEGY: It is necessary to select the lightest column that satisfies Eq. (10.55). This involves a trial-and-error process, which can be shortened if the first 8-in. wide-flange shape selected is close to the final solution. This is done by using the allowable-stress method, Eq. (10.53), with an approximate allowable stress.

MODELING and ANALYSIS:

So that we can select a trial section, we use the allowable-stress method with $\sigma_{all} = 22$ ksi and write

$$\sigma_{all} = \frac{P}{A} + \frac{Mc}{I_x} = \frac{P}{A} + \frac{Mc}{Ar_x^2} \tag{1}$$

(continued)

P = 85 kips
z
y
5 in.
C
x
=
z
P = 85 kips
y
C
x
M = (85 kips)(5 in.)
= 425 kip · in.

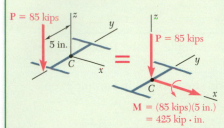

Fig. 1 Eccentric loading replaced by equivalent force-couple at column's centroid.

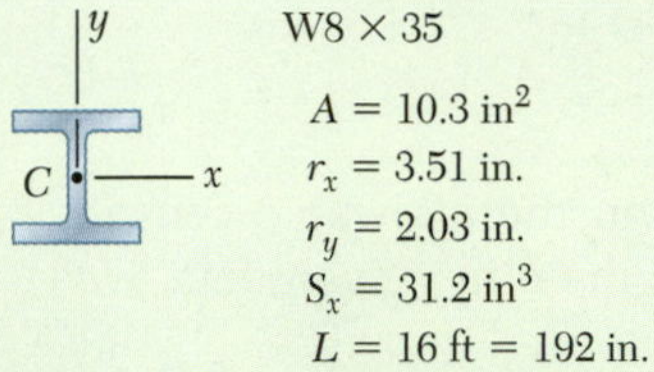

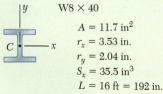

y
C
x
W8 × 35
A = 10.3 in²
r_x = 3.51 in.
r_y = 2.03 in.
S_x = 31.2 in³
L = 16 ft = 192 in.

Fig. 2 Section properties for W8 × 35.

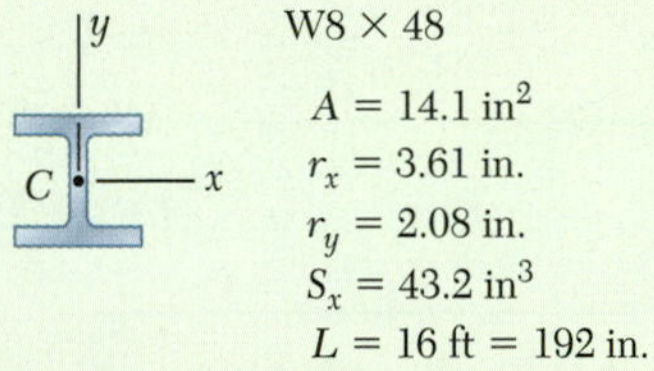

y
C
x
W8 × 48
A = 14.1 in²
r_x = 3.61 in.
r_y = 2.08 in.
S_x = 43.2 in³
L = 16 ft = 192 in.

Fig. 3 Section properties for W8 × 48.

y
C
x
W8 × 40
A = 11.7 in²
r_x = 3.53 in.
r_y = 2.04 in.
S_x = 35.5 in³
L = 16 ft = 192 in.

Fig. 4 Section properties for W8 × 40.

From Appendix C, we observe that for shapes of 8-in. nominal depth $c \approx 4$ in. and $r_x \approx 3.5$ in. Using Fig. 1 and substituting into Eq. (1),

$$22 \text{ ksi} = \frac{85 \text{ kips}}{A} + \frac{(425 \text{ kip·in.})(4 \text{ in.})}{A(3.5 \text{ in.})^2} \qquad A \approx 10.2 \text{ in}^2$$

For a first trial shape, select W8 × 35.

Trial 1: W8 × 35 (Fig. 2). The allowable stresses are

Allowable Bending Stress: (see data) $(\sigma_{\text{all}})_{\text{bending}} = 22$ ksi

Allowable Concentric Stress: The largest slenderness ratio of the column is $L/r_y = (192 \text{ in.})/(2.03 \text{ in.}) = 94.6$. Using Eq. (10.41) with $E = 29 \times 10^6$ psi and $\sigma_Y = 36$ ksi, the slenderness ratio at the junction between the two equations for σ_{cr} is $L/r = 133.7$. Thus, use Eqs. (10.38) and (10.39) and find $\sigma_{cr} = 22.5$ ksi. Using Eq. (10.42), the allowable stress is

$$(\sigma_{\text{all}})_{\text{centric}} = 22.5/1.67 = 13.46 \text{ ksi}$$

For the W8 × 35 trial shape,

$$\frac{P}{A} = \frac{85 \text{ kips}}{10.3 \text{ in}^2} = 8.25 \text{ ksi} \qquad \frac{Mc}{I} = \frac{M}{S_x} = \frac{425 \text{ kip·in.}}{31.2 \text{ in}^3} = 13.62 \text{ ksi}$$

With this data, the left-hand member of Eq. (10.55) is

$$\frac{P/A}{(\sigma_{\text{all}})_{\text{centric}}} + \frac{Mc/I}{(\sigma_{\text{all}})_{\text{bending}}} = \frac{8.25 \text{ ksi}}{13.46 \text{ ksi}} + \frac{13.62 \text{ ksi}}{22 \text{ ksi}} = 1.232$$

Since $1.232 > 1.000$, the requirement expressed by the interaction formula is not satisfied. Select a larger trial shape.

Trial 2: W8 × 48 (Fig.3). Following the procedure used in trial 1 gives

$$\frac{L}{r_y} = \frac{192 \text{ in.}}{2.08 \text{ in.}} = 92.3 \qquad (\sigma_{\text{all}})_{\text{centric}} = 13.76 \text{ ksi}$$

$$\frac{P}{A} = \frac{85 \text{ kips}}{14.1 \text{ in}^2} = 6.03 \text{ ksi} \qquad \frac{Mc}{I} = \frac{M}{S_x} = \frac{425 \text{ kip·in.}}{43.2 \text{ in}^3} = 9.84 \text{ ksi}$$

Substituting into Eq. (10.55) gives

$$\frac{P/A}{(\sigma_{\text{all}})_{\text{centric}}} + \frac{Mc/I}{(\sigma_{\text{all}})_{\text{bending}}} = \frac{6.03 \text{ ksi}}{13.76 \text{ ksi}} + \frac{9.82 \text{ ksi}}{22 \text{ ksi}} = 0.885 < 1.000$$

The W8 × 48 shape is satisfactory but may be unnecessarily large.

Trial 3: W8 × 40 (Fig.4). Following the same procedure, the interaction formula is not satisfied.

Selection of Shape. The shape to be used is W8 × 48 ◄

Problems

10.89 An eccentric load is applied at a point 22 mm from the geometric axis of a 60-mm-diameter rod made of a steel for which $\sigma_Y = 250$ MPa and $E = 200$ GPa. Using the allowable-stress method, determine the allowable load **P**.

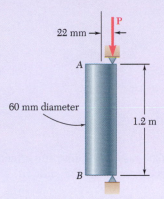

Fig. P10.89

10.90 Solve Prob. 10.89, assuming that the load is applied at a point 40 mm from the geometric axis and that the effective length is 0.9 m.

10.91 A sawn-lumber column of 5.0×7.5-in. cross section has an effective length of 8.5 ft. The grade of wood used has an adjusted allowable stress for compression parallel to the grain $\sigma_C = 1180$ psi and an adjusted modulus $E = 440 \times 10^3$ psi. Using the allowable-stress method, determine the largest eccentric load **P** that can be applied when (*a*) $e = 0.5$ in., (*b*) $e = 1.0$ in.

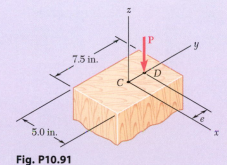

Fig. P10.91

10.92 Solve Prob. 10.91 using the interaction method and an allowable stress in bending of 1300 psi.

10.93 A column of 5.5-m effective length is made of the aluminum alloy 2014-T6 for which the allowable stress in bending is 220 MPa. Using the interaction method, determine the allowable load **P**, knowing that the eccentricity is (*a*) $e = 0$, (*b*) $e = 40$ mm.

10.94 Solve Prob. 10.93, assuming that the effective length of the column is 3.0 m.

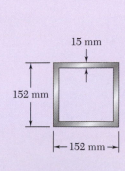

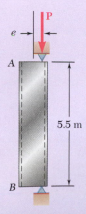

Fig. P10.93

10.95 A steel compression member of 9-ft effective length supports an eccentric load as shown. Using the allowable-stress method, determine the maximum allowable eccentricity e if (*a*) $P = 30$ kips, (*b*) $P = 18$ kips. Use $\sigma_Y = 36$ ksi and $E = 29 \times 10^6$ psi.

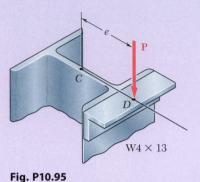

Fig. P10.95

10.96 Solve Prob. 10.95, assuming that the effective length of the column is increased to 12 ft and that (*a*) $P = 20$ kips, (*b*) $P = 15$ kips.

10.97 Two L4 $\times$ 3 $\times$ $\frac{3}{8}$-in. steel angles are welded together to form the column AB. An axial load **P** of magnitude 14 kips is applied at point D. Using the allowable-stress method, determine the largest allowable length L. Assume $\sigma_Y = 36$ ksi and $E = 29 \times 10^6$ psi.

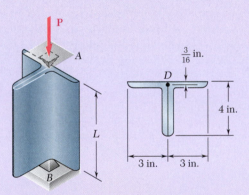

Fig. P10.97

10.98 Solve Prob. 10.97 using the interaction method with $P = 18$ kips and an allowable stress in bending of 22 ksi.

10.99 A rectangular column is made of a grade of sawn wood that has an adjusted allowable stress for compression parallel to the grain $\sigma_C = 8.3$ MPa and an adjusted modulus of elasticity $E = 11.1$ GPa. Using the allowable-stress method, determine the largest allowable effective length L that can be used.

10.100 Solve Prob. 10.99, assuming that $P = 105$ kN.

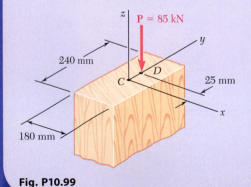

Fig. P10.99

10.101 An eccentric load $P = 48$ kN is applied at a point 20 mm from the geometric axis of a 50-mm-diameter rod made of the aluminum alloy 6061-T6. Using the interaction method and an allowable stress in bending of 145 MPa, determine the largest allowable effective length L that can be used.

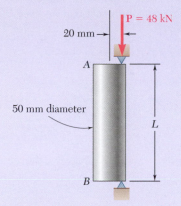

P = 48 kN

20 mm

A

50 mm diameter

L

B

Fig. P10.101

10.102 Solve Prob. 10.101, assuming that the aluminum alloy used is 2014-T6 and that the allowable stress in bending is 180 MPa.

10.103 A compression member made of steel has a 720-mm effective length and must support the 198-kN load **P** as shown. For the material used $\sigma_Y = 250$ MPa and $E = 200$ GPa. Using the interaction method with an allowable bending stress equal to 150 MPa, determine the smallest dimension d of the cross section that can be used.

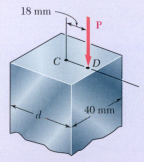

18 mm

P

C   D

d   40 mm

Fig. P10.103

10.104 Solve Prob. 10.103, assuming that the effective length is 1.62 m and that the magnitude of P of the eccentric load is 128 kN.

10.105 A steel tube of 80-mm outer diameter is to carry a 93-kN load **P** with an eccentricity of 20 mm. The tubes available for use are made with wall thicknesses in increments of 3 mm from 6 mm to 15 mm. Using the allowable-stress method, determine the lightest tube that can be used. Assume $E = 200$ GPa and $\sigma_Y = 250$ MPa.

10.106 Solve Prob. 10.105, using the interaction method with $P = 165$ kN, $e = 15$ mm, and an allowable stress in bending of 150 MPa.

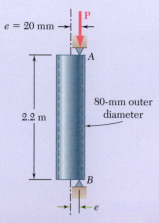

e = 20 mm

P

A

2.2 m

80-mm outer diameter

B

e

Fig. P10.105

10.107 A sawn lumber column of rectangular cross section has a 2.2-m effective length and supports a 41-kN load as shown. The sizes available for use have b equal to 90 mm, 140 mm, 190 mm, and 240 mm. The grade of wood has an adjusted allowable stress for compression parallel to the grain $\sigma_C = 8.1$ MPa and an adjusted modulus $E = 8.3$ GPa. Using the allowable-stress method, determine the lightest section that can be used.

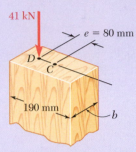

Fig. P10.107

10.108 Solve Prob. 10.107, assuming that $e = 40$ mm.

10.109 A compression member of rectangular cross section has an effective length of 36 in. and is made of the aluminum alloy 2014-T6 for which the allowable stress in bending is 24 ksi. Using the interaction method, determine the smallest dimension d of the cross section that can be used when $e = 0.4$ in.

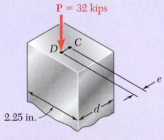

Fig. P10.109

10.110 Solve Prob. 10.109, assuming that $e = 0.2$ in.

10.111 An aluminum tube of 3-in. outside diameter is to carry a load of 10 kips having an eccentricity $e = 0.6$ in. Knowing that the stock of tubes available for use are made of alloy 2014-T6 and have wall thicknesses in increments of $\frac{1}{16}$ in. up to $\frac{1}{2}$ in., determine the lightest tube that can be used. Use the allowable-stress method.

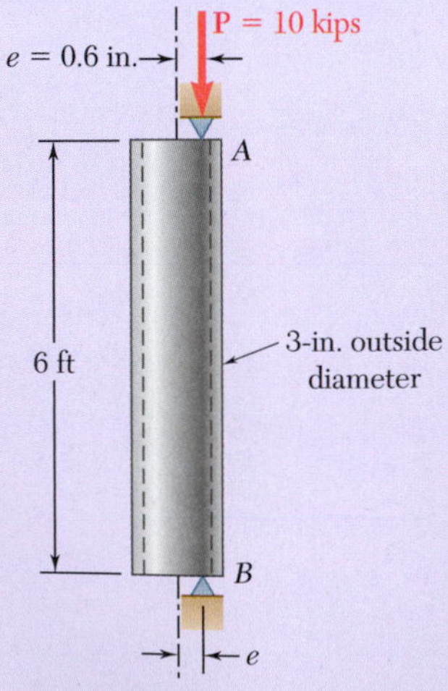

10.112 Solve Prob. 10.111, using the interaction method of design with an allowable stress in bending of 25 ksi.

Fig. P10.111

10.113 A steel column having a 24-ft effective length is loaded eccentrically as shown. Using the allowable-stress method, select the wide-flange shape of 14-in. nominal depth that should be used. Use $\sigma_Y = 36$ ksi and $E = 29 \times 10^6$ psi.

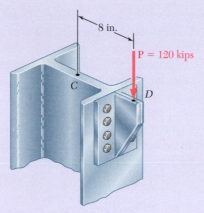

8 in.

P = 120 kips

C

D

Fig. P10.113

10.114 Solve Prob. 10.113 using the interaction method, assuming that $\sigma_Y = 50$ ksi and the allowable stress in bending is 30 ksi.

10.115 A steel compression member of 5.8-m effective length is to support a 296-kN eccentric load **P**. Using the interaction method, select the wide-flange shape of 200-mm nominal depth that should be used. Use $E = 200$ GPa, $\sigma_Y = 250$ MPa, and $\sigma_{all} = 150$ MPa in bending.

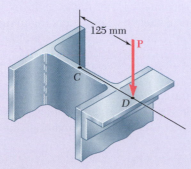

125 mm

P

C

D

Fig. P10.115

10.116 A steel column of 7.2-m effective length is to support an 83-kN eccentric load **P** at a point D, located on the x axis as shown. Using the allowable-stress method, select the wide-flange shape of 250-mm nominal depth that should be used. Use $E = 200$ GPa and $\sigma_Y = 250$ MPa.

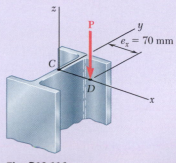

z

P

y

$e_x = 70$ mm

C

D

x

Fig. P10.116

Review and Summary

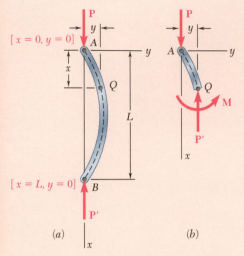

Fig. 10.38 Free-body diagrams of (a) buckled column and (b) portion AQ.

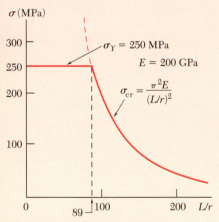

Fig. 10.39 Plot of critical stress.

Critical Load

The design and analysis of columns (i.e., prismatic members supporting axial loads), is based on the determination of the *critical load*. Two equilibrium positions of the column model are possible: the original position with zero transverse deflections and a second position involving deflections that could be quite large. The first equilibrium position is unstable for $P > P_{cr}$ and stable for $P < P_{cr}$, since in the latter case it was the only possible equilibrium position.

We considered a pin-ended column of length L and constant flexural rigidity EI subjected to an axial centric load P. Assuming that the column buckled (Fig. 10.38), the bending moment at point Q is equal to $-Py$. Thus,

$$\frac{d^2y}{dx^2} = \frac{M}{EI} = -\frac{P}{EI}y \tag{10.4}$$

Euler's Formula

Solving this differential equation, subject to the boundary conditions corresponding to a pin-ended column, we determined the smallest load P for which buckling can take place. This load, known as the *critical load* and denoted by P_{cr}, is given by *Euler's formula*:

$$P_{cr} = \frac{\pi^2 EI}{L^2} \tag{10.11a}$$

where L is the length of the column. For this or any larger load, the equilibrium of the column is unstable, and transverse deflections will occur.

Slenderness Ratio

Denoting the cross-sectional area of the column by A and its radius of gyration by r, the critical stress σ_{cr} corresponding to the critical load P_{cr} is

$$\sigma_{cr} = \frac{\pi^2 E}{(L/r)^2} \tag{10.13a}$$

The quantity L/r is the *slenderness ratio*. The critical stress σ_{cr} is plotted as a function of L/r in Fig. 10.39. Since the analysis was based on stresses remaining below the yield strength of the material, the column will fail by yielding when $\sigma_{cr} > \sigma_Y$.

Effective Length

The critical load of columns with various end conditions is written as

$$P_{cr} = \frac{\pi^2 EI}{L_e^2} \tag{10.11b}$$

where L_e is the *effective length* of the column, i.e., the length of an equivalent pin-ended column. The effective lengths of several columns with various end conditions were calculated and shown in Fig. 10.18 on page 700.

Eccentric Axial Load

For a pin-ended column subjected to a load P applied with an eccentricity e, the load can be replaced with a centric axial load and a couple of moment $M_A = Pe$ (Fig. 10.40). The maximum transverse deflection is

$$y_{max} = e\left[\sec\left(\sqrt{\frac{P}{EI}}\frac{L}{2}\right) - 1\right] \qquad (10.28)$$

Secant Formula

The maximum stress in a column supporting an eccentric axial load can be found using the *secant formula*:

$$\frac{P}{A} = \frac{\sigma_{max}}{1 + \dfrac{ec}{r^2}\sec\left(\dfrac{1}{2}\sqrt{\dfrac{P}{EA}}\dfrac{L_e}{r}\right)} \qquad (10.36)$$

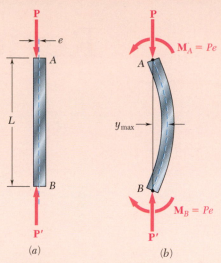

Fig. 10.40 (a) Column with an eccentric load (b) modeled as a column with an equivalent centric force-couple load.

This equation can be solved for the force per unit area P/A, which causes a specified maximum stress σ_{max} in a pin-ended or other column of effective slenderness ratio L_e/r.

Design of Real Columns

Since imperfections exist in all columns, the *design of real columns* is done with empirical formulas based on laboratory tests, set forth in specifications and codes issued by professional organizations. For *centrically loaded columns* made of steel, aluminum, or wood, design is based on equations for the allowable stress as a function of the slenderness ratio L/r. For structural steel, the *Load and Resistance Factor Design* method also can be used.

Design of Eccentrically Loaded Columns

Two methods can be used for the design of columns under an *eccentric* load. The first method is the *allowable-stress method*. This conservative method assumes that the allowable stress is the same as if the column were centrically loaded. The allowable-stress method requires that the following inequality to be satisfied:

$$\frac{P}{A} + \frac{Mc}{I} \leq \sigma_{all} \qquad (10.53)$$

The second method is the *interaction method,* which is the basis of most modern specifications. In this method, the allowable stress for a centrically loaded column is used for the portion of the total stress due to the axial load, and the allowable stress in bending is used for the stress due to bending. Thus, the inequality to be satisfied is

$$\frac{P/A}{(\sigma_{all})_{centric}} + \frac{Mc/I}{(\sigma_{all})_{bending}} \leq 1 \qquad (10.55)$$

Review Problems

10.117 Determine (*a*) the critical load for the steel strut, (*b*) the dimension *d* for which the aluminum strut will have the same critical load. (*c*) Express the weight of the aluminum strut as a percent of the weight of the steel strut.

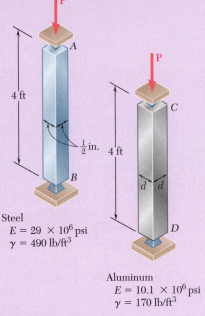

4 ft

$\frac{1}{2}$ in.

4 ft

d *d*

Steel
$E = 29 \times 10^6$ psi
$\gamma = 490$ lb/ft^3

Aluminum
$E = 10.1 \times 10^6$ psi
$\gamma = 170$ lb/ft^3

Fig. P10.117

10.118 The rigid rod *AB* is attached to a hinge at *A* and to two springs, each of constant *k*. If *h* = 450 mm, *d* = 300 mm, and *m* = 200 kg, determine the range of values of *k* for which the equilibrium of rod *AB* is stable in the position shown. Each spring can act in either tension or compression.

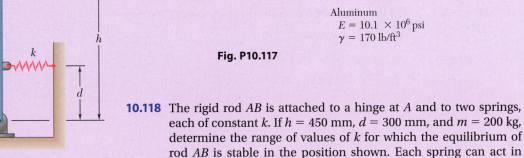

Fig. P10.118

10.119 A column of 3-m effective length is to be made by welding together two C130 × 13 rolled-steel channels. Using *E* = 200 GPa, determine for each arrangement shown the allowable centric load if a factor of safety of 2.4 is required.

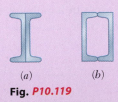

(*a*)   (*b*)

Fig. *P10.119*

10.120 (*a*) Considering only buckling in the plane of the structure shown and using Euler's formula, determine the value of θ between 0 and 90° for which the allowable magnitude of the load **P** is maximum. (*b*) Determine the corresponding maximum value of *P* knowing that a factor of safety of 3.2 is required. Use $E = 29 \times 10^6$ psi.

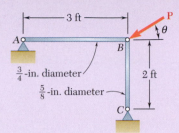

Fig. P10.120

10.121 Member *AB* consists of a single C130 × 10.4 steel channel of length 2.5 m. Knowing that the pins *A* and *B* pass through the centroid of the cross section of the channel, determine the factor of safety for the load shown with respect to buckling in the plane of the figure when $\theta = 30°$. Use $E = 200$ GPa.

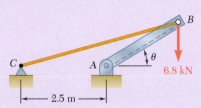

Fig. P10.121

10.122 The line of action of the 75-kip axial load is parallel to the geometric axis of the column *AB* and intersects the *x* axis at $x = 0.6$ in. Using $E = 29 \times 10^6$ psi, determine (*a*) the horizontal deflection of the midpoint *C* of the column, (*b*) the maximum stress in the column.

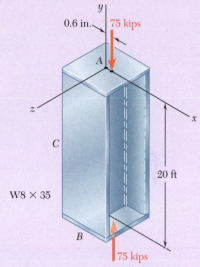

Fig. P10.122

10.123 Supports *A* and *B* of the pin-ended column shown are at a fixed distance *L* from each other. Knowing that at a temperature T_0 the force in the column is zero and that buckling occurs when the temperature is $T_1 = T_0 + \Delta T$, express ΔT in terms of *b*, *L* and the coefficient of thermal expansion α.

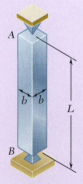

Fig. P10.123

10.124 A column is made from half of a W360 × 216 rolled-steel shape, with the geometric properties as shown. Using allowable stress design, determine the allowable centric load if the effective length of the column is (*a*) 4.0 m, (*b*) 6.5 m. Use $\sigma_Y = 345$ MPa and $E = 200$ GPa.

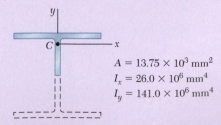

$A = 13.75 \times 10^3 \text{ mm}^2$
$I_x = 26.0 \times 10^6 \text{ mm}^4$
$I_y = 141.0 \times 10^6 \text{ mm}^4$

Fig. P10.124

10.125 A rectangular column with a 4.4-m effective length is made of glued laminated wood. Knowing that for the grade of wood used the adjusted allowable-stress for compression parallel to the grain is $\sigma_C = 8.3$ MPa and the adjusted modulus $E = 4.6$ GPa, determine the maximum allowable centric load for the column.

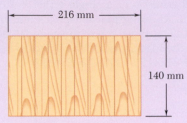

Fig. P10.125

10.126 A column of 4.5-m effective length must carry a centric load of 900 kN. Knowing that $\sigma_Y = 345$ MPa and $E = 200$ GPa, use allowable-stress design to select the wide-flange shape of 250-mm nominal depth that should be used.

10.127 An 11-kip vertical load **P** is applied at the midpoint of one edge of the square cross section of the steel compression member AB, which is free at its top A and fixed at its base B. Knowing that for the grade of steel used $\sigma_Y = 36$ ksi and $E = 29 \times 10^6$ psi and using the allowable-stress method, determine the smallest allowable dimension d.

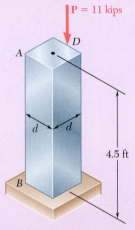

Fig. P10.127

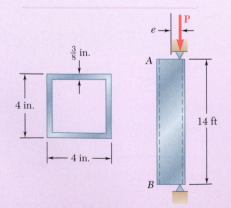

Fig. P10.128

10.128 A column of 14-ft effective length consists of a section of steel tubing having the cross section shown. Using the allowable-stress method, determine the maximum allowable eccentricity e if (*a*) $P = 55$ kips, (*b*) $P = 35$ kips. Use $\sigma_Y = 36$ ksi and $E = 29 \times 10^6$ psi.

Computer Problems

The following problems are designed to be solved with a computer.

10.C1 A solid steel rod having an effective length of 500 mm is to be used as a compression strut to carry a centric load **P**. For the grade of steel used, $E = 200$ GPa and $\sigma_Y = 245$ MPa. Knowing that a factor of safety of 2.8 is required and using Euler's formula, write a computer program and use it to calculate the allowable centric load P_{all} for values of the radius of the rod from 6 mm to 24 mm, using 2-mm increments.

10.C2 An aluminum bar is fixed at end A and supported at end B so that it is free to rotate about a horizontal axis through the pin. Rotation about a vertical axis at end B is prevented by the brackets. Knowing that $E = 10.1 \times 10^6$ psi, use Euler's formula with a factor of safety of 2.5 to determine the allowable centric load **P** for values of b from 0.75 in. to 1.5 in., using 0.125-in. increments.

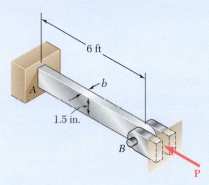

Fig. C10.C2

10.C3 The pin-ended members AB and BC consist of sections of aluminum pipe of 120-mm outer diameter and 10-mm wall thickness. Knowing that a factor of safety of 3.5 is required, determine the mass m of the largest block that can be supported by the cable arrangement shown for values of h from 4 m to 8 m, using 0.25-m increments. Use $E = 70$ GPa and consider only buckling in the plane of the structure.

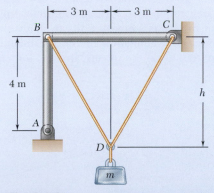

Fig. C10.C3

10.C4 An axial load **P** is applied at a point located on the x axis at a distance $e = 0.5$ in. from the geometric axis of the W8 × 40 rolled-steel column AB. Using $E = 29 \times 10^6$ psi, write a computer program and use it to calculate for values of P from 25 to 75 kips, using 5-kip increments, (a) the horizontal deflection at the midpoint C, (b) the maximum stress in the column.

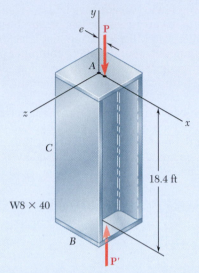

Fig. C10.C4

10.C5 A column of effective length L is made from a rolled-steel shape and carries a centric axial load **P**. The yield strength for the grade of steel used is denoted by σ_Y, the modulus of elasticity by E, the cross-sectional area of the selected shape by A, and its smallest radius of gyration by r. Using the AISC design formulas for allowable stress design, write a computer program that can be used with either SI or U.S. customary units to determine the allowable load **P**. Use this program to solve (a) Prob. 10.57, (b) Prob. 10.58, (c) Prob. 10.124.

10.C6 A column of effective length L is made from a rolled-steel shape and is loaded eccentrically as shown. The yield strength of the grade of steel used is denoted by σ_Y, the allowable stress in bending by σ_{all}, the modulus of elasticity by E, the cross-sectional area of the selected shape by A, and its smallest radius of gyration by r. Write a computer program that can be used with either SI or U.S. customary units to determine the allowable load **P**, using either the allowable-stress method or the interaction method. Use this program to check the given answer for (a) Prob. 10.113, (b) Prob. 10.114.

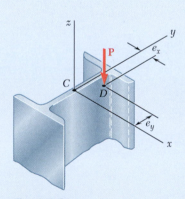

Fig. C10.C6

11

Energy Methods

When the diver jumps on the board, the potential energy due to his elevation above the board is converted into strain energy as the board bends.

Learning Objectives

In this chapter, you will:

- **Compute** the strain energy due to axial, bending, and torsion loading
- **Determine** the effect of impact loading on members
- **Define** the work done by a force or couple
- **Determine** displacements from a single load using the work-energy method
- **Apply** Castigliano's theorem to determine displacements due to multiple loads
- **Solve** statically indeterminate problems using Castigliano's theorem

Introduction

In the previous chapter we were concerned with the relations existing between forces and deformations under various loading conditions. Our analysis was based on two fundamental concepts, the concept of stress (Chap. 1) and the concept of strain (Chap. 2). A third important concept, the concept of *strain energy*, will now be introduced.

The *strain energy* of a member is the increase in energy associated with the deformation of the member. The strain energy is equal to the work done by a slowly increasing load applied to the member. The *strain-energy density* of a material is the strain energy per unit volume; this it is equal to the area under the stress-strain diagram of the material (Sec. 11.1B). From the stress-strain diagram of a material, two additional properties are the *modulus of toughness* and the *modulus of resilience* of the material.

In Sec. 11.2 the elastic strain energy associated with *normal stresses* is discussed for members under axial loading and in bending. The elastic strain energy associated with shearing stresses (such as in torsional loadings of shafts and in transversely loaded beams) is discussed. Strain energy for a *general state of stress* is considered in Sec. 11.3, where the *maximum-distortion-energy criterion* for yielding is derived.

The effect of *impact loading* on members is considered in Sec. 11.4. The *maximum stress* and the *maximum deflection* caused by a moving mass impacting a member are calculated. Properties that increase the ability of a structure to withstand impact loads effectively are discussed in Sec. 11.4B.

In Sec. 11.5A the elastic strain of a member subjected to a *single concentrated load* is calculated, and in Sec. 11.5B the deflection at the point of application of a single load is determined.

The last portion of the chapter is devoted to the strain energy of structures subjected to *multiple loads* (Sec. 11.6). *Castigliano's theorem* is derived (Sec. 11.7) and used (Sec. 11.8) to determine the deflection at a given point of a structure subjected to several loads. Indeterminate structures are analyzed using Castigliano's theorem (Sec. 11.9).

11.1   STRAIN ENERGY

11.1A   Strain Energy Concepts

Consider a rod BC with a length of L and uniform cross-sectional area A that is attached at B to a fixed support and subjected at C to a *slowly increasing* axial load $\mathbf{P}$ (Fig. 11.1). By plotting the magnitude P of the load against the deformation x of the rod (Sec. 2.1), a load-deformation diagram is obtained (Fig. 11.2) that is characteristic of rod BC.

Fig. 11.2   Load-deformation diagram for axial loading.

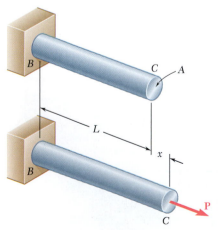

Fig. 11.1   Axially loaded rod

Now consider the work dU done by the load $\mathbf{P}$ as the rod elongates by a small amount dx. This *elementary work* is equal to the product of the magnitude P of the load and of the small elongation dx

$$dU = P\,dx \qquad (11.1)$$

Note that this equation is equal to the element with an area of width dx located under the load-deformation diagram (Fig. 11.3). The *total work* U done by the load as the rod undergoes a deformation x_1 is

$$U = \int_0^{x_1} P\,dx$$

and is equal to the area under the load-deformation diagram between $x = 0$ and $x = x_1$.

The work done by load $\mathbf{P}$ as it is slowly applied to the rod results in the increase of energy associated with the deformation of the rod. This energy is the *strain energy* of the rod.

$$\text{Strain energy} = U = \int_0^{x_1} P\,dx \qquad (11.2)$$

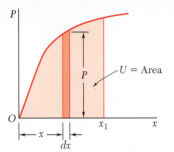

Fig. 11.3   Work due to load $\mathbf{P}$ is equal to the area under the load-deformation diagram.

Recall that work and energy should be obtained by multiplying units of length by units of force. So, when SI units are used, work and energy are expressed in N·m, which is called a *joule* (J). When U.S. customary units are used, work and energy are in ft·lb or in·lb.

For a linear elastic deformation, the portion of the load-deformation diagram involved can be represented by a straight line with equation $P = kx$ (Fig. 11.4). Substituting for P in Eq. (11.2) gives

$$U = \int_0^{x_1} kx\,dx = \tfrac{1}{2}kx_1^2$$

or

$$U = \tfrac{1}{2}P_1 x_1 \qquad (11.3)$$

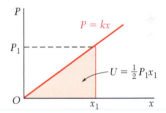

Fig. 11.4   Work due to linear elastic deformation.

where P_1 is the load corresponding to the deformation x_1.

Strain energy can be used to determine the effects of impact loadings on structures or machine components. For example, a body of mass m moving with a velocity $\mathbf{v}_0$ strikes the end B of a rod AB (Fig. 11.5a). Neglecting the inertia of the rod and assuming no dissipation of energy during the impact, the maximum strain energy U_m acquired by the rod (Fig. 11.5b) is equal to the original kinetic energy $T = \tfrac{1}{2}mv_0^2$ of the moving body. If we then determine P_m of the static load (which would have produced the same strain energy in the rod), we can obtain σ_m of the largest stress occurring in the rod by dividing P_m by the cross-sectional area.

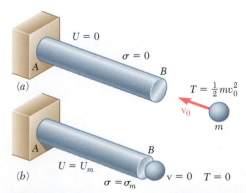

Fig. 11.5   Rod subject to impact loading.

11.1B Strain-Energy Density

As noted in Sec. 2.1, the load-deformation diagram for a rod BC depends upon the length L and the cross-sectional area A. The strain energy U defined by Eq. (11.2), therefore, will also depend on the dimensions of the rod. To eliminate the effect of size and direct attention to the properties of the material, the strain energy per unit volume will be considered. Dividing the strain energy U by the volume $V = AL$ of the rod (Fig. 11.1) and using Eq. (11.2) gives

$$\frac{U}{V} = \int_0^{x_1} \frac{P}{A} \frac{dx}{L}$$

Recalling that P/A represents the normal stress σ_x and x/L the normal strain ϵ_x,

$$\frac{U}{V} = \int_0^{\epsilon_1} \sigma_x \, d\epsilon_x$$

where ϵ_1 denotes the value of the strain corresponding to the elongation x_1. The strain energy per unit volume U/V is called the *strain-energy density*, denoted by u. Therefore,

> **Strain-energy density** = $u = \displaystyle\int_0^{\epsilon_1} \sigma_x \, d\epsilon_x$ **(11.4)**

The strain-energy density u is expressed in units of energy divided by units of volume. Thus, when SI metric units are used, the strain-energy density is in J/m^3 or its multiples kJ/m^3 and MJ/m^3. When U.S. customary units are used, they are in $in\cdot lb/in^3$.[†]

Referring to Fig. 11.6, the strain-energy density u is equal to the area under the stress-strain curve, which is measured from $\epsilon_x = 0$ to $\epsilon_x = \epsilon_1$. If the material is unloaded, the stress returns to zero. However, there is a permanent deformation represented by the strain ϵ_p, and only the part of the strain energy per unit volume corresponding to the triangular area is recovered. The remainder of the energy spent deforming the material is dissipated in the form of heat.

The strain-energy density obtained by setting $\epsilon_1 = \epsilon_R$ in Eq. (11.4), where ϵ_R is the strain at rupture, is called the *modulus of toughness* of the material. It is equal to the area under the entire stress-strain diagram (Fig. 11.7) and represents the energy per unit volume required for the material to rupture. The toughness of a material is related to its ductility, as well as its ultimate strength (Sec. 2.1B), and the capacity of a structure

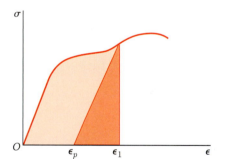

Fig. 11.6 Strain-energy density is the area under the stress-strain curve between $\epsilon_x = 0$ and $\epsilon_x = \epsilon_1$. If loaded into the plastic region, only the energy associated with elastic unloading is recovered.

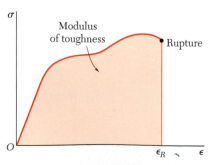

Fig. 11.7 Modulus of toughness is the area under the stress-strain curve to rupture.

[†]Note that $1\ J/m^3$ and $1\ Pa$ are both equal to $1\ N/m^2$, while $1\ in.\cdot lb/in^3$ and 1 psi are both equal to $1\ lb/in^2$. Thus, strain-energy density and stress are dimensionally equal and can be expressed in the same units.

to withstand an impact load depends upon the toughness of the material used (Photo 11.1).

If the stress σ_x remains within the proportional limit of the material, Hooke's law applies:

$$\sigma_x = E\epsilon_x \tag{11.5}$$

Substituting for σ_x from Eq. (11.5) into Eq. (11.4) gives

$$u = \int_0^{\epsilon_1} E\epsilon_x \, d\epsilon_x = \frac{E\epsilon_1^2}{2} \tag{11.6}$$

or using Eq. (11.5) to express ϵ_1 in terms of the corresponding stress σ_1 gives

$$u = \frac{\sigma_1^2}{2E} \tag{11.7}$$

Photo 11.1 The railroad coupler is made of a ductile steel that has a large modulus of toughness.

The strain-energy density obtained by setting $\sigma_1 = \sigma_Y$ in Eq. (11.7), where σ_Y is the yield strength, is called the *modulus of resilience* of the material, and is denoted by u_Y. So,

$$u_Y = \frac{\sigma_Y^2}{2E} \tag{11.8}$$

The modulus of resilience is equal to the area under the straight-line portion OY of the stress-strain diagram (Fig. 11.8) and represents the energy per unit volume that the material can absorb without yielding. A structure's ability to withstand an impact load without being permanently deformed depends on the resilience of the material used.

Since the modulus of toughness and the modulus of resilience represent characteristic values of the strain-energy density of the material considered, they are both expressed in J/m^3 or its multiples if SI units are used, and in in·lb/in^3 if U.S. customary units are used.[†]

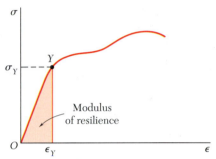

Fig. 11.8 Modulus of resilience is the area under the stress-strain curve to yield.

11.2 ELASTIC STRAIN ENERGY

11.2A Normal Stresses

Since the rod considered in the preceding section was subjected to uniformly distributed stresses σ_x, the strain-energy density was constant throughout the rod and could be defined as the ratio U/V of the strain energy U and the volume V of the rod. In a structural element or machine part with a nonuniform stress distribution, the strain-energy density u can be defined by considering the strain energy of a small element of material of volume ΔV. So,

$$u = \lim_{\Delta V \to 0} \frac{\Delta U}{\Delta V}$$

or

$$u = \frac{dU}{dV} \tag{11.9}$$

[†]Note that the modulus of toughness and the modulus of resilience could be expressed in the same units as stress.

The expression previously obtained for u (Sec. 11.1B) in terms of σ_x and ϵ_x is valid, so

$$u = \int_0^{\epsilon_x} \sigma_x \, d\epsilon_x \qquad (11.10)$$

which allows for variations in the stress σ_x, the strain ϵ_x, and the strain-energy density u from point to point.

For values of σ_x within the proportional limit, $\sigma_x = E\epsilon_x$ in Eq. (11.10) and

$$u = \frac{1}{2}E\epsilon_x^2 = \frac{1}{2}\sigma_x\epsilon_x = \frac{1}{2}\frac{\sigma_x^2}{E} \qquad (11.11)$$

The strain energy U of a body subjected to uniaxial normal stresses can be obtained by substituting for u from Eq. (11.11) into Eq. (11.9) and integrating both members.

$$U = \int \frac{\sigma_x^2}{2E} \, dV \qquad (11.12)$$

This equation is valid only for elastic deformations and is called the *elastic strain energy* of the body.

Strain Energy under Axial Loading.   Recall from Sec. 2.10 that when a rod is subjected to a centric axial load, it can be assumed that the normal stresses σ_x are uniformly distributed in any given cross section. Using the area of the section A located at a distance x from end B of the rod (Fig. 11.9) and the internal force P in that section, we write $\sigma_x = P/A$. Substituting for σ_x into Eq. (11.12) gives

$$U = \int \frac{P^2}{2EA^2} \, dV$$

or setting $dV = A \, dx$,

$$U = \int_0^L \frac{P^2}{2AE} \, dx \qquad (11.13)$$

For a rod of uniform cross section with equal and opposite forces of magnitude P at its ends (Fig. 11.10), Eq. (11.13) yields

$$U = \frac{P^2L}{2AE} \qquad (11.14)$$

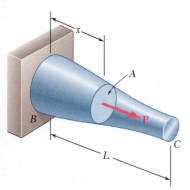

Fig. 11.9   Rod with centric axial load.

Fig. 11.10   Prismatic rod with centric axial load.

Concept Application 11.1

A rod consists of two portions BC and CD of the same material, same length, but of different cross sections (Fig. 11.11). Determine the strain energy of the rod when it is subjected to a centric axial load **P**, expressing the result in terms of P, L, E, the cross-sectional area A of portion CD, and the ratio n of the two diameters.

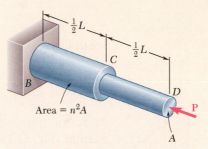

Fig. 11.11 Axially loaded stepped rod.

Use Eq. (11.14) for the strain energy of each of the two portions, and add the expressions obtained:

$$U_n = \frac{P^2(\frac{1}{2}L)}{2AE} + \frac{P^2(\frac{1}{2}L)}{2(n^2A)E} = \frac{P^2L}{4AE}\left(1 + \frac{1}{n^2}\right)$$

or

$$U_n = \frac{1 + n^2}{2n^2}\frac{P^2L}{2AE} \tag{1}$$

Check that, for $n = 1$,

$$U_1 = \frac{P^2L}{2AE}$$

which is the same as Eq. (11.14) for a rod of length L and uniform cross section of area A. Also note that for $n > 1$, $U_n < U_1$. As an example, when $n = 2$, $U_2 = (\frac{5}{8})U_1$. Since the maximum stress occurs in portion CD of the rod and is equal to $\sigma_{\max} = P/A$, then for a given allowable stress, increasing the diameter of portion BC of the rod results in a *decrease* of the overall energy-absorbing capacity. Unnecessary changes in cross-sectional area should be avoided in the design of members subjected to loads (such as impact loadings) where the energy-absorbing capacity of the member is critical.

Concept Application 11.2

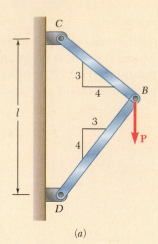

(a)

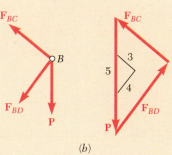

(b)

Fig. 11.12 (a) Frame *CBD* supporting a vertical force *P*. (b) Free-body diagram of joint *B* and corresponding force triangle.

A load **P** is supported at *B* by two rods of the same material and of the same uniform cross section of area *A* (Fig. 11.12*a*). Determine the strain energy of the system.

Using the forces F_{BC} and F_{BD} in members *BC* and *BD* and recalling Eq. (11.14), the strain energy of the system is

$$U = \frac{F_{BC}^2(BC)}{2AE} + \frac{F_{BD}^2(BD)}{2AE} \qquad (1)$$

From Fig. 11.12*a*,

$$BC = 0.6l \qquad BD = 0.8l$$

From the free-body diagram of pin *B* and the corresponding force triangle (Fig. 11.12*b*),

$$F_{BC} = +0.6P \qquad F_{BD} = -0.8P$$

Substituting into Eq. (1) gives

$$U = \frac{P^2 l[(0.6)^3 + (0.8)^3]}{2AE} = 0.364 \frac{P^2 l}{AE}$$

Strain Energy in Bending.

Consider a beam *AB* subjected to a given loading (Fig. 11.13), and let *M* be the bending moment at a distance *x* from end *A*. Neglecting the effect of shear and taking into account only the normal stresses $\sigma_x = My/I$, substitute this expression into Eq. (11.12) and write

$$U = \int \frac{\sigma_x^2}{2E} dV = \int \frac{M^2 y^2}{2EI^2} dV$$

Setting $dV = dA\, dx$, where dA represents an element of the cross-sectional area, and recalling that $M^2/2EI^2$ is a function of *x* alone gives

$$U = \int_0^L \frac{M^2}{2EI^2} \left(\int y^2\, dA \right) dx$$

Recall that the integral within the parentheses represents the moment of inertia *I* of the cross section about its neutral axis. Thus,

$$U = \int_0^L \frac{M^2}{2EI} dx \qquad (11.15)$$

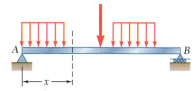

Fig. 11.13 Transversely loaded beam.

Concept Application 11.3

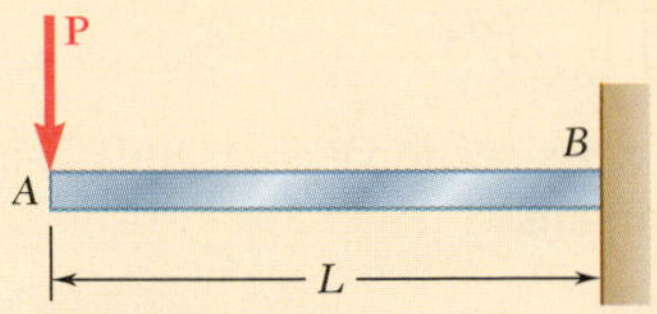

Determine the strain energy of the prismatic cantilever beam AB (Fig. 11.14), taking into account only the effect of the normal stresses. The bending moment at a distance x from end A is $M = -Px$. Substitute this expression into Eq. (11.15) to obtain

$$U = \int_0^L \frac{P^2 x^2}{2EI} dx = \frac{P^2 L^3}{6EI}$$

Fig. 11.14 Cantilever beam with load P.

11.2B   Shearing Stresses

When a material is subjected to plane shearing stresses τ_{xy}, the strain-energy density at a given point is

$$u = \int_0^{\gamma_{xy}} \tau_{xy} d\gamma_{xy} \tag{11.16}$$

where γ_{xy} is the shearing strain corresponding to τ_{xy} (Fig. 11.15a). Note that the strain-energy density u is equal to the area under the shearing-stress-strain diagram (Fig. 11.15b).

For values of τ_{xy} within the proportional limit, $\tau_{xy} = G\gamma_{xy}$, where G is the modulus of rigidity of the material. Substituting for τ_{xy} into Eq. (11.16) and integrating gives

$$u = \frac{1}{2} G\gamma_{xy}^2 = \frac{1}{2}\tau_{xy}\gamma_{xy} = \frac{\tau_{xy}^2}{2G} \tag{11.17}$$

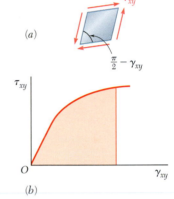

(a)

(b)

Fig. 11.15   (a) Shearing strain corresponding to τ_{xy}. (b) Strain-energy density u is the area under the stress-strain diagram.

The value of the strain energy U of a body subjected to plane shearing stresses can be obtained by recalling from Sec. 11.2A that

$$u = \frac{dU}{dV} \tag{11.9}$$

Substituting for u from Eq. (11.17) into Eq. (11.9) and integrating both members gives

$$U = \int \frac{\tau_{xy}^2}{2G} dV \tag{11.18}$$

This equation defines the elastic strain energy associated with the shear deformations of the body. Like the similar expression in Sec. 11.2A for uniaxial normal stresses, it is only valid for elastic deformations.

Strain Energy in Torsion.   Consider a shaft BC of non-uniform circular cross section with a length of L subjected to one or several twisting couples. Using the polar moment of inertia J of the cross section located

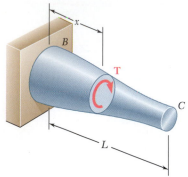

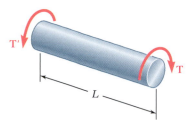

Fig. 11.16 Shaft subject to torque.

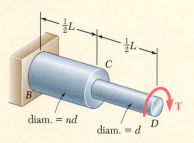

Fig. 11.17 Prismatic shaft subject to torque.

a distance x from B (Fig. 11.16) and the internal torque T, recall that the shearing stresses are $\tau_{xy} = T\rho/J$. Substituting for τ_{xy} into Eq. (11.18),

$$U = \int \frac{\tau_{xy}^2}{2G} dV = \int \frac{T^2\rho^2}{2GJ^2} dV$$

Set $dV = dA\,dx$, where dA is the element cross-sectional area, and observe that $T^2/2GJ^2$ is a function of x alone to obtain

$$U = \int_0^L \frac{T^2}{2GJ^2}\left(\int \rho^2\,dA\right)dx$$

Recall that the integral within the parentheses represents the polar moment of inertia J of the cross section, giving

$$U = \int_0^L \frac{T^2}{2GJ}\,dx \tag{11.19}$$

For a shaft of uniform cross section subjected at its ends to equal and opposite couples of magnitude T (Fig. 11.17), Eq. (11.19) yields

$$U = \frac{T^2 L}{2GJ} \tag{11.20}$$

Concept Application 11.4

A circular shaft consists of two portions BC and CD of the same material and length, but of different cross sections (Fig. 11.18). Determine the strain energy of the shaft when it is subjected to a twisting couple $\mathbf{T}$ at end D. Express the results in terms of T, L, G, the polar moment of inertia J of the smaller cross section, and the ratio n of the two diameters.

Use Eq. (11.20) to compute the strain energy of each of the two portions of shaft, and add the expressions obtained. Note that the polar moment of inertia of portion BC is equal to $n^4 J$ giving

$$U_n = \frac{T^2(\tfrac{1}{2}L)}{2GJ} + \frac{T^2(\tfrac{1}{2}L)}{2G(n^4 J)} = \frac{T^2 L}{4GJ}\left(1 + \frac{1}{n^4}\right)$$

or

$$U_n = \frac{1 + n^4}{2n^4}\frac{T^2 L}{2GJ} \tag{1}$$

For $n = 1$,

$$U_1 = \frac{T^2 L}{2GJ}$$

which is the expression given in Eq. (11.20) for a shaft of length L and uniform cross section. Note that, for $n > 1$, $U_n < U_1$. For example, when $n = 2$, $U_2 = (\tfrac{17}{32})U_1$. Since the maximum shearing stress occurs in segment CD of the shaft and is proportional to the torque T, increasing the diameter of segment BC results in a *decrease* of the overall energy-absorbing capacity of the shaft.

Fig. 11.18 Stepped shaft subject to torque T.

Strain Energy under Transverse Loading. In Sec. 11.2A an equation for the strain energy of a beam subjected to a transverse loading was obtained. However, in that expression, only the effect of the normal stresses due to bending is taken into account and the effect of the shearing stresses is neglected. In Concept Application 11.5, both types of stresses will be taken into account.

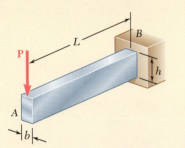

Fig. 11.19 Rectangular cantilever beam with load P.

Concept Application 11.5

Determine the strain energy of the rectangular cantilever beam AB (Fig. 11.19), taking into account the effect of both normal and shearing stresses.

Recall from Concept Application 11.3 that the strain energy due to the normal stresses σ_x is

$$U_\sigma = \frac{P^2 L^3}{6EI}$$

To determine the strain energy U_τ due to the shearing stresses τ_{xyw}, recall Eq. (6.9) and find that, for a beam with a rectangular cross section of width b and depth h,

$$\tau_{xy} = \frac{3}{2}\frac{V}{A}\left(1 - \frac{y^2}{c^2}\right) = \frac{3}{2}\frac{P}{bh}\left(1 - \frac{y^2}{c^2}\right)$$

Substituting for τ_{xy} into Eq. (11.18),

$$U_\tau = \frac{1}{2G}\left(\frac{3}{2}\frac{P}{bh}\right)^2\int\left(1 - \frac{y^2}{c^2}\right)^2 dV$$

or setting $dV = b\,dy\,dx$, and after reductions,

$$U_\tau = \frac{9P^2}{8Gbh^2}\int_{-c}^{c}\left(1 - 2\frac{y^2}{c^2} + \frac{y^4}{c^4}\right)dy\int_0^L dx$$

Performing integrations and recalling that $c = h/2$,

$$U_\tau = \frac{9P^2 L}{8Gbh^2}\left[y - \frac{2}{3}\frac{y^3}{c^2} + \frac{1}{5}\frac{y^5}{c^4}\right]_{-c}^{+c} = \frac{3P^2 L}{5Gbh} = \frac{3P^2 L}{5GA}$$

The total strain energy of the beam is

$$U = U_\sigma + U_\tau = \frac{P^2 L^3}{6EI} + \frac{3P^2 L}{5GA}$$

or with $I/A = h^2/12$ and factoring the expression for U_σ,

$$U = \frac{P^2 L^3}{6EI}\left(1 + \frac{3Eh^2}{10GL^2}\right) = U_\sigma\left(1 + \frac{3Eh^2}{10GL^2}\right) \qquad (1)$$

Recall from Sec. 2.7 that $G \geq E/3$. Considering the parenthesis, this equation is less than $1 + 0.9(h/L)^2$ and the relative error is less than $0.9(h/L)^2$ when the effect of shear is neglected. For a beam with a ratio h/L less than $\frac{1}{10}$, the percentage error is less than 0.9%. It is therefore customary in engineering practice to neglect the effect of shear to compute the strain energy of shallow beams.

11.3 STRAIN ENERGY FOR A GENERAL STATE OF STRESS

In the preceding sections, we determined the strain energy of a body in a state of uniaxial stress (Sec. 11.2A) and in a state of plane shearing stress (Sec. 11.2B). In a body in a general state of stress characterized by the six stress components σ_x, σ_y, σ_z, τ_{xy}, τ_{yz}, and τ_{zx} (Fig. 2.35), the strain-energy density is obtained by adding the expressions given in Eqs. (11.10) and (11.16), as well as the four other expressions obtained through a permutation of the subscripts.

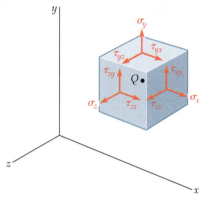

Fig. 2.35 (repeated) Positive stress component at point Q for a general state of stress.

In the elastic deformation of an isotropic body, each of the six stress-strain relationships involved is linear, and the strain-energy density is

$$u = \tfrac{1}{2}(\sigma_x\epsilon_x + \sigma_y\epsilon_y + \sigma_z\epsilon_z + \tau_{xy}\gamma_{xy} + \tau_{yz}\gamma_{yz} + \tau_{zx}\gamma_{zx}) \quad \textbf{(11.21)}$$

Recalling Eq. (2.29) and substituting for the strain components into Eq. (11.21), the most general state of stress at a given point of an elastic isotropic body is

$$u = \frac{1}{2E}\left[\sigma_x^2 + \sigma_y^2 + \sigma_z^2 - 2\nu(\sigma_x\sigma_y + \sigma_y\sigma_z + \sigma_z\sigma_x)\right]$$

$$+ \frac{1}{2G}(\tau_{xy}^2 + \tau_{yz}^2 + \tau_{zx}^2) \quad \textbf{(11.22)}$$

If the principal axes at the given point are used as coordinate axes, the shearing stresses become zero and Eq. (11.22) reduces to

$$u = \frac{1}{2E}\left[\sigma_a^2 + \sigma_b^2 + \sigma_c^2 - 2\nu(\sigma_a\sigma_b + \sigma_b\sigma_c + \sigma_c\sigma_a)\right] \quad \textbf{(11.23)}$$

where σ_a, σ_b, and σ_c are the principal stresses at the given point.

Now recall from Sec. 7.5A that one criterion used to predict whether a given state of stress causes a ductile material to yield is the maximum-distortion-energy criterion, which is based on the energy per unit volume associated with the distortion (or change in shape) of that material.

Separating the strain-energy density u at a given point into two parts, u_v associated with a change in volume of the material at that point and u_d associated with a distortion of the material at the same point,

$$u = u_v + u_d \tag{11.24}$$

In order to determine u_v and u_d, we introduce the *average value* $\overline{\sigma}$ of the principal stresses at the point considered as

$$\overline{\sigma} = \frac{\sigma_a + \sigma_b + \sigma_c}{3} \tag{11.25}$$

and set

$$\sigma_a = \overline{\sigma} + \sigma_a' \qquad \sigma_b = \overline{\sigma} + \sigma_b' \qquad \sigma_c = \overline{\sigma} + \sigma_c' \tag{11.26}$$

Thus, the given state of stress (Fig. 11.20*a*) can be obtained by superposing the states of stress shown in Fig. 11.20*b* and *c*. The state of stress described in Fig. 11.20*b* tends to change the volume of the element but not its shape, since all of the faces of the element are subjected to the same stress $\overline{\sigma}$. On the other hand, it follows from Eqs. (11.25) and (11.26) that

$$\sigma_a' + \sigma_b' + \sigma_c' = 0 \tag{11.27}$$

This indicates that some of the stresses shown in Fig. 11.20*c* are tensile and others compressive. Thus, this state of stress tends to change the shape of the element. However, it does not tend to change its volume. Recall from Eq. (2.22) that the dilatation e (i.e., the change in volume per unit volume) caused by this state of stress is

$$e = \frac{1 - 2\nu}{E} (\sigma_a' + \sigma_b' + \sigma_c')$$

or $e = 0$ in view of Eq. (11.27). Thus, the portion u_v of the strain-energy density must be associated with the state of stress shown in Fig. 11.20*b*, while the portion u_d must be associated with the state of stress shown in Fig. 11.20*c*.

The portion u_v of the strain-energy density corresponding to a change in volume of the element can be obtained by substituting $\overline{\sigma}$ for each of the principal stresses in Eq. (11.23). Thus,

$$u_v = \frac{1}{2E} [3\overline{\sigma}^2 - 2\nu(3\overline{\sigma}^2)] = \frac{3(1 - 2\nu)}{2E} \overline{\sigma}^2$$

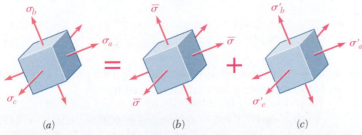

(a) (b) (c)

Fig. 11.20   *(a)* Element subject to multiaxial state of stress expressed as the superposition of *(b)* stresses tending to cause volume change, *(c)* stresses tending to cause distortion.

or recalling Eq. (11.25),

$$u_v = \frac{1 - 2\nu}{6E}(\sigma_a + \sigma_b + \sigma_c)^2 \tag{11.28}$$

The portion of the strain-energy density corresponding to the distortion of the element is found by solving Eq. (11.24) for u_d and substituting for u and u_v from Eqs. (11.23) and (11.28), respectively.

$$u_d = u - u_v = \frac{1}{6E}[3(\sigma_a^2 + \sigma_b^2 + \sigma_c^2) - 6\nu(\sigma_a\sigma_b + \sigma_b\sigma_c + \sigma_c\sigma_a)$$
$$- (1 - 2\nu)(\sigma_a + \sigma_b + \sigma_c)^2]$$

Expanding the square and rearranging terms,

$$u_d = \frac{1 + \nu}{6E}[(\sigma_a^2 - 2\sigma_a\sigma_b + \sigma_b^2) + (\sigma_b^2 - 2\sigma_b\sigma_c + \sigma_c^2)$$
$$+ (\sigma_c^2 - 2\sigma_c\sigma_a + \sigma_a^2)]$$

Noting that each of the parentheses inside the bracket is a perfect square, and recalling from Eq. (2.34) that the coefficient in front of the bracket is equal to $1/12G$, the portion u_d of the strain-energy density (i.e., the distortion energy per unit volume) is

$$u_d = \frac{1}{12G}[(\sigma_a - \sigma_b)^2 + (\sigma_b - \sigma_c)^2 + (\sigma_c - \sigma_a)^2] \tag{11.29}$$

In *plane stress*, assuming that the c axis is perpendicular to the plane of stress, $\sigma_c = 0$ and Eq. (11.29) reduces to

$$u_d = \frac{1}{6G}(\sigma_a^2 - \sigma_a\sigma_b + \sigma_b^2) \tag{11.30}$$

Considering a tensile-test specimen, at yield $\sigma_a = \sigma_Y$, $\sigma_b = 0$, and $(u_d)_Y = \sigma_Y^2/6G$. The maximum-distortion-energy criterion for plane stress indicates that a given state of stress is safe when $u_d < (u_d)_Y$, or by substituting for u_d from Eq. (11.30), it is safe as long as

$$\sigma_a^2 - \sigma_a\sigma_b + \sigma_b^2 < \sigma_Y^2 \tag{7.26}$$

which is the condition in Sec. 7.5A and represented graphically by the ellipse of Fig. 7.32. For a general state of stress, Eq. (11.29) obtained for u_d should be used. The maximum-distortion-energy criterion is then found by the condition:

$$(\sigma_a - \sigma_b)^2 + (\sigma_b - \sigma_c)^2 + (\sigma_c - \sigma_a)^2 < 2\sigma_Y^2 \tag{11.31}$$

which indicates that a given state of stress is safe if the point of coordinates σ_a, σ_b, σ_c is located within the surface defined by

$$(\sigma_a - \sigma_b)^2 + (\sigma_b - \sigma_c)^2 + (\sigma_c - \sigma_a)^2 = 2\sigma_Y^2 \tag{11.32}$$

This surface is a circular cylinder with a radius of $\sqrt{2/3}\,\sigma_Y$ and an axis of symmetry forming equal angles with the three principal axes of stress.

Sample Problem 11.1

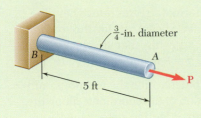

During a routine manufacturing operation, rod AB must acquire an elastic strain energy of 120 in·lb. Using $E = 29 \times 10^6$ psi, determine the required yield strength of the steel if the factor of safety is 5 with respect to permanent deformation.

STRATEGY: Use the specified factor of safety to determine the required strain-energy density, and then use Eq. (11.8) to determine the yield strength.

MODELING and ANALYSIS:

Factor of Safety. Since a factor of safety of 5 is required, the rod should be designed for a strain energy of

$$U = 5(120 \text{ in·lb}) = 600 \text{ in·lb}$$

Strain-Energy Density. The volume of the rod is

$$V = AL = \frac{\pi}{4}(0.75 \text{ in.})^2(60 \text{ in.}) = 26.5 \text{ in}^3$$

Since the rod has a uniform cross section, the required strain-energy density is

$$u = \frac{U}{V} = \frac{600 \text{ in·lb}}{26.5 \text{ in}^3} = 22.6 \text{ in·lb/in}^3$$

Yield Strength. Recall that the modulus of resilience is equal to the strain-energy density when the maximum stress is equal to σ_Y (Fig. 1). Using Eq. (11.8),

$$u = \frac{\sigma_Y^2}{2E}$$

$$22.6 \text{ in·lb/in}^3 = \frac{\sigma_Y^2}{2(29 \times 10^6 \text{ psi})} \qquad \sigma_Y = 36.2 \text{ ksi} \blacktriangleleft$$

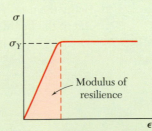

Fig. 1 The modulus of resilience equals the strain-energy density up to yield.

REFLECT and THINK: Since energy loads are not linearly related to the stresses they produce, factors of safety associated with energy loads should be applied to the energy loads and not to the stresses.

Sample Problem 11.2

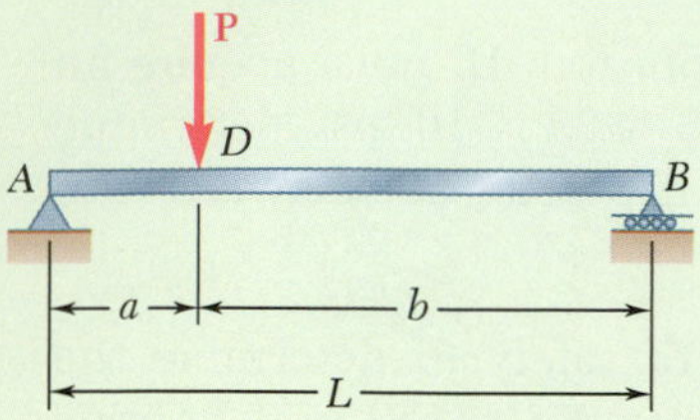

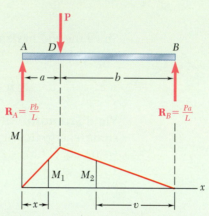

(*a*) Taking into account only the effect of normal stresses due to bending, determine the strain energy of the prismatic beam *AB* for the loading shown. (*b*) Evaluate the strain energy, knowing that the beam is a W10 × 45, $P = 40$ kips, $L = 12$ ft, $a = 3$ ft, $b = 9$ ft, and $E = 29 \times 10^6$ psi.

STRATEGY: Use a free-body diagram to determine the reactions, and write equations for the moment as a function of the coordinate along the beam. The strain energy required for part *a* is then determined from Eq. (11.15). Use this with the data to numerically evaluate the strain energy for part *b*.

MODELING:

Bending Moment. Using the free-body diagram of the entire beam (Fig. 1), determine the reactions

$$\mathbf{R}_A = \frac{Pb}{L} \uparrow \qquad \mathbf{R}_B = \frac{Pa}{L} \uparrow$$

Fig. 1 Free-body and bending-moment diagrams.

Using the free-body diagram in Fig. 2, the bending moment for portion *AD* of the beam is

$$M_1 = \frac{Pb}{L} x$$

From *A* to *D*:

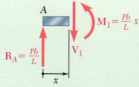

Fig. 2 Free-body diagram, taking a section within portion *AD*.

(continued)

Similarly using the free-body diagram in Fig. 3, the bending moment for portion DB at a distance v from end B is

$$M_2 = \frac{Pa}{L}v$$

From B to D:

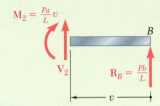

Fig. 3 Free-body diagram, taking a section within portion DB.

ANALYSIS:

a. Strain Energy. Since strain energy is a scalar quantity, add the strain energy of segment AD to that of DB to obtain the total strain energy of the beam. Using Eq. (11.15),

$$U = U_{AD} + U_{DB}$$

$$= \int_0^a \frac{M_1^2}{2EI}\,dx + \int_0^b \frac{M_2^2}{2EI}\,dv$$

$$= \frac{1}{2EI}\int_0^a \left(\frac{Pb}{L}x\right)^2 dx + \frac{1}{2EI}\int_0^b \left(\frac{Pa}{L}v\right)^2 dv$$

$$= \frac{1}{2EI}\frac{P^2}{L^2}\left(\frac{b^2a^3}{3} + \frac{a^2b^3}{3}\right) = \frac{P^2a^2b^2}{6EIL^2}(a+b)$$

or since $(a+b) = L$,

$$U = \frac{P^2a^2b^2}{6EIL} \blacktriangleleft$$

b. Evaluation of the Strain Energy. The moment of inertia of a W10 × 45 rolled-steel shape is obtained from Appendix C, and the given data is restated using units of kips and inches.

$P = 40$ kips	$L = 12$ ft $= 144$ in.
$a = 3$ ft $= 36$ in.	$b = 9$ ft $= 108$ in.
$E = 29 \times 10^6$ psi $= 29 \times 10^3$ ksi	$I = 248$ in^4

Substituting into the expression for U,

$$U = \frac{(40\text{ kips})^2(36\text{ in.})^2(108\text{ in.})^2}{6(29 \times 10^3\text{ ksi})(248\text{ in}^4)(144\text{ in.})} \qquad U = 3.89\text{ in}\cdot\text{kips} \blacktriangleleft$$

Problems

11.1 Determine the modulus of resilience for each of the following grades of structural steel:
  (*a*) ASTM A709 Grade 50: $\sigma_Y = 50$ ksi
  (*b*) ASTM A913 Grade 65: $\sigma_Y = 65$ ksi
  (*c*) ASTM A709 Grade 100: $\sigma_Y = 100$ ksi

11.2 Determine the modulus of resilience for each of the following aluminum alloys:
  (*a*) 1100-H14: $E = 70$ GPa: $\sigma_Y = 55$ MPa
  (*b*) 2014-T6: $E = 72$ GPa: $\sigma_Y = 220$ MPa
  (*c*) 6061-T6: $E = 69$ GPa: $\sigma_Y = 150$ MPa

11.3 Determine the modulus of resilience for each of the following metals:
  (*a*) Stainless steel
     AISI 302 (annealed): $E = 190$ GPa $\sigma_Y = 260$ MPa
  (*b*) Stainless steel
     AISI 302 (cold-rolled): $E = 190$ GPa $\sigma_Y = 520$ MPa
  (*c*) Malleable cast iron: $E = 165$ GPa $\sigma_Y = 230$ MPa

11.4 Determine the modulus of resilience for each of the following alloys:
  (*a*) Titanium: $E = 16.5 \times 10^6$ psi $\sigma_Y = 120$ ksi
  (*b*) Magnesium: $E = 6.5 \times 10^6$ psi $\sigma_Y = 29$ ksi
  (*c*) Cupronickel (annealed): $E = 20 \times 10^6$ psi $\sigma_Y = 16$ ksi

11.5 The stress-strain diagram shown has been drawn from data obtained during a tensile test of a specimen of structural steel. Using $E = 29 \times 10^6$ psi, determine (*a*) the modulus of resilience of the steel, (*b*) the modulus of toughness of the steel.

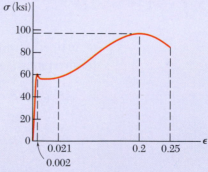

Fig. P11.5

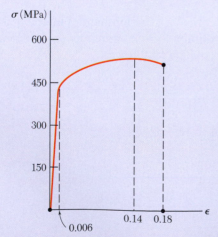

Fig. P11.6

11.6 The stress-strain diagram shown has been drawn from data obtained during a tensile test of an aluminum alloy. Using $E = 72$ GPa, determine (*a*) the modulus of resilience of the alloy, (*b*) the modulus of toughness of the alloy.

11.7 The load-deformation diagram shown has been drawn from data obtained during a tensile test of a specimen of an aluminum alloy. Knowing that the cross-sectional area of the specimen was 600 mm² and that the deformation was measured using a 400-mm gage length, determine by approximate means (*a*) the modulus of resilience of the alloy, (*b*) the modulus of toughness of the alloy.

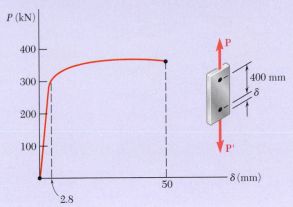

Fig. P11.7

11.8 The load-deformation diagram shown has been drawn from data obtained during a tensile test of a $\frac{5}{8}$-in.-diameter rod of structural steel. Knowing that the deformation was measured using an 18-in. gage length, determine by approximate means (*a*) the modulus of resilience of the steel, (*b*) the modulus of toughness of the steel.

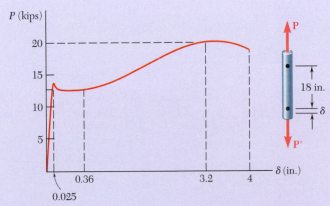

Fig. P11.8

11.9 Using $E = 29 \times 10^6$ psi, determine (*a*) the strain energy of the steel rod *ABC* when *P* = 8 kips, (*b*) the corresponding strain energy density in portions *AB* and *BC* of the rod.

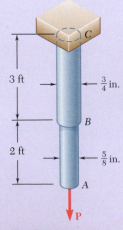

Fig. P11.9

11.10 Using $E = 200$ GPa, determine (*a*) the strain energy of the steel rod *ABC* when $P = 25$ kN, (*b*) the corresponding strain-energy density in portions *AB* and *BC* of the rod.

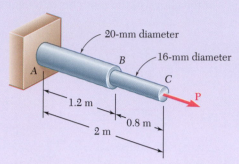

Fig. P11.10

11.11 A 30-in. length of aluminum pipe of cross-sectional area 1.85 in² is welded to a fixed support *A* and to a rigid cap *B*. The steel rod *EF*, of 0.75-in. diameter, is welded to cap *B*. Knowing that the modulus of elasticity is 29×10^6 psi for the steel and 10.6×10^6 psi for the aluminum, determine (*a*) the total strain energy of the system when $P = 8$ kips, (b) the corresponding strain-energy density of the pipe *CD* and in the rod *EF*.

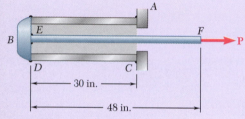

Fig. P11.11

11.12 A single 6-mm-diameter steel pin *B* is used to connect the steel strip *DE* to two aluminum strips, each of 20-mm width and 5-mm thickness. The modulus of elasticity is 200 GPa for the steel and 70 GPa for the aluminum. Knowing that for the pin at *B* the allowable shearing stress is $\tau_{\text{all}} = 85$ MPa, determine, for the loading shown, the maximum strain energy that can be acquired by the assembled strips.

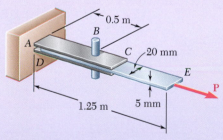

Fig. P11.12

11.13 Rods *AB* and *BC* are made of a steel for which the yield strength is $\sigma_Y = 300$ MPa and the modulus of elasticity is $E = 200$ GPa. Determine the maximum strain energy that can be acquired by the assembly without causing any permanent deformation when the length *a* of rod *AB* is (*a*) 2 m, (*b*) 4 m.

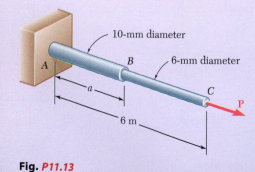

Fig. P11.13

11.14 Rod *BC* is made of a steel for which the yield strength is $\sigma_Y = 300$ MPa and the modulus of elasticity is $E = 200$ GPa. Knowing that a strain energy of 10 J must be acquired by the rod when the axial load **P** is applied, determine the diameter of the rod for which the factor of safety with respect to permanent deformation is six.

11.15 The assembly *ABC* is made of a steel for which $E = 200$ GPa and $\sigma_Y = 320$ MPa. Knowing that a strain energy of 5 J must be acquired by the assembly as the axial load **P** is applied, determine the factor of safety with respect to permanent deformation when (*a*) $x = 300$ mm, (*b*) $x = 600$ mm.

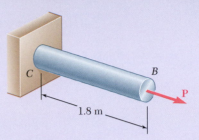

Fig. P11.14

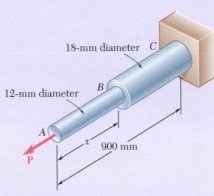

Fig. P11.15

11.16 Show by integration that the strain energy of the tapered rod *AB* is

$$U = \frac{1}{4}\frac{P^2 L}{EA_{min}}$$

where A_{min} is the cross-sectional area at end *B*.

11.17 Using $E = 10.6 \times 10^6$ psi, determine by approximate means the maximum strain energy that can be acquired by the aluminum rod shown if the allowable normal stress is $\sigma_{all} = 22$ ksi.

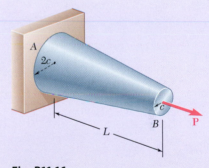

Fig. P11.16

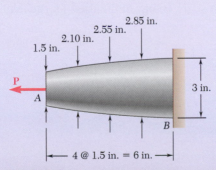

Fig. P11.17

11.18 through 11.21 In the truss shown, all members are made of the same material and have the uniform cross-sectional area indicated. Determine the strain energy of the truss when the load **P** is applied.

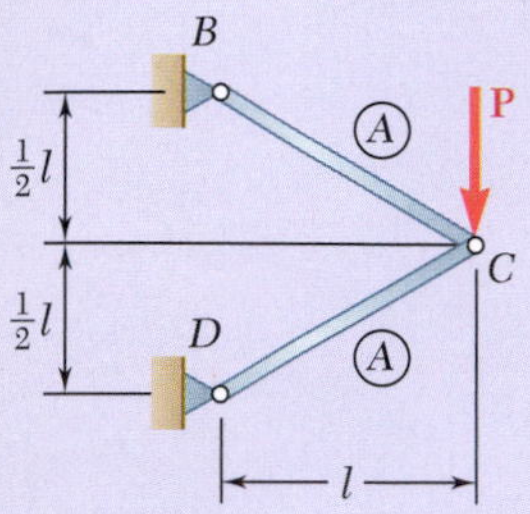

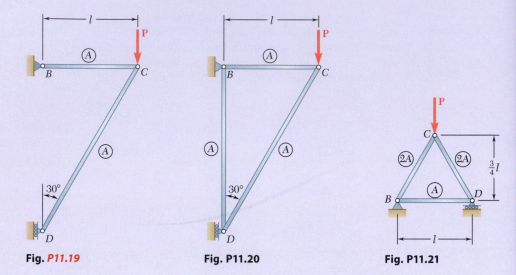

Fig. P11.18

Fig. *P11.19*

Fig. **P11.20**

Fig. **P11.21**

11.22 Each member of the truss shown is made of aluminum and has the cross-sectional area shown. Using $E = 72$ GPa, determine the strain energy of the truss for the loading shown.

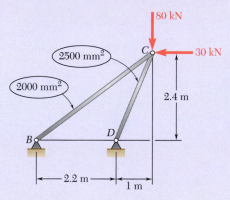

Fig. *P11.22*

11.23 Each member of the truss shown is made of aluminum and has the cross-sectional area shown. Using $E = 10.5 \times 10^6$ psi, determine the strain energy of the truss for the loading shown.

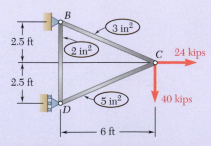

Fig. P11.23

11.24 through 11.27 Taking into account only the effect of normal stresses, determine the strain energy of the prismatic beam AB for the loading shown.

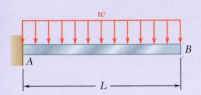

Fig. P11.24

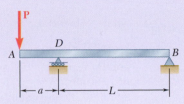

Fig. P11.25

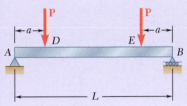

Fig. *P11.26*

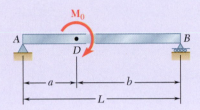

Fig. P11.27

11.28 and *11.29* Using $E = 29 \times 10^6$ psi , determine the strain energy due to bending for the steel beam and loading shown. (Neglect the effect of shearing stresses.)

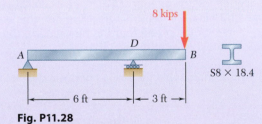

Fig. P11.28

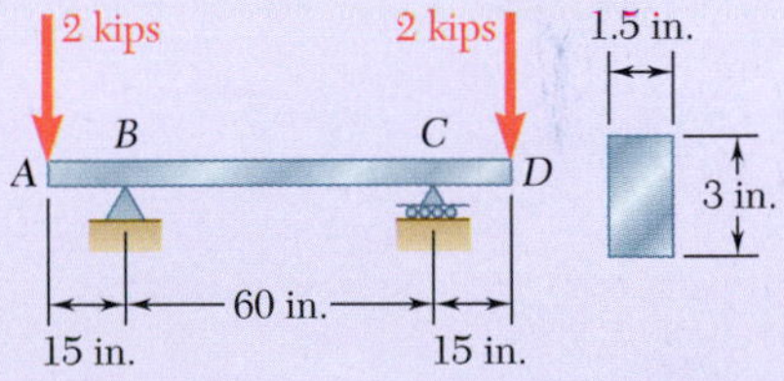

Fig. *P11.29*

11.30 and *11.31* Using $E = 200$ GPa, determine the strain energy due to bending for the steel beam and loading shown. (Neglect the effect of shearing stresses.)

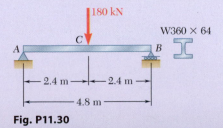

Fig. P11.30

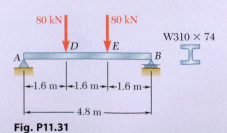

Fig. P11.31

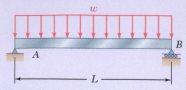

Fig. P11.32

11.32 Assuming that the prismatic beam AB has a rectangular cross section, show that for the given loading the maximum value of the strain-energy density in the beam is

$$u_{max} = \frac{45}{8}\frac{U}{V}$$

where U is the strain energy of the beam and V is its volume.

11.33 In the assembly shown, torques $\mathbf{T}_A$ and $\mathbf{T}_B$ are exerted on disks A and B, respectively. Knowing that both shafts are solid and made of aluminum ($G = 73$ GPa), determine the total strain energy acquired by the assembly.

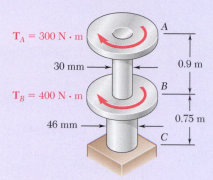

Fig. P11.33

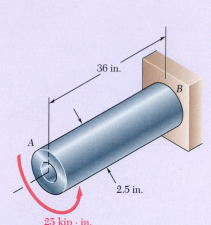

Fig. *P11.34*

11.34 The design specifications for the steel shaft AB require that the shaft acquire a strain energy of 400 in · lb as the 25-kip · in. torque is applied. Using $G = 11.2 \times 10^6$ psi, determine (a) the largest inner diameter of the shaft that can be used, (b) the corresponding maximum shearing stress in the shaft.

11.35 Show by integration that the strain energy in the tapered rod AB is

$$U = \frac{7}{48}\frac{T^2 L}{GJ_{min}}$$

where J_{min} is the polar moment of inertia of the rod at end B.

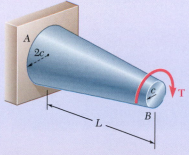

Fig. P11.35

11.36 The state of stress shown occurs in a machine component made of a brass for which $\sigma_Y = 160$ MPa. Using the maximum-distortion-energy criterion, determine the range of values of σ_z for which yield does not occur.

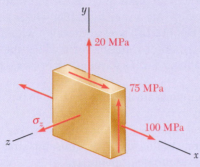

Fig. P11.36 and P11.37

11.37 The state of stress shown occurs in a machine component made of a brass for which $\sigma_Y = 160$ MPa. Using the maximum-distortion-energy criterion, determine whether yield occurs when (a) $\sigma_z = +45$ MPa, (b) $\sigma_z = -45$ MPa.

11.38 The state of stress shown occurs in a machine component made of a grade of steel for which $\sigma_Y = 65$ ksi. Using the maximum-distortion-energy criterion, determine the range of values of σ_y for which the factor of safety associated with the yield strength is equal to or larger than 2.2.

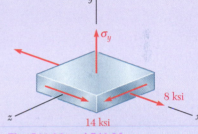

Fig. P11.38 and P11.39

11.39 The state of stress shown occurs in a machine component made of a grade of steel for which $\sigma_Y = 65$ ksi. Using the maximum-distortion-energy criterion, determine the factor of safety associated with the yield strength when (a) $\sigma_y = +16$ ksi, (b) $\sigma_y = -16$ ksi.

11.40 Determine the strain energy of the prismatic beam AB, taking into account the effect of both normal and shearing stresses.

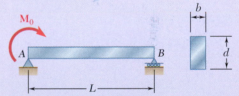

Fig. P11.40

*11.41 A vibration isolation support is made by bonding a rod A, of radius R_1, and a tube B, of inner radius R_2, to a hollow rubber cylinder. Denoting by G the modulus of rigidity of the rubber, determine the strain energy of the hollow rubber cylinder for the loading shown.

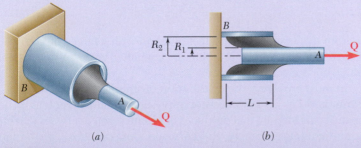

(a)   (b)

Fig. P11.41

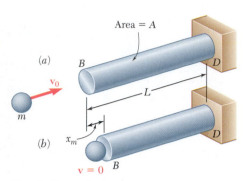

Area = A

(a)

B

D

v_0

L

m

(b)

x_m

D

$v = 0$ B

Fig. 11.21   Rod subject to impact loading.

Photo 11.2   Steam alternately lifts a weight inside the pile driver and then propels it downward. This delivers a large impact load to the pile that is being driven into the ground.

11.4   IMPACT LOADS

11.4A   Analysis

Consider a rod *BD* with an uniform cross section that is hit at end *B* by a body of mass *m* moving with a velocity $\mathbf{v}_0$ (Fig. 11.21*a*). As the rod deforms under the impact (Fig. 11.21*b*), stresses develop within the rod and reach a maximum value σ_m. After vibrating for a while, the rod comes to rest, and all stresses disappear. Such a sequence of events is called *impact loading* (Photo 11.2).

Several assumptions are made to determine the maximum value σ_m of the stress at a given point of a structure subjected to an impact load.

First, the kinetic energy $T = \frac{1}{2}mv_0^2$ of the striking body is assumed to be transferred entirely to the structure. Thus, the strain energy U_m corresponding to the maximum deformation x_m is

$$U_m = \tfrac{1}{2}mv_0^2 \tag{11.33}$$

This assumption leads to the following requirements.

1. No energy should be dissipated during the impact.
2. The striking body should not bounce off the structure and retain part of its energy. This, in turn, necessitates that the inertia of the structure be negligible, compared to the inertia of the striking body.

In practice, neither of these requirements is satisfied, and only part of the kinetic energy of the striking body is actually transferred to the structure. Thus, assuming that all of the kinetic energy of the striking body is transferred to the structure leads to a conservative design.

The stress-strain diagram obtained from a static test of the material is also assumed to be valid under impact loads. So, for an elastic deformation of the structure, the maximum value of the strain energy is

$$U_m = \int \frac{\sigma_m^2}{2E}\, dV \tag{11.34}$$

For the uniform rod in Fig. 11.21, the maximum stress σ_m has the same value throughout the rod, and $U_m = \sigma_m^2\, V/2E$. Solving for σ_m and substituting for U_m from Eq. (11.33) gives

$$\sigma_m = \sqrt{\frac{2U_m E}{V}} = \sqrt{\frac{mv_0^2 E}{V}} \tag{11.35}$$

Note from this equation that selecting a rod with a large volume *V* and a low modulus of elasticity *E* results in a smaller value of the maximum stress σ_m for a given impact load.

In most problems, the distribution of stresses in the structure is not uniform, and Eq. (11.35) does not apply. It is then convenient to determine the static load $\mathbf{P}_m$ that produces the same strain energy as the impact load and compute from P_m the corresponding value σ_m of the largest stress in the structure.

Concept Application 11.6

A body of mass m moving with a velocity $\mathbf{v}_0$ hits the end B of the non-uniform rod BCD (Fig. 11.22). Knowing that the diameter of segment BC is twice the diameter of portion CD, determine the maximum value σ_m of the stress in the rod.

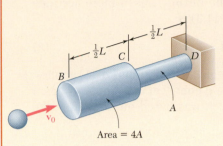

Fig. 11.22 Stepped rod impacted by a body of mass m.

Making $n = 2$ in Eq. (1) from Concept Application 11.1, when rod BCD is subjected to a static load P_m, its strain energy is

$$U_m = \frac{5P_m^2 L}{16AE} \tag{1}$$

where A is the cross-sectional area of segment CD. Solving Eq. (1) for P_m, the static load that produces the same strain energy as the given impact load is

$$P_m = \sqrt{\frac{16}{5}\frac{U_m AE}{L}}$$

where U_m is given by Eq. (11.33). The largest stress occurs in segment CD. Dividing P_m by the area A of that portion,

$$\sigma_m = \frac{P_m}{A} = \sqrt{\frac{16}{5}\frac{U_m E}{AL}} \tag{2}$$

or substituting for U_m from Eq. (11.33) gives

$$\sigma_m = \sqrt{\frac{8}{5}\frac{mv_0^2 E}{AL}} = 1.265\sqrt{\frac{mv_0^2 E}{AL}}$$

Comparing this with the value obtained for σ_m in the uniform rod of Fig. 11.21 and making $V = AL$ in Eq. (11.35), note that the maximum stress in the rod of variable cross section is 26.5% larger than in the lighter uniform rod. Thus, as in our discussion of Concept Application 11.1, increasing the diameter of segment BC results in a *decrease* of the energy-absorbing capacity of the rod.

Concept Application 11.7

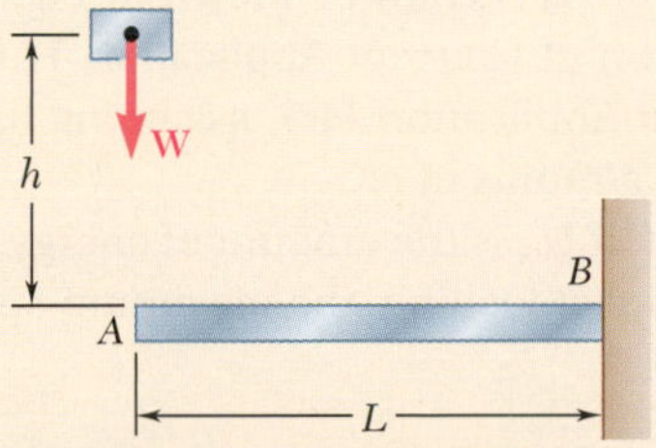

A block of weight W is dropped from a height h onto the free end of the cantilever beam AB (Fig. 11.23). Determine the maximum value of the stress in the beam.

As it falls through the distance h, the potential energy Wh of the block is transformed into kinetic energy. As a result of the impact, the kinetic energy is transformed into strain energy. Therefore,

$$U_m = Wh \tag{1}$$

Fig. 11.23 Weight W falling on cantilever beam.

(continued)

The total distance the block drops is actually $h + y_m$, where y_m is the maximum deflection of the end of the beam. Thus, a more accurate expression for U_m (see Sample Prob. 11.3) is

$$U_m = W(h + y_m) \tag{2}$$

However, when $h \gg y_m$, y_m may be neglected, and thus Eq. (1) applies.

Recalling the equation for the strain energy of the cantilever beam AB in Concept Application 11.3, which was based on neglecting the effect of shear,

$$U_m = \frac{P_m^2 L^3}{6EI}$$

Solving this equation for P_m, the static force that produces the same strain energy in the beam is

$$P_m = \sqrt{\frac{6U_m EI}{L^3}} \tag{3}$$

The maximum stress σ_m occurs at the fixed end B and is

$$\sigma_m = \frac{|M|c}{I} = \frac{P_m L c}{I}$$

Substituting for P_m from Eq. (3),

$$\sigma_m = \sqrt{\frac{6U_m E}{L(I/c^2)}} \tag{4}$$

or recalling Eq. (1),

$$\sigma_m = \sqrt{\frac{6WhE}{L(I/c^2)}}$$

11.4B   Design

Now compare the values from the preceding section for the maximum stress σ_m: (*a*) in the rod of uniform cross section of shown in Fig. 11.21, (*b*) in the rod of variable cross section of Concept Application 11.6, and (*c*) in the cantilever beam of Concept Application 11.7, assuming that the last has a circular cross section with a radius of c.

(*a*) Equation (11.35) shows that, if U_m is the amount of energy transferred to the rod as a result of the impact loading, the maximum stress in the rod of uniform cross section is

$$\sigma_m = \sqrt{\frac{2U_m E}{V}} \tag{11.36a}$$

where V is the volume of the rod.

(b) Considering the rod of Concept Application 11.6 and observing that the volume of the rod is

$$V = 4A(L/2) + A(L/2) = 5AL/2$$

substitute $AL = 2V/5$ into Eq. (2) of Concept Application 11.6 and write

$$\sigma_m = \sqrt{\frac{8U_mE}{V}} \qquad\qquad \textbf{(11.36b)}$$

(c) Finally, recalling that $I = \frac{1}{4}\pi c^4$ for a beam of circular cross section,

$$L(I/c^2) = L(\tfrac{1}{4}\pi c^4/c^2) = \tfrac{1}{4}(\pi c^2 L) = \tfrac{1}{4}V$$

where V is the volume of the beam. Substituting into Eq. (4) of Concept Application 11.7, the maximum stress in the cantilever beam is

$$\sigma_m = \sqrt{\frac{24U_mE}{V}} \qquad\qquad \textbf{(11.36c)}$$

In each case, the maximum stress σ_m is proportional to the square root of the modulus of elasticity of the material and inversely proportional to the square root of the volume of the member. Assuming that all three members have the same volume and are of the same material, we note that for a given value of the absorbed energy, the uniform rod experiences the lowest maximum stress and the cantilever beam the highest.

This is explained by the fact that the distribution of stresses is uniform in case a, and the strain energy is uniformly distributed throughout the rod. In case b, on the other hand, the stresses in segment BC of the rod are only 25% as large as the stresses in segment CD. This uneven distribution of the stresses and strain energy results in a maximum stress σ_m that is twice as large as the corresponding stress in the uniform rod. Finally, in case c, where the cantilever beam is subjected to a transverse impact load, the stresses vary linearly along the beam as well as through a transverse section. This uneven distribution of strain energy causes the maximum stress σ_m to be 3.46 times larger than in the same member loaded axially (as in case a).

The properties discussed in this section are quite general and can be observed in all types of structures subject to impact loads. Thus, a structural member designed to most effectively withstand an impact load should

1. Have a large volume.
2. Be made of a material with a low modulus of elasticity and a high yield strength.
3. Be shaped so that the stresses are distributed as evenly as possible throughout the member.

11.5 SINGLE LOADS

11.5A Energy Formulation

The concept of strain energy introduced at the beginning of this chapter considered the work done by an axial load **P** applied to the end of a rod of uniform cross section (Fig. 11.1). The strain energy of the rod for an elongation x_1 was defined as the work of the load **P** as it is slowly increased from 0 to P_1 corresponding to x_1. Thus,

$$\text{Strain energy} = U = \int_0^{x_1} P\,dx \tag{11.2}$$

For an elastic deformation, the work of load **P** and the strain energy of the rod is

$$U = \tfrac{1}{2}P_1 x_1 \tag{11.3}$$

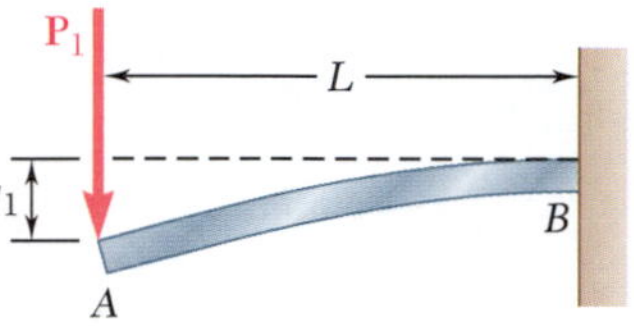

The strain energy of structural members under various load conditions was determined in Sec. 11.2 using the strain-energy density u at every point of the member and integrating u over the entire member.

When a structure or member is subjected to a *single concentrated load,* Eq. (11.3) can be used to evaluate its elastic strain energy, provided that the relationship between the load and the resulting deformation is known. For instance, the cantilever beam of Concept Application 11.3 (Fig. 11.24) has

$$U = \tfrac{1}{2}P_1 y_1$$

and substituting the value from the table of *Beam Deflections and Slopes* of Appendix D for y_1 gives

$$U = \frac{1}{2}P_1\left(\frac{P_1 L^3}{3EI}\right) = \frac{P_1^2 L^3}{6EI} \tag{11.37}$$

A similar approach can be used to determine the strain energy of a structure or member subjected to a *single couple.* Recall that the elementary work of a couple of moment M is $M\,d\theta$, where $d\theta$ is a small angle. Since M and θ are linearly related, the elastic strain energy of a cantilever beam AB subjected to a single couple $\mathbf{M}_1$ at its end A (Fig. 11.25) is

$$U = \int_0^{\theta_1} M\,d\theta = \tfrac{1}{2}M_1\theta_1 \tag{11.38}$$

where θ_1 is the slope of the beam at A. Substituting the value obtained from Appendix D for θ_1 gives

$$U = \frac{1}{2}M_1\left(\frac{M_1 L}{EI}\right) = \frac{M_1^2 L}{2EI} \tag{11.39}$$

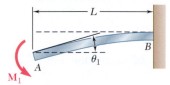

Fig. 11.24 Cantilever beam with load **P**₁.

Fig. 11.25 Cantilever beam with couple **M**₁.

Likewise, the elastic strain energy of a uniform circular shaft AB with a length of L subjected at its end B to a single torque $\mathbf{T}_1$ (Fig. 11.26) is

$$U = \int_0^{\phi_1} T \, d\phi = \tfrac{1}{2} T_1 \phi_1 \qquad (11.40)$$

Substituting for the angle of twist ϕ_1 from Eq. (3.15) gives

$$U = \frac{1}{2} T_1 \left(\frac{T_1 L}{JG} \right) = \frac{T_1^2 L}{2JG}$$

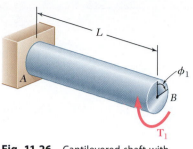

Fig. 11.26 Cantilevered shaft with torque $\mathbf{T}_1$.

The method presented in this section may simplify the solution of many impact-loading problems. In Concept Application 11.8, the crash of an automobile into a barrier (Photo 11.3) is analyzed by using a simplified model consisting of a block and a simple beam.

Photo 11.3 As the automobile crashed into the barrier, considerable energy is dissipated as heat during the permanent deformation of the automobile and the barrier.

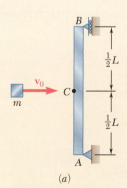

(a)

Fig. 11.27 (a) Simply supported beam having block propelled into its midpoint.

Concept Application 11.8

A block of mass m moving with a velocity $\mathbf{v}_0$ hits the prismatic member AB squarely at its midpoint C (Fig. 11.27a). Determine (a) the equivalent static load P_m, (b) the maximum stress σ_m in the member, and (c) the maximum deflection x_m at point C.

a. Equivalent Static Load. The maximum strain energy of the member is equal to the kinetic energy of the block before impact.

$$U_m = \tfrac{1}{2} m v_0^2 \qquad (1)$$

On the other hand, U_m can be given as the work of the equivalent horizontal static load as it is slowly applied at the midpoint C

$$U_m = \tfrac{1}{2} P_m x_m \qquad (2)$$

(continued)

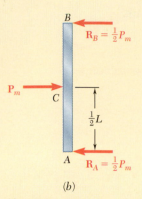

Fig. 11.27 (continued) (b) Free-body diagram of beam.

where x_m is the deflection of C corresponding to the static load P_m. From the table of *Beam Deflections and Slopes* of Appendix D,

$$x_m = \frac{P_m L^3}{48EI} \tag{3}$$

Substituting for x_m from Eq. (3) into Eq. (2),

$$U_m = \frac{1}{2}\frac{P_m^2 L^3}{48EI}$$

Solving for P_m and recalling Eq. (1), the static load equivalent to the given impact loading is

$$P_m = \sqrt{\frac{96U_m EI}{L^3}} = \sqrt{\frac{48mv_0^2 EI}{L^3}} \tag{4}$$

b. Maximum Stress. Drawing the free-body diagram of the member (Fig. 11.27b), the maximum value of the bending moment occurs at C and is $M_{max} = P_m L/4$. The maximum stress occurs in a transverse section through C and is equal to

$$\sigma_m = \frac{M_{max}c}{I} = \frac{P_m L c}{4I}$$

Substituting for P_m from Eq. (4),

$$\sigma_m = \sqrt{\frac{3mv_0^2 EI}{L(I/c)^2}}$$

c. Maximum Deflection. Substituting into Eq. (3) the expression obtained for P_m in Eq. (4):

$$x_m = \frac{L^3}{48EI}\sqrt{\frac{48mv_0^2 EI}{L^3}} = \sqrt{\frac{mv_0^2 L^3}{48EI}}$$

11.5B Deflections

The preceding section showed that, if the deflection x_1 of a structure or member under a single concentrated load $\mathbf{P}_1$ is known, the corresponding strain energy U is

$$U = \tfrac{1}{2}P_1 x_1 \tag{11.3}$$

A similar equation for the strain energy of a structural member under a single couple $\mathbf{M}_1$ is

$$U = \tfrac{1}{2}M_1 \theta_1 \tag{11.38}$$

If the strain energy U of a structure or member subjected to a single concentrated load $\mathbf{P}_1$ or couple $\mathbf{M}_1$ is known, Eq. (11.3) or (11.38) can be used to determine the corresponding deflection x_1 or angle θ_1. In order to find the deflection under a single load applied to a structure with several components, rather than use one of the methods of Chap. 9, it is often easier to first compute the strain energy of the structure by integrating the strain-energy density over its various parts, as was done in Sec. 11.2, and then use either Eq. (11.3) or Eq. (11.38) for the desired deflection. Similarly, the angle of twist ϕ_1 of a composite shaft can be obtained by integrating the strain-energy density over various parts of the shaft and solving Eq. (11.40) for ϕ_1.

The method in this section can be used *only if the given structure is subjected to a single concentrated load or couple*. The strain energy of a structure subjected to several loads *cannot* be determined by computing the work of each load as if applied independently to the structure (see Sec. 11.6). Even if it is possible to determine the strain energy of the structure in this manner, only one equation is available for the deflections corresponding to various loads. In Secs. 11.7 and 11.8, another method based on the concept of strain energy is developed that can be used to find the deflection or slope at a given point—even when that structure is subjected to several concentrated loads, distributed loads, or couples simultaneously.

Concept Application 11.9

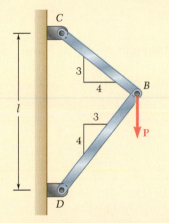

Fig. 11.28 Frame *CBD* with vertical load **P**.

A load **P** is supported at B by two uniform rods of the same cross-sectional area A (Fig. 11.28). Determine the vertical deflection of point B.

The strain energy of the system under the given load was determined in Concept Application 11.2. Equating U to the work of the load, write

$$U = 0.364 \frac{P^2 l}{AE} = \frac{1}{2} P y_B$$

and solving for the vertical deflection of B,

$$y_B = 0.728 \frac{Pl}{AE}$$

Remark. Once the forces in the two rods have been obtained (see Concept Application 11.2), the deformations $\delta_{B/C}$ and $\delta_{B/D}$ can be obtained using the method in Chap. 2. However, determining the vertical deflection of point B from these deformations requires a careful geometric analysis of the various displacements. The strain-energy method used here makes such an analysis unnecessary.

Concept Application 11.10

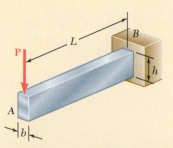

Fig. 11.29 Cantilevered rectangular beam with load **P**.

Determine the deflection of end A of the cantilever beam AB (Fig. 11.29), taking into account the effect of (*a*) the normal stresses only, (*b*) the normal and shearing stresses.

a. Effect of Normal Stresses. The work of the force **P** as it is slowly applied to A is

$$U = \tfrac{1}{2} P y_A$$

(continued)

Substituting for U the strain energy of the beam in Concept Application 11.3, where only the effect of the normal stresses was considered, write

$$\frac{P^2L^3}{6EI} = \frac{1}{2}Py_A$$

and solving for y_A,

$$y_A = \frac{PL^3}{3EI}$$

b. Effect of Normal and Shearing Stresses. Now substitute for U the expression for the total strain energy of the beam obtained in Concept Application 11.5, where the effects of both the normal and shearing stresses were taken into account. Thus,

$$\frac{P^2L^3}{6EI}\left(1 + \frac{3Eh^2}{10GL^2}\right) = \frac{1}{2}Py_A$$

and solving for y_A,

$$y_A = \frac{PL^3}{3EI}\left(1 + \frac{3Eh^2}{10GL^2}\right)$$

Note that the relative error when the effect of shear is neglected is the same as that obtained in Concept Application 11.5, (i.e., less than $0.9(h/L)^2$). This is less than 0.9% for a beam with a ratio h/L less than $\frac{1}{10}$.

Concept Application 11.11

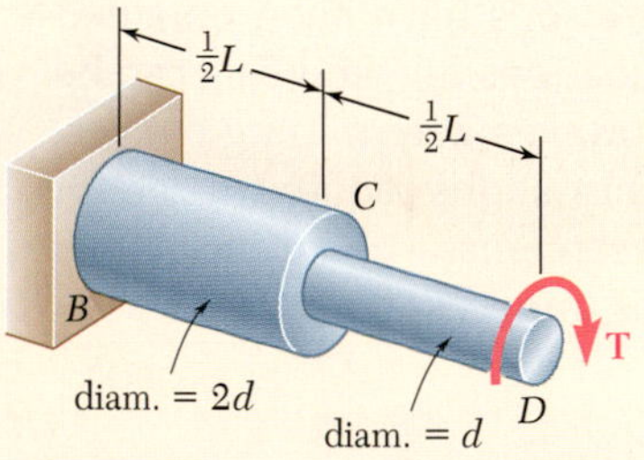

Fig. 11.30 Stepped shaft *BCD* with torque **T**.

A torque **T** is applied at the end D of shaft BCD (Fig. 11.30). Knowing that both portions of the shaft are of the same material and length, but that the diameter of BC is twice the diameter of CD, determine the angle of twist for the entire shaft.

In Concept Application 11.4, the strain energy of a similar shaft was determined by breaking the shaft into its component parts BC and CD. Making $n = 2$ in Eq. (1) of that Concept Application gives

$$U = \frac{17}{32}\frac{T^2L}{2GJ}$$

where G is the modulus of rigidity of the material and J is the polar moment of inertia of segment CD. Making U equal to the work of the torque as it is slowly applied to end D and recalling Eq. (11.40), write

$$\frac{17}{32}\frac{T^2L}{2GJ} = \frac{1}{2}T\phi_{D/B}$$

and solving for the angle of twist $\phi_{D/B}$,

$$\phi_{D/B} = \frac{17TL}{32GJ}$$

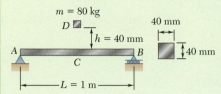

Sample Problem 11.3

The block *D* of mass *m* is released from rest and falls a distance *h* before it strikes the midpoint *C* of the aluminum beam *AB*. Using *E* = 73 GPa, determine (*a*) the maximum deflection of point *C*, (*b*) the maximum stress in the beam.

STRATEGY: Calculate the strain energy of the beam in terms of the deflection and equate this to the work done by the block. This then can be used with the data to solve part *a*. Using the relation between the applied load and deflection (Appendix D), obtain the equivalent static load and use this to get the normal stress due to bending.

MODELING:

Principle of Work and Energy.   The block is released from rest (Fig. 1, position *1*). Note that in this position both the kinetic and strain energy are zero. In position *2* (Fig. 1), where the maximum deflection y_m occurs, the kinetic energy is also zero. Use to the table of *Beam Deflections and Slopes* in Appendix D to find the expression for y_m shown in Fig. 2. The strain energy of the beam in position *2* is

$$U_2 = \frac{1}{2}P_m y_m = \frac{1}{2}\frac{48EI}{L^3}y_m^2 \qquad U_2 = \frac{24EI}{L^3}y_m^2$$

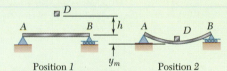

Fig. 1   Block released from rest (position *1*) and maximum deflection of beam (position *2*).

The work done by the weight **W** of the block is $W(h + y_m)$. Equating the strain energy of the beam to the work done by **W** gives

$$\frac{24EI}{L^3}y_m^2 = W(h + y_m) \tag{1}$$

ANALYSIS:

a. Maximum Deflection of Point C.   From the given data,

$$EI = (73 \times 10^9 \text{ Pa})\tfrac{1}{12}(0.04 \text{ m})^4 = 15.573 \times 10^3 \text{ N·m}^2$$

$$L = 1 \text{ m} \qquad h = 0.040 \text{ m} \qquad W = mg = (80 \text{ kg})(9.81 \text{ m/s}^2) = 784.8 \text{ N}$$

Substituting *W* into Eq. (1), we obtain a quadratic equation that can be solved for the deflection:

$$(373.8 \times 10^3)y_m^2 - 784.8y_m - 31.39 = 0 \qquad \qquad y_m = 10.27 \text{ mm} \blacktriangleleft$$

(continued)

b. Maximum Stress.   The value of P_m (Fig. 2) is

$$P_m = \frac{48EI}{L^3} y_m = \frac{48(15.573 \times 10^3 \text{ N·m})}{(1 \text{ m})^3}(0.01027 \text{ m}) \qquad P_m = 7677 \text{ N}$$

Recalling that $\sigma_m = M_{\max}c/I$ and $M_{\max} = \frac{1}{4}P_mL$, write

$$\sigma_m = \frac{(\frac{1}{4}P_mL)c}{I} = \frac{\frac{1}{4}(7677 \text{ N})(1 \text{ m})(0.020 \text{ m})}{\frac{1}{12}(0.040 \text{ m})^4} \qquad \sigma_m = 179.9 \text{ MPa} \blacktriangleleft$$

From Appendix D

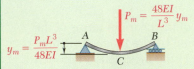

$$y_m = \frac{P_mL^3}{48EI}$$

$$P_m = \frac{48EI}{L^3}y_m$$

Fig. 2   Equivalent static force to cause deflection y_m.

REFLECT and THINK: An approximation for the work done by the weight of the block is obtained by omitting y_m from the expression for the work and from the right-hand member of Eq. (1), as was done in Concept Application 11.7. If this approximation is used here, $y_m = 9.16$ mm, and the error is 10.8%. However, if an 8-kg block is dropped from a height of 400 mm (producing the same value for Wh), omitting y_m from the right-hand member of Eq. (1) results in an error of only 1.2%.

Sample Problem 11.4

Members of the truss shown consist of sections of aluminum pipe with the cross-sectional areas indicated. Using $E = 73$ GPa, determine the vertical deflection of point E caused by load **P**.

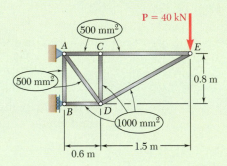

STRATEGY: Draw a free-body diagram of the truss to determine the reactions and then use free-body diagrams at each joint to find the member forces. Eq. (11.14) can then be used to determine the strain energy in each member. Equate the total strain energy in the members to the work done by the load **P** to determine the vertical deflection at the load.

MODELING:

Axial Forces in Truss Members.   The reactions are found by using the free-body diagram of the entire truss (Fig. 1a). Consider the equilibrium of joints E, C, D, and B in sequence (Fig. 1b through 1e). At each joint, determine the forces indicated by dashed lines. At joint B, the equation $\Sigma F_x = 0$ provides a check of the computations.

(continued)

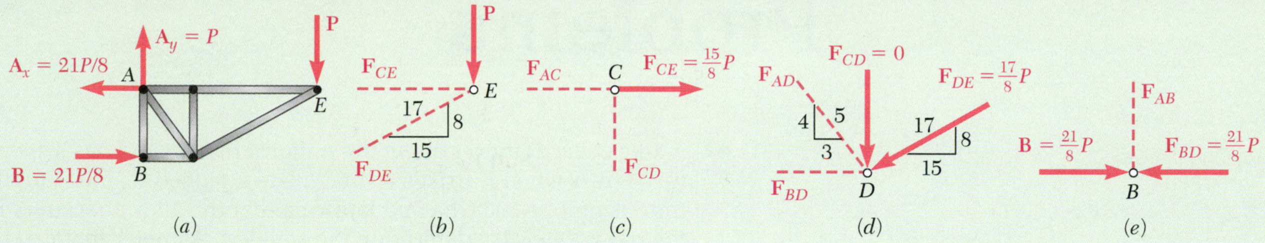

Fig. 1 (a) Free-body diagram of truss. (b-e) Force diagrams at joints.

$\Sigma F_y = 0$: $F_{DE} = -\frac{17}{8}P$ $\Sigma F_x = 0$: $F_{AC} = +\frac{15}{8}P$ $\Sigma F_y = 0$: $F_{AD} = +\frac{5}{4}P$ $\Sigma F_y = 0$: $F_{AB} = 0$

$\Sigma F_x = 0$: $F_{CE} = +\frac{15}{8}P$ $\Sigma F_y = 0$: $F_{CD} = 0$ $\Sigma F_x = 0$: $F_{BD} = -\frac{21}{8}P$ $\Sigma F_x = 0$: (Checks)

ANALYSIS:

Strain Energy. Noting that E is the same for all members, the strain energy of the truss is

$$U = \sum \frac{F_i^2 L_i}{2A_i E} = \frac{1}{2E} \sum \frac{F_i^2 L_i}{A_i} \tag{1}$$

where F_i is the force in a given member indicated in the following table and where the summation is extended over all members of the truss.

Member	F_i	L_i, m	A_i, m^2	$\dfrac{F_i^2 L_i}{A_i}$
AB	0	0.8	500×10^{-6}	0
AC	$+15P/8$	0.6	500×10^{-6}	$4\,219P^2$
AD	$+5P/4$	1.0	500×10^{-6}	$3\,125P^2$
BD	$-21P/8$	0.6	1000×10^{-6}	$4\,134P^2$
CD	0	0.8	1000×10^{-6}	0
CE	$+15P/8$	1.5	500×10^{-6}	$10\,547P^2$
DE	$-17P/8$	1.7	1000×10^{-6}	$7\,677P^2$

$$\sum \frac{F_i^2 L_i}{A_i} = 29\,700P^2$$

Returning to Eq. (1),

$$U = (1/2E)(29.7 \times 10^3 P^2).$$

Principle of Work-Energy. The work done by the load **P** as it is gradually applied is $\frac{1}{2}Py_E$. Equating the work done by **P** to the strain energy U and recalling that $E = 73$ GPa and $P = 40$ kN,

$$\frac{1}{2}Py_E = U \qquad \frac{1}{2}Py_E = \frac{1}{2E}(29.7 \times 10^3 P^2)$$

$$y_E = \frac{1}{E}(29.7 \times 10^3 P) = \frac{(29.7 \times 10^3)(40 \times 10^3)}{73 \times 10^9}$$

$$y_E = 16.27 \times 10^{-3}\,\text{m} \qquad y_E = 16.27\,\text{mm}\downarrow \quad \blacktriangleleft$$

Problems

11.42 A 5-kg collar D moves along the uniform rod AB and has a speed $v_0 = 6$ m/s when it strikes a small plate attached to end A of the rod. Using $E = 200$ GPa and knowing that the allowable stress in the rod is 250 MPa, determine the smallest diameter that can be used for the rod.

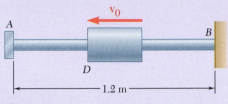

Fig. P11.42

11.43 The 18-lb cylindrical block E has a horizontal velocity v_0 when it strikes squarely the yoke BD that is attached to the $\frac{7}{8}$-in.-diameter rods AB and CD. Knowing that the rods are made of a steel for which $\sigma_Y = 50$ ksi and $E = 29 \times 10^6$ psi, determine the maximum allowable speed v_0 if the rods are not to be permanently deformed.

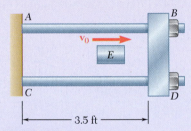

Fig. P11.43 and P11.44

11.44 The cylindrical block E has a speed $v_0 = 16$ ft/s when it strikes squarely the yoke BD that is attached to the $\frac{7}{8}$-in.-diameter rods AB and CD. Knowing that the rods are made of a steel for which $\sigma_Y = 50$ ksi and $E = 29 \times 10^6$ psi, determine the weight of block E for which the factor of safety is five with respect to permanent deformation of the rods.

11.45 The 35-kg collar D is released from rest in the position shown and is stopped by a plate attached at end C of the vertical rod ABC. Knowing that $E = 200$ GPa for both portions of the rod, determine the distance h for which the maximum stress in the rod is 250 MPa.

11.46 The 15-kg collar D is released from rest in the position shown and is stopped by a plate attached at end C of the vertical rod ABC. Knowing that $E = 200$ GPa for both portions of the rod, determine (a) the maximum deflection of end C, (b) the equivalent static load, (c) the maximum stress that occurs in the rod.

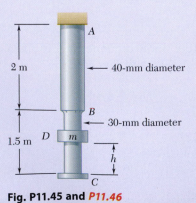

Fig. P11.45 and *P11.46*

11.47 The 48-kg collar G is released from rest in the position shown and is stopped by plate BDF that is attached to the 20-mm-diameter steel rod CD and to the 15-mm-diameter steel rods AB and EF. Knowing that for the grade of steel used $\sigma_{all} = 180$ MPa and $E = 200$ GPa, determine the largest allowable distance h.

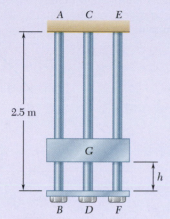

Fig. P11.47

11.48 A 25-lb block C moving horizontally with at velocity $\mathbf{v}_0$ hits the post AB squarely as shown. Using $E = 29 \times 10^6$ psi, determine the largest speed v_0 for which the maximum normal stress in the post does not exceed 18 ksi.

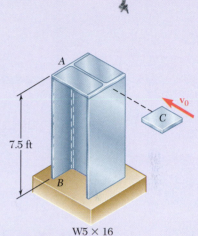

W5 × 16

Fig. P11.48

11.49 Solve Prob. 11.48, assuming that the post AB has been rotated 90° about its longitudinal axis.

11.50 An aluminum tube having the cross section shown is struck squarely in its midsection by a 6-kg block moving horizontally with a speed of 2 m/s. Using $E = 70$ GPa, determine (a) the equivalent static load, (b) the maximum stress in the beam, (c) the maximum deflection at the midpoint C of the beam.

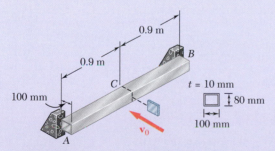

Fig. P11.50

11.51 Solve Prob. 11.50, assuming that the tube has been replaced by a solid aluminum bar with the same outside dimensions as the tube.

11.52 The 2-kg block D is dropped from the position shown onto the end of a 16-mm-diameter rod. Knowing that $E = 200$ GPa, determine (a) the maximum deflection of end A, (b) the maximum bending moment in the rod, (c) the maximum normal stress in the rod.

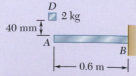

Fig. P11.52

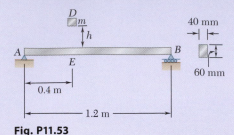

Fig. P11.53

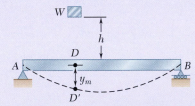

Fig. P11.55

11.53 The 10-kg block D is dropped from a height $h = 450$ mm onto the aluminum beam AB. Knowing that $E = 70$ GPa, determine (*a*) the maximum deflection of point E, (*b*) the maximum stress in the beam.

11.54 The 4-lb block D is dropped from the position shown onto the end of a $\frac{5}{8}$-in.-diameter rod. Knowing that $E = 29 \times 10^6$ psi, determine (*a*) the maximum deflection at point A, (*b*) the maximum bending moment in the rod, (*c*) the maximum normal stress in the rod.

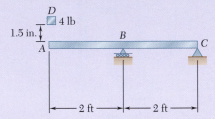

Fig. P11.54

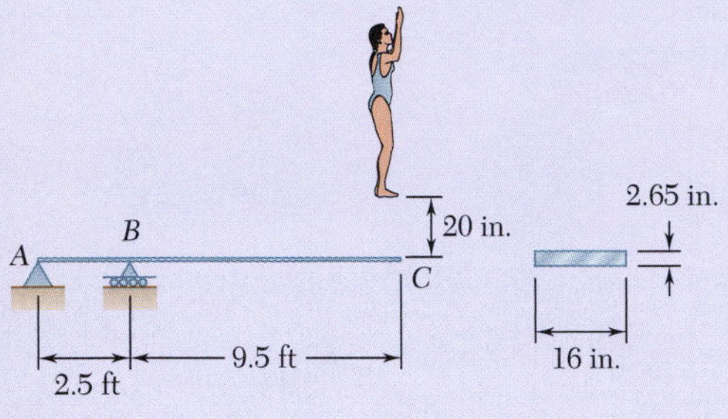

11.55 A 160-lb diver jumps from a height of 20 in. onto end C of a diving board having the uniform cross section shown. Assuming that the diver's legs remain rigid and using $E = 1.8 \times 10^6$ psi, determine (*a*) the maximum deflection at point C, (*b*) the maximum normal stress in the board, (*c*) the equivalent static load.

11.56 A block of weight W is dropped from a height h onto the horizontal beam AB and hits it at point D. (*a*) Show that the maximum deflection y_m at point D can be expressed as

$$y_m = y_{st}\left(1 + \sqrt{1 + \frac{2h}{y_{st}}}\right)$$

where y_{st} represents the deflection at D caused by a static load W applied at that point and where the quantity in parenthesis is referred to as the *impact factor*. (*b*) Compute the impact factor for the beam of Prob. 11.52.

Fig. P11.56 and P11.57

11.57 A block of weight W is dropped from a height h onto the horizontal beam AB and hits point D. (*a*) Denoting by y_m the exact value of the maximum deflection at D and by y'_m the value obtained by neglecting the effect of this deflection on the change in potential energy of the block, show that the absolute value of the relative error is $(y'_m - y_m)/y_m$, never exceeding $y'_m/2h$. (*b*) Check the result obtained in part *a* by solving part *a* of Prob. 11.52 without taking y_m into account when determining the change in potential energy of the load, and comparing the answer obtained in this way with the exact answer to that problem.

11.58 and 11.59 Using the method of work and energy, determine the deflection at point D caused by the load **P**.

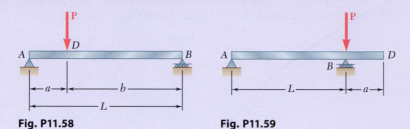

Fig. P11.58   Fig. P11.59

11.60 and 11.61 Using the method of work and energy, determine the slope at point D caused by the couple $\mathbf{M}_0$.

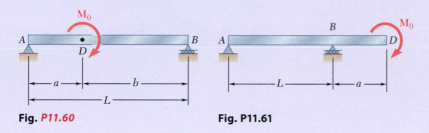

Fig. *P11.60*   Fig. P11.61

11.62 and 11.63 Using the method of work and energy, determine the deflection at point C caused by the load **P**.

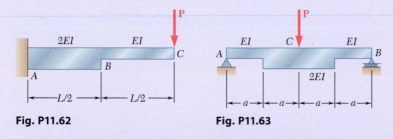

Fig. P11.62   Fig. P11.63

11.64 Using the method of work and energy, determine the slope at point B caused by the couple $\mathbf{M}_0$.

11.65 Using the method of work and energy, determine the slope at point D caused by the couple $\mathbf{M}_0$.

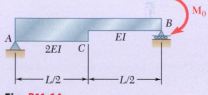

Fig. *P11.64*

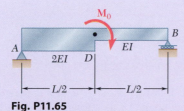

Fig. P11.65

11.66 The 20-mm diameter steel rod BC is attached to the lever AB and to the fixed support C. The uniform steel lever is 10 mm thick and 30 mm deep. Using the method of work and energy, determine the deflection of point A when $L = 600$ mm. Use $E = 200$ GPa and $G = 77.2$ GPa.

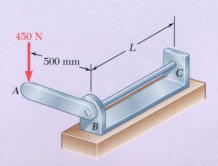

Fig. P11.66

11.67 Torques of the same magnitude T are applied to the steel shafts AB and CD. Using the method of work and energy, determine the length L of the hollow portion of shaft CD for which the angle of twist at C is equal to 1.25 times the angle of twist at A.

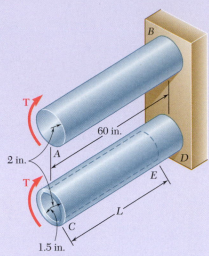

Fig. P11.67

11.68 Two steel shafts, each of 0.75-in.-diameter, are connected by the gears shown. Knowing that $G = 11.2 \times 10^6$ psi and that shaft DF is fixed at F, determine the angle through which end A rotates when a 750-lb·in. torque is applied at A. (Neglect the strain energy due to the bending of the shafts.)

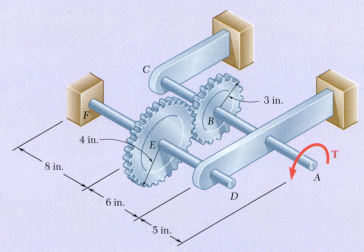

Fig. P11.68

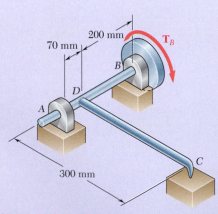

Fig. P11.69

11.69 The 20-mm-diameter steel rod CD is welded to the 20-mm-diameter steel shaft AB as shown. End C of rod CD is touching the rigid surface shown when a couple $\mathbf{T}_B$ is applied to a disk attached to shaft AB. Knowing that the bearings are self aligning and exert no couples on the shaft, determine the angle of rotation of the disk when $T_B = 400$ N·m. Use $E = 200$ GPa and $G = 77.2$ GPa. (Consider the strain energy due to both bending and twisting in shaft AB and to bending in arm CD.)

11.70 The thin-walled hollow cylindrical member *AB* has a noncircular cross section of nonuniform thickness. Using the expression given in Eq. (3.50) of Sec. 3.10 and the expression for the strain-energy density given in Eq. (11.17), show that the angle of twist of member *AB* is

$$\phi = \frac{TL}{4\mathfrak{a}^2 G} \oint \frac{ds}{t}$$

where *ds* is the length of a small element of the wall cross section and $\mathfrak{a}$ is the area enclosed by center line of the wall cross section.

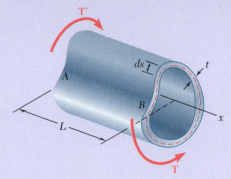

Fig. P11.70

11.71 and 11.72 Each member of the truss shown has a uniform cross-sectional area *A*. Using the method of work and energy, determine the horizontal deflection of the point of application of the load **P**.

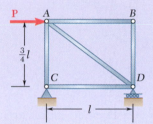
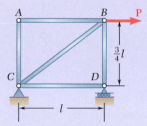

Fig. *P11.71*   **Fig. *P11.72***

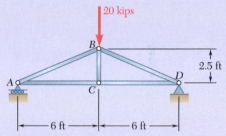

Fig. P11.73

11.73 Each member of the truss shown is made of steel and has a uniform cross-sectional area of 5 in². Using $E = 29 \times 10^6$ psi, determine the vertical deflection of joint *B* caused by the application of the 20-kip load.

11.74 Each member of the truss shown is made of steel. The cross-sectional area of member *BC* is 800 mm², and for all other members the cross-sectional area is 400 mm². Using $E = 200$ GPa, determine the deflection of point *D* caused by the 60-kN load.

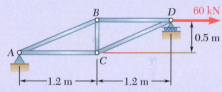

Fig. *P11.74*

11.75 Each member of the truss shown is made of steel and has a cross-sectional area of 5 in². Using $E = 29 \times 10^6$ psi, determine the vertical deflection of point *C* caused by the 15-kip load.

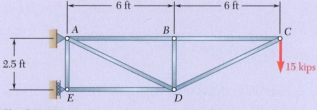

Fig. P11.75

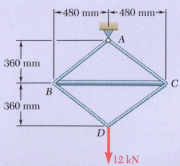

11.76 The steel rod *BC* has a 24-mm diameter and the steel cable *ABDCA* has a 12-mm diameter. Using $E = 200$ GPa, determine the deflection of joint *D* caused by the 12-kN load.

Fig. P11.76

*11.6 WORK AND ENERGY UNDER MULTIPLE LOADS

In this section, the strain energy of a structure subjected to several loads is considered and expressed in terms of the loads and resulting deflections.

Consider an elastic beam AB subjected to two concentrated loads $\mathbf{P}_1$ and $\mathbf{P}_2$. The strain energy of the beam is equal to the work of $\mathbf{P}_1$ and $\mathbf{P}_2$ as they are slowly applied to the beam at C_1 and C_2, respectively (Fig. 11.31). However, in order to evaluate this work, the deflections x_1 and x_2 must be expressed in terms of the loads $\mathbf{P}_1$ and $\mathbf{P}_2$.

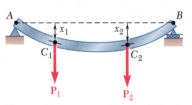

Fig. 11.31 Beam with multiple loads.

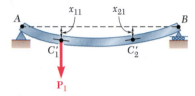

Fig. 11.32 Beam deflections at C_1 and C_2 due to single load $\mathbf{P}_1$.

Assume that only $\mathbf{P}_1$ is applied to the beam (Fig. 11.32). Both C_1 and C_2 are deflected, and their deflections are proportional to the load P_1. Denoting these deflections by x_{11} and x_{21}, respectively, write

$$x_{11} = \alpha_{11}P_1 \qquad x_{21} = \alpha_{21}P_1 \qquad \textbf{(11.41)}$$

where α_{11} and α_{21} are constants called *influence coefficients*. These constants represent the deflections of C_1 and C_2 when a unit load is applied at C_1 and are characteristics of the beam.

Now assume that only $\mathbf{P}_2$ is applied to the beam (Fig. 11.33). The resulting deflections of C_1 and C_2 are denoted by x_{12} and x_{22}, respectively, so

$$x_{12} = \alpha_{12}P_2 \qquad x_{22} = \alpha_{22}P_2 \qquad \textbf{(11.42)}$$

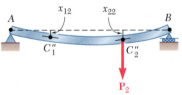

Fig. 11.33 Beam deflections at C_1 and C_2 due to single load $\mathbf{P}_2$.

where α_{12} and α_{22} are the influence coefficients representing the deflections of C_1 and C_2 when a unit load is applied at C_2. Applying the principle of superposition, the deflections x_1 and x_2 of C_1 and C_2 when both loads are applied (Fig. 11.31) are

$$x_1 = x_{11} + x_{12} = \alpha_{11}P_1 + \alpha_{12}P_2 \qquad \textbf{(11.43)}$$

$$x_2 = x_{21} + x_{22} = \alpha_{21}P_1 + \alpha_{22}P_2 \qquad \textbf{(11.44)}$$

To compute the work done by $\mathbf{P}_1$ and $\mathbf{P}_2$ and thus the strain energy of the beam, assume that $\mathbf{P}_1$ is first applied slowly at C_1 (Fig. 11.34a). Recalling the first of Eqs. (11.41), the work of $\mathbf{P}_1$ is

$$\tfrac{1}{2}P_1 x_{11} = \tfrac{1}{2}P_1(\alpha_{11}P_1) = \tfrac{1}{2}\alpha_{11}P_1^2 \tag{11.45}$$

Note that $\mathbf{P}_2$ does no work while C_2 moves through x_{21}, since it has not yet been applied to the beam.

Now slowly apply $\mathbf{P}_2$ at C_2 (Fig. 11.34b). Recalling the second of Eqs. (11.42), the work of $\mathbf{P}_2$ is

$$\tfrac{1}{2}P_2 x_{22} = \tfrac{1}{2}P_2(\alpha_{22}P_2) = \tfrac{1}{2}\alpha_{22}P_2^2 \tag{11.46}$$

But, as $\mathbf{P}_2$ is slowly applied at C_2, the point of application at $\mathbf{P}_1$ moves through x_{12} from C_1' to C_1, and load $\mathbf{P}_1$ does work. Since $\mathbf{P}_1$ is *fully applied* during this displacement (Fig. 11.35), its work is equal to $P_1 x_{12}$, or recalling the first of Eqs. (11.42),

$$P_1 x_{12} = P_1(\alpha_{12}P_2) = \alpha_{12}P_1 P_2 \tag{11.47}$$

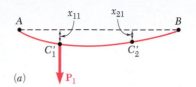

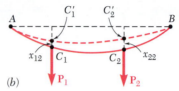

(a)

(b)

Fig. 11.34 (a) Deflection due to P_1 only. (b) Additional deflection due to subsequent application of P_2.

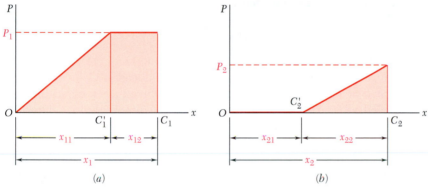

(a)

(b)

Fig. 11.35 Load-displacement diagrams for application of P_1 followed by P_2. (a) Load-displacement diagram for C_1. (b) Load-displacement diagram for C_2.

Adding the expressions in Eq. (11.45), (11.46), and (11.47), the strain energy of the beam under the loads $\mathbf{P}_1$ and $\mathbf{P}_2$ is

$$U = \tfrac{1}{2}(\alpha_{11}P_1^2 + 2\alpha_{12}P_1 P_2 + \alpha_{22}P_2^2) \tag{11.48}$$

If load $\mathbf{P}_2$ had been applied first to the beam (Fig. 11.36a), followed by load $\mathbf{P}_1$ (Fig. 11.36b), the work done by each load would have been as

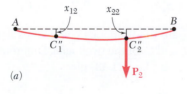

(a)

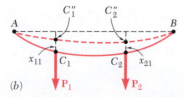

(b)

Fig. 11.36 (a) Deflection due to P_2 only. (b) Additional deflection due to subsequent application of P_1.

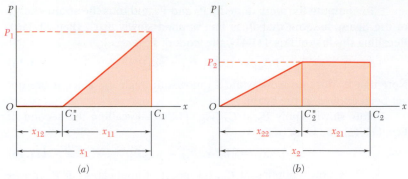

Fig. 11.37 Load-displacement diagrams for application of P_2 followed by P_1. (a) Load-displacement diagram for C_1. (b) Load-displacement diagram for C_2.

shown in Fig. 11.37. Similar calculations would lead to an alternative expression for the strain energy of the beam:

$$U = \tfrac{1}{2}(\alpha_{22}P_2^2 + 2\alpha_{21}P_2P_1 + \alpha_{11}P_1^2) \tag{11.49}$$

Equating the right-hand members of Eqs. (11.48) and (11.49), $\alpha_{12} = \alpha_{21}$, and we thus conclude that the deflection produced at C_1 by a unit load applied at C_2 is equal to the deflection produced at C_2 by a unit load applied at C_1. This is known as *Maxwell's reciprocal theorem,* after the British physicist James Clerk Maxwell (1831–1879).

While we are now able to express the strain energy U of a structure subjected to several loads as a function of these loads, the work energy method of Sec. 11.5B cannot be used determine the deflections of such a structure. Computing the strain energy U by integrating the strain-energy density u over the structure and substituting the expression obtained into Eq. (11.48) yields only one equation, which clearly can not be solved for the multiple coefficients α.

*11.7 CASTIGLIANO'S THEOREM

Recall from the previous section that the strain energy of an elastic structure subjected to two loads $\mathbf{P}_1$ and $\mathbf{P}_2$ is

$$U = \tfrac{1}{2}(\alpha_{11}P_1^2 + 2\alpha_{12}P_1P_2 + \alpha_{22}P_2^2) \tag{11.48}$$

where α_{11}, α_{12}, and α_{22} are the influence coefficients associated with the points of application C_1 and C_2 of the two loads. Differentiating Eq. (11.48) with respect to P_1 and using Eq. (11.43) gives

$$\frac{\partial U}{\partial P_1} = \alpha_{11}P_1 + \alpha_{12}P_2 = x_1 \tag{11.50}$$

Differentiating Eq. (11.48) with respect to P_2, using Eq. (11.44), and keeping in mind that $\alpha_{12} = \alpha_{21}$, we have

$$\frac{\partial U}{\partial P_2} = \alpha_{12}P_1 + \alpha_{22}P_2 = x_2 \tag{11.51}$$

More generally if an elastic structure is subjected to n loads $\mathbf{P}_1$, $\mathbf{P}_2, \ldots, \mathbf{P}_n$, the deflection x_j of the point of application of $\mathbf{P}_j$, and measured along the line of action of $\mathbf{P}_j$ is expressed as the partial derivative of the strain energy of the structure with respect to the load $\mathbf{P}_j$. Thus,

$$x_j = \frac{\partial U}{\partial P_j} \tag{11.52}$$

This is *Castigliano's theorem*, named after the Italian engineer Alberto Castigliano (1847–1884) who first stated it.[†]

Recall that the work of a couple $\mathbf{M}$ is $\frac{1}{2}M\theta$, where θ is the angle of rotation at the point where the couple is slowly applied. Castigliano's theorem can be used to determine the slope of a beam at the point of application of a couple $\mathbf{M}_j$. Thus,

$$\theta_j = \frac{\partial U}{\partial M_j} \tag{11.55}$$

Similarly, the angle of twist ϕ_j in a section of a shaft where a torque $\mathbf{T}_j$ is slowly applied is obtained by differentiating the strain energy of the shaft with respect to T_j:

$$\phi_j = \frac{\partial U}{\partial T_j} \tag{11.56}$$

[†]For an elastic structure subjected to n loads $\mathbf{P}_1$, $\mathbf{P}_2, \ldots, \mathbf{P}_n$, the deflection of the point of application of $\mathbf{P}_j$, measured along the line of action of $\mathbf{P}_j$, is

$$x_j = \sum_k \alpha_{jk} P_k \tag{11.53}$$

and the strain energy of the structure is

$$U = \tfrac{1}{2} \sum_i \sum_k \alpha_{ik} P_i P_k \tag{11.54}$$

Differentiating U with respect to $\mathbf{P}_j$ and observing that $\mathbf{P}_j$ is found in terms corresponding to either $i = j$ or $k = j$ gives

$$\frac{\partial U}{\partial P_j} = \frac{1}{2} \sum_k \alpha_{jk} P_k + \frac{1}{2} \sum_i \alpha_{ij} P_i$$

or since $\alpha_{ij} = \alpha_{ji}$,

$$\frac{\partial U}{\partial P_j} = \frac{1}{2} \sum_k \alpha_{jk} P_k + \frac{1}{2} \sum_i \alpha_{ji} P_i = \sum_k \alpha_{jk} P_k$$

Recalling Eq. (11.53), we verify that

$$x_j = \frac{\partial U}{\partial P_j} \tag{11.52}$$

*11.8 DEFLECTIONS BY CASTIGLIANO'S THEOREM

We saw in the preceding section that the deflection x_j of a structure at the point of application of a load $\mathbf{P}_j$ can be determined by computing the partial derivative $\partial U/\partial P_j$ of the strain energy U of the structure. As we recall from Secs. 11.2A and 11.2B, the strain energy U is obtained by integrating or summing over the structure the strain energy of each element of the structure. The calculation by Castigliano's theorem of the deflection x_j is simplified if the differentiation with respect to the load P_j is carried out before the integration or summation.

For the beam from Sec. 11.2A, the strain energy was found to be

$$U = \int_0^L \frac{M^2}{2EI} dx \tag{11.15}$$

and the deflection x_j of the point of application of the load $\mathbf{P}_j$ is then

$$x_j = \frac{\partial U}{\partial P_j} = \int_0^L \frac{M}{EI} \frac{\partial M}{\partial P_j} dx \tag{11.57}$$

For a truss of n uniform members with a length L_i, cross-sectional area A_i, and internal force F_i, Eq. (11.14) can be used for the strain energy U to write

$$U = \sum_{i=1}^n \frac{F_i^2 L_i}{2A_i E} \tag{11.58}$$

The deflection x_j of the point of application of the load $\mathbf{P}_j$ is obtained by differentiating each term of the sum with respect to P_j. Thus,

$$x_j = \frac{\partial U}{\partial P_j} = \sum_{i=1}^n \frac{F_i L_i}{A_i E} \frac{\partial F_i}{\partial P_j} \tag{11.59}$$

Concept Application 11.12

The cantilever beam AB supports a uniformly distributed load w and a concentrated load $\mathbf{P}$ (Fig. 11.38). Knowing that $L = 2$ m, $w = 4$ kN/m, $P = 6$ kN, and $EI = 5$ MN·m², determine the deflection at A.

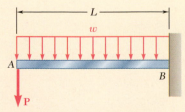

Fig. 11.38 Cantilever beam loaded as shown.

(continued)

The deflection y_A of point A where load **P** is applied is obtained from Eq. (11.57). Since **P** is vertical and directed downward, y_A represents a vertical deflection and is positive downward.

$$y_A = \frac{\partial U}{\partial P} = \int_0^L \frac{M}{EI} \frac{\partial M}{\partial P} dx \tag{1}$$

The bending moment M at a distance x from A is

$$M = -(Px + \tfrac{1}{2}wx^2) \tag{2}$$

and its derivative with respect to P is

$$\frac{\partial M}{\partial P} = -x$$

Substituting for M and $\partial M/\partial P$ into Eq. (1),

$$y_A = \frac{1}{EI} \int_0^L \left(Px^2 + \frac{1}{2} wx^3 \right) dx$$

$$y_A = \frac{1}{EI} \left(\frac{PL^3}{3} + \frac{wL^4}{8} \right) \tag{3}$$

Substituting the given data,

$$y_A = \frac{1}{5 \times 10^6 \text{ N·m}^2} \left[\frac{(6 \times 10^3 \text{ N})(2 \text{ m})^3}{3} + \frac{(4 \times 10^3 \text{ N/m})(2 \text{ m})^4}{8} \right]$$

$$y_A = 4.8 \times 10^{-3} \text{ m} \qquad y_A = 4.8 \text{ mm} \downarrow$$

Note that the computation of the partial derivative $\partial M/\partial P$ *could not have been carried out* if the numerical value of P had been substituted for P in Eq. (2) for the bending moment.

The deflection x_j of a structure at a given point C_j can be obtained using the direct application of Castigliano's theorem only if load $\mathbf{P}_j$ is applied at C_j in the direction for which x_j is to be determined. When no load is applied at C_j or a load is applied in a direction other than the desired one, the deflection x_j still can be found using Castigliano's theorem if we use the following procedure. First, a fictitious or "dummy" load $\mathbf{Q}_j$ at C_j is applied in the direction in which the deflection x_j is to be determined. Then Castigliano's theorem is used to obtain the deflection

$$x_j = \frac{\partial U}{\partial Q_j} \tag{11.60}$$

due to $\mathbf{Q}_j$ and the actual loads. Making $Q_j = 0$ in Eq. (11.60) yields the deflection at C_j in the desired direction under the given load.

The slope θ_j of a beam at a point C_j can be found by applying a fictitious couple $\mathbf{M}_j$ at C_j, computing the partial derivative $\partial U/\partial M_j$, and making $M_j = 0$ in the expression obtained.

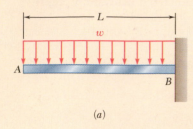

(a)

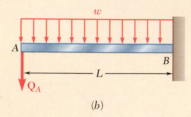

(b)

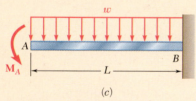

(c)

Fig. 11.39 *(a)* Cantilever beam supporting a uniformly distributed load. *(b)* Dummy load Q_A applied to determine deflection at *A*. *(c)* Dummy load M_A applied to determine the slope at *A*.

Concept Application 11.13

The cantilever beam *AB* supports a uniformly distributed load *w* (Fig. 11.39*a*). Determine the deflection and slope at *A*.

Deflection at A. Apply a dummy downward load $\mathbf{Q}_A$ at *A* (Fig. 11.39*b*) and write

$$y_A = \frac{\partial U}{\partial Q_A} = \int_0^L \frac{M}{EI} \frac{\partial M}{\partial Q_A} dx \quad (1)$$

The bending moment *M* at a distance *x* from *A* is

$$M = -Q_A x - \tfrac{1}{2} w x^2 \quad (2)$$

and its derivative with respect to Q_A is

$$\frac{\partial M}{\partial Q_A} = -x \quad (3)$$

Substituting for *M* and $\partial M/\partial Q_A$ from Eq. (2) and (3) into Eq. (1) and making $Q_A = 0$, the deflection at *A* for the given load is:

$$y_A = \frac{1}{EI} \int_0^L (-\tfrac{1}{2} w x^2)(-x)\, dx = +\frac{wL^4}{8EI}$$

Since the dummy load was directed downward, the positive sign indicates that

$$y_A = \frac{wL^4}{8EI} \downarrow$$

Slope at A. Apply a dummy counterclockwise couple $\mathbf{M}_A$ at *A* (Fig. 11.39*c*) and write

$$\theta_A = \frac{\partial U}{\partial M_A}$$

Recalling Eq. (11.15),

$$\theta_A = \frac{\partial}{\partial M_A} \int_0^L \frac{M^2}{2EI} dx = \int_0^L \frac{M}{EI} \frac{\partial M}{\partial M_A} dx \quad (4)$$

The bending moment *M* at a distance *x* from *A* is

$$M = -M_A - \tfrac{1}{2} w x^2 \quad (5)$$

and its derivative with respect to M_A is

$$\frac{\partial M}{\partial M_A} = -1 \quad (6)$$

Substituting for *M* and $\partial M/\partial M_A$ from Eq. (5) and (6) into Eq. (4) and making $M_A = 0$, the slope at *A* for the given load is:

$$\theta_A = \frac{1}{EI} \int_0^L (-\tfrac{1}{2} w x^2)(-1)\, dx = +\frac{wL^3}{6EI}$$

Since the dummy couple was counterclockwise, the positive sign indicates that the angle θ_A is also counterclockwise:

$$\theta_A = \frac{wL^3}{6EI} \ \measuredangle$$

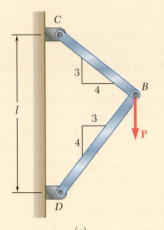

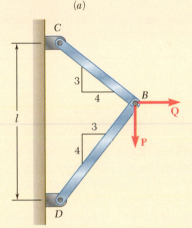

(a)

(b)

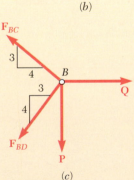

(c)

Fig. 11.40 (a) Frame *CBD* supporting vertical load *P*. (b) Frame *CBD* with horizontal dummy load *Q* applied. (c) Free-body diagram of joint *B* for finding member forces in terms of loads *P* and *Q*.

Concept Application 11.14

A load **P** is supported at *B* by two rods of the same material and the same cross-sectional area *A* (Fig. 11.40*a*). Determine the horizontal and vertical deflection of point *B*.

We apply a dummy horizontal load **Q** at *B* (Fig. 11.40*b*). From Castigliano's theorem,

$$x_B = \frac{\partial U}{\partial Q} \qquad y_B = \frac{\partial U}{\partial P}$$

Using Eq. (11.14) to obtain the strain energy for the rods

$$U = \frac{F_{BC}^2 (BC)}{2AE} + \frac{F_{BD}^2 (BD)}{2AE}$$

where F_{BC} and F_{BD} represent the forces in *BC* and *BD*, respectively. Therefore,

$$x_B = \frac{\partial U}{\partial Q} = \frac{F_{BC} (BC)}{AE} \frac{\partial F_{BC}}{\partial Q} + \frac{F_{BD} (BD)}{AE} \frac{\partial F_{BD}}{\partial Q} \qquad \textbf{(1)}$$

and

$$y_B = \frac{\partial U}{\partial P} = \frac{F_{BC} (BC)}{AE} \frac{\partial F_{BC}}{\partial P} + \frac{F_{BD} (BD)}{AE} \frac{\partial F_{BD}}{\partial P} \qquad \textbf{(2)}$$

From the free-body diagram of pin *B* (Fig. 11.40*c*),

$$F_{BC} = 0.6P + 0.8Q \qquad F_{BD} = -0.8P + 0.6Q \qquad \textbf{(3)}$$

Differentiating these equations with respect to *Q* and *P*, write

$$\frac{\partial F_{BC}}{\partial Q} = 0.8 \qquad \frac{\partial F_{BD}}{\partial Q} = 0.6$$

$$\frac{\partial F_{BC}}{\partial P} = 0.6 \qquad \frac{\partial F_{BD}}{\partial P} = -0.8 \qquad \textbf{(4)}$$

Substituting from Eqs. (3) and (4) into both Eqs. (1) and (2), making $Q = 0$, and noting that $BC = 0.6l$ and $BD = 0.8l$, the horizontal and vertical deflections of point *B* under the given load **P** are

$$x_B = \frac{(0.6P)(0.6l)}{AE}(0.8) + \frac{(-0.8P)(0.8l)}{AE}(0.6)$$

$$= -0.096 \frac{Pl}{AE}$$

$$y_B = \frac{(0.6P)(0.6l)}{AE}(0.6) + \frac{(-0.8P)(0.8l)}{AE}(-0.8)$$

$$= +0.728 \frac{Pl}{AE}$$

Referring to the directions of the loads **Q** and **P**, we conclude that

$$x_B = 0.096 \frac{Pl}{AE} \leftarrow \qquad y_B = 0.728 \frac{Pl}{AE} \downarrow$$

We check that the expression found for the vertical deflection of *B* is the same as obtained in Concept Application 11.9.

*11.9 STATICALLY INDETERMINATE STRUCTURES

The reactions at the supports of a statically indeterminate elastic structure can be determined using Castigliano's theorem. For example, in a structure indeterminate to the first degree, designate one of the reactions as redundant and eliminate or modify accordingly the corresponding support. The redundant reaction is treated like an unknown load that, together with the other loads, must produce deformations compatible with the original supports. First calculate the strain energy U of the structure due to the combined action of the loads and the redundant reaction. Observing that the partial derivative of U with respect to the redundant reaction represents the deflection (or slope) at the support that has been eliminated or modified, we then set this derivative equal to zero and solve for the redundant reaction.[†] The remaining reactions are found using the equations of statics.

[†]This is in the case of a rigid support allowing no deflection. For other types of support, the partial derivative of U should be set equal to the allowed deflection.

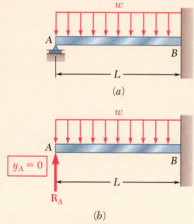

Fig. 11.41 (a) Beam statically indeterminate to first degree. (b) Redundant reaction at A and zero displacement boundary condition.

Concept Application 11.15

Determine the reactions at the supports for the prismatic beam and load shown (Fig. 11.41a).

The beam is statically indeterminate to the first degree. The reaction at A is redundant and the beam is released from that support. The reaction $\mathbf{R}_A$ is considered to be an unknown load (Fig. 11.41b) and will be determined under the condition that the deflection y_A at A must be zero. By Castigliano's theorem, $y_A = \partial U / \partial R_A$, where U is the strain energy of the beam under the distributed load and the redundant reaction. Recalling Eq. (11.57),

$$y_A = \frac{\partial U}{\partial R_A} = \int_0^L \frac{M}{EI} \frac{\partial M}{\partial R_A} \, dx \qquad (1)$$

The bending moment M for the load of Fig. 11.41b at a distance x from A is

$$M = R_A x - \tfrac{1}{2} w x^2 \qquad (2)$$

and its derivative with respect to R_A is

$$\frac{\partial M}{\partial R_A} = x \qquad (3)$$

Substituting for M and $\partial M / \partial R_A$ from Eqs. (2) and (3) into Eq. (1), write

$$y_A = \frac{1}{EI} \int_0^L \left(R_A x^2 - \frac{1}{2} w x^3 \right) dx = \frac{1}{EI} \left(\frac{R_A L^3}{3} - \frac{w L^4}{8} \right)$$

(continued)

Set $y_A = 0$ and solve for R_A:

$$R_A = \tfrac{3}{8}wL \qquad \mathbf{R}_A = \tfrac{3}{8}wL \uparrow$$

From the conditions of equilibrium for the beam, the reaction at B consists of the force and couple:

$$\mathbf{R}_B = \tfrac{5}{8}wL \uparrow \qquad \mathbf{M}_B = \tfrac{1}{8}wL^2 \downarrow$$

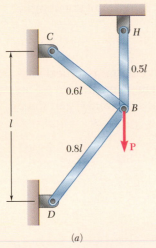

(a)

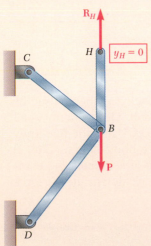

(b)

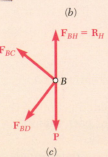

(c)

Fig. 11.42 (a) Statically indeterminate frame supporting a vertical load P. (b) Redundant reaction at H and zero displacement boundary condition. (c) Free-body diagram of joint B.

Concept Application 11.16

A load $\mathbf{P}$ is supported at B by three rods of the same material and the same cross-sectional area A (Fig. 11.42a). Determine the force in each rod.

The structure is statically indeterminate to the first degree. The reaction at H is choosen as the redundant. Thus rod BH is released from its support at H. The reaction $\mathbf{R}_H$ is now considered to be an unknown load (Fig. 11.42b) and will be determined under the condition that the deflection y_H of point H must be zero. By Castigliano's theorem, $y_H = \partial U/\partial R_H$, where U is the strain energy of the three-rod system under load $\mathbf{P}$ and the redundant reaction $\mathbf{R}_H$. Recalling Eq. (11.59),

$$y_H = \frac{F_{BC}(BC)}{AE}\frac{\partial F_{BC}}{\partial R_H} + \frac{F_{BD}(BD)}{AE}\frac{\partial F_{BD}}{\partial R_H} + \frac{F_{BH}(BH)}{AE}\frac{\partial F_{BH}}{\partial R_H} \tag{1}$$

Note that the force in rod BH is equal to R_H, or

$$F_{BH} = R_H \tag{2}$$

Then, from the free-body diagram of pin B (Fig. 11.42c),

$$F_{BC} = 0.6P - 0.6R_H \qquad F_{BD} = 0.8R_H - 0.8P \tag{3}$$

Differentiating with respect to R_H the force in each rod gives

$$\frac{\partial F_{BC}}{\partial R_H} = -0.6 \qquad \frac{\partial F_{BD}}{\partial R_H} = 0.8 \qquad \frac{\partial F_{BH}}{\partial R_H} = 1 \tag{4}$$

Substituting from Eq. (2), (3), and (4) into Eq. (1) and noting that the lengths BC, BD, and BH are equal to $0.6l$, $0.8l$, and $0.5l$, respectively,

$$y_H = \frac{1}{AE}\big[(0.6P - 0.6R_H)(0.6l)(-0.6) + (0.8R_H - 0.8P)(0.8l)(0.8) + R_H(0.5l)(1)\big]$$

Setting $y_H = 0$ gives

$$1.228R_H - 0.728P = 0$$

and solving for R_H:

$$R_H = 0.593P$$

Carrying this value into Eqs. (2) and (3), the forces in the three rods are

$$F_{BC} = +0.244P \qquad F_{BD} = -0.326P \qquad F_{BH} = +0.593P$$

Sample Problem 11.5

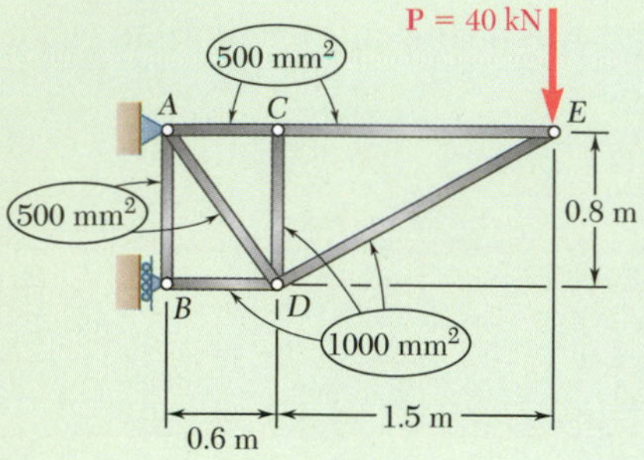

P = 40 kN

(500 mm²)

(500 mm²)

(1000 mm²)

0.8 m

0.6 m   1.5 m

For the truss and loading of Sample Prob. 11.4, determine the vertical deflection of joint C.

STRATEGY: Add a dummy load associated with the desired vertical deflection at joint C. The truss is then analyzed to determine the member forces, first by drawing a free-body diagram of the truss to find the reactions and then by using equilibrium at each joint to find the member forces. Use Eq. (11.59) to get the deflection in terms of the dummy load $\mathbf{Q}$.

MODELING and ANALYSIS:

Castigliano's Theorem. We introduce the dummy vertical load $\mathbf{Q}$ as shown in Fig. 1. Using Castigliano's theorem where the force F_i in a given member i is caused by the combined load of $\mathbf{P}$ and $\mathbf{Q}$ and since $E =$ constant,

$$y_C = \sum \left(\frac{F_i L_i}{A_i E} \right) \frac{\partial F_i}{\partial Q} = \frac{1}{E} \sum \left(\frac{F_i L_i}{A_i} \right) \frac{\partial F_i}{\partial Q} \quad \text{(1)}$$

Fig. 1 Dummy load $\mathbf{Q}$ applied to joint C used to determine vertical deflection at C.

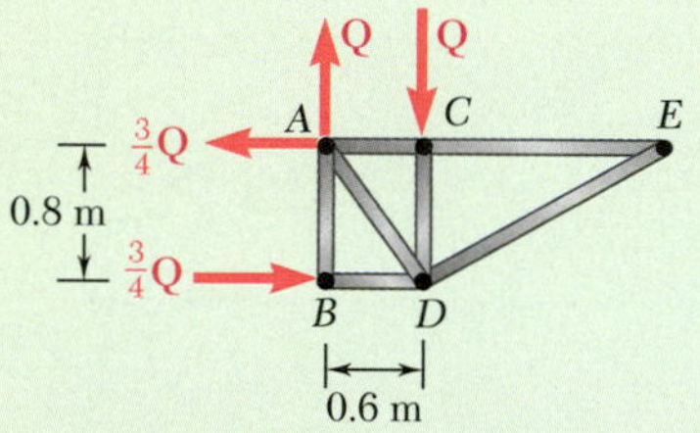

Fig. 2 Free-body diagram of truss with only dummy load Q.

Force in Members. Since the force in each member caused by the load $\mathbf{P}$ was previously found in Sample Prob. 11.4, we only need to determine the force in each member due to $\mathbf{Q}$. Using the free-body diagram of the truss with load $\mathbf{Q}$, we draw a free-body diagram (Fig. 2) to determine the reactions. Then, considering in sequence the equilibrium of joints E, C, B and D and using Fig. 3, we determine the force in each member caused by load $\mathbf{Q}$.

Joint E: $F_{CE} = F_{DE} = 0$

Joint C: $F_{AC} = 0$; $F_{CD} = -Q$

Joint B: $F_{AB} = 0$; $F_{BD} = -\frac{3}{4}Q$

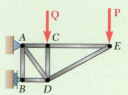

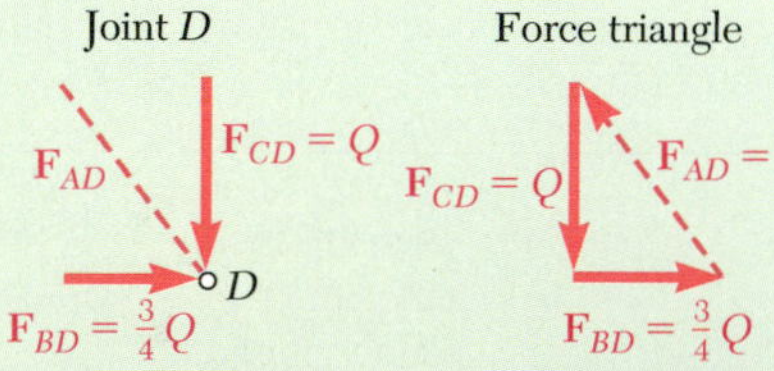

Fig. 3 Force analysis diagrams for joint D.

The total force in each member under the combined action of $\mathbf{Q}$ and $\mathbf{P}$ is shown in the following table. Form $\partial F_i / \partial Q$ for each member, then compute $(F_i L_i / A_i)(\partial F_i / \partial Q)$, as indicated.

(continued)

Member	F_i	$\partial F_i/\partial Q$	L_i, m	A_i, m^2	$\left(\dfrac{F_iL_i}{A_i}\right)\dfrac{\partial F_i}{\partial Q}$
AB	0	0	0.8	500×10^{-6}	0
AC	$+15P/8$	0	0.6	500×10^{-6}	0
AD	$+5P/4 + 5Q/4$	$\frac{5}{4}$	1.0	500×10^{-6}	$+3125P + 3125Q$
BD	$-21P/8 - 3Q/4$	$-\frac{3}{4}$	0.6	1000×10^{-6}	$+1181P + 338Q$
CD	$-Q$	-1	0.8	1000×10^{-6}	$+ 800Q$
CE	$+15P/8$	0	1.5	500×10^{-6}	0
DE	$-17P/8$	0	1.7	1000×10^{-6}	0

$$\sum \left(\frac{F_iL_i}{A_i}\right)\frac{\partial F_i}{\partial Q} = 4306P + 4263Q$$

Deflection of C. Substituting into Eq. (1), we have

$$y_C = \frac{1}{E} \sum \left(\frac{F_iL_i}{A_i}\right)\frac{\partial F_i}{\partial Q} = \frac{1}{E}(4306P + 4263Q)$$

Since load **Q** is not part of the original load, set $Q = 0$. Substituting $P = 40$ kN and $E = 73$ GPa gives

$$y_C = \frac{4306\,(40 \times 10^3\,\text{N})}{73 \times 10^9\,\text{Pa}} = 2.36 \times 10^{-3}\,\text{m} \qquad y_C = 2.36\ \text{mm} \downarrow \blacktriangleleft$$

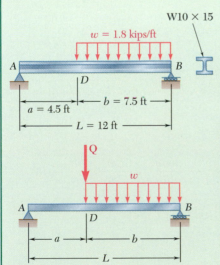

Fig. 1 Dummy load **Q** used to determine vertical deflection at point D.

Sample Problem 11.6

For the beam and loading shown, determine the deflection at point D. Use $E = 29 \times 10^6$ psi.

STRATEGY: Add a dummy load associated with the desired vertical deflection at joint D. Use a free-body diagram to determine the reactions due to both the dummy load and the distributed load. The moments in each segment are then written as a function of the coordinate along the beam. Eq. (11.57) is used to determine the deflection.

MODELING and ANALYSIS:

Castigliano's Theorem. We introduce a dummy load **Q** as shown in Fig. 1. Using Castigliano's theorem and noting that EI is constant, write

$$y_D = \int \frac{M}{EI}\left(\frac{\partial M}{\partial Q}\right)dx = \frac{1}{EI}\int M\left(\frac{\partial M}{\partial Q}\right)dx \qquad (1)$$

The integration will be performed separately for segments AD and DB.

(continued)

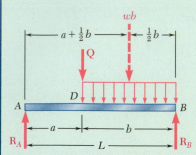

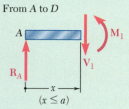

Fig. 2 Free-body diagram of beam.

From A to D

Fig. 3 Free-body diagram of left portion (in AD).

From B to D

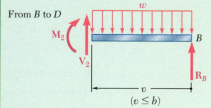

Fig. 4 Free-body diagram of right portion (in BD).

Reactions. Using the free-body diagram of the entire beam (Fig. 2) gives

$$\mathbf{R}_A = \frac{wb^2}{2L} + Q\frac{b}{L} \uparrow \qquad \mathbf{R}_B = \frac{wb(a + \frac{1}{2}b)}{L} + Q\frac{a}{L} \uparrow$$

Portion AD of Beam. Using the free-body diagram shown in Fig. 3,

$$M_1 = R_A x = \left(\frac{wb^2}{2L} + Q\frac{b}{L}\right) x \qquad \frac{\partial M_1}{\partial Q} = +\frac{bx}{L}$$

Substituting into Eq. (1) and integrating from A to D gives

$$\frac{1}{EI} \int M_1 \frac{\partial M_1}{\partial Q}\, dx = \frac{1}{EI} \int_0^a R_A x \left(\frac{bx}{L}\right) dx = \frac{R_A a^3 b}{3EIL}$$

Then substitute for R_A and set the dummy load Q equal to zero.

$$\frac{1}{EI} \int M_1 \frac{\partial M_1}{\partial Q}\, dx = \frac{wa^3 b^3}{6EIL^2} \qquad (2)$$

Portion DB of Beam. Using the free-body diagram shown in Fig. 4, the bending moment at a distance v from end B is

$$M_2 = R_B v - \frac{wv^2}{2} = \left[\frac{wb(a + \frac{1}{2}b)}{L} + Q\frac{a}{L}\right] v - \frac{wv^2}{2} \qquad \frac{\partial M_2}{\partial Q} = +\frac{av}{L}$$

Substitute into Eq. (1) and integrate from point B (where $v = 0$) to point D (where $v = b$) for

$$\frac{1}{EI} \int M_2 \frac{\partial M_2}{\partial Q}\, dv = \frac{1}{EI} \int_0^b \left(R_B v - \frac{wv^2}{2}\right)\left(\frac{av}{L}\right) dv = \frac{R_B ab^3}{3EIL} - \frac{wab^4}{8EIL}$$

Substituting for R_B and setting $Q = 0$,

$$\frac{1}{EI} \int M_2 \frac{\partial M_2}{\partial Q}\, dv = \left[\frac{wb(a + \frac{1}{2}b)}{L}\right]\frac{ab^3}{3EIL} - \frac{wab^4}{8EIL} = \frac{5a^2 b^4 + ab^5}{24EIL^2} w \qquad (3)$$

Deflection at Point D. Recalling Eqs. (1), (2), and (3),

$$y_D = \frac{wab^3}{24EIL^2}(4a^2 + 5ab + b^2) = \frac{wab^3}{24EIL^2}(4a + b)(a + b) = \frac{wab^3}{24EIL}(4a + b)$$

From Appendix C, $I = 68.9$ in^4 for a W10 $\times$ 15 beam. Substituting the numerical values for I, w, a, b, and L,

$$y_D = 0.262 \text{ in.} \downarrow \blacktriangleleft$$

Sample Problem 11.7

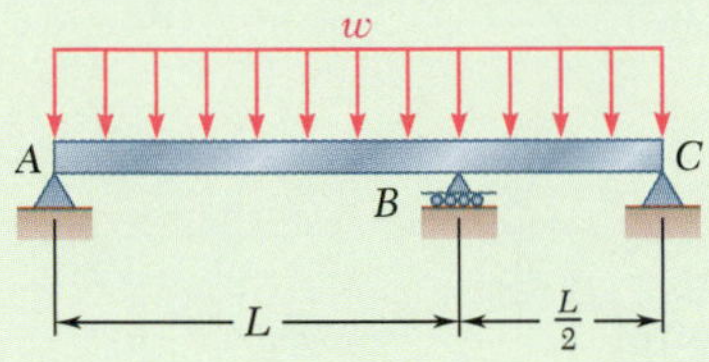

For the uniform beam and loading shown, determine the reactions at the supports.

STRATEGY: The beam is indeterminate to the first degree, and we must choose one of the reactions as a redundant. We then use a free-body diagram to solve for the reactions due to the distributed load and the redundant reaction. Using free-body diagrams of the segments, we obtain the moments as a function of the coordinate along the beam. Using Eq. (11.57), we write Castigliano's theorem for deflection associated with the redundant reaction. We set this deflection equal to zero, and solve for the redundant reaction. Equilibrium can then be used to find the other reactions.

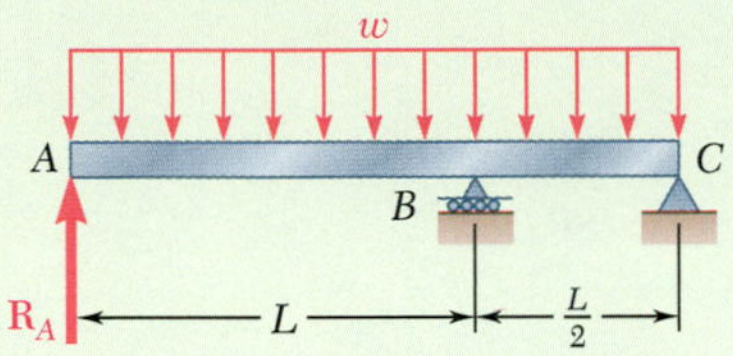

Fig. 1 Released beam, replacing support at A with redundant reaction R_A.

MODELING and ANALYSIS:

Castigliano's Theorem. Choose the reaction $\mathbf{R}_A$ as the redundant one (Fig. 1). Using Castigliano's theorem, determine the deflection at A due to the combined action of $\mathbf{R}_A$ and the distributed load. Since EI is constant,

$$y_A = \int \frac{M}{EI}\left(\frac{\partial M}{\partial R_A}\right)dx = \frac{1}{EI}\int M\frac{\partial M}{\partial R_A}\,dx \qquad (1)$$

The integration will be performed separately for portions AB and BC of the beam. $\mathbf{R}_A$ is then obtained by setting y_A equal to zero.

Free Body: Entire Beam. Using Fig. 2, the reactions at B and C in terms of R_A and the distributed load are

$$R_B = \tfrac{9}{4}\,wL - 3R_A \qquad R_C = 2R_A - \tfrac{3}{4}\,wL \qquad (2)$$

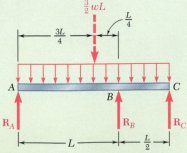

Fig. 2 Free-body diagram of beam.

From A to B

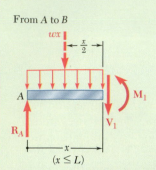

Fig. 3 Free-body diagram of left portion showing internal shear and moment.

From C to B

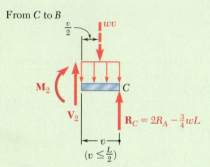

Fig. 4 Free-body diagram of right portion showing internal shear and moment.

Portion AB of Beam. Using the free-body diagram shown in Fig. 3, find

$$M_1 = R_A x - \frac{wx^2}{2} \qquad \frac{\partial M_1}{\partial R_A} = x$$

Substituting into Eq. (1) and integrating from A to B gives

$$\frac{1}{EI} \int M_1 \frac{\partial M}{\partial R_A}\, dx = \frac{1}{EI} \int_0^L \left(R_A x^2 - \frac{wx^3}{2} \right) dx = \frac{1}{EI} \left(\frac{R_A L^3}{3} - \frac{wL^4}{8} \right) \quad \textbf{(3)}$$

Portion BC of Beam. Using the free-body diagram shown in Fig. 4, find

$$M_2 = \left(2R_A - \frac{3}{4}wL \right) v - \frac{wv^2}{2} \qquad \frac{\partial M_2}{\partial R_A} = 2v$$

Substituting into Eq. (1) and integrating from C (where $v = 0$) to B (where $v = \frac{1}{2}L$) gives

$$\frac{1}{EI} \int M_2 \frac{\partial M_2}{\partial R_A}\, dv = \frac{1}{EI} \int_0^{L/2} \left(4R_A v^2 - \frac{3}{2}wLv^2 - wv^3 \right) dv$$

$$= \frac{1}{EI} \left(\frac{R_A L^3}{6} - \frac{wL^4}{16} - \frac{wL^4}{64} \right) = \frac{1}{EI} \left(\frac{R_A L^3}{6} - \frac{5wL^4}{64} \right) \quad \textbf{(4)}$$

Reaction at A. Adding the expressions from Eqs. (3) and (4), we obtain y_A and set it equal to zero:

$$y_A = \frac{1}{EI} \left(\frac{R_A L^3}{3} - \frac{wL^4}{8} \right) + \frac{1}{EI} \left(\frac{R_A L^3}{6} - \frac{5wL^4}{64} \right) = 0$$

Thus, $\qquad\qquad R_A = \frac{13}{32}wL \qquad\qquad\qquad R_A = \frac{13}{32}wL \uparrow \quad \blacktriangleleft$

Reactions at B and C. Substituting for R_A into Eqs. (2), we obtain

$$R_B = \frac{33}{32}wL \uparrow \qquad\qquad\qquad R_C = \frac{wL}{16} \uparrow \qquad \blacktriangleleft$$

Problems

11.77 and 11.78 Using the information in Appendix D, compute the work of the loads as they are applied to the beam (*a*) if the load **P** is applied first, (*b*) if the couple **M** is applied first.

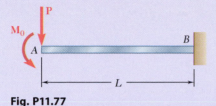

Fig. P11.77

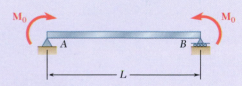

Fig. P11.78

11.79 through 11.82 For the beam and loading shown, (*a*) compute the work of the loads as they are applied successively to the beam, using the information provided in Appendix D, (*b*) compute the strain energy of the beam by the method of Sec. 11.2A and show that it is equal to the work obtained in part *a*.

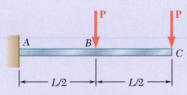

Fig. *P11.79*

Fig. P11.80

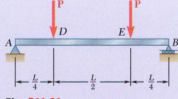

Fig. *P11.81*

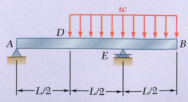

Fig. P11.82

11.83 through 11.85 For the prismatic beam shown, determine the deflection of point *D*.

11.86 through 11.88 For the prismatic beam shown, determine the slope at point *D*.

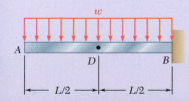

Fig. P11.83 and P11.86

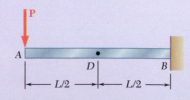

Fig. *P11.84* and *P11.87*

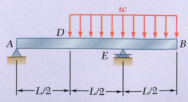

Fig. P11.85 and P11.88

11.89 For the prismatic beam shown, determine the slope at point A.

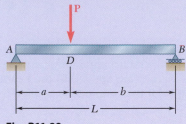

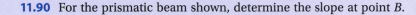

Fig. P11.89

11.90 For the prismatic beam shown, determine the slope at point B.

11.91 For the beam and loading shown, determine the deflection of point B. Use $E = 29 \times 10^6$ psi.

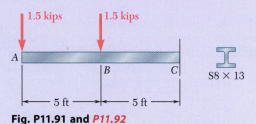

Fig. P11.91 and *P11.92*

11.92 For the beam and loading shown, determine the deflection of point A. Use $E = 29 \times 10^6$ psi.

11.93 and 11.94 For the beam and loading shown, determine the deflection at point B. Use $E = 200$ GPa.

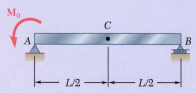

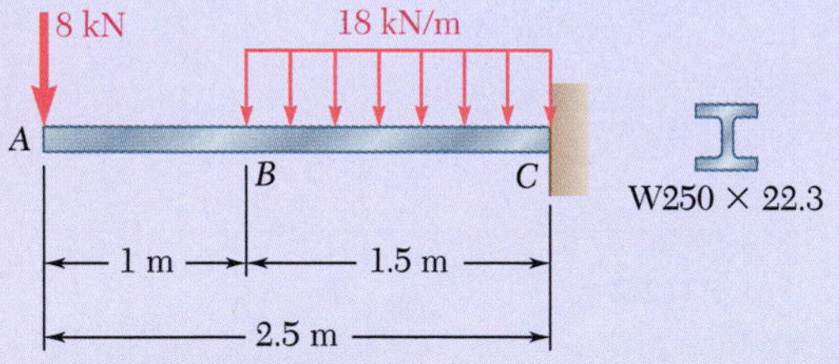

Fig. P11.90

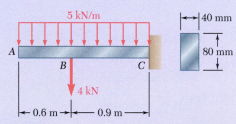

Fig. P11.94

11.95 For the beam and loading shown, determine the slope at end A. Use $E = 200$ GPa.

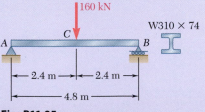

Fig. P11.95

Fig. P11.93

11.96 For the beam and loading shown, determine the deflection at point D. Use $E = 200$ GPa.

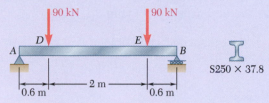

Fig. P11.96

11.97 For the beam and loading shown, determine the slope at end A. Use $E = 29 \times 10^6$ psi.

11.98 For the beam and loading shown, determine the deflection at point C. Use $E = 29 \times 10^6$ psi.

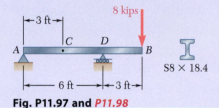

Fig. P11.97 and *P11.98*

11.99 and 11.100 For the truss and loading shown, determine the horizontal and vertical deflection of joint C.

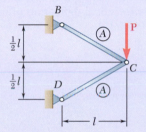

Fig. P11.99

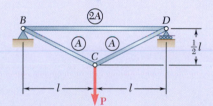

Fig. *P11.100*

11.101 and 11.102 Each member of the truss shown is made of steel and has the cross-sectional area shown. Using $E = 29 \times 10^6$ psi, determine the deflection indicated.

11.101 Vertical deflection of joint C.

11.102 Horizontal deflection of joint C.

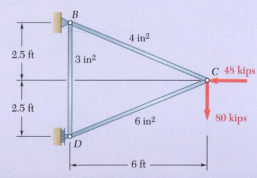

Fig. P11.101 and P11.102

11.103 and 11.104 Each member of the truss shown is made of steel and has a cross-sectional area of 500 mm². Using $E = 200$ GPa, determine the deflection indicated.

11.103 Vertical deflection of joint B.

11.104 Horizontal deflection of joint B.

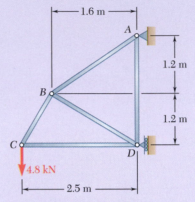

Fig. P11.103 and *P11.104*

11.105 A uniform rod of flexural rigidity EI is bent and loaded as shown. Determine (a) the vertical deflection of point A, (b) the horizontal deflection of point A.

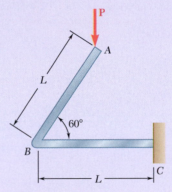

Fig. P11.105

11.106 For the uniform rod and loading shown and using Castigliano's theorem, determine the deflection of point B.

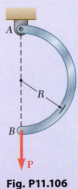

Fig. P11.106

11.107 For the beam and loading shown and using Castigliano's theorem, determine (a) the horizontal deflection of point B, (b) the vertical deflection of point B.

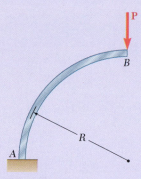

Fig. P11.107

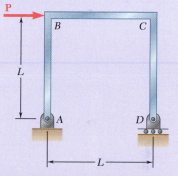

Fig. P11.108

11.108 Two rods AB and BC of the same flexural rigidity EI are welded together at B. For the loading shown, determine (a) the deflection of point C, (b) the slope of member BC at point C.

11.109 Three rods, each of the same flexural rigidity EI, are welded to form the frame ABCD. For the loading shown, determine the deflection of point D.

11.110 Three rods, each of the same flexural rigidity EI, are welded to form the frame ABCD. For the loading shown, determine the angle formed by the frame at point D.

11.111 through 11.115 Determine the reaction at the roller support and draw the bending-moment diagram for the beam and loading shown.

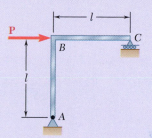

Fig. P11.109 and P11.110

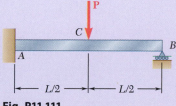

Fig. P11.111

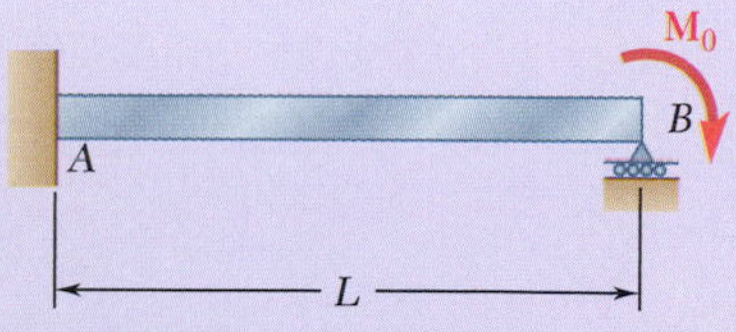

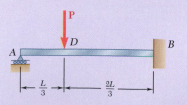

Fig. P11.112

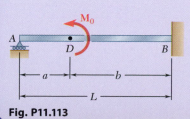

Fig. P11.113

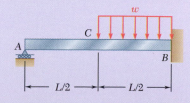

Fig. P11.114

Fig. P11.115

821

11.116 For the uniform beam and loading shown, determine the reaction at each support.

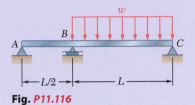

Fig. _P11.116_

11.117 through 11.120 Three members of the same material and same cross-sectional area are used to support the load **P**. Determine the force in member *BC*.

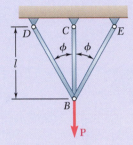

Fig. P11.117

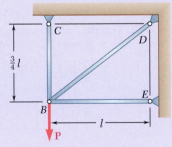

Fig. P11.118

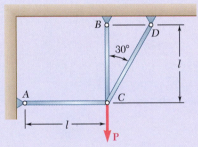

Fig. P11.119

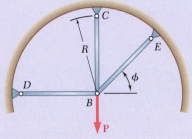

Fig. _P11.120_

11.121 and 11.122 Knowing that the eight members of the indeterminate truss shown have the same uniform cross-sectional area, determine the force in member *AB*.

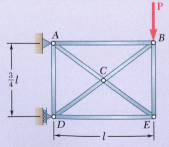

Fig. P11.121

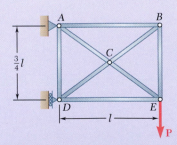

Fig. _P11.122_

Strain Energy

We considered a uniform rod subjected to a slowly increasing axial load **P** (Fig. 11.43). The area under the load-deformation diagram (Fig. 11.44) represents the work done by **P**. This work is equal to the *strain energy* of the rod associated with the deformation caused by load **P**:

$$\text{Strain energy} = U = \int_0^{x_1} P\, dx \qquad \textbf{(11.2)}$$

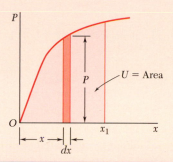

Fig. 11.44 Work due to load P is equal to the area under the load-deformation diagram.

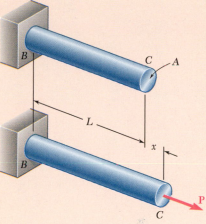

Fig. 11.43 Axially loaded rod.

Strain-Energy Density

Since the stress is uniform throughout the rod shown in Fig. 11.43, the strain energy can be divided by the volume of the rod to obtain the strain energy per unit volume. This is the *strain-energy density* of the material.

$$\text{Strain-energy density} = u = \int_0^{\epsilon_1} \sigma_x\, d\epsilon_x \qquad \textbf{(11.4)}$$

The strain-energy density is equal to the area under the stress-strain diagram of the material (Fig. 11.45). Equation (11.4) remains valid when the stresses are not uniformly distributed, but the strain-energy density now varies from point to point. If the material is unloaded, there is a permanent strain ϵ_p, and only the strain-energy density corresponding to the triangular area is recovered. The remainder of the energy is dissipated in the form of heat during the deformation of the material.

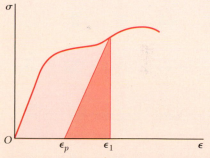

Fig. 11.45 Strain-energy density is the area under the stress-strain curve between $\epsilon_x = 0$ and $\epsilon_x = \epsilon_1$. If loaded into the plastic region, only the energy associated with elastic unloading is recovered.

Modulus of Toughness

The area under the entire stress-strain diagram (from zero to rupture) is called the *modulus of toughness* and is a measure of the total energy that can be acquired by the material.

Modulus of Resilience

If the normal stress σ remains within the proportional limit of the material, the strain-energy density u is

$$u = \frac{\sigma^2}{2E}$$

The area under the stress-strain curve from zero strain to the strain ϵ_Y at yield (Fig. 11.46) is the *modulus of resilience* of the material. It represents the energy per unit volume that the material can absorb without yielding:

$$u_Y = \frac{\sigma_Y^2}{2E} \tag{11.8}$$

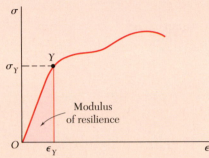

Fig. 11.46 Modulus of resilience is the area under the stress-strain curve to yield.

Strain Energy Under Axial Load

The strain energy under axial load is associated with *normal stresses*. If a rod of length L and *variable cross-sectional area A* is subjected to a centric axial load **P** at its end, the strain energy of the rod is

$$U = \int_0^L \frac{P^2}{2AE}\, dx \tag{11.13}$$

If the rod has a *uniform cross section* with an area A, the strain energy is

$$U = \frac{P^2L}{2AE} \tag{11.14}$$

Strain Energy Due to Bending

For a beam subjected to transverse loads (Fig. 11.47), the strain energy associated with the normal stresses is

$$U = \int_0^L \frac{M^2}{2EI}\, dx \tag{11.15}$$

where M is the bending moment and EI is the flexural rigidity of the beam.

Strain Energy Due to Shearing Stresses

The strain energy also can be associated with *shearing stresses*. The strain-energy density for a material in pure shear is

$$u = \frac{\tau_{xy}^2}{2G} \tag{11.17}$$

where τ_{xy} is the shearing stress and G is the modulus of rigidity of the material.

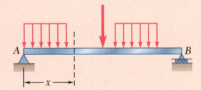

Fig. 11.47 Transversely loaded beam.

Strain Energy Due to Torsion

For a shaft with a length of L and uniform cross section subjected to couples of magnitude T at its ends (Fig. 11.48), the strain energy is

$$U = \frac{T^2 L}{2GJ} \qquad \textbf{(11.20)}$$

where J is the polar moment of inertia of the cross-sectional area of the shaft.

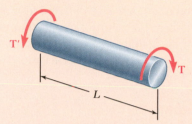

Fig. 11.48 Prismatic shaft subjected to torque.

General State of Stress

The strain energy of an elastic isotropic material under a general state of stress was considered where the strain-energy density at a given point is expressed in terms of the principal stresses σ_a, σ_b, and σ_c at that point:

$$u = \frac{1}{2E} \left[\sigma_a^2 + \sigma_b^2 + \sigma_c^2 - 2\nu(\sigma_a\sigma_b + \sigma_b\sigma_c + \sigma_c\sigma_a) \right] \qquad \textbf{(11.23)}$$

The strain-energy density at a given point is divided into two parts: u_v, which is associated with a change in volume of the material at that point, and u_d, which is associated with a distortion of the material at the same point. Thus, $u = u_v + u_d$, where

$$u_v = \frac{1 - 2\nu}{6E} (\sigma_a + \sigma_b + \sigma_c)^2 \qquad \textbf{(11.28)}$$

and

$$u_d = \frac{1}{12G} \left[(\sigma_a - \sigma_b)^2 + (\sigma_b - \sigma_c)^2 + (\sigma_c - \sigma_a)^2 \right] \qquad \textbf{(11.29)}$$

This equation for u_d is used to derive the maximum-distortion-energy criterion to predict whether or not a ductile material yields under a given state of plane stress.

Impact Loads

For the *impact loading* of an elastic structure being hit by a mass moving with a given velocity, it is assumed that the kinetic energy of the mass is transferred entirely to the structure. The *equivalent static load* is the load that causes the same deformations and stresses as the impact load.

A structural member designed to withstand an impact load effectively should be shaped so that the stresses are evenly distributed throughout the member. Also, the material used should have a low modulus of elasticity and a high yield strength.

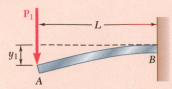

Fig. 11.49 Cantilever beam with load $\mathbf{P}_1$.

Members Subjected to a Single Load

The strain energy of structural members subjected to a *single load* was considered for the beam and loading of Fig. 11.49. The strain energy of the beam is

$$U = \frac{P_1^2 L^3}{6EI} \tag{11.37}$$

Observing that the work done by load $\mathbf{P}$ is equal to $\frac{1}{2}P_1 y_1$, the work of the load and the strain energy of the beam are equal and can be equated to determine the deflection y_1 at the point of application of the load.

The method just described is of limited value, since it is restricted to structures subjected to a single concentrated load and to the determination of the deflection at the point of application of that load. In the remaining sections of the chapter, we presented a more general method, which can be used to determine deflections at various points of structures subjected to several loads.

Castigliano's Theorem

Castigliano's theorem states that the deflection x_j of the point of application of a load $\mathbf{P}_j$ measured along the line of action of $\mathbf{P}_j$ is equal to the partial derivative of the strain energy of the structure with respect to the load $\mathbf{P}_j$. Thus,

$$x_j = \frac{\partial U}{\partial P_j} \tag{11.52}$$

Castigliano's theorem also can be used to determine the *slope* of a beam at the point of application of a couple $\mathbf{M}_j$ by writing

$$\theta_j = \frac{\partial U}{\partial M_j} \tag{11.55}$$

Similarly the *angle of twist* is determined in a section of a shaft where a torque $\mathbf{T}_j$ is applied by writing

$$\phi_j = \frac{\partial U}{\partial T_j} \tag{11.56}$$

Castigliano's theorem can be applied to determine the deflections and slopes at various points of a given structure. Dummy loads are used to determine displacements at points where no actual load is applied. The calculation of a deflection x_j is simpler if the differentiation with respect to load P_j is carried out before the integration. For a beam,

$$x_j = \frac{\partial U}{\partial P_j} = \int_0^L \frac{M}{EI} \frac{\partial M}{\partial P_j} \, dx \tag{11.57}$$

For a truss consisting of n members, the deflection x_j at the point of application of the load $\mathbf{P}_j$ is

$$x_j = \frac{\partial U}{\partial P_j} = \sum_{i=1}^{n} \frac{F_i L_i}{A_i E} \frac{\partial F_i}{\partial P_j} \tag{11.59}$$

Indeterminate Structures

Castigliano's theorem can also be used in the analysis of statically indeterminate structures, as shown in Sec. 11.9.

Review Problems

11.123 Rod AB is made of a steel for which the yield strength is $\sigma_Y = 450$ MPa and $E = 200$ GPa; rod BC is made of an aluminum alloy for which $\sigma_Y = 280$ MPa and $E = 73$ GPa. Determine the maximum strain energy that can be acquired by the composite rod ABC without causing any permanent deformations.

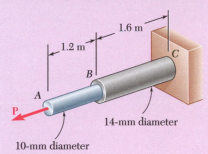

Fig. P11.123

11.124 Each member of the truss shown is made of steel and has the cross-sectional area shown. Using $E = 29 \times 10^6$ psi, determine the strain energy of the truss for the loading shown.

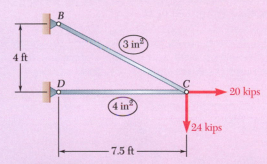

Fig. P11.124

11.125 The ship at A has just started to drill for oil on the ocean floor at a depth of 5000 ft. The steel drill pipe has an outer diameter of 8 in. and a uniform wall thickness of 0.5 in. Knowing that the top of the drill pipe rotates through two complete revolutions before the drill bit at B starts to operate and using $G = 11.2 \times 10^6$ psi, determine the maximum strain energy acquired by the drill pipe.

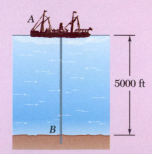

Fig. P11.125

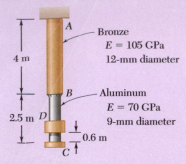

Fig. P11.126

11.126 Collar D is released from rest in the position shown and is stopped by a small plate attached at end C of the vertical rod ABC. Determine the mass of the collar for which the maximum normal stress in portion BC is 125 MPa.

11.127 Each member of the truss shown is made of steel and has a cross-sectional area of 400 mm². Using E = 200 GPa, determine the deflection of point D caused by the 16-kN load.

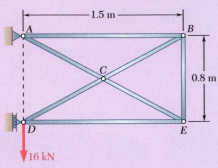

Fig. *P11.127*

11.128 A block of weight W is placed in contact with a beam at some given point D and released. Show that the resulting maximum deflection at point D is twice as large as the deflection due to a static load W applied at D.

11.129 Two solid steel shafts are connected by the gears shown. Using the method of work and energy, determine the angle through which end D rotates when $T = 820$ N · m. Use $G = 77.2$ GPa.

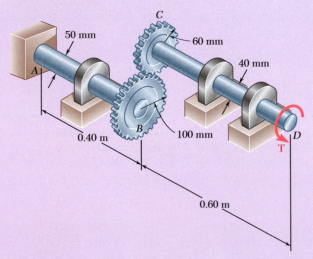

Fig. P11.129

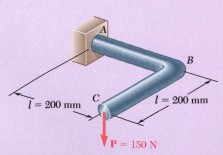

Fig. P11.130

11.130 The 12-mm-diameter steel rod ABC has been bent into the shape shown. Knowing that $E = 200$ GPa and $G = 77.2$ GPa, determine the deflection of end C caused by the 150-N force.

11.131 For the prismatic beam shown, determine the slope at point D.

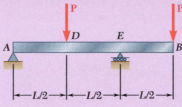

Fig. P11.131

11.132 A disk of radius a has been welded to end B of the solid steel shaft AB. A cable is then wrapped around the disk and a vertical force **P** is applied to end C of the cable. Knowing that the radius of the shaft is L and neglecting the deformations of the disk and of the cable, show that the deflection of point C caused by the application of **P** is

$$\delta_c = \frac{PL^3}{3EI}\left(1 + 1.5\frac{Ea^2}{GL^2}\right)$$

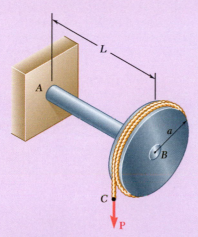

Fig. P11.132

11.133 A uniform rod of flexural rigidity EI is bent and loaded as shown. Determine (a) the horizontal deflection of point D, (b) the slope at point D.

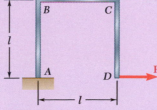

Fig. P11.133

11.134 The steel bar ABC has a square cross section of side 0.75 in. and is subjected to a 50-lb load **P**. Using $E = 29 \times 10^6$ psi for rod BD and the bar, determine the deflection of point C.

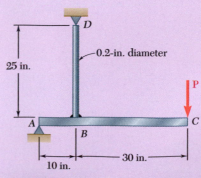

Fig. P11.134

Computer Problems

The following problems are designed to be solved with a computer.

11.C1 A rod consisting of n elements, each of which is homogeneous and of uniform cross section, is subjected to a load **P** applied at its free end. The length of element i is denoted by L_i and its diameter by d_i. (*a*) Denoting by E the modulus of elasticity of the material used in the rod, write a computer program that can be used to determine the strain energy acquired by the rod and the deformation measured at its free end. (*b*) Use this program to determine the strain energy and deformation for the rods of Probs. 11.9 and 11.10.

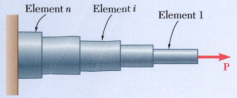

Fig. P11.C1

11.C2 Two 0.75×6-in. cover plates are welded to a W8 $\times$ 18 rolled-steel beam as shown. The 1500-lb block is to be dropped from a height $h = 2$ in. onto the beam. (*a*) Write a computer program to calculate the maximum normal stress on transverse sections just to the left of D and at the center of the beam for values of a from 0 to 60 in. using 5-in. increments. (*b*) From the values considered in part *a*, select the distance a for which the maximum normal stress is as small as possible. Use $E = 29 \times 10^6$ psi.

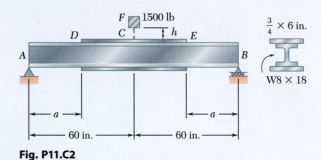

Fig. P11.C2

11.C3 The 16-kg block D is dropped from a height h onto the free end of the steel bar AB. For the steel used $\sigma_{all} = 120$ MPa and $E = 200$ GPa. (*a*) Write a computer program to calculate the maximum allowable height h for values of the length L from 100 mm to 1.2 m, using 100-mm increments. (*b*) From the values considered in part *a*, select the length corresponding to the largest allowable height.

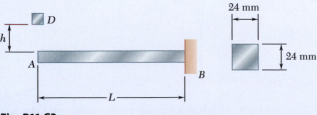

Fig. P11.C3

11.C4 The block D of mass $m = 8$ kg is dropped from a height $h = 750$ mm onto the rolled-steel beam AB. Knowing that $E = 200$ GPa, write a computer program to calculate the maximum deflection of point E and the maximum normal stress in the beam for values of a from 100 to 900 mm 900 mm, using 100-mm increments.

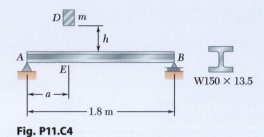

Fig. P11.C4

11.C5 The steel rods AB and BC are made of a steel for which $\sigma_Y = 300$ MPa and $E = 200$ GPa. (*a*) Write a computer program to calculate for values of a from 0 to 6 m, using 1-m increments, the maximum strain energy that can be acquired by the assembly without causing any permanent deformation. (*b*) For each value of a considered, calculate the diameter of a uniform rod of length 6 m and of the same mass as the original assembly, and the maximum strain energy that could be acquired by this uniform rod without causing permanent deformation.

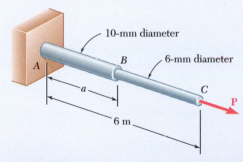

Fig. P11.C5

11.C6 A 160-lb diver jumps from a height of 20 in. onto end C of a diving board having the uniform cross section shown. Write a computer program to calculate for values of a from 10 to 50 in., using 10-in. increments, (*a*) the maximum deflection of point C, (*b*) the maximum bending moment in the board, (*c*) the equivalent static load. Assume that the diver's legs remain rigid and use $E = 1.8 \times 10^6$ psi.

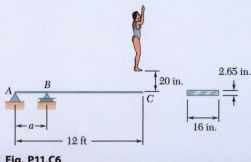

Fig. P11.C6

Appendices

[†]Courtesy of the American Institute of Steel Construction, Chicago, Illinois.

A Moments of Areas

A.1 First Moment of an Area and Centroid of an Area

Consider an area A located in the xy plane (Fig. A.1). Using x and y as the coordinates of an element of area dA, the *first moment of the area A with respect to the x axis* is the integral

$$Q_x = \int_A y \, dA \tag{A.1}$$

Similarly, the *first moment of the area A with respect to the y axis* is the integral

$$Q_y = \int_A x \, dA \tag{A.2}$$

Note that each of these integrals may be positive, negative, or zero, depending on the position of the coordinate axes. When SI units are used, the first moments Q_x and Q_y are given in m^3 or mm^3. When U.S. customary units are used, they are given in ft^3 or in^3.

The *centroid of the area A* is the point C of coordinates $\bar{x}$ and $\bar{y}$ (Fig. A.2), which satisfy the relationship

$$\int_A x \, dA = A\bar{x} \quad \int_A y \, dA = A\bar{y} \tag{A.3}$$

Comparing Eqs. (A.1) and (A.2) with Eqs. (A.3), the first moments of the area A can be expressed as the products of the area and the coordinates of its centroid:

$$Q_x = A\bar{y} \quad Q_y = A\bar{x} \tag{A.4}$$

When an area possesses an *axis of symmetry,* the first moment of the area with respect to that axis is zero. Considering area A of Fig. A.3, which is symmetric with respect to the y axis, every element of area dA of

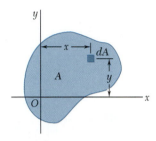

Fig. A.1 General area A with infinitesimal area dA referred to xy coordinate system.

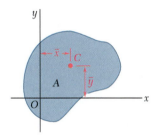

Fig. A.2 Centroid of area A.

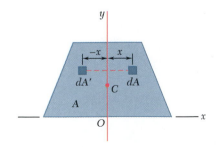

Fig. A.3 Area having axis of symmetry.

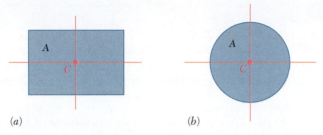

(a)   *(b)*

Fig. A.4 Areas having two axes of symmetry have the centroid at their intersection.

abscissa x corresponds to an element of area dA' of abscissa $-x$. Therefore, the integral in Eq. (A.2) is zero, and $Q_y = 0$. From the first of the relationships in Eq. (A.3), $\bar{x} = 0$. Thus, if an area A possesses an axis of symmetry, its centroid C is located on that axis.

Since a rectangle possesses two axes of symmetry (Fig. A.4a), the centroid C of a rectangular area coincides with its geometric center. Similarly, the centroid of a circular area coincides with the center of the circle (Fig. A.4b).

When an area possesses a *center of symmetry O*, the first moment of the area about any axis through O is zero. Considering the area A of Fig. A.5, every element of area dA with coordinates x and y corresponds to an element of area dA' with coordinates $-x$ and $-y$. It follows that the integrals in Eqs. (A.1) and (A.2) are both zero, and $Q_x = Q_y = 0$. From Eqs. (A.3), $\bar{x} = \bar{y} = 0$, so the centroid of the area coincides with its center of symmetry.

When the centroid C of an area can be located by symmetry, the first moment of that area with respect to any given axis can be obtained easily from Eqs. (A.4). For example, for the rectangular area of Fig. A.6,

$$Q_x = A\bar{y} = (bh)(\tfrac{1}{2}h) = \tfrac{1}{2}bh^2$$

and

$$Q_y = A\bar{x} = (bh)(\tfrac{1}{2}b) = \tfrac{1}{2}b^2h$$

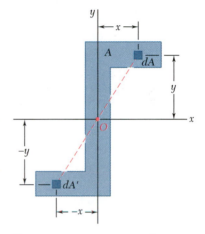

Fig. A.5 Area with center of symmetry has its centroid at the origin.

In most cases, it is necessary to perform the integrations indicated in Eqs. (A.1) through (A.3) to determine the first moments and the centroid of a given area. While each of the integrals is actually a double integral, it is possible to select elements of area dA in the shape of thin horizontal or vertical strips and to reduce the equations to integrations in a single variable. This is illustrated in Concept Application A.1. Centroids of common geometric shapes are given in a table inside the back cover.

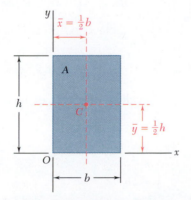

Fig. A.6 Centroid of a rectangular area.

Concept Application A.1

For the triangular area of Fig. A.7a, determine (a) the first moment Q_x of the area with respect to the x axis, (b) the ordinate $\bar{y}$ of the centroid of the area.

a. First Moment Q_x.   We choose to select as an element of area a horizontal strip with a length of u and thickness dy. Note that all of the points within the element are at the same distance y from the x axis (Fig. A.7b). From similar triangles,

$$\frac{u}{b} = \frac{h - y}{h} \quad u = b\frac{h - y}{h}$$

and

$$dA = u\,dy = b\frac{h - y}{h}\,dy$$

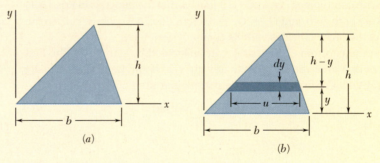

Fig. A.7   (a) Triangular area. (b) Horizontal element used in integration to find centroid.

The first moment of the area with respect to the x axis is

$$Q_x = \int_A y\,dA = \int_0^h yb\frac{h - y}{h}\,dy = \frac{b}{h}\int_0^h (hy - y^2)\,dy$$

$$= \frac{b}{h}\left[h\frac{y^2}{2} - \frac{y^3}{3}\right]_0^h \quad Q_x = \tfrac{1}{6}bh^2$$

b. Ordinate of Centroid.   Recalling the first of Eqs. (A.4) and observing that $A = \tfrac{1}{2}bh$,

$$Q_x = A\bar{y} \quad \tfrac{1}{6}bh^2 = (\tfrac{1}{2}bh)\bar{y}$$

$$\bar{y} = \tfrac{1}{3}h$$

A.2   The First Moment and Centroid of a Composite Area

Consider area A of the quadrilateral area shown in Fig. A.8, which can be divided into simple geometric shapes. The first moment Q_x of the area with respect to the x axis is represented by the integral $\int y\,dA$, which extends over the entire area A. Dividing A into its component parts A_1, A_2, A_3, write

$$Q_x = \int_A y\,dA = \int_{A_1} y\,dA + \int_{A_2} y\,dA + \int_{A_3} y\,dA$$

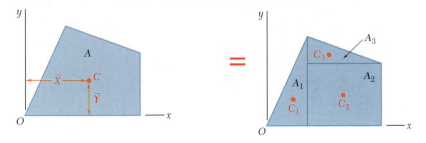

Fig. A.8   Quadrilateral area divided into simple geometric shapes.

or recalling the second of Eqs. (A.3),

$$Q_x = A_1\bar{y}_1 + A_2\bar{y}_2 + A_3\bar{y}_3$$

where $\bar{y}_1, \bar{y}_2$, and $\bar{y}_3$ represent the ordinates of the centroids of the component areas. Extending this to an arbitrary number of component areas and noting that a similar expression for Q_y may be obtained, write

$$Q_x = \sum A_i\bar{y}_i \quad Q_y = \sum A_i\bar{x}_i \tag{A.5}$$

To obtain the coordinates $\bar{X}$ and $\bar{Y}$ of the centroid C of the composite area A, substitute $Q_x = A\bar{Y}$ and $Q_y = A\bar{X}$ into Eqs. (A.5):

$$A\bar{Y} = \sum_i A_i\bar{y}_i \quad A\bar{X} = \sum_i A_i\bar{x}_i$$

Solving for $\bar{X}$ and $\bar{Y}$ and recalling that the area A is the sum of the component areas A_i,

$$\bar{X} = \frac{\sum_i A_i\bar{x}_i}{\sum_i A_i} \quad \bar{Y} = \frac{\sum_i A_i\bar{y}_i}{\sum_i A_i} \tag{A.6}$$

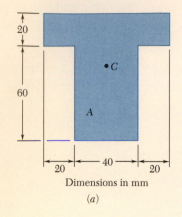

Dimensions in mm

(a)

Concept Application A.2

Locate the centroid C of the area A shown in Fig. A.9a.

Selecting the coordinate axes shown in Fig. A.9b, note that the centroid C must be located on the y axis, since this is an axis of symmetry. Thus, $\overline{X} = 0$.

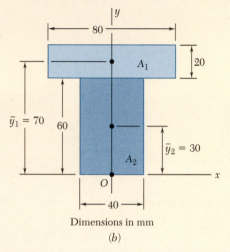

Dimensions in mm

(b)

Fig. A.9   (a) Area A. (b) Composite areas A_1 and A_2 used to determine overall centroid.

Divide A into its component parts A_1 and A_2 and use the second of Eqs. (A.6) to determine the ordinate $\overline{Y}$ of the centroid. The actual computation is best carried out in tabular form:

	Area, mm^2	$\overline{y}_i$, mm	$A_i\overline{y}_i$, mm^3
A_1	$(20)(80) = 1600$	70	112×10^3
A_2	$(40)(60) = 2400$	30	72×10^3
	$\sum_i A_i = 4000$		$\sum_i A_i\overline{y}_i = 184 \times 10^3$

$$\overline{Y} = \frac{\sum_i A_i\overline{y}_i}{\sum_i A_i} = \frac{184 \times 10^3 \text{ mm}^3}{4 \times 10^3 \text{ mm}^2} = 46 \text{ mm}$$

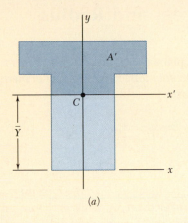

(a)

Concept Application A.3

Referring to the area A of Concept Application A.2, consider the horizontal x' axis through its centroid C (called a *centroidal axis*). The portion of A located above that axis is A' (Fig. A.10a). Determine the first moment of A' with respect to the x' axis.

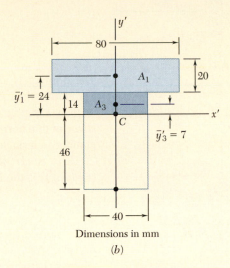

Dimensions in mm

(b)

Solution.   Divide the area A' into its components A_1 and A_3 (Fig. A.10b). Recall from Concept Application A.2 that C is located 46 mm above the lower edge of A. The ordinates $\bar{y}'_1$ and $\bar{y}'_3$ of A_1 and A_3 and the first moment $Q'_{x'}$ of A' with respect to x' are

$$Q'_{x'} = A_1\bar{y}'_1 + A_3\bar{y}'_3$$
$$= (20 \times 80)(24) + (14 \times 40)(7) = 42.3 \times 10^3\,\text{mm}^3$$

Alternative Solution.   Since the centroid C of A is located on the x' axis, the first moment $Q_{x'}$ of the *entire area* A with respect to that axis is zero:

$$Q_{x'} = A\bar{y}' = A(0) = 0$$

Using A'' as the portion of A located below the x' axis and $Q''_{x'}$ as its first moment with respect to that axis,

$$Q_{x'} = Q'_{x'} + Q''_{x'} = 0 \quad \text{or} \quad Q'_{x'} = -Q''_{x'}$$

This shows that the first moments of A' and A'' have the same magnitude and opposite signs. Referring to Fig. A.10c, write

$$Q''_{x'} = A_4\bar{y}'_4 = (40 \times 46)(-23) = -42.3 \times 10^3\,\text{mm}^3$$

and

$$Q'_{x'} = -Q''_{x'} = +42.3 \times 10^3\,\text{mm}^3$$

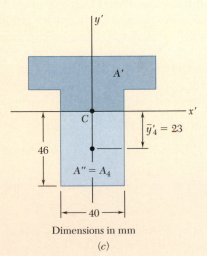

Dimensions in mm

(c)

Fig. A.10 (a) Area A with centroidal $x'y'$ axes, highlighting portion A'. (b) Areas used to determine the first moment of area A' with respect to the x' axis. (c) Alternative solution using the other portion A'' of the total area A.

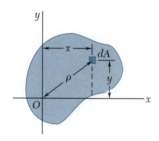

Fig. A.1   (repeated)

A.3   Second Moment, or Moment of Inertia of an Area, and Radius of Gyration

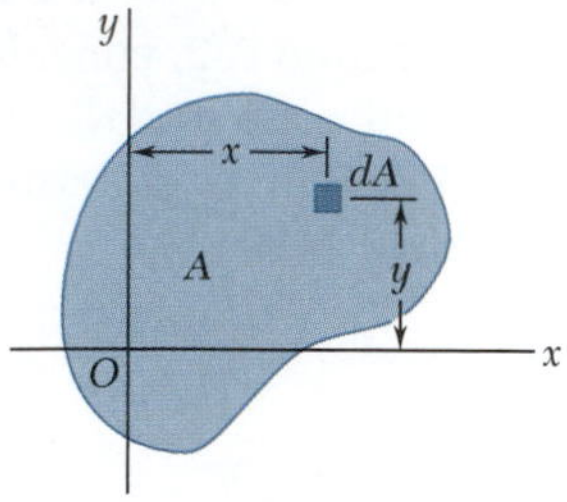

Consider an area A located in the xy plane (Fig. A.1) and the element of area dA of coordinates x and y. The *second moment,* or *moment of inertia,* of area A with respect to the x and y axes is

$$I_x = \int_A y^2\, dA \quad I_y = \int_A x^2\, dA \tag{A.7}$$

These integrals are called *rectangular moments of inertia,* since they are found from the rectangular coordinates of element dA. While each integral is actually a double integral, it is possible to select elements of area dA in the shape of thin horizontal or vertical strips and to reduce the equations to integrations in a single variable. This is illustrated in Concept Application A.4.

The *polar moment of inertia* of area A with respect to point O (Fig. A.11) is the integral

$$J_O = \int_A \rho^2\, dA \tag{A.8}$$

Fig. A.11   Area dA located by distance ρ from point O.

where ρ is the distance from O to the element dA. While this integral is also a double integral, for circular areas it is possible to select elements of area dA in the shape of thin circular rings and to reduce the equation of J_O to a single integration (see Concept Application A.5).

Note from Eqs. (A.7) and (A.8) that the moments of inertia of an area are positive quantities. When SI units are used, moments of inertia are given in m^4 or mm^4. When U.S. customary units are used, they are given in ft^4 or in^4.

An important relationship can be established between the polar moment of inertia J_O of a given area and the rectangular moments of inertia I_x and I_y. Noting that $\rho^2 = x^2 + y^2$,

$$J_O = \int_A \rho^2\, dA = \int_A (x^2 + y^2)\, dA = \int_A y^2\, dA + \int_A x^2\, dA$$

or

$$J_O = I_x + I_y \tag{A.9}$$

The *radius of gyration* of an area A with respect to the x axis is r_x, which satisfies the relationship

$$I_x = r_x^2 A \tag{A.10}$$

where I_x is the moment of inertia of A with respect to the x axis. Solving Eq. (A.10) for r_x,

$$r_x = \sqrt{\frac{I_x}{A}} \tag{A.11}$$

The radii of gyration with respect to the y axis and the origin O are

$$I_y = r_y^2 A \qquad r_y = \sqrt{\frac{I_y}{A}} \tag{A.12}$$

$$J_O = r_O^2 A \qquad r_O = \sqrt{\frac{J_O}{A}} \tag{A.13}$$

Substituting for J_O, I_x, and I_y in terms of the corresponding radii of gyration in Eq. (A.9),

$$r_O^2 = r_x^2 + r_y^2 \tag{A.14}$$

The results obtained in the following two Concept Applications are included in the table for moments of inertias of common geometric shapes, located inside the back cover of this book.

Concept Application A.4

For the rectangular area of Fig. A.12a, determine (a) the moment of inertia I_x of the area with respect to the centroidal x axis, (b) the corresponding radius of gyration r_x.

a. Moment of Inertia I_x. We choose to select a horizontal strip of length b and thickness dy (Fig. A.12b). Since all of the points within the strip are at the same distance y from the x axis, the moment of inertia of the strip with respect to that axis is

$$dI_x = y^2\, dA = y^2(b\, dy)$$

Integrating from $y = -h/2$ to $y = +h/2$,

$$I_x = \int_A y^2\, dA = \int_{-h/2}^{+h/2} y^2(b\, dy) = \tfrac{1}{3}b[y^3]_{-h/2}^{+h/2}$$

$$= \tfrac{1}{3}b\left(\frac{h^3}{8} + \frac{h^3}{8}\right)$$

or

$$I_x = \tfrac{1}{12}bh^3$$

b. Radius of Gyration r_x. From Eq. (A.10),

$$I_x = r_x^2 A \quad \tfrac{1}{12}bh^3 = r_x^2(bh)$$

and solving for r_x gives

$$r_x = h/\sqrt{12}$$

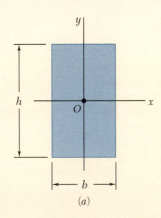

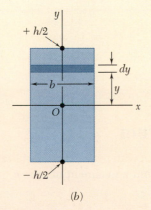

Fig. A.12 (a) Rectangular area. (b) Horizontal strip used to determine moment of inertia I_x.

Concept Application A.5

For the circular area of Fig. A.13a, determine (a) the polar moment of inertia J_O, (b) the rectangular moments of inertia I_x and I_y.

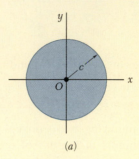

(a)

a. Polar Moment of Inertia. We choose to select as an element of area a ring of radius ρ with a thickness $d\rho$ (Fig. A.13b). Since all of the points within the ring are at the same distance ρ from the origin O, the polar moment of inertia of the ring is

$$dJ_O = \rho^2 \, dA = \rho^2 (2\pi\rho \, d\rho)$$

Integrating in ρ from 0 to c,

$$J_O = \int_A \rho^2 \, dA = \int_0^c \rho^2 (2\pi\rho \, d\rho) = 2\pi \int_0^c \rho^3 \, d\rho$$

$$J_O = \tfrac{1}{2}\pi c^4$$

b. Rectangular Moments of Inertia. Because of the symmetry of the circular area, $I_x = I_y$. Recalling Eq. (A.9), write

$$J_O = I_x + I_y = 2I_x \quad \tfrac{1}{2}\pi c^4 = 2I_x$$

and,

$$I_x = I_y = \tfrac{1}{4}\pi c^4$$

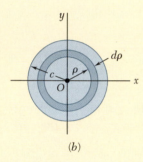

(b)

Fig. A.13 (a) Circular area. (b) Annular strip used to determine polar moment of inertia J_O.

A.4 Parallel-Axis Theorem

Consider the moment of inertia I_x of an area A with respect to an arbitrary x axis (Fig. A.14). Using y as the distance from an element of area dA to that axis, recall from Sec. A.3 that

$$I_x = \int_A y^2 \, dA$$

We now draw the *centroidal x′ axis,* which is the axis parallel to the x axis that passes through the centroid C. Using $y′$ as the distance from element dA

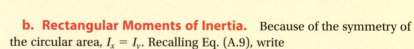

Fig. A.14 General area with centroidal $x′$ axis, parallel to arbitrary x axis.

to that axis, $y = y' + d$, where d is the distance between the two axes. Substituting for y in the integral representing I_x gives

$$I_x = \int_A y^2 \, dA = \int_A (y' + d)^2 dA$$

$$I_x = \int_A y'^2 \, dA + 2d \int_A y' \, dA + d^2 \int_A dA \qquad \textbf{(A.15)}$$

The first integral in Eq. (A.15) represents the moment of inertia $\bar{I}_{x'}$ of the area with respect to the centroidal x' axis. The second integral represents the first moment $Q_{x'}$ of the area with respect to the x' axis and is equal to zero, since the centroid C of the area is located on that axis. In other words, recalling from Sec. A.1 we write

$$Q_{x'} = A\bar{y}' = A(0) = 0$$

The last integral in Eq. (A.15) is equal to the total area A. Therefore,

$$I_x = \bar{I}_{x'} + Ad^2 \qquad \textbf{(A.16)}$$

This equation shows that the moment of inertia I_x of an area with respect to an arbitrary x axis is equal to the moment of inertia $\bar{I}_{x'}$ of the area with respect to the centroidal x' axis parallel to the x axis *plus* the product Ad^2 of the area A and of the square of the distance d between the two axes. This result is known as the *parallel-axis theorem*. With this theorem, the moment of inertia of an area with respect to a given axis can be determined when its moment of inertia with respect to a centroidal axis of the same direction is known. Conversely, it makes it possible to determine the moment of inertia $\bar{I}_{x'}$ of an area A with respect to a centroidal axis x' when the moment of inertia I_x of A with respect to a parallel axis is known. This is done by *subtracting* from I_x the product Ad^2. Note that the parallel-axis theorem may be used *only if one of the two axes involved is a centroidal axis.*

A similar formula relates the polar moment of inertia J_O of an area with respect to an arbitrary point O and the polar moment of inertia $\bar{J}_C$ of the same area with respect to its centroid C. Using d as the distance between O and C,

$$J_O = \bar{J}_C + Ad^2 \qquad \textbf{(A.17)}$$

A.5 Moment of Inertia of a Composite Area

Consider a composite area A made of several component parts A_1, A_2, and so forth. Since the integral for the moment of inertia of A can be subdivided into integrals extending over A_1, A_2, etc. the moment of inertia of A with respect to a given axis is obtained by adding the moments of inertia of the areas A_1, A_2, etc. with respect to the same axis. The moment of inertia of an area made of several common shapes may be found by using the formulas shown in the inside back cover of this book. Before adding the moments of inertia of the component areas, the parallel-axis theorem should be used to transfer each moment of inertia to the desired axis. This is shown in Concept Application A.6.

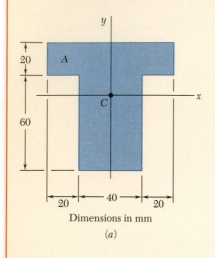

Dimensions in mm

(a)

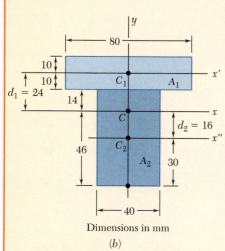

Dimensions in mm

(b)

Fig. A.15 (a) Area A. (b) Composite areas and centroids.

Concept Application A.6

Determine the moment of inertia $\bar{I}_x$ of the area shown with respect to the centroidal x axis (Fig. A.15a).

Location of Centroid. The centroid C of the area has been located in Concept Application A.2 for the given area. From this, C is located 46 mm above the lower edge of area A.

Computation of Moment of Inertia. Area A is divided into two rectangular areas A_1 and A_2 (Fig. A.15b), and the moment of inertia of each area is found with respect to the x axis.

Rectangular Area A_1. To obtain the moment of inertia $(I_x)_1$ of A_1 with respect to the x axis, first compute the moment of inertia of A_1 *with respect to its own centroidal axis x'*. Recalling the equation in part *a* of Concept Application A.4 for the centroidal moment of inertia of a rectangular area,

$$(\bar{I}_{x'})_1 = \tfrac{1}{12}bh^3 = \tfrac{1}{12}(80 \text{ mm})(20 \text{ mm})^3 = 53.3 \times 10^3 \text{ mm}^4$$

Using the parallel-axis theorem, transfer the moment of inertia of A_1 from its centroidal axis x' to the parallel axis x:

$$(I_x)_1 = (\bar{I}_{x'})_1 + A_1 d_1^2 = 53.3 \times 10^3 + (80 \times 20)(24)^2$$
$$= 975 \times 10^3 \text{ mm}^4$$

Rectangular Area A_2. Calculate the moment of inertia of A_2 with respect to its centroidal axis x'' and use the parallel-axis theorem to transfer it to the x axis to obtain

$$(\bar{I}_{x''})_2 = \tfrac{1}{12}bh^3 = \tfrac{1}{12}(40)(60)^3 = 720 \times 10^3 \text{ mm}^4$$

$$(I_x)_2 = (\bar{I}_{x''})_2 + A_2 d_2^2 = 720 \times 10^3 + (40 \times 60)(16)^2$$

$$= 1334 \times 10^3 \text{ mm}^4$$

Entire Area A. Add the values for the moments of inertia of A_1 and A_2 with respect to the x axis to obtain the moment of inertia $\bar{I}_x$ of the entire area:

$$\bar{I}_x = (I_x)_1 + (I_x)_2 = 975 \times 10^3 + 1334 \times 10^3$$

$$\bar{I}_x = 2.31 \times 10^6 \text{ mm}^4$$

Appendix B   Typical Properties of Selected Materials Used in Engineering[1,5]

(U.S. Customary Units)

| Material | Specific Weight, lb/in³ | Ultimate Strength | | | Yield Strength[3] | | Modulus of Elasticity, 10⁶ psi | Modulus of Rigidity, 10⁶ psi | Coefficient of Thermal Expansion, 10⁻⁶/°F | Ductility, Percent Elongation in 2 in. |
		Tension, ksi	Compression,[2] ksi	Shear, ksi	Tension, ksi	Shear, ksi				
Steel										
Structural (ASTM-A36)	0.284	58			36	21	29	11.2	6.5	21
High-strength-low-alloy										
ASTM-A709 Grade 50	0.284	65			50		29	11.2	6.5	21
ASTM-A913 Grade 65	0.284	80			65		29	11.2	6.5	17
ASTM-A992 Grade 50	0.284	65			50		29	11.2	6.5	21
Quenched & tempered										
ASTM-A709 Grade 100	0.284	110			100		29	11.2	6.5	18
Stainless, AISI 302										
Cold-rolled	0.286	125			75		28	10.8	9.6	12
Annealed	0.286	95			38	22	28	10.8	9.6	50
Reinforcing Steel										
Medium strength	0.283	70			40		29	11	6.5	
High strength	0.283	90			60		29	11	6.5	
Cast Iron										
Gray Cast Iron										
4.5% C, ASTM A-48	0.260	25	95	35			10	4.1	6.7	0.5
Malleable Cast Iron										
2% C, 1% Si,										
ASTM A-47	0.264	50	90	48	33		24	9.3	6.7	10
Aluminum										
Alloy 1100-H14										
(99% Al)	0.098	16		10	14	8	10.1	3.7	13.1	9
Alloy 2014-T6	0.101	66		40	58	33	10.9	3.9	12.8	13
Alloy 2024-T4	0.101	68		41	47		10.6		12.9	19
Alloy 5456-H116	0.095	46		27	33	19	10.4		13.3	16
Alloy 6061-T6	0.098	38		24	35	20	10.1	3.7	13.1	17
Alloy 7075-T6	0.101	83		48	73		10.4	4	13.1	11
Copper										
Oxygen-free copper										
(99.9% Cu)										
Annealed	0.322	32		22	10		17	6.4	9.4	45
Hard-drawn	0.322	57		29	53		17	6.4	9.4	4
Yellow Brass										
(65% Cu, 35% Zn)										
Cold-rolled	0.306	74		43	60	36	15	5.6	11.6	8
Annealed	0.306	46		32	15	9	15	5.6	11.6	65
Red Brass										
(85% Cu, 15% Zn)										
Cold-rolled	0.316	85		46	63		17	6.4	10.4	3
Annealed	0.316	39		31	10		17	6.4	10.4	48
Tin bronze	0.318	45			21		14		10	30
(88 Cu, 8Sn, 4Zn)										
Manganese bronze	0.302	95			48		15		12	20
(63 Cu, 25 Zn, 6 Al, 3 Mn, 3 Fe)										
Aluminum bronze	0.301	90	130		40		16	6.1	9	6
(81 Cu, 4 Ni, 4 Fe, 11 Al)										

(Table continued on page A14)

Appendix B   Typical Properties of Selected Materials Used in Engineering[1,5]

(SI Units)

Material	Density kg/m^3	Ultimate Strength			Yield Strength[3]		Modulus of Elasticity, GPa	Modulus of Rigidity, GPa	Coefficient of Thermal Expansion, 10^{-6}/°C	Ductility, Percent Elongation in 50 mm
		Tension, MPa	Compression,[2] MPa	Shear, MPa	Tension, MPa	Shear, MPa				
Steel										
Structural (ASTM-A36)	7860	400			250	145	200	77.2	11.7	21
High-strength-low-alloy										
ASTM-A709 Grade 345	7860	450			345		200	77.2	11.7	21
ASTM-A913 Grade 450	7860	550			450		200	77.2	11.7	17
ASTM-A992 Grade 345	7860	450			345		200	77.2	11.7	21
Quenched & tempered										
ASTM-A709 Grade 690	7860	760			690		200	77.2	11.7	18
Stainless, AISI 302										
Cold-rolled	7920	860			520		190	75	17.3	12
Annealed	7920	655			260	150	190	75	17.3	50
Reinforcing Steel										
Medium strength	7860	480			275		200	77	11.7	
High strength	7860	620			415		200	77	11.7	
Cast Iron										
Gray Cast Iron										
4.5% C, ASTM A-48	7200	170	655	240			69	28	12.1	0.5
Malleable Cast Iron										
2% C, 1% Si,										
ASTM A-47	7300	345	620	330	230		165	65	12.1	10
Aluminum										
Alloy 1100-H14										
(99% Al)	2710	110		70	95	55	70	26	23.6	9
Alloy 2014-T6	2800	455		275	400	230	75	27	23.0	13
Alloy-2024-T4	2800	470		280	325		73		23.2	19
Alloy-5456-H116	2630	315		185	230	130	72		23.9	16
Alloy 6061-T6	2710	260		165	240	140	70	26	23.6	17
Alloy 7075-T6	2800	570		330	500		72	28	23.6	11
Copper										
Oxygen-free copper										
(99.9% Cu)										
Annealed	8910	220		150	70		120	44	16.9	45
Hard-drawn	8910	390		200	265		120	44	16.9	4
Yellow-Brass										
(65% Cu, 35% Zn)										
Cold-rolled	8470	510		300	410	250	105	39	20.9	8
Annealed	8470	320		220	100	60	105	39	20.9	65
Red Brass										
(85% Cu, 15% Zn)										
Cold-rolled	8740	585		320	435		120	44	18.7	3
Annealed	8740	270		210	70		120	44	18.7	48
Tin bronze	8800	310			145		95		18.0	30
(88 Cu, 8Sn, 4Zn)										
Manganese bronze	8360	655			330		105		21.6	20
(63 Cu, 25 Zn, 6 Al, 3 Mn, 3 Fe)										
Aluminum bronze	8330	620	900		275		110	42	16.2	6
(81 Cu, 4 Ni, 4 Fe, 11 Al)										

(Table continued on page A15

Appendix B   Typical Properties of Selected Materials Used in Engineering[1,5]

(U.S. Customary Units)

Continued from page A14

Material	Specific Weight, lb/in³	Ultimate Strength			Yield Strength[3]		Modulus of Elasticity, 10⁶ psi	Modulus of Rigidity, 10⁶ psi	Coefficient of Thermal Expansion, 10⁻⁶/°F	Ductility, Percent Elongation in 2 in.
		Tension, ksi	Compression,[2] ksi	Shear, ksi	Tension, ksi	Shear, ksi				
Magnesium Alloys										
Alloy AZ80 (Forging)	0.065	50		23	36		6.5	2.4	14	6
Alloy AZ31 (Extrusion)	0.064	37		19	29		6.5	2.4	14	12
Titanium										
Alloy (6% Al, 4% V)	0.161	130			120		16.5		5.3	10
Monel Alloy 400(Ni-Cu)										
Cold-worked	0.319	98			85	50	26		7.7	22
Annealed	0.319	80			32	18	26		7.7	46
Cupronickel										
(90% Cu, 10% Ni)										
Annealed	0.323	53			16		20	7.5	9.5	35
Cold-worked	0.323	85			79		20	7.5	9.5	3
Timber, air dry										
Douglas fir	0.017	15	7.2	1.1			1.9	.1	Varies	
Spruce, Sitka	0.015	8.6	5.6	1.1			1.5	.07	1.7 to 2.5	
Shortleaf pine	0.018		7.3	1.4			1.7			
Western white pine	0.014		5.0	1.0			1.5			
Ponderosa pine	0.015	8.4	5.3	1.1			1.3			
White oak	0.025		7.4	2.0			1.8			
Red oak	0.024		6.8	1.8			1.8			
Western hemlock	0.016	13	7.2	1.3			1.6			
Shagbark hickory	0.026		9.2	2.4			2.2			
Redwood	0.015	9.4	6.1	0.9			1.3			
Concrete										
Medium strength	0.084		4.0				3.6		5.5	
High strength	0.084		6.0				4.5		5.5	
Plastics										
Nylon, type 6/6, (molding compound)	0.0412	11	14		6.5		0.4		80	50
Polycarbonate	0.0433	9.5	12.5		9		0.35		68	110
Polyester, PBT (thermoplastic)	0.0484	8	11		8		0.35		75	150
Polyester elastomer	0.0433	6.5		5.5			0.03			500
Polystyrene	0.0374	8	13		8		0.45		70	2
Vinyl, rigid PVC	0.0520	6	10		6.5		0.45		75	40
Rubber	0.033	2							90	600
Granite (Avg. values)	0.100	3	35	5			10	4	4	
Marble (Avg. values)	0.100	2	18	4			8	3	6	
Sandstone (Avg. values)	0.083	1	12	2			6	2	5	
Glass, 98% silica	0.079		7				9.6	4.1	44	

[1]Properties of metals vary widely as a result of variations in composition, heat treatment, and mechanical working.
[2]For ductile metals the compression strength is generally assumed to be equal to the tension strength.
[3]Offset of 0.2 percent.
[4]Timber properties are for loading parallel to the grain.
[5]See also *Marks' Mechanical Engineering Handbook,* 10th ed., McGraw-Hill, New York, 1996; *Annual Book of ASTM,* American Society for Testing Materials, Philadelphia, Pa.; *Metals Handbook,* American Society for Metals, Metals Park, Ohio; and *Aluminum Design Manual,* The Aluminum Association, Washington, DC.

Appendix B   Typical Properties of Selected Materials Used in Engineering[1,5]

(SI Units)
Continued from page A15

Material	Density kg/m³	Ultimate Strength			Yield Strength[3]		Modulus of Elasticity, GPa	Modulus of Rigidity, GPa	Coefficient of Thermal Expansion, 10⁻⁶/°C	Ductility, Percent Elongation in 50 mm
		Tension, MPa	Compression,[2] MPa	Shear, MPa	Tension, MPa	Shear, MPa				
Magnesium Alloys										
Alloy AZ80 (Forging)	1800	345		160	250		45	16	25.2	6
Alloy AZ31 (Extrusion)	1770	255		130	200		45	16	25.2	12
Titanium										
Alloy (6% Al, 4% V)	4730	900			830		115		9.5	10
Monel Alloy 400(Ni-Cu)										
Cold-worked	8830	675			585	345	180		13.9	22
Annealed	8830	550			220	125	180		13.9	46
Cupronickel										
(90% Cu, 10% Ni)										
Annealed	8940	365			110		140	52	17.1	35
Cold-worked	8940	585			545		140	52	17.1	3
Timber, air dry										
Douglas fir	470	100	50	7.6			13	0.7	Varies	
Spruce, Sitka	415	60	39	7.6			10	0.5	3.0 to 4.5	
Shortleaf pine	500		50	9.7			12			
Western white pine	390		34	7.0			10			
Ponderosa pine	415	55	36	7.6			9			
White oak	690		51	13.8			12			
Red oak	660		47	12.4			12			
Western hemlock	440	90	50	10.0			11			
Shagbark hickory	720		63	16.5			15			
Redwood	415	65	42	6.2			9			
Concrete										
Medium strength	2320		28				25		9.9	
High strength	2320		40				30		9.9	
Plastics										
Nylon, type 6/6, (molding compound)	1140	75	95		45		2.8		144	50
Polycarbonate	1200	65	85		35		2.4		122	110
Polyester, PBT (thermoplastic)	1340	55	75		55		2.4		135	150
Polyester elastomer	1200	45		40			0.2			500
Polystyrene	1030	55	90		55		3.1		125	2
Vinyl, rigid PVC	1440	40	70		45		3.1		135	40
Rubber	910	15							162	600
Granite (Avg. values)	2770	20	240	35			70	4	7.2	
Marble (Avg. values)	2770	15	125	28			55	3	10.8	
Sandstone (Avg. values)	2300	7	85	14			40	2	9.0	
Glass, 98% silica	2190		50				65	4.1	80	

[1]Properties of metals very widely as a result of variations in composition, heat treatment, and mechanical working.

[2]For ductile metals the compression strength is generally assumed to be equal to the tension strength.

[3]Offset of 0.2 percent.

[4]Timber properties are for loading parallel to the grain.

[5]See also *Marks' Mechanical Engineering Handbook,* 10th ed., McGraw-Hill, New York, 1996; *Annual Book of ASTM,* American Society for Testing Materials, Philadelphia, Pa.; *Metals Handbook,* American Society of Metals, Metals Park, Ohio; and *Aluminum Design Manual,* The Aluminum Association, Washington, DC.

Appendix C Properties of Rolled-Steel Shapes
(U.S. Customary Units)

W Shapes
(Wide-Flange Shapes)

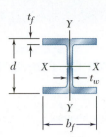

Designation[†]	Area A, in²	Depth d, in.	Flange Width b_f, in.	Flange Thickness t_f, in.	Web Thickness t_w, in.	Axis X-X I_x, in⁴	Axis X-X S_x, in³	Axis X-X r_x, in.	Axis Y-Y I_y, in⁴	Axis Y-Y S_y, in³	Axis Y-Y r_y, in.
W36 × 302	88.8	37.3	16.7	1.68	0.945	21100	1130	15.4	1300	156	3.82
135	39.7	35.6	12.0	0.790	0.600	7800	439	14.0	225	37.7	2.38
W33 × 201	59.2	33.7	15.7	1.15	0.715	11600	686	14.0	749	95.2	3.56
118	34.7	32.9	11.5	0.740	0.550	5900	359	13.0	187	32.6	2.32
W30 × 173	51.0	30.4	15.0	1.07	0.655	8230	541	12.7	598	79.8	3.42
99	29.1	29.7	10.50	0.670	0.520	3990	269	11.7	128	24.5	2.10
W27 × 146	43.1	27.4	14.0	0.975	0.605	5660	414	11.5	443	63.5	3.20
84	24.8	26.70	10.0	0.640	0.460	2850	213	10.7	106	21.2	2.07
W24 × 104	30.6	24.1	12.8	0.750	0.500	3100	258	10.1	259	40.7	2.91
68	20.1	23.7	8.97	0.585	0.415	1830	154	9.55	70.4	15.7	1.87
W21 × 101	29.8	21.4	12.3	0.800	0.500	2420	227	9.02	248	40.3	2.89
62	18.3	21.0	8.24	0.615	0.400	1330	127	8.54	57.5	14.0	1.77
44	13.0	20.7	6.50	0.450	0.350	843	81.6	8.06	20.7	6.37	1.26
W18 × 106	31.1	18.7	11.2	0.940	0.590	1910	204	7.84	220	39.4	2.66
76	22.3	18.2	11.0	0.680	0.425	1330	146	7.73	152	27.6	2.61
50	14.7	18.0	7.50	0.570	0.355	800	88.9	7.38	40.1	10.7	1.65
35	10.3	17.7	6.00	0.425	0.300	510	57.6	7.04	15.3	5.12	1.22
W16 × 77	22.6	16.5	10.3	0.760	0.455	1110	134	7.00	138	26.9	2.47
57	16.8	16.4	7.12	0.715	0.430	758	92.2	6.72	43.1	12.1	1.60
40	11.8	16.0	7.00	0.505	0.305	518	64.7	6.63	28.9	8.25	1.57
31	9.13	15.9	5.53	0.440	0.275	375	47.2	6.41	12.4	4.49	1.17
26	7.68	15.7	5.50	0.345	0.250	301	38.4	6.26	9.59	3.49	1.12
W14 × 370	109	17.9	16.5	2.66	1.66	5440	607	7.07	1990	241	4.27
145	42.7	14.8	15.5	1.09	0.680	1710	232	6.33	677	87.3	3.98
82	24.0	14.3	10.1	0.855	0.510	881	123	6.05	148	29.3	2.48
68	20.0	14.0	10.0	0.720	0.415	722	103	6.01	121	24.2	2.46
53	15.6	13.9	8.06	0.660	0.370	541	77.8	5.89	57.7	14.3	1.92
43	12.6	13.7	8.00	0.530	0.305	428	62.6	5.82	45.2	11.3	1.89
38	11.2	14.1	6.77	0.515	0.310	385	54.6	5.87	26.7	7.88	1.55
30	8.85	13.8	6.73	0.385	0.270	291	42.0	5.73	19.6	5.82	1.49
26	7.69	13.9	5.03	0.420	0.255	245	35.3	5.65	8.91	3.55	1.08
22	6.49	13.7	5.00	0.335	0.230	199	29.0	5.54	7.00	2.80	1.04

[†]A wide-flange shape is designated by the letter W followed by the nominal depth in inches and the weight in pounds per foot.

(Table continued on page A18)

Appendix C   Properties of Rolled-Steel Shapes
(SI Units)

W Shapes
(Wide-Flange Shapes)

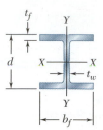

Designation[†]	Area A, mm²	Depth d, mm.	Flange Width b_f, mm	Flange Thickness t_f, mm	Web Thickness t_w mm	Axis X-X I_x 10^6 mm⁴	S_x 10^3 mm³	r_x mm	Axis Y-Y I_y 10^6 mm⁴	S_y 10^3 mm³	r_y mm
W920 × 449	57300	947	424	42.7	24.0	8780	18500	391	541	2560	97.0
201	25600	904	305	20.1	15.2	3250	7190	356	93.7	618	60.5
W840 × 299	38200	856	399	29.2	18.2	4830	11200	356	312	1560	90.4
176	22400	836	292	18.8	14.0	2460	5880	330	77.8	534	58.9
W760 × 257	32900	772	381	27.2	16.6	3430	8870	323	249	1310	86.9
147	18800	754	267	17.0	13.2	1660	4410	297	53.3	401	53.3
W690 × 217	27800	696	356	24.8	15.4	2360	6780	292	184	1040	81.3
125	16000	678	254	16.3	11.7	1190	3490	272	44.1	347	52.6
W610 × 155	19700	612	325	19.1	12.7	1290	4230	257	108	667	73.9
101	13000	602	228	14.9	10.5	762	2520	243	29.3	257	47.5
W530 × 150	19200	544	312	20.3	12.7	1010	3720	229	103	660	73.4
92	11800	533	209	15.6	10.2	554	2080	217	23.9	229	45.0
66	8390	526	165	11.4	8.89	351	1340	205	8.62	104	32.0
W460 × 158	20100	475	284	23.9	15.0	795	3340	199	91.6	646	67.6
113	14400	462	279	17.3	10.8	554	2390	196	63.3	452	66.3
74	9480	457	191	14.5	9.02	333	1460	187	16.7	175	41.9
52	6650	450	152	10.8	7.62	212	944	179	6.37	83.9	31.0
W410 × 114	14600	419	262	19.3	11.6	462	2200	178	57.4	441	62.7
85	10800	417	181	18.2	10.9	316	1510	171	17.9	198	40.6
60	7610	406	178	12.8	7.75	216	1060	168	12.0	135	39.9
46.1	5890	404	140	11.2	6.99	156	773	163	5.16	73.6	29.7
38.8	4950	399	140	8.76	6.35	125	629	159	3.99	57.2	28.4
W360 × 551	70300	455	419	67.6	42.2	2260	9950	180	828	3950	108
216	27500	376	394	27.7	17.3	712	3800	161	282	1430	101
122	15500	363	257	21.7	13.0	367	2020	154	61.6	480	63.0
101	12900	356	254	18.3	10.5	301	1690	153	50.4	397	62.5
79	10100	353	205	16.8	9.40	225	1270	150	24.0	234	48.8
64	8130	348	203	13.5	7.75	178	1030	148	18.8	185	48.0
57.8	7230	358	172	13.1	7.87	160	895	149	11.1	129	39.4
44	5710	351	171	9.78	6.86	121	688	146	8.16	95.4	37.8
39	4960	353	128	10.7	6.48	102	578	144	3.71	58.2	27.4
32.9	4190	348	127	8.51	5.84	82.8	475	141	2.91	45.9	26.4

[†]A wide-flange shape is designated by the letter W followed by the nominal depth in millimeters and the mass in kilograms permeter.

(Table continued on page A19)

Appendix C   Properties of Rolled-Steel Shapes

(U.S. Customary Units)
Continued from page A18

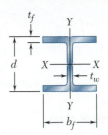

W Shapes
(Wide-Flange Shapes)

Designation[†]	Area A, in²	Depth d, in.	Flange Width b_f, in.	Flange Thickness t_f, in.	Web Thickness t_w, in.	Axis X-X I_x, in⁴	Axis X-X S_x, in³	Axis X-X r_x, in.	Axis Y-Y I_y, in⁴	Axis Y-Y S_y, in³	Axis Y-Y r_y, in.
W12 × 96	28.2	12.7	12.2	0.900	0.550	833	131	5.44	270	44.4	3.09
72	21.1	12.3	12.0	0.670	0.430	597	97.4	5.31	195	32.4	3.04
50	14.6	12.2	8.08	0.640	0.370	391	64.2	5.18	56.3	13.9	1.96
40	11.7	11.9	8.01	0.515	0.295	307	51.5	5.13	44.1	11.0	1.94
35	10.3	12.5	6.56	0.520	0.300	285	45.6	5.25	24.5	7.47	1.54
30	8.79	12.3	6.52	0.440	0.260	238	38.6	5.21	20.3	6.24	1.52
26	7.65	12.2	6.49	0.380	0.230	204	33.4	5.17	17.3	5.34	1.51
22	6.48	12.3	4.03	0.425	0.260	156	25.4	4.91	4.66	2.31	0.848
16	4.71	12.0	3.99	0.265	0.220	103	17.1	4.67	2.82	1.41	0.773
W10 × 112	32.9	11.4	10.4	1.25	0.755	716	126	4.66	236	45.3	2.68
68	20.0	10.4	10.1	0.770	0.470	394	75.7	4.44	134	26.4	2.59
54	15.8	10.1	10.0	0.615	0.370	303	60.0	4.37	103	20.6	2.56
45	13.3	10.1	8.02	0.620	0.350	248	49.1	4.32	53.4	13.3	2.01
39	11.5	9.92	7.99	0.530	0.315	209	42.1	4.27	45.0	11.3	1.98
33	9.71	9.73	7.96	0.435	0.290	171	35.0	4.19	36.6	9.20	1.94
30	8.84	10.5	5.81	0.510	0.300	170	32.4	4.38	16.7	5.75	1.37
22	6.49	10.2	5.75	0.360	0.240	118	23.2	4.27	11.4	3.97	1.33
19	5.62	10.2	4.02	0.395	0.250	96.3	18.8	4.14	4.29	2.14	0.874
15	4.41	10.0	4.00	0.270	0.230	68.9	13.8	3.95	2.89	1.45	0.810
W8 × 58	17.1	8.75	8.22	0.810	0.510	228	52.0	3.65	75.1	18.3	2.10
48	14.1	8.50	8.11	0.685	0.400	184	43.2	3.61	60.9	15.0	2.08
40	11.7	8.25	8.07	0.560	0.360	146	35.5	3.53	49.1	12.2	2.04
35	10.3	8.12	8.02	0.495	0.310	127	31.2	3.51	42.6	10.6	2.03
31	9.12	8.00	8.00	0.435	0.285	110	27.5	3.47	37.1	9.27	2.02
28	8.24	8.06	6.54	0.465	0.285	98.0	24.3	3.45	21.7	6.63	1.62
24	7.08	7.93	6.50	0.400	0.245	82.7	20.9	3.42	18.3	5.63	1.61
21	6.16	8.28	5.27	0.400	0.250	75.3	18.2	3.49	9.77	3.71	1.26
18	5.26	8.14	5.25	0.330	0.230	61.9	15.2	3.43	7.97	3.04	1.23
15	4.44	8.11	4.01	0.315	0.245	48.0	11.8	3.29	3.41	1.70	0.876
13	3.84	7.99	4.00	0.255	0.230	39.6	9.91	3.21	2.73	1.37	0.843
W6 × 25	7.34	6.38	6.08	0.455	0.320	53.4	16.7	2.70	17.1	5.61	1.52
20	5.87	6.20	6.02	0.365	0.260	41.4	13.4	2.66	13.3	4.41	1.50
16	4.74	6.28	4.03	0.405	0.260	32.1	10.2	2.60	4.43	2.20	0.967
12	3.55	6.03	4.00	0.280	0.230	22.1	7.31	2.49	2.99	1.50	0.918
9	2.68	5.90	3.94	0.215	0.170	16.4	5.56	2.47	2.20	1.11	0.905
W5 × 19	5.56	5.15	5.03	0.430	0.270	26.3	10.2	2.17	9.13	3.63	1.28
16	4.71	5.01	5.00	0.360	0.240	21.4	8.55	2.13	7.51	3.00	1.26
W4 × 13	3.83	4.16	4.06	0.345	0.280	11.3	5.46	1.72	3.86	1.90	1.00

†A wide-flange shape is designated by the letter W followed by the nominal depth in inches and the weight in pounds per foot.

Appendix C   Properties of Rolled-Steel Shapes
(SI Units)
Continued from page A19

W Shapes
(Wide-Flange Shapes)

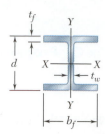

Designation[†]	Area A, mm²	Depth d, mm	Flange Width b_f, mm	Flange Thickness t_f, mm	Web Thickness t_w, mm	Axis X-X I_x 10^6 mm⁴	Axis X-X S_x 10^3 mm³	Axis X-X r_x mm	Axis Y-Y I_y 10^6 mm⁴	Axis Y-Y S_y 10^3 mm³	Axis Y-Y r_y mm
W310 × 143	18200	323	310	22.9	14.0	347	2150	138	112	728	78.5
107	13600	312	305	17.0	10.9	248	1600	135	81.2	531	77.2
74	9420	310	205	16.3	9.40	163	1050	132	23.4	228	49.8
60	7550	302	203	13.1	7.49	128	844	130	18.4	180	49.3
52	6650	318	167	13.2	7.62	119	747	133	10.2	122	39.1
44.5	5670	312	166	11.2	6.60	99.1	633	132	8.45	102	38.6
38.7	4940	310	165	9.65	5.84	84.9	547	131	7.20	87.5	38.4
32.7	4180	312	102	10.8	6.60	64.9	416	125	1.94	37.9	21.5
23.8	3040	305	101	6.73	5.59	42.9	280	119	1.17	23.1	19.6
W250 × 167	21200	290	264	31.8	19.2	298	2060	118	98.2	742	68.1
101	12900	264	257	19.6	11.9	164	1240	113	55.8	433	65.8
80	10200	257	254	15.6	9.4	126	983	111	42.9	338	65.0
67	8580	257	204	15.7	8.89	103	805	110	22.2	218	51.1
58	7420	252	203	13.5	8.00	87.0	690	108	18.7	185	50.3
49.1	6260	247	202	11.0	7.37	71.2	574	106	15.2	151	49.3
44.8	5700	267	148	13.0	7.62	70.8	531	111	6.95	94.2	34.8
32.7	4190	259	146	9.14	6.10	49.1	380	108	4.75	65.1	33.8
28.4	3630	259	102	10.0	6.35	40.1	308	105	1.79	35.1	22.2
22.3	2850	254	102	6.86	5.84	28.7	226	100	1.20	23.8	20.6
W200 × 86	11000	222	209	20.6	13.0	94.9	852	92.7	31.3	300	53.3
71	9100	216	206	17.4	10.2	76.6	708	91.7	25.3	246	52.8
59	7550	210	205	14.2	9.14	60.8	582	89.7	20.4	200	51.8
52	6650	206	204	12.6	7.87	52.9	511	89.2	17.7	174	51.6
46.1	5880	203	203	11.0	7.24	45.8	451	88.1	15.4	152	51.3
41.7	5320	205	166	11.8	7.24	40.8	398	87.6	9.03	109	41.1
35.9	4570	201	165	10.2	6.22	34.4	342	86.9	7.62	92.3	40.9
31.3	3970	210	134	10.2	6.35	31.3	298	88.6	4.07	60.8	32.0
26.6	3390	207	133	8.38	5.84	25.8	249	87.1	3.32	49.8	31.2
22.5	2860	206	102	8.00	6.22	20.0	193	83.6	1.42	27.9	22.3
19.3	2480	203	102	6.48	5.84	16.5	162	81.5	1.14	22.5	21.4
W150 × 37.1	4740	162	154	11.6	8.13	22.2	274	68.6	7.12	91.9	38.6
29.8	3790	157	153	9.27	6.60	17.2	220	67.6	5.54	72.3	38.1
24	3060	160	102	10.3	6.60	13.4	167	66.0	1.84	36.1	24.6
18	2290	153	102	7.11	5.84	9.20	120	63.2	1.24	24.6	23.3
13.5	1730	150	100	5.46	4.32	6.83	91.1	62.7	0.916	18.2	23.0
W130 × 28.1	3590	131	128	10.9	6.86	10.9	167	55.1	3.80	59.5	32.5
23.8	3040	127	127	9.14	6.10	8.91	140	54.1	3.13	49.2	32.0
W100 × 19.3	2470	106	103	8.76	7.11	4.70	89.5	43.7	1.61	31.1	25.4

[†]A wide-flange shape is designated by the letter W followed by the nominal depth in millimeters and the mass in kilograms per meter.

Appendix C Properties of Rolled-Steel Shapes
(U.S. Customary Units)

S Shapes
(American Standard Shapes)

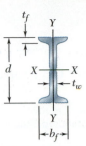

Designation[†]	Area A, in²	Depth d, in.	Flange Width b_f, in.	Flange Thickness t_f, in.	Web Thickness t_w, in.	Axis X-X I_x, in⁴	Axis X-X S_x, in³	Axis X-X r_x, in.	Axis Y-Y I_y, in⁴	Axis Y-Y S_y, in³	Axis Y-Y r_y, in.
S24 × 121	35.5	24.5	8.05	1.09	0.800	3160	258	9.43	83.0	20.6	1.53
106	31.1	24.5	7.87	1.09	0.620	2940	240	9.71	76.8	19.5	1.57
100	29.3	24.0	7.25	0.870	0.745	2380	199	9.01	47.4	13.1	1.27
90	26.5	24.0	7.13	0.870	0.625	2250	187	9.21	44.7	12.5	1.30
80	23.5	24.0	7.00	0.870	0.500	2100	175	9.47	42.0	12.0	1.34
S20 × 96	28.2	20.3	7.20	0.920	0.800	1670	165	7.71	49.9	13.9	1.33
86	25.3	20.3	7.06	0.920	0.660	1570	155	7.89	46.6	13.2	1.36
75	22.0	20.0	6.39	0.795	0.635	1280	128	7.62	29.5	9.25	1.16
66	19.4	20.0	6.26	0.795	0.505	1190	119	7.83	27.5	8.78	1.19
S18 × 70	20.5	18.0	6.25	0.691	0.711	923	103	6.70	24.0	7.69	1.08
54.7	16.0	18.0	6.00	0.691	0.461	801	89.0	7.07	20.7	6.91	1.14
S15 × 50	14.7	15.0	5.64	0.622	0.550	485	64.7	5.75	15.6	5.53	1.03
42.9	12.6	15.0	5.50	0.622	0.411	446	59.4	5.95	14.3	5.19	1.06
S12 × 50	14.6	12.0	5.48	0.659	0.687	303	50.6	4.55	15.6	5.69	1.03
40.8	11.9	12.0	5.25	0.659	0.462	270	45.1	4.76	13.5	5.13	1.06
35	10.2	12.0	5.08	0.544	0.428	228	38.1	4.72	9.84	3.88	0.980
31.8	9.31	12.0	5.00	0.544	0.350	217	36.2	4.83	9.33	3.73	1.00
S10 × 35	10.3	10.0	4.94	0.491	0.594	147	29.4	3.78	8.30	3.36	0.899
25.4	7.45	10.0	4.66	0.491	0.311	123	24.6	4.07	6.73	2.89	0.950
S8 × 23	6.76	8.00	4.17	0.425	0.441	64.7	16.2	3.09	4.27	2.05	0.795
18.4	5.40	8.00	4.00	0.425	0.271	57.5	14.4	3.26	3.69	1.84	0.827
S6 × 17.2	5.06	6.00	3.57	0.359	0.465	26.2	8.74	2.28	2.29	1.28	0.673
12.5	3.66	6.00	3.33	0.359	0.232	22.0	7.34	2.45	1.80	1.08	0.702
S5 × 10	2.93	5.00	3.00	0.326	0.214	12.3	4.90	2.05	1.19	0.795	0.638
S4 × 9.5	2.79	4.00	2.80	0.293	0.326	6.76	3.38	1.56	0.887	0.635	0.564
7.7	2.26	4.00	2.66	0.293	0.193	6.05	3.03	1.64	0.748	0.562	0.576
S3 × 7.5	2.20	3.00	2.51	0.260	0.349	2.91	1.94	1.15	0.578	0.461	0.513
5.7	1.66	3.00	2.33	0.260	0.170	2.50	1.67	1.23	0.447	0.383	0.518

†An American Standard Beam is designated by the letter S followed by the nominal depth in inches and the weight in pounds per foot.

Appendix C Properties of Rolled-Steel Shapes
(SI Units)
S Shapes
(American Standard Shapes)

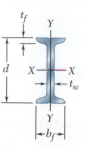

Designation[†]	Area A, mm²	Depth d, mm	Flange		Web Thickness t_w, mm	Axis X-X			Axis Y-Y		
			Width b_f, mm	Thickness t_f, mm		I_x 10⁶ mm⁴	S_x 10³ mm³	r_x mm	I_y 10⁶ mm⁴	S_y 10³ mm³	r_y mm
S610 × 180	22900	622	204	27.7	20.3	1320	4230	240	34.5	338	38.9
158	20100	622	200	27.7	15.7	1220	3930	247	32.0	320	39.9
149	18900	610	184	22.1	18.9	991	3260	229	19.7	215	32.3
134	17100	610	181	22.1	15.9	937	3060	234	18.6	205	33.0
119	15200	610	178	22.1	12.7	874	2870	241	17.5	197	34.0
S510 × 143	18200	516	183	23.4	20.3	695	2700	196	20.8	228	33.8
128	16300	516	179	23.4	16.8	653	2540	200	19.4	216	34.5
112	14200	508	162	20.2	16.1	533	2100	194	12.3	152	29.5
98.2	12500	508	159	20.2	12.8	495	1950	199	11.4	144	30.2
S460 × 104	13200	457	159	17.6	18.1	384	1690	170	10.0	126	27.4
81.4	10300	457	152	17.6	11.7	333	1460	180	8.62	113	29.0
S380 × 74	9480	381	143	15.8	14.0	202	1060	146	6.49	90.6	26.2
64	8130	381	140	15.8	10.4	186	973	151	5.95	85.0	26.9
S310 × 74	9420	305	139	16.7	17.4	126	829	116	6.49	93.2	26.2
60.7	7680	305	133	16.7	11.7	112	739	121	5.62	84.1	26.9
52	6580	305	129	13.8	10.9	94.9	624	120	4.10	63.6	24.9
47.3	6010	305	127	13.8	8.89	90.3	593	123	3.88	61.1	25.4
S250 × 52	6650	254	125	12.5	15.1	61.2	482	96.0	3.45	55.1	22.8
37.8	4810	254	118	12.5	7.90	51.2	403	103	2.80	47.4	24.1
S200 × 34	4360	203	106	10.8	11.2	26.9	265	78.5	1.78	33.6	20.2
27.4	3480	203	102	10.8	6.88	23.9	236	82.8	1.54	30.2	21.0
S150 × 25.7	3260	152	90.7	9.12	11.8	10.9	143	57.9	0.953	21.0	17.1
18.6	2360	152	84.6	9.12	5.89	9.16	120	62.2	0.749	17.7	17.8
S130 × 15	1890	127	76.2	8.28	5.44	5.12	80.3	52.1	0.495	13.0	16.2
S100 × 14.1	1800	102	71.1	7.44	8.28	2.81	55.4	39.6	0.369	10.4	14.3
11.5	1460	102	67.6	7.44	4.90	2.52	49.7	41.7	0.311	9.21	14.6
S75 × 11.2	1420	76.2	63.8	6.60	8.86	1.21	31.8	29.2	0.241	7.55	13.0
8.5	1070	76.2	59.2	6.60	4.32	1.04	27.4	31.2	0.186	6.28	13.2

[†]An American Standard Beam is designated by the letter S followed by the nominal depth in millimeters and the mass in kilograms per meter.

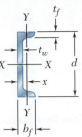

Appendix C   Properties of Rolled-Steel Shapes
(U.S. Customary Units)

C Shapes
(American Standard Channels)

Designation[†]	Area A, in^2	Depth d, in	Flange Width b_f, in	Flange Thickness t_f, in	Web Thickness t_w, in	Axis X-X I_x, in^4	Axis X-X S_x, in^3	Axis X-X r_x, in	Axis Y-Y I_y, in^4	Axis Y-Y S_y, in^3	Axis Y-Y r_y, in	x, in
C15 × 50	14.7	15.0	3.72	0.650	0.716	404	53.8	5.24	11.0	3.77	0.865	0.799
40	11.8	15.0	3.52	0.650	0.520	348	46.5	5.45	9.17	3.34	0.883	0.778
33.9	10.0	15.0	3.40	0.650	0.400	315	42.0	5.62	8.07	3.09	0.901	0.788
C12 × 30	8.81	12.0	3.17	0.501	0.510	162	27.0	4.29	5.12	2.05	0.762	0.674
25	7.34	12.0	3.05	0.501	0.387	144	24.0	4.43	4.45	1.87	0.779	0.674
20.7	6.08	12.0	2.94	0.501	0.282	129	21.5	4.61	3.86	1.72	0.797	0.698
C10 × 30	8.81	10.0	3.03	0.436	0.673	103	20.7	3.42	3.93	1.65	0.668	0.649
25	7.34	10.0	2.89	0.436	0.526	91.1	18.2	3.52	3.34	1.47	0.675	0.617
20	5.87	10.0	2.74	0.436	0.379	78.9	15.8	3.66	2.80	1.31	0.690	0.606
15.3	4.48	10.0	2.60	0.436	0.240	67.3	13.5	3.87	2.27	1.15	0.711	0.634
C9 × 20	5.87	9.00	2.65	0.413	0.448	60.9	13.5	3.22	2.41	1.17	0.640	0.583
15	4.41	9.00	2.49	0.413	0.285	51.0	11.3	3.40	1.91	1.01	0.659	0.586
13.4	3.94	9.00	2.43	0.413	0.233	47.8	10.6	3.49	1.75	0.954	0.666	0.601
C8 × 18.7	5.51	8.00	2.53	0.390	0.487	43.9	11.0	2.82	1.97	1.01	0.598	0.565
13.7	4.04	8.00	2.34	0.390	0.303	36.1	9.02	2.99	1.52	0.848	0.613	0.554
11.5	3.37	8.00	2.26	0.390	0.220	32.5	8.14	3.11	1.31	0.775	0.623	0.572
C7 × 12.2	3.60	7.00	2.19	0.366	0.314	24.2	6.92	2.60	1.16	0.696	0.568	0.525
9.8	2.87	7.00	2.09	0.366	0.210	21.2	6.07	2.72	0.957	0.617	0.578	0.541
C6 × 13	3.81	6.00	2.16	0.343	0.437	17.3	5.78	2.13	1.05	0.638	0.524	0.514
10.5	3.08	6.00	2.03	0.343	0.314	15.1	5.04	2.22	0.860	0.561	0.529	0.500
8.2	2.39	6.00	1.92	0.343	0.200	13.1	4.35	2.34	0.687	0.488	0.536	0.512
C5 × 9	2.64	5.00	1.89	0.320	0.325	8.89	3.56	1.83	0.624	0.444	0.486	0.478
6.7	1.97	5.00	1.75	0.320	0.190	7.48	2.99	1.95	0.470	0.372	0.489	0.484
C4 × 7.2	2.13	4.00	1.72	0.296	0.321	4.58	2.29	1.47	0.425	0.337	0.447	0.459
5.4	1.58	4.00	1.58	0.296	0.184	3.85	1.92	1.56	0.312	0.277	0.444	0.457
C3 × 6	1.76	3.00	1.60	0.273	0.356	2.07	1.38	1.08	0.300	0.263	0.413	0.455
5	1.47	3.00	1.50	0.273	0.258	1.85	1.23	1.12	0.241	0.228	0.405	0.439
4.1	1.20	3.00	1.41	0.273	0.170	1.65	1.10	1.17	0.191	0.196	0.398	0.437

[†]An American Standard Channel is designated by the letter C followed by the nominal depth in inches and the weight in pounds per foot.

Appendix C   Properties of Rolled-Steel Shapes
(SI Units)

C Shapes
(American Standard Channels)

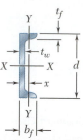

Designation[†]	Area A, mm^2	Depth d, mm	Flange		Web Thickness t_w, mm	Axis X-X			Axis Y-Y			
			Width b_f, mm	Thickness t_f, mm		I_x 10^6 mm^4	S_x 10^3 mm^3	r_x mm	I_y 10^6 mm^4	S_y 10^3 mm^3	r_y mm	x mm
C380 × 74	9480	381	94.5	16.5	18.2	168	882	133	4.58	61.8	22.0	20.3
60	7610	381	89.4	16.5	13.2	145	762	138	3.82	54.7	22.4	19.8
50.4	6450	381	86.4	16.5	10.2	131	688	143	3.36	50.6	22.9	20.0
C310 × 45	5680	305	80.5	12.7	13.0	67.4	442	109	2.13	33.6	19.4	17.1
37	4740	305	77.5	12.7	9.83	59.9	393	113	1.85	30.6	19.8	17.1
30.8	3920	305	74.7	12.7	7.16	53.7	352	117	1.61	28.2	20.2	17.7
C250 × 45	5680	254	77.0	11.1	17.1	42.9	339	86.9	1.64	27.0	17.0	16.5
37	4740	254	73.4	11.1	13.4	37.9	298	89.4	1.39	24.1	17.1	15.7
30	3790	254	69.6	11.1	9.63	32.8	259	93.0	1.17	21.5	17.5	15.4
22.8	2890	254	66.0	11.1	6.10	28.0	221	98.3	0.945	18.8	18.1	16.1
C230 × 30	3790	229	67.3	10.5	11.4	25.3	221	81.8	1.00	19.2	16.3	14.8
22	2850	229	63.2	10.5	7.24	21.2	185	86.4	0.795	16.6	16.7	14.9
19.9	2540	229	61.7	10.5	5.92	19.9	174	88.6	0.728	15.6	16.9	15.3
C200 × 27.9	3550	203	64.3	9.91	12.4	18.3	180	71.6	0.820	16.6	15.2	14.4
20.5	2610	203	59.4	9.91	7.70	15.0	148	75.9	0.633	13.9	15.6	14.1
17.1	2170	203	57.4	9.91	5.59	13.5	133	79.0	0.545	12.7	15.8	14.5
C180 × 18.2	2320	178	55.6	9.30	7.98	10.1	113	66.0	0.483	11.4	14.4	13.3
14.6	1850	178	53.1	9.30	5.33	8.82	100	69.1	0.398	10.1	14.7	13.7
C150 × 19.3	2460	152	54.9	8.71	11.1	7.20	94.7	54.1	0.437	10.5	13.3	13.1
15.6	1990	152	51.6	8.71	7.98	6.29	82.6	56.4	0.358	9.19	13.4	12.7
12.2	1540	152	48.8	8.71	5.08	5.45	71.3	59.4	0.286	8.00	13.6	13.0
C130 × 13	1700	127	48.0	8.13	8.26	3.70	58.3	46.5	0.260	7.28	12.3	12.1
10.4	1270	127	44.5	8.13	4.83	3.11	49.0	49.5	0.196	6.10	12.4	12.3
C100 × 10.8	1370	102	43.7	7.52	8.15	1.91	37.5	37.3	0.177	5.52	11.4	11.7
8	1020	102	40.1	7.52	4.67	1.60	31.5	39.6	0.130	4.54	11.3	11.6
C75 × 8.9	1140	76.2	40.6	6.93	9.04	0.862	22.6	27.4	0.125	4.31	10.5	11.6
7.4	948	76.2	38.1	6.93	6.55	0.770	20.2	28.4	0.100	3.74	10.3	11.2
6.1	774	76.2	35.8	6.93	4.32	0.687	18.0	29.7	0.0795	3.21	10.1	11.1

[†]An American Standard Channel is designated by the letter C followed by the nominal depth in millimeters and the mass in kilograms per meter.

Appendix C   Properties of Rolled-Steel Shapes
(U.S. Customary Units)

Angles
Equal Legs

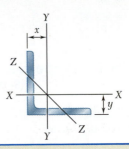

Size and Thickness, in.	Weight per Foot, lb/ft	Area, in²	Axis X-X and Axis Y-Y				Axis Z-Z r_z, in.
			I, in⁴	S, in³	r, in.	x or y, in.	
L8 × 8 × 1	51.0	15.0	89.1	15.8	2.43	2.36	1.56
¾	38.9	11.4	69.9	12.2	2.46	2.26	1.57
½	26.4	7.75	48.8	8.36	2.49	2.17	1.59
L6 × 6 × 1	37.4	11.0	35.4	8.55	1.79	1.86	1.17
¾	28.7	8.46	28.1	6.64	1.82	1.77	1.17
⅝	24.2	7.13	24.1	5.64	1.84	1.72	1.17
½	19.6	5.77	19.9	4.59	1.86	1.67	1.18
⅜	14.9	4.38	15.4	3.51	1.87	1.62	1.19
L5 × 5 × ¾	23.6	6.94	15.7	4.52	1.50	1.52	0.972
⅝	20.0	5.86	13.6	3.85	1.52	1.47	0.975
½	16.2	4.75	11.3	3.15	1.53	1.42	0.980
⅜	12.3	3.61	8.76	2.41	1.55	1.37	0.986
L4 × 4 × ¾	18.5	5.44	7.62	2.79	1.18	1.27	0.774
⅝	15.7	4.61	6.62	2.38	1.20	1.22	0.774
½	12.8	3.75	5.52	1.96	1.21	1.18	0.776
⅜	9.80	2.86	4.32	1.50	1.23	1.13	0.779
¼	6.60	1.94	3.00	1.03	1.25	1.08	0.783
L3½ × 3½ × ½	11.1	3.25	3.63	1.48	1.05	1.05	0.679
⅜	8.50	2.48	2.86	1.15	1.07	1.00	0.683
¼	5.80	1.69	2.00	0.787	1.09	0.954	0.688
L3 × 3 × ½	9.40	2.75	2.20	1.06	0.895	0.929	0.580
⅜	7.20	2.11	1.75	0.825	0.910	0.884	0.581
¼	4.90	1.44	1.23	0.569	0.926	0.836	0.585
L2½ × 2½ × ½	7.70	2.25	1.22	0.716	0.735	0.803	0.481
⅜	5.90	1.73	0.972	0.558	0.749	0.758	0.481
¼	4.10	1.19	0.692	0.387	0.764	0.711	0.482
³⁄₁₆	3.07	0.900	0.535	0.295	0.771	0.687	0.482
L2 × 2 × ⅜	4.70	1.36	0.476	0.348	0.591	0.632	0.386
¼	3.19	0.938	0.346	0.244	0.605	0.586	0.387
⅛	1.65	0.484	0.189	0.129	0.620	0.534	0.391

Appendix C   Properties of Rolled-Steel Shapes
(SI Units)

Angles
Equal Legs

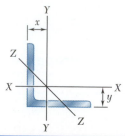

Size and Thickness, mm	Mass per Meter, kg/m	Area, mm²	Axis X-X				Axis Z-Z
			I 10^6 mm⁴	S 10^3 mm³	r mm	x or y mm	r_z mm
L203 × 203 × 25.4	75.9	9680	37.1	259	61.7	59.9	39.6
19	57.9	7350	29.1	200	62.5	57.4	39.9
12.7	39.3	5000	20.3	137	63.2	55.1	40.4
L152 × 152 × 25.4	55.7	7100	14.7	140	45.5	47.2	29.7
19	42.7	5460	11.7	109	46.2	45.0	29.7
15.9	36.0	4600	10.0	92.4	46.7	43.7	29.7
12.7	29.2	3720	8.28	75.2	47.2	42.4	30.0
9.5	22.2	2830	6.41	57.5	47.5	41.1	30.2
L127 × 127 × 19	35.1	4480	6.53	74.1	38.1	38.6	24.7
15.9	29.8	3780	5.66	63.1	38.6	37.3	24.8
12.7	24.1	3060	4.70	51.6	38.9	36.1	24.9
9.5	18.3	2330	3.65	39.5	39.4	34.8	25.0
L102 × 102 × 19	27.5	3510	3.17	45.7	30.0	32.3	19.7
15.9	23.4	2970	2.76	39.0	30.5	31.0	19.7
12.7	19.0	2420	2.30	32.1	30.7	30.0	19.7
9.5	14.6	1850	1.80	24.6	31.2	28.7	19.8
6.4	9.80	1250	1.25	16.9	31.8	27.4	19.9
L89 × 89 × 12.7	16.5	2100	1.51	24.3	26.7	26.7	17.2
9.5	12.6	1600	1.19	18.8	27.2	25.4	17.3
6.4	8.60	1090	0.832	12.9	27.7	24.2	17.5
L76 × 76 × 12.7	14.0	1770	0.916	17.4	22.7	23.6	14.7
9.5	10.7	1360	0.728	13.5	23.1	22.5	14.8
6.4	7.30	929	0.512	9.32	23.5	21.2	14.9
L64 × 64 × 12.7	11.4	1450	0.508	11.7	18.7	20.4	12.2
9.5	8.70	1120	0.405	9.14	19.0	19.3	12.2
6.4	6.10	768	0.288	6.34	19.4	18.1	12.2
4.8	4.60	581	0.223	4.83	19.6	17.4	12.2
L51 × 51 × 9.5	7.00	877	0.198	5.70	15.0	16.1	9.80
6.4	4.70	605	0.144	4.00	15.4	14.9	9.83
3.2	2.40	312	0.0787	2.11	15.7	13.6	9.93

Appendix C   Properties of Rolled-Steel Shapes
(U.S. Customary Units)

Angles
Unequal Legs

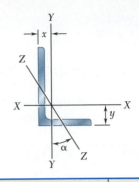

Size and Thickness, in.	Weight per Foot, lb/ft	Area, in²	Axis X-X				Axis Y-Y				Axis Z-Z	
			I_x, in⁴	S_x, in³	r_x, in.	y, in.	I_y, in⁴	S_y, in³	r_y, in.	x, in.	r_z, in.	tan α
L8 × 6 × 1	44.2	13.0	80.9	15.1	2.49	2.65	38.8	8.92	1.72	1.65	1.28	0.542
¾	33.8	9.94	63.5	11.7	2.52	2.55	30.8	6.92	1.75	1.56	1.29	0.550
½	23.0	6.75	44.4	8.01	2.55	2.46	21.7	4.79	1.79	1.46	1.30	0.557
L6 × 4 × ¾	23.6	6.94	24.5	6.23	1.88	2.07	8.63	2.95	1.12	1.07	0.856	0.428
½	16.2	4.75	17.3	4.31	1.91	1.98	6.22	2.06	1.14	0.981	0.864	0.440
⅜	12.3	3.61	13.4	3.30	1.93	1.93	4.86	1.58	1.16	0.933	0.870	0.446
L5 × 3 × ½	12.8	3.75	9.43	2.89	1.58	1.74	2.55	1.13	0.824	0.746	0.642	0.357
⅜	9.80	2.86	7.35	2.22	1.60	1.69	2.01	0.874	0.838	0.698	0.646	0.364
¼	6.60	1.94	5.09	1.51	1.62	1.64	1.41	0.600	0.853	0.648	0.652	0.371
L4 × 3 × ½	11.1	3.25	5.02	1.87	1.24	1.32	2.40	1.10	0.858	0.822	0.633	0.542
⅜	8.50	2.48	3.94	1.44	1.26	1.27	1.89	0.851	0.873	0.775	0.636	0.551
¼	5.80	1.69	2.75	0.988	1.27	1.22	1.33	0.585	0.887	0.725	0.639	0.558
L3½ × 2½ × ½	9.40	2.75	3.24	1.41	1.08	1.20	1.36	0.756	0.701	0.701	0.532	0.485
⅜	7.20	2.11	2.56	1.09	1.10	1.15	1.09	0.589	0.716	0.655	0.535	0.495
¼	4.90	1.44	1.81	0.753	1.12	1.10	0.775	0.410	0.731	0.607	0.541	0.504
L3 × 2 × ½	7.70	2.25	1.92	1.00	0.922	1.08	0.667	0.470	0.543	0.580	0.425	0.413
⅜	5.90	1.73	1.54	0.779	0.937	1.03	0.539	0.368	0.555	0.535	0.426	0.426
¼	4.10	1.19	1.09	0.541	0.953	0.980	0.390	0.258	0.569	0.487	0.431	0.437
L2½ × 2 × ⅜	5.30	1.55	0.914	0.546	0.766	0.826	0.513	0.361	0.574	0.578	0.419	0.612
¼	3.62	1.06	0.656	0.381	0.782	0.779	0.372	0.253	0.589	0.532	0.423	0.624

Appendix C   Properties of Rolled-Steel Shapes
(SI Units)

Angles
Unequal Legs

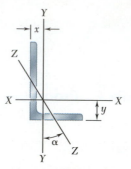

Size and Thickness, mm	Mass per Meter kg/m	Area mm²	Axis X-X				Axis Y-Y				Axis Z-Z	
			I_x 10⁶ mm⁴	S_x 10³ mm³	r_x mm	y mm	I_y 10⁶ mm⁴	S_y 10³ mm³	r_y mm	x mm	r_z mm	tan α
L203 × 152 × 25.4	65.5	8390	33.7	247	63.2	67.3	16.1	146	43.7	41.9	32.5	0.542
19	50.1	6410	26.4	192	64.0	64.8	12.8	113	44.5	39.6	32.8	0.550
12.7	34.1	4350	18.5	131	64.8	62.5	9.03	78.5	45.5	37.1	33.0	0.557
L152 × 102 × 19	35.0	4480	10.2	102	47.8	52.6	3.59	48.3	28.4	27.2	21.7	0.428
12.7	24.0	3060	7.20	70.6	48.5	50.3	2.59	33.8	29.0	24.9	21.9	0.440
9.5	18.2	2330	5.58	54.1	49.0	49.0	2.02	25.9	29.5	23.7	22.1	0.446
L127 × 76 × 12.7	19.0	2420	3.93	47.4	40.1	44.2	1.06	18.5	20.9	18.9	16.3	0.357
9.5	14.5	1850	3.06	36.4	40.6	42.9	0.837	14.3	21.3	17.7	16.4	0.364
6.4	9.80	1250	2.12	24.7	41.1	41.7	0.587	9.83	21.7	16.5	16.6	0.371
L102 × 76 × 12.7	16.4	2100	2.09	30.6	31.5	33.5	0.999	18.0	21.8	20.9	16.1	0.542
9.5	12.6	1600	1.64	23.6	32.0	32.3	0.787	13.9	22.2	19.7	16.2	0.551
6.4	8.60	1090	1.14	16.2	32.3	31.0	0.554	9.59	22.5	18.4	16.2	0.558
L89 × 64 × 12.7	13.9	1770	1.35	23.1	27.4	30.5	0.566	12.4	17.8	17.8	13.5	0.485
9.5	10.7	1360	1.07	17.9	27.9	29.2	0.454	9.65	18.2	16.6	13.6	0.495
6.4	7.30	929	0.753	12.3	28.4	27.9	0.323	6.72	18.6	15.4	13.7	0.504
L76 × 51 × 12.7	11.5	1450	0.799	16.4	23.4	27.4	0.278	7.70	13.8	14.7	10.8	0.413
9.5	8.80	1120	0.641	12.8	23.8	26.2	0.224	6.03	14.1	13.6	10.8	0.426
6.4	6.10	768	0.454	8.87	24.2	24.9	0.162	4.23	14.5	12.4	10.9	0.437
L64 × 51 × 9.5	7.90	1000	0.380	8.95	19.5	21.0	0.214	5.92	14.6	14.7	10.6	0.612
6.4	5.40	684	0.273	6.24	19.9	19.8	0.155	4.15	15.0	13.5	10.7	0.624

Appendix D   Beam Deflections and Slopes

Beam and Loading	Elastic Curve	Maximum Deflection	Slope at End	Equation of Elastic Curve
1		$-\dfrac{PL^3}{3EI}$	$-\dfrac{PL^2}{2EI}$	$y = \dfrac{P}{6EI}(x^3 - 3Lx^2)$
2		$-\dfrac{wL^4}{8EI}$	$-\dfrac{wL^3}{6EI}$	$y = -\dfrac{w}{24EI}(x^4 - 4Lx^3 + 6L^2x^2)$
3		$-\dfrac{ML^2}{2EI}$	$-\dfrac{ML}{EI}$	$y = -\dfrac{M}{2EI}x^2$
4		$-\dfrac{PL^3}{48EI}$	$\pm\dfrac{PL^2}{16EI}$	For $x \le \frac{1}{2}L$: $y = \dfrac{P}{48EI}(4x^3 - 3L^2x)$
5		For $a > b$: $-\dfrac{Pb(L^2 - b^2)^{3/2}}{9\sqrt{3}\,EIL}$ at $x_m = \sqrt{\dfrac{L^2 - b^2}{3}}$	$\theta_A = -\dfrac{Pb(L^2 - b^2)}{6EIL}$ $\theta_B = +\dfrac{Pa(L^2 - a^2)}{6EIL}$	For $x < a$: $y = \dfrac{Pb}{6EIL}[x^3 - (L^2 - b^2)x]$ For $x = a$: $y = -\dfrac{Pa^2b^2}{3EIL}$
6		$-\dfrac{5wL^4}{384EI}$	$\pm\dfrac{wL^3}{24EI}$	$y = -\dfrac{w}{24EI}(x^4 - 2Lx^3 + L^3x)$
7		$\dfrac{ML^2}{9\sqrt{3}\,EI}$	$\theta_A = +\dfrac{ML}{6EI}$ $\theta_B = -\dfrac{ML}{3EI}$	$y = -\dfrac{M}{6EIL}(x^3 - L^2x)$

E Fundamentals of Engineering Examination

Engineers are required to be licensed when their work directly affects the public health, safety, and welfare. The intent is to ensure that engineers have met minimum qualifications, involving competence, ability, experience, and character. The licensing process involves an initial exam, called the *Fundamentals of Engineering Examination,* professional experience, and a second exam, called the *Principles and Practice of Engineering.* Those who successfully complete these requirements are licensed as a *Professional Engineer.* The exams are developed under the auspices of the *National Council of Examiners for Engineering and Surveying.*

The first exam, the *Fundamentals of Engineering Examination,* can be taken just before or after graduation from a four-year accredited engineering program. The exam stresses subject material in a typical undergraduate engineering program, including *Mechanics of Materials.* The topics included in the exam cover much /of the material in this book. The following is a list of the main topic areas with references to appropriate sections in this book. Also included are problems that can be solved to review this material.

Stresses (1.2–1.4)
Problems: 1.4, 1.10, 1.30, 1.37

Strains (2.1–2.4; 2.7–2.8)
Problems: 2.4, 2.19, 2.41, 2.47, 2.61, 2.68

Torsion (3.1–3.3; 3.9–3.10)
Problems: 3.6, 3.27, 3.35, 3.53, 3.129, 3.137

Bending (4.1–4.4; 4.7)
Problems: 4.9, 4.22, 4.33, 4.49, 4.103, 4.107

Shear and Bending–Moment Diagrams (5.1–5.2)
Problems: 5.5, 5.11, 5.39, 5.43

Normal Stresses in Beams (5.3)
Problems: 5.18, 5.21, 5.56, 5.60

Shear (6.1; 6.3–6.4)
Problems: 6.1, 6.11, 6.32, 6.38

Transformation of Stresses and Strains (7.1–7.2; 7.5–7.6)
Problems: 7.5, 7.15, 7.32, 7.43, 7.81, 7.87, 7.100, 7.104

Deflection of Beams (9.1; 9.4)
Problems: 9.5, 9.13, 9.71, 9.77

Columns (10.1)
Problems: 10.11, 10.19, 10.22

Strain Energy (11.1–11.2)
Problems: 11.9, 11.15, 11.21

Answers to Problems

Answers to problems with a number set in straight type are given on this and the following pages. Answers to problems with a number set in italic and red are not listed.

CHAPTER 1

1.1 (*a*) 84.9 MPa. (*b*) −96.8 MPa.
1.2 $d_1 = 22.6$ mm, $d_2 = 40.2$ mm.
1.3 (*a*) 17.93 ksi. (*b*) 22.6 ksi.
1.4 6.75 kips.
1.7 (*a*) 101.6 MPa. (*b*) −21.7 MPa.
1.8 1084 ksi.
1.9 285 mm^2.
1.10 (*a*) 11.09 ksi. (*b*) −12.00 ksi.
1.13 −4.97 MPa.
1.14 (*a*) 12.73 MPa. (*b*) −4.77 MPa.
1.15 43.4 mm.
1.16 2.25 kips.
1.17 889 psi.
1.18 67.9 kN.
1.20 29.4 mm.
1.21 (*a*) 3.33 MPa. (*b*) 525 mm.
1.23 (*a*) 1.030 in. (*b*) 38.8 ksi.
1.24 8.31 kN.
1.25 (*a*) 11.45 mm. (*b*) 134.9 MPa. (*c*) 90.0 MPa.
1.28 (*a*) 9.94 ksi. (*b*) 6.25 ksi.
1.29 (*a*) $\sigma = 489$ kPa; $\tau = 489$ kPa.
1.30 (*a*) 13.95 kN. (*b*) 620 kPa.
1.31 $\sigma = 70.0$ psi; $\tau = 40.4$ psi.
1.32 (*a*) 1.500 kips. (*b*) 43.3 psi.
1.35 $\sigma = -21.6$ MPa; $\tau = 7.87$ MPa.
1.36 833 kN.
1.37 3.09 kips.
1.40 (*a*) 181.3 mm^2. (*b*) 213 mm^2.
1.41 (*a*) 3.97. (*b*) 265 mm^2.
1.42 0.268 in^2.
1.45 2.87.
1.46 0.798 in.
1.47 10.25 kN.
1.48 (*a*) 2.92. (*b*) $b = 40.3$ mm, $c = 97.2$ mm.
1.50 3.24.
1.52 283 lb.
1.53 2.42.
1.54 2.05.
1.55 3.72 kN.
1.56 3.97 kN.
1.57 (*a*) 362 kg. (*b*) 1.718.
1.58 (*a*) 629 lb. (*b*) 1.689.
1.59 195.3 MPa.
1.60 (*a*) 14.64 ksi. (*b*) −9.96 ksi.
1.62 25.2 mm.
1.64 (*a*) −640 psi. (*b*) −320 psi.
1.65 (*a*) 444 psi. (*b*) 7.50 in. (*c*) 2400 psi.
1.67 3.45.
1.68 $\sigma_{all}\, d/4\, \tau_{all}$.

1.70 $21.3° < \theta < 32.3°$.
1.C2 (*c*) 16 mm ≤ *d* ≤ 22 mm. (*d*) 18 mm ≤ *d* ≤ 22 mm.
1.C3 (*c*) 0.70 in. ≤ *d* ≤ 1.10 in. (*d*) 0.85 in. ≤ *d* ≤ 1.25 in.
1.C4 (*b*) For $\beta = 38.66°$, $\tan \beta = 0.8$; *BD* is perpendicular to *BC*.
   (*c*) F.S. = 3.58 for $\alpha = 26.6°$; **P** is perpendicular to line *AC*.
1.C5 (*b*) *Member of Fig. P 1.29, for* $\alpha = 60°$:
   (1) 70.0 psi; (2) 40.4 psi; (3) 2.14; (4) 5.30; (5) 2.14.
   Member of Fig. P 1.31, for $\alpha = 45°$:
   (1) 489 kPa; (2) 489 kPa; (3) 2.58; (4) 3.07; (5) 2.58.
1.C6 (*d*) $P_{all} = 5.79$ kN; stress in links is critical.

CHAPTER 2

2.1 (*a*) 0.546 mm. (*b*) 36.3 MPa.
2.2 (*a*) 0.0303 in. (*b*) 15.28 ksi.
2.3 (*a*) 9.82 kN. (*b*) 500 MPa.
2.4 (*a*) 81.8 MPa. (*b*) 1.712.
2.6 (*a*) 5.32 mm. (*b*) 1.750 m.
2.7 (*a*) 0.381 in. (*b*) 17.58 ksi.
2.9 9.21 mm.
2.11 48.4 kips.
2.13 0.429 in.
2.14 1.988 kN.
2.15 0.868 in.
2.17 (*a*) 25.5×10^{-3} in. (*b*) 15.56×10^{-3} in.
2.19 (*a*) 32.8 kN. (*b*) 0.0728 mm ↓.
2.20 (*a*) 0.0189 mm ↑. (*b*) 0.0919 mm ↓.
2.21 (*a*) $\delta_{AB} = -2.11$ mm; $\delta_{AC} = 2.03$ mm.
2.23 (*a*) 0.1767 in. (*b*) 0.1304 in.
2.24 50.4 kN.
2.25 14.74 kN.
2.26 (*a*) −0.0302 mm. (*b*) 0.01783 mm.
2.27 4.71×10^{-3} in. ↓.
2.29 (*a*) $\rho g l^2/2E$. (*b*) $W/2$.
2.30 $Ph/\pi Eab$ ↓.
2.33 (*a*) 140.6 MPa. (*b*) 93.8 MPa.
2.34 (*a*) 15.00 mm. (*b*) 288 kN.
2.35 $\sigma_s = -8.34$ ksi; $\sigma_c = -1.208$ ksi.
2.36 695 kips.
2.39 (*a*) $R_A = 2.28$ kips ↑; $R_C = 9.72$ kips ↑. (*b*) $\sigma_{AB} = +1.857$ ksi;
   $\sigma_{BC} = -3.09$ ksi.
2.41 (*a*) 62.8 kN ← at *A*; 37.2 kN ← at *E*. (*b*) 46.3 μm →.
2.42 (*a*) 45.5 kN ← at *A*; 54.5 kN ← at *E*. (*b*) 48.8 μm →.
2.43 0.536 mm ↓.
2.44 (*a*) $P_{BE} = 205$ lb; $P_{CF} = 228$ lb. (*b*) 0.0691 in. ↓.
2.45 $P_A = 0.525\, P$; $P_B = 0.200\, P$; $P_C = 0.275\, P$.
2.47 −8.15 MPa.
2.48 −56.2 MPa.
2.50 $\sigma_s = -1.448$ ksi; σ_C 54.2 psi.
2.51 142.6 kN.
2.52 (*a*) $\sigma_{AB} = -5.25$ ksi; $\sigma_{BC} = -11.82$ ksi. (*b*) 6.57×10^{-3} in. →.

2.54 (a) −98.3 MPa. (b) −38.3 MPa.

2.55 (a) 21.4°C. (b) 3.67 MPa.

2.56 5.70 kN.

2.58 (a) 201.6°F. (b) 18.0107 in.

2.59 (a) 52.3 kips. (b) 9.91×10^{-3}in.

2.61 29×10^3 psi; 10.03×10^3 psi; 0.444.

2.63 0.399.

2.64 (a) 0.0358 mm. (b) −0.00258 mm. (c) −0.000344 mm. (d) −0.00825 mm^2.

2.66 94.9 kips.

2.67 (a) −0.0724 mm. (b) −0.01531 mm.

2.68 (a) 0.00312 in. (b) 0.00426 in. (c) 0.00505 in.

2.69 (a) 352×10^{-6} in. (b) 82.8×10^{-6} in. (c) 307×10^{-6} in.

2.70 (a) −63.0 MPa. (b) −13.50 mm^2. (c) −540 mm^3.

2.77 $a = 42.9$ mm; $b = 160.7$ mm.

2.78 75.0 kN; 40.0 mm.

2.79 (a) 10.42 in. (b) 0.813 in.

2.80 $\tau = 62.5$ psi; $G = 156.3$ psi.

2.81 16.67 MPa.

2.82 19.00×10^3 kN/m.

2.83 (a) 588×10^{-6} in. (b) 33.2×10^{-3} in^3. (c) 0.0294%.

2.84 (a) −0.0746 mm; −143.9 mm^3.
(b) −0.0306 mm; −521 mm^3.

2.85 (a) 193.2×10^{-6}; 1.214×10^{-3} in^3.
(b) 396×10^{-6}; 2.49×10^{-3} in^3.

2.88 3.00.

2.91 (a) 0.0303 mm. (b) $\sigma_x = 40.6$ MPa, $\sigma_y = \sigma_z = 5.48$ MPa.

2.92 (a) $\sigma_x = 44.6$ MPa; $\sigma_y = 0$; $\sigma_z = 3.45$ MPa. (b) −0.0129 mm.

2.93 (a) 58.3 kN. (b) 64.3 kN.

2.94 (a) 87.0 MPa. (b) 75.2 MPa. (c) 73.9 MPa.

2.97 (a) 11.4 mm. (b) 28.8 kN.

2.98 36.7 mm.

2.99 (a) 12.02 kips. (b) 108.0%.

2.100 23.9 kips.

2.101 (a) 15.90 kips; 0.1745 in. (b) 15.90 kips; 0.274 in.

2.102 (a) 44.2 kips; 0.0356 in. (b) 44.2 kips; 0.1606 in.

2.105 176.7 kN; 3.84 mm.

2.106 176.7 kN; 3.16 mm.

2.107 (a) 0.292 mm. (b) $\sigma_{AC} = 250$ MPa; $\sigma_{BC} = -307$ MPa.
(c) 0.0272 mm.

2.108 (a) 990kN. (b) $\sigma_{AC} = 250$ MPa; $\sigma_{BC} = -316$ MPa.
(c) 0.0313 mm.

2.111 (a) 112.1 kips. (b) 50 ksi in low-strength steel;
82.9 ksi in high-strength steel. (c) 0.00906 in.

2.112 (a) 0.0309 in. (b) 64.0 ksi. (c) 0.00387 in.

2.113 (a) $\sigma_{AD} = 250$ MPa. (b) $\sigma_{BE} = 124.3$ MPa. (c) 0.622 mm ↓.

2.114 (a) $\sigma_{AD} = 233$ MPa; $\sigma_{BE} = 250$ MPa. (b) 1.322 mm ↓.

2.115 (a) $\sigma_{AD} = -4.70$ MPa; $\sigma_{BE} = 19.34$ MPa. (b) 0.0967 mm ↓.

2.116 (a) −36.0 ksi. (b) 15.84 ksi.

2.117 (a) $\sigma_{AC} = -150.0$ MPa; $\sigma_{CB} = -250$ MPa.
(b) −0.1069 mm →.

2.118 (a) $\sigma_{AC} = 56.5$ MPa; $\sigma_{CB} = 9.41$ MPa. (b) 0.0424 mm →.

2.121 (a) 915°F. (b) 1759°F.

2.122 (a) 0.1042 mm. (b) $\sigma_{AC} = \sigma_{CB} = -65.2$ MPa.

2.123 (a) 0.00788 mm. (b) $\sigma_{AC} = \sigma_{CB} = -6.06$ MPa.

2.125 1.219 in.

2.127 4.67°C.

2.128 (a) 9.53 kips. (b) 1.254×10^{-3} in.

2.130 (steel) − 15.80 ksi; (concrete) − 1.962 ksi.

2.131 (a) 9.73 kN. (b) 2.02 mm ←.

2.133 0.01870 in.

2.135 (a) $A\,\sigma_Y/\mu g$. (b) $E\,A/L$.

2.C1 Prob. 2.126: (a) 11.90×10^{-3} in. ↓. (b) 5.66×10^{-3} in. ↑.

2.C3 Prob. 2.60: (a) −116.2 MPa. (b) 0.363 mm.

2.C5 $r = 0.25$ in.: 3.89 kips;
$r = 0.75$ in.: 2.78 kips.

2.C6 (a) −0.40083. (b) −0.10100. (c) −0.00405.

CHAPTER 3

3.1 641 N·m.

3.2 87.3 MPa.

3.3 (a) 5.17 kN·m. (b) 87.2 MPa.

3.4 (a) 7.55 ksi. (b) 7.64 ksi.

3.6 (a) 70.7 MPa. (b) 35.4 MPa. (c) 6.25%.

3.7 (a) 19.21 kip·in. (b) 2.01 in.

3.9 (a) 8.35 ksi. (b) 5.94 ksi.

3.10 (a) 1.292 in. (b) 1.597 in.

3.11 (a) shaft CD. (b) 85.8 MPa.

3.13 (a) 77.6 MPa. (b) 62.8 MPa. (c) 20.9 MPa.

3.15 9.16 kip·in.

3.16 (a) 1.503 in. (b) 1.853 in.

3.18 (a) $d_{AB} = 52.9$ mm. (b) $d_{BC} = 33.3$ mm.

3.20 3.18 kN·m.

3.21 (a) 59.6 mm. (b) 43.9 mm.

3.22 (a) 72.5 MPa. (b) 68.7 MPa.

3.23 1.189 kip·in.

3.25 4.30 kip·in.

3.27 73.6 N·m.

3.28 (a) $d_{AB} = 38.6$ mm. (b) $d_{CD} = 52.3$ mm. (c) 75.5 mm.

3.29 1.0; 1.025; 1.120; 1.200; 1.0.

3.31 11.87 mm.

3.32 9.38 ksi.

3.33 (a) 1.390°. (b) 1.482°.

3.35 (a) 1.384°. (b) 3.22°.

3.37 (a) 14.43°. (b) 46.9°.

3.38 6.02°.

3.40 1.140°.

3.41 3.77°.

3.42 3.78°.

3.44 53.8°.

3.45 36.1 mm.

3.46 0.837 in.

3.47 1.089 in.

3.48 62.9 mm.

3.49 42.0 mm.

3.50 22.5 mm.

3.53 (a) 4.72 ksi. (b) 7.08 ksi. (c) 4.35°.

3.54 7.37°.

3.56 (a) $T_A = 1090$ N·m; $T_C = 310$ N·m.
(b) 47.4 MPa. (c) 28.8 MPa.

3.57 $\tau_{AB} = 68.9$ MPa; $\tau_{CD} = 14.70$ MPa.

3.59 12.24 MPa.

3.61 0.241 in.

3.63 $T/2\pi t r_1^2$ at r_1.

3.64 (a) 9.51 ksi. (b) 4.76 ksi.

3.65 (a) 46.9 MPa. (b) 23.5 MPa.

3.66 (a) 0.893 in. (b) 0.709 in.

3.67 (a) 20.1 mm. (b) 15.94 mm.

3.68 25.6 kW.

3.69 2.64 mm.

3.71 (a) 51.7 kW. (b) 6.17°.

3.73 (a) 47.5 MPa. (b) 30.4 mm.

3.76 (a) 4.08 ksi. (b) 6.79 ksi.

3.77 (*a*) 0.799 in. (*b*) 0.947 in.
3.78 (*a*) 16.02 Hz. (*b*) 27.2 Hz.
3.79 1917 rpm.
3.80 50.0 kW.
3.81 36.1 mm.
3.84 10.8 mm.
3.86 (*a*) 5.36 ksi. (*b*) 5.02 ksi.
3.87 63.5 kW.
3.88 42.6 Hz.
3.89 (*a*) 2.61 ksi. (*b*) 2.01 ksi.
3.90 (*a*) 203 N·m. (*b*) 165.8 N·m.
3.92 (*a*) 9.64 kN·m. (*b*) 9.91 kN·m.
3.93 2230 lb·in.
3.94 (*a*) 18.86 ksi; 1.500 in. (*b*) 21.0 ksi; 0.916 in.
3.95 (*a*) 113.3 MPa; 15.00 mm.
  (*b*) 145.0 MPa; 6.90 mm.
3.98 (*a*) 6.72°. (*b*) 18.71°.
3.99 (*a*) 2.47°. (*b*) 4.34°.
3.100 (*a*) 977 N·m. (*b*) 8.61 mm.
3.101 (*a*) 52.1 kip·in. (*b*) 80.8 kip·in.
3.102 $\tau_{max} = 145.0$ MPa; $\phi = 19.70°$.
3.104 (*a*) 8.17 mm. (*b*) 42.1°.
3.106 (*a*) 8.02°. (*b*) 14.89 kN·m.
3.107 (*a*) 11.71 kN·m; 3.44°. (*b*) 14.12 kN·m; 4.81°.
3.110 (*a*) 5.24 kip·in. (*b*) 6.88°.
3.111 (*a*) 1.322 kip·in. (*b*) 12.60 ksi.
3.112 2.32 kN·m.
3.113 2.26 kN·m.
3.114 5.63 ksi.
3.115 14.62°.
3.118 68.0 MPa at inner surface.
3.119 20.2°.
3.120 (*a*) $c_0 = 0.1500c$. (*b*) $T_0 = 0.221\tau_y c^3$.
3.121 0.0505 in.
3.122 68.2 in.
3.123 (*a*) 189.2 N·m; 9.05°. (*b*) 228 N·m; 7.91°.
3.124 (*a*) 74.0 MPa; 9.56°. (*b*) 61.5 MPa; 6.95°.
3.127 5.07 MPa.
3.128 59.2 MPa.
3.129 0.944.
3.131 0.198.
3.132 0.883.
3.133 (*a*) 1.193 in. (*b*) 1.170 in. (*c*) 0.878 in.
3.135 (*a*) 157.0 kN·m. (*b*) 8.70°.
3.136 (*a*) 7.52 ksi. (*b*) 4.61°.
3.137 (*a*) 1007 N·m. (*b*) 9.27°.
3.138 (*a*) 4.55 ksi. (*b*) 2.98 ksi. (*c*) 2.56°.
3.139 (*a*) 5.82 ksi. (*b*) 2.91 ksi.
3.142 (*a*) 16.85 N·m.
3.143 8.45 N·m.
3.144 $\tau_a = 4.73$ MPa, $\tau_b = 9.46$ MPa.
3.146 0.894 in.
3.147 (*a*) 12.76 MPa. (*b*) 5.40 kN·m.
3.149 (*a*) $3c/t$. (*b*) $3c^2/t^2$.
3.150 (*b*) 0.25% , 1.000%, 4.00%.
3.151 637 kip·in.
3.153 12.22°.
3.155 1.285 in.
3.156 (*a*) 73.7 MPa. (*b*) 34.4 MPa. (*c*) 5.06°.
3.157 4.12 kN·m.
3.158 (*a*) 18.80 kW. (*b*) 24.3 MPa.
3.160 7.34 kip·ft.

3.162 (*a*) 0.347 in. (*b*) 37.2°.
3.C2 *Prob. 3.44:* 2.21°.
3.C5 (*a*) −3.282%. (*b*) −0.853%.
  (*c*) −0.138%. (*d*) −0.00554%.
3.C6 (*a*) −1.883%. (*b*) −0.484%.
  (*c*) −0.078%. (*d*) −0.00313%.

CHAPTER 4

4.1 (*a*) −61.6 MPa. (*b*) 91.7 MPa.
4.2 (*a*) −2.38 ksi. (*b*) −0.650 ksi.
4.3 80.2 kN·m.
4.4 24.8 kN·m.
4.5 (*a*) 1.405 kip·in. (*b*) 3.19 kip·in.
4.6 (*a*) −39.3 MPa. (*b*) 26.2 MPa.
4.9 −14.71 ksi; 8.82 ksi.
4.10 −10.38 ksi; 15.40 ksi.
4.11 −102.4 MPa; 73.2 MPa.
4.12 61.3 kN.
4.15 20.4 kip·in.
4.16 106.1 N·m.
4.18 4.63 kip·in.
4.19 3.79 kN·m.
4.21 (*a*) 96.5 MPa. (*b*) 20.5 N·m.
4.22 (*a*) 0.602 mm. (*b*) 0.203 N·m.
4.23 (*a*) 145.0 ksi. (*b*) 384 lb·in.
4.24 (*a*) $\sigma = 75.0$ MPa, $\rho = 26.7$ m.
  (*b*) $\sigma = 125.0$ MPa, $\rho = 9.60$ m.
4.25 (*a*) 9.17 kN·m (*b*) 10.24 kN·m.
4.26 (*a*) 45.1 kip·in. (*b*) 49.7 kip·in.
4.29 (*a*) $(8/9) h_0$. (*b*) 0.949.
4.30 (*a*) 1007 in. (*b*) 3470 in. (*c*) 0.01320°.
4.31 (*a*) 139.1 m. (*b*) 480m.
4.32 (*a*) $[(\sigma_x)_{max} /2 \rho c](y^2 − c^2)$. (*b*) $− (\sigma_x)_{max} c/2\rho$.
4.33 1.240 kN·m.
4.34 887 N·m.
4.37 689 kip·in.
4.38 335 kip·in.
4.39 (*a*) −56.0 MPa. (*b*) 66.4 MPa.
4.40 (*a*) −56.0 MPa. (*b*) 68.4 MPa.
4.41 (*a*) −1.979 ksi. (*b*) 16.48 ksi.
4.43 8.70 m.
4.44 8.59 m.
4.45 625 ft.
4.47 3.87 kip·ft.
4.48 2.88 kip·ft.
4.49 (*a*) 212 MPa. (*b*) −15.59 MPa.
4.50 (*a*) 210 MPa. (*b*) −14.08 MPa.
4.54 (*a*) 1674 mm². (*b*) 90.8 kN·m.
4.55 (*a*) $\sigma_A = 6.86$ ksi; $\sigma_B = 6.17$ ksi; $\sigma_S = 4.11$ ksi. (*b*) 151.9 ft.
4.57 (*a*) −22.5 ksi. (*b*) 17.78 ksi.
4.59 (*a*) 6.15 MPa. (*b*) −8.69 MPa.
4.61 (*a*) 219 MPa. (*b*) 176.0 MPa.
4.63 (*a*) 6.79 kip·in. (*b*) 5.59 kip·in.
4.64 (*a*) 4.71 ksi. (*b*) 5.72 ksi.
4.65 (*a*) 147.0 MPa. (*b*) 119.0 MPa.
4.67 (*a*) 38.4 N·m. (*b*) 52.8 N·m.
4.68 (*a*) 57.6 N·m. (*b*) 83.2 N·m.
4.69 2460 lb·in.
4.71 (*a*) 5.87 mm. (*b*) 2.09 m.
4.72 (*a*) 21.9 mm. (*b*) 7.81 m.
4.75 (*a*) 1759 kip·in. (*b*) 2650 kip·in.

4.77 (*a*) 29.2 kN·m. (*b*) 1.500.

4.78 (*a*) 27.5 kN·m. (*b*) 1.443.

4.79 (*a*) 2840 kip·in. (*b*) 1.611.

4.80 (*a*) 4820 kip·in. (*b*) 1.443.

4.81 1.866 kN·m.

4.82 19.01 kN·m.

4.84 22.8 kip·in.

4.86 212 kip·in.

4.87 120 MPa.

4.88 106.4 MPa.

4.91 (*a*) 106.7 MPa. (*b*) $y_0 = -31.2$ mm, 0, 31.2 mm. (*c*) 24.1 m.

4.92 (*a*) 13.36 ksi. (*b*) $y_0 = -1.517$ in., 0, 1.517 in. (*c*) 168.8 ft.

4.94 (*a*) $0.707\rho_Y$. (*b*) $6.09\rho_Y$.

4.96 (*a*) 4.69 m. (*b*) 7.29 kN·m.

4.99 (*a*) -102.8 MPa. (*b*) 80.6 MPa.

4.100 (*a*) -212 psi. (*b*) -637 psi. (*c*) -1061 psi.

4.101 (*a*) $-2P/\pi r^2$. (*b*) $-5P/\pi r^2$.

4.103 (*a*) -37.8 MPa. (*b*) -38.6 MPa.

4.105 (*a*) 288 lb. (*b*) 209 lb.

4.106 1.994 kN.

4.107 14.40 kN.

4.108 16.04 mm.

4.109 43.0 kips.

4.110 0.500*d*.

4.113 7.86 kips ↓; 9.15 kips ↑.

4.114 5.32 kips ↓; 10.79 kips ↑.

4.115 (*a*) 47.6 MPa. (*b*) -49.4 MPa. (*c*) 9.80 mm below top of section.

4.116 (*a*) $-P/2at$. (*b*) $-2P/at$. (*c*) $-P/2at$.

4.117 (*a*) 1125 kN. (*b*) 817 kN.

4.121 (*a*) 30.0 mm. (*b*) 94.5 kN.

4.122 (*a*) 5.00 mm. (*b*) 243 kN.

4.124 $P = 44.2$ kips; $Q = 57.3$ kips.

4.125 (*a*) 152.3 kips. (*b*) $x = 0.59$ in. (*c*) 300 μ.

4.127 (*a*) -3.37 MPa. (*b*) -18.60 MPa. (*c*) 3.37 MPa.

4.128 (*a*) 9.86 ksi. (*b*) -2.64 ksi. (*c*) -9.86 ksi.

4.129 (*a*) -29.3 MPa. (*b*) -144.8 MPa. (*c*) -125.9 MPa.

4.130 (*a*) 0.321 ksi. (*b*) -0.107 ksi. (*c*) 0.427 ksi.

4.133 (*a*) 57.8 MPa. (*b*) -56.8 MPa. (*c*) 25.9 MPa.

4.134 (*a*) 57.4°. (*b*) 75.7 MPa.

4.135 (*a*) 18.29°. (*b*) 13.74 ksi.

4.137 (*a*) 10.03°. (*b*) 54.2 MPa.

4.138 (*a*) 27.5°. (*b*) 5.07 ksi.

4.139 (*a*) 32.9°. (*b*) 61.4 MPa.

4.141 113.0 MPa.

4.143 10.46 ksi.

4.144 (*a*) $\sigma_A = 31.5$ MPa; $\sigma_B = -10.39$ MPa. (*b*) 94.0 mm above point *A*.

4.145 (*a*) 17.11 mm.

4.146 0.1638 in.

4.147 53.9 kips.

4.150 29.1 kip·in.

4.151 29.1 kip·in.

4.152 733 N·m.

4.153 1.323 kN·m.

4.155 900 N·m.

4.161 (*a*) -77.3 MPa. (*b*) -55.7 MPa.

4.162 $\sigma_A = -65.1$ MPa; $\sigma_B = 39.7$ MPa.

4.163 (*a*) 12.19 ksi. (*b*) 11.15 ksi.

4.164 $\sigma_A = 10.77$ ksi; $\sigma_B = -3.22$ ksi.

4.167 655 lb.

4.169 73.2 mm.

4.170 (*a*) -82.4 MPa. (*b*) 36.6 MPa.

4.171 (*a*) 3.06 ksi. (*b*) -2.81 ksi. (*c*) 0.529 ksi.

4.173 13.80 kN·m.

4.174 8.49 kN·m.

4.175 (*a*) 16.05 ksi. (*b*) -9.84 ksi.

4.177 (*a*) 41.8 MPa. (*b*) -20.4 MPa.

4.178 27.2 mm.

4.179 107.8 N·m.

4.180 (*a*) -32.5 MPa. (*b*) 34.2 MPa.

4.181 (*a*) -3.65 ksi. (*b*) 3.72 ksi.

4.183 (*a*) -5.96 ksi. (*b*) 3.61 ksi.

4.184 (*a*) -6.71 ksi. (*b*) 3.24 ksi.

4.185 (*a*) 63.9 MPa. (*b*) -52.6 MPa.

4.191 -0.536 ksi.

4.192 67.8 MPa; -81.8 MPa

4.194 (*a*) $\sigma_{\max} = 6\,M/a^3$, $1/\rho = 12M/Ea^4$. (*b*) $\sigma_{\max} = 8.49\,M/a^3$, $1/\rho = 12M/Ea^4$.

4.195 48.6 kN·m.

4.196 (*a*) 46.9 MPa. (*b*) 18.94 MPa. (*c*) 55.4 m.

4.198 (*a*) -20.9 ksi. (*b*) -22.8 ksi.

4.199 60.9 mm.

4.201 (*a*) 56.7 kN·m. (*b*) 20.0 mm.

4.202 $P = 75.6$ kips ↓; $Q = 87.1$ kips ↓.

4.203 (*a*) $\sigma_A = -\frac{1}{2}\,\sigma$; $\sigma_B = \sigma_1$; $\sigma_C = -\sigma_1$; $\sigma_D = \frac{1}{2}\,\sigma_1$. (*b*) $4/3\,\rho_1$.

4.C1 $a = 4$ mm: $\sigma_a = 50.6$ MPa, $\sigma_s = 107.9$ MPa. $a = 14$ mm: $\sigma_a = 89.7$ MPa, $\sigma_s = 71.8$ MPa. (*a*) 1 16.6 MPa. (*b*) 6.61 mm.

4.C2 $y_Y = 65$ mm, $M = 52.6$ kN.m, $\rho = 43.3$; $y_Y = 45$ mm, $M = 55.6$ kN·m, $\rho = 30.0$ m.

4.C3 $\beta = 30°$: $\sigma_A = -7.83$ ksi, $\sigma_B = -5.27$ ksi, $\sigma_C = 7.19$ ksi, $\sigma_D = 5.91$ ksi; $\beta = 120°$: $\sigma_A = 1.557$ ksi, $\sigma_B = 6.01$ ksi, $\sigma_C = -2.67$ ksi, $\sigma_D = -4.89$ ksi.

4.C4 $r_1/h = 0.529$ for 50% increase in $\sigma_{\max}$.

4.C5 *Prob. 4.10:* -102.4 MPa; 73.2 MPa.

4.C6 $y_Y = 0.8$ *in.*: 76.9 kip·in., 552 in.; $y_Y = 0.2$ *in.*: 95.5 kip·in., 138.1 in.

4.C7 $a = 0.2$ *in.*: -7.27 ksi, $a = 0.8$ *in.*: -6.61 ksi. For $a = 0.625$ in., $\sigma = -6.51$ ksi.

CHAPTER 5

5.1 (*b*) $V = w(L/2 - x)$; $M = wx(L - x)/2$.

5.2 (*b*) *A* to *B*: $V = \dfrac{Pb}{L}$; $M = Pbx/L$. *B* to *C*: $V = Pa/L$; $M = Pa\,(L - x)/L$.

5.3 (*b*) $V = w_0 L/2 - w_0\,x^2/2L$; $M = -w_0 L^2/3 + w_0 Lx/2 - w_0 x^3/6L$

5.4 (*b*) $V = w(L - x)$; $M = w/2\,(L - x)^2$.

5.5 (*b*) $(0 < x < a)$: $V = -P$; $M = Px$. $(a < x < 2a)$: $V = 2P$; $M = -2Px + Pa$.

5.6 (*b*) *A* to *B*: $V = w(a - x)$; $M = w(ax - x^2/2)$. *B* to *C*: $V = 0$; $M = wa^2/2$. *C* to *D*: $V = w(L - x - a)$; $M = w[a(L - x) - (L - x)^2/2]$.

5.7 (*a*) 3.00 kN. (*b*) 0.800 kN·m.

5.8 (*a*) 150.0 lb. (*b*) 1500 lb·in.

5.9 (*a*) 62.5 kN. (*b*) 47.6 kN·m.

5.11 (*a*) 3.45 kN. (*b*) 1125 N·m.

5.12 (*a*) 2000 lb. (*b*) 19200 lb·in.

5.13 (*a*) 900 N. (*b*) 112.5 N·m.

5.15 10.89 MPa.

5.16 950 psi.

5.18 139.2 MPa.

5.20 9.90 ksi.

5.21 14.17 ksi.

5.23 $|V|_{max} = 342$ N; $|M|_{max} = 51.6$ N·m; $\sigma = 17.19$ MPa.

5.25 10.34 ksi.

5.26 $|V|_{max} = 6.00$ kN; $|M|_{max} = 4.00$ kN·m; $\sigma_{max} = 14.29$ MPa.

5.27 (a) 10.67 kN. (b) 9.52 MPa.

5.29 (a) 866 mm. (b) 99.2 MPa.

5.30 (a) 819 mm. (b) 89.5 MPa.

5.31 (a) 3.09 ft. (b) 12.95 ksi.

5.32 1.021 in.

5.33 (a) 33.3 mm. (b) 6.66 mm.

5.34 See 5.1.

5.35 See 5.2.

5.36 See 5.3.

5.37 See 5.4.

5.38 See 5.5.

5.39 See 5.6.

5.40 See 5.7.

5.41 See 5.8.

5.42 See 5.9.

5.43 See 5.10.

5.46 See 5.15.

5.47 See 5.16.

5.48 See 5.18.

5.49 See 5.20.

5.52 (a) $V = (w_0 L/\pi) \cos(\pi x/L)$; $M = (w_0 L^2/\pi^2) \sin(\pi x/L)$. (b) $w_0 L^2/\pi^2$.

5.53 (a) $V = w_0(L^2 - 3x^2)/6L$; $M = w_0(Lx - x^3/L)/6$. (b) $0.0642\, w_0 L^2$.

5.54 $|V|_{max} = 15.75$ kips; $|M|_{max} = 27.8$ kip·ft; $\sigma = 13.58$ ksi.

5.55 $|V|_{max} = 16.80$ kN; $|M|_{max} = 8.82$ kN·m; $\sigma = 73.5$ MPa.

5.56 $|V|_{max} = 20.7$ kN; $|M|_{max} = 9.75$ kN·m; $\sigma = 60.2$ MPa.

5.58 $|V|_{max} = 1400$ lb; $|M|_{max} = 19.20$ kip·in; $\sigma = 6.34$ ksi.

5.59 $|V|_{max} = 76.0$ kN; $|M|_{max} = 67.3$ kN·m; $\sigma = 68.5$ MPa.

5.60 $|V|_{max} = 48.0$ kN; $|M|_{max} = 12.00$ kN·m; $\sigma = 62.2$ MPa.

5.61 $|V|_{max} = 30.0$ lb; $|M|_{max} = 24.0$ lb·ft; $\sigma = 6.95$ ksi.

5.63 (a) $|V|_{max} = 24.5$ kips; $|M|_{max} = 36.3$ kip·ft; (b) 15.82 ksi.

5.64 $|V|_{max} = 1150$ N; $|M|_{max} = 221$ N·m; $P = 500$ N; $Q = 250$ N.

5.65 $h > 173.2$ mm.

5.68 $b > 6.20$ in.

5.69 $h > 203$ mm.

5.70 $b > 48.0$ mm.

5.71 W21 × 62.

5.72 W27 × 84.

5.73 W530 × 92.

5.74 W250 × 28.4.

5.76 S15 × 42.9.

5.77 S510 × 98.2.

5.79 9 mm.

5.80 C180 × 14.6.

5.81 C9 × 15.

5.82 3/8 in.

5.83 W610 × 101.

5.84 W24 × 68.

5.85 176.8 kN/m.

5.86 108.8 kN/m.

5.89 (a) 1.485 kN/m. (b) 1.935 m.

5.91 (a) S15 × 42.9. (b) W27 × 84.

5.92 (a) 6.49 ft. (b) W16 × 31.

5.94 383 mm.

5.95 336 mm.

5.96 W27 × 84.

5.97 +23.2%.

5.98 (a) $V = -w_0 x + w_0 \langle x - a \rangle^1$; $M = w_0 x^2/2 + (w_0/2) \langle x - a \rangle^2$. (b) $-3\, w_0 a^2/2$.

5.100 (a) $V = -w_0 x + w_0 x^2/2a - (w_0/2a)\langle x - a \rangle^2$; $M = -w_0 x^2/2 + w_0 x^3/6a - (w_0/6a)\langle x - a \rangle^3$. (b) $-5\, w_0 a^2/6$.

5.102 (a) $V = -w_0 \langle x - a \rangle^1 - 3w_0 a/4 + (15\, w_0 a/4)\langle x - 2a \rangle^0$; $M = -(w_0/2)\langle x - a \rangle^2 - 3w_0 ax/4 + (15\, w_0 a/4)\langle x - 2a \rangle^1$. (b) $-w_0 a^2/2$.

5.103 (a) $V = 1.25\, P - P\langle x - a \rangle^0 - P\langle x - 2a \rangle^0$; $M = 1.25\, Px - P\langle x - a \rangle^1 - P\langle x - 2a \rangle^1$. (b) $0.750\, Pa$.

5.104 (a) $V = -P/2 - P\langle x - a \rangle^0$; $M = Px/2 - P\langle x - a \rangle^1 + Pa + Pa\langle x - a \rangle^0$. (b) $3\, Pa/2$.

5.105 (a) $V = -P\langle x - a \rangle^0$; $M = -P\langle x - a \rangle^1 - Pa\langle x - a \rangle^0$. (b) $-Pa$:

5.106 (a) $V = 40 - 48\langle x - 1.5 \rangle^0 - 60\langle x - 3.0 \rangle^0 + 60\langle x - 3.6 \rangle^0$ kN; $M = 40x - 48\langle x - 1.5 \rangle^1 - 60\langle x - 3.0 \rangle^1 + 60\langle x - 3.6 \rangle^1$ kN·m. (b) 60.0 kN·m.

5.107 (a) $V = -3 + 9.75\langle x - 3 \rangle^0 - 6\langle x - 7 \rangle^0 - 6\langle x - 11 \rangle^0$ kips; $M = -3x + 9.75\langle x - 3 \rangle^1 - 6\langle x - 7 \rangle^1 - 6\langle x - 11 \rangle^1$ kip·ft. (b) 21.0 kip·ft.

5.108 (a) $V = 62.5 - 25\langle x - 0.6 \rangle^1 + 25\langle x - 2.4 \rangle^1 - 40\langle x - 0.6 \rangle^0 - 40\langle x - 2.4 \rangle^0$ kN; $M = 62.5x - 12.5\langle x - 0.6 \rangle^2 + 12.5\langle x - 2.4 \rangle^2 - 40\langle x - 0.6 \rangle^1 - 40\langle x - 2.4 \rangle^1$ kN·m (b) 47.6 kN·m.

5.109 (a) $V = 13 - 3x + 3\langle x - 3 \rangle^1 - 8\langle x - 7 \rangle^0 - 3\langle x - 11 \rangle^1$ kips; $M = 13x - 1.5x^2 + 1.5\langle x - 3 \rangle^2 - 8\langle x - 7 \rangle^1 - 1.5\langle x - 11 \rangle^2$ kip·ft. (b) 41.5 kip·ft.

5.110 (a) $V = 30 - 24\langle x - 0.75 \rangle^0 - 24\langle x - 1.5 \rangle^0 - 24\langle x - 2.25 \rangle^0 + 66\langle x - 3 \rangle^0$ kN; $M = 30x - 24\langle x - 0.75 \rangle^1 - 24\langle x - 1.5 \rangle^1 - 24\langle x - 2.25 \rangle^1 + 66\langle x - 3 \rangle^1$ kN·m. (b) 87.7 MPa.

5.114 (a) 122.7 kip·ft at $x = 6.50$ ft. (b) W16 × 40.

5.115 (a) 121.5 kip·ft at $x = 6.00$ ft. (b) W16 × 40.

5.118 $|V|_{max} = 35.6$ kN; $|M|_{max} = 25.0$ kN·m.

5.119 $|V|_{max} = 89.0$ kN; $|M|_{max} = 178.0$ kN·m.

5.120 $|V|_{max} = 15.30$ kips; $|M|_{max} = 38.0$ kip·ft.

5.122 (a) $|V|_{max} = 13.80$ kN; $|M|_{max} = 16.16$ kN·m. (b) 83.8 MPa.

5.123 (a) $|V|_{max} = 40.0$ kN; $|M|_{max} = 30.0$ kN·m. (b) 40.0 MPa.

5.124 (a) $|V|_{max} = 3.84$ kips; $|M|_{max} = 3.80$ kip·ft. (b) 0.951 ksi.

5.126 (a) $h = h_0 [(x/L)(1 - x/L)]^{1/2}$. (b) 4.44 kip·in.

5.127 (a) $h = h_0 (x/L)^{1/2}$. (b) 20.0 kips.

5.128 (a) $h = h_0 (x/L)^{3/2}$. (b) 167.7 mm.

5.130 (a) $h = h_0 \sqrt{2x/L}$. (b) 60.0 kN.

5.132 $l_2 = 6.00$ ft; $l_2 = 4.00$ ft.

5.134 1.800 m.

5.135 1.900 m.

5.136 $d = d_0 (2x/L)^{1/3}$ for $0 \le x \le L/2$; $d = d_0 [2(L - x)/L]^{1/3}$ for $L/2 \le x \le L$.

5.139 (a) $b = b_0 (1 - x/L)^2$. (b) 160.0 lb/in.

5.140 (a) 155.2 MPa. (b) 143.3 MPa.

5.141 (a) 25.0 ksi. (b) 18.03 ksi.

5.143 193.8 kN.

5.144 (a) 152.6 MPa. (b) 133.6 MPa.

5.145 (a) 4.49 m. (b) 211. mm.

5.147 (a) 11.16 ft. (b) 14.31 in.

5.149 (a) 240 mm. (b) 150.0 MPa.

5.150 (a) 15.00 in. (b) 320 lb/in.

5.151 (a) 30.0 in. (b) 12.80 kips.

5.152 (a) 85.0 N. (b) 21.3 N·m.

5.154 (a) 1.260 ft. (b) 7.24 ksi.

5.156 $|V|_{max} = 200$ kN; $|M|_{max} = 300$ kN·m; 136.4 MPa.

5.158 $h > 14.27$ in.

5.159 W27 × 84.

5.161 (a) 225.6 kN·m at $x = 3.63$ m.
(b) 60.6 MPa.

5.163 (a) $b_0 (1 - x/L)$. (b) 20.8 mm.

5.C4 *For x = 13.5 ft:* $M_1 = 131.25$ kip·ft;
$M_2 = 156.25$ kip·ft; $M_C = 150.0$ kip·ft.

5.C6 *Prob. 5.112:* $V_A = 29.5$ kN, $M_{max} = 28.3$ kN·m,
at 1.938 m from A.

CHAPTER 6

6.1 60.0 mm.

6.2 2.00 kN.

6.3 326 lb.

6.4 (a) 155.8 N. (b) 329 kPa.

6.5 193.5 kN.

6.7 10.56 ksi.

6.9 (a) 8.97 MPa. (b) 8.15 MPa.

6.11 (a) 13.15 ksi. (b) 11.16 ksi.

6.12 (a) 3.17 ksi. (b) 2.40 ksi.

6.13 114.0 kN.

6.15 1733 lb.

6.17 (a) 84.2 kips. (b) 60.2 kips.

6.18 87.3 mm.

6.19 (b) $h = 320$ mm; $b = 97.7$ mm.

6.21 (a) 1.745 ksi. (b) 2.82 ksi.

6.22 (a) 31.0 MPa. (b) 23.2 MPa.

6.23 3.21 ksi.

6.24 32.7 MPa.

6.26 (a) Line at mid-height. (b) 2.00.

6.28 (a) Line at mid-height. (b) 1.500.

6.29 1.672 in.

6.30 10.79 kN.

6.31 (a) 59.9 psi. (b) 79.8 psi.

6.32 (a) 379 kPa. (b) 0.

6.35 (a) 95.2 MPa. (b) 112.8 MPa.

6.36 (a) 101.6 MPa. (b) 79.9 MPa.

6.37 $\tau_a = 33.7$ MPa; $\tau_b = 75.0$ MPa; $\tau_c = 43.5$ MPa.

6.38 (a) 40.5 psi. (b) 55.2 psi.

6.40 $\tau_a = 0$; $\tau_b = 1.262$ ksi; $\tau_c = 3.30$ ksi;
$\tau_d = 6.84$ ksi; $\tau_e = 7.86$ ksi.

6.42 (a) 18.23 MPa. (b) 14.59 MPa. (c) 46.2 MPa.

6.43 7.19 ksi.

6.44 9.05 mm.

6.45 0.371 in.

6.46 83.3 MPa.

6.48 $\tau_a = 10.76$ MPa; $\tau_b = 0$; $\tau_c = 11.21$ MPa;
$\tau_d = 22.0$ MPa; $\tau_e = 9.35$ MPa.

6.49 (a) 50.9 MPa. (b) 62.4 MPa.

6.51 1.4222 in.

6.52 10.53 ksi.

6.56 (a) 6.73 MPa. (b) 1.515 MPa.

6.57 (a) 23.3 MPa. (b) 109.7 MPa.

6.58 (a) 1.323 ksi. (b) 1.329 ksi.

6.59 (a) 0.888 ksi. (b) 1.453 ksi.

6.61 0.345a.

6.62 0.714a.

6.63 1.250a.

6.64 $3(b^2 - a^2)/[6(a + b) + h]$.

6.65 (a) 19.06 mm. (b) $\tau_A = 0$; $(\tau_B)_{AB} = 50.5$ MPa;
$(\tau_B)_{BD} = 25.3$ MPa; $\tau_c = 59.0$ MPa.

6.68 (a) 10.22 mm. (b) 0 at B and E; 41.1 MPa at A;
68.5 MPa just above D; 13.71 MPa just to the right of D;
77.7 MPa just below D; 81.8 MPa at center of DF.

6.69 0.727 in.

6.70 20.2 mm.

6.71 1.265 in.

6.72 6.14 mm.

6.75 2.37 in.

6.76 21.7 mm.

6.77 40.0 mm.

6.78 3.75 in.

6.81 (maximum) P/at.

6.82 (maximum) 1.333 P/at.

6.83 (a) 144.6 N·m. (b) 65.9 MPa.

6.84 (a) 144.6 N·m. (b) 106.6 MPa.

6.87 (maximum) 0.428 ksi at B'.

6.88 (maximum) 1.287 ksi at C'.

6.89 738 N.

6.90 (a) 17.63 MPa. (b) 13.01 MPa.

6.91 143.3 kips.

6.93 189.6 lb.

6.94 (a) 41.4 MPa. (b) 41.4 MPa.

6.96 53.9 kips.

6.97 (a) 146.1 kN/m. (b) 19.99 MPa.

6.99 40.0 mm.

6.100 0.433 in.

6.C1 (a) $h = 173.2$ mm. (b) $h = 379$ mm.

6.C2 (a) $L = 37.5$ in.; $b = 1.250$ in.
(b) $L = 70.3$ in.; $b = 1.172$ in.
(c) $L = 59.8$ in.; $b = 1.396$ in.

6.C4 (a) $\tau_{max} = 2.03$ ksi; $\tau_B = 1.800$ ksi. (b) 194 psi.

6.C5 *Prob. 6.66:* (a) 2.67 in. (b) $\tau_B = 0.917$ ksi;
$\tau_D = 3.36$ ksi; $\tau_{max} = 4.28$ ksi.

CHAPTER 7

7.1 $\sigma = 9.46$ ksi; $\tau = 1.013$ ksi.

7.2 $\sigma = 32.9$ MPa; $\tau = 71.0$ MPa.

7.3 $\sigma = 10.93$ ksi; $\tau = 0.536$ ksi.

7.4 $\sigma = -0.521$ MPa; $\tau = 56.4$ MPa.

7.5 (a) −37.0°, 53.0°. (b) −13.60 MPa. (c) −86.4 MPa.

7.7 (a) − 26.6°; 63.4°. (b) 190.0 MPa, −10.00 MPa.

7.9 (a) 8.0°, 98.0°. (b) 36.4 MPa. (c) −50.0 MPa.

7.10 (a) −26.6°, 63.4°. (b) 5.00 ksi. (c) 6.00 ksi.

7.11 (a) 18.4°, 108.4°. (b) 100.0 MPa. (c) 90.0 MPa.

7.12 (a) −31.0°, 59.0° (b) 17.00 ksi. (c) 3.00 ksi.

7.13 (a) $\sigma_{x'} = -2.40$ ksi; $\tau_{x'y'} = 0.1498$ ksi; $\sigma_{y'} = 10.40$ ksi.
(b) $\sigma_{x'} = 1.951$ ksi; $\tau_{x'y'} = 6.07$ ksi; $\sigma_{y'} = 6.05$ ksi.

7.15 (a) $\sigma_{x'} = 9.02$ ksi; $\tau_{x'y'} = 3.80$ ksi; $\sigma_{y'} = -13.02$ ksi.
(b) $\sigma_{x'} = 5.34$ ksi; $\tau_{x'y'} = -9.06$ ksi; $\sigma_{y'} = -9.34$ ksi.

7.17 (a) 217 psi. (b) −125.0 psi.

7.18 (a) −0.300 MPa. (b) −2.92 MPa.

7.19 16.58 kN.

7.21 (a) 18.4°. (b) 16.67 ksi.

7.23 (a) 18.9°, 108.9°, 18.67 MPa, −158.5 MPa. (b) 88.6 MPa.

7.24 (a) 25.1 ksi, −0.661 ksi, 12.88 ksi.

7.25 5.12 ksi, −1.640 ksi, 3.38 ksi.

7.26 12.18 MPa, −48.7 MPa; 30.5 MPa.

7.27 205 MPa.

7.29 (a) −2.89 MPa. (b) 12.77 MPa, 1.226 MPa.

7.53 (a) −8.66 MPa. (b) 17.00 MPa, −3.00 MPa.

7.55 33.8°, 123.8°; 168.6 MPa, 6.42 MPa.

7.56 0°, 90°; σ_0, $-\sigma_0$.

7.57 −30°, 60°; $-\sqrt{3}\tau_0$, $\sqrt{3}\tau_0$.

7.58 −120.0 MPa ≤ τ_{xy} ≤ 120.0 MPa.

7.59 −141.4 MPa ≤ τ_{xy} ≤ 141.1 MPa.

7.61 16.5° ≤ θ ≤ 110.1°.

7.62 −5.1° ≤ θ ≤ 132.0°.

7.63 (a) 33.7°, 123.7°. (b) 18.00 ksi. (c) 6.50 ksi.

7.65 (b) $|\tau_{xy}| = \sqrt{\sigma_x\sigma_y - \sigma_{max}\sigma_{min}}$.

7.66 (a) 13.00 ksi. (b) 15.00 ksi.

7.68 (a) 94.3 MPa. (b) 105.3 MPa.

7.69 (a) 100.0 MPa. (b) 110.0 MPa.

7.70 (a) 91.0 MPa. (b) 91.0 MPa. (c) 108.0 MPa.

7.71 (a) 113.0 MPa. (b) 91.0 MPa. (c) 143.0 MPa.

7.73 (a) 18.5 ksi. (b) 13.00 ksi. (c) 11.00 ksi.

7.74 (a) ±6.00 ksi. (b) ±11.24 ksi.

7.75 ±60.0 MPa.

7.77 2.00 ksi; 9.33 ksi.

7.79 −40.0 MPa; 130.0 MPa.

7.80 (a) 45.7 MPa. (b) 92.9 MPa.

7.81 (a) 1.228. (b) 1.098. (c) Yielding occurs.

7.82 (a) 1.083. (b) Yielding occurs. (c) Yielding occurs.

7.83 (a) 1.287. (b) 1.018. (c) Yielding occurs.

7.84 (a) 1.119. (b) Yielding occurs. (c) Yielding occurs.

7.87 8.19 kip·in.

7.88 9.46 kip·in.

7.89 Rupture will occur.

7.90 Rupture will occur.

7.91 No Rupture.

7.92 Rupture will occur.

7.94 ±8.49 MPa.

7.95 50.0 MPa.

7.96 196.9 N·m.

7.98 (a) 1.290 MPa. (b) 0.852 mm.

7.100 5.49.

7.102 10.25 ksi; 5.12 ksi.

7.103 2.94 MPa.

7.104 12.76 m.

7.105 σ_{max} = 113.7 MPa; τ_{max} = 56.8 MPa.

7.106 σ_{max} = 136.0 MPa; τ_{max} = 68.0 MPa.

7.108 σ_{max} = 78.5 MPa; τ_{max} = 39.3 MPa.

7.109 251 psi.

7.111 0.307 in.

7.112 3.29 MPa.

7.113 3.80 MPa.

7.114 (a) 44.2 MPa. (b) 15.39 MPa.

7.115 56.8°.

7.117 (a) 3750 psi. (b) 1079 psi.

7.118 387 psi.

7.120 (a) 3.15 ksi. (b) 1.9993 ksi.

7.121 (a) 1.486 ksi. (b) 3.16 ksi.

7.122 σ_{max} = 68.6 MPa; τ_{max} = 34.3 MPa.

7.124 σ_{max} = 77.4 MPa; τ_{max} = 38.7 MPa.

7.126 (a) 5.64 ksi. (b) 282 psi.

7.127 (a) 2.28 ksi. (b) 228 psi.

7.128 $\epsilon_{x'}$ = −653 μ; $\epsilon_{y'}$ = 303 μ; $\gamma_{x'y'}$ = −829 μ.

7.129 $\epsilon_{x'}$ = 115.0 μ; $\epsilon_{y'}$ = 285 μ; $\gamma_{x'y'}$ = −5.72 μ.

7.131 $\epsilon_{x'}$ = 36.7 μ; $\epsilon_{y'}$ = 283 μ; $\gamma_{x'y'}$ = 227 μ.

7.132 $\epsilon_{x'}$ = −653 μ; $\epsilon_{y'}$ = 303 μ; $\gamma_{x'y'}$ = −829 μ.

7.133 $\epsilon_{x'}$ = 115.0 μ; $\epsilon_{y'}$ = 285 μ; $\gamma_{x'y'}$ = −5.72 μ.

7.135 $\epsilon_{x'}$ = 36.7 μ; $\epsilon_{y'}$ = 283 μ; $\gamma_{x'y'}$ = 227 μ.

7.136 (a) −33.7°, 56.3°; −420 μ, 100 μ, 160 μ. (b) 520 μ. (c) 580 μ.

7.137 (a) −30.1°, 59.9°; −702 μ, −298 μ, 500 μ. (b) 403 μ. (c) 1202 μ.

7.139 (a) −26.6°, 64.4°; −150.0 μ, 750 μ, −300 μ. (b) 900 μ. (c) 1050 μ.

7.140 (a) 7.8°, 97.8°; 56.6 μ, 243 μ, 0. (b) 186.8 μ.(c) 243 μ.

7.141 (a) 121.0°, 31.0°; 513 μ, 87.5 μ, 0. (b) 425 μ.(c) 513 μ.

7.143 (a) 127.9°, 37.9°; −383 μ, −57.5 μ, 0. (b) 325 μ.(c) 383 μ.

7.146 (a) −300 × 10^{-6} in./in. (b) 435 × 10^{-6} in./in., −315 × 10^{-6} in./in.; 750 × 10^{-6} in./in.

7.149 (a) 30.0°, 120.0°; 560 × 10^{-6} in./in, −140.0 × 10^{-6} in./in. (b) 700 × 10^{-6} in./in.

7.150 P = 69.6 kips; Q = 30.3 kips.

7.151 P = 34.8 kips; Q = 38.4 kips.

7.154 1.421 MPa.

7.155 1.761 MPa.

7.156 −22.5°, 67.5°; 426 μ, −952 μ, −224 μ.

7.157 −32.1°, 57.9°; −70.9 MPa, −29.8 MPa.

7.158 −4.76 ksi; −0.467 ksi.

7.159 (a) 47.9 MPa. (b) 102.7 MPa.

7.161 θ/2, (θ + π)/2; $\sigma_0 + \sigma_0 \cos θ$, $\sigma_0 -\sigma_0 \cos θ$.

7.163 (a) 40.0 MPa. (b) 72.0 MPa.

7.164 (a) 1.286. (b) 1.018. (c) Yielding occurs.

7.165 σ_{max} = 45.1 MPa; τ_{max} (in-plane) = 9.40 MPa.

7.167 3.43 ksi (compression).

7.169 415 × 10^{-6} in./in.

7.C1 *Prob. 7.14:* (a) −56.2 MPa, 86.2 MPa, −38.2 MPa. (b) −45.2 MPa, 75.2 MPa, 53.8 MPa.
Prob. 7.16: (a) 24.0 MPa, −104.0 MPa, −1.50 MPa. (b) −19.51 MPa, −60.5 MPa, −60.7 MPa.

7.C4 *Prob. 7.93:* Rupture occurs at τ_0 = 3.67 ksi.

7.C6 *Prob. 7.138:* (a) −21.6°, 68.4°; 279μ, −599μ, 160.0μ. (b) 877μ. (c) 877μ.

7.C7 *Prob. 7.142:* (a) 11.3°, 101.3°; 310μ, 50.0μ, 0. (b) 260μ. (b) 310μ.

7.C8 *Prob. 7.144:* ϵ_x = 253μ; ϵ_y = 307; γ_{xy} = −893. ϵ_a = 727μ; ϵ_b = −167.2; γ_{max} = −894.
Prob. 7.145: ϵ_x = 725μ; ϵ_y = −75.0; γ_{xy} = 173.2. ϵ_a = 734μ; ϵ_b = −84.3; γ_{max} = 819.

CHAPTER 8

8.1 (a) 10.69 ksi. (b) 19.18 ksi. (c) Not acceptable.

8.2 (a) 10.69 ksi. (b) 13.08 ksi. (c) Acceptable.

8.3 (a) 94.6 MPa. (b) 93.9 MPa. (c) Acceptable.

8.4 (a) 91.9 MPa. (b) 95.1 MPa. (c) Acceptable.

8.5 (a) W 310 × 38.7. (b) 147.8 MPa; 18.18 MPa; 140.2 MPa.

8.6 (a) W 690 × 125. (b) 118.2 MPa; 34.7 MPa; 122.3 MPa.

8.9 (a) 137.5 MPa. (b) 129.5 MPa.

8.11 (a) 17.90 ksi. (b) 17.08 ksi.

8.12 (a) 19.39 ksi. (b) 20.7 ksi.

8.13 (a) 131.3 MPa. (b) 135.5 MPa.

8.15 41.2 mm.

8.19 873 lb.

8.20 1.578 in.

8.22 (a) H : 6880 psi; K : 6760 psi. (b) H : 7420 psi; K : 7010 psi.

8.23 57.7 mm.

8.24 54.3 mm.

8.27 37.0 mm.

8.28 43.9 mm.

8.29 1.822 in.

8.30 1.792 in.

8.31 (a) 11.06 ksi; 0. (b) −0.537 ksi; 1.610 ksi. (c) −12.13 ksi; 0.

8.32 (a) −12.34 ksi; 0. (b) −1.073 ksi; 0.805 ksi. (c) 10.20 ksi; 0.

8.35 (a) −37.9 MPa; 14.06 MPa. (b) −131.6 MPa; 0.

8.36 (a) −32.5 MPa; 14.06 MPa. (b) −126.2 MPa; 0.

8.37 (a) 0; 3.34 ksi. (b) −8.80 ksi; 2.93 ksi.

8.38 (a) 20.4 MPa; 14.34 MPa. (b) −21.5 MPa; 19.98 MPa.

8.39 (a) 4.79 ksi; 3.07 ksi. (b) −2.57 ksi; 3.07 ksi.

8.40 −14.98 MPa; 17.29 MPa.

8.42 55.0 MPa, −55.0 MPa; −45.0°, 45.0°; 55.0 MPa.

8.43 (a) 4.30 MPa, −93.4 MPa; 12.1°, 102.1°. (b) 48.9 MPa.

8.46 (a) 3.47 ksi; 1.042 ksi. (b) 7.81 ksi; 0.781 ksi. (c) 12.15 ksi; 0.

8.47 (a) 18.39 MPa; 0.391 MPa. (b) 21.3 MPa; 0.293 MPa.
(c) 24.1 MPa; 0.

8.48 (a) −7.98 MPa; 0.391 MPa. (b) −5.11 MPa; 0.293 MPa.
(c) −2.25 MPa; 0.

8.49 1506 psi, −4150 psi; 31.1°, 121.1°; 2830 psi.

8.51 25.2 MPa, −0.870 MPa; 13.06 MPa.

8.52 34.6 MPa, −10.18 MPa; 22.4 MPa.

8.53 (a) 86.5 MPa; 0. (b) 57.0 MPa; 9.47 MPa.

8.55 12.94 MPa, −1.328 MPa; 7.13 MPa.

8.57 4.05 ksi, −0.010 ksi; 2.03 ksi.

8.58 1.468 ksi, −3.90 ksi; 2.68 ksi.

8.60 (a) 51.0 kN. (b) 39.4 kN.

8.61 12.2 MPa, −12.2 MPa; 12.2 MPa.

8.62 (a) 12.90 ksi, −0.32 ksi; −8.9°, 81.1°.; 6.61 ksi.
(b) 6.43 ksi, −6.43 ksi; ± 45.0°; 6.43 ksi.

8.64 0.48 ksi, −44.7 ksi; 22.6 ksi.

8.65 (a) W14 × 22. (b) 23.6 ksi, 4.89 ksi; 22.4 ksi.

8.66 BC: 21.7 mm; CD: 33.4 mm.

8.68 46.5 mm.

8.69 (a) −11.07 ksi; 0. (b) 2.05 ksi; 2.15 ksi. (c) 15.17 ksi; 0.

8.71 $P(2R + 4r/3)/\pi r^3$.

8.74 30.1 MPa, −0.62 MPa; −8.2°, 81.8°; 15.37 MPa.

8.75 (a) −16.41 ksi; 0. (b) −15.63 ksi; 0.0469 ksi.
(c) −7.10 ksi; 1.256 ksi.

8.76 (a) 7.50 MPa. (b) 11.25 MPa. (c) 56.3°; 13.52 MPa.

8.C3 *Prob. 8.18*: 37.3 mm.

8.C5 *Prob. 8.45*: σ = 6.00 ksi; τ = 0.781 ksi.

CHAPTER 9

9.1 (a) $y = -(Px^2/6EI)(3L - x)$. (b) $PL^3/3EI \downarrow$.
(c) $PL^2/2EI \searrow$.

9.2 (a) $y = (M_0/2EI)(L - x)^2$. (b) $M_0L^2/2EI \uparrow$.
(c) $M_0 L/EI \searrow$.

9.3 (a) $y = -(w/24EI)(x^4 - 4L^3x + 3L^4)$. (b) $wL^4/8EI \downarrow$.
(c) $wL^3/6EI \searrow$.

9.4 (a) $y = -(w_0/120EIL)(x^5 - 5L^4x)$. (b) $w_0L^4/30 EI \downarrow$.
(c) $w_0L^3/24 EI \searrow$.

9.5 (a) $y = (w/72 EI)(3x^4 - 16ax^3)$. (b) $10 wa^4/9 EI \downarrow$.
(c) $4 wa^3/3EI \searrow$.

9.7 (a) $y = (w_0/EIL)(L^2x^3/48 - x^5/120 - L^4x/80)$.
(b) $w_0 L^4/256 EI \downarrow$. (c) $w_0 L^3/120 EI \searrow$.

9.9 (a) 3.92×10^{-3} rad $\searrow$. (b) 0.1806 in. $\downarrow$.

9.10 (a) 2.79×10^{-3} rad $\searrow$. (b) 1.859 mm $\downarrow$.

9.11 (a) $0.00652w_0L^4/EI \downarrow$; $0.481L$. (b) 0.229 in. $\downarrow$.

9.12 (a) $0.211L$; $0.01604M_0L^2/EI$. (b) 6.08 m.

9.13 0.398 in. $\downarrow$.

9.16 (a) $(P/EI)(ax^2/2 - aLx/2 + a^3/6)$. (b) 1.976 mm $\downarrow$.

9.17 (a) $y = -(w_0/EIL^2)(L^2x^4/24 - LX^5/30 + X^6/120 - L^4x^2/24)$.
(b) $w_0 L^4/40 EI \downarrow$.

9.18 (a) $y = w_0(x^6 - 15 L^2x^4 + 25L^3x^3 - 11 L^5x)/360 EIL^2$.
(b) $11w_0L^3/360 EI \searrow$. (c) $0.00916 w_0L^4/EI \downarrow$.

9.19 $3wL/8 \uparrow$.

9.20 $3M_0/2L \uparrow$.

9.23 4.00 kips $\uparrow$.

9.24 9.75 kN $\uparrow$.

9.25 $R_B = 5P/16 \uparrow$; $M_A = -3PL/16$, $M_C = 5PL/32$, $M_B = 0$.

9.26 $R_B = 9M_0/8L \uparrow$; $M_A = M_0/8$, $M_{C-} = -7M_0/16$,
$M_{C+} = 9M_0/16$.

9.27 $R_A = 9w_0L/640 \uparrow$; $M_M = 0.00814 w_0L^2$, $M_B = -0.0276 w_0L^2$.

9.28 $R_A = 7wL/128 \uparrow$; $M_C = 0.0273 wL^2$, $M_B = -0.0703 wL^2$,
$M = 0.0288 wL^2$ at $x = 0.555L$.

9.30 $R_B = 17wL/64 \uparrow$; $y_C = wL^4/1024 EI \downarrow$.

9.32 $R_B = 5M_0/6L \downarrow$; $y_D = 7M_0L^2/486 EI \uparrow$.

9.33 $wL/2\uparrow$, $wL^2/12 \searrow$; $M = w[6x(L - x) - L^2]/12$.

9.34 $R_A = w_0L/4 \uparrow$, $M_A = 0.0521 w_0L^2\searrow$; $M_C = 0.0313 w_0L_2$.

9.35 (a) $y = w\{Lx^3/48 - \langle x - L/2\rangle^4/24 - 7L^3x/384\}/EI$.
(b) $7wL^3/384 EI \searrow$. (c) $5wL^4/768 EI \downarrow$.

9.36 (a) $y = (M_0/6EIL)\{x^3 - 3L\langle x - a\rangle^2 + (3b^2 - L^2) x\}$
(b) $M_0 (3b^2 - L^2)/6EIL \searrow$. (c) $M_0 ab (b - a)/3 EIL \uparrow$.

9.37 (a) $9Pa^3/4EI \downarrow$. (b) $19Pa^3/6EI \downarrow$. (c) $9Pa^3/4EI\downarrow$.

9.38 (a) $5Pa^3/2EI \downarrow$. (b) $49Pa^3/6EI \downarrow$. (c) $15Pa^3/EI$.

9.41 (a) $y = w\{ax^3/6 - \langle x - a\rangle^4/24 + \langle x - 3a\rangle^4/24 - 11 a^3x/6\}/EI$.
(b) $19 wa^4/8EI \downarrow$.

9.43 (a) $y = w_0\{-5L^3x^2/48 + L^2 x^3/24 - \langle x - L/2\rangle^5/60\}/EIL$.
(b) $w_0L^4/48 EI \downarrow$. (c) $121 w_0L^4/1920 EI \downarrow$.

9.44 (a) $y = (w/24 EI)\{-x^4 + \langle x - L/2\rangle^4 - \langle x - L\rangle^4 + Lx^3 + 3L \langle x - L\rangle^3 - L^3x/16\}$.
(b) $wL^4/768 EI \uparrow$. (c) $5wL^4/256 EI$.

9.45 (a) 9.51×10^{-3} rad $\searrow$. (b) 5.80 mm $\downarrow$.

9.46 (a) 8.66×10^{-3} rad $\searrow$. (b) 0.1503 in. $\downarrow$.

9.48 (a) 5.40×10^{-3} rad $\searrow$. (b) 3. 06 mm $\downarrow$.

9.49 (a) $5P/16 \uparrow$. (b) $7 PL^3/168EI \downarrow$.

9.50 (a) $9M_0/8L \uparrow$. (b) $M_0L^2/128EI \downarrow$.

9.51 (a) $2P/3 \uparrow$. (b) $5PL^3/486EI \downarrow$.

9.53 (a) 11.54 kN $\uparrow$. (b) 4.18 mm $\downarrow$.

9.54 (a) 41.3 kN $\uparrow$. (b) 0.705 mm $\downarrow$.

9.56 (a) 7.38 kips $\uparrow$. (b) 0.0526 in. $\downarrow$.

9.57 (a) $1.280wa \uparrow$; $1.333wa^2\searrow$. (b) $0.907 wa^4/EI \downarrow$.

9.58 (a) $20P/27 \uparrow$; $4PL/27\searrow$. (b) $5PL^3/1296 EI \downarrow$.

9.59 5.80 mm $\downarrow$ at $x = 0.991$ m.

9.60 0.1520 in. $\downarrow$ at $x = 26.4$ in.

9.61 0.281 in. $\downarrow$ at $x = 8.40$ ft.

9.62 3.07 mm $\downarrow$ at $x = 0.942$ m.

9.65 $wL^3/48EI \measuredangle$; $wL^4/384EI \uparrow$.

9.66 $PL^2/24 EI \searrow$; $PL^3/48 EI$.

9.67 $3PL^2/4 EI \measuredangle$; $13 PL^3/24 EI \downarrow$.

9.68 $Pa (2L - a)/2 EI \searrow$; $Pa (3L^2 - 3aL + a^2)/6 EI \uparrow$.

9.71 (a) $wL^4/128 EI$. (b) $wL^3/72 EI$.

9.72 (a) $PL^3/486 EI$. (b) $PL^2/81 EI \searrow$.

9.73 6.32×10^{-3} rad $\searrow$; 5.55 mm $\downarrow$.

9.75 7.91×10^{-3} rad $\measuredangle$; 0.340 in. $\downarrow$.

9.76 6.98×10^{-3} rad $\measuredangle$; 0.1571 in. $\downarrow$.

9.77 (a) 0.601×10^{-3} rad $\searrow$. (b) 3.67 mm $\downarrow$.

9.79 (a) $4P/3 \uparrow$; $PL/3 \searrow$. (b) $2P/3 \uparrow$.

9.80 (a) $41 wL/128 \uparrow$. (b) $23 wL/128 \uparrow$; $7wL^2/128 \downarrow$.

9.82 $R_A = 2M_0/L \uparrow$; $R_B = 3M_0/L \downarrow$; $R_C = M_C/L \uparrow$.
9.84 $wL/2 \uparrow$, $wL^2/2 \downarrow$.
9.85 (a) 5.94 mm ↓. (b) 6.75 mm ↓.
9.86 $y_B = 0.210$ in. ↓; $y_c = 0.1709$ in. ↓.
9.87 (a) 5.06×10^{-3} rad ↰. (b) 0.0477 in. ↓.
9.88 121.5 N/m.
9.90 5.63 kN.
9.91 (a) 0.00937 mm ↓. (b) 229 N ↑.
9.93 0.278 in. ↓.
9.94 9.31 mm ↓.
9.95 (a) M_0L/EI ↰. (b) $M_0L^2/2EI$ ↑.
9.96 (a) $PL^2/2EI$ ↲. (b) $PL^3/3\,EI$ ↓.
9.97 (a) $wL^3/6EI$ ↲. (b) $wL^4/8EI$ ↓.
9.98 (a) $w_0L^3/24EI$ ↲. (b) $w_0L^4/30\,EI$ ↓.
9.101 (a) 4.24×10^{-3} rad ↰. (b) 0.0698 in. ↓.
9.102 (a) 5.20×10^{-3} rad ↲. (b) 10.85 mm ↓.
9.103 (a) 5.84×10^{-3} rad ↰. (b) 0.300 in. ↓.
9.104 (a) 7.15×10^{-3} rad ↲. (b) 17.67 mm ↓.
9.105 (a) $wL^3/16EI$ ↲. (b) $47wL^4/1152\,EI$ ↓.
9.107 (a) 3.43×10^{-3} rad ↲. (b) 6.66 mm ↓.
9.109 (a) $PL^2/16EI$ ↰. (b) $PL^3/48EI$ ↓.
9.110 (a) $Pa(L - a)/2EI$ ↰. (b) $Pa\,(3L^2 - 4a^2)/24EI$ ↓.
9.111 (a) $PL^2/32\,EI$ ↰. (b) $PL^3/128EI$ ↓.
9.112 (a) $wa^2\,(3L - 2a)/12EI$ ↰. (b) $wa^2\,(3L^2 - 2a^2/48EI)$ ↓.
9.113 (a) $M_0\,(L - 2a)/2EI$ ↰. (b) $M_0\,(L^2 - 4a^2)/8EI$ ↓.
9.115 (a) $5Pa^2/8EI$ ↰. (b) $3Pa^3/4EI$ ↓.
9.118 (a) 4.71×10^{-3} rad ↰. (b) 5.84 mm ↓.
9.119 (a) 4.50×10^{-3} rad ↰. (b) 8.26 mm ↓.
9.120 (a) 5.21×10^{-3} rad ↰. (b) 21.2 mm ↓.
9.122 3.84 kN/m.
9.123 $0.211L$.
9.124 $0.223L$.
9.125 (a) $4PL^3/243EI$ ↑. (b) $14\,PL^2/81\,EI$ ↲.
9.126 (a) $5PL^3/768EI$ ↓. (b) $3PL^2/128\,EI$ ↰.
9.128 (a) $5w_0L^4/768\,EI$ ↓. (b) $7w_0\,L^3/360\,EI$ ↰.
9.129 (a) 8.74×10^{-3} rad ↰. (b) 15.10 mm ↓.
9.130 (a) 7.48×10^{-3} rad ↰. (b) 5.35 mm ↓.
9.132 (a) 5.31×10^{-3} rad ↰. (b) 0.204 in. ↓.
9.133 (a) $M_0\,(L + 3a)/3EI$ ↲. (b) $M_0a\,(2L + 3a)/6\,EI$ ↓.
9.135 (a) 5.33×10^{-3} rad ↲. (b) 0.01421 in. ↓.
9.136 (a) 3.61×10^{-3} rad ↰. (b) 0.960 mm ↑.
9.137 (a) 2.34×10^{-3} rad ↰. (b) 0.1763 in. ↓.
9.139 (a) $9wL^3/256\,EI$ ↰. (b) $7wL^3/256\,EI$ ↲.
(c) $5wL^4/512EI$ ↓.
9.140 (a) $17PL^3/972\,EI$ ↓. (b) $19PL^3/972\,EI$ ↓.
9.142 $0.00652\,w_0L^4/EI$ at $x = 0.519L$.
9.144 0.212 in. at $x = 5.15$ ft.
9.145 0.1049 in.
9.146 1.841 mm.
9.147 $5P/16$ ↑.
9.148 $9M_0/8L$.
9.150 $7wL/128$ ↑.
9.152 $R_A = 3P/32 \downarrow$; $R_B = 13P/32 \uparrow$; $R_C = 11P/16 \uparrow$.
9.153 (a) 6.87 mm ↑. (b) 46.3 kN ↑.
9.154 10.18 kips ↑; $M_A = -87.9$ kip·ft; $M_D = 46.3$ kip·ft; $M_B = 0$.
9.155 $48\,EI/7L^3$.
9.156 $144\,EI/L^3$.
9.157 (a) $y = (w_0/EIL)\,(L^3x^2/6 - Lx^4/12 + x^5/120)$.
(b) $11w_0L^4/120\,EI$ ↓. (c) $w_0L^3/8EI$ ↰.
9.158 (a) $0.0642\,M_0L^2/EI$ a+ $x = 0.423\,L$. (b) 45.3 kN·m.
9.160 $R_A = R_B = P/2 \uparrow$, $M_A = PL/8$ ↱; $M_B = PL/8 \downarrow$; $M_C = PL/8$.
9.161 (a) 2.49×10^{-3} rad ↰. (b) 1.078 mm ↓.

9.163 0.210 in. ↓.
9.165 (a) 2.55×10^{-3} rad ↰. (b) 6.25 mm ↓.
9.166 (a) 5.86×10^{-3} rad ↲. (b) 0.0690 in. ↑.
9.168 (a) 65.2 kN ↑; $M_A = 0°$; $M_D = 58.7$ kN·m;
$M_B = -82.8$ kN·m.
9.C1 *Prob. 9.74:* 5.56×10^{-3} rad ↰; 2.50 mm ↓.
9.C2 *a = 6 ft:* (a) 3.14×10^{-3} rad ↰, 0.292 in. ↓;
(b) 0.397 in. ↓ at 11.27 ft to the right of A.
9.C3 *x = 1.6 m:* (a) 7.90×10^{-3} rad ↰, 8.16 mm ↓;
(b) 6.05×10^{-3} rad ↰, 5.79 mm ↓;
(c) 1.021×10^{-3} rad ↰, 0.314 mm ↓.
9.C5 (a) *a = 3 ft:* 1.586×10^{-3} rad ↰; 0.1369 in. ↓;
(b) *a = 1.0 m:* 0.293×10^{-3} rad ↰, 0.479 mm ↓.
9.C7 *x = 2.5 m:* 5.31 mm ↓; *x = 5.0 m:* 11.2.28 mm ↓.

CHAPTER 10

10.1 kL.
10.2 k/L.
10.3 $kL/4$.
10.4 $2\,kL/9$.
10.6 120.0 kips.
10.7 $ka^2/2l$.
10.9 305 kN.
10.10 8.37 lb.
10.11 1.421.
10.13 14.10 mm; round strut; 61.4 kN; square strut: 64.3 kN.
10.15 70.2 kips.
10.16 467 kN.
10.17 335 kips.
10.19 2.27.
10.21 (a) 0.500. (b) 2.46.
10.22 (a) $L_{BC} = 4.20$ ft; $L_{CD} = 1.050$ ft. (b) 4.21 kips.
10.24 657 mm.
10.25 (a) 0.500. (b) $b = 14.15$ mm; $d = 28.3$ mm.
10.27 (a) 2.55. (b) $d_2 = 28.3$ mm; $d_3 = 14.14$ mm;
$d_4 = 16.72$ mm; $d_5 = 20.0$ mm.
10.28 (1) 319 kg; (2) 79.8 kg; (3) 319 kg; (4) 653 kg.
10.29 (a) 4.32 mm. (b) 44.4 MPa.
10.30 (a) 1.658 mm. (b) 78.9 MPa.
10.32 (a) 0.0399 in. (b) 19.89 ksi.
10.34 (a) 0.247 in. (b) 12.95 ksi.
10.35 (a) 13.29 kips. (b) 15.50 ksi.
10.36 (a) 235 kN. (b) 149.6 MPa.
10.37 (a) 151.6 kN. (b) 109.5 MPa.
10.39 (a) 370 kN. (b) 104.6 MPa.
10.40 (a) 224 kN. (b) 63.3 MPa.
10.41 58.9°F.
10.43 (a) 189.0 kN. (b) 229 kN.
10.44 (a) 147.0 kN. (b) 174.0 kN.
10.45 (a) 49.6 kips. (b) 0.412.
10.47 1.302 m.
10.49 (a) 26.8 ft. (b) 8.40 ft.
10.50 (a) 4.54 m. (b) 2.41 m.
10.51 W 200 × 26.6.
10.53 2.125 in.
10.54 2.625 in.
10.56 3.09.
10.57 (a) 220 kN. (b) 841 kN.
10.58 (a) 86.6 kips. (b) 88.1 kips.
10.59 414 kN.
10.60 35.9 kN.

10.62 (a) 26.6 kN. (b) 33.0 kN.
10.64 76.8 kips.
10.65 76.3 kips.
10.66 1596 kN.
10.68 173.5 kips.
10.69 (a) 66.3 kN. (b) 243 kN.
10.71 123.1 mm.
10.72 6.53 in.
10.74 1.615 in.
10.75 22.3 mm.
10.77 W200 × 46.1.
10.78 W14 × 82.
10.79 W10 × 54.
10.80 (a) 30.1 mm. (b) 33.5 mm.
10.83 L89 × 64 × 12.7.
10.84 56.1 kips.
10.86 (a) $P_D = 433$ kN; $P_L = 321$ kN.
　　　(b) $P_D = 896$ kN; $P_L = 664$ kN.
10.87 W310 × 74.
10.88 5/16 in.
10.89 76.7 kN.
10.91 (a) 18.26 kips. (b) 14.20 kips.
10.92 (a) 21.1 kips. (b) 18.01 kips.
10.93 (a) 329 kN. (b) 280 kN.
10.95 (a) 0.698 in. (b) 2.11 in.
10.97 16.44 ft.
10.99 5.48 m.
10.100 4.81 m.
10.101 1.021 m.
10.102 1.175 m.
10.103 83.4 mm.
10.104 87.2 mm.
10.105 12.00 mm.
10.106 15.00 mm.
10.107 140.0 mm.
10.109 1.882 in.
10.110 1.735 in.
10.111 $t = \frac{1}{4}$ in.
10.113 W14 × 145.
10.114 W14 × 68.
10.116 W250 × 58.
10.117 (a) 647 lb. (b) 0.651 in. (c) 58.8%.
10.118 $k > 4.91$ kN/m.
10.120 (a) 47.2°. (b) 1.582 kips.
10.121 2.44.
10.123 $\Delta T = \pi^2 b^2/12L^2\alpha$.
10.125 107.7 kN.
10.126 W250 × 67.
10.128 (a) 0.0987 in. (b) 0.787 in.
10.C1 $r = 8$ mm: 9.07 kN. $r = 16$ mm: 70.4 kN.
10.C2 $b = 1.0$ in.: 3.85 kips. $b = 1.375$ in.: 6.07 kips.
10.C3 $h = 5.0$ m: 9819 kg. $h = 7.0$ m: 13,255 kg.
10.C4 $P = 35$ kips: (a) 0.086 in.; (b) 4.69 ksi.
　　　$P = 55$ kips: (a) 0.146 in.; (b) 7.65 ksi.
10.C6 Prob. 10.113: $P_{all} = 282.6$ kips.
　　　Prob. 10.114: $P_{all} = 139.9$ kips.

CHAPTER 11

11.1 (a) 43.1 in · lb/in³. (b) 72.8 in · lb/in³.
　　　(c) 172.4 in · lb/in³.
11.2 (a) 21.6 kJ/m³. (b) 336 kJ/m³, (c) 163.0 kJ/m³.

11.3 (a) 177.9 kJ/m³. (b) 712 kJ/m³. (c) 160.3 kJ/m³.
11.5 (a) 58.0 in · lb/in³. (b) 20.0 in · kip/in³.
11.6 (a) 1296 kJ/m³. (b) 90.0 MJ/m³.
11.7 (a) 1.750 MJ/m³. (b) 71.2 MJ/m³.
11.9 (a) 176.2 in · lb.
　　　(b) $u_{AB} = 11.72$ in · lb/in³; $u_{BC} = 5.65$ in · lb/in³.
11.10 (a) 12.18 J. (b) $u_{AB} = 15.83$kJ/m³; $u_{BC} = 38.6$ kJ/m³.
11.11 (a) 168.8 in · lb.
　　　(b) $u_{CD} = 0.882$ in · lb/in³; $u_{EF} = 5.65$ in · lb/in³.
11.12 0.846 J.
11.15 (a) 3.28. (b) 4.25.
11.17 102.7 in · lb.
11.18 1.398 P^2l/EA.
11.20 2.37 P^2l/EA.
11.21 0.233 P^2l/EA.
11.23 6.68 kip · in.
11.24 $W^2L^5/40$ EI.
11.25 $(P^2a^2/6$ EI$)(a + L)$.
11.27 $(M_0^2/6$ EIL²$)(a^3 + b^3)$.
11.28 89.5 in · lb.
11.30 1048 J.
11.31 670 J.
11.33 12.70 J.
11.37 (a) No yield. (b) Yield occurs.
11.39 (a) 2.33. (b) 2.02.
11.40 $(2M_0^2 L / Ebd^3)(1 + 3Ed^2/10GL^2)$.
11.41 $(Q^2/4\pi GL) \ln (R_2 / R_1)$.
11.42 24.7 mm.
11.43 25.5 ft/sec.
11.44 9.12 lb.
11.45 841 mm.
11.48 11.09 ft/s.
11.50 (a) 7.54 kN. (b) 41.3 MPa. (c) 3.18 mm.
11.51 (a) 9.60 kN. (b) 32.4 MPa. (c) 2.50 mm.
11.52 (a) 15.63 mm. (b) 83.8 N·m. (c) 208 MPa.
11.53 (a) 7.11 mm. (b) 140.1 MPa.
11.54 (a) 0.903 in. (b) 511 lb·in. (c) 21.3 ksi.
11.56 (b) 7.12.
11.57 (b) 0.152.
11.58 $Pa^2b^2/3EI \downarrow$.
11.59 $Pa^2 (a + L)/3EI \uparrow$.
11.61 $M_0 (L + 3a)/3EI \searrow$.
11.62 $3PL^3/16$ EI $\downarrow$.
11.63 $3Pa^3/4$ EI $\downarrow$.
11.65 $M_0L/16$ EI $\searrow$.
11.66 59.8 mm $\downarrow$.
11.67 32.4 in.
11.68 3.12°.
11.72 2.38$Pl/EA \rightarrow$.
11.73 0.650 in. $\downarrow$.
11.75 0.366 in. $\downarrow$.
11.76 1.111 mm $\downarrow$.
11.77 (a) and (b) $P^2L^3/6$ EI $+ PM_0L^2/2EI + M_0^2L/2$ EI.
11.78 (a) and (b) $P^2L^3/48$ EI $+ M_0PL^2/8EI + M_0^2L/2$ EI.
11.80 (a) and (b) $5 M_0^2L/4$ EI.
11.82 (a) and (b) $M_0^2L/2$ EI.
11.83 $0.0443wL^4/EI \downarrow$.
11.85 $wL^4/768$ EI $\uparrow$.
11.86 $7wL^3/48$ EI $\measuredangle$.
11.88 $wL^3/384$ EI $\measuredangle$.
11.89 $(Pab/6EIL^2)(3La + 2a^2 + 2b^2) \searrow$.
11.90 $M_0L/6$ EI $\searrow$.

11.91 0.329 in. $\downarrow$.

11.93 5.12 mm $\downarrow$.

11.94 7.25 mm $\downarrow$.

11.95 7.07×10^{-3} rad $\searrow$.

11.96 3.80 mm $\downarrow$.

11.97 2.07×10^{-3} rad $\measuredangle$.

11.99 $x_C = 0$, $y_C = 2.80 \, P_L/EA \downarrow$.

11.101 0.1613 in. $\downarrow$.

11.102 0.01034 in. $\leftarrow$.

11.103 0.1459 mm $\downarrow$.

11.105 (a) $PL^3/6EI \downarrow$. (b) $0.1443 \, PL^3/EI$.

11.106 $\pi PR^3/2 \, EI \downarrow$.

11.107 (a) $PR^3/2 \, EI \rightarrow$. (b) $\pi PR^3/4 \, EI \downarrow$.

11.109 $5PL^3/6EI$.

11.111 $5P/16 \uparrow$.

11.112 $3M_0/2L \uparrow$.

11.113 $3M_0 b(L + a) /2L^3 \uparrow$.

11.114 $7wL/128 \uparrow$.

11.117 $P/(1 + 2 \cos^3 \phi)$.

11.118 $7P/8$.

11.119 $0.652P$.

11.121 $2P/3$.

11.123 136.6 J.

11.124 1.767 in · kip.

11.126 4.76 kg.

11.129 2.55°.

11.130 11.57 mm $\downarrow$.

11.134 0.807 in . $\downarrow$.

11.C2 (a) $a = 15$ in.: $\sigma_D = 17.19$ ksi, $\sigma_C = 21.0$ ksi;
$a = 45$ in.: $\sigma_D = 36.2$ ksi, $\sigma_C = 14.74$ ksi.
(b) $a = 18.34$ in., $\sigma = 20.67$ ksi.

11.C3 (a) $L = 200$ mm: $h = 2.27$ mm;
$L = 800$ mm: $h = 1.076$ mm.
(b) $L = 440$ mm: $h = 3.23$ mm.

11.C4 $a = 300$ mm: 1.795 mm, 179.46 MPa;
$a = 600$ mm: 2.87 mm, 179.59 MPa.

11.C5 $a = 2$ m: (a) 30.0 J; (b) 7.57 mm, 60.8 J.
$a = 4$ m: (a) 21.9 J; (b) 8.87 mm, 83.4 J.

11.C6 $a = 20$ in: (a) 13.26 in.; (b) 99.5 kip · in.; (c) 803 lb.
$a = 50$ in: (a) 9.46 in.; (b) 93.7 kip · in.; (c) 996 lb.

Photo Credits

Image Research by Danny Meldung/Photo Affairs, Inc.

CHAPTER 1
Opener: © Pete Ryan/Getty Images RF; 1.1: © David R. Frazier/Science Source; 1.2: © Walter Bibikow/agefotostock; 1.3: © John DeWolf.

CHAPTER 2
Opener: © Sylvain Grandadam/agefotostock; 2.1: © John DeWolf; 2.2: Courtesy of Tinius Olsen Testing Machine Co., Inc.; 2.3-2.5: © John DeWolf; 2.6: © John Fisher.

CHAPTER 3
Opener: © incamerastock/Alamy; 3.1: © 2008 Ford Motor Company; 3.2: © John DeWolf; 3.3: Courtesy of Tinius Olsen Testing Machine Co., Inc.; 3.4: © koi88/Alamy RF.

CHAPTER 4
Opener: © Mel Curtis/Getty Images RF; 4.1: Courtesy of Flexifoil; 4.2: © Tony Freeman/PhotoEdit; 4.3: © Hisham Ibrahim/Getty Images RF; 4.4: © Bohemian Nomad Picturemakers/Corbis; 4.5: © Tony Freeman/PhotoEdit; 4.6: © John DeWolf.

CHAPTER 5
Opener: © Digital Vision/Getty Images RF; 5.1: © Huntstock/agefotostock RF; 5.2: © David Nunuk/Science Source.

CHAPTER 6
Opener: © Aurora Photos/Alamy; 6.1: © John DeWolf; 6.2: © Jake Wyman/Getty Images; 6.3: © Rodho/shutterstock.com.

CHAPTER 7
Opener: NASA; 7.1: © Walter G. Allgöwer Image Broker/Newscom; 7.2: © FELLOW/agefotostock; 7.3: © Clair Dunn/Alamy; 7.4: © Spencer C. Grant/PhotoEdit.

CHAPTER 8
Opener: © S. Meltzer/PhotoLink/Getty Images RF.

CHAPTER 9
Opener: © Jetta Productions/Getty Images RF; 9.1: © Royalty-Free/Corbis; 9.2–9.3: © John DeWolf; 9.4: © FOTOG/Getty Images RF; 9.5: © Royalty-Free/Corbis.

CHAPTER 10
Opener: © Jose Manuel/Getty Images RF; 10.1: Courtesy of Fritz Engineering Laboratory, Lehigh University; 10.2a: © Steve Photo/Alamy RF; 10.2b: © Peter Marlow/Magnum Photos.

CHAPTER 11
Opener: © Corbis Super RF/Alamy; 11.1: © Israel Antunes/Alamy; 11.2: © Tony Freeman/PhotoEdit; 11.3: © TRL Ltd./Science Source.

Index

Centroids of Common Shapes of Areas and Lines

Shape		$\bar{x}$	$\bar{y}$	Area
Triangular area			$\dfrac{h}{3}$	$\dfrac{bh}{2}$
Quarter-circular area		$\dfrac{4r}{3\pi}$	$\dfrac{4r}{3\pi}$	$\dfrac{\pi r^2}{4}$
Semicircular area		0	$\dfrac{4r}{3\pi}$	$\dfrac{\pi r^2}{2}$
Semiparabolic area		$\dfrac{3a}{8}$	$\dfrac{3h}{5}$	$\dfrac{2ah}{3}$
Parabolic area		0	$\dfrac{3h}{5}$	$\dfrac{4ah}{3}$
Parabolic spandrel		$\dfrac{3a}{4}$	$\dfrac{3h}{10}$	$\dfrac{ah}{3}$
Circular sector		$\dfrac{2r\sin\alpha}{3\alpha}$	0	αr^2
Quarter-circular arc		$\dfrac{2r}{\pi}$	$\dfrac{2r}{\pi}$	$\dfrac{\pi r}{2}$
Semicircular arc		0	$\dfrac{2r}{\pi}$	πr
Arc of circle		$\dfrac{r\sin\alpha}{\alpha}$	0	$2\alpha r$